Design and Analysis of Experiments

Classical and Regression Approaches with SAS

STATISTICS: Textbooks and Monographs

D. B. Owen

Founding Editor, 1972–1991

STATISTICS: Textbooks and Monographs

Recent Titles

Visualizing Statistical Models and Concepts, *R. W. Farebrother and Michaël Schyns*

Financial and Actuarial Statistics: An Introduction, *Dale S. Borowiak*

Nonparametric Statistical Inference, Fourth Edition, Revised and Expanded, *Jean Dickinson Gibbons and Subhabrata Chakraborti*

Computer-Aided Econometrics, *edited by David E.A. Giles*

The EM Algorithm and Related Statistical Models, *edited by Michiko Watanabe and Kazunori Yamaguchi*

Multivariate Statistical Analysis, Second Edition, Revised and Expanded, *Narayan C. Giri*

Computational Methods in Statistics and Econometrics, *Hisashi Tanizaki*

Applied Sequential Methodologies: Real-World Examples with Data Analysis, *edited by Nitis Mukhopadhyay, Sujay Datta, and Saibal Chattopadhyay*

Handbook of Beta Distribution and Its Applications, *edited by Arjun K. Gupta and Saralees Nadarajah*

Item Response Theory: Parameter Estimation Techniques, Second Edition, *edited by Frank B. Baker and Seock-Ho Kim*

Statistical Methods in Computer Security, *edited by William W. S. Chen*

Elementary Statistical Quality Control, Second Edition, *John T. Burr*

Data Analysis of Asymmetric Structures, *Takayuki Saito and Hiroshi Yadohisa*

Mathematical Statistics with Applications, *Asha Seth Kapadia, Wenyaw Chan, and Lemuel Moyé*

Advances on Models, Characterizations and Applications, *N. Balakrishnan, I. G. Bairamov, and O. L. Gebizlioglu*

Survey Sampling: Theory and Methods, Second Edition, *Arijit Chaudhuri and Horst Stenger*

Statistical Design of Experiments with Engineering Applications, *Kamel Rekab and Muzaffar Shaikh*

Quality by Experimental Design, Third Edition, *Thomas B. Barker*

Handbook of Parallel Computing and Statistics, *Erricos John Kontoghiorghes*

Statistical Inference Based on Divergence Measures, *Leandro Pardo*

A Kalman Filter Primer, *Randy Eubank*

Introductory Statistical Inference, *Nitis Mukhopadhyay*

Handbook of Statistical Distributions with Applications, *K. Krishnamoorthy*

A Course on Queueing Models, *Joti Lal Jain, Sri Gopal Mohanty, and Walter Böhm*

Univariate and Multivariate General Linear Models: Theory and Applications with SAS, Second Edition, *Kevin Kim and Neil Timm*

Randomization Tests, Fourth Edition, *Eugene S. Edgington and Patrick Onghena*

Design and Analysis of Experiments: Classical and Regression Approaches with SAS, *Leonard C. Onyiah*

Analytical Methods for Risk Management: A Systems Engineering Perspective, *Paul R. Garvey*

Confidence Intervals in Generalized Regression Models, *Esa Uusipaikka*

Design and Analysis of Experiments

Classical and Regression Approaches with SAS

Leonard C. Onyiah

St. Cloud State University

Minnesota, U.S.A.

CRC Press is an imprint of the
Taylor & Francis Group, an informa business
A CHAPMAN & HALL BOOK

CRC Press
Taylor & Francis Group
6000 Broken Sound Parkway NW, Suite 300
Boca Raton, FL 33487-2742

First issued in paperback 2019

ISBN-13: 978-1-4200-6054-6 (hbk)
ISBN-13: 978-0-367-38707-5 (pbk)

Library of Congress Cataloging-in-Publication Data

Onyiah, Leonard C.
Design and analysis of experiments : classical and regression approaches with SAS / by Leonard Onyiah.
p. cm.
Includes bibliographical references and index.
ISBN-13: 978-1-4200-6054-6 (alk. paper)
ISBN-10: 1-4200-6054-6 (alk. paper)
1. Regression analysis--Data processing. 2. Anaylsis of variance--Data processing. 3. SAS (Computer file) I. Title.

QA278.2.O49 2008
519.5'36--dc22 2007052021

Visit the Taylor & Francis Web site at
http://www.taylorandfrancis.com

and the CRC Press Web site at
http://www.crcpress.com

Contents

Preface

This book is intended for students who have had a mix of 100 and 200 level classes of statistics, covering exploratory data analysis, basic descriptive data analysis, basic probability, statistical distributions, confidence intervals, and hypothesis testing. Knowledge of algebra of matrices, determinants, solutions of simultaneous linear equations, and quadratic equations will be helpful. This book contains the required computations. Graduate students in biology, psychology, computer science, the physical sciences (physics and chemistry), and the different fields of engineering have felt comfortable in my classes of STAT 440/540, which were taught from material in this book.

Books written on design and analysis of experiments have hitherto mostly emphasized the classical approaches to the modeling and analysis of experimental data. *Design and Analysis of Experiments: Classical and Regression Approaches with SAS* not only includes classical experimental design theory but also emphasizes regression approaches while teaching data analysis using the invaluable tool, SAS, which is the widely used software in statistical data analysis.

With the growth of statistical computing and availability of cutting-edge software, the time has come for a new approach for the design and analysis of experiments. Consequently, in addition to all the classical manual analyses in this book, explicit SAS programs have been written to analyze the responses of each experiment, in addition to presenting outputs that result from the executions of the programs. This book presents explicit SAS codes for the analyses of responses in the examples of designs presented in the text. The SAS codes cover classical analyses and analyses based on regression models. The example SAS codes presented in this book are aimed at providing help for students and any user in writing their own SAS programs. The purpose of presenting these codes is to help students/users become versatile in their ability to analyze data, acquire the skills to do it manually or by using SAS, and writing their own programs and interpreting the outputs. Exercises provided at the end of each chapter enable students/users to have hands-on SAS programming experience in analyzing data for both classical and regression approaches. This book, in addition to teaching design and analysis of experiments, is a useful tool for learning statistical computing with SAS, which will prepare the student/user for future work environments in which SAS is the software used routinely for statistical data analysis. Chapters dealing with classical designs are followed by the regression equivalents of the chapter. For instance,

Chapter 4 treats the regression approaches to classical designs treated in Chapter 3.

SAS analyses in this text mostly use PROC GLM, PROC ANOVA, PROC MIXED, and PROC REG. For ANOVA using SAS, this text uses PROC GLM for the more general and complicated analyses. PROC GLM is used to carry out the bulk of the analyses, but occasionally uses PROC ANOVA for the straightforward analyses of variance for balanced data. PROC MIXED is employed mostly for the mixed randomized complete block designs and repeated measures designs.

This book treats most of the different designs covered in a typical experimental design course. It provides full coverage of the material needed for undergraduate classes on experimental design and analysis and more. In some institutions, the course that uses this text could be considered as Experimental Design and Regression, which could be run in two semesters for in-depth treatment. This book introduces general regression in the first chapter. In the other chapters, the treatment of regression emphasizes aspects of regression concerned with modeling the designed experiments. The text covers enough material for the first semester of a masters-level course in experimental design. For institutions where experimental design is taught to graduate students of disciplines other than statistics, this book contains material that would cover more than a first course.

In using this text to teach STAT 440/540, the design and analysis of experiments, I usually invite students to read Chapter 1 to review relevant basic statistical concepts. We spend reasonable time on Chapter 2, which treats the requirements for good experimentation, the completely randomized design (CRD), use of orthogonal contrast to test hypotheses, and model adequacy check. The model adequacy check is applicable to all designs treated in this book; consequently, it is thoroughly treated, dealing with the theory to be used for diagnostics in other designs treated later in the text. In Chapter 3, we quickly work our way through two-factor factorial experiments learning about fixed, random, mixed effects models, repeated measures, and their analyses. We treat all these models and others from the general linear model and regression approaches, providing examples of manual analyses complemented by the corresponding SAS analyses. Designs with randomization restrictions (randomized complete block, Latin and Graeco-Latin square designs) are briefly treated before we delve into the special cases of the 2^k and 3^k factorial designs including fractional replication and confounding. We introduce response surfaces, and thereafter, treat balanced incomplete block or hierarchical designs. Some of the designs in the text never get included in a typical undergraduate course. Keeping in mind a class that is predominantly made up of graduate students, the material is treated in more detail, and content is varied to cover things like the MANOVA equivalents of the univariate designs and ANCOVA. Material not thoroughly treated in class is fortified by assignments, exercises, projects, and directed reading, which treat either the design or the analytical aspects of the designs. If material in the text is followed in detail, it should require two one-semester courses to treat each experimental design in this text thoroughly.

I have provided a solutions manual for all the exercises in this book. This manual can be purchased with the book to aid in understanding the material. Exercises included in the book provide a mix of questions requiring manual and SAS analyses. The problems cover classical and regression approaches. Doing the exercises should help to induce a well-rounded understanding of the theory.

Acknowledgments

The accomplishment of work of this magnitude uses the help of many. I am therefore grateful to everybody who contributed to the successful completion of this book.

I wish to thank my wife Constance Onyiah for support throughout the process of writing this book. I am also grateful to all my children (Comfort C. Onyiah, MD, Joseph C. Onyiah, MD, Leonard C. Onyiah, Jr, and Constance I. Onyiah) whose support and encouragement formed a large part of the inspiration needed to complete it. I am particularly very indebted to Comfort C. Onyiah, MD, for taking time to read through the material and making helpful suggestions and comments. I also thank Leonard C. Onyiah, Jr, for designing the cover of this book.

I am very indebted to Ene I. Ette, PhD, FCP, FCCP, formerly of FDA and Vertex Corp., now CEO Anoixis Corp, for friendship, encouragement, support, and inspiration. I thank David G. Kleinbaum, PhD, Department of Epidemiology, RSPH, Emory University for helpful material. I thank my various students of STAT 440 at St. Cloud State University, St. Cloud, Minnesota, who were taught from and read the earlier versions of the materials in this book. I appreciate their helpful suggestions.

I thank my colleagues at the Department of Statistics and Computer Networking of St. Cloud State University for supporting the use of the material to teach Experimental Design and Analysis, helping the material as it metamorphosed into a text on the subject. I appreciate the kind comments of colleagues in the department who had read earlier versions of this book. I thank Dr. Hui Xu who proofread the material.

I wish to thank Karin Duncan, Dennis Murphy, and Subigya Shagya of the Learning Resources and Technology Services (LRTS) of St. Cloud State University for their help with the production of quality graphics for this book.

The outputs, codes, and data analyses for this text were generated using SAS software, which is a copyright of SAS Institute Inc. SAS and all other SAS Inc. product or service names are registered trademarks of SAS Institute Inc., Cary, North Carolina. I would also like to thank SAS Inc. for allowing the use of their copyrighted material in this book.

I thank the various people in the Taylor & Francis Group who helped to get this book out. In particular, I thank David Grubbs, editor for mathematics and statistics; Jessica Vakili, project coordinator; and Richard Tressider, the project editor for their invaluable contributions to the production of this book.

Lastly, but above all, I thank Jesus Christ, my Lord, for providing me with health, strength, and knowledge to complete the writing and production of this book.

Author

Leonard C. Onyiah is currently a professor of statistics and the director of the statistics program in the Department of Statistics and Computer Networking of St. Cloud State University, St. Cloud, Minnesota. He has several years of experience in teaching statistics at the University of Northumbria, Newcastle upon Tyne, England; The Queen's College, Glasgow (now, Glasgow Caledonian University, Glasgow, Scotland); and Anambra State University of Technology, Enugu, Nigeria. He holds the degrees of MPhil and PhD in statistics from University of Strathclyde, Glasgow, Scotland, Great Britain. He also holds the degrees of BSc in statistics from University of Nigeria, Nsukka, Nigeria, and MSc in statistics from the University of Ibadan, Ibadan, Nigeria.

He has published some papers in peer-reviewed journals and written *SAS Programming and Data Analysis: A Theory and Program-Driven Approach* (2005), University Press of America, Lanham. He contributed to the text: *Pharmacometrics: The Science of Quantitative Pharmacology*, edited by Ette and Williams.

Chapter 1

Introductory Statistical Inference and Regression Analysis

1.1 Elementary Statistical Inference

Generally, when statisticians try to draw some conclusions about a population of interest, it is often cheaper and more convenient to draw these conclusions using a sample drawn from the population. Some of the samples used may arise from experiments that are deliberately planned and executed to elicit information of interest. Others may arise from pure observational studies in which the statistician does not control the process but simply observes the responses of the process. Statistical inference, a major area of statistical endeavor, is built around a practical way of obtaining information about populations, namely, sampling. One of the problems commonly dealt with in inferential statistics is the estimation of population values called parameters. A statistic can be used to estimate parameters like mean, median, and variance of a population. Population proportions and the rth moments about the mean are also examples of parameters.

Population. The entire collection of attributes that are under study is called a population. The attributes may refer to characteristics of objects such as pencils, steel rods, cars, etc. They may also refer to characteristics of human or animal subjects.

Sample. The subset of the population that is used in the actual study is called a sample.

Statistical inference. The general field of statistical endeavor that uses samples and sample characteristics to make decisions (make inference) about populations and population characteristics is called statistical inference.

Statistic. A value derived from a sample is called a statistic.

For clarity, we distinguish between an estimator and an estimate. If we wish to estimate θ, the mean of the normal distribution, it is known that by maximum likelihood estimation (see Hogg and Craig, 1978, p. 202), the solution for θ that maximizes the likelihood function is $\sum_{i=1}^{n} x_i/n$. The estimator of θ is the statistic $\hat{\theta} = \sum_{i=1}^{n} X_i/n$. The observed value of $\hat{\theta}$, say in an experiment, $\sum_{i=1}^{n} x_i/n$ is an estimate of θ.

1.1.1 Unbiased and Efficient Estimators

Most of the inference made about a parameter is based on a statistic, which is deemed to be an estimate of the parameter. Statisticians, therefore, endeavor to choose the best statistic for inferential purposes. The way to arrive at the best choice would depend on the qualities of the statistic. Two important qualities that are generally considered in choosing the best statistic are unbiasedness and efficiency of the statistic. Let us look closely at the definitions for an unbiased estimator:

Random variable. A function that associates a unique numeric value with every outcome of an experiment is a random variable. Thus, the value of the random variable will vary from one trial to another as the experiment is repeated. For example, if we toss a coin six times, the random variable Y, which is the number of heads, can take the values 0, 1, 2, ..., 6. A similar definition of a random variable is found in Hogg and Craig (1978, p. 200) in which they state "We shall refer to a random variable X as the outcome of a random experiment and the space of X as the sample space."

Unbiased estimator. Let θ be a parameter of the random variable Y. If, $E(\hat{\theta} - \theta) = 0$, then $\hat{\theta}$ is an unbiased estimator of the parameter θ. Otherwise, $\hat{\theta}$ is biased for estimating θ.

For instance, from statistical theory we know that the sample variance $s^2 = (\sum_{i=1}^{n}(y_i - \bar{y})^2)/n - 1$ is an unbiased estimator of the parameter σ^2, the population variance. We also know that the approximation (large sample definition) of sample variance $s^2 = (\sum_{i=1}^{n}(y_i - \bar{y})^2)/n$ is biased for estimating the population variance σ^2.

Example 1.1

Show that $s^2 = \sum_{i=1}^{n}(y_i - \bar{y})^2/n$ is a biased estimator of σ^2.

Solution:

$$E(s^2) = E\left(\frac{\sum_{i=1}^{n}(y_i - \bar{y})^2}{n}\right) = \left(\frac{1}{n}\right)\left(E\sum_{i=1}^{n} y_i^2 - E(n\bar{y}^2)\right)$$

Note: $E(\bar{y}) = \mu;\quad E(\bar{y})^2 = \dfrac{\sigma^2}{n} \Rightarrow E(\bar{y}^2) = \mu^2 + \dfrac{\sigma^2}{n} = \dfrac{n\mu^2 + \sigma^2}{n}$

Similarly,

$$E\left(\sum_{i=1}^{n} y_i^2\right) - E\left(\sum_{i=1}^{n} y_i\right)^2 = n\sigma^2 \Rightarrow E\left(\sum_{i=1}^{n} y_i^2\right)$$

$$= n\sigma^2 + n^2\mu^2 \Rightarrow E\left(\frac{1}{n}\right)\left(\sum_{i=1}^{n} y_i^2\right) = \sigma^2 + n\mu^2$$

$$E(s^2) = E\frac{1}{n}\sum_{i=1}^{n} y_i^2 - E(\bar{y}^2) = \sigma^2 + n\mu^2 - \frac{n\mu^2 + \sigma^2}{n}$$

$$= \frac{n\sigma^2 + n\mu^2 - n\mu^2 - \sigma^2}{n} = \frac{(n-1)\sigma^2}{n} \neq \sigma^2$$

The conclusion reached above shows that $s^2 = \sum_{i=1}^{n} (y_i - \bar{y})^2/n$ is a biased estimator of σ^2.

Efficient estimator. An efficient estimator for a parameter is an unbiased estimator with the minimum variance.

A statistic could be an unbiased estimator for a parameter but may not be an efficient estimator.

Most efficient estimator. The estimator with the least variance of all possible estimators of a parameter is termed the most efficient estimator for that parameter.

It is desirable that an unbiased estimator of a parameter should be the most efficient estimator. □

Example 1.2

Show that for a sample of size n taken from a normal population, the sample mean $\bar{y}$ is a more efficient estimator of the population mean μ than the sample median $\bar{\bar{y}}$.

Solution:

There are several unbiased estimators of the population mean for the normal distribution. Apart from the sample mean, another statistic that is unbiased for the population mean is the median if sample data are normal.

$$E(\bar{y}) = \mu \quad (\mu \text{ is the population mean})$$

$$E(\bar{\bar{y}}) = \mu \quad (\bar{\bar{y}} \text{ is the sample median})$$

It is known from statistical theory (see Wackerly et al., 2002, p. 417) that

$$\text{Var}(\bar{y}) = \frac{\sigma^2}{n} \quad (\bar{y} \text{ is the sample mean, and } \sigma^2 \text{ is the population variance})$$

$$\text{Var}(\bar{\bar{y}}) = \frac{(1.253)^2\sigma^2}{n} \quad (\bar{\bar{y}} \text{ is the sample median}).$$

(Proof is beyond the scope of this section)

The variance of the median is greater than the variance of the sample mean. Therefore, the sample mean is a more efficient estimator of the population mean than the sample median. (We can see that the sample mean $\bar{y}$ and the sample median $\bar{\bar{y}}$ have the same mean μ, the population mean. This means that both are unbiased estimators of the population mean μ. We have just shown that the variance of the sample median is greater than the variance for the sample mean, with the result that the sample mean is an efficient estimator of the population mean, while the sample median is not an efficient estimator of the mean.)

Therefore, the estimator with the smaller variance of two possible estimators of the population mean would be considered to be the more efficient estimator of the mean. In fact, the sample mean is a more efficient estimator of the population mean μ than the median. □

1.1.2 Point and Interval Estimation

We can estimate a parameter by a single number statistic called a point estimator. Alternatively, we can also indicate that the parameter lies in an interval, for example, the length equal to 7.8 ± 0.7 cm, where the length lies between 7.1 and 8.5 cm. This type of estimate is termed an interval estimate. Interval estimates are often preferred to point estimates because the former contains a statement of error that indicates the reliability of the estimate.

1.1.3 Confidence Intervals for Parameters of a Population

An interval estimator usually depends on the values of the sample measurements. And this is the rule that helps us to specify two numbers which are the end points of the interval. The upper and lower ends of the interval are called its upper and lower limits, respectively. The probability that an interval will enclose the parameter of interest is called the confidence coefficient. The confidence coefficient is a measure of the proportion of the times that if we take repeated sampling measurements from the population, the parameter of interest would be contained in the interval. This probability is an indication of the confidence we can attach to that interval as an estimate of the parameter. The interval is called a confidence interval (CI) estimate. The CIs are obtained using the sampling distribution of the statistic which was chosen to estimate the parameter. From sampling theory and large sample theory, we know that most of the statistics used would be normally distributed. A table of such statistics and their distributions has been presented in Spiegel (1974, p. 178). Typically, the CI estimate for a parameter θ would be written as a $100(1-\alpha)\%$ CI, converting the probability attached to the CI into a percentage. Provided that the estimator of θ is normally distributed, then the $100(1-\alpha)\%$ CI for θ is $\hat{\theta} \pm Z_{\alpha/2}\sigma_{\hat{\theta}}$ where $\sigma_{\hat{\theta}}$ is the standard error of the statistic $\hat{\theta}$. Hence, for the 95% CI for θ, $\alpha = 5\%$, and the CI would be written as $100(1-0.05)\%$

confidence limits for θ is $\hat{\theta} \pm Z_{0.025}\sigma_{\hat{\theta}}$. These limits represent the lower and upper ends of the CI $(\hat{\theta} - Z_{0.025}\sigma_{\hat{\theta}}, \hat{\theta} + Z_{0.025}\sigma_{\hat{\theta}})$. The lower limit is obtained when the value of the standard normal variate $Z = -1.96$, corresponding to a probability of 0.025 (half of α). The upper limit is obtained when $Z = 1.96$ is the value of the standard normal variate corresponding to a probability of 0.975 (equivalent of $1 - \alpha/2$) (see Appendices A1 and A2 for table of the standard normal distribution).

To obtain the CI for a parameter, we need to address four points (1) the sampling distribution of its estimator, (2) the values of the variate at extreme ends of the interval based on the confidence of $100(1 - \alpha)\%$, (3) the computed value of the estimate from a sample of the population, and (4) the computed value of standard error for the estimate, if it is known. These values are substituted in the formula for $100(1 - \alpha)\%$ CI to obtain the desired limits for the interval. Because the intervals have limits, it is often said that we are looking for the $100(1 - \alpha)\%$ confidence limits. This is exactly the same thing as CIs, for once we know the upper and lower limits, we know the interval.

1.1.4 Confidence Intervals for the Means

We stated earlier that if a random variable Y is normally distributed with mean μ and variance σ^2, the sample mean $\bar{y}$ is an unbiased and efficient estimator of the population mean μ. We also know from sampling theory that the sample mean is normally distributed, so that

$$\bar{y} \sim N\left(\mu, \frac{\sigma^2}{n}\right)$$

We can easily set up the $100(1 - \alpha)\%$ confidence limits for the mean of the population based on the sample mean as

$$\bar{y} \pm Z_{\alpha/2}\frac{\sigma}{\sqrt{n}}$$

so that the corresponding interval is

$$\left(\bar{y} - Z_{\alpha/2}\frac{\sigma}{\sqrt{n}}, \bar{y} + Z_{\alpha/2}\frac{\sigma}{\sqrt{n}}\right)$$

Sometimes, when the underlying distribution of the data is normal with mean μ and variance σ^2, which are unknown, we have to estimate such variance from the sample data. Then, by sampling theory, the sample mean is not normally distributed, but the standardized sample mean is $(\bar{y} - \mu)/(s/\sqrt{n}) \sim t(n-1)$, where n is the number of observations in the sample or is the sample size. In such situations, σ is replaced by s in the formula for the $100(1 - \alpha)\%$ CI. The resulting $100(1 - \alpha)\%$ confidence limits are: $\bar{y} \pm t(n - 1, \alpha/2)(s/\sqrt{n})$. The corresponding interval is

$(\bar{y} - t(n-1, \alpha/2)(s/\sqrt{n}), \bar{y} + t(n-1, \alpha/2)(s/\sqrt{n}))$. Note that sample variance is $s^2 = \sum_{i=1}^{n} (y_i - \bar{y})^2/(n-1)$.

We then need to refer to the table of the t-distribution in Appendix A3 to obtain the values of t to be used in the above formula. For a small sample size, we can use small sample theory to obtain the $100(1-\alpha)\%$ CI for the mean as:

$$\left(\bar{y} - t(n-1, \alpha/2)\frac{s}{\sqrt{n}}, \bar{y} + t(n-1, \alpha/2)\frac{s}{\sqrt{n}}\right)$$

The central limit theorem states that if $Y_1, Y_2, \ldots, Y_n$ are independent, identically distributed random variables with $E(Y_i) = \mu$ and variance of Y_i is $\sigma^2 < \infty$, then $\bar{y}$ is asymptotically normally distributed with mean μ and variance σ^2/n.

The conclusions of the central limit theorem can be applied to any random sample $Y_1, Y_2, \ldots, Y_n$ drawn from any population as long as the mean $E(Y_i) = \mu$ and variance $\mathrm{Var}(Y_i) = \sigma^2$ are finite, and the sample size n is large.

Therefore, if the sample size is large, we can obtain the interval by using the standard normal distribution as

$$\left(\bar{y} - Z_{\alpha/2}\frac{s}{\sqrt{n}}, \bar{y} + Z_{\alpha/2}\frac{s}{\sqrt{n}}\right)$$

Example 1.3

The variance of the birth-weights of premature babies born in a large metropolitan hospital is known to be 0.56 kg. Samples of 190 premature babies were taken to estimate the population mean birth-weight for premature babies in the metropolis, and a sample mean of 3.02 was found. Obtain (1) the 95% and (2) the 99% CIs for mean birth-weights of premature babies born in the metropolis.

Solution:

Let the birth-weights be Y. We know that by central limit theorem, the sample mean $\bar{y} \sim N\left(\mu, \frac{\sigma^2}{n}\right)$:

$$\bar{y} = 3.02; \quad \hat{\sigma}_{\bar{y}} = \sqrt{\frac{0.56}{190}} = 0.0543$$

1. $$\begin{aligned} 95\% \text{ confidence limits} &= 100(1-0.05)\% \text{ CI} \\ &= 3.02 \pm Z_{(0.05/2)}(\sigma/\sqrt{n}) \\ &= 3.02 \pm Z_{(0.025)}(0.0543) \\ &= 3.02 \pm 0.1064 \end{aligned}$$

   The 95% CI is (2.9136, 3.1264).

2. 99% confidence limits $= 100(1 - 0.01)\%$ CI

$$= 3.02 \pm Z_{(0.01/2)}(\sigma/\sqrt{n})$$
$$= 3.02 \pm Z_{(0.005)}(\sigma/\sqrt{n})$$
$$= 3.02 \pm 2.57(0.0543)$$
$$= 3.02 \pm 0.1396$$

99% CI is (2.8804, 3.1596). ☐

Example 1.4

The following data represent the scores of 20 students in a general education statistics course involving hundreds of students. Obtain a 95% CI for the mean of the marks of students in this course. Assume that the scores in this class are normally distributed. What is the 90% CI for the mean score?

75 45 67 88 90 78 83 89 74 68 44 73 84 91 77 56 64 55 79 72

Solution:

$$\bar{y} = \frac{1452}{20} = 72.6;$$
$$s^2 = \left(\frac{1}{20-1}\right)\left[(75^2 + 45^2 + \cdots + 72^2) - \frac{1452^2}{20}\right] = 198.7 \Rightarrow s = 14.095$$
$$\frac{s}{\sqrt{n}} = \frac{14.095}{\sqrt{20}} = 3.15$$

$$100(1-\alpha)\%\ \text{CI} = \bar{y} \pm t(n-1, \alpha/2)\frac{s}{\sqrt{n}}$$

1. 95% CI $= 100(1 - 0.05)\%$ CI

$$= 72.6 \pm t(19{,}0.025)(3.15)$$
$$= 72.6 \pm 2.09(3.15)$$
$$= 72.6 \pm 6.584$$

95% CI is (66.02, 79.18).

2. 90% CI $= 100(1 - 0.10)\%$ CI

$$= 72.6 \pm t(19{,}0.05)(3.15)$$
$$= 72.6 \pm 1.73(3.15)$$
$$= 72.6 \pm 5.45$$

90% CI is (67.15, 78.05). ☐

1.1.5 Confidence Intervals for Differences between Two Means

Sometimes, we need to establish the CI for the difference between the means of two independent samples represented by Y_1 and Y_2. In this case, we are looking for the CI for the difference in means $\mu_1 - \mu_2$. If we have a large sample ($n \geq 30$), then we can assume that the difference between the means of the samples $\bar{y}_1 - \bar{y}_2 \sim N(\mu_1 - \mu_2, (\sigma_1^2/n_1) + (\sigma_2^2/n_2))$ and thus the $100(1-\alpha)\%$ CI for $\mu_1 - \mu_2$ is

$$(\bar{y}_1 - \bar{y}_2) \pm Z_{\alpha/2}\sigma_{\bar{y}_1 - \bar{y}_2}$$

where $\sigma_{\bar{y}_1 - \bar{y}_2} = \sqrt{(\sigma_1^2/n_1) + (\sigma_2^2/n_2)}$ if the two samples are truly independent. Unfortunately, we often do not know the values of the population variances σ_1^2 and σ_2^2, but when the sample sizes are large, we can substitute the sample variances for the population variances, so that

$$\hat{\sigma}_{\bar{y}_1 - \bar{y}_2} = \sqrt{\frac{s_1^2}{n_1} + \frac{s_2^2}{n_2}}$$

Let us consider the following example:

Example 1.5

A survey of the heights of 18-year-old students (260 male and 200 female students) in the North Jade County was made. The mean height for the males was 1.68 m with variance 0.076 m. The mean height for the females was 1.57 m with variance 0.065 m. Obtain the 94% confidence limits for the difference in the mean heights of 18-year-old males and females from this county.

Solution:

Since the sample sizes are large, each being greater than 30, we can use large sample theory. This means that the applicable CI is obtained by the formula

$$100(1-\alpha)\%\,\text{CI} = (\bar{y}_1 - \bar{y}_2) \pm Z_{\alpha/2}\sqrt{\frac{s_1^2}{n_1} + \frac{s_2^2}{n_2}}$$

$$\Rightarrow 94\%\,\text{CI} = 100(1-0.04)\,\text{CI} = (1.68 - 1.57) \pm 1.88\sqrt{\frac{0.076}{260} + \frac{0.065}{200}}$$

$$= 0.11 \pm 0.0467$$

The 94% CI is (0.0633, 0.1567).

When the sample size is not large, the large sample theory does not apply and we cannot obtain a CI based on the assumption used above. We have to resort to small sample theory. The difference between the means has a different distribution, which is not normal. Suppose that sample Y_1 contains n_1 and sample Y_2 contains n_2 observations, then there are two possibilities. If we have reason to believe that the two small samples might have come from normal populations with equal variance, and we know this common variance, we can establish a CI for the difference between the two means by using the formula

$$100(1-\alpha)\%\,\text{CI} = (\bar{y}_1 - \bar{y}_2) \pm Z_{\alpha/2}\sqrt{\sigma^2\left(\frac{1}{n_1} + \frac{1}{n_2}\right)}$$

However, if the variance is unknown, we can pool the variances of the two samples to obtain an estimate of the common variance. Then the pooled variance is estimated by

$$s_{\mathrm{p}}^2 = \frac{(n_1 - 1)s_1^2 + (n_2 - 1)s_2^2}{n_1 + n_2 - 2}$$

The standard error for the difference in the two means is obtained as

$$\hat{\sigma}_{(\bar{y}_1 - \bar{y}_2)} = s_{\mathrm{p}}\sqrt{\frac{1}{n_1} + \frac{1}{n_2}}$$

Under the above conditions, the difference in means is distributed as t, so that

$$\frac{(\bar{y}_1 - \bar{y}_2)}{s_{\mathrm{p}}\sqrt{\dfrac{1}{n_1} + \dfrac{1}{n_2}}} \sim t(n_1 + n_2 - 2)$$

The $100(1-\alpha)\%$ CI for the difference in the two means is obtained by using the formula

$$(\bar{y}_1 - \bar{y}_2) \pm t(n_1 + n_2 - 2, \alpha/2)s_{\mathrm{p}}\sqrt{\frac{1}{n_1} + \frac{1}{n_2}}$$

The second case arises when we have no reason to believe that the two populations have equal variance. Here, we are comparing two independent samples, just like the one considered earlier, but the sample sizes are small (each of n_1 and n_2, is less than 30). We will deal with this case later in this chapter. To find the CI for the difference between the two means when this condition arises, refer to the method used in Example 1.16.

When we believe that the variances of the two populations are equal, we use the formulas presented above. We illustrate with an example. □

Example 1.6

The skills of repairmen trained by managers A and B who use different styles for training repairmen are being compared. A sample of 12 repairmen trained by manager A was chosen, and the time taken by each repairmen to complete a task in the repair process was obtained. The times of completion of the same task by a sample of 14 repairmen trained by manager B were obtained. The completion times they turned out (in minutes) are as follows:

A: 10, 20, 30, 14, 16, 25, 15, 18, 12, 16, 20, 23
B: 12, 22, 22, 27, 31, 16, 20, 24, 30, 10, 26, 24, 28, 31

Obtain a 99% CI for the difference in the mean performance times for the repairmen trained by managers A and B.

Solution:

$$\sum_{i=1}^{12} y_{1i} = 219 \quad \sum_{i=1}^{12} y_{1i}^2 = 4355 \quad s_1^2 = \left(\frac{1}{12-1}\right)\left[4355 - \frac{219^2}{12}\right] = 32.568$$

$$\sum_{i=1}^{14} y_{2i} = 323 \quad \sum_{i=1}^{14} y_{2i}^2 = 8031 \quad s_2^2 = \left(\frac{1}{14-1}\right)\left[8031 - \frac{323^2}{14}\right] = 44.53$$

$$s_{\text{p}}^2 = \frac{(14-1)(44.53) + (12-1)(32.568)}{14+12-2} = 39.048$$

$$\sigma_{(\bar{y}_1 - \bar{y}_2)} = \sqrt{39.048\left(\frac{1}{14} + \frac{1}{12}\right)} = 2.4583$$

The $100(1-0.01)\%$ CI for the difference between the means for the repairmen trained by managers B and A is obtained using the limits

$$\left(\frac{323}{14} - \frac{219}{12}\right) \pm t(24, 0.005)\sigma_{\bar{y}_1 - \bar{y}_2} = 4.82 \pm 2.797(2.4583) = 4.82 \pm 6.88$$

The 99% CI between the means is (−2.055, 11.7). □

1.1.6 Confidence Intervals for Proportions

Sometimes, we may sample from a binomial process in which p is the probability of success and $q = 1 - p$ is the probability of failure. We can obtain the CI for the proportion of successes, p, by using the proportion of successes in our sample, so that the $100(1-\alpha)\%$ confidence limit for p is $\hat{p} \pm Z_{\alpha/2}\sigma_{\hat{p}}$.

Provided that we are sampling from a very large or an infinite population, or that we are sampling from a finite population with replacement, then the CI can be written as [$100(1-\alpha)\%$ confidence limits]

$$\hat{p} \pm Z_{\alpha/2}\sqrt{\frac{\hat{p}(1-\hat{p})}{n}}$$

1.1.6.1 Confidence Interval for Difference between Two Proportions

Often our interest is in the CI for the difference between two proportions. If n_1 and n_2, the sizes of the samples from which estimates of the proportions were obtained, are large (i.e., n_1, $n_2 \geq 30$), we can establish CIs for the difference in the two proportions. The standard error for the difference is

$$\sigma_{\hat{p}_1-\hat{p}_2} = \sqrt{\frac{\hat{p}_1(1-\hat{p}_1)}{n_1} + \frac{\hat{p}_2(1-\hat{p}_2)}{n_2}}$$

Since the sample sizes are large, the corresponding CI for the difference in the two proportions depends on the normal distribution. Hence, the $100(1-\alpha)\%$ CI is

$$100(1-\alpha)\%\,\text{CI} = (\hat{p}_1 - \hat{p}_2) \pm Z_{\alpha/2}\sigma_{\hat{p}_1-\hat{p}_2} \Rightarrow 100(1-\alpha)\%\,\text{CI}$$
$$= (\hat{p}_1 - \hat{p}_2) \pm Z_{\alpha/2}\sqrt{\frac{\hat{p}_1(1-\hat{p}_1)}{n_1} + \frac{\hat{p}_2(1-\hat{p}_2)}{n_2}}$$

Example 1.7

"The country is not yet ready for a female president." In two surveys of this political question in 1995 and 1999, the following figures were obtained from the respondents:

Year	Yes	No	Total
1995	125	348	473
1999	270	1905	2175

Obtain the 95% CI for (1) the proportion of those who said "no" in 1995 and (2) the difference in proportions of those who said "yes" in 1995 and 1999.

Solution:

To establish the CI, we first calculate the estimates of the proportions from the samples and then apply the large sample theory since it is relevant in this situation.

1. $\hat{p}_{\text{no}} = \dfrac{348}{473} = 0.735729 \quad \sigma_{\hat{p}_{\text{no}}} = \sqrt{\dfrac{0.735729(1-0.735729)}{473}} = 0.020275$

$$\begin{aligned}100(1-0.05)\%\,\text{CI} &= 0.735729 \pm Z_{0.025} \times 0.020275 \\ &= 0.735729 \pm 1.96 \times 0.020275 \\ &= 0.735729 \pm 0.039738 \\ &= (0.695991, 0.775468).\end{aligned}$$

2. $\hat{p}_1 = \frac{125}{473} = 0.2642 \quad \hat{p}_2 = \frac{270}{2175} = 0.1241$

$$\begin{aligned} 100(1-0.05)\%\,\text{CI} &= (0.2642 - 0.1241) \pm Z_{0.025} \\ &\quad \times \sqrt{\frac{0.2642(1-0.2642)}{473} + \frac{0.1241(1-0.1241)}{2175}} \\ &= 0.14 \pm 1.96\sqrt{0.000411 + 0.0000499} \\ &= 0.14 \pm 0.042 \end{aligned}$$

The 95% CI for difference between the proportions is (0.098, 0.182). ▯

1.1.7 Tests of Hypotheses

The statistician is frequently required to confirm or disprove a belief about a population. Such beliefs are usually formulated into a hypothesis. An assumption, a guess, a statement, or a postulate about the population, which is to be proved or disproved. Since it is not often that the entire population is available, a random sample is usually taken from the population, and the data from the sample is used to prove or disprove the hypothesis. The decision is usually arrived at by carrying out a test of the hypothesis to establish whether the hypothesis is true. To test a statistical hypothesis, we state the main hypothesis to be tested and refer to it as the null hypothesis. Against this hypothesis, we state an alternative hypothesis. Based on a statistic (a sample value) whose distribution is known, a test statistic is calculated and compared with some tabulated values; thereafter, a decision is made to reject or not to reject the null hypothesis.

1.1.7.1 The Null Hypothesis

Although the statement to be tested is called the main hypothesis, we sometimes formulate this hypothesis to reject or nullify it. For instance, if we wish to compare the effects of two different fertilizer treatments on the yield of the same crop, we can state a null hypothesis that says "there is no difference between the mean yields of the crop under the two fertilizer treatments." If we take samples of yields after the application of these two fertilizers, any difference we observe should only be attributable to random fluctuations. We usually denote the null hypothesis by H_0. Once we reject this null hypothesis, we accept the alternative. However, there may be borderline situations in which we may reserve decisions until more sample information is available. This happens when the value of the calculated test statistics falls on the line bordering the rejection and acceptance regions.

1.1.7.2 The Alternative Hypothesis

This is the hypothesis that we intend to accept if the null hypothesis is rejected. Often, this is the complement of the statement that appears in the null hypothesis. For instance, we could say that the alternative hypothesis to the null hypothesis stated above is "there is a difference in the mean yield of the crop under the two fertilizer treatments." We usually denote the alternative hypothesis by H_1 or H_{A}.

1.1.7.3 Type I and Type II Errors

There are two types of errors that we could make in the process of testing a hypothesis. These errors occur when wrong decisions are made. They are called the type I and type II errors.

Type I error. This error occurs when we reject the null hypothesis when it should have been accepted. In other words, type I error is made when we fail to accept the null hypothesis when it is true.

Type II error. If we accept the null hypothesis when it should have been rejected (or when the alternative is true), we make a type II error. A good statistical test of hypothesis should be designed in such a way as to minimize the sizes of the two errors. However, in practice, it is often discovered that as we seek to minimize one error, the other error is increased. A practical decision is usually reached by determining which of the errors is more serious and this error is then minimized at the expense of the less serious error, thereby achieving a compromise.

1.1.7.4 Level of Significance

Along with the null hypothesis, we set up a level of significance usually denoted by α, representing the level of error we make if we wrongly decide to reject the null hypothesis when it ought to be accepted, a type I error. The level of significance represents the maximum probability with which we are prepared to risk making a type I error. Since this error could be serious, we make its risk as small as possible. Values of α ranging from 1% to 5% are frequently used. Above 5%, the risk begins to be progressively less acceptable, although values as high as 10% are sometimes used.

1.1.7.5 Test of Hypothesis, Parametric Tests

As mentioned earlier (Section 1.1.7), we take a sample from the population for which we wish to test a hypothesis. Based on sampling theory, we calculate a test statistic. There is always a test statistic for every hypothesis to be tested. The test statistic is compared with some tabulated values. Normally, the tabulated values obtained for the test are based on the distribution of the statistic and on the level of significance α. From the comparison, we can decide whether the findings from our sample data differ significantly from what is postulated for the population. In the earlier example regarding two fertilizer treatments, we determined whether the observed mean yields differ significantly from one another. The process that enables us to arrive at this decision is called a test of hypothesis, or test of significance. Hypothesis testing is part of a subject area in statistics called inferential statistics or statistical inference.

1.1.8 Tests for a Single Parameter (Mean) Involving the Normal Distribution

In statistical theory, a parameter (a population value, a constant) may be estimated by a statistic (a variable obtained from a sample). Some of these statistics are known to follow the normal distribution. In such cases, assuming a normal distribution, if Y_s denotes the statistic, the mean will be μ_s and variance will be σ_s^2. If the variance of the normal population σ^2 is known, we use it to obtain σ_s^2. A hypothesis may be tested that its mean is, for example, μ_0, against an alternative that it is not, requiring that a random sample of size n be taken from the normal population to obtain the value of a test statistic. The hypothesis is stated as follows:

$$H_0 : \mu_s = \mu_0 \quad \text{vs} \quad H_1 : \mu_s \neq \mu_0$$

The statistic associated with this test is the sample mean $\bar{y}$

$$y_s = \bar{y}$$

We calculate the test statistic Z_c, which is used in the test as follows:

$$Z_c = \frac{\bar{y} - \mu_0}{\sigma/\sqrt{n}}$$

The above test has a two-sided or two-tailed alternative. We can also have tests with one-sided or one-tailed alternatives. The test statistic obtained for this test is compared with appropriate tabulated values of Z. This Z has the standard normal distribution with mean 0 and variance 1. The value read from the table depends on the chosen level of significance or size of type I error, α.

Decision rules. There are three versions of possible hypotheses that can be stated for the above problem (Section 1.1.8), and the three sets of hypotheses can be tested using the same test statistic Z_c. Two one-sided alternatives are available along with the two-sided alternative stated earlier. The choice of alternative hypotheses determines the differences in the sets of hypotheses that can be tested. The choice is made according to the interest of the researcher. Although the same test statistic would do for the three versions of the test, the decision rules are different for the three versions of the hypotheses.

Case 1: One-sided; H_0: $\mu = \mu_0$ vs H_1: $\mu > \mu_0$

Decision: If $Z_c > Z_\alpha$ reject H_0 and accept H_1

Case 2: One-sided; H_0: $\mu = \mu_0$ vs H_1: $\mu < \mu_0$

Decision: If $Z_c < -Z_\alpha$ reject H_0 and accept H_1

Case 3: Two-sided; H_0: $\mu = \mu_0$ vs H_1: $\mu \neq \mu_0$

Decision: If $Z_c < -Z_{\alpha/2}$ or if $Z_c > Z_{\alpha/2}$ reject H_0 and accept H_1

The critical or rejection regions and their sizes for the three cases mentioned above are illustrated in Figures 1.1 through 1.3. We use an example to illustrate the test of hypothesis under the normal distribution.

Example 1.8
It is known that the resting heart rates of a large number of students after a given period on a treadmill are normally distributed with a variance of 53. To test the hypothesis that the mean rate is 77/min, the following random

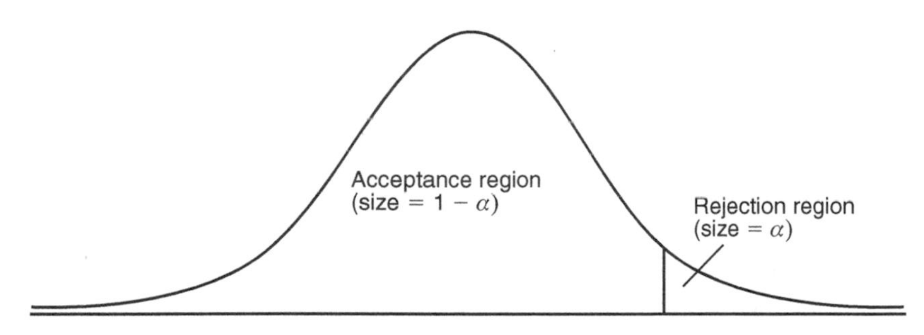

FIGURE 1.1: One-sided test with rejection region to the right.

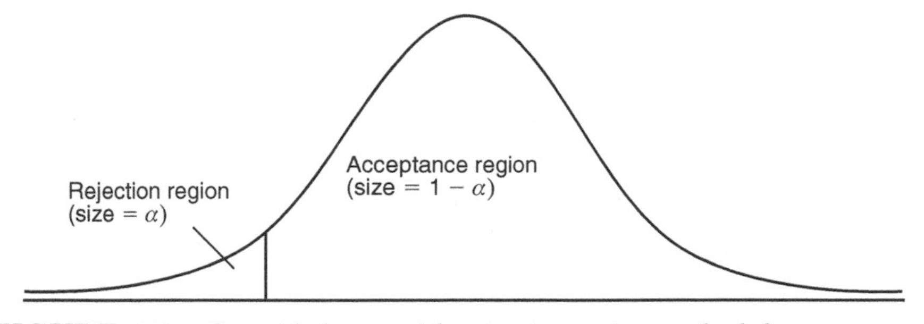

FIGURE 1.2: One-sided test with rejection region to the left.

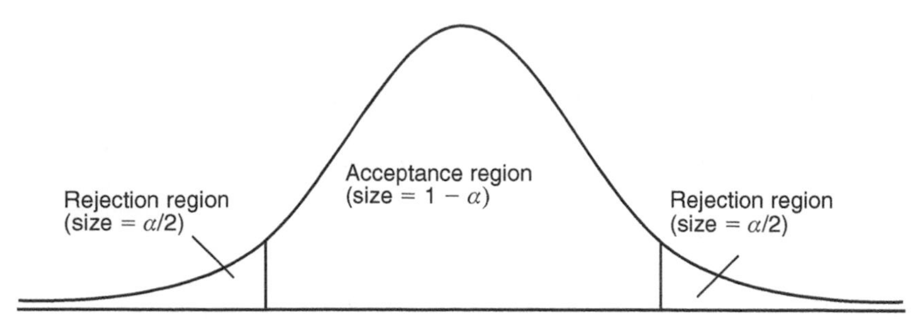

FIGURE 1.3: Two-sided test with rejection regions to the right and left.

sample of 44 students was chosen. After running at a given rate on a treadmill, the heart rates of the students were recorded as follows:

66	75	92	72	72	92
68	62	87	75	62	76
75	90	79	76	80	60
83	75	92	65	73	65
74	83	68	65	70	
88	80	68	66	62	
68	73	60	82	91	
58	62	53	58	86	

Carry out this test against an alternative that the mean rate is less than 77, using a level of significance $\alpha = 5\%$.

Solution:
First, we set up the null hypothesis and the alternative as

$$H_0 : \mu = 77 \quad \text{vs} \quad H_1 : \mu < 77$$

We know from statistical theory that $\bar{y} \sim N\left(\mu, \frac{\sigma^2}{n}\right)$.

The required test statistic Z_c is distributed as standard normal variable with mean 0 and variance 1 as

$$Z_c = \frac{\bar{y} - \mu}{\sigma/\sqrt{n}} \sim N(0,1)$$

The values of Z associated with a given level of significance α can be read from the tables of the normal distribution.

Now,

$$\sum_{i=1}^{44} y_i = 3227 \Rightarrow \bar{y} = \frac{1}{44}\sum_{i=1}^{44} y_i = \frac{3227}{44} = 73.34091$$

$$Z_c = \frac{73.34091 - 77}{\sqrt{(53/44)}} = -3.33397$$

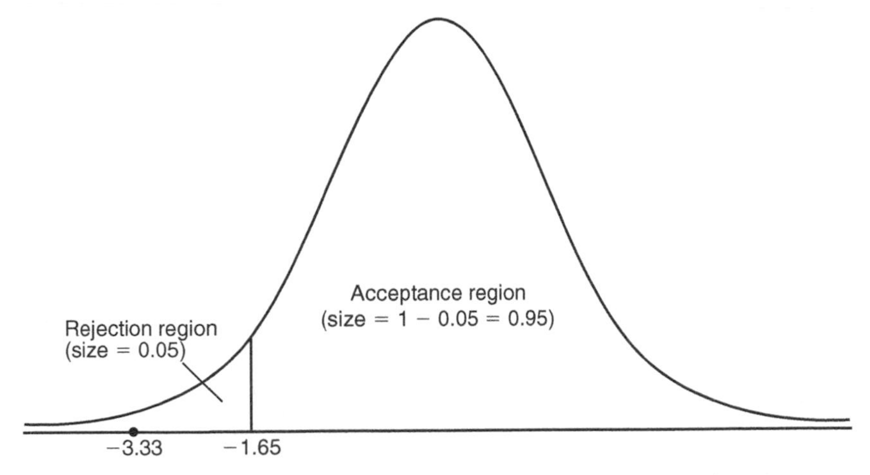

FIGURE 1.4: Rejection and acceptance regions for Example 1.8.

Using $\alpha = 5\%$, we read from the table of $N(0,1)$ that $Z_\alpha = 1.65$, so that $-Z_\alpha = -1.65$.

Since our test corresponds to Case 2 above, we compare Z_c with $-Z_{0.05}$. Clearly, $Z_c = -3.33397 < -Z_\alpha = -1.65$, so we reject the null hypothesis and accept the alternative that the mean heart rate is less than 77/min. Alternatively, we can make our decision by drawing the figure of the normal distribution and identifying the rejection and acceptance regions, so that if Z_c falls into the rejection region, we reject the null hypothesis; otherwise, we accept it. This is illustrated in Figure 1.4. □

Example 1.9

It is thought that a laboratory population of fruit flies is made up of flies of about equal number of gray and black hues. A random sample of 300 flies yielded 163 black flies. Can we conclude that the proportion of flies are equal for the two colors? Use levels of significance (1) 5% and (2) 1%.

Solution:

To set up null and alternative hypotheses, we know that number of black flies follows the binomial distribution. For large n, the number of trials, the outcome is approximately normally distributed with mean np and variance npq if p = probability of a black fly and q = probability of a gray fly. Since $np = (1/2)(300) = 150$ and $npq = (1/2)(1/2)(300) = 75$, the resulting hypothesis is

$$H_0 : p = 0.50 \quad \text{vs} \quad H_1 : p \neq 0.50$$

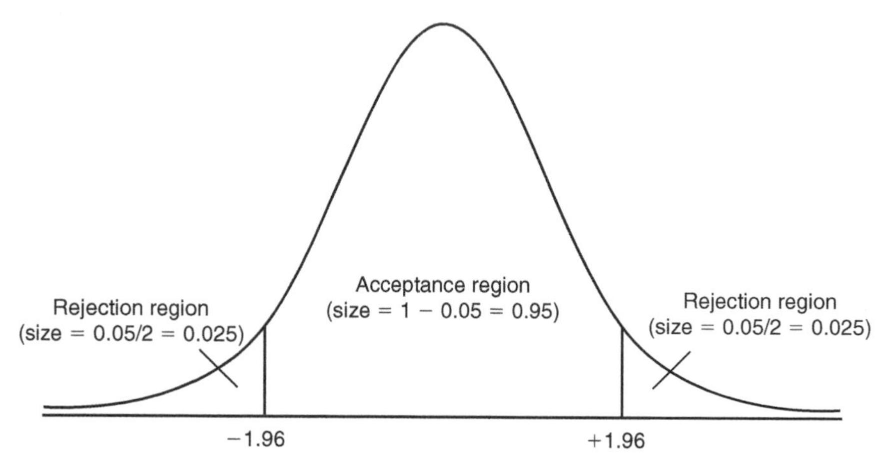

FIGURE 1.5: Rejection and acceptance regions for Example 1.9.

The number of outcomes of the experiment is used to obtain the test statistic Z_c as follows:

$$Z_c = \frac{y - np}{\sqrt{npq}} = \frac{163 - 150}{\sqrt{75}} = 1.50$$

From the tables of the standard normal distribution (Appendices A1 and A2) for $\alpha = 5\%$, $Z_{\alpha/2} = 1.96$ and $-Z_{\alpha/2} = -1.96$. By the decision rule, Case 3, we see that Z_c lies between -1.96 and $+1.96$, so we conclude that the null hypothesis H_0 should not be rejected. This means that the proportions of black and gray flies are probably equal. The decision rule applied here is depicted in Figure 1.5. □

Example 1.10

A biotech company claims that the time it takes for a unit biomass to degrade is normally distributed with mean of 1500 h and variance of $(110\,\text{h})^2$. This is disputed by a rival company, which believes that a unit biomass degrades in less time. The lifetimes of a random sample of 20 units of these biomasses were studied, and their mean degrading times was found to be 1494 h. Can we support the claim of the producer at 5% level of significance?

Solution:

The null and alternative hypotheses to be tested are

$$H_0 : \mu = 1500 \quad \text{vs} \quad H_1 : \mu < 1500$$

The test statistic is

$$Z_c = \frac{\bar{y} - \mu}{\sigma/\sqrt{n}} = \frac{1494 - 1500}{\frac{110}{\sqrt{20}}} = -0.2439$$

From the tables (Appendices A1 and A2), $Z_{0.05} = -1.645$, so we do not reject the null hypothesis. The decision rule applied in this case is represented in Figure 1.6. □

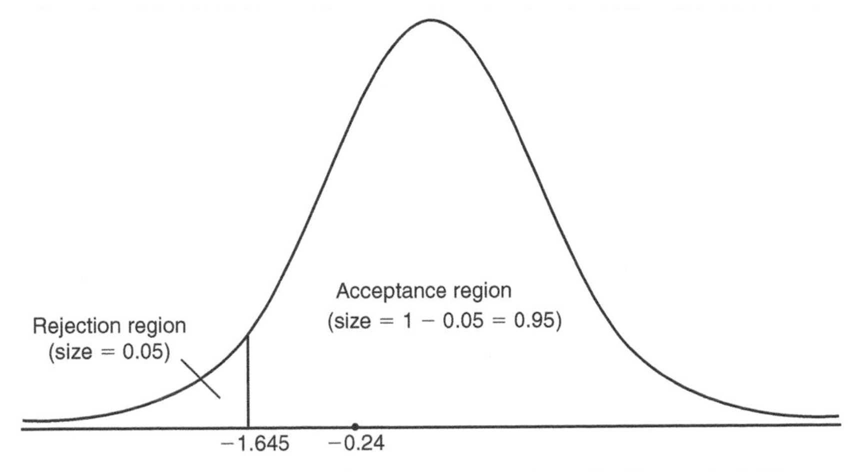

FIGURE 1.6: Rejection and acceptance regions for Example 1.10.

1.1.9 Tests of Hypotheses for Single Means Involving the Student's t-Distribution

If a large sample ($n \geq 30$) is taken from most distributions in statistical theory, the distribution of many statistics obtained can be approximated by the normal distribution. The larger we make the sample size, the better the approximation we obtain if we represent the data by the normal distribution. Problems arise when small samples are taken, which cannot be approximated by the normal distribution. In fact, the smaller the sample size the less useful is any attempt to use the normal distribution to approximate the distribution of any statistic from the sample. Instead of using the normal distribution for small samples, statisticians have found that the Student's t-distribution gives better results when used to represent the distributions of statistics calculated from small samples drawn from a normal population. The results obtained could equally be applied to large samples. This is because as the sample size gets larger the t-distribution is a closer approximation of the standard normal distribution.

For the tests in the previous section, regardless of the sample size, when the distribution seems normal and population standard deviation is unknown, the best test statistic to use has a t-distribution with $n-1$ degrees of freedom where $n =$ sample size. Further, if the sample size is small, then the t-statistic is the best to use. A small sample of size n may be taken from a population that is believed to be normal with unknown standard deviation. The sample standard deviation with some modifications could then be used in place of the population standard deviation in the calculation of the test statistic. To test the following hypotheses (represented by Cases 1–3)

Case 1: $H_0 : \mu = \mu_0 \quad \text{vs} \quad H_1 : \mu > \mu_0$
or

Case 2: $H_0 : \mu = \mu_0$ vs $H_1 : \mu < \mu_0$
or
Case 3: $H_0 : \mu = \mu_0$ vs $H_1 : \mu \neq \mu_0$

we use

$$t_c = \frac{\bar{y} - \mu_0}{s/\sqrt{n}}$$

where $s^2 = \sum_{i=1}^{n}(y_i - \bar{y})^2/n - 1$.

Under H_0, t_c is t distributed with $n-1$ degrees of freedom. The above statistic will be used for testing the 2 one-sided and the 1 two-sided tests stated above in Cases 1–3.

Decision rule. Let $\alpha\%$ be the level of significance for the tests. Then, here are the decision rules for the tests:

Case 1: If $t_c > t(n-1,\, \alpha)$ reject null hypothesis, accept the alternative.

Case 2: If $t_c < -t(n-1,\, \alpha)$ reject null hypothesis, accept the alternative.

Case 3: If $t_c > t(n-1,\, \alpha/2)$ or $t_c < -t(n-1,\, \alpha/2)$ reject null hypothesis, accept the alternative.

Example 1.11

A manager of a market garden chain was planning for inventory purposes and believed that a particular shop in the chain sold less than 60 plants per day on average. He took a random sample of plant sales for 21 days and obtained the following figures:

71	50	76
62	43	22
63	66	93
68	63	10
81	25	61
61	66	32
59	20	67

Assuming that a daily sale of plants is normally distributed, test the claim of the manager using a level of significance of 5%.

Solution:
We state the hypothesis to be tested as follows:

$$H_0 : \mu = 60 \quad \text{vs} \quad H_1 : \mu < 60$$

Now, we calculate all the statistics required for the test $\bar{y}$, s^2, and t_c as follows:

$$\bar{y} = \sum_{i=1}^{21} y_i = 1159 \Rightarrow \bar{y} = \frac{1159}{21} = 55.19948$$

$$s^2 = \frac{1}{21-1}\left[\sum_{i=1}^{21} y_i^2 - 21(\bar{y})^2\right] = \frac{1}{20}[73539 - 21(55.19948)^2] = 478.6619$$

$$t_c = \frac{\bar{y}-\mu}{s/\sqrt{n}} = \frac{55.19948 - 60}{\sqrt{478.6619}/\sqrt{21}} = -1.00739$$

From the tables of the t-distribution (see Appendix A3), we see that $t(n-1, \alpha) = t(20, 0.05) = 1.725$. Since $t_c > -t(20, 0.05)$, we do not reject the null hypothesis. The shop probably sells 60 plants per day on average. □

Example 1.12

It is claimed that the containers of fuel made from agricultural waste produced by the JTC Company weigh 1900 lb each. A manager suspects that the afternoon shift workers are not producing according to specification. So the manager took a sample of 20 fuel containers produced by the shift and measured their weights. The results were as follows: 1895, 1924, 1891, 1889, 1915, 1910, 1904, 1894, 1889, 1906, 1903, 1913, 1922, 1886, 1893, 1993, 1914, 1897, 1895, and 1912. Assume that the weights are normally distributed. Test the hypothesis that the shift was working within specifications using levels of significance of (1) 1% and (2) 5%.

Solution:

In this case, we can carry out a two-sided test. We set up the hypothesis to be tested as follows:

$$H_0 : \mu = 1900 \quad \text{vs} \quad H_1 : \mu \neq 1900$$

$$\sum_{i=1}^{20} y_i = 36250 \Rightarrow \bar{y} = \frac{36250}{20} = 1907.25$$

$$s^2 = \frac{1}{20-1}[72762267 - 20(1907.25)^2] = 537.6711 \Rightarrow s = 23.18773$$

Now we calculate the test statistic as

$$t_c = \frac{\bar{y}-\mu}{s/\sqrt{n}} = \frac{1907.25 - 1900}{23.18773/\sqrt{20}} = 1.3983$$

From the tables, we read $t(19, 0.025) = 2.093$, $t(19, 0.005) = 2.861$. In both cases, we do not reject the null hypothesis and reject the alternative. The shift is producing according to specification. □

1.1.10 Comparing Two Populations Using t-Distribution

Frequently, we need to compare two samples taken from what appears to be two different populations to ascertain whether the populations are actually different from each other or the same. This type of comparison can arise in two ways. We may have made observations from the same individuals before and after applying treatment to them, and we wish to compare the measurements to see if there has been a treatment effect. This type of comparison of pairs of observations is called paired comparison test. However, there are cases where the two samples clearly come from different individuals and from what appears to be different populations. Here, the same attributes are being measured from two distinct individuals. In this case, we are testing whether there is significant difference between the means of the populations from which they were drawn. There are two types of tests available for this type of comparison. One of these tests applies when we have reason to suspect that the variances of the two populations are the same. The test used is the pooled variance t-test. When the variances are not known and we have reason to believe they are not equal, we shall refer to the test used as the two-sample t-test with unknown variances. We discuss each of these three tests in the following sections.

1.1.10.1 Paired Comparison or Matched Pair t-Test

Suppose that the heart rates of n students were measured before they were asked to run around the school building 10 times, and after this exercise, their heart rates were measured again. We should expect the heart rates to be faster just after the exercise—that is what is believed theoretically. But is there a real increase in heart rate or is the observed increase due to random fluctuations only? We can test this claim by using the paired comparison test. The same could be applied to medical treatments in which the level of a disease could be measured before and after the treatment to ascertain whether there is an improvement in the patient's condition after the treatment. There are several other situations which give rise to matched pairs of observations. The test procedure we discuss below illustrates this. Consider n matched pairs of observations; if we take the differences between pairs i ($i = 1, 2, \ldots, n$), we obtain the difference d_i (see the table below).

Pairs (i)	Before treatment	After treatment	Differences (d_i)
1	Y_{11}	Y_{12}	$d_1 = Y_{12} - Y_{11}$
2	Y_{21}	Y_{22}	$d_2 = Y_{22} - Y_{21}$
3	Y_{31}	Y_{32}	$d_3 = Y_{32} - Y_{31}$
4	Y_{41}	Y_{42}	$d_4 = Y_{42} - Y_{41}$
$\vdots$	$\vdots$	$\vdots$	$\vdots$
$n-1$	Y_{n-11}	Y_{n-12}	$d_{n-1} = Y_{n-12} - Y_{n-11}$
n	Y_{n1}	Y_{n2}	$d_n = Y_{n2} - Y_{n1}$

We now use the differences d_i in our inference as follows:

We set up the null and alternative hypotheses as

1. $H_0 : \mu_d = 0$ vs $H_1 : \mu_d > 0$ (one-sided test)

   or

2. $H_0 : \mu_d = 0$ vs $H_1 : \mu_d < 0$ (one-sided test)

or

3. $H_0 : \mu_d = 0$ vs $H_1 : \mu_d \neq 0$ (two-sided test)

The test statistic required for this test is

$$t_c = \frac{\bar{d}}{s_d/\sqrt{n}}$$

where $s_d^2 = \sum_{i=1}^{n} (d_i - \bar{d})^2/n - 1$ and $s_d^2 = \sum_{i=1}^{n} d_i/n$

Decision rule. The decision rule for Case 1 requires that t_c be compared with the tabulated value $t(\alpha,\, n-1)$ and if $t_c > t(\alpha,\, n-1)$, we reject the null hypothesis and accept the alternative. For Case 2, if $t_c < -t(\alpha, n-1)$, we reject the null hypothesis and accept the alternative. For Case 3, if $t_c < -t(\alpha/2, n-1)$ or $t_c > t(\alpha/2, n-1)$, we reject the null hypothesis and accept the alternative.

Example 1.13

A drug company claims that a new baby formula it has prepared could improve the growth of malnourished babies, more than a generic brand. The weight gains for 14 babies in the month before they were given the new formula and the month after they were given the new formula were recorded as follows:

Babies	Weight Gain a Month before Being Fed on Generic Brand	Weight Gain a Month after Being Fed on New Formula	Differences
1	175	142	−33
2	132	211	79
3	218	337	119
4	151	262	111
5	200	302	102
6	219	195	−24
7	234	253	19
8	199	149	−50
9	236	187	−49
10	248	211	−37
11	206	176	−30
12	179	214	35
13	214	206	−8
14	249	179	−70

Is there any improvement in the babies' growth because of the use of the new formula? Test at $\alpha = 5\%$.

Solution:
We set up the null and alternative hypotheses as follows:

$$H_0 : \mu_d = 0 \quad \text{vs} \quad H_1 : \mu_d > 0$$

$$\sum_{i=1}^{n} d_i = 164; \Rightarrow \bar{d} = \frac{164}{14} = 11.71429;$$

$$\sum_{i=1}^{n} d_i^2 = 58512; \Rightarrow s_d^2 = \frac{1}{14-1}[58512 - 14(11.71429)^2]$$

$$= 4352.143 \Rightarrow s_d = 65.97835$$

$$t_c = \frac{\bar{d}}{s_d/\sqrt{n}} = \frac{11.71429}{65.97835/\sqrt{14}} = 0.664332$$

From the t tables (Appendix A3), $t(0.05, 13) = 1.771$; hence, $t_c < t(0.05, 13)$, so we do not reject the null hypothesis. The new baby formula has not significantly improved the growth of babies. □

1.1.10.2 Pooled Variance t-Test

If we have n_1 observations from population 1 and n_2 observations from population 2, which come from normal populations with unknown variances, but we have reason to believe that these variances are equal, we can compare their means by using the pooled variance t-test. The pooled variance serves as an estimate of the common variance for the two populations. The hypotheses that can be tested are as follows:

1. $H_0 : \mu_1 = \mu_2$ vs $H_1 : \mu_1 > \mu$ (one-sided test)

   or

2. $H_0 : \mu_1 = \mu_2$ vs $H_1 : \mu_1 < \mu_2$ (one-sided test)

   or

3. $H_0 : \mu_1 = \mu_2$ vs $H_1 : \mu_1 \neq \mu_2$ (two-sided test)

The test statistic to be used for all the above tests is

$$t_c = \frac{\bar{y}_1 - \bar{y}_2}{s_p \sqrt{\left[\dfrac{1}{n_1} + \dfrac{1}{n_2}\right]}}$$

$$s_p^2 = \frac{(n_1 - 1)s_1^2 + (n_2 - 1)s_2^2}{n_1 + n_2 - 2} \quad \text{and} \quad s_1^2 = \frac{1}{n_1 - 1}\left[\sum_{i=1}^{n_1} y_1^2 - n_1\bar{y}_1^2\right] \quad \text{and}$$

$$s_2^2 = \frac{1}{n_2 - 1}\left[\sum_{i=1}^{n_2} y_2^2 - n_2\bar{y}_2^2\right]$$

Decision rule. The decision rule for Case 1 requires that t_c be compared with the tabulated value $t(\alpha, n_1+n_2-2)$ and if $t_c > t(\alpha, n_1+n_2-2)$, we reject the null hypothesis and accept the alternative. For Case 2, if $t_c < -t(\alpha, n_1+n_2-2)$, we reject the null hypothesis and accept the alternative. For Case 3, if $t_c < -t(\alpha/2, n_1+n_2-2)$ or $t_c > t(\alpha/2, n_1+n_2-2)$, we reject the null hypothesis and accept the alternative.

Example 1.14

In a pilot project in the textile industry, two similar machines were being compared for speed of production. The time taken by the machines to produce the same piece of cloth was recorded 15 times. Compare the mean production times for both machines. Use level of significance of 5%.

Machine 1 (t_1)	Machine 2 (t_2)
1603	1602
1604	1597
1605	1596
1605	1601
1602	1599
1601	1603
1596	1604
1598	1602
1599	1601
1602	1607
1614	1600
1612	1596
1607	1595
1593	1606
1604	1597

Solution:

To simplify matters, we can subtract an arbitrary number that appears to be common to all observations, which in this case is 1600, from all the numbers

(t_1) and (t_2), so that we are left with the numbers in the table, reducing our calculations without affecting our decision in any way.

$y_1 = t_1 - 1600$	$y_2 = t_2 - 1600$	y_1^2	y_2^2
3	2	9	4
4	−3	16	9
5	−4	25	16
5	1	25	1
2	−1	4	1
1	3	1	9
−4	4	16	16
−2	2	4	4
−1	1	1	1
2	7	4	49
14	0	196	0
12	−4	144	16
7	−5	49	25
−7	6	49	36
4	−3	16	9
45	6	559	196

The last row represents the column totals. We set up the hypothesis to be tested as

$$H_0 : \mu_1 = \mu_2 \quad \text{vs} \quad H_1 : \mu_1 > \mu_2$$

We can assume that production time for both machines have equal variance. From the above data we estimate the pooled variance.

$$\sum_{i=1}^{15} y_{1i} = 45 \Rightarrow \bar{y}_1 = \frac{45}{15} = 3$$

$$\sum_{i=1}^{15} y_{2i} = 6 \Rightarrow \bar{y}_2 = \frac{6}{15} = 0.40$$

$$\sum_{i=1}^{15} y_{1i}^2 = 559 \Rightarrow s_1^2 = \frac{1}{15-1}\left[559 - 15(3)^2\right] = 30.28571$$

$$\sum_{i=1}^{15} y_{2i}^2 = 196 \Rightarrow s_2^2 = \frac{1}{15-1}\left[196 - 15(0.4)^2\right] = 13.82857$$

$$s_\text{p}^2 = \frac{(15-1)(30.28571) + (15-1)(13.82857)}{15+15-2} = 22.05714$$

$$s_\text{p} = 4.696503$$

$$t_\text{c} = \frac{\bar{y}_1 - \bar{y}_2}{s_\text{p}\sqrt{\dfrac{1}{n_1} + \dfrac{1}{n_2}}} = \frac{3 - 0.4}{4.696503\sqrt{\dfrac{1}{15} + \dfrac{1}{15}}} = 1.5161$$

From the tables, $t(0.08, 28) = 1.701$. We therefore do not reject the null hypothesis and reject the alternative.

We conclude that the mean production times for both machines are the same. □

1.1.10.3 Two-Sample t-Test with Unknown Variances

This test involves two samples from different populations, and we wish to compare the means. In this case, we have no reason to believe that the variances are equal. In addition, if the two variances are unknown, then we need a modified form of the test used in the previous section. In fact, the exact distribution of the test statistic is unknown, and several tests have been suggested. The earliest and perhaps the most controversial is the Fisher (1935) and Behrens (1929) Test, which is based on fiducial probability. Tables to be used in connection with this test were prepared by Sukhatme (1938). Other tests include those by Bartlett (1936, 1939), Welch (1938), Cochran and Cox (1950), and Smith (1936), which were further expanded by Satterthwaite (1946) and Dixon and Massey (1951). The hypotheses to be tested are as follows:

1. $H_0 : \mu_1 = \mu_2$ vs $H_1 : \mu_1 > \mu_2$ (one-sided test)

   or

2. $H_0 : \mu_1 = \mu_2$ vs $H_1 : \mu_1 < \mu_2$ (one-sided test)

   or

3. $H_0 : \mu_1 = \mu_2$ vs $H_1 : \mu_1 \neq \mu_2$ (two-sided test)

For these tests, Smith–Satterthwaite and Dixon–Massey appear to be the most popular, and both of them use the same test statistic. The test statistic used is not exactly t-distributed; however, its distribution can usefully be approximated by the t-distribution. While the two versions agree regarding the test statistic, they differ slightly on the method for calculating the degree of freedom for the test statistics. In both cases, if we have two samples of Y_1 of size n_1 and Y_2 of size n_2, then the test statistic required for testing the above set of hypotheses is

$$t_c = \frac{\bar{y}_1 - \bar{y}_2}{\sqrt{\dfrac{s_1^2}{n_1} + \dfrac{s_2^2}{n_2}}}$$

The Smith–Satterthwaite version suggests that the degree of freedom of t_c is

$$\nu = \frac{\left[\dfrac{s_1^2}{n_1} + \dfrac{s_2^2}{n_2}\right]^2}{\left[\left(\dfrac{1}{n_1 - 1}\right)\left(\dfrac{s_1^2}{n_1}\right)^2 + \left(\dfrac{1}{n_2 - 1}\right)\left(\dfrac{s_2^2}{n_2}\right)^2\right]}$$

while Dixon and Massey prefer to use

$$\nu = \frac{\left[\dfrac{s_1^2}{n_1} + \dfrac{s_2^2}{n_2}\right]^2}{\left[\left(\dfrac{1}{n_1+1}\right)\left(\dfrac{s_1^2}{n_1}\right) + \left(\dfrac{1}{n_2+1}\right)\left(\dfrac{s_2^2}{n_2}\right)\right]} - 2$$

Both versions are popular and are being regularly used.

Example 1.15

The times taken by a sample of 18 chemistry students from School A and a sample of 15 chemistry students from School B to complete a titration experiment is as follows:

$$\begin{aligned}&\mathrm{A}: 2.3, 6.7, 3.8, 5, 4.9, 6.1, 4.4, 5.2, 3.9, 4.8, 4.6, 5.7, 5.3, 4.7, 4.2, 5.7, 4.8, 4.7\\&\mathrm{B}: 6.7, 7.3, 4.4, 8.3, 6.2, 4.3, 5.5, 3.2, 8.4, 7.3, 5.5, 4.8, 4.9, 6.7, 7.5\end{aligned}$$

Are the perceived differences significant? Use level of significance $\alpha = 5\%$.

Solution:

$$H_0 : \mu_1 = \mu_2 \quad \text{vs} \quad H_1 : \mu_1 > \mu_2$$

Since we are not sure about the background of students, it is reasonable not to suppose that the variances of the samples are the same. We therefore use the two-sample t-test with unknown variances.

$$\sum_{i=1}^{18} y_{1i} = 86.8 \quad \sum_{i=1}^{15} y_{2i} = 91 \Rightarrow \bar{y}_1 = 4.8222 \quad \bar{y}_2 = 6.0667$$

$$\sum_{i=1}^{18} y_{1i}^2 = 434.58 \quad \sum_{i=1}^{15} y_{2i}^2 = 586.14$$

$$s_1^2 = \frac{1}{18-1}\left[434.58 - 18\left(\frac{86.8}{18}\right)^2\right] = 0.9418$$

$$s_2^2 = \frac{1}{15-1}\left[586.14 - 15\left(\frac{91}{15}\right)^2\right] = 2.43381$$

The test statistic is

$$t_c = \frac{4.8222 - 6.0667}{\sqrt{\dfrac{0.9418}{18} + \dfrac{2.43381}{15}}} = -2.6866$$

with the degree of freedom of

$$\nu = \frac{\left[\frac{0.9418}{18} + \frac{2.43381}{15}\right]^2}{\left[\left(\frac{1}{18-1}\right)\left(\frac{0.9418}{18}\right)^2 + \left(\frac{1}{15-1}\right)\left(\frac{2.43381}{15}\right)^2\right]} = 22.5383 \approx 23$$

From the tables, $t(0.05, 23) = 1.714$. We therefore reject the null hypothesis. ◻

Example 1.16

Refer to Example 1.15. Obtain the 95% CI for the difference in the population means of Y_1 and Y_2.

$$\text{The } 100(1-\alpha)\%\,\text{CI} = \bar{y}_1 - \bar{y}_2 \pm t(\nu, \alpha/2)\sqrt{\frac{s_1^2}{n_1} + \frac{s_2^2}{n_2}}$$

$$\nu = \frac{\left[\frac{0.9418}{18} + \frac{2.43381}{15}\right]^2}{\left[\left(\frac{1}{18-1}\right)\left(\frac{0.9418}{18}\right)^2 + \left(\frac{1}{15-1}\right)\left(\frac{2.43381}{15}\right)^2\right]} = 22.5383 \approx 23$$

so that

$$\begin{aligned} 100(1-0.05)\text{CI} &= (4.8222 - 6.06667) \pm t(23, 0.025)\sqrt{\frac{0.9418}{18} + \frac{2.43381}{15}} \\ &= -1.24 \pm 2.07(0.4632) \\ &= -1.24 \pm 0.9588 \end{aligned}$$

The 95% CI for $\mu_1 - \mu_2 = (-2.198, -0.28)$. ◻

1.1.11 Operating Characteristic Curves

As we mentioned earlier, we can limit the type I error by choosing the level of significance, the α value, before beginning the experiment. The problem of dealing with type II error is not as simple as that. In fact, the only way we are going to avoid making type II error is if we refrain from accepting hypotheses. Since this is virtually impossible in many practical situations, it is usual to construct and use the operating characteristic (OC) curves, graphs that relate the different hypotheses that can be true, with the size of type II error that is associated with it. These curves show us how well a given test will be able to help us to minimize the size of type II error. Thus, the OC curves give us the power of a test. Since these curves depend on the sample sizes, they also enable us to decide what sample size to use for an experiment.

Example 1.17

Consider the problem in Example 1.10; obtain the OC curve for the test.

Solution:

We note that there was a sample of size 20, and the null and alternative hypotheses were

$$H_0 : \mu = 1500 \quad \text{vs} \quad H_1 : \mu < 1500$$

We can see that $Z = (\bar{y} - \mu)/(\sigma/\sqrt{n}) \Rightarrow \bar{y} = \mu + Z\sigma/\sqrt{n}$. With respect to the two hypotheses tested in Example 1.10, we reject H_0 if $\bar{y} < 1500 - 1.65 \times 110/\sqrt{20} = 1459.4154$. Suppose we are more specific and wish to consider the set of hypotheses $H_0 : \mu = 1500$ against $H_1 : \mu = 1450$, we will only accept H_1 if $Z < (1459.4154 - 1450/110/\sqrt{20}) = 0.38$. This can happen with $P[Z < 0.38] = 0.648$. This is the value of β, the type II error for this test. Using this argument, we can calculate the values of β for different values of $\bar{y}$.

$\bar{y}$	1350	1375	1400	1425	1450	1475	1500	1525	1550
β	1.00	0.9997	0.9918	0.9177	0.648	0.2643	0.05	0.0039	0.0001

□

Example 1.18

Repeat the above problem; this time, assume that $n = 30$ and use the same data from Example 1.10 and compare the two OC curves with that in Example 1.9.

Solution:

In this case, we assume that $n = 30, \bar{y} = 1494$, and $\sigma = 110$; we calculate Z_c as follows:

$$Z_c = \frac{\bar{y} - \mu}{\sigma/\sqrt{n}} = \frac{1494 - 1500}{110/\sqrt{30}} = -0.2988$$

We come to the same conclusion as was reached earlier that null hypothesis should be accepted and the alternative rejected.

The rule for rejecting the null hypothesis at $\alpha = 5\%$ remains; we will accept H_1 only if

$$\bar{y} = \mu + Z\left(\frac{\sigma}{\sqrt{n}}\right) = 1500 - 1.65\left(\frac{110}{\sqrt{30}}\right) = 1466.8628$$

If we take the specific hypotheses considered in Example 1.10, $H_0 : \mu = 1500$ vs H_1: $\mu < 1450$, then we will accept H_1 only if

$$Z_c = \frac{1466.8628 - 1450}{110/\sqrt{30}} = 0.84 = P(Z_c < 0.84) = 0.7996 = \beta$$

Following the same procedure as in Example 1.10, we obtain the values of β corresponding to different values of $\bar{y}$ as follows:

$\bar{y}$	1350	1375	1400	1425	1450	1475	1500	1525	1550
β	1.00	1.00	0.9995	0.9812	0.7996	0.3427	0.05	0.0019	0.0

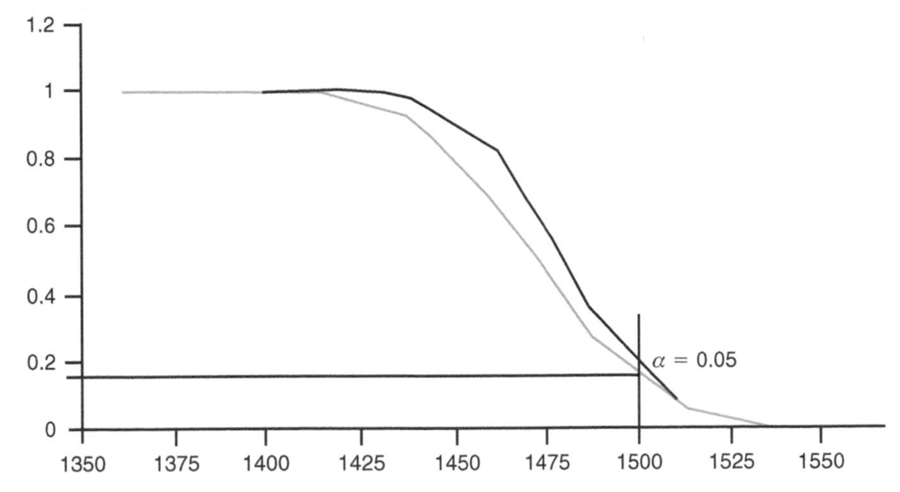

FIGURE 1.7: Sample means and OC curves for $n = 20$ and $n = 30$.

We now combine the results in Examples 1.9 and 1.10 to produce two OC curves as in Figure 1.7. □

1.1.12 The p-Value Approach to Decisions in Statistical Inference

All the methods we adopted for rejecting or accepting the hypotheses we have tested so far depended on the division of the area under the distribution of the test statistic (when the null hypotheses is true) into rejection and acceptance regions, using the level of significance. We rejected or did not reject the null hypotheses depending on where the value of the test statistic fell within the rejection or acceptance regions. This approach is the classical test of hypotheses. The more modern and equivalent approach uses the p-value. To understand the p-value, we need to understand two concepts: extreme values and the direction of extreme.

Extreme value. An observation is said to be an extreme value if (when H_0 is true) it is more likely under the alternative hypothesis than under the null hypothesis. An extreme value has more probability of occurring under the alternative hypothesis than under the null hypothesis. An extreme value lends more support to the alternative hypothesis than the null hypothesis.

Direction of extreme. This is the set of all values which is more likely under the alternative hypothesis than under the null hypothesis. This set makes up the direction of extreme. The direction of extreme can be either to the right or to the left depending on whether the alternative hypothesis is to the right (greater than) or to the left (less than). The direction of extreme can also be both to the right and to the left if the alternative hypothesis is stated as "not equal." In general, the direction of extreme consists of all values which lend more support to the alternative hypothesis than the null hypothesis. From the above definitions it is clear that some values are certainly more extreme than others. We can eventually arrive at the most extreme value for any set of hypotheses.

The p*-value.* The p-value is defined as the chance (probability) under H_0 of observing what one observed (connoting a sample-dependent value, a statistic), or something more extreme.

Thus, suppose without loss of generality, that we are dealing with a discrete distribution, and that y_1 is less extreme than y_2, and y_2 is less extreme than $y_3, \ldots,$ and y_{n-1} is less extreme than y_n for a given hypothesis. If the value of the test statistic one observed for the hypothesis is y_3, and the alternative hypothesis is to the left, this means that $y_1 < y_2 < y_3 < \cdots < y_n$ and the p-value $= P(Y \leq y_3)$ (under H_0).

If the alternative hypothesis is to the right, this means that $y_1 > y_2 > y_3 > \cdots > y_n$ and the p-value $= P(Y \geq y_3)$ (under H_0), if the observed test statistic is y_3.

If the alternative hypothesis is both to the left and right (a two-sided test), this means that $y_1 < y_2 < y_3 < \cdots < y_n$ and $-y_1 > -y_2 > -y_3 > \cdots > -y_n$, and the p-value $= P(Y \geq y_3) + P(Y \leq -y_3)$ (under H_0), if the observed test statistic is y_3.

1.1.13 Making a Decision in Hypothesis Testing with p-Value

To make a decision with a p-value, we simply compare the p-value with the level of significance we are willing to allow for the test. Most researchers simply state the p-value and allow the reader to assume their own levels of significance and see whether they would agree that the data are significant.

Generally, however, we reject H_0 if the p-value is less than or equal to α, the level of significance. We do not reject the null hypothesis if the p-value is greater than α. Thus, if we choose an acceptable size of type I error, we compute p-value for our test, compare it with the size of type I error and make a decision accordingly.

1.1.13.1 Applications to the Decisions Made in Previous Examples

For Example 1.8, the hypothesis is

$$H_0 : \mu = 77 \quad \text{vs} \quad H_1 : \mu < 77$$

The direction of extreme is to the left since the alternative hypothesis is stated as "less than." The test statistic obtained is -3.33397. Therefore, under the assumption that H_0 is true, p-value $= P(Z < -3.33397) = 0.000428$. Since we were testing at 5% level of significance, we see that the p-value is $\ll 0.05$; therefore, we reject the null hypothesis. We arrive at the same conclusion reached with the classical approach.

For Example 1.9, the hypothesis states $H_0 : p = 0.50$ vs $H_1 : p \neq 0.50$. The alternative hypothesis states "not equal." This means that the direction of extreme is both to the left and the right. The observed test statistic is 1.50. Thus, under the assumption that H_0 is true, the p-value $= P(Z > 1.50) + P(Z < -1.50) = 0.068807 + 0.068807 = 0.133614$.

The level of significance for this test was 5%, and we see that $0.133614 > 0.05$, so we do not reject the null hypothesis. Again, we come to the same conclusion reached under the classical approach.

For Example 1.12, the hypothesis was $H_0 : \mu = 1900$ vs $H_1 : \mu \neq 1900$. This is another two-sided test with direction of extreme to the right and to the left. The observed test statistic is 1.3983. The distribution of the test statistic is t with 19 degrees of freedom, so we find the p-value via the t-distribution. Thus, under the assumption that H_0 is true, p-value $= P[t(19) > 1.3983] + P[t(19) < -1.3983] = 0.089066 + 0.089066 = 0.178133$. The p-value is much greater than the level of significance for this test, which is 1%, so we do not reject the null hypothesis. We reached the same conclusion as we were led to reach under the classical approach.

Thus, if a test statistic has a χ^2, or F-distributions when the null hypothesis is true, we obtain the p-value using the relevant χ^2 or F-distributions. In general, if the test statistic has any distribution whatsoever when the null hypothesis is true, we find the p-value under that distribution.

1.2 Regression Analysis

In experimental or pure observational studies, we may have reasons to believe that a relationship exists between two or more variables. Sometimes, we explore these relationships with a view to predicting one of the variables if we know others. This process of prediction involves both interpolation and extrapolation, that is, prediction within the range of data on the basis of which the relationship was deduced, as well as beyond this range. A chemist suspects that there is a relationship between the catalyst concentration and the rate of completion of a reaction. He would like to find out this relationship to use it to predict what the rate of completion of reaction would be for a given level of catalyst concentration.

Suppose that a response variable y is known to depend on m independent variables $x_1, x_2, \ldots, x_m$, then we can obtain the relationship between y and $x_1, x_2, \ldots, x_m$ called the regression equation; in this case, y is called the dependent, response or uncontrolled variable, while $x_1, x_2, \ldots, x_m$ are called the independent or controlled variables. The model which should represent the relationship between y and $x_1, x_2, \ldots, x_m$ is a function

$$y = f(x_1, x_2, \ldots, x_m) + \varepsilon \tag{1.1}$$

Usually, the true functional relationship in Equation 1.1 is unknown to the investigator. In such situations, a polynomial of some order is chosen to approximate the functional relationship. The choice of the polynomial equation may depend on the initial exploratory analysis of the data. One exploratory method which is often used when $m = 1$, involves the plotting of the response variable y against the independent variable x called a scatterplot. Such a plot may indicate that a linear, quadratic, or even cubic relationship may be adequate. The scatterplot may indicate that the relationship is linear, giving a simple linear regression. Figures 1.8 through 1.10 show the

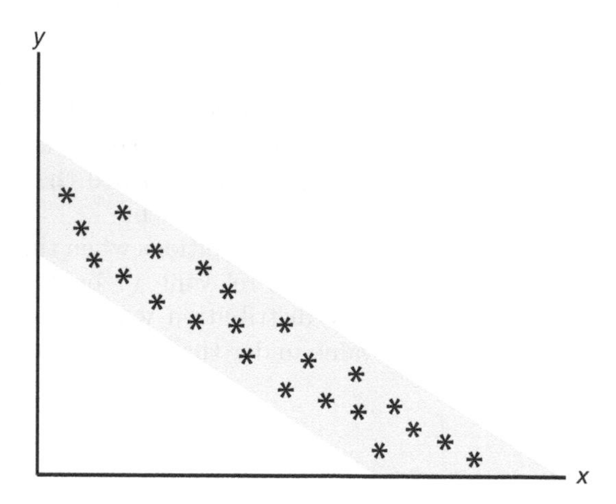

FIGURE 1.8: Scatterplot of response variable y against independent variable x showing a linear relationship between x and y.

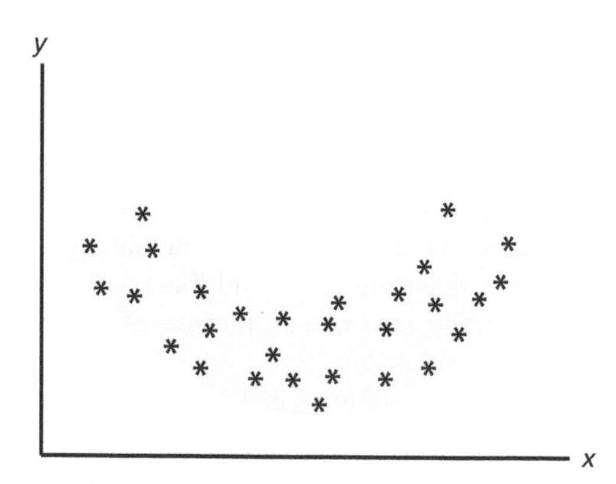

FIGURE 1.9: Scatterplot of response variable y against independent variable x showing a quadratic relationship between x and y.

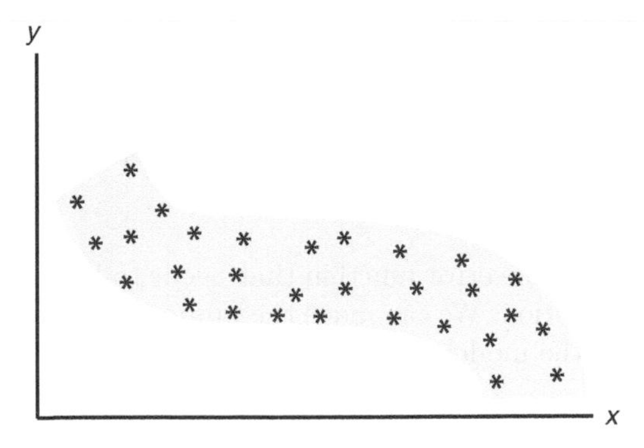

FIGURE 1.10: Scatterplot of response variable y against independent variable x showing a cubic relationship between x and y.

scatterplots that suggest linear, quadratic, and cubic relationships between y and x.

1.2.1 Simple Linear Regression

When there is only one variable x (the independent or predictor variable), whose relationship with the response variable y we wish to explore, then the experiment is carried out, and for different settings of x, we observe y. If it is a pure observational study, we simply observe changes in y as the values of x change. For n pairs of observations (x, y), we obtain the scatterplot. If the scatterplot (of y vs x) indicates that a straight line could adequately represent the data, then the model to be fitted is of the form

$$y = \beta_0 + \beta_1 x + \varepsilon \tag{1.2}$$

The parameters β_0 and β_1 are unknown quantities, which we shall estimate using the responses of our experiments or pure outcomes of observational studies. The random error $\varepsilon \sim \text{NID}(0, \sigma^2)$ (NID means normal, independently distributed). The model 1.2 further assumes that the errors are uncorrelated. In this case, Equation 1.2 is the simple linear regression model, because there is only one controlled variable x. It is clear from Equation 1.2 that

$$E(y \mid x) = \beta_0 + \beta_1 x \tag{1.3}$$

We can fit a straight line to the n pairs of observations $(x_1, y_1), (x_2, y_2), \ldots,$ (x_n, y_n) obtained from the experiments or pure observational studies using the least squares method. This is a mathematical technique which seeks to minimize the sum of the squares of the errors ε_i that occur between the actual observations x_i and the fitted straight line. We note that if we rewrite Equation 1.2 in terms of the ith observation, then

$$y_i = \beta_0 + \beta_1 x_i + \varepsilon_i \quad i = 1, 2, \ldots, n \tag{1.4}$$

The function to be minimized is

$$g(\varepsilon) = \sum_{i=1}^{n} \varepsilon_i^2 = \sum_{i=1}^{n} (y_i - \beta_0 - \beta_1 x_i)^2 \tag{1.5}$$

Equation 1.5 shows the error function that needs to be minimized to satisfy the least squares criterion. We can make the minimization of the function $g(\varepsilon)$ easier by rewriting the model of Equation 1.2 as

$$y_i = \beta_0 + \beta_1(x_i - \bar{x}) + \varepsilon_i \tag{1.6a}$$

where $\bar{x} = \sum_{i=1}^{n} x_i/n$ and

$$\beta_0' = \beta_0 + \beta_1\bar{x} \tag{1.6b}$$

With this transformed model, the least squares function $g(\varepsilon)$ can be rewritten as

$$g(\varepsilon) = \sum_{i=1}^{n} [y_i - \beta_0' - \beta_1(x_i - \bar{x})]^2 \tag{1.7a}$$

Minimizing, with respect to β_0', β_1 we obtain Equations 1.7b and 1.7c as

$$\frac{\partial g(\varepsilon)}{\partial \beta_0'} = -2\sum_{i=1}^{n} [y_i - \beta_0' - \beta_1(x_i - \bar{x})] = 0 \Rightarrow n\beta_0' = \sum_{i=1}^{n} y_i \tag{1.7b}$$

$$\frac{\partial g(\varepsilon)}{\partial \beta_1} = -2\sum_{i=1}^{n} [y_i - \beta_0' - \beta_1(x_i - \bar{x})](x_i - \bar{x}) = 0$$

$$\Rightarrow \beta_1 \sum (x_i - \bar{x})^2 = \beta_1 \sum y_i(x_i - \bar{x}) \tag{1.7c}$$

Solving Equations 1.7b and 1.7c simultaneously, we obtain the estimates of the parameters of the transformed model as

$$\hat{\beta}_0' = \bar{y} \tag{1.8a}$$

$$\hat{\beta}_1 = \frac{\sum_{i=1}^{n} y_i(x_i - \bar{x})}{\sum_{i=1}^{n} (x_i - \bar{x})^2} \tag{1.8b}$$

Equations 1.8a and 1.8b lead to the fitted model of the form

$$\hat{y}_i = \hat{\beta}'_0 - \hat{\beta}_1(x_i - \bar{x}) \tag{1.9}$$

for the ith observation, y_i.

To obtain the intercept of the original model 1.4 from Equation 1.9, we see that

$$\beta_0 = \beta'_0 - \beta_1\bar{x}$$

Consequently, the equivalent fitted model for the original model 1.4 is

$$\hat{y}_i = \hat{\beta}_0 + \hat{\beta}_1 x_i \tag{1.10}$$

The result of Equation 1.10 is obtained by substituting for β_0 in Equation 1.9 using Equation 1.6b. From this, we see that the fitted model is

$$\hat{y}_i = \hat{\beta}_0 - \beta_1\bar{x} + \hat{\beta}_1(x_i - \bar{x}) \Rightarrow \hat{y}_i = \hat{\beta}_0 + \hat{\beta}_1 x_i$$

From Equations 1.8a and 1.8b, we note that if we define

$$S_{xy} = \sum_{i=1}^{n} y_i(x_i - \bar{x}) = \sum_{i=1}^{n} x_i y_i - n\overline{xy} \tag{1.11}$$

$$S_{xx} = \sum_{i=1}^{n} (x_i - \bar{x})^2 = \sum_{i=1}^{n} x_i^2 - n\bar{x}^2 \tag{1.12}$$

$$\hat{\beta}_1 = \frac{S_{xy}}{S_{xx}} \tag{1.13}$$

S_{xy} is the corrected sum of cross products of x and y while S_{xx} is the corrected sum of squares of x.

Since $\hat{\beta}'_0$ and $\hat{\beta}_1$ are estimators of the parameters β'_0 and β_1, they are random variables so that we can find their means and variances. Since $\hat{\beta}'_0$ depends on $\hat{\beta}_1$, let us find the mean and variance of $\hat{\beta}_1$ first, and from those values find the means and variances of $\hat{\beta}'_0$.

$$\begin{aligned}
E(\hat{\beta}_1) &= E\left[\frac{S_{xy}}{S_{xx}}\right] = \left(\frac{1}{S_{xx}}\right) E\left[\sum_{i=1}^{n} y_i(x_i - \bar{x})\right] \\
&= \left(\frac{1}{S_{xx}}\right) E\left\{\sum_{i=1}^{n} [\beta'_0 + \beta_1(x_i - \bar{x}) + \varepsilon_i](x_i - \bar{x})\right\} \\
&= \left(\frac{1}{S_{xx}}\right) E\left\{\sum_{i=1}^{n} \beta'_0(x_i - \bar{x}) + \beta_1 \sum_{i=1}^{n}(x_i - \bar{x})^2 + \sum_{i=1}^{n} \varepsilon_i(x_i - \bar{x})\right\} = \beta_1
\end{aligned}$$

because $\sum_{i=1}^{n} \beta'_0(x_i - \bar{x})$ and $\varepsilon_i \sim \text{NID}(0, \sigma^2)$.

From Equation 1.8a, we have

$$E(\hat{\beta}_0') = E\left(\frac{1}{n}\right)\sum_{i=1}^{n}(y_i) = E\left(\frac{1}{n}\right)\sum_{i=1}^{n}[\beta_0' + \beta_1(x_i - \bar{x}) + \varepsilon_i] = \frac{1}{n}n\beta_0' = \beta_0'$$

$$\text{Var}(\hat{\beta}_0') = \text{Var}\left(\frac{1}{n}\right)\sum_{i=1}^{n}(y_i) = \left(\frac{1}{n}\right)^2 n\sigma^2 = \frac{\sigma^2}{n}$$

$$E(\hat{\beta}_0) = E(\bar{y} - \hat{\beta}_1\bar{x}) = E\left\{\left(\frac{1}{n}\right)\sum_{i=1}^{n}[\beta_0' + \beta_1(x_i - \bar{x}) + \varepsilon_i] - \beta_1\bar{x}\right\}$$

$$= E(\beta_0 - \beta_1\bar{x}) = \beta_0 + \beta_1\bar{x} - \beta_1\bar{x} = \beta_0$$

$$\text{Var}(\hat{\beta}_0) = \text{Var}(\bar{y} - \hat{\beta}_1\bar{x}) = \text{Var}(\bar{y}) + \text{Var}(\hat{\beta}_1\bar{x}) - 2\,\text{Cov}(\bar{y}, \hat{\beta}_1\bar{x})$$

$$= \frac{\sigma^2}{n} + \bar{x}^2\text{Var}(\hat{\beta}_1) - 2\bar{x}\,\text{Cov}(\bar{y}, \hat{\beta}_1) = \frac{\sigma^2}{n} + \frac{\bar{x}^2\sigma^2}{S_{xx}} - 2\bar{x}\,\text{Cov}(\bar{y}, \hat{\beta}_1)$$

But

$$2\bar{x}\,\text{Cov}(\bar{y}, \hat{\beta}_1) = -2\bar{x}\left(\bar{y}, \frac{S_{xy}}{S_{xx}}\right) = -2\bar{x}\,\text{Cov}\left[\frac{1}{n}\sum_{i=1}^{n} y_i, \frac{1}{S_{xx}}\sum_{i=1}^{n} y_i(x_i - \bar{x})\right]$$

$$= \frac{-2\bar{x}}{nS_{xx}}\text{Cov}\{y_1(x_i - \bar{x}), \ldots, y_n(x_i - \bar{x})\}$$

$$= \frac{-2\bar{x}}{nS_{xx}}\sum_{i=1}^{n}(x_i - \bar{x})\sigma^2 = 0$$

Consequently,

$$\text{Var}(\hat{\beta}_0) = \frac{\sigma^2}{n} + \frac{\bar{x}^2\sigma^2}{S_{xx}}$$

Example 1.19

In a chemical experiment, for different catalyst concentrations x, the yields y of the process were recorded:

Observation	x	y
1	1.5	28
2	2.5	34.5
3	3	42.5
4	3	40.5
5	5	46
6	7	52
7	8	56
8	9	54.5
9	10	55
10	11	58
11	12	62
12	14	66
13	17	72
Total	103	667

Find the relationship between concentration x and yield y. It seems that the relationship between x and y could be linear as shown in the scatterplots of Figure 1.11. We will use the least squares method previously described to fit a straight line to the data.

Observation	x	y	xy	x^2
1	1.5	28	42	2.25
2	2.5	34.5	86.25	6.25
3	3	42.5	127.5	9
4	3	40.5	121.5	9
5	5	46	230	25
6	7	52	364	49
7	8	56	448	64
8	9	54.5	490.5	81
9	10	55	550	100
10	11	58	638	121
11	12	62	744	144
12	14	66	924	196
13	17	72	1224	289
Total	103	667	5989.75	1095.5

$$\sum_{i=1}^{13} x_i = 103 \quad \sum_{i=1}^{13} y_i = 667 \quad \sum_{i=1}^{13} x_i y_i = 5989.75 \quad \sum_{i=1}^{13} x_i^2 = 1095.5$$

$$n = 13 \quad \bar{x} = 7.9231 \quad \bar{y} = 51.308$$

$$S_{xy} = 5989.75 - 13(7.9231)\,(51.308) = 705.0577$$

$$S_{xx} = 1095.5 - 13(7.9231)^2 = 279.4231$$

$$S_{yy} = 28^2 + 34.5^2 + \cdots + 72^2 - 13(51.308)^2 = 1897.769$$

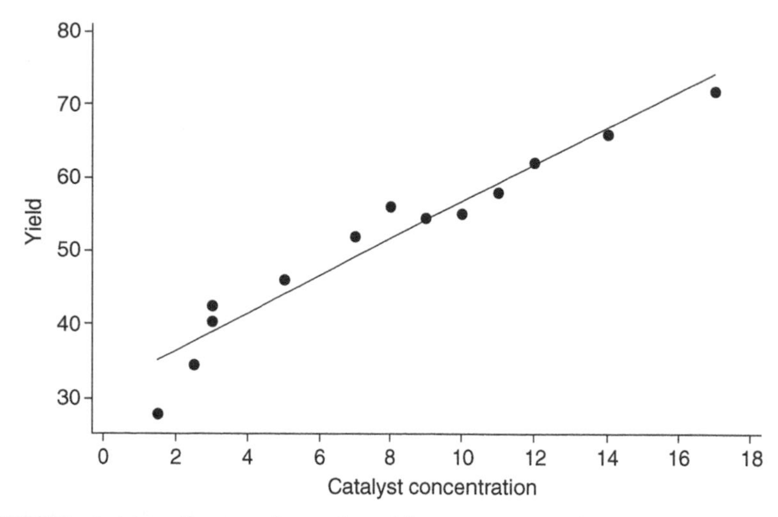

FIGURE 1.11: Scatterplot of yield against catalyst concentration for Example 1.19.

$$\hat{\beta}_1 = \frac{705.0577}{279.4231} = 2.523262 \quad \hat{\beta}'_0 = \bar{y} = 51.308$$

so that

$$\hat{y} = 51.308 + 2.523262(x - 7.9231) \tag{1.14a}$$

or

$$\hat{y} = 31.3157 + 2.523262x \tag{1.14b}$$

□

1.2.2 Checking Model Adequacy—Diagnosis by Residual Plots

We usually need to check for nonviolation of the basic model assumptions. We also need to check the adequacy of the regression models we fitted to data. This is even more necessary given that regression models are often fitted to sets of data when their true functional relationship is unknown.

As usual, residual plots against the fitted values and regressors provide bases for judging whether the fit is adequate. Generally, interpretation of residual plots should follow some patterns (we shall discuss the patterns in Chapter 2). Apart from the plots suggested above, insight into the adequacy of the model could be gained by plotting the residuals against any variable that was omitted in the analysis. It is possible for such a plot to indicate that inclusion of such a variable in the analysis will improve the model.

TABLE 1.1: Fitted Values and Residuals for Example 1.19

x	y	$\hat{y}_i$	$y_i - \hat{y}_i$
1.5	28	35.10059	−7.10059
2.5	34.5	37.62386	−3.12386
3	42.5	38.88549	3.614514
3	40.5	38.88549	1.614514
5	46	43.93201	2.06799
7	52	48.97853	3.021466
8	56	51.5018	4.498204
9	54.5	54.02506	0.474942
10	55	56.54832	−1.54832
11	58	59.07158	−1.07158
12	62	61.59484	0.405156
14	66	66.64137	−0.64137
17	72	74.21115	−2.21115
103	667	667.0001	−8.6E−05

A normal probability plot of the residuals of a regression analysis will indicate whether the assumption that $\varepsilon_i \sim \text{NID}(0, \sigma^2)$ is grossly violated. On the other hand, if there is a trend of some sort in the data or some variable of interest has been omitted in the analysis, this is likely to be thrown up by a plot of residuals against time sequence or fitted variables. The residuals are obtained as

$$\hat{\varepsilon}_i = y_i - \hat{y}_i \tag{1.14c}$$

where $\hat{y}_i$ is the fitted value for the ith observation y_i.

Using either Equation 1.14a or 1.14b, we obtain the fitted values and the ensuing residuals for the above fitted linear relationship between the yield (y) and the catalyst concentration (x) as presented in Table 1.1.

We present a plot of residuals against fitted variables in Figure 1.12. From the above plot, it seems that a quadratic rather than a linear model would be a better fit for the data.

This is further confirmed by plotting the same data and joining the points as shown in Figure 1.13.

It seems from the pattern of the scatterplots that a quadratic term is needed in the regression model. We will return to this problem after treating polynomial regression. We stated earlier (Section 1.2.1) that it is necessary to plot the data and to check whether a straight line would be adequate before fitting one to the data set. This is important because with the least squares method, one can fit a straight line to any set of observations. A straight line fitted when such a line is not an adequate representation of the actual relationship between the variables x and y cannot serve as a good predictor of this relationship. There are analytic methods for determining how well a straight line fits our data points. One of such methods is the coefficient of determination,

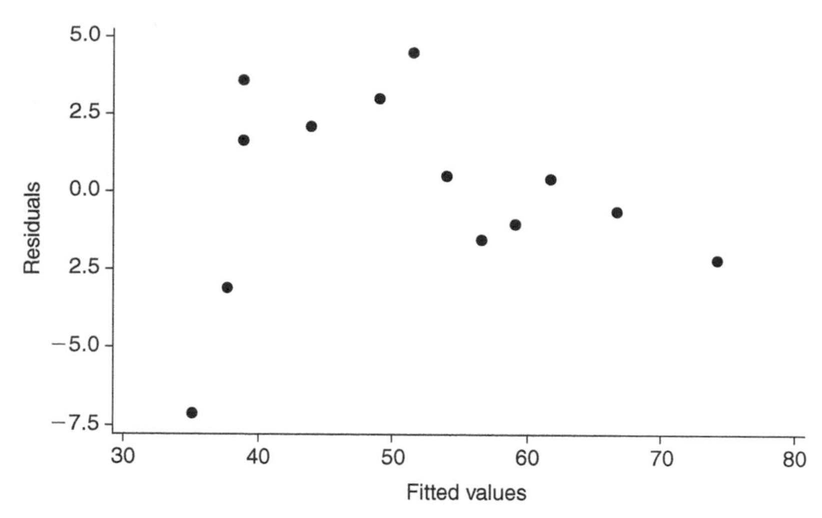

FIGURE 1.12: Scatterplot of residuals against fitted values for Example 1.19.

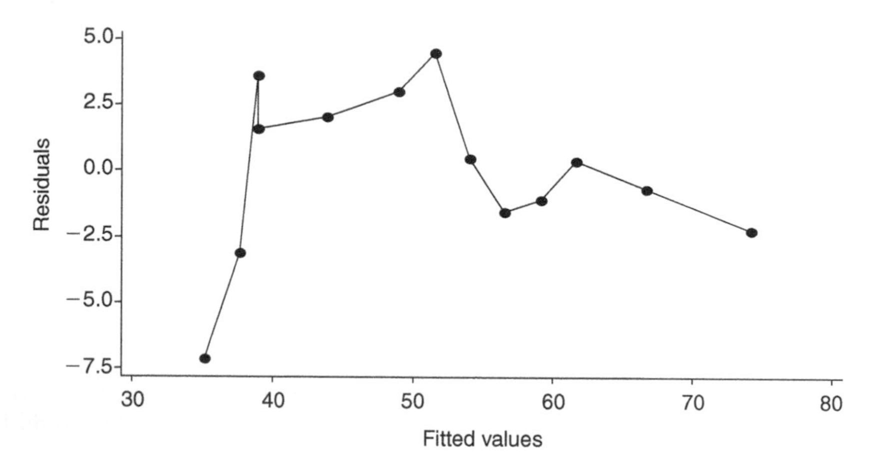

FIGURE 1.13: Scatterplot of residuals against fitted values for Example 1.19 with the points joined.

while the other uses analysis of variance (ANOVA) to test whether the variability in y is properly explained by a linear relationship with the variable x. Although we can always obtain r, which is the measure of the linear correlation between x and y, it is not always easy to explain in practical terms what r means. The coefficient of determination r^2 is more useful in this regard, because it enables us to interpret all values of r in addition to those that are easily explained ($r = -1, r = 0$, and $r = +1$). It is also useful because it is associated with both regression and correlation and provides a link between them. A detailed discussion of the correlation coefficient r can be found in most basic texts on statistics. The value of r can easily be calculated using statistical packages such as MINITAB, SPSS, SAS, and others. In addition to S_{xx} and S_{xy} defined in Section 1.2 (see Equations 1.12 and 1.13), we define another measure S_{yy}. While S_{xy} is a measure of the covariance between X and Y, S_{xx} and S_{yy} are the respective measures of variations in X and Y or the corrected sums of squares of X and Y.

$$S_{yy} = \sum_{i=1}^{n} y_i^2 - n\bar{y}^2 \tag{1.15}$$

The coefficient of correlation is r, where

$$r = \frac{\sum_{i=1}^{n} x_i y_i - \left(\sum_{i=1}^{n} x_i\right)\left(\sum_{i=1}^{n} y_i\right)}{\sqrt{\left[n\sum_{i=1}^{n} x_i^2 - \left(\sum_{i=1}^{n} x_i\right)^2\right]\left[n\sum_{i=1}^{n} y_i^2 - \left(\sum_{i=1}^{n} y_i\right)^2\right]}} \tag{1.16}$$

and this can be rewritten in terms of the measures S_{xx}, S_{xy}, and S_{yy} as

$$r = \frac{S_{xy}}{\sqrt{S_{xx} S_{yy}}} \tag{1.17}$$

By using Equations 1.15 and 1.17, we can easily show that

$$r = \frac{\beta_1 \sqrt{S_{xx}}}{\sqrt{S_{yy}}} \tag{1.18}$$

Consider the error sum of squares SS_{E} and from Equation 1.7a, we see that

$$\begin{aligned} SS_{\text{E}} &= \sum_{i=1}^{n} (y_i - \hat{y}_i)^2 = \sum_{i=1}^{n} [y_i - \beta_0' - \beta_1(x_i - \bar{x})]^2 \\ &= \sum_{i=1}^{n} \left[(y_i - \bar{y}) - \left(\frac{S_{xy}}{S_{xx}}\right)(x_i - \bar{x})\right]^2 \Rightarrow SS_{\text{E}} \\ &= \sum_{i=1}^{n} (y_i - \bar{y})^2 - \frac{2S_{xy}^2}{S_{xx}} + \frac{S_{xy} S_{xx}}{S_{xx}^2} = S_{yy} - \beta_1 S_{xy} \end{aligned} \tag{1.19}$$

Dividing throughout by S_{yy}, we obtain

$$\frac{SS_{\mathrm{E}}}{S_{yy}} = \frac{S_{yy}}{S_{yy}} - \frac{\beta_1 S_{xy}}{S_{yy}} = 1 - \frac{S_{xy}^2}{S_{xx}S_{xy}}$$

$$= 1 - r^2 \quad \text{because } r = \frac{\beta_1\sqrt{S_{xy}}}{\sqrt{S_{yy}}}$$

$$SS_{\mathrm{E}} = (1 - r^2)S_{yy}$$

so that

$$r^2 = \frac{S_{yy} - SS_{\mathrm{E}}}{S_{yy}} \tag{1.20}$$

Clearly, Equation 1.20 shows that r^2 is the ratio of the variation in Y due to linearity and the total variation in Y. This means that r^2 (often written as R^2) represents the proportion of the total variation in Y that could be attributed to linearity. Further, if r^2 is expressed in percentage, it will indicate the percentage of the variation responses Y, which is explained by linearity. The coefficient of determination, r^2 better explains the linear relationship between Y and X than does r alone. A high r^2 should indicate a high linear association between X and Y and in turn indicates that predictions obtained by using the fitted relationship should be good. In any regression analysis, linear or not, r^2 represents the ratio of variation due to regression model to total variation. Another way of testing whether a significant amount of the variability in the response variable y is explained by a straight line is to carry out an ANOVA, which involves the partitioning of the total variation in y into two parts (1) the part explained by regression of Y on X and (2) the unexplained variation (the residual variation). The partitions are obtained as

$$S_{yy} = SS_{\mathrm{R}} + SS_{\mathrm{E}} \tag{1.21}$$

If linear regression adequately explains the variations in the response, then (SS_{R} the regression sum of squares $SS_{\mathrm{R}} = \beta_1 S_{xy}$) will be high and hence significant in ANOVA. From Equation 1.4, the regression model, with n observations, we can see that the total degrees of freedom is $n-1$. The regression process estimated two parameters; therefore, the regression sum of squares should have 1 degree of freedom, while the error sum of squares has $n-2$ degrees of freedom. In terms of the model 1.4, we can express the hypothesis of no linear regression (relationship) as

$$H_0 : \beta_1 = 0 \quad \text{vs} \quad H_1 : \beta_1 \neq 0$$

The typical ANOVA for least squares linear regression is presented in Table 1.2.

TABLE 1.2: ANOVA for Typical Linear Regression

Source	**Sum of squares**	**Degrees of freedom**	**Mean square**	***F*-ratio**
Regression	$\beta_1 S_{xy}$	1	$SS_{\mathrm{R}}/1$	$MS_{\mathrm{R}}/MS_{\mathrm{E}}$
Error	$S_{yy} - \beta_1 S_{xy}$	$n-2$	$SS_{\mathrm{E}}/(n-2)$	
Total	S_{yy}	$n-1$		

F-ratio for the regression is compared with the tabulated $F(\alpha, 1, n-2)$. If the ratio is greater than the tabulated value, the null hypothesis is rejected. To illustrate both the coefficient of determination and the technique of ANOVA in linear regression, we calculate the coefficient of determination and carry out the test on data of Example 1.19 (Table 1.3).

$$S_{yy} = 1897.769 \quad SS_{\mathrm{R}} = 1779.045 \quad SS_{\mathrm{E}} = 118.7238$$

$$r^2 = \frac{S_{yy} - SS_{\mathrm{E}}}{S_{yy}} = \frac{1897.769 - 118.7238}{1897.769} = 0.93744$$

$$\Rightarrow r = \sqrt{0.93744} = 0.968215$$

TABLE 1.3: ANOVA for Regression in Example 1.1

Source	**Sum of squares**	**Degrees of freedom**	**Mean square**	***F*-ratio**	***p*-Value**
Regression	1779.045	1	1779.045	164.8322	5.79199×10^{-8}
Error	118.7238	11	10.79307		
Total	1897.769	12			

$F(1, 11, 0.01) = 9.65$. The regression effect is significant. The linear regression explained a high percentage of the total variation in the response variable Y $(r^2 = 0.93744)$.

1.2.3 Checking Model Adequacy—Lack of Fit Test

In fitting straight lines or in exploratory fitting of polynomial models, we would often like to test whether the order of the curve fitted at any stage is correct. Lack of fit (LOF) of our model renders it inadequate for prediction purposes and defeats our basic aim for fitting the model to the data. The "lack of fit" or "goodness of fit" test we present here will be applied in the exploration of response surfaces, which we will discuss in Chapter 11. Although we will describe the test for one independent variable, note that even when $k (k \geq 2)$ independent variables are being studied, the results can be generalized.

The hypothesis to be tested is of the form

H_0 : The model adequately fits the data vs H_1 : The model does not fit the data

To carry out this test, we partition the residual sum of squares into two components:

$$SS_{\text{residual}} = SS_{\text{LOF}} + SS_{\text{PE}} \tag{1.22}$$

SS_{PE} is the sum of squares arising from pure experimental error. SS_{LOF} is the sum of squares due to LOF of the model. As usual, we need a set of repeated measurements or responses y, at the same value or level of the control variable x, to estimate pure experimental error. In factorial experiments, we are unable to estimate error unless we replicated the experiment. Suppose the responses were obtained as follows:

$$\begin{array}{ll} y_{11}, y_{12}, \ldots, y_{1n_1} & n_1 \text{ repeated observations at } x_1 \\ y_{21}, y_{22}, \ldots, y_{2n_2} & n_2 \text{ repeated observations at } x_2 \\ \vdots & \vdots \\ y_{m1}, y_{m2}, \ldots, y_{mn_m} & n_m \text{ repeated observations at } x_m \end{array}$$

There are n_i repeated observations for each level of x_i, so that pure error sum of squares is calculated as

$$SS_{\text{PE}} = \sum_{i=1}^{m} \sum_{j=1}^{n_i} (y_{ij} - \bar{y}_i)^2 \tag{1.23}$$

with $\upsilon_{\text{PE}} = \sum_{i=1}^{m} (n_i - 1)$ degrees of freedom. We can obtain the LOF sum of squares by subtraction from the residual sum of squares:

$$SS_{\text{LOF}} = SS_{\text{residual}} - SS_{\text{PE}} \tag{1.24}$$

with the LOF degrees of freedom of $\upsilon_{\text{LOF}} = n - 2 - \upsilon_{\text{PE}}$. The ratio of $SS_{\text{LOF}}/\upsilon_{\text{LOF}}$ and $SS_{\text{PE}}/\upsilon_{\text{PE}}$ follows the F-distribution with υ_{LOF} and υ_{PE} degrees of freedom. We can, therefore, use ANOVA to carry out the LOF test.

Example 1.20

To test for LOF in Example 1.19, we use the values of the predictor variable at which repeated measurements have been made. Repeated measurements were made only on $X = 3$, so the data are not suitable for LOF test as it is.

Suppose in the problem of Example 1.19, we make repeated measurements at different values of x and then can carry out a LOF test for the data. We present the data for repeated experiments in the table below.

i	x_i	y_i
1	1.5	28
2	2.5	34.5
3	3	42.5
4	3	40.5
5	5	46
6	7	53
7	7	52
8	7	49
9	8	56
10	9	54.6
11	10	55
12	11	58
13	12	62
14	14	66
15	17	72
16	17	69.5
17	17	72.6

$$\sum_{i=1}^{17} x_i = 151 \quad \sum_{i=1}^{17} x_i^2 = 1771.5 \quad \sum_{i=1}^{17} y_i = 911.2$$

$$\sum_{i=1}^{17} y_i^2 = 51441.92 \quad \sum_{i=1}^{17} x_i y_i = 9120.35$$

$$\bar{y} = 53.6 \quad \bar{x} = 8.882353;$$

For this problem, $n = 17$; $n_1 = 2$; $n_2 = n_3 = 3$; $m = 3$;

$$S_{xx} = 1771.5 - 17(8.882353)^2 = 430.2647$$
$$S_{yy} = 51441.92 - 17(53.6)^2 = 2601.60$$
$$S_{xy} = 9120.35 - (8.882353)(53.6) = 1026.75$$

$$\hat{\beta}_1 = \frac{S_{xy}}{S_{xx}} = \frac{1026.75}{430.2647} = 2.386322 \quad \hat{\beta}_0' = \bar{y} = 53.6$$

The fitted line is

$$\hat{y} = 53.6 + 2.386322(x - 8.882353) \quad \text{or} \quad \hat{y} = 32.4038 + 2.386322x$$

$$S_{yy} = SS_{\text{regression}} + SS_{\text{residual}}$$

$$SS_{\text{regression}} = \hat{\beta}_1 S_{xy} = 2.386322(1026.75) = 2450.156$$

$$SS_{\text{residual}} = 2601.6 - 2450.156 = 151.4442$$

To obtain the SS_{PE}, we note that responses were repeated twice, when $x=3$: (42.5, 41.5), thrice when $x=7$: (53, 52, 49), and thrice when $x=17$: (72, 72.6, 69.5). Using the formula

$$SS_{\text{PE}} = \sum_{i=1}^{3}\sum_{j=1}^{n_i}(y_{ij}-\bar{y}_i)^2$$

$$\begin{aligned} SS_{\text{PE}} &= (42.5-41.5)^2+(40.5-41.5)^2+(53-51.33)^2+(52-51.33)^2 \\ &\quad +(49-51.33)^2+(72-71.367)^2+(72.6-71.367)^2 \\ &\quad +(69.5-71.367)^2 \\ &= 16.29527 \end{aligned}$$

$$SS_{\text{LOF}} = SS_{\text{residual}} - SS_{\text{PE}} = 151.4442 - 16.29527 = 135.1516$$

We now present the ANOVA for the regression as well as the LOF test for the replicated regression (Table 1.4).

TABLE 1.4: ANOVA for Replicated Regression

Source	Sum of squares	Degrees of freedom	Mean square	*F*-ratio	*p*-Value
Regression	2450.156	1	2450.156	242.6791	1.13686×10^{-10}
Residual	151.4442	15	10.09628		
Lack of fit	135.1516	10	13.51516	4.14696	0.064945596
Pure error	16.29527	5	3.259054		
Total	2601.60	16			

From Table 1.4, we observe that $F(1,\ 15,\ 0.01)=8.68$ while $F(10,\ 5,\ 0.01)=10.05$. We cannot reject the hypothesis as the model adequately fits the data. □

1.2.4 Multiple Linear Regression

If the number of independent variables $m \geq 2$, the regression is called multiple regression. In this case, these m variables are related to the response variable by an equation of the form

$$y_j = \beta_0 + \beta_1 x_{1j} + \cdots + \beta_m x_{mj} + \varepsilon_j \tag{1.25}$$

We fit the linear equation by using the least squares method. We are required to minimize the function

$$L(\varepsilon, x) = \sum_{j=1}^{m} \varepsilon_j^2 \tag{1.26}$$

where

$$\varepsilon_j = y_j - \beta_0 - \beta_1 x_{1j} - \cdots - \beta_m x_{mj} \tag{1.27}$$

By partial differentiation of $L(\varepsilon, x)$ with respect to each of the parameters $\beta_0, \beta_1, \ldots, \beta_m$ and equating to 0, that is, $\partial L(\varepsilon, x)/\partial \beta_i = 0$, we obtain the normal equations, and by rearranging the normal equations, we obtain

$$\begin{aligned}
\sum_{j=1}^{n} y_j &= n\beta_0 + \beta_1 \sum_{j=1}^{n} x_{1j} + \beta_2 \sum_{j=1}^{n} x_{2j} + \cdots + \beta_m \sum_{j=1}^{n} x_{mj} \\
\sum_{j=1}^{n} x_{1j} y_j &= \beta_0 \sum_{j=1}^{n} x_{1j} + \beta_1 \sum_{j=1}^{n} x_{1j}^2 + \beta_2 \sum_{j=1}^{n} x_{2j} x_{1j} + \cdots + \beta_m \sum_{j=1}^{n} x_{mj} x_{1j} \\
\sum_{j=1}^{n} x_{2j} y_j &= \beta_0 \sum_{j=1}^{n} x_{2j} + \beta_1 \sum_{j=1}^{n} x_{2j} x_{1j} + \beta_2 \sum_{j=1}^{n} x_{2j}^2 + \cdots + \beta_m \sum_{j=1}^{n} x_{mj} x_{2j} \\
&\vdots \\
\sum_{j=1}^{n} x_{mj} y_j &= \beta_0 \sum_{j=1}^{n} x_{mj} + \beta_1 \sum_{j=1}^{n} x_{mj} x_{1j} + \beta_2 \sum_{j=1}^{n} x_{2j} x_{mj} + \cdots + \beta_m \sum_{j=1}^{n} x_{mj}^2
\end{aligned} \tag{1.28}$$

Solving these simultaneous normal equations, we obtain estimates of the parameters of the model as follows:

$$\begin{bmatrix} \hat{\beta}_0 \\ \hat{\beta}_1 \\ \vdots \\ \hat{\beta}_m \end{bmatrix} = \begin{bmatrix} n & \sum_{j=1}^{n} x_{1j} & \cdots & \sum_{j=1}^{n} x_{mj} \\ \sum_{j=1}^{n} x_{1j} & \sum_{j=1}^{n} x_{1j}^2 & \cdots & \sum_{j=1}^{n} x_{1j} x_{mj} \\ \vdots & \vdots & \cdots & \vdots \\ \sum_{j=1}^{n} x_{mj} & \sum_{j=1}^{n} x_{1j} x_{mj} & \cdots & \sum_{j=1}^{n} x_{mj}^2 \end{bmatrix}^{-1} \begin{bmatrix} \sum_{j=1}^{n} y_j \\ \sum_{j=1}^{n} x_{1j} y_j \\ \vdots \\ \sum_{j=1}^{n} x_{mj} y_j \end{bmatrix} \tag{1.29}$$

Example 1.21

In a large company that is into the catering business, the cooking of a type of beans depends on quantity, temperature, and pressure. A pilot study was carried out 15 times under 15 different cooking conditions, and the following data on the cooking times of the beans were obtained. Determine the equation of the linear relationship between time and the three variables, quantity, temperature, and pressure.

Quantity (lb)	Temperature (°C)	Pressure (gm)	Time (min)
11	31	6	46
13	35	6	53
6	35	11	26
18	34	4	66
9	40	16	20
7	50	11	21
13	36	6	56
11	34	6	71
19	45	14	25
24	35	4	91
17	37	6	62
21	40	3	81
16	45	9	69
7	48	12	28
6	52	9	25

Find the relationship between these variables and the response.

Let $x_1 = \text{quantity}$, $x_2 = \text{temperature}$, $x_3 = \text{pressure}$, $y = \text{time}$, $m = 3$, and $n = 15$.

$$\sum_{j=1}^{15} x_{1j} = 198 \quad \sum_{j=1}^{15} x_{2j} = 597 \quad \sum_{j=1}^{15} x_{3j} = 123 \quad \sum_{j=1}^{15} y_j = 740$$

$$\sum_{j=1}^{15} x_{1j}^2 = 3078 \quad \sum_{j=1}^{15} x_{2j}^2 = 24371 \quad \sum_{j=1}^{15} x_{3j}^2 = 1221 \quad \sum_{j=1}^{15} x_{1j}x_{2j} = 7702$$

$$\sum_{j=1}^{15} x_{1j}x_{3j} = 1456 \quad \sum_{j=1}^{15} x_{2j}x_{3j} = 5088 \quad \sum_{j=1}^{15} x_{1j}y_j = 11239$$

$$\sum_{j=1}^{15} x_{2j}y_j = 28308 \quad \sum_{j=1}^{15} x_{3j}y_j = 4968$$

The normal equations for the above data are

$$740 = 15\beta_0 + 198\beta_1 + 597\beta_2 + 123\beta_3$$

$$11239 = 198\beta_0 + 3078\beta_1 + 7702\beta_2 + 1456\beta_3$$

$$28308 = 597\beta_0 + 7702\beta_1 + 24371\beta_2 + 5088\beta_3$$

$$4968 = 123\beta_0 + 1456\beta_1 + 5088\beta_2 + 1221\beta_3$$

We obtain the parameters $\beta_0, \beta_1, \ldots, \beta_m$ as follows:

$$\begin{bmatrix} \hat{\beta}_0 \\ \hat{\beta}_1 \\ \hat{\beta}_2 \\ \hat{\beta}_3 \end{bmatrix} = \begin{bmatrix} 15 & 198 & 597 & 123 \\ 198 & 3078 & 7702 & 11239 \\ 597 & 7702 & 24371 & 5088 \\ 123 & 1456 & 5088 & 1221 \end{bmatrix}^{-1} \begin{bmatrix} 740 \\ 11239 \\ 28308 \\ 4968 \end{bmatrix}$$

$$= \begin{bmatrix} 4.20279 & -0.06545 & -0.07824 & -0.01930 \\ -0.06545 & 0.003025 & 0.000183 & 0.002221 \\ -0.07824 & 0.000183 & 0.002306 & -0.00195 \\ -0.01930 & 0.002221 & -0.00195 & 0.008225 \end{bmatrix} \begin{bmatrix} 740 \\ 11239 \\ 28308 \\ 4968 \end{bmatrix}$$

The fitted model is

$$y = 63.80119 + 1.797404x_1 - 0.2373x_2 - 3.5545x_3$$

We can test hypothesis about the significance of the fitted regression model. The hypothesis to be tested is

$$H_0 : \beta_1 = \beta_2 = \beta_3 = 0 \quad \text{vs} \quad H_1 : \text{at least one } \beta_i \neq 0$$

To test this, we use the fact that

$$S_{yy} = SS_{\mathrm{R}} + SS_{\mathrm{E}}$$

$$SS_{\mathrm{E}} = \sum_{i=1}^{n} (y_i - \hat{y}_i) = \sum_{i=1}^{n} y_i^2 - [\beta_0 \quad \beta_1 \quad \beta_2 \quad \beta_3] \begin{bmatrix} \sum_{j=1}^{n} y_j \\ \sum_{j=1}^{n} x_1 y_j \\ \sum_{j=1}^{n} x_2 y_j \\ \sum_{j=1}^{n} x_3 y_j \end{bmatrix}$$

$$= 44456 - [63.80119 \quad 1.797404 \quad -0.2273 \quad -3.5545] \begin{bmatrix} 740 \\ 11239 \\ 28308 \\ 4968 \end{bmatrix}$$

$$= 1135.260244$$

$$S_{yy} = \sum_{i=1}^{n} y_i^2 - \frac{\left(\sum_{i=1}^{15} y_i\right)^2}{15} = 44456 - \frac{740^2}{15} = 7949.33$$

$$SS_{\text{regression}} = 7949.33 - 1135.260244 = 6814.069756$$

The ANOVA for this test is presented in Table 1.5.

TABLE 1.5: ANOVA for the Beans Problem

Source	Sum of squares	Degrees of freedom	Mean square	F-ratio
Regression	6814.069756	3	2271.3565	22.0081
Residual	1135.260244	11	103.20547	
Total	7949.33	14		

Clearly, the regression effects are significant at 1% because $F(3, 11, 0.01) = 6.22$. Now, we calculate another quantity, R^2, which (as stated earlier) is the ratio of the regression sum of squares to the total sum of squares. That is, $R^2 = (S_{yy} - SS_{\text{E}})/S_{yy}$ is a measure of how well the predicted model fits the dependent variable. It has values between 0 and 1. Thus, if the R^2 in a given regression analysis is 1, then we say that the model is perfect. The quantity R^2 is often called the coefficient of determination.

For the above regression example, we see that

$$R^2 = \frac{7949.33 - 1135.260244}{7949.33} = 0.85289$$

We can also obtain the adjusted R^2 denoted by R^2_{adj} which is obtained by rescaling R^2 by the degrees of freedom so that its adjusted form depends only on the mean squares rather than the sums of squares. The adjusted form of coefficient of determination is obtained as

$$R^2_{\text{adj}} = 1 - \frac{MS_{\text{residual}}}{MS_{\text{total}}} = 1 - \frac{(1 - R^2)(n - 1)}{n - m}$$

$$MS_{\text{residual}} = 103.20547 \quad MS_{\text{total}} = \frac{7949.33}{14} = 567.8092857$$

$$R^2_{\text{adj}} = 1 - \frac{103.20547}{567.8092857} = 1 - 0.181708 = 0.8182392$$ □

1.2.5 Polynomial Regression

If the relationship between a single independent variable x and an outcome variable y is a straight line, we usually determine the relationship by using the model 1.4 as

$$y_i = \beta_0 + \beta_1 x_i + \varepsilon_i$$

where $\varepsilon_i \sim \text{NID}(0, \sigma^2)$. Sometimes, the linear relationship may not be adequate to represent the relationship between the variables x and y, so that a second-order term needs to be added to the model 1.4. The model that may be adequate would be of the form

$$y_i = \beta_0 + \beta_1 x_i + \beta_2 x_i^2 + \varepsilon_i \tag{1.30}$$

The model 1.30 is said to be parabolic. It may be possible that the parabola represented by model 1.30 may not be adequate; in that case, a higher order model is required. Such higher order models are called polynomials and can generally be represented in the following model for $k \geq 3$:

$$y_i = \beta_0 + \beta_1 x_i + \beta_2 x_i^2 + \cdots + \beta_k x_i^k + \varepsilon_i \tag{1.31}$$

The model 1.31 represents a polynomial of order k, since the highest exponent of x in the model is k.

1.2.5.1 Fitting a Parabola

If we have the independent variable x and the response variable y, one could start by first fitting a linear relationship, testing for adequacy (LOF), and then fitting the second order or parabolic model if indicated by the LOF test. The fitting of the parabola is based on the method of least squares. Based on the model 1.31, we usually minimize the sum of the squares of the errors

$$L(x, \varepsilon) = \sum (y_i - \beta_0 - \beta_1 x_i - \beta_2 x_i^2)^2 \tag{1.32}$$

with respect to the parameters of the model β_0, β_1, and β_2. By minimizing we obtain

$$\frac{\partial L(x,\varepsilon)}{\partial \beta_0} = -2\sum \left(y_i - \beta_0 - \beta_1 x_i - \beta_2 x_i^2\right) = 0$$

$$\frac{\partial L(x,\varepsilon)}{\partial \beta_1} = -2\sum x_i \left(y_i - \beta_0 - \beta_1 x_i - \beta_2 x_i^2\right) = 0$$

$$\frac{\partial L(x,\varepsilon)}{\partial \beta_2} = -2\sum x_i^2 \left(y_i - \beta_0 - \beta_1 x_i - \beta_2 x_i^2\right) = 0$$

These partial differentials lead to the normal equations

$$n\beta_0 + \beta_1 \sum_{i=1}^{n} x_i + \beta_2 \sum_{i=1}^{n} x_i^2 = \sum_{i=1}^{n} y_i$$

$$\beta_0 \sum_{i=1}^{n} x_i + \beta_1 \sum_{i=1}^{n} x_i^2 + \beta_2 \sum_{i=1}^{n} x_i^3 = \sum_{i=1}^{n} x_i y_i$$

$$\beta_0 \sum_{i=1}^{n} x_i^2 + \beta_1 \sum_{i=1}^{n} x_i^2 + \beta_2 \sum_{i=1}^{n} x_i^4 = \sum_{i=1}^{n} x_i^2 y_i$$

and subsequently lead to

$$\begin{bmatrix} \hat{\beta}_0 \\ \hat{\beta}_1 \\ \hat{\beta}_2 \end{bmatrix} = \begin{bmatrix} n & \sum_{i=1}^{n} x_i & \sum_{i=1}^{n} x_i^2 \\ \sum_{i=1}^{n} x_i & \sum_{i=1}^{n} x_i^2 & \sum_{i=1}^{n} x_i^3 \\ \sum_{i=1}^{n} x_i^2 & \sum_{i=1}^{n} x_i^3 & \sum_{i=1}^{n} x_i^4 \end{bmatrix}^{-1} \begin{bmatrix} \sum_{i=1}^{n} y_i \\ \sum_{i=1}^{n} x_i y_i \\ \sum_{i=1}^{n} x_i^2 y_i \end{bmatrix} \tag{1.33}$$

To carry out ANOVA to test for significance of the fitted model, we obtain the following:

$$SS_{\text{regression}} = S_{yy} - SS_{\text{E}} \tag{1.34}$$

$$SS_{\text{E}} = \sum_{i=1}^{n} y_i^2 - \begin{bmatrix} \hat{\beta}_0 \\ \hat{\beta}_1 \\ \hat{\beta}_2 \end{bmatrix}' \begin{bmatrix} \sum_{i=1}^{n} y_i \\ \sum_{i=1}^{n} x_i y_i \\ \sum_{i=1}^{n} x_i^2 y_i \end{bmatrix} \tag{1.35}$$

$$S_{yy} = \sum_{i=1}^{n} y_i^2 - \frac{\left(\sum_{i=1}^{n} y_i\right)^2}{n} \tag{1.36}$$

With this, we can carry out ANOVA to test for the significance of the model. The typical ANOVA table for fitting a parabola is shown in Table 1.6.

Example 1.22
While carrying out diagnostics in Example 1.19, we discovered that the residual plot indicated that a quadratic term might be needed in the data. Let us

TABLE 1.6: ANOVA for Typical Polynomial Regression (Fitting a Parabola)

Source	**Sum of squares**	**Degrees of freedom**	**Mean square**	***F*-ratio**
Regression	$S_{yy} - SS_{\text{E}}$	2	$SS_{\text{R}}/2$	$MS_{\text{R}}/MS_{\text{E}}$
Error	SS_{E}	$n-3$	$SS_{\text{E}}/(n-2)$	
Total	S_{yy}	$n-1$		

fit a parabola to the data set in Example 1.19 and test for the significance of the model.

Solution:
The model to be fitted is

$$y_i = \beta_0 + \beta_1 x_i + \beta_2 x_i^2 + \varepsilon_i \tag{1.37}$$

To fit this, we need all the quantities in the normal equations of Equation 1.33. We calculate the quantities in the table below.

	x	x^2	x^3	x^4	y	y^2	xy	x^2y
	1.5	2.25	3.375	5.0625	28	784	42	63
	2.5	6.25	15.625	39.0625	34.5	1190.25	86.25	215.625
	3	9	27	81	42.5	1806.25	127.5	382.5
	3	9	27	81	40.5	1640.25	121.5	364.5
	5	25	125	625	46	2116	230	1150
	7	49	343	2401	52	2704	364	2548
	8	64	512	4096	56	3136	448	3584
	9	81	729	6561	54.5	2970.25	490.5	4414.5
	10	100	1000	10000	55	3025	550	5500
	11	121	1331	14641	58	3364	638	7018
	12	144	1728	20736	62	3844	744	8928
	14	196	2744	38416	66	4356	924	12936
	17	289	4913	83521	72	5184	1224	20808
Total	103	1095.5	13498	181203.1	667	36120	5989.75	67912.13

From the above data, we see that

$$n = 13 \quad \sum_{i=1}^{13} y_i = 667 \quad \sum_{i=1}^{13} y_i^2 = 36120 \quad \sum_{i=1}^{13} x_i y_i = 5989.75$$

$$\sum_{i=1}^{13} x_i^2 y_i = 67912.13 \quad \sum_{i=1}^{13} x_i = 103 \quad \sum_{i=1}^{n} x_i^2 = 1095.5 \quad \sum_{i=1}^{n} x_i^3 = 13498$$

$$\sum_{i=1}^{n} x_i^4 = 181203.1$$

$$\begin{bmatrix} \hat{\beta}_0 \\ \hat{\beta}_1 \\ \hat{\beta}_2 \end{bmatrix} = \begin{bmatrix} 13 & 103 & 1095.5 \\ 103 & 1095.5 & 13498 \\ 1095.5 & 13498 & 181203.1 \end{bmatrix}^{-1} \begin{bmatrix} 667 \\ 5989.75 \\ 67912.13 \end{bmatrix}$$

$$= \begin{bmatrix} 0.774011 & -0.183959 & 0.0090238 \\ -0.183959 & 0.054830 & -0.0029722 \\ 0.009024 & -0.002972 & 0.0001724 \end{bmatrix} \begin{bmatrix} 667 \\ 5989.75 \\ 67912.13 \end{bmatrix} = \begin{bmatrix} 27.2261 \\ 3.8703 \\ -0.0781 \end{bmatrix}$$

The fitted model is

$$\hat{y} = 27.2261 + 3.8703x - 0.0781x^2$$

TABLE 1.7: ANOVA for the Fitted Parabola

Source	Sum of squares	Degrees of freedom	Mean square	F-ratio	p-Value
Regression	1814.4396	2	907.2198	108.882	1.63145×10^{-7}
Residual	83.3214	10	8.33214		
Total	1879.769	12			

Next, we test for the significance of the regression model. To do this, we need to obtain the quantities S_{yy}, $S_{\text{regression}}$, and SS_{E}. They are obtained as follows (Table 1.7):

$$S_{yy} = \sum_{i=1}^{13} y_i^2 - \frac{\left(\sum_{i=1}^{13} y_i\right)^2}{13} = 36120 - \frac{667^2}{13} = 1879.769$$

$$SS_{\text{E}} = \sum_{i=1}^{13} y_i^2 - \begin{bmatrix} \hat{\beta}_0 & \hat{\beta}_1 & \hat{\beta}_2 \end{bmatrix} \begin{bmatrix} \sum_{i=1}^{13} y_i \\ \sum_{i=1}^{13} x_i y_i \\ \sum_{i=1}^{13} x_i^2 y_i \end{bmatrix}$$

$$= 36120 - \begin{bmatrix} 27.2261 & 3.8703 & -0.0781 \end{bmatrix} \begin{bmatrix} 667 \\ 5989.75 \\ 67912.13 \end{bmatrix} = 83.3214$$

$$SS_{\text{regression}} = S_{yy} - SS_{\text{E}} = 1897.761 - 83.3214 = 1814.4396$$

From the tables, $F(2, 10, 0.01) = 7.56$. This implies that the fitted model is significant. However, we are more interested in finding out whether the added simple polynomial is significant. If we want to find out whether the simple polynomial x^2 which we fitted is significant, we can use the results we obtained by fitting the linear model and so obtain the extra sum of squares contributed by this simple polynomial.

Source	Sum of squares	Degrees of freedom	Mean square	F-ratio	p-Value
x	1779.045	1	1779.045	213.5159	4.49589×10^{-8}
$x^2 \mid x$	35.3946	1	35.3946	4.248	0.066268539
Residual	83.3214	10	8.33214		
Total	1879.769	12			

We see that $F(1, 10, 0.10) = 3.29$ and $F(1, 10, 0.05) = 4.96$, so we can say that the model is only significant at 10% based on our limited tables for the

F-distribution. Actually, the p-value for the addition of $x^2|x$ is 0.066269. We leave the decision to the reader as it depends on the size of type I error that the researcher is willing to allow. □

1.2.6 Fitting Higher Order Polynomials

To fit higher order polynomials using natural polynomials, all we need is to extend the results for the parabola so that the model to be fitted corresponds to model 1.30. In that case, the results of minimization and solving of the least squares equations are the following normal equations:

$$\begin{aligned}
&n\beta_0 + \beta_1\sum_{i=1}^{n} x_i + \cdots + \beta_k\sum_{i=1}^{n} x_i^k = \sum_{i=1}^{n} y_i \\
&\beta_0\sum_{i=1}^{n} x_i + \beta_1\sum_{i=1}^{n} x_i^2 + \cdots + \beta_k\sum_{i=1}^{n} x_i^{k+1} = \sum_{i=1}^{n} x_i y_i \\
&\vdots \\
&\beta_0\sum_{i=1}^{n} x_i^k + \beta_1\sum_{i=1}^{n} x_i^{k+1} + \cdots + \beta_2\sum_{i=1}^{n} x_i^{2k} = \sum_{i=1}^{n} x_i^k y_i
\end{aligned}$$

Estimates of the parameters are obtained by solving the following equations:

$$\begin{bmatrix} \hat{\beta}_0 \\ \hat{\beta}_1 \\ \vdots \\ \hat{\beta}_k \end{bmatrix} = \begin{bmatrix} n & \sum_{i=1}^{n} x_i & \cdots & \sum_{i=1}^{n} x_i^k \\ \sum_{i=1}^{n} x_i & \sum_{i=1}^{n} x_i^2 & \cdots & \sum_{i=1}^{n} x_i^{k+1} \\ \vdots & \vdots & \cdots & \vdots \\ \sum_{i=1}^{n} x_i^k & \sum_{i=1}^{n} x_i^{k+1} & \cdots & \sum_{i=1}^{n} x_i^{2k} \end{bmatrix}^{-1} \begin{bmatrix} \sum_{i=1}^{n} y_i \\ \sum_{i=1}^{n} x_i y_i \\ \vdots \\ \sum_{i=1}^{n} x_i^k y_i \end{bmatrix} \tag{1.38}$$

$$SS_{\text{regression}} = S_{yy} - SS_{\text{E}} \tag{1.39}$$

$$SS_{\text{E}} = \sum_{i=1}^{n} y_i^2 - \begin{bmatrix} \hat{\beta}_0 \\ \hat{\beta}_1 \\ \vdots \\ \hat{\beta}_k \end{bmatrix}^T \begin{bmatrix} \sum_{i=1}^{n} y_i \\ \sum_{i=1}^{n} x_i y_i \\ \vdots \\ \sum_{i=1}^{n} x_i^k y_i \end{bmatrix} \tag{1.40}$$

$$S_{yy} = \sum_{i=1}^{n} y_i^2 - \frac{\left(\sum_{i=1}^{n} y_i\right)^2}{n} \tag{1.41}$$

The above equations will enable us to fit polynomials of order $k \geq 3$ to the responses when required. The method used for fitting the parabola should be followed.

1.2.7 Orthogonal Polynomials

The methods we discussed in the previous sections relate to how to fit natural polynomials to response data. When we fit natural polynomials to responses of an experiment, we may run into the problem of collinearity. Collinearity arises in a situation where there is a close relationship between the predictors or the predictors are highly correlated. Collinearity leads to inaccurate estimates of the model parameters and therefore should be avoided. A more in-depth treatment of collinearity—what it is and how to deal with it—can be found in Chapter 14 of Kleinbaum et al. (2008). The problem of collinearity involves natural polynomial regression terms, and they mostly occur as scaling problems. One way of combating collinearity in fitting polynomial models is by using the orthogonal polynomials to fit the desired kth-order polynomial to the data. Orthogonal polynomials are coded forms of the simple polynomials $x, x^2, x^3 \ldots$. The use of the orthogonal polynomials enables us to achieve the following two objectives:

1. To maintain exactly the same information that is in the simple polynomials and
2. To create variables that are uncorrelated with one another

Thus, when we use orthogonal polynomials to fit the desired model to the data, we eliminate the occurrence of collinearity. Also, any model based on orthogonal polynomials will provide the same information that we sought to elicit from the data by fitting the model with simple polynomials. A typical natural polynomial model of order k is written as

$$y = \beta_0 + \beta_1 x + \beta_2 x^2 + \cdots + \beta_k x^k$$

The simple polynomials used in the above polynomial of order k are $x, x^2, x^3, \ldots, x^k$. Orthogonal polynomials were obtained as linear combinations of these simple polynomials. If we denote orthogonal polynomials as

$z_1, z_2, \ldots, z_k$ then

$$\begin{aligned} z_1 &= c_{01} + c_{11}x \\ z_2 &= c_{02} + c_{12}x + c_{22}x^2 \\ &\vdots \\ z_k &= c_{0k} + c_{1k}x + \cdots + c_{kk}x^k \end{aligned}$$

Thus, each of the zs is a polynomial made up of simple polynomials. Conversely, it is also possible to convert a simple polynomial into a linear combination of the orthogonal polynomials, so that

$$\begin{aligned} x &= \alpha_{01} + \alpha_{11}z_1 \\ x^2 &= \alpha_{02} + \alpha_{12}z_1 + \alpha_{22}z_2 \\ &\vdots \\ x^k &= \alpha_{0k} + \alpha_{1k}z_1 + \cdots + \alpha_{kk}z_k \end{aligned}$$

Thus, no information is lost if we write the model as either

$$y = \beta_0 + \beta_1 x + \beta_2 x^2 + \cdots + \beta_k x^k + \varepsilon$$

or

$$y = \alpha_0 + \alpha_1 z_1 + \alpha_2 z_2 + \cdots + \alpha_k z_k + \varepsilon$$

It should be noted that partial F-test for $H_0 : \alpha_i = 0$ under the orthogonal polynomial model is equivalent to the partial F-test for $H_0 : \beta_i = 0$ for the natural polynomial model. We can see that the orthogonal polynomial model is useful in eliminating some of the undesirable features of natural polynomial model, especially the propensity for highly correlated relationships that leads to collinearity.

1.2.7.1 Fitting Orthogonal Polynomials

In Appendix A10, we present orthogonal polynomial coefficients. These coefficients can be used to set up an orthogonal polynomial model if and only if the levels of the predictor variable x are equally spaced for the data and an equal number of observations y were made for each value of x. If the data do not conform to these restrictions, then we need to generate the relevant orthogonal polynomial. One way of generating such orthogonal polynomials is to use the ORPOL statement within PROC IML syntax in SAS. The generated orthogonal polynomials are used as the predictor variables in what essentially becomes like the fitting of a multiple regression model. With orthogonal polynomials, we are able to fit a polynomial of order $k-1$ to a data set that contains k measurements of the responses for k settings of the predictor variable. Thus, usually $k-1$ sets of orthogonal polynomials would be defined for the purpose of fitting polynomials of order up to $k-1$. We can obviously fit polynomial of order less than $k-1$; in that case, we use the number of orthogonal polynomials we need, utilizing the theories of multiple regression discussed previously to fit the desired polynomial. We illustrate with the following example:

Example 1.23

The yields of a chemical process are dependent on the percentage of reactant added. For such an experiment, reactant concentrations and the yields (x and y) are given below.

x	y	x	y
1.5	14.9	1.5	14.81
2.5	15.08	2.5	14.99
3.5	14.6	3.5	14.75
4.5	14.49	4.5	14.32
5.5	14.53	5.5	14.45
6.5	14.59	6.5	14.64

Use orthogonal polynomials to fit a quadratic relationship between reactant concentrations and yields, and test for significance of the fitted model.

Solution:

Since there were two measurements at each value of x, the data qualify for the use of the orthogonal polynomials listed in the table in Appendix A10. There are six settings of reactant concentrations; therefore, the highest order of the polynomial that can be fitted is of order five. So from the tables, we provide all the orthogonal polynomials (z_1, z_2, z_3, z_4, z_5) and show how they are used to replace the original values of x, and then proceed to fit a polynomial of order three. The relevant orthogonal coefficients for fitting up to the fifth-order polynomial are given in Table 1.8.

TABLE 1.8: Orthogonal Contrast Values for Example 1.23

z_1	z_2	z_3	z_4	z_5	y
−5	5	−5	1	−1	14.9
−3	−1	7	−3	5	15.08
−1	−4	4	2	−10	14.6
1	−4	−4	2	10	14.49
3	−1	−7	−3	−5	14.53
5	5	5	1	1	14.59
−5	5	−5	1	−1	14.81
−3	−1	7	−3	5	14.99
−1	−4	4	2	−10	14.75
1	−4	−4	2	10	14.32
3	−1	−7	−3	−5	14.45
5	5	5	1	1	14.64

TABLE 1.9: Functions for Fitting Polynomial (Example 1.23) by Orthogonal Polynomial Method

z_1	z_2	z_3	z_1^2	z_2^2	z_3^2	z_1z_2	z_1z_3	z_2z_3	y	z_1y	z_2y	z_3y
−5	5	−5	25	25	25	−25	25	−25	14.9	−74.5	74.5	−74.5
−3	−1	7	9	1	49	3	−21	−7	15.08	−45.24	−15.08	105.56
−1	−4	4	1	16	16	4	−4	−16	14.6	−14.6	−58.4	58.4
1	−4	−4	1	16	16	−4	−4	16	14.49	14.49	−57.96	−57.96
3	−1	−7	9	1	49	−3	−21	7	14.53	43.59	−14.53	−101.71
5	5	5	25	25	25	25	25	25	14.59	72.95	72.95	72.95
−5	5	−5	25	25	25	−25	25	−25	14.81	−74.05	74.05	−74.05
−3	−1	7	9	1	49	3	−21	−7	14.99	−44.97	−14.99	104.93
−1	−4	4	1	16	16	4	−4	−16	14.75	−14.75	−59	59
1	−4	−4	1	16	16	−4	−4	16	14.32	14.32	−57.28	−57.28
3	−1	−7	9	1	49	−3	−21	7	14.45	43.35	−14.45	−101.15
5	5	5	25	25	25	25	25	25	14.64	73.2	73.2	73.2
0	0	0	140	168	360	0	0	0	176.15	−6.21	3.01	7.39

To fit the third-order polynomial, we use only z_1, z_2, z_3 as shown in Table 1.9.

Using the results of Equations 1.38 through 1.41, we see that the third-order polynomial model would be of the form

$$y = \alpha_0 + \alpha_1 z_1 + \alpha_2 z_2 + \alpha_3 z_3 + \varepsilon \tag{1.42}$$

$$\begin{aligned}
\begin{bmatrix} \hat{\alpha}_0 \\ \hat{\alpha}_1 \\ \hat{\alpha}_2 \\ \hat{\alpha}_3 \end{bmatrix} &= \begin{bmatrix} n & \sum z_1 & \sum z_2 & \sum z_3 \\ \sum z_1 & \sum z_1^2 & \sum z_1 z_2 & \sum z z_3 \\ \sum z_2 & \sum z_1 z_2 & \sum z_2^2 & \sum z_2 z_3 \\ \sum z_3 & \sum z_1 z_3 & \sum z_2 z_3 & \sum z_3^2 \end{bmatrix}^{-1} \begin{bmatrix} \sum y \\ \sum z_1 y \\ \sum z_2 y \\ \sum z_3 y \end{bmatrix} \\
&= \begin{bmatrix} 12 & 0 & 0 & 0 \\ 0 & 140 & 0 & 0 \\ 0 & 0 & 168 & 0 \\ 0 & 0 & 0 & 360 \end{bmatrix}^{-1} \begin{bmatrix} 176.15 \\ -6.21 \\ 3.01 \\ 7.39 \end{bmatrix} \\
&= \begin{bmatrix} 0.166667 & 0.0000000 & 0.0000000 & 0.0000000 \\ 0.000000 & 0.0071429 & 0.0000000 & 0.0000000 \\ 0.000000 & 0.0000000 & 0.0059524 & 0.0000000 \\ 0.000000 & 0.0000000 & 0.0000000 & 0.0027778 \end{bmatrix} \begin{bmatrix} 176.15 \\ -6.21 \\ 3.01 \\ 7.39 \end{bmatrix} \\
&= \begin{bmatrix} 14.6792 \\ -0.0444 \\ 0.0179 \\ 0.0205 \end{bmatrix}
\end{aligned}$$

Thus the fitted model is

$$y = 14.6792 - 0.0444 z_1 + 0.0179 z_2 + 0.0205 z_3 \tag{1.43}$$

This can be written in terms of the natural polynomials as

$$y = 14.6792 - 0.0444x + 0.0179x^2 + 0.0205x^3 \tag{1.44}$$

To perform ANOVA and test for significance of the model, we first obtain

$$S_{yy} = \sum_{i=1}^{12} y_i^2 - \frac{\left(\sum_{i=1}^{12} y_i\right)^2}{12} = 2586.319 - \frac{(176.15)^2}{12} = 0.583492$$

$$SS_{\text{E}} = \sum_{i=1}^{12} y_i^2 - \begin{bmatrix} \hat{\alpha}_0 & \hat{\alpha}_1 & \hat{\alpha}_2 & \hat{\alpha}_3 \end{bmatrix} \begin{bmatrix} \sum y \\ \sum z_1 y \\ \sum z_2 y \\ \sum z_3 y \end{bmatrix}$$

$$= 2586.319 - \begin{bmatrix} 14.6792 & -0.444 & 0.179 & 0.0205 \end{bmatrix} \begin{bmatrix} 176.15 \\ -6.21 \\ 3.01 \\ 7.39 \end{bmatrix} = 0.1027$$

$$SS_{\text{regression}} = S_{yy} - SS_{\text{E}} = 0.583492 - 0.1027 = 0.480792$$

The ANOVA for the fitted model is presented in Table 1.10.

TABLE 1.10: Testing for Significance of Fitted Model by ANOVA

Source	**Sum of squares**	**Degrees of freedom**	**Mean square**	***F*-ratio**	***p*-Value**
Regression	0.480792	3	0.160264	12.48404	0.002189
Error	0.1027	8	0.012838		
Total	0.583492	11			

From Table 1.9 for the F-distribution, $F(3, 8, 0.05) = 4.06618$. Thus, the fitted model is highly significant at 5% level of significance. □

1.2.7.2 SAS Analysis of the Data of Example 1.4

Using the orthogonal polynomials from Table 1.10 we can write a SAS program to analyze the above data. There are two ways of writing the SAS program: one can enter the orthogonal contrasts as the predictors and use the REG procedure to fit the polynomial of order three or one can define the orthogonal contrasts in a CONTRAST statement in the GLM procedure in SAS to achieve the same purpose. We present both programs.

SAS PROGRAM

```
data poly1;
input x1-x5 y @@;
cards;
```

```
-5  5  -5  1  -1  14.90  -3  -1  7  -3  5  15.08
-1  -4  4  2  -10  14.6  1  -4  -4  2  10  14.49
3  -1  -7  -3  -5  14.53  5  5  5  1  1  14.59
-5  5  -5  1  -1  14.81  -3  -1 7  -3 5  14.99
-1  -4  4  2  -10  14.75  1  -4  -4  2  10  14.32
3  -1  -7  -3  -5  14.45  5  5  5  1  1  14.64
;
proc reg data=poly1;
model y=x1-x3; run;
```

SAS OUTPUT

```
                          The REG Procedure
                           Model: MODEL1
                       Dependent Variable: y

              Number of Observations Read          12
              Number of Observations Used          12
                        Analysis of Variance

                                   Sum of        Mean
Source                   DF       Squares      Square    F Value    Pr > F

Model                     3       0.48109     0.16036      12.53    0.0022
Error                     8       0.10240     0.01280
Corrected Total          11       0.58349

          Root MSE                0.11314    R-Square     0.8245
          Dependent Mean         14.67917    Adj R-Sq     0.7587
          Coeff Var               0.77075
                       Parameter Estimates

                             Parameter       Standard
   Variable       DF          Estimate          Error    t Value    Pr > |t|

   Intercept       1          14.67917        0.03266     449.45      <.0001
   x1              1          -0.04436        0.00956      -4.64      0.0017
   x2              1           0.01792        0.00873       2.05      0.0742
   x3              1           0.02053        0.00596       3.44      0.0088
```

Another SAS program that would produce the above result is given below:

```
data poly2;
input z y @@;
cards;
1.5 14.90 2.5 15.08 3.5 14.6 4.5 14.49 5.5 14.53 6.5 14.59
1.5 14.81 2.5 14.99 3.5 14.75 4.5 14.32 5.5 14.45 6.5 14.64
;
```

```
proc glm data=poly2;
class z;
model y=z;
contrast 'linear'  z-5 -3 -1 1 3 5;
contrast 'quadratic' z  5 -1 -4 -4 -1 5;
contrast 'cubic' z -5 7 4 -4 -7 5;
/*contrast 'quartic' z 1 -3 2 2 -3 1;
contrast 'quintic' z -1 5 -10 10 -5 1*/;run;
```

We have commented out the contrasts for quartic and quintic so that we can have SAS fit only a polynomial model of order three. To fit a full polynomial model of order five, just remove the comment, that is (/* and */), then SAS will return output for fitting a fifth-order polynomial.

1.2.8 Use of Dummy Variables in Regression Models

In the regression analyses, we have encountered hitherto, we have been dealing with only continuous response variables and continuous predictor variables. Sometimes, we need to incorporate predictor variables which are categorical, in our regression analysis. For instance, some survey questions are designed to elicit only categorical responses such as "yes" or "no"; "male" or "female"; "republican," "democrat," or "independent." To account for such predictor variables in a regression analysis, we need to find a way to include them in the model. This is achieved by defining an "indicator" or "dummy" variable that assigns a value of a nominal variable to each of the categories of a categorical variable. Thus, when a dummy variable is defined, values such as (0, 1, −1) are assigned to a categorical variable to indicate its different categories. Typical examples of definitions of dummy variables are

$$X_1 = \begin{cases} 0 & \text{if gender is male} \\ 1 & \text{if gender is female} \end{cases}$$

$$X_2 = \begin{cases} 1 & \text{if party affiliation is republican} \\ -1 & \text{if party affiliation is democratic} \\ 0 & \text{if party affiliation is independent} \end{cases}$$

$$X_3 = \begin{cases} 1 & \text{if response is ``yes''} \\ 0 & \text{if response is ``no''} \end{cases}$$

With these types of dummy variables, we can set up regression models that contain both continuous and categorical variables. We demonstrate this in the next example.

Example 1.24
The data below shows the ages and weights of 10 females and 18 males.

Gender	Age	Weight	Gender	Age	Weight
F	23	133	M	19	146
M	41	186	M	33	154
M	57	176	F	57	116
M	40	167	F	45	144
F	57	134	M	21	150
M	45	187	M	20	142
F	19	126	F	42	156
M	21	120	F	30	136
M	47	176	M	20	124
F	57	156	M	20	146
F	42	132	M	34	150
M	21	142	M	34	170
M	19	140	M	34	173
M	45	180	F	58	160

Our aim is to obtain a regression equation with age and gender as predictors of weight. First, let gender be X_1, age be X_2, and weight be Y. We represent X_1 as a dummy variable as

$$X_1 = \begin{cases} 1 & \text{if gender is male} \\ 0 & \text{if gender is female} \end{cases}$$

The regression model required is (Equation 1.45)

$$y = \beta_0 + \beta_1 x_1 + \beta_2 x_2 + \varepsilon \tag{1.45}$$

Table 1.11 below contains the values of the functions of the response and predictor variables needed to fit the model described in Equation 1.45 to the data of Example 1.24.

TABLE 1.11: Functions for Fitting a Model Using Dummy Variables (Example 1.25)

x_1	x_2	y	x_1^2	x_2^2	x_1x_2	x_1y	x_2y
0	23	133	0	529	0	0	3059
1	41	186	1	1681	41	186	7626
1	57	176	1	3249	57	176	10032
1	40	167	1	1600	40	167	6680
0	57	134	0	3249	0	0	7638
1	45	187	1	2025	45	187	8415
0	19	126	0	361	0	0	2394
1	21	120	1	441	21	120	2520

Continued

TABLE 1.11: Continued

x_1	x_2	y	x_1^2	x_2^2	x_1x_2	x_1y	x_2y
1	47	176	1	2209	47	176	8272
0	57	156	0	3249	0	0	8892
0	42	132	0	1764	0	0	5544
1	21	142	1	441	21	142	2982
1	19	140	1	361	19	140	2660
1	45	180	1	2025	45	180	8100
1	19	146	1	361	19	146	2774
1	33	154	1	1089	33	154	5082
0	57	116	0	3249	0	0	6612
0	45	144	0	2025	0	0	6480
1	21	150	1	441	21	150	3150
1	20	142	1	400	20	142	2840
0	42	156	0	1764	0	0	6552
0	30	136	0	900	0	0	4080
1	20	124	1	400	20	124	2480
1	20	146	1	400	20	146	2920
1	34	150	1	1156	34	150	5100
1	34	170	1	1156	34	170	5780
1	34	173	1	1156	34	173	5882
0	58	160	0	3364	0	0	9280
18	1001	4222	18	41045	571	2829	153826

$$\begin{pmatrix}\hat{\beta}_0\\ \hat{\beta}_1\\ \hat{\beta}_1\end{pmatrix} = \begin{bmatrix} n & \sum x_1 & \sum x_2 \\ \sum x_1 & \sum x_1^2 & \sum x_1x_2 \\ \sum x_2 & \sum x_1x_2 & \sum x_2^2 \end{bmatrix}^{-1} \begin{bmatrix}\sum y\\ \sum x_1y\\ \sum x_2y\end{bmatrix}$$

$$= \begin{bmatrix} 28 & 18 & 1001 \\ 18 & 18 & 571 \\ 1001 & 571 & 41045 \end{bmatrix}^{-1} \begin{bmatrix} 4222 \\ 2829 \\ 153826 \end{bmatrix}$$

$$= \begin{bmatrix} 0.516290 & -0.209182 & -0.0096812 \\ -0.209182 & 0.184191 & 0.0025391 \\ -0.009681 & 0.002539 & 0.0002251 \end{bmatrix} \begin{bmatrix} 4222 \\ 2829 \\ 153826 \end{bmatrix} = \begin{bmatrix} 98.7859 \\ 28.4925 \\ 0.9422 \end{bmatrix}$$

The fitted model is

$$\hat{y} = 98.7859 + 28.4925x_1 + 0.9422x_2$$

Because $x_1 = 1$ when the subject is male, the fitted model for males is

$$\hat{y} = 127.2784 + 0.9422x_2 \tag{1.46}$$

When the subject is female, $x_1 = 0$, so that

$$\hat{y} = 98.7859 + 0.9422x_2 \tag{1.47}$$

TABLE 1.12: ANOVA for Dummy Variable Regression of Example 1.25

Source	Sum of squares	Degrees of freedom	Mean square	*F*-ratio	*p*-Value
Regression	2	5995.015	2997.507	15.04864	5.13007×10^{-5}
Residual	25	4979.7	199.188		
Total	27	10974.71			

We can carry out ANOVA for the above regression as follows (Table 1.12):

$$SS_{\mathrm{E}} = 647592 - \begin{bmatrix} 98.7859 & 28.49246 & 0.942188 \end{bmatrix} \begin{bmatrix} 4222 \\ 2829 \\ 153826 \end{bmatrix} = 4979.75$$

$$SS_{\mathrm{regression}} = S_{yy} - SS_{\mathrm{E}} = 10974.71 - 4979.75 = 5994.965$$

$$S_{yy} = 647592 - \frac{4222^2}{28} = 10974.71$$

There are other ways of defining dummy variables than the ones we have mentioned and used so far. We could have achieved the same results by defining the dummy variable differently. One of the ways we could redefine the dummy variable for gender is

$$X_1 = \begin{cases} 1 & \text{if gender is male} \\ -1 & \text{if gender is female} \end{cases}$$

In this case, we obtain

$$\hat{y} = 113.0321 + 14.24623x_1 + 0.942188x_2 \tag{1.48}$$

Because $x_1 = 1$ when the subject is male, the fitted model for males is

$$\hat{y} = 127.2784 + 0.9422x_2$$

When the subject is female, $x_1 = -1$, so that

$$\hat{y} = 98.7859 + 0.9422x_2$$

In other words, the definition of the dummy variable for gender in these two different ways made no difference to the actual fitted models for the relationship between weights of the males and their ages, as well as weights of females and ages.

Dummy variables are used very extensively in regression models. For instance, dummy variables are largely behind the regression approaches to ANOVA in modeling and analyzing the relationships between explanatory variables and responses in experimental designs. The above definitions and examples do not reflect all the variety in their definitions and uses in regression models. Such models will be used in our consideration of regression approaches to ANOVA for the different experimental designs in this book. We will discuss and define the relevant indicator variables to be used in appropriate sections in the book. □

1.3 Exercises

Exercise 1.1

The number of routine tasks (y) that a machine can accomplish in a time t appears to be linearly dependent on the length of t. The following data show the number of tasks performed by the machine for given values of t. Obtain a scatterplot of data and linear relationship between y and t. Test for significance of regression.

t	2.9	1.9	2.3	3.0	3.6	3.5	3.4	3.7	2.9	2.8	2.8	2.6	2.7	3.0	2.1
y	21	15	19	23	24	28	29	22	23	20	19	17	21	21	16

□

Exercise 1.2

The following data show the systolic blood pressures (SBP) of 12 men as well as their weights.

Weight (lb)	185	198	196	188	203	195	193	198	190	194	189	200
SBP (mmHg)	133	144	142	135	150	140	141	143	138	139	136	147

Obtain by least squares regression the linear relationship between the weights and the SBP and test for the parameters of the regression equation. Use ANOVA to test whether the fitted model is significant. Obtain the estimated values of SBP and hence, using the residuals, check that the assumptions underlying linear regression are satisfied. □

Exercise 1.3

The yield y of a chemical process is thought to depend on the catalyst concentration (x_1) and duration of reaction (x_2). Fifteen runs of the process under different combinations of catalyst concentrations and reaction times yielded the following data.

x_1	x_2	y	x_1	x_2	y
13	19	3.2	9	18	3.6
8	14	2.4	7	12	2.8
10	10	2.8	14	19	2.7
12	19	2.6	16	16	3.0
14	18	3.6	12	11	2.7
11	13	2.9	15	15	3.0
15	13	3.0	16	19	3.3
13	17	3.4			

By multiple regression, obtain the relationship among x_1, x_2, and the yield y. □

Exercise 1.4

It is believed that there is a linear relationship between the scores of students in the final year in statistical theory and their scores in a prerequisite course. Obtain the least square regression line that represents this relationship. Test for the significance of the parameters of the regression line.

Prerequisite	82	71	86	79	61	80	67	58	86	64	71	80	90	79	55	93
Statistical theory	76	77	88	76	68	81	68	61	89	71	66	76	91	77	59	94

□

Exercise 1.5

Consider the following data:

x_1	x_2	x_3	x_4	y	x_1	x_2	x_3	x_4	y
10	5	7	4	50	0	4	15	5	43
9	7	8	3	48	9	3	3	4	36
6	10	12	4	45	4	6	10	5	34
11	4	6	3	51	8	14	9	3	63
7	4	5	6	68	6	5	7	6	44
14	6	4	5	59	14	4	5	4	58
20	12	12	7	91	8	7	4	3	45
7	9	15	4	60	20	6	6	5	81
2	12	3	5	68	22	12	12	8	99

Fit a multiple regression model of the form $y = \beta_1 x_1 + \beta_2 x_2 + \beta_3 x_3 + \beta_4 x_4 + \varepsilon$ to the data. Test the hypothesis $H_0 : \beta_i = 0$. □

Exercise 1.6

The following data are the scores of two groups of students of the same class in statistics. Assume that both sets of scores are from populations that follow normal distributions with equal variance. Test the hypothesis that the means are the same. Use the level of significance $\alpha = 0.05$.

A	B	A	B
63	55	66	43
68	47	61	93
64	44	63	60
57	77	69	66
60	70	65	48
64	52	72	51
69	64	70	56
56	63	58	63
65	61	55	67
59	74	68	70

□

Exercise 1.7

A new smart card is being studied for placement into a machine to improve the time it takes to perform various tasks. The following data show the time (in minutes) taken by the machine to perform a random sample of tasks before and after the placement of the smart card. Does the card improve the performance of the machine?

Test at $\alpha = 0.01$.

Before	After	Before	After
27	20	27	25
26	25	38	33
25	23	34	33
30	30	34	36
37	29	40	38
30	29	38	35
28	30	21	25
26	29		

□

Exercise 1.8

In a physiotherapy study, two instruments (a Goniometer and a tape) were used in turn to measure the elbow flexions of 12 manikins (in degrees). Can the two methods be said to turn out the same values on average or is the tape measurement lower on average? Use $\alpha = 0.05$.

Tape	Goniometer	Tape	Goniometer
47.6	46.9	45.3	46.6
47.4	50.1	44.3	46.8
45.5	48.3	48.1	50.2
44.1	46.5	47.8	46.8
44.3	46.1	47.3	47.9
45.9	45.5	46.4	48.2

□

Exercise 1.9

In a clothing factory, a therapist was brought in to design and provide suitable seats for 10 machinists in a trial evaluation to provide for better ergonomics and productivity. Average weekly production of a particular dress by these machinists before and after changes of seats is shown below.

Machinist	On old seats	On new seats
1	59	65
2	63	70
3	51	48
5	57	53
6	66	69
7	50	58
8	52	60
9	53	54
10	62	65

Have the new seats improved the productivity of the machinists? Test at 5% level of significance. □

Exercise 1.10

A new diet is believed to reduce the weights of its adherents. The weights of 12 women (in lb) before and after they had been on the diet for 1 week are shown below.

Weight before	Weight after
130	128
145	142
134	136
165	158
148	147
137	137
178	168
126	121
124	122
144	140
135	130
123	123

Is the diet effective for reducing weights? Use $\alpha = 0.01$. □

Exercise 1.11

An experiment was performed to investigate whether there is deterioration in the performance of a car if it uses unleaded fuel rather than leaded fuel. Ten cars filled with leaded fuel were run on the same route using the same driver, and the performance of each car in miles per gallon was noted. The same cars were filled with unleaded fuel and run on the same route by the same driver. The results of the experiment are presented as follows:

Cars	1	2	3	4	5	6	7	8	9	10
Leaded	37.6	39.4	39.5	37.1	35.3	35.9	36.3	38.4	37.3	39.1
Unleaded	36.8	40.0	38.2	36.4	36.0	35.4	36.5	38.0	36.4	40.1

Compare the performances at $\alpha = 0.05$ by (1) classical method and (2) p-value approach if you have access to computer software (EXCEL, MINITAB) to obtain the p-value. ▯

Exercise 1.12

Two machines used in manufacturing textiles were set to perform a task in 16 min. The performance times of nine such tasks were recorded for each machine as follows:

Machine 1	16.03	16.04	16.05	16.05	16.02	16.01	15.96	15.98	16.02
Machine 2	16.02	15.96	16.01	15.99	16.03	16.04	16.02	16.01	16.00

Can the machines be said to perform the tasks in significantly different mean time from each other? Use $\alpha = 0.05$ and assume that $\sigma_1^2 = \sigma_2^2$. Carry out analysis by (1) classical method and (2) p-value approach if you have access to computer software (EXCEL, MINITAB) to obtain the p-value. ▯

Exercise 1.13

A shop that stocks ladies/gents clothing is about to open a new wing and is carrying out a small study to make the most efficient use of space; it has decided that it will predominantly stock the new wing with ladies' or gents' clothes, whichever is proved to attract more customers. Sales figures in thousands of dollars of gents/ladies clothes per week in the shop in 14 weeks chosen at random are as follows:

Ladies	117	106	98	131	119	94	89	105	108	112	96	121	108	89
Gents	109	94	97	86	105	87	111	91	98	104	98	96	113	76

Which of the two types of clothes would you suggest should predominantly be stocked in the new wing? Use $\alpha = 0.05$. ▯

Exercise 1.14

Use p-value approach to make decisions about the hypothesis tested in Examples 1.13 and 1.14. ▯

References

Bartlett, M.S. (1936) The information available in small samples. *Proceedings of Cambridge Philosophical Society*, 32, 560.

Bartlett, M.S. (1939) Complete simultaneous fiducial distribution. *Annals of Mathematical Statistics*, 10, 129.

Behrens, W.U. (1929) Ein Beitrag zur Fehlen-Berechnung bei wenigen Beobachtungen. *Landw. Jb*, 68, 807–837.

Cochran, W.G. and Cox, G.M. (1950) *Experimental Designs*, J Wiley, New York.

Dixon, W.J. and Massey, F.J. (1951) *Introduction to Statistical Analysis*, McGraw-Hill, New York.

Fisher, R.A. (1935) The fiducial argument statistical inference. *Annals Eugenic*, 6, 391–398.

Hogg, R.V. and Craig, T.C. (1978) *Introduction to Mathematical Statistics*, 4th ed., Collier Macmillan International Inc., New York.

Kleinbaum, D.G., Kupper, L.L., Nizam, A., and Muller, K.E. (2008) *Applied Regression Analysis and Other Multivariable Methods*, 4th ed., Brooks/Cole, Belmont, CA.

Satterthwaite, F.E. (1946) An approximate distribution of estimates of variance components. *Biometrics Bulletin*, 2, 110–112.

Smith, H.F. (1936) The problem of comparing the results of two experiments with unequal errors. *Journal of Scientific and Industrial Research*, 9, 211–212.

Spiegel, M.R. (1974) *Schaum Outline of Theory and Problems of Probability and Statistics. Schaum Outlines Series*, McGraw-Hill, New York.

Sukhatme, P.V. (1938) On the Fisher-Behrens test of significance for the difference in means of two normal samples. *Sankya*, 4, 39.

Wackerly, D.D., Mendenhall III, W., and Scheaffer, R.L. (2002) *Mathematical Statistics with Applications*, 6th ed., Duxbury, Belmont, CA.

Welch, B.L. (1938) The significance of the difference between two means when the population variances are unequal. *Biometrika*, 39, 350–361.

Chapter 2

Experiments, the Completely Randomized Design—Classical and Regression Approaches

2.1 Experiments

What is an experiment? What are the requirements for a good experiment? These are appropriate questions to ask at the beginning of this study of "Design and Analysis of Experiments." Answers to these crucial questions will help us understand how to perform good experiments. To perform experiments to study a process, we need the right combinations of tools (experimental and analysis tools). If we use the right tools, our methods will be right, and our analysis will be accurate, leading us to make valid deductions. The use of tools, which we have described in this chapter and intend to discuss in the rest of the text, should enable us to be more informed about the processes we investigate.

Investigators in different fields carry out studies which generally have the objective of either discovering some new phenomena or finding out the effect of imposition of certain conditions on some processes. Studies in which the investigator controls some or all the conditions that are imposed on the study units are called experiments. For instance, a microbiologist could study the growth of algae in a given medium. The study may focus on the effect of varying concentrations of a particular substance in the medium on the growth of the algae under certain conditions. An alternative focus could be on studying factors (internal or external to the medium) that, if varied, are likely to influence the growth of algae. Another possible investigation may involve a case where there are, say, two media in which the algae could be grown and the focus of the study is on the differential patterns of growth of the algae which may be observed for the two media.

The advent of computers has simplified the analysis as well as presentation of experimental results to a large extent. Consequently, virtually all experimental results in the sciences are presented and summarized in some statistical form or another. While this is good, problems could arise concerning how the data being summarized were collected. If the experiment that yielded the data was poorly designed, the conclusions of the analysis would be unreliable.

Sometimes nonstatisticians request help with the analysis of data obtained from poorly designed experiments. Most often, the statistician discovers that due to the poor designs of the experiments that led to the data, it is impossible to use proper methods of analysis for studying the responses of such experiments. In such circumstances, only salvage work can be attempted; more often than not, such analyses often do not lead to reliable inferences. It is therefore important to ask for help with the design of our experiments, or better still, to take a class of design and analysis of experiments to be able to design and carry out experiments, and analyze their responses.

We need to make the right conclusions about any process we investigate or study. We cannot overemphasize the need that any scientific experiment should be properly designed. As a part of that process, take for example, the study (mentioned earlier), in which the microbiologist was interested in the differential growth of a particular algae in two media. If the experimenter's objective was to find out which of the two media encourages the fastest growth of the algae, then to design the experiments properly, some questions should be asked before proceeding with the study. Such questions should include the following:

1. Are there some other media in which the algae are likely to thrive better than in the two under consideration?

2. Are there some factors, internal or external to the media which should be studied, controlled, or investigated?

3. Given the budget, how many days are allowed for each study under a given set of conditions?

4. Are there some other ways of reducing the number of replications and obtaining results of comparable precision?

5. How are the data going to be collected?

6. What method of statistical analysis of the responses of the experiment is going to be employed?

7. How much difference in growth of algae in the media could be considered significant?

Of course, before asking these questions, the experimenter should find out what attributes should be measured to carry out the desired investigation. It is clear from the discussion that, for this particular study, the experimenter should already have settled that the response variable in which the experimenter is interested is the amount of growth experienced by the algae. There could possibly be more questions which the investigator should answer before setting out on his investigations. In statistics, we generally examine some numbers and try to make inferences on the population from which it was collected. Since the conclusions that we reach on any phenomenon we are investigating depend on the numbers obtained from it, the way the numbers were collected could make a difference to the conclusions we draw. There are four categories of data whose distinguishing attributes are the methods by which they were collected:

1. Data relating to experiments in which a lot of control is exerted.
2. Data relating to pure observational studies.
3. Data relating to controlled prospective surveys. Here, individuals are chosen but what kind of measurements to be made on them are determined *a priori.*
4. Data related to controlled retrospective surveys. Individuals are chosen and the measurements to be obtained from them are determined *a posteriori.*

These four categories represent the entire spectrum of control exerted in collecting data; the first two are the extremes, while the last two are intermediate forms of controlled investigation.

2.2 Experiments to Compare Treatments

Suppose that at XYZ Teaching Hospital, a medical team is trying to decide on which of two treatments is better for a disease. Let these treatments be T_1 and T_2. They could look at hospital records of those who have already received the treatments, collect the records of 30 patients with treatment T_1 and 30 patients with treatment T_2. They, then count the number in each group surviving after 1 year. On the other hand, they could toss a coin and assign a treatment to each patient as follows:

Heads: treatment T_1; Tails: treatment T_2.

Again, they check the number in each group surviving after a specified period, say, 1 year. It is possible to obtain the same result for both experiments. However, with randomization that was applied to the second design, bias could be eliminated to a high degree, and confidence in the results obtained is enhanced. In these circumstances, any observed differences in results could then be attributed to the treatments and not to other extraneous factors. In the first design of this study, if there is any significant difference in the responses to the two treatments T_1 and T_2, the medical team is not protected against the possibility that it is due to

1. Measured factors confounded with (not distinguishable from) treatment, for example, if treatment T_1 is always given to men and treatment T_2 is always given to women or
2. Unconscious bias on the part of the medical team

This example further emphasizes that the important part of any investigation should be the design of the experiment or the design of the study procedure.

2.2.1 An Illustrative Example

To further underscore the importance of good design, we present an example: A company which manufactures tennis shoes is interested in comparing two materials A and B for making soles of tennis shoes. They are interested in the measures of wear. The wear is measured after a man has put on the shoes and has played with it for a specified number of hours. Suppose that the company could only afford to use 24 shoes for the experiment (12 with soles made of material A and 12 with soles made of material B) for the study, then a number of ideas for performing the experiments are likely to be considered. We think of the following three:

2.2.1.1 Idea I

One could easily take 24 men tennis players, put the shoes made of materials A and B on their right feet. Alternatively, 12 men out of the 24 can be chosen at random to receive shoes made of material A while the rest receive shoes made of soles of material B. After the specified number of hours of play, measurements of wear are taken: Let Y_1 be measurements of wear made for material A and Y_2 be measurements of wear related to material B; where Y_1 is $(y_{11}, y_{12}, \ldots, y_{112})$ and Y_2 is $(y_{21}, y_{22}, \ldots, y_{212})$. Appropriate test statistic for the comparison of the two treatments has a t-distribution and is often referred to as the pooled variance t-test. So that we can obtain the test statistic, t_c as

$$t_c = \frac{\overline{y}_1 - \overline{y}_2}{\sqrt{s_{\mathrm{p}}^2\left(\frac{1}{n_1} + \frac{1}{n_2}\right)}} \tag{2.1}$$

where

$$s_{\mathrm{p}}^2 = \frac{(n_1 - 1)s_1^2 + (n_2 - 1)s_2^2}{n_1 + n_2 - 2}$$

$$\overline{y}_1 = \frac{1}{n_1}\sum_{i=1}^{n_1} y_{1i} \quad \overline{y}_2 = \frac{1}{n_2}\sum_{j=1}^{n_2} y_{2j}$$

$$s_1^2 = \frac{1}{n_1 - 1}\left[\sum_{i=1}^{n_1} y_{1i}^2 - n_1\overline{y}_1^2\right]$$

$$s_2^2 = \frac{1}{n_2 - 1}\left[\sum_{j=1}^{n_2} y_{2j}^2 - n_2\overline{y}_2^2\right]$$

The hypothesis of no difference between the means of A and B is rejected if $t_c < -t(\alpha/2,\ n_1 + n_2 - 2)$ or $t_c > t(\alpha/2,\ n_1 + n_2 - 2)$. In this particular experiment, $n_1 = n_2 = 12$.

A possibility in this type of design is that the differences between men tennis players could swamp the differences between the materials A and B from which the soles are made. Although the experiment has the appearance of being well designed, the inter-tennis player differences or variations may mask any differences in soles. The conclusions we make on the soles based on this design would be incorrect as there is an extraneous factor affecting our observations.

2.2.1.2 Idea II

In this case, 12 men tennis players are chosen at random. Sole A is put on their right feet while sole B is put on their left feet. The test statistic for the model has a t-distribution which is usually employed for paired comparison. After the measurements, we calculated the differences between the observations obtained for each tennis players' legs as: $d_1 = y_{11} - y_{21}, d_2 = y_{12} - y_{22}, \ldots, d_{12} = y_{112} - y_{212}$.

The test statistic for this model (that is paired comparison) is

$$t_c = \frac{\overline{d}}{\sqrt{\frac{(s_d^2)}{n}}} \tag{2.2}$$

where

$$\overline{d} = (1/n)\sum_{i=1}^{n} d_i \quad s_d^2 = (1/(n-1))\left(\sum_{i=1}^{n} d_i^2 - n\overline{d}^2\right)$$

In this case, n is 12 and in the test of no difference between the populations, we reject the hypothesis if $t_c < -t(\alpha/2, n-1)$ or $t_c > t(\alpha/2, n-1)$.

This idea leads to a bad design because although tennis player differences are eliminated, differences between soles are *confounded* with differences between right and left feet. This model also limits our ability to detect what we are looking for—differences in sole materials.

2.2.1.3 Idea III

The test is now conducted as in Idea II, but this time, a coin is tossed for each tennis player to determine which foot receives which sole material. For instance, for each player, a coin is tossed so that if head turns up, he wears shoe with sole material A on the right while he puts the shoe with sole material B on the left foot. The test statistic remains the same as was used for Idea II; that is,

$$t_c = \frac{\overline{d}}{\sqrt{\frac{(s_d^2)}{n}}}$$

Although we have lost 11 degrees of freedom in this test as compared to Idea I, and on the surface this appears to be undesirable, it will be seen that $s_d^2 \ll s_p^2$, and this more than compensates for the losses in degrees of freedom. We indeed eliminated a source of variability—differences between the players' feet and enhanced the confidence that could be placed on our conclusions.

Example 2.1

In an experiment carried out on two materials A and B by a company to compare the durability of the materials for soles of shoes, the following observations were made:

Material A	Material B
13	13.8
8	8.6
10.7	11
14.1	14
10.5	11.6
6.4	6.2
9.3	9.6
10.6	11.1
8.6	9.1
13.1	13.4
12.2	10.4
11.4	10.4

The responses represent the percentage wear of the materials. If we assume that the results of these experiments were obtained according to Idea III, the model we employed is the paired comparison design with randomization. The pairwise differences are presented in the following table:

i	Material A	Material B	Differences (d_i)
1	13	13.8	$13-13.8=-0.8$
2	8	8.6	$8-8.6=-0.6$
3	10.7	11	$10.7-11=-0.30$
4	14.1	14	$14.1-14=0.10$
5	10.5	11.6	$10.5-11.6=-1.10$
6	6.4	6.2	$6.4-6.2=0.2$
7	9.3	9.6	$9.3-9.6=-0.30$
8	10.6	11.1	$10.6-11.1=-0.50$
9	8.6	9.1	$8.6-9.1=-0.50$
10	13.1	13.4	$13.1-13.4=-0.30$
11	12.2	11.8	$12.2-11.8=0.40$
12	11.4	10.8	$13.3-13.6=0.60$

To make inference on the wear of materials A and B, the null hypothesis to be tested is "there is no difference in treatment means (or the mean wear of materials A and B are the same)." Initially, we assume the two-sided alternative hypothesis, that is,

$$H_0\colon \mu_d = 0 \quad \text{vs} \quad H_1\colon \mu_d \neq 0$$

This means that if $|t_c| > t(\alpha/2, 9)$, we would reject the null hypothesis and accept the alternative. Now $s_d^2 = 0.249924$; $\overline{d} = -0.25833$; so that

$$t_c = \frac{\overline{d}}{\frac{s_d}{\sqrt{n}}}$$

$$= \frac{-0.25833}{\frac{0.499924}{\sqrt{12}}} = -3.58066$$

Because $t(0.025, 11) = 2.201$ and $t_c = -1.79006$, we do not accept the null hypothesis. In other words, we accept that there is no significant difference between sole materials A and B. Alternatively, we could, for the purpose of illustration, consider the data as having been independently obtained as in Idea I. Then, the analysis gives the following results (The test applied here is the pooled-variance t-test. Each of the observations is treated as an independent unit. Here, the null hypothesis is the same as that stated earlier.):

$\overline{y}_1 = 10.65833$ $\overline{y}_2 = 10.91667$ $s_1^2 = 5.229924$ $s_2^2 = 5.26697$, so that the pooled variance becomes

$$s_{\text{p}}^2 = \frac{(12-1)(5.229924) + (12-1)(5.26697)}{12+12-2} = 5.248447$$

Consequently, we obtain the test statistic,

$$t_c = \frac{10.65833 - 10.91666}{\sqrt{(5.248447)\left(\frac{1}{12} + \frac{1}{12}\right)}} = \frac{-0.25833}{0.935276} = -0.27621$$

Again, because $t(0.025,\ 22) = 2.074$ and $t_c = -0.27621$, we accept the null hypothesis. Notable differences arise when we construct the confidence intervals for the means. *Here the differences between the tennis players' legs (block differences) has swelled the error and made it hard for us to detect that the materials A and B are significantly different from each other.* Under Idea I, the 95% confidence interval for the difference in means of wear of materials A and B is

$$\overline{y}_1 - \overline{y}_2 \pm t(\alpha/2, n_1 + n_2 - 2) \times \sqrt{s_{\text{p}}^2\left(\frac{1}{n_1} + \frac{1}{n_2}\right)}$$

$$\text{i.e.,}\quad -0.25833 \pm 2.074 \times \sqrt{(5.248447)\left(\frac{1}{12} + \frac{1}{12}\right)}$$

$$-0.25833 \pm 1.9398$$

However, under Idea III, the 95% confidence interval for the differences in the means is

$$-0.25833 \pm 2.201 \times \frac{0.249924}{\sqrt{12}}$$

$$-0.25833 \pm 0.31769$$

The confidence interval under Idea III (paired comparison) is much narrower than that obtained under the pooled-variance independent analysis (Idea I).

What we have done (Idea III) is the paired comparison design. In the process, we have introduced the idea of *blocking*. In this case, the two legs of each tennis player constituted a block. We have carried out a special case of the *randomized complete block* experiment. The blocking so applied has reduced the variation and increased the confidence in the conclusions of our analysis.

It should be noted that if the within block variability was equal to the between block variability, the variance of $\overline{y}_1 - \overline{y}_2$ would have been the same regardless of which design was used (Idea I or Idea III). Blocking in such situations would have led to a loss of some degrees of freedom ($n-1$ of them if there were n observations from each material). This loss in degrees of freedom would have led to wider confidence interval on the differences between means, which would have been undesirable. We return later to blocking as discussed in this section in Chapter 5. □

2.3 Some Basic Ideas

We have used terms in our discussions so far that need to be properly understood. Terms such as response, factor, blocking, randomizing, confounding, and replications need to be defined. We proceed to define each of them.

Block. A block is a set of *experimental units* which are expected to be more homogeneous than the whole aggregate. In our example, the left and right feet of a tennis player, in the experiment represented by Idea III, together constitute a block. While there may be variations in the effect which either the left or right foot of a tennis player would have on materials A and B, there is likely to be more homogeneity between the two legs of a tennis player than with the legs of all the other tennis players put together.

Experimental units. An experimental unit is the smallest subdivision of experimental material such that any two units may receive different treatments in the actual experiment.

Randomization. This is a deliberate, impersonal way of converting uncontrolled variation in an experiment (regardless of source) to random variation. Randomization helps to justify some basic assumptions and provide the statistical basis for the analysis of responses. It protects against possible biases which if present, make the conclusions reached less dependable.

Confounding. This is said to take place when it becomes impossible to distinguish between two effects. We discussed this previously in relation to Idea II in our example.

Replicate. A repetition of the whole designed experiment with the same assignment of treatments. This leads to higher accuracy. However, it costs more in terms of time and money.

Observation or response. This is the quantity which an experiment is designed to account for (in our example above, wear of soles of shoes is the response).

Factor. This is an explanatory variable; a quantity which is expected to account for the response obtained in an experiment.

If we denote a response from an experiment by y, and the explanatory variables as $x_1, x_2, \ldots, x_k$ then, we usually assume that

$$y = f(x_1, x_2, \ldots, x_k) + \text{error}$$

Our experiments are usually designed to investigate f and show that all or some of $x_1, x_2, \ldots, x_k$ do explain adequately the response. A factor can be qualitative (such as type of soil, kind of herbicide, sole A, and sole B, specific fertilizer, leg of a tennis player used for our experiment) or quantitative (temperature, humidity, and rainfall).

Treatment. The list of values or settings of explanatory variables $x_1, x_2, \ldots, x_k$ applied to a particular experimental unit.

2.4 Requirements of a Good Experiment

From the foregone discussions, it can be seen that there are some basic requirements for a good experiment. A good discussion of these requirements can be found in Cox (1958), as well as Cochran and Cox (1957). The requirements include the following:

1. Use your nonstatistical knowledge of the problem. For instance, a microbiologist studying the growth of algae should use his knowledge of biology of algae to determine what factors are likely to influence growth which should be modeled into the design.

2. Systematic error should be absent. In the previously discussed example (Idea II), if we always put the sole material A on the right foot of each tennis player, and sole material B on the left foot, then even if we increased the number of tennis players and so the tests, any differences observed in the wear of the soles cannot be clearly distinguished from the differences between right and left legs. Only Idea III eliminates this systematic error through randomization.

3. The precision of the experiment should be high. The emphasis here is that the method of measurement of responses should be accurate; there should also be high variability in materials used for the experiment. As

mentioned earlier, replications will ensure that the number of *experimental units* to which treatment is applied is high, thereby increasing accuracy. All information available to the experimenter should be used to make the standard error small, but not too small because too small standard error suggests that the experiment was unnecessary. Too high standard error implies that the experiment was not useful.

4. The conclusions should have a wide range of validity. Usually conclusions reached in an experiment apply in the main to the units used under the conditions they were studied. However, it is possible to plan an experiment so that results could be extrapolated beyond the immediate units and conditions. This is achieved by widening the range of conditions investigated in the experiment. In particular, in factorial designs, when large effects of the same class are to be investigated, a few are chosen and assumed to be random as opposed to fixed. The results of such experiments when analyzed can be extrapolated beyond the immediate effects.

5. Keep the design and analysis as simple as possible. Do not be overzealous and use complex and sophisticated statistical techniques. Most efficient designs are simple and lead to simple analysis. Complicated analysis, which often arise from complicated designs, lead to waste. Analysis of such designs lead to difficulties in interpreting the outcome.

6. It should be possible to calculate the uncertainties as well as the estimates of treatment differences from the data. For this to be achieved *experimental units* ought to respond in a completely independent fashion to different treatments. Randomization as, we discussed earlier, ensures this and gives validity to the statistical basis of the analysis which is subsequently carried out.

7. Experiments are usually iterative. Do not design a very comprehensive experiment at the onset of a study. Exploratory designs enable us to determine which factors are important and which should be eliminated. This favors an iterative, sequential approach.

2.5 One-Way Experimental Layout or the Completely Randomized Design: Design and Analysis

We have so far discussed the comparison of the means of two treatments or the means of two groups of observations. Recall the comparison of sole materials for shoes in Section 2.2. Often, however, the need arises for the means of more than two treatments or more than two levels of a factor to be compared. For instance, in industrial research, managers often come up with three or more methods of production of the same material and have to make choices.

The new methods may arise from innovative reappraisal of methods already in use. Piecewise testing of these methods by choosing two at a time and comparing them until one arrives at the most efficient method (particularly, when the number of methods is large) is both time-consuming and wasteful. It is therefore more desirable to construct an efficient design to provide one simple, less expensive experiment for making such a decision. A tool devised for this purpose which over the years has proved useful, is the analysis of variance (ANOVA) in one-way experimental layouts. We now discuss this design and the method of analysis which is appropriate for it. When interest is in the comparison of several treatments or effects of different levels of a factor, there should be one variable (the response variable) in which we should be primarily interested. Once we identify this variable, we ought to determine the most efficient method for measuring it. This has the benefit of enhancing precision of the results of any further analysis we carry out on the data. If we have k treatments or a factor with k levels, denoted by $B_1, B_2, \ldots, B_k$ which we wish to test their effects on a response variable y, the usual procedure under the one-way classification is to perform the experiments, making n observations when B_1 is present, the same number of observations when B_2 is present, and so on, until n observations are made when B_k is switched on. Each of these observations is made for each treatment when all other treatments are absent. This is akin to making n observations from each of the k groups represented by $B_1, \ldots, B_k$ and seeking to compare the means for the k groups.

The mathematical model suitable for the kind of design we have adopted expresses the responses as a linear function of the grand mean, the treatment effects, and the error term as follows:

$$y_{ij} = \mu + \alpha_i + \varepsilon_{ij} \quad i = 1, 2, \ldots, k \;\; j = 1, 2, \ldots, n \tag{2.3}$$

The responses of the experiment could be presented in a one-way layout as indicated in Table 2.1.

In Table 2.1,

$$y_{..} = \sum_{i=1}^{k}\sum_{j=1}^{n} y_{ij} \quad y_{i.} = \sum_{j=1}^{n} y_{ij} \quad \overline{y}_{i.} = \frac{1}{n}\sum_{j=1}^{n} y_{ij} \quad \overline{y}_{..} = \frac{1}{kn}\sum_{i=1}^{k}\sum_{j=1}^{n} y_{ij} \tag{2.4}$$

TABLE 2.1: Observations in One-Way Experimental Layouts

Treatments or levels of factor	Replicates						
	1	**2**	...	$n-1$	n	**Total**	**Mean**
B_1	y_{11}	y_{12}	...	y_{1n-1}	y_{1n}	$y_{1.}$	$\overline{y}_{1.}$
B_2	y_{21}	y_{22}	...	y_{2n-1}	y_{2n}	$y_{2.}$	$\overline{y}_{2.}$
$\vdots$	$\vdots$	$\vdots$	$\vdots$	$\vdots$	$\vdots$	$\vdots$	$\vdots$
B_{k-1}	$y_{(k-1)1}$	$y_{(k-1)2}$	...	$y_{(k-1)(n-1)}$	$y_{(k-1)n}$	$y_{k-1.}$	$\overline{y}_{k-1.}$
B_k	y_{k1}	y_{k2}	...	$y_{k(n-1)}$	y_{kn}	y_k	$\overline{y}_{k.}$

In our assumed model (Equation 2.3), μ is a parameter which is common to all observations y_{ij} (called the overall mean), while α_i is the treatment effect of the ith level of the factor (or the effect of the ith treatment). To carry out the analysis of the responses of this study (using the procedure called ANOVA), we need to state the objective of our analysis as well as the assumptions we make about the responses of the experiment. This objective is usually stated in the form of a hypothesis; in this case, the hypothesis we are to test is that there is no treatment effect. In other words, the means of all the observations made for each of the treatments (or levels of factor) are the same. The error component, ε_{ij} is assumed to be normally and independently distributed with mean zero and variance, σ^2 [i.e., $\varepsilon_{ij} \sim \text{NID}(0, \sigma^2)$] which means that σ^2 is a constant at all levels of the factor or for all treatments or groups being studied. The order in which these treatments are applied to experimental units within each level of the factor or treatment groups should be random in order to ensure the uniform treatment of the units and the elimination of any systematic error. This is a completely randomized design (CRD). The model we have developed is called the one-way experimental layout because only one factor at different levels is tested at a time. Although groups which have received different treatments may be compared under this design, the basic hypothesis assumes that there is no difference between their means, that is, there is no treatment effect on the k different groups being compared.

The model we stated in Equation 2.3 represents two basic assumptions about the treatments. If in the course of planning the experiment, the investigator had chosen the k treatment groups with the intent to test hypothesis about treatment means without extrapolating the conclusions of the study to similar treatments not specifically included in the study, then we have the fixed effects model. In a different approach to the design, the experimenter could have chosen the treatments as a sample from a population containing a large number of similar treatments with a view to extending the deductions of the investigations to the rest of the treatments in the population not necessarily considered in the design. In such circumstances, primary interest would be in the testing of the variability of α_i and in attempting to estimate this variability. This is the random effects model. It is also referred to as the components of variance model. For the fixed effects model, it is assumed that $\sum_{i=1}^{k} \alpha_i = 0$. For the components of variance model, $\alpha_i \sim \text{NID}(0, \sigma_\alpha^2)$. In some experiments where more than one factor is being studied, it is possible to have a mixed model in which some effects are fixed and others are random.

2.6 Analysis of Experimental Data (Fixed Effects Model)

To analyze the data obtained from our experiments (when we assume the fixed effects model), we have to bear in mind that the implications of having fixed effects is that treatment effects are defined as departures from the overall mean, μ. That is

$$\sum_{i=1}^{k} \alpha_i = 0 \tag{2.5}$$

Since $y_{i.}$ represents the total of all the observations for the ith treatment, then $\overline{y}_{i.}$ is the average of the observations for the same treatment. The grand total (sum of all observations for all treatments or all levels of a factor) is $y_{..}$ while $\overline{y}_{..}$ is the grand average for all observations. We proceed to test the hypothesis that there is no treatment effect; that is, treatment effects in our model are all zero; against the alternative that there is at least one which is not zero. The hypothesis is stated as:

$$H_0 : \alpha_1 = \alpha_2 = \cdots = \alpha_k = 0 \quad \text{vs} \quad H_1 : \alpha_i \neq 0 \quad \textit{for at least one } i$$

We now partition the total *sum of squares* (*SS*) arising from the least squares analysis of our model, which represents the total variability in the model, to obtain its component parts and use them in the analysis of variance. The total sum of squares is obtained as the sum of the squares of all the deviations from the grand average:

$$SS_{\text{total}} = \sum_{i=1}^{k} \sum_{j=1}^{n} [y_{ij} - \overline{y}_{..}]^2 \tag{2.6}$$

The total sum of squares can be rewritten as

$$\begin{aligned} SS_{\text{total}} &= \sum_{i=1}^{k} \sum_{j=1}^{n} [(y_{ij} - \overline{y}_{i.}) + (\overline{y}_{i.} - \overline{y}_{..})]^2 \\ &= \sum_{i=1}^{k} \sum_{j=1}^{n} (y_{ij} - \overline{y}_{i.})^2 + \sum_{i=1}^{k} \sum_{j=1}^{n} (\overline{y}_{i.} - \overline{y}_{..})^2 - 2\sum_{i=1}^{k} \sum_{j=1}^{n} (y_{ij} - \overline{y}_{i.})(\overline{y}_{i.} - \overline{y}_{..}) \end{aligned} \tag{2.7}$$

But

$$\sum_{i=1}^{k} \sum_{j=1}^{n} (y_{ij} - \overline{y}_{i.})(\overline{y}_{i.} - \overline{y}_{..}) = \sum_{i=1}^{k} [(n\overline{y}_{i.} - n\overline{y}_{i.})(\overline{y}_{i.} - \overline{y}_{..})] = 0$$

$$\Rightarrow SS_{\text{total}} = \sum_{i=1}^{k} \sum_{j=1}^{n} (y_{ij} - \overline{y}_{i.})^2 + \sum_{i=1}^{k} \sum_{j=1}^{n} (\overline{y}_{i.} - \overline{y}_{..})^2 \tag{2.8}$$

The term containing the squares of the differences between treatment averages and the grand average is called the treatment sum of squares. The other term which contains the squares of the differences between observations within a treatment and the average for that treatment could only arise due to residual fluctuations or experimental error. We therefore call this the residual sum of squares. If we rewrite Equation 2.8 in terms of these components, we have

$$SS_{\text{total}} = SS_{\text{treatment}} + SS_{\text{residual}} \tag{2.9}$$

Furthermore (since we have assumed equal variances for all treatments), it is possible to rewrite SS_{residual} as

$$SS_{\text{residual}} = \sum_{i=1}^{k}\left[\sum_{j=1}^{n}(y_{ij} - \overline{y}_{i.})^2\right] = k(n-1)s^2 \tag{2.10}$$

where

$$s^2 = \frac{\sum_{j=1}^{n}[y_{ij} - \overline{y}_{i.}]^2}{n-1}$$

Consequently, an estimate of the common variance becomes

$$MS_{\text{residual}} = \frac{SS_{\text{residual}}}{k(n-1)} = \frac{SS_{\text{residual}}}{N-k} \tag{2.11}$$

where $N = kn$ and MS is the mean square.

2.7 Expected Values for the Sums of Squares

Under our hypothesis, in which we assume there is no difference between treatments, we could use the squares of the differences between treatment averages and the grand average to estimate the common variance, σ^2. Since there are k treatments or k levels of the factor, it can easily be seen that

$$\frac{\sum_{i=1}^{k}(\overline{y}_{i.} - \overline{y}_{..})^2}{k-1}$$

is an estimate of σ^2/n, the variance of $\overline{y}_{i.}$, so that

$$\frac{n\sum_{i=1}^{k}(\overline{y}_{i.} - \overline{y}_{..})^2}{k-1}$$

is an estimate of σ^2. We therefore obtain a second estimate of σ^2 as

$$MS_{\text{treatment}} = \frac{SS_{\text{treatment}}}{k-1} \tag{2.12}$$

The next step is to find the expected values of $MS_{\text{treatment}}$ and MS_{residual} which will justify the intuitive approach we have adopted in assuming that both are estimates of the common variance, σ^2.

$$EMS_{\text{residual}} = E\left(\frac{SS_{\text{residual}}}{N-k}\right) = E\left[\frac{\sum_{i=1}^{k}\sum_{j=1}^{n}(y_{ij}-\overline{y}_{i.})^2}{N-k}\right]$$

$$= E\left[\frac{\sum_{i=1}^{k}\sum_{j=1}^{n}(y_{ij}^2 - 2\overline{y}_{i.}y_{ij} + \overline{y}_{i.}^2)}{N-k}\right]$$

$$= E\left[\frac{\sum_{i=1}^{k}\left(\sum_{j=1}^{n} y_{ij}^2 - n\overline{y}_{i.}^2\right)}{N-k}\right]$$

By substituting for y_{ij} (using Equation 2.3) in

$$MS_{\text{residual}} = \frac{\left[\sum_{i=1}^{k}\left(\sum_{j=1}^{n} y_{ij}^2 - n\overline{y}_{i.}^2\right)\right]}{N-k}$$

we get

$$E(MS_{\text{residual}}) = E\frac{\left[\sum_{i=1}^{k}\sum_{j=1}^{n} y_{ij}^2 - \sum_{i=1}^{k}\left(\frac{1}{n}\sum_{j=1}^{n} y_{ij}\right)^2\right]}{k(n-1)}$$

$$= E\frac{\left[\sum_{i=1}^{k}\sum_{j=1}^{n}(\mu + \alpha_i + \varepsilon_{ij})^2\right]}{k(n-1)}$$

$$- E\frac{\left[\sum_{i=1}^{k}\left(n\mu + n\alpha_i + \sum_{j=1}^{n}\varepsilon_{ij}\right)^2\right]}{kn(n-1)}$$

But we know that

$$\sum_{i=1}^{k}\alpha_i = 0; \quad \varepsilon_{ij} \sim \text{NID}(0,\sigma^2) \Rightarrow E\varepsilon_{ij}^2 = \sigma^2; \quad E\varepsilon_{ij} = 0, \text{and so on.}$$

so that

$$E(MS_{\text{residual}})$$
$$= E\frac{\left[\sum_{i=1}^{k}\sum_{j=1}^{n}(\mu^2+\alpha_i^2+\varepsilon_{ij}^2+2\mu\varepsilon_{ij}+2\alpha_i\varepsilon_{ij}+2\mu\alpha_i)\right]}{k(n-1)}$$
$$-E\frac{\left\{\sum_{i=1}^{k}\left[n^2\mu^2+n^2\alpha_i^2+\left(\sum_{j=1}^{n}\varepsilon_{ij}\right)^2+2n^2\mu\alpha_i + 2n\alpha_i\sum_{j=1}^{n}\varepsilon_{ij}+2n\mu\sum_{j=1}^{n}\varepsilon_{ij}\right]\right\}}{kn(n-1)}$$
$$= \frac{\left(kn\mu^2+n\sum_{i=1}^{k}\alpha_i^2+kn\sigma^2\right)}{k(n-1)} - \frac{\left(kn\mu^2+n\sum_{i=1}^{k}\alpha_i^2+k\sigma^2\right)}{k(n-1)}$$
$$= \frac{k(n-1)\sigma^2}{k(n-1)} = \sigma^2$$

Similarly,

$$SS_{\text{treatment}} = \sum_{i=1}^{k}\frac{y_{i.}^2}{n} - \frac{y_{..}^2}{nk}$$

so that

$$ESS_{\text{treatment}} = E\sum_{i=1}^{k}\frac{y_{i.}^2}{n} - E\frac{y_{..}^2}{nk}.$$

Consider

$$E\sum_{i=1}^{k}\frac{y_{i.}^2}{n} = E\sum_{i=1}^{k}\left(\frac{\sum_{j=1}^{n_i}y_{ij}}{n}\right)^2 = E\sum_{i=1}^{k}\frac{1}{n}\left[\sum_{j=1}^{n}(\mu+\alpha_i+\varepsilon_{ij})\right]^2$$
$$= E\sum_{i=1}^{k}\left(\frac{1}{n}\right)(n^2\mu^2+n^2\varepsilon_i^2+n\sigma^2+2n^2\mu\alpha_i)$$
$$= E\sum_{i=1}^{k}\left(n\mu^2+n\varepsilon_i^2+\sigma^2+2\sum\mu n\alpha_i\right) = \mu^2\sum_{i=1}^{k}n+\sum_{i=1}^{k}n\alpha_i^2+k\sigma^2$$

Now consider

$$E\frac{\left(\sum_{i=1}^{k}\sum_{j=1}^{n_i} y_{ij}\right)^2}{kn} = \frac{\left[\sum_{i=1}^{k}\sum_{j=1}^{n_i}(\mu+\alpha_i+\varepsilon_{ij})\right]^2}{kn}$$

$$= E\frac{\left(\mu\sum_{i=1}^{k} n + \sum_{i=1}^{k} n\alpha_i + \sum_{i=1}^{k}\sum_{j=1}^{n_i}\varepsilon_{ij}\right)^2}{kn}$$

$$\left(\text{since } \sum_{i=1}^{k}\sum_{j=1}^{n_i}\alpha_i = n\sum_{i=1}^{k}\alpha_i = 0\right)$$

$$= E\frac{\left[\mu^2(kn)^2 + \left(\sum_{i=1}^{k}\sum_{j=1}^{n_i}\varepsilon_{ij}\right)^2 + 2\mu kn\sum_{i=1}^{k}\sum_{j=1}^{n_i}\varepsilon_{ij}\right]}{kn}$$

$$= \frac{k^2n^2\mu^2 + kn\sigma^2}{kn} = kn\mu^2 + \sigma^2$$

$$EMS_{\text{treatment}} = \frac{SS_{\text{treatment}}}{k-1}$$

$$EMS_{\text{treatment}} = \frac{\left(kn\mu^2 + n\sum_{i=1}^{k}\alpha_i^2 + k\sigma^2 - kn\mu^2 - \sigma^2\right)}{k-1}$$

$$= \frac{(k-1)\sigma^2 + \sum_{i=1}^{k} n\alpha_i^2}{k-1} = \sigma^2 + \frac{n\sum_{i=1}^{k}\alpha_i^2}{k-1} \quad (2.13)$$

It should be noted that $SS_{\text{treatment}}$ is the sum of the squares of normal variables and therefore it can be shown that $SS_{\text{treatment}}/\sigma^2$ is χ^2 distributed with $k-1$ degrees of freedom. Similarly, it can be shown that $SS_{\text{residual}}/\sigma^2$ is χ^2 distributed with $k(n-1)$ degrees of freedom. It follows that the ratio $[SS_{\text{treatment}}/(k-1)]/[SS_{\text{residual}}/k(n-1)]$ is distributed as F with $k-1$ and $k(n-1)$ degrees of freedom. In the test of hypothesis called ANOVA which follows, the above ratio is obtained from the responses of the experiment and is compared with some critical value of F with $k-1$ and $k(n-1)$ degrees of freedom. This critical value usually depends on α, the level of significance or the size of Type I error.

Another important principle underlying the test of significance (ANOVA) has to do with $E(MS_{\text{treatment}})$ and $E(MS_{\text{residual}})$. Under the null hypothesis, $E(MS_{\text{treatment}})$ as well as $E(MS_{\text{residual}})$ are estimates of σ^2, the common variance of all the groups. If the null hypothesis is true, then H_0: $\alpha_1=\alpha_2=\cdots=\alpha_k=0$, and the ratio $MS_{\text{treatment}}/MS_{\text{residual}}$ would be small and therefore insignificant, ensuring that the null hypothesis is not rejected.

2.7.1 Estimating the Parameters of the Model

The model is

$$y_{ij} = \mu + \alpha_i + \varepsilon_{ij} \quad i = 1, 2, \ldots, k \;\; j = 1, 2, \ldots, n$$

This implies that $\varepsilon_{ij} = y_{ij} - \mu - \alpha_i$. To estimate the parameters of the model, we use least squares method, so that the values that minimize L with respect to μ and α_i where

$$L = \sum_{i=1}^{k}\sum_{j=1}^{n} \varepsilon_{ij}^2 = \sum_{i=1}^{k}\sum_{j=1}^{n} (y_{ij} - \mu - \alpha_i)^2$$

are the estimates. We solve the following simultaneous equations:

$$\left.\frac{\partial L}{\partial \mu}\right|_{\hat{\mu},\hat{\alpha}_i} = 0$$

$$\left.\frac{\partial L}{\partial \alpha_i}\right|_{\hat{\mu},\hat{\alpha}_i} = 0$$

That is

$$-2\sum_{i=1}^{k}\sum_{j=1}^{n} (y_{ij} - \hat{\mu} - \hat{\alpha}_i) = 0$$

$$-2\sum_{j=1}^{n} (y_{ij} - \hat{\mu} - \hat{\alpha}_i) \quad i = 1, 2, \ldots, k$$

Solving, we see that the above equations lead to the following $k + 1$ normal equations:

$$kn\hat{\mu} + n\hat{\alpha}_1 + n\hat{\alpha}_2 + \cdots + n\hat{\alpha}_k = \sum_{i=1}^{k}\sum_{j=1}^{n} y_{ij} = y_{..}$$

$$n\hat{\mu} + n\hat{\alpha}_1 + 0 + 0 + \cdots + 0 = \sum_{j=1}^{n} y_{1j} = y_{1.}$$

$$n\hat{\mu} + 0 + n\hat{\alpha}_2 + 0 + 0 + \cdots + 0 = y_{2.}$$

$$\vdots$$

$$n\hat{\mu} + 0 + 0 + \cdots + 0 + n\hat{\alpha}_k = y_{k.}$$

The above equations are not linearly independent so that no unique solution exists. However, if we impose the constraint that $\sum_{i=1}^{k} \alpha_i = 0$ (this condition is often called the fixed effects model) then the solutions for the normal equations are:

$$\hat{\mu} = \overline{y}_{..}$$

$$\hat{\alpha}_i = \overline{y}_{i.} - \overline{y}_{..}$$

It is clear that if we had imposed another constraint on the normal equations, we would have obtained different estimates for μ and α_i. This means that the estimates depend on the constraints assumed by the experimenter and for the same data, estimates would vary according to the constraint imposed on the analysis. However, there are some functions which depend on the model parameters that can be estimated as the same no matter what constraint we impose on the normal equations.

One example is μ_i, the mean of the ith treatment, $\mu_i = \mu + \alpha_i$, which is estimated as $\hat{\mu}_i = \hat{\mu} + \hat{\alpha}_i = \overline{y}_{i.}$. Another is the difference between two treatment means $\mu_i - \mu_j$, which is estimated as $\hat{\alpha}_i - \hat{\alpha}_j = \overline{y}_{i.} - \overline{y}_{j.}$

2.8 Analysis of Variance Table

It is usual to display the different estimates, MS_{residual} and $MS_{\text{treatment}}$ in what is referred to as ANOVA table. It presents all the characteristics of the statistical analysis of the responses of our experiment in a clear concise manner. For the one-way classification we discussed in the previous section, Table 2.2 presents the ANOVA for the typical CRD.

For ease of calculation, it turns out that algebraic manipulations of the sum of squares lead to the following more easily manageable formulae:

$$SS_{\text{total}} = \sum_{i=1}^{k}\sum_{j=1}^{n} y_{ij}^2 - \frac{y_{..}^2}{kn} \tag{2.14}$$

$$SS_{\text{treatment}} = \frac{1}{n}\sum_{i=1}^{k} y_{i.}^2 - \frac{y_{..}^2}{kn} \tag{2.15}$$

$$SS_{\text{residual}} = SS_{\text{E}} = \sum_{i=1}^{k}\sum_{j=1}^{n} y_{ij}^2 - \frac{1}{n}\sum_{i=1}^{k} y_{i.}^2 = SS_{\text{total}} - SS_{\text{E}} \tag{2.16}$$

TABLE 2.2: Analysis of Variance Table

Source of variation	**Sum of squares**	**Degrees of freedom**	**Mean square**	***F*-ratio**
Between groups (between treatments)	$SS_{\text{treatment}}$	$k-1$	$MS_{\text{treatment}} = \frac{SS_{\text{treatments}}}{k-1}$	$\frac{MS_{\text{treatment}}}{MS_{\text{residual}}}$
Within groups (residual or within treatments)	SS_{residual}	$N-k=k(n-1)$	$MS_{\text{residual}} = \frac{MS_{\text{residual}}}{k(n-1)}$	
Total	SS_{total}	$N-1=kn-1$		

The total sum of squares should be equal to the sum of the between and within treatments sums of squares. The sum of degrees of freedom for both the within and between groups sums of squares is equal to the total number of degrees of freedom. We have deliberately restricted our discussions to cases where the observations made for all the treatments or levels of a factor are equal. In practice, this is not always the case. For the result, when the number of observations are not equal, see Exercise 2.2 at the end of this chapter.

The hypothesis usually tested is

$$H_0 : \alpha_1 = \alpha_2 = \cdots = \alpha_k = 0 \quad \text{vs} \quad H_1 : \alpha_i \neq 0 \quad \textit{for at least one } i.$$

The null hypothesis is rejected if $F > F(k-1, k(n-1), \alpha)$. $F(k-1, k(n-1), \alpha)$ is read from the table of the F-distribution in Appendices A5 through A7; α is the level of significance.

The one-way layout can be used to compare k populations, by an arrangement in which we choose independent random samples from each of the populations of interest (see Wackerly et al., 2002, p. 620). We now consider an example to illustrate the ANOVA for the responses of an experiment carried out according to the one-way classification.

Example 2.2

Five different methods proposed for extracting a nutrient from a fruit were being compared. Fifty extractions were carried out in random order on an equal amount of the same properly mixed fruit pulp, ten using each method. The amount of nutrient extracted in milligrams is shown below. Test the hypothesis that the mean nutrient extracted for each method are the same, against that the means are different for at least one of the methods. Use a level of significance of 1%.

	Sample measurement number (replications)										
Extraction method	**1**	**2**	**3**	**4**	**5**	**6**	**7**	**8**	**9**	**10**	**Total ($y_{i.}$)**
I	10	12	10	11	12	9	11	13	10	9	107
II	14	14	14	12	16	10	14	15	10	15	134
III	17	13	15	12	16	10	14	15	10	15	142
IV	19	15	19	14	23	12	17	17	15	19	170
V	20	19	23	16	20	18	21	22	18	20	197

Analysis:

For this study, the underlying model is

$$y_{ij} = \mu + \alpha_i + \varepsilon_{ij} \quad i = 1, 2, 3, 4, 5; \; j = 1, 2, \ldots, 10$$
$$\varepsilon_{ij} \sim \text{NID}(0, \sigma^2)$$

This is a fixed effects model, so that

$$\sum_{i=1}^{5} \alpha_i = 0$$

which means that the sum of the effects for the five different methods of nutrient extraction in the model is zero.

$$SS_{\text{total}} = \sum_{i=1}^{k}\sum_{j=1}^{n} y_{ij}^2 - \frac{y_{..}^2}{kn} = 11952 - \frac{750^2}{5(10)} = 702$$

$$SS_{\text{extraction method}} = \frac{\sum_{i=1}^{k} y_{i.}^2}{n} - \frac{y_{..}^2}{kn} = \frac{107^2 + 134^2 + 142^2 + 170^2 + 197^2}{10} - \frac{750^2}{5(10)} = 477.80$$

$$SS_{\text{E}} = SS_{\text{total}} - SS_{\text{extraction method}} = 702 - 477.80 = 224.2$$

We perform ANOVA (Table 2.3) to test the following hypothesis (which can be stated in two possible ways):

$$H_0 : \mu_1 = \mu_2 = \mu_3 = \mu_4 = \mu_5 \quad \text{vs} \quad H_1 : \textit{at least one } \mu_i \textit{ is different}$$

or

$$H_0 : \alpha_1 = \alpha_2 = \alpha_3 = \alpha_4 = \alpha_5 = 0 \quad \text{vs} \quad H_1 : \textit{at least one } \alpha_i \neq 0$$

From the tables of the F-distribution in Appendix A10, $F(4, 45, 0.01) = 3.77$. Since $F_{\text{ratio}} > F(4, 45, 0.01)$, the tabulated value, we reject H_0 and accept the alternative, H_1. Therefore the method of extraction effect is significant at 1% level of significance. We have found that there is a significant

TABLE 2.3: ANOVA for Comparing Means for the Extraction Method Study

Source	**Sum of squares**	**Degrees of freedom**	**Mean square**	***F*-ratio**	***p*-Value**
Between extraction methods (treatment)	477.80	4	119.45	23.97525	1.14625E−10
Within the extraction methods(error)	224.2	45	4.98222		
Total	702	49			

difference between the mean nutrients extracted by the five methods from the fruit. Since the ANOVA has found that the means are significantly different, we may be interested in finding out which method mean is greater or lesser than the other. This is addressed in the follow-up analysis. □

2.9 Follow-Up Analysis to Check for Validity of the Model

Rejection of the hypothesis indicates that there is significant difference between the methods. A number of questions about the result of the analysis give rise to some follow-up studies of the methods. These questions are

1. Is there any method that is better than others?
2. Is there any method that is worse than others?
3. Is it possible to divide the methods into homogeneous groups, each consisting of nutrient extraction methods, which are not significantly different from one another?

2.9.1 Least Significant Difference Method

We can answer the above questions using a test for comparing means called the least significant difference (LSD). Since we assumed that our model is

$$y_{ij} = \mu + \alpha_i + \varepsilon_{ij}$$

with $\varepsilon_{ij} \sim N(0, \sigma^2)$, then the mean of the ith extraction method is $\mu_i = \mu + \alpha_i$. We could then estimate μ by using the fact that

$$\begin{aligned} \hat{\mu} &= \overline{y}_{..}; \quad \hat{\alpha}_i = \overline{y}_{i.} - \overline{y}_{..} \\ \hat{\mu}_i &= \hat{\mu} + \hat{\alpha}_i = \overline{y}_{i.} \end{aligned} \tag{2.17}$$

But $\overline{y}_{i.} \sim N(\mu + \alpha_i, \sigma^2/n)$, this means that for any pair of extraction method levels, i, j; $i \neq j, i, j = 1, 2, \ldots, k$,

$$\overline{y}_{i.} - \overline{y}_{j.} \sim N\left(0, \frac{2\sigma^2}{n}\right) \tag{2.18}$$

The statement in Equation 2.18 is true if $\mu_i = \mu_j$.

Often, however, we do not know σ^2 so that the appropriate distribution to use is the t with σ^2 replaced by its estimate, s^2. The LSD at α level of significance is

$$\text{LSD} = t\left[\frac{\alpha}{2}, k(n-1)\right] s\sqrt{\left(\frac{2}{n}\right)} \tag{2.19}$$

Returning to Example 2.2, since we rejected the hypothesis of equality of means, at $\alpha = 1\%$, we use the same α to calculate the LSD with a view to investigating further the relationships between the means for the different methods for extracting the nutrient from the fruit. MS_{E} is the best estimate of s^2 so that

$$s^2 = MS_{\mathrm{E}} = 4.98222 \quad t(0.005, 45) = 2.693$$

$$s = \sqrt{4.98222} = 2.232089$$

$$\begin{aligned}\mathrm{LSD} &= t\left[\frac{\alpha}{2}, k(n-1)\right] s\sqrt{\left(\frac{2}{n}\right)} \\ &= t(0.025, 45)(2.232089)\sqrt{\frac{2}{10}} \\ &= (2.693)(2.232089)(0.4472) = 2.688126\end{aligned}$$

For the methods of extraction, the mean yields of the nutrient are given as follows:

Means				
Method 1	**Method 2**	**Method 3**	**Method 4**	**Method 5**
10.7	13.4	14.2	17	19.7
$\overline{y}_{1.}$	$\overline{y}_{2.}$	$\overline{y}_{3.}$	$\overline{y}_{4.}$	$\overline{y}_{5.}$

We now arrange the treatment means (means of methods for extraction of nutrient from the fruit) according to their magnitude as

$$\overline{y}_{1.} < \overline{y}_{2.} < \overline{y}_{3.} < \overline{y}_{4.} < \overline{y}_{5.}$$

The differences are as follows:

$$\begin{aligned}
\overline{y}_{1.} - \overline{y}_{2.} &= 10.7 - 13.4 = -2.7 \text{ (significant)} \\
\overline{y}_{1.} - \overline{y}_{3.} &= 10.7 - 14.2 = -3.5 \text{ (significant)} \\
\overline{y}_{1.} - \overline{y}_{4.} &= 10.7 - 17 = -6.3 \text{ (significant)} \\
\overline{y}_{1.} - \overline{y}_{5.} &= 10.7 - 19.7 = -9.0 \text{ (significant)}
\end{aligned}$$

$$\begin{aligned}
\overline{y}_{2.} - \overline{y}_{3.} &= 13.4 - 14.2 = -0.80 \text{ (not significant)} \\
\overline{y}_{2.} - \overline{y}_{4.} &= 13.4 - 17 = -3.6 \text{ (significant)} \\
\overline{y}_{2.} - \overline{y}_{5.} &= 13.4 - 19.7 = -6.3 \text{ (significant)}
\end{aligned}$$

$$\begin{aligned}
\overline{y}_{3.} - \overline{y}_{4.} &= 14.2 - 17 = -2.80 \text{ (significant)} \\
\overline{y}_{3.} - \overline{y}_{5.} &= 14.2 - 19.7 = -5.50 \text{ (significant)}
\end{aligned}$$

$$\overline{y}_{4.} - \overline{y}_{5.} = 17 - 19.7 = -2.70 \text{ (significant)}$$

Since LSD is 2.688126, our conclusion is that μ_1 and μ_2 are significantly different from each other; μ_2 and μ_3 are not significantly different from each other, while μ_4 and μ_5 are significantly different from each other and from others. Thus, the method 5 turned out the highest mean yield of the nutrient of all the five methods studied. Method 5 is followed by 4 and both 2 and 3, with the least mean yield coming from method 1. In short, the means can be grouped according to means that are not significantly different as follows: (μ_1), (μ_2, μ_3), (μ_4), (μ_5).

2.9.2 Duncan's Multiple Range Test

Another method for comparing all pairs of means is the multiple range test (Duncan, 1955). To use this method when samples are of equal size, the treatment averages are arranged in ascending order of magnitude and the standard error of each average is calculated so that

$$s_{\overline{y}} = \sqrt{\frac{MS_{\mathrm{E}}}{n}}$$

When the samples are of unequal sizes so that there are k groups with sizes n_1, $n_2, \ldots, n_k$, respectively, then we replace n in the above formula by n_u where

$$n_u = \frac{k}{\sum_{i=1}^{k} \left(\frac{1}{n_i}\right)}$$

From the Duncan's table of significant ranges (see Appendices A8 and A9), we can obtain the tabulated $r_\alpha(p, f)$ for $p = 2, 3, \ldots, k$ where α is the significance level; f is the degree of freedom with which MS_{residual} was calculated. These ranges are then converted to $k-1$ least significant ranges (LSR) where $(\mathrm{LSR})_p = r_\alpha(p, f).s_{\overline{y}}$. The observed difference between means are then compared with LSR ($MS_{\mathrm{residual}} = MS_{\mathrm{E}}$).

To use LSR for Example 2.2, we arrange the averages in ascending order as

$$\begin{aligned}
\overline{y}_{1.} &= 10.70 \\
\overline{y}_{2.} &= 13.40 \\
\overline{y}_{3.} &= 14.20 \\
\overline{y}_{4.} &= 17.00 \\
\overline{y}_{5.} &= 19.70
\end{aligned}$$

Since the MS_{residual} was calculated using 45 degrees of freedom, then at 1% significance level, we calculate, the $r_{0.01}(p, f)$ as well as the LSR as follows:

$$\begin{aligned}
r_{0.01}(2, 45) &= 3.805 \quad r_{0.01}(3, 45) = 3.9725 \\
r_{0.01}(4, 45) &= 4.0825 \quad r_{0.01}(5, 45) = 4.1575
\end{aligned}$$

$$\text{LSR}_2 = 3.805\sqrt{\frac{4.982222}{10}} = 2.685754$$

$$\text{LSR}_3 = 3.9725\sqrt{\frac{4.982222}{10}} = 2.803983$$

$$\text{LSR}_4 = 4.0825\sqrt{\frac{4.982222}{10}} = 2.881627$$

$$\text{LSR}_5 = 4.1575\sqrt{\frac{4.982222}{10}} = 2.934565$$

Having obtained the LSRs, we compare the treatments and obtain the following results:

$$\begin{array}{lll}
\overline{y}_{5.} \text{ vs } \overline{y}_{1.} = 19.70 - 10.70 = 9.00 > 2.934565(\text{LSR}_5) & \text{(significant)} \\
\overline{y}_{5.} \text{ vs } \overline{y}_{2.} = 19.70 - 13.40 = 6.30 > 2.881627(\text{LSR}_4) & \text{(significant)} \\
\overline{y}_{5.} \text{ vs } \overline{y}_{3.} = 19.70 - 14.20 = 5.50 > 2.803983(\text{LSR}_3) & \text{(significant)} \\
\overline{y}_{5.} \text{ vs } \overline{y}_{4.} = 19.70 - 17.00 = 2.70 > 2.685754(\text{LSR}_2) & \text{(significant)}
\end{array}$$

$$\begin{array}{ll}
\overline{y}_{4.} \text{ vs } \overline{y}_{1.} = 17.00 - 10.70 = 6.30 > 2.881627(\text{LSR}_4) & \text{(significant)} \\
\overline{y}_{4.} \text{ vs } \overline{y}_{2.} = 17.00 - 13.40 = 3.60 > 2.803983(\text{LSR}_3) & \text{(significant)} \\
\overline{y}_{4.} \text{ vs } \overline{y}_{3.} = 17.00 - 14.20 = 2.80 > 2.685754(\text{LSR}_2) & \text{(significant)}
\end{array}$$

$$\begin{array}{ll}
\overline{y}_{3.} \text{ vs } \overline{y}_{1.} = 14.20 - 10.70 = 3.50 > 2.803983(\text{LSR}_3) & \text{(significant)} \\
\overline{y}_{3.} \text{ vs } \overline{y}_{2.} = 14.20 - 13.40 = 0.80 < 2.685754(\text{LSR}_2) & \text{(not significant)}
\end{array}$$

$$\overline{y}_{2.} \text{ vs } \overline{y}_{1.} = 13.40 - 10.70 = 2.70 > 2.685754(\text{LSR}_2) \quad \text{(significant)}$$

Using the Duncan's multiple range test, we have come to exactly the same conclusions that we reached under the LSD. That is, μ_1 and μ_2 are significantly different from each other; μ_2 and μ_3 are not significantly different from each other, while μ_4 and μ_5 are significantly different from each other and from others. Thus, the extraction method 5 turned out the highest mean yield of all the five extraction methods studied, followed by μ_4, then by both μ_2 and μ_3, with the least mean yield coming from method 1.

2.9.3 Tukey's Studentized Range Test

There is yet another test that could be used to study the differences between the means which have been found significantly different in the ANOVA of a typical CRD to classify them into groups with homogeneous means. This method is often called the Tukey's studentized range (Tukey, 1953) or the Tukey's method. The method is used for all pairwise comparisons of means, and utilizes the differences in the means

$$\mu_i - \mu_j \quad i \neq j$$

If the sample sizes in group i and j are equal, then the confidence attached to the range of the set is $100(1-\alpha)\%$. If the sample sizes are not equal, then the confidence is greater than $100(1-\alpha)\%$. There is a generalized version of the Tukey's method for working with unequal sample sizes which is often called the Tukey–Kramer method. Since the data we have for balanced CRD have equal number of observations per group, this method would be useful for comparing the means after they have been found to be significantly different in ANOVA.

Tukey's method is based on the distribution of the studentized range. Typically, we draw n independent observations from a population Y, where $Y \sim N(\mu, \sigma^2)$. Suppose we have an estimate σ^2 denoted by s^2 which is based on, say, δ degrees of freedom, and this estimate is independent of the sample values, y_i then we define the studentized range as

$$q_{(n,\delta)} = \frac{r}{s} \tag{2.20}$$

where r is the range of the data obtained as *maximum–minimum*. In order to use Tukey's method for testing the hypothesis that

$$\mu_i - \mu_j = 0 \quad i \neq j$$

in a typical CRD, we simply need to obtain the quantity

$$\overline{y}_{i.} - \overline{y}_{j.} \pm q_{[k,k(n-1),1-\alpha]}\sqrt{\frac{MS_{\mathrm{E}}}{n}} \tag{2.21}$$

where there are k groups being compared in the CRD, each made up of n observations. This quantity gives the $100(1-\alpha)\%$ confidence interval for the difference between the pair of means. So we should be able to reject the hypothesis above, if the interval does not include zero. The quantities $q_{[k,k(n-1),1-\alpha]}$ are tabulated and presented in Appendices A11 through A13. Tukey's method is similar to the Duncan's multiple range tests because one has to first arrange the k means to be compared from the largest to the least. Thereafter, the largest is compared to the least, by computing the interval in Equation 2.21; after which the largest is compared to the next least; and so on; comparison is stopped if either the largest is compared to the second largest, or nonsignificance occurs. Next, the second largest is compared to the least, then the next largest is compared to the second smallest, until a nonsignificant result is obtained, or the second largest is compared to the third largest. The comparison is carried out in this manner, stopping at each stage if a nonsignificant result is obtained. At the end of the comparisons, we should graphically represent the relationships between the means. We will demonstrate Tukey's method using data of Example 2.2.

Arranging the data we see that

$$\overline{y}_{5.} > \overline{y}_{4.} > \overline{y}_{3.} \geq \overline{y}_{2.} \geq \overline{y}_{1.}$$

$k = 5$, $n = 10$, $nk - k = 45$, and from the tables, Appendices A11 through A13, we read $q_{(5,45,0.99)} = 4.9025$ so that

$$q_{(5,45,0.99)}\sqrt{\frac{MS_{\text{E}}}{n}} = 4.9025\sqrt{\frac{4.982222}{10}} = 3.46042$$

$$\Rightarrow (\overline{y}_{i.} - \overline{y}_{j}) \pm 3.46042 \text{ by Equation 2.21}$$

Thus, we compare two means μ_i and μ_j and by comparing $\overline{y}_{i.}$ and $\overline{y}_{j.}$ we have the following results:

First, we compare μ_5 and μ_1

$$(19.7 - 10.7) \pm 3.46042 = (5.53958, 12.46042)$$

The interval does not contain zero, so there is significant difference between μ_5 and μ_1. Next, we compare μ_5 and μ_2 yielding

$$(19.7 - 13.4) \pm 3.46042 = (2.83958, 9.76042)$$

The interval does not contain zero, so there is significant difference between μ_5 and μ_2. Next, we compare μ_5 and μ_3 yielding

$$(19.7 - 14.2) \pm 3.46042 = (2.03958, 8.96042)$$

The interval does not contain zero, so there is significant difference between μ_5 and μ_3. Next, we compare μ_5 and μ_4 yielding

$$(19.7 - 17.0) \pm 3.46042 = (-0.76042, 6.16042)$$

The interval contains zero, so there is no significant difference between μ_5 and μ_4.

Next, we compare μ_4 and μ_1

$$(17.0 - 10.7) \pm 3.46042 = (2.83958, 9.76042)$$

The interval does not contain zero, so there is significant difference between μ_4 and μ_1. Next, we compare μ_4 and μ_2 yielding

$$(17.0 - 13.4) \pm 3.46042 = (0.13958, 7.06042)$$

The interval does not contain zero, so there is significant difference between μ_4 and μ_2. Next, we compare μ_4 and μ_3 yielding

$$(17.0 - 14.2) \pm 3.46042 = (-0.66042, 6.26042)$$

This interval contains zero, so there is no significant difference between μ_4 and μ_3.

Next, we compare μ_2 and μ_1

$$(13.4 - 10.7) \pm 3.46042 = (-0.76042, 6.16042)$$

This interval contains zero, so there is no significant difference between μ_2 and μ_1.

The Tukey's analysis indicates the following pairs of means which are not significantly different from one another:

$$(\mu_5, \mu_4), (\mu_4, \mu_3), (\mu_3, \mu_2), (\mu_2, \mu_1).$$

Tukey's studentized range gives more precise results because the intervals are narrower than those of the LSD. However, the Tukey's test is more conservative than the Duncan's test. It has a Type I error rate of α for all pairwise comparisons on an experimentwise basis.

2.9.4 SAS Program for Analysis of Responses of the Experiment in Example 2.2

Next, we write a SAS program to analyze the data. In the program, we perform ANOVA and invoke the Tukey's, LSD and Duncan's tests to compare the treatment means, that is the means of extraction methods used to extract nutrients. The outcomes of the analyses agree with results obtained by manual analysis.

SAS PROGRAM

```
data crdone;
input method nutrient@@;
datalines;
1 10 1 12 1 10 1 11 1 12 1 9  1 11 1 13 1 10 1 9
2 14 2 14 2 14 2 12 2 16 2 10 2 14 2 15 2 10 2 15
3 17 3 13 3 15 3 12 3 16 3 10 3 15 3 15 3 13 3 16
4 19 4 15 4 19 4 14 4 23 4 12 4 17 4 17 4 15 4 19
5 20 5 19 5 23 5 16 5 20 5 18 5 21 5 22 5 18 5 20
;
proc print data=crdone;
proc anova data=crdone;
class method;
model nutrient=method;
means method/tukey alpha=0.01;
means method/lsd alpha=0.01;
means method/Duncan alpha=0.01;run;
```

SAS OUTPUT FOR EXAMPLE 2.2

Obs	method	nutrient
1	1	10
2	1	12
3	1	10
4	1	11
5	1	12
6	1	9
7	1	11
8	1	13

```
Obs    method    nutrient
  9      1          10
 10      1           9
 11      2          14
 12      2          14
 13      2          14
 14      2          12
 15      2          16
 16      2          10
 17      2          14
 18      2          15
 19      2          10
 20      2          15
 21      3          17
 22      3          13
 23      3          15
 24      3          12
 25      3          16
 26      3          10
 27      3          15
 28      3          15
 29      3          13
 30      3          16
 31      4          19
 32      4          15
 33      4          19
 34      4          14
 35      4          23
 36      4          12
 37      4          17
 38      4          17
 39      4          15
 40      4          19
 41      5          20
 42      5          19
 43      5          23
 44      5          16
 45      5          20
 46      5          18
 47      5          21
 48      5          22
 49      5          18
 50      5          20

The SAS System   14:26 Tuesday, October 17, 2006  2

            The ANOVA Procedure

          Class Level Information

          Class      Levels  Values

          method        5   1 2 3 4 5

      Number of Observations Read     50
      Number of Observations Used     50
```

```
                                 The ANOVA Procedure

Dependent Variable: nutrient

                                      Sum of
   Source               DF           Squares   Mean Square  F Value  Pr > F

   Model                 4       477.8000000   119.4500000    23.98  <.0001

   Error                45       224.2000000     4.9822222

   Corrected Total      49       702.0000000

           R-Square    Coeff Var    Root MSE   nutrient Mean

           0.680627     14.88059    2.232089        15.00000

   Source               DF          Anova SS   Mean Square  F Value  Pr > F

   method                4       477.8000000   119.4500000    23.98  <.0001

           The SAS System   14:26 Tuesday, October 17, 2006  4

                            The ANOVA Procedure

            Tukey's Studentized Range (HSD) Test for nutrient

NOTE: This test controls the Type I experimentwise error rate, but it
      generally has a higher Type II error rate than REGWQ.

           Alpha                                    0.01
           Error Degrees of Freedom                45
           Error Mean Square                        4.982222
           Critical Value of Studentized Range      4.89269
           Minimum Significant Difference           3.4535

      Means with the same letter are not significantly different.

             Tukey   Grouping        Mean      N  method

                          A       19.7000     10       5
                          A
                 B        A       17.0000     10       4
                 B
                 B        C       14.2000     10       3
                          C
                 D        C       13.4000     10       2
                 D
                 D                10.7000     10       1

                       The ANOVA Procedure

                     t Tests (LSD) for nutrient

NOTE: This test controls the Type I comparisonwise error rate, not the
      experimentwise error rate.
```

```
            Alpha                             0.01
            Error Degrees of Freedom        45
            Error Mean Square                 4.982222
            Critical Value of t               2.68959
            Least Significant Difference      2.6848

    Means with the same letter are not significantly different.

        t Grouping          Mean      N    method

                A        19.7000     10         5

                B        17.0000     10         4

                C        14.2000     10         3
                C
                C        13.4000     10         2

                D        10.7000     10         1

               The ANOVA Procedure

          Duncan's Multiple Range Test for nutrient

NOTE: This test controls the Type I comparisonwise error rate, not the
      experimentwise error rate.

              Alpha                          0.01
              Error Degrees of Freedom     45
              Error Mean Square              4.982222

        Number of Means    2        3        4        5
        Critical Range     2.685    2.799    2.877    2.934

        Means with the same letter are not significantly different.

        Duncan Grouping          Mean      N    method

                      A       19.7000     10         5

                      B       17.0000     10         4

                      C       14.2000     10         3
                      C
                      C       13.4000     10         2

                      D       10.7000     10         1
```

The SAS output mirrors the manual analysis. The Tukey's test output contains the same ambiguities we found by manual analysis. Duncan's and LSD tests present clear classifications and differences.

2.10 Checking Model Assumptions

The essential parts of any mathematical model include the basic assumptions on which the model has been built. If any of these assumptions is violated, conclusions reached on the basis of the model would be suspect. In the one-way ANOVA, we assumed that

$$y_{ij} = \mu + \alpha_i + \varepsilon_{ij}$$

and $\varepsilon_{ij} \sim N(0, \sigma^2)$. Explicitly, the assumptions are that the effect of each treatment is additive and that this model adequately describes the data. Further, there is no other explanatory variable (factors) whose influence contributes to the observed response which was omitted from the design; errors are normally and independently distributed (NID) random variables with mean, zero, and variance, σ^2, that is, $\varepsilon_{ij} \sim$ NID(0, σ^2). It is prudent after ANOVA, to carry out a check to ensure that these basic assumptions are satisfied.

One way of checking for violations of the assumptions is by looking at the residuals. The residuals are obtained after ANOVA as

$$\varepsilon_{ij} = y_{ij} - \hat{y}_{ij}$$

but $\hat{y}_{ij} = \hat{\mu}_i$ (see Equation 2.17), so that the residuals are obtained as the difference between each observation and its group mean as in the following equation:

$$\varepsilon_{ij} = y_{ij} - \overline{y}_{i.} \qquad (2.22)$$

Examination of residuals is the first step that should be taken after ANOVA to find out whether there are any violations of the assumptions of the model. Such a check on the residuals could indicate a number of violations. If the residuals have a regular pattern, it may be due to an uncontrolled factor systematically portrayed in the residuals. In such circumstances, randomization or blocking could be used to correct the violation. Another violation that could be detected is nonnormality of the errors (or responses). Although the F-distribution used in ANOVA as well as the t-distribution (used in Chapter 1) are robust to departures from normality, and this reduces the seriousness of some of the violations of this assumption, it is important that any serious departures be investigated and properly dealt with. Sometimes, this process may lead to transformation of the data. Further, it may be discovered that error variances change from one treatment to another, violating one of the basic assumptions, that is, the assumption of equality of variances across levels of factors or treatment groups. This can be corrected by either transforming the data (where possible); otherwise, it would be better to compare the treatment means using individual estimates of treatment variances rather than a combined estimate. How to treat these departures from basic assumptions was treated in Cochran and Cox (1957), Wetherill (1981), Keppel and Wickens (2004), Montgomery (2005), Tabachnick and Fidell (2007). A discussion of graphical diagnostic methods using residuals follows in the next section.

2.10.1 Analysis of Residuals

In Section 2.10, we mentioned that some of the violations of basic assumptions of our model could be detected by studying the resulting residuals of the ANOVA (usually obtained by using Equation 2.22). As the procedure for examining residuals which apply in one-way ANOVA is basically the same for most of the other models we shall study, we present some of the methods here. Indeed, checking the adequacy of the models ought to be a regular part of all the analysis of data we obtain from the experiments we perform under different designs. In any study, it is advisable to carry out residual analysis; otherwise, it is possible to make conclusions which are not valid because some of the underlying assumptions of the basic model have been violated. Diagnostic treatment of residuals involves carrying out different plots of these residuals against some variables in the experiment. As a general rule, the residuals obtained after analysis of the experimental observations can be plotted against any variable in the model such as treatments, blocks, fitted variables, time sequence at which observations were made, and so on. Plots against time sequence has the special advantage of detecting departures from independence for residuals. It is, however, not proper to plot residuals against the response variables. This is because the least squares curve fitting on which the models we use depend leaves the residuals correlated with the response variables. Typical pictures, some, indicating departures from assumptions which could be obtained when plots of residuals are carried out against some variables are shown in Figures 2.1 through 2.5.

Interpretations of the underlying violations of the model assumptions in any of the plots depend on the variables against which the residuals are plotted. In Figure 2.1, there is no violation of any of the assumptions since the residuals appear random and most of the residuals lie between $\pm 2\sqrt{(MS_{\text{residual}})}$. A good residual plot should have most residuals in the interval $\pm 2\sqrt{(MS_{\text{residual}})}$. Figure 2.2 indicates heteroscedasticity (meaning nonhomogeneity of variances for the different groups), regardless of whether the plots are against the fitted values or any other quantity in the model. If this is observed in the course of analysis of the responses of an experiment, it may be helpful to transform the variables. Figure 2.3 indicates trend and if it arises as a result of a plot of residuals against time, a linear term in time may need to be added to the

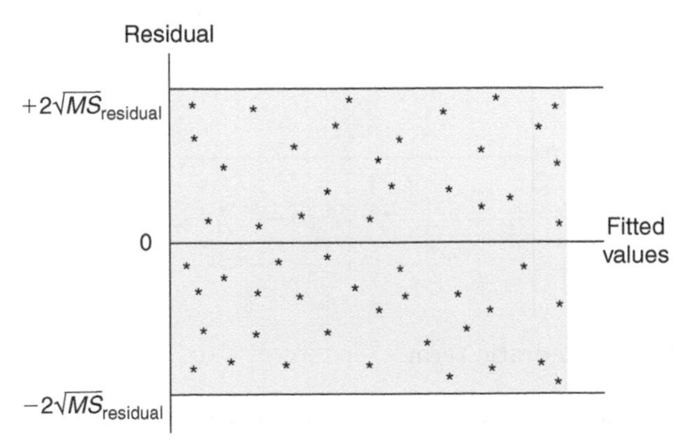

FIGURE 2.1: Random error, no violation of model assumption.

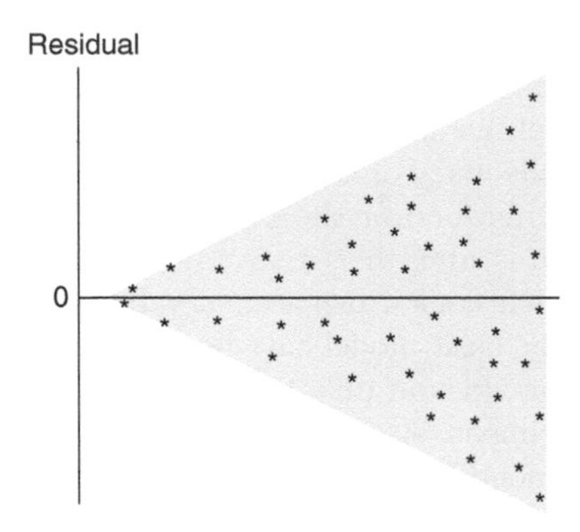

FIGURE 2.2: Inequality of variances.

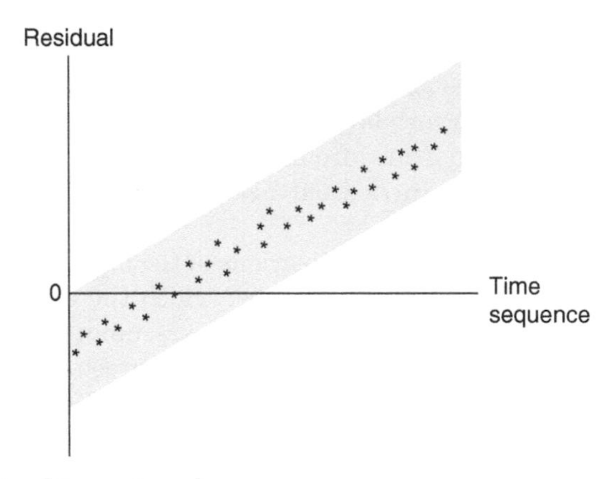

FIGURE 2.3: Linear trend.

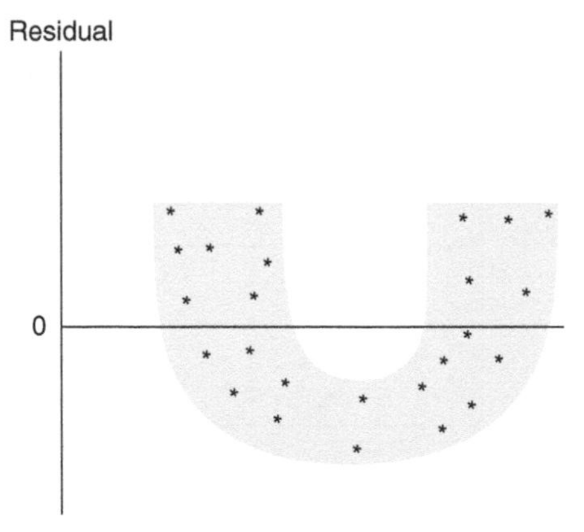

FIGURE 2.4: Quadratic term.

design. The diagnosis will be different if the plot was carried out against fitted variables; if this indicates trend, then there may be some error of calculation or analysis which should be corrected. If Figure 2.4 is a plot of residuals against a variable in the model, it indicates that a quadratic term should be added to

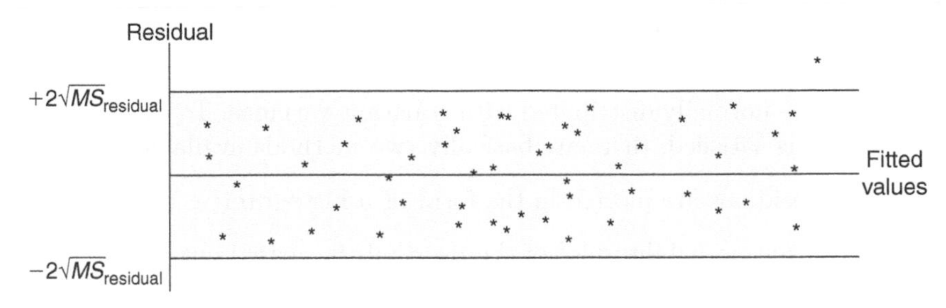

FIGURE 2.5: An outlier.

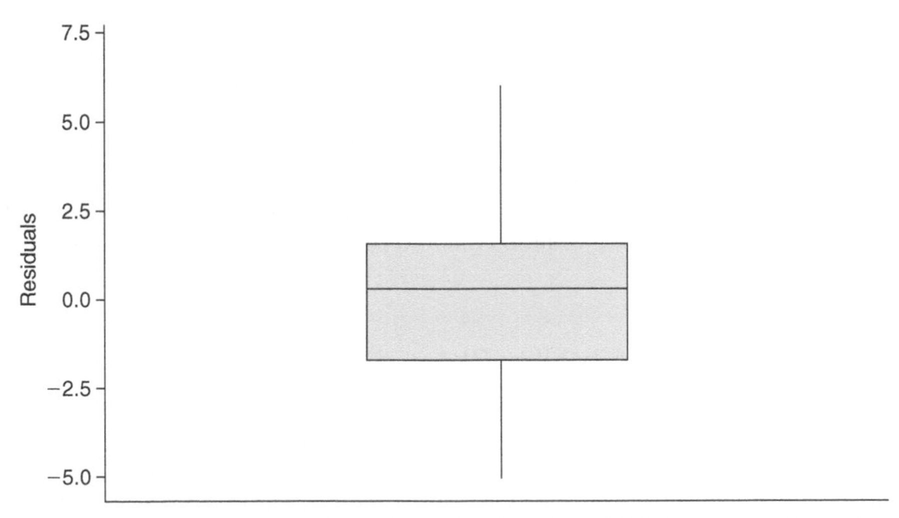

FIGURE 2.6: Boxplot of residuals for fitted model (Equation 2.3) to data for Example 2.2 showing no outlier.

the model or indeed another explanatory variable. Figure 2.5 indicates that there is an outlier. When such an outlier is detected, calculations, entries, and records should be checked to ensure that no mistake has been made. Barnett and Lewis (1994), John and Prescott (1975), and Stefansky (1972) discuss methods of detecting outliers. Another method of detecting outliers is to carry out a boxplot of the residuals. Methods for obtaining a boxplot of a dataset is readily available in most statistical texts that deal with elements of statistics. A boxplot is used as part of exploratory data analysis. If any reader is unfamiliar with boxplots, we suggest consulting a text on the elements of statistics. A boxplot would usually indicate whether a value in a dataset is an outlier. Here, we present the boxplot for the residuals of the fitted model (Equation 2.3) for the data of Example 2.2.

The boxplot in Figure 2.6 indicates that none of the residuals is an outlier. Customarily, Figure 2.6 would indicate the location of an outlier with an asterisk. In the next few sections, we discuss further some of the violations that have been depicted by these plots (Figures 2.1 through 2.5). Apart from initial inspection of residuals, possible methods of detection of departures from model assumptions are treated.

2.10.2 Checking for Normality

One of the assumptions underlying our model is $\varepsilon_{ij} \sim \text{NID}(0,\sigma^2)$, that is, the errors are normally distributed with constant variance. To check that this assumption is satisfied, there are basically two methods available:

1. The residuals are plotted in the form of a histogram
2. A normal probability plot of the residuals is carried out

2.10.2.1 Histogram

The residuals are plotted in the form of a histogram from which it is possible to deduce whether there is any departure from normality. However, this method is fraught with problems since the sample sizes used in the experiments are not often large enough to ensure that any departures observed is a serious violation of this assumption. With small samples, if small departures from the expected distribution occur, they are not to be taken seriously; only gross violations should be viewed seriously and be further investigated.

2.10.2.2 Normal Probability Plot

More often, it is better to use the normal probability plot of the residuals as a way of checking whether the normality assumption is violated. The plot also helps in the detection of any outliers whose contributions might adversely affect the conclusions. The plots are usually carried out on a normal probability paper. The normal probability paper is a graph paper specially designed so that plots of cumulative normal distribution appears on it as a straight line. Most computer packages for statistical analysis contain procedures for obtaining the normal probability plot. These could be used for normal probability plots instead of carrying them out on normal probability paper. In addition, normal probability paper is becoming difficult to obtain; soon, computer-based plots may well be the only way to carry out this investigation. However, to understand the foundations for the plot, we discuss how to carry out the plot on a normal probability paper. To obtain a normal probability plot, we arrange the residuals in ascending order of their magnitude so that if we plot the jth order residual against the cumulative probability, $F_j = (j - 0.5)/N$ (where $N = kn$ is the total number of observations), on the normal probability paper, a straight line will result if the underlying distribution is normal. We shall illustrate the normal probability plot by constructing one using the residuals of the nutrient extraction experiment in Table 2.4. In the same table, we show the responses (y), the estimated responses ($\hat{y}$), residuals, ordered residuals, calculated by using Equation 2.9, and the corresponding cumulative probabilities. As a result of our analysis, the ordered residuals are shown in Table 2.4.

TABLE 2.4: Responses, Fitted Responses, Ordered Residuals and Their Cumulative Probabilities (Extraction Method Experiment)

Order (j)	y	$\hat{y}$	Residuals	Ordered residuals	$\frac{(j-0.5)}{N}$
1	10	10.7	−0.7	−5.0	0.01
2	12	10.7	1.3	−4.2	0.03
3	10	10.7	−0.7	−3.7	0.05
4	11	10.7	0.3	−3.4	0.07
5	12	10.7	1.3	−3.4	0.09
6	9	10.7	−1.7	−3.0	0.11
7	11	10.7	0.3	−2.2	0.13
8	13	10.7	2.3	−2.0	0.15
9	10	10.7	−0.7	−2.0	0.17
10	9	10.7	−1.7	−1.7	0.19
11	14	13.4	0.6	−1.7	0.21
12	14	13.4	0.6	−1.7	0.23
13	14	13.4	0.6	−1.7	0.25
14	12	13.4	−1.4	−1.4	0.27
15	16	13.4	2.6	−1.2	0.29
16	10	13.4	−3.4	−1.2	0.31
17	14	13.4	0.6	−0.7	0.33
18	15	13.4	1.6	−0.7	0.35
19	10	13.4	−3.4	−0.7	0.37
20	15	13.4	1.6	−0.7	0.39
21	17	14.2	2.8	0.0	0.41
22	13	14.2	−1.2	0.0	0.43
23	15	14.2	0.8	0.3	0.45
24	12	14.2	−2.2	0.3	0.47
25	16	14.2	1.8	0.3	0.49
26	10	14.2	−4.2	0.3	0.51
27	15	14.2	0.8	0.3	0.53
28	15	14.2	0.8	0.6	0.55
29	13	14.2	−1.2	0.6	0.57
30	16	14.2	1.8	0.6	0.59
31	19	17	2	0.6	0.61
32	15	17	−2	0.8	0.63
33	19	17	2	0.8	0.65
34	14	17	−3	0.8	0.67
35	23	17	6	1.3	0.69
36	12	17	−5	1.3	0.71
37	17	17	0	1.3	0.73
38	17	17	0	1.6	0.75
39	15	17	−2	1.6	0.77
40	19	17	2	1.8	0.79
41	20	19.7	0.3	1.8	0.81
42	19	19.7	−0.7	2.0	0.83
43	23	19.7	3.3	2.0	0.85
44	16	19.7	−3.7	2.0	0.87
45	20	19.7	0.3	2.3	0.89
46	18	19.7	−1.7	2.3	0.91
47	21	19.7	1.3	2.6	0.93
48	22	19.7	2.3	2.8	0.95
49	18	19.7	−1.7	3.3	0.97
50	20	19.7	0.3	6.0	0.99

TABLE 2.5: The Nutrient Contents of the Fruit Varieties and Their Variances

	Method 1	**Method 2**	**Method 3**	**Method 4**	**Method 5**
	10	14	17	19	20
	12	14	13	15	19
	10	14	15	19	23
	11	12	12	14	16
	12	16	16	23	20
	9	10	10	12	18
	11	14	15	17	21
	13	15	15	17	22
	10	10	13	15	18
	9	15	16	19	20
Variance(s_i^2)	1.788889	4.266667	4.622222	10	4.233333

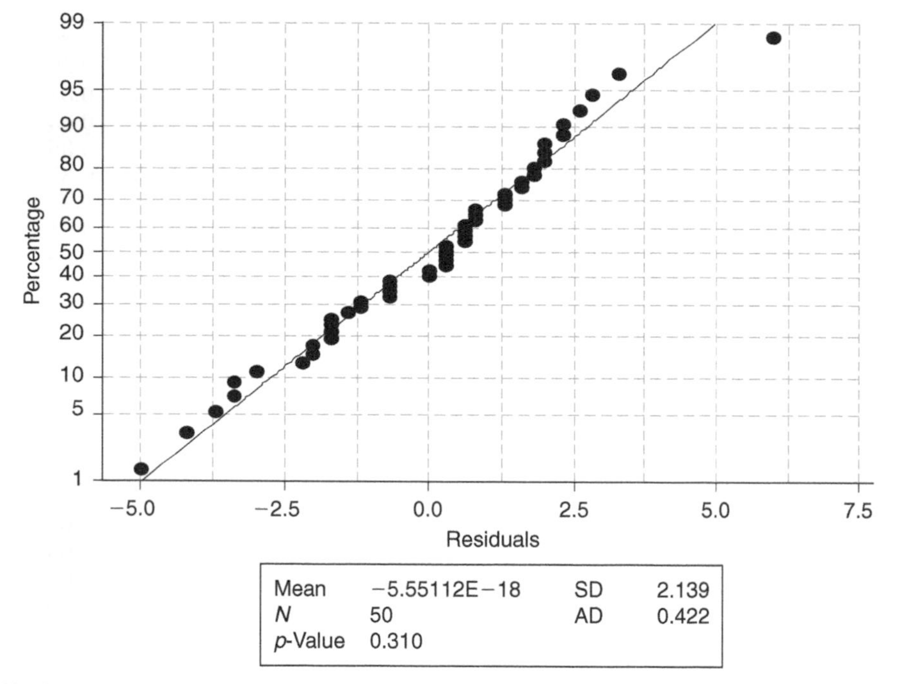

FIGURE 2.7: Normal probability plot of residuals for the fruit variety problem.

As can be seen in the plot of the residuals (Figure 2.7), there is not much departure from the expected straight line, when fluctuations due to the smallness of the size of the sample used in the experiment are taken into consideration. This would indicate that the normality assumption is satisfied. Also if we look at the Anderson–Darling test for normality, the test statistic is 0.422 and its p-value is 0.310. Thus, the data are not significant, leading us to accept that the residuals are normally distributed. We stated earlier that normal probability plots could help in the detection of an outlier; that is, a residual which stands out from the others. It could arise because the corresponding observation from the experiment is either by far larger or smaller than all other observations obtained in the experiment for the particular treatment from which it arose. When this is detected, it would be routine to scrutinize the particular observation that gave rise to the residual to find out whether it was wrongly entered in the calculation. Further, the experiment that gave rise to the observation should be checked to make sure that every thing was properly carried out. Once all is known to be in order, the outlier is accepted. Sometimes, the outliers may be more informative than the residuals of other observations and should not be discarded unnecessarily.

Apart from the plots, there are a number of tests which could be performed to check whether the normality assumption is satisfied. Apart from the Kolmogorov–Smirnov test (Kolmogorov, 1933; Smirnov, 1939) and χ^2 tests, one other test which is highly recommended is the W-test developed by Shapiro and Wilks (1965). For a comparison of the performances of these tests, see Shapiro et al. (1968).

2.10.3 Nonhomogeneity of Variances

Residual plots could indicate that variances of treatments in our model may not be equal. As pointed out earlier, nonhomogeneity of variances is one of the ways the basic assumptions underlying our analysis of variance could be violated. If we suspect that this is the case, then we could, apart from checking this by plots of residuals, carry out tests for the nonhomogeneity of variances. A number of tests have been developed for this phenomenon. These tests were developed by Bartlett (1934), Bartlett and Kendall (1946), and Burr and Foster (1972), among others. A good review of suitable tests was carried out by Anderson and McLean (1974). We discuss a test which is widely used for this purpose due to Bartlett (1934). Tests show that Bartlett's test is sensitive to nonnormality and should not be applied if there is reason to suspect that the data are not normally distributed. In such situations, the Bartlett's test often gives significant results. Bartlett's method is used to test the hypothesis:

$$H_0 : \sigma_1^2 = \sigma_2^2 = \cdots = \sigma_k^2 \quad \text{vs} \quad H_1 : \sigma_i^2 \neq \sigma_j^2 \quad for\ some\ i, j$$

To test the hypothesis, we shall compute the statistic B, whose distribution is approximately χ^2 with $k-1$ degrees of freedom, if there are k treatments or indeed k random samples, and if these k random samples are from independent normal populations. The statistic B is calculated as

$$B = \frac{2.3026A}{D} \tag{2.23}$$

where

$$A = (N-k)\log_{10} s_c^2 - \sum_{i=1}^{k}(n_i - 1)\log_{10} s_i^2 \quad (2.24a)$$

$$D = 1 + \left[\frac{1}{3(k-1)}\right]\left[\sum_{i=1}^{k}\left(\frac{1}{n_i - 1}\right) - \frac{1}{N-k}\right] \quad (2.24b)$$

$$s_c^2 = \sum_{i=1}^{k}\left[\frac{(n_i - 1)s_i^2}{N-k}\right] \quad (2.24c)$$

while s_i^2 is the sample variance of the ith population.

In applying the Bartlett's test here, we treat each set of observations obtained for each treatment as a sample from an independent normal population. We shall now apply the test to the data on the study of the nutrient contents of the varieties of the methods in nutrient extraction of fruit (Example 2.2). The data as we recall were

$$s_c^2 = \frac{\begin{array}{c}(10-1)(1.788889) + (10-1)(4.2666667) \\ + (10-1)(4.622222) + (10-1)(10) + (10-1)(4.233333)\end{array}}{50-5} = 4.982222$$

$$A = 45\log_{10}(4.982222) - 9[\log_{10}(1.788889) + \log_{10}(4.266667)$$

$$+ \log_{10}(4.622222) + \log_{10}(10) + \log_{10}(4.23333)] = 2.816191$$

$$D = 1 + \frac{1}{3(4)}\left(\frac{5}{9} - \frac{1}{45}\right) = 1.04444$$

$$B = \frac{2.3026(2.816191)}{1.04444} = 6.208625$$

We carry out the test at 1% level of significance. From the tables, $\chi^2_{(0.01,4)} = 13.277 > B$. We do not reject the null hypothesis of homogeneity of variances.

2.10.4 Modified Levene's Test for Homogeneity of Variances

Because of concern about the sensitivity of the Bartlett's test to nonnormality, one of the tests proposed is the modified Levene's test (Levene, 1960) as modified by Brown and Forsythe (1974). A good survey of the method can be found in Conover et al. (1981). The test uses the absolute deviation of each observation from its group median. Thus in a typical one-way ANOVA with model

$$y_{ij} = \mu + \alpha_i + \varepsilon_{ij} \quad i = 1, 2, \ldots, k; \; j = 1, 2, \ldots, n_i$$

for each observation y_{ij}, we define d_{ij}, its absolute deviations from its group median as

$$d_{ij} = |y_{ij} - \tilde{y}_{i.}| \quad i = 1, 2, \ldots, k; \ j = 1, 2, \ldots, n_i$$

The usual one-way ANOVA, is used to determine whether there is significant difference between the means of the deviations, d_{ij} for the k groups. If this happens, then the variances of the k groups are considered not equal.

2.10.4.1 Application of Data from Example 2.2 on Varieties of Nutrient Extractions of Fruit

From the data of Example 2.2, the medians for the five methods for extraction of the fruit nutrients are shown here in the table:

Medians for the methods of extraction

Method 1	**Method 2**	**Method 3**	**Method 4**	**Method 5**
10.7	13.4	14.2	17	19.7
$\tilde{y}_{1.}$	$\tilde{y}_{2.}$	$\tilde{y}_{3.}$	$\tilde{y}_{4.}$	$\tilde{y}_{5.}$

The deviations from the medians are as shown in the following table:

d_{1j}	d_{2j}	d_{3j}	d_{4j}	d_{5j}
0.5	0	2.5	2	0
1.5	0	1.5	2	1
0.5	0	0.5	2	3
0.5	2	2.5	3	4
1.5	2	1.5	6	0
1.5	4	4.5	5	2
0.5	0	0.5	0	1
2.5	1	0.5	0	2
0.5	4	4.5	2	2
1.5	1	0.5	2	0
11	**14**	**19**	**24**	**15**

$$SS_{\text{total}} = 0.5^2 + 1.5^2 + \cdots + 2^2 + 0^2 - \frac{83^2}{50} = 108.22$$

$$SS_{\text{groups}} = \frac{11^2 + 14^2 + 19^2 + 24^2 + 15^2}{10} - \frac{83^2}{50} = 10.12$$

$$SS_{\text{E}} = 108.22 - 10.12 = 98.10$$

From the analysis (Table 2.6), it is clear that with a p-value of 0.340849, we should conclude that the data are not significant; the variances are therefore homogeneous.

TABLE 2.6: ANOVA for Testing for Homogeneity (Example 2.2)

Source	Sum of squares	Degrees of freedom	Mean square	F-ratio	p-Value
Groups	10.12	4	2.53	1.16055	0.340849
Error	98.1	45	2.18		
Total	108.22	49			

2.10.5 Dealing with Heteroscedasticity

If the test indicates heteroscedasticity (that is nonhomogeneity of variances), it is usual to transform the data to equalize the variances before carrying out the ANOVA. The appropriate transformation to be used depends on which distribution is perceived by the experimenter to be the underlying distribution of the observations. This can be achieved by inspecting the residuals. If the underlying distribution is the Poisson distribution, then the appropriate transformation may be $y^*_{ij} = \sqrt{y_{ij}}$ or $y^*_{ij} = \sqrt{(1 + y_{ij})}$. When the binomial distribution is considered to be the appropriate distribution of the responses, then $y^*_{ij} = \sin^{-1}(y_{ij})$ while the transformation when the underlying distribution is log-normal is $y^*_{ij} = \ln(y_{ij})$. Sometimes when there are no obvious transformations, the experimenter will have to decide empirically what transformations may be appropriate. Whatever transformations are eventually adopted, it should be noted that the conclusions of the analysis apply only to the transformed data and not to the original responses before transformation. A number of possible transformations were treated in Bartlett (1936, 1947). Box and Cox (1964) discuss some ways to determine the formula for the transformation when it is unknown. Another way may be to use the Welch F-test (Welch, 1951) if the variances are not equal.

2.10.5.1 Variance-Stabilizing Transformations

When a violation of the constant variance assumption is found, the goal of the experimenter would be to find a transformation which will make the variances a function of the responses constant, thus enabling the model to be applied and analysis to be carried out. These are called variance, stabilizing transformations. The following can help us to obtain adequate transformations of the original responses.

Let the mean of the original responses y be μ; suppose we find that the standard deviation σ_y, of y is proportional to some power of the mean, so that

$$\sigma_y = \mu^\alpha$$

If we set the transformation as a power of the original data, so that

$$y^* = y^\lambda$$

Box et al. (2005, p. 329) show that for the variance of y^* to be constant, λ must be equal to $1-\alpha$. Thus some appropriate transformations for various types of data are given in the following table:

Relationship between σ_y and μ	α	$\lambda = 1-\alpha$	**Transformation**	**Example**
$\sigma_y \propto$ constant	0	1	No transformation	
$\sigma_y \propto \mu^{\frac{3}{2}}$	3/2	−1/2	Reciprocal square root	
$\sigma_y \propto \mu^{\frac{1}{2}}$	1/2	1/2	Square root	Poisson
$\sigma_y \propto \mu$	1	0	Log	Sample variance
$\sigma_y \propto \mu^2$	2	−1	Reciprocal	

2.10.5.2 Use of Welch *F*-Test to Deal with Responses with Unequal Variances

Welch (1951) developed a test for comparing groups which have equal means but their variances could be equal or unequal. The test statistic is:

$$W = \frac{\sum_i w_i(y_{i.} - \bar{y}_{..})^2/(k-1)}{\left[1 + \frac{2(k-2)}{(k^2-1)}\sum_i \frac{(1-w_i/u)^2}{(n_i-1)}\right]}$$

where $w_i = n_i/s_i^2$; $u = \sum_i w_i$; $\bar{y}_{..} = \sum_i w_i y_{i.}$ is approximately F-distributed if the means are equal (even if variances are unequal) with $k-1$, and f degrees of freedom. f is obtained implicitly from the relation

$$\frac{1}{f} = \frac{3}{k^2-1}\sum_i \frac{(1-w_i/u)^2}{(n_i-1)}$$

2.10.6 One-Way ANOVA with Unequal Observations per Group

When the number of observations made per group, or per level of factor, are not equal, we have the unbalanced design. In this case, n_1 observations are made at level 1, n_2 at level 2, ..., and n_k at level k. The fixed effects model for the design (the equivalent of model 2.3) when the design is unbalanced is

$$y_{ij} = \mu + \alpha_i + \varepsilon_{ij} \quad i = 1, 2, \ldots, k \;\; j = 1, 2, \ldots, n_i$$

It is easy to show the results we obtained for the balanced design of model 2.3 for this unbalanced form. We illustrate the analysis of the responses of this form of the design by an example.

Example 2.2a

In a one-way experimental layout, there are k treatments with n_1 observations for treatment 1, n_2 observations for treatment 2, ..., and n_k observations for treatment k. Show that the partitions of the sums of squares for ANOVA for this design are represented by the following:

$$SS_{\text{total}} = \sum_{i=1}^{k}\sum_{j=1}^{n} y_{ij}^2 - \frac{y_{..}^2}{\sum_{i=1}^{k} n_i}$$

$$SS_{\text{treatment}} = \sum_{i=1}^{k} \frac{y_{i.}^2}{n_i} - \frac{y^2}{\sum_{i=1}^{k} n_i}$$

$$SS_{\text{Error}} = \sum_{i=1}^{k}\sum_{j=1}^{n} y_{ij}^2 - \sum_{i=1}^{k} \frac{y_{i.}^2}{n_i}$$

Use the above formulae to carry out the analysis of the following data, comparing means of treatments at 1% level of significance.

Treatment	Responses
1	64 55 65 46 66 48 58 67 54 58 60
2	86 64 75 75 69 87 60 69 71
3	89 97 96 68 90 82 80 88 76 64 71 90 78 69
4	57 78 98 76 88 59 77 65

Solution:

The equivalent of Equation 2.8 for unequal observations per level of factor is

$$SS_{\text{total}} = \sum_{i=1}^{k}\sum_{j=1}^{n_i} (y_{ij} - \overline{y}_{i.})^2 + \sum_{i=1}^{k}\sum_{j=1}^{n_i} (\overline{y}_{i.} - \overline{y}_{..})^2$$

$$= SS_{\text{E}} + SS_{\text{treatment}}; \quad \sum_{i=1}^{k} y_{i.} = y_{..}$$

We carry out the proofs required as follows:

$$SS_{\text{treatment}} = \sum_{i=1}^{k}\sum_{j=1}^{n_i} (\overline{y}_{i.} - \overline{y}_{..})^2 = \sum_{i=1}^{k}\sum_{j=1}^{n_i} \left(\frac{y_{i.}}{n_i} - \frac{y_{..}}{\sum_{i=1}^{k} n_i} \right)^2$$

$$= \sum_{i=1}^{k}\sum_{j=1}^{n_i} \left[\left(\frac{y_{i.}}{n_i}\right)^2 - 2\frac{y_{i.}}{n_i}\frac{y_{..}}{\sum_{i=1}^{k} n_i} + \frac{y_{..}^2}{\left(\sum_{i=1}^{k} n_i\right)^2} \right]$$

$$= \sum_{i=1}^{k} \left[n_i \left(\frac{y_{i.}}{n_i} \right)^2 - 2n_i \frac{y_{i.}}{n_i} \frac{y_{..}}{\sum_{i=1}^{k} n_i} + \frac{n_i y_{..}^2}{\left(\sum_{i=1}^{k} n_i \right)^2} \right]$$

$$= \sum_{i=1}^{k} \frac{y_{i.}^2}{n_i} - 2 \frac{y_{..}}{\sum_{i=1}^{k} n_i} \sum_{i=1}^{k} y_{i.} + y_{..}^2 \frac{\sum_{i=1}^{k} n_i}{\left(\sum_{i=1}^{k} n_i \right)^2}$$

$$= \sum_{i=1}^{k} \frac{y_{i.}^2}{n_i} - \frac{y_{..}^2}{\sum_{i=1}^{k} n_i}$$

$$SS_{\text{total}} = \sum_{i=1}^{k} \sum_{j=1}^{n_i} (y_{ij} - \overline{y}_{..})^2 = \sum_{i=1}^{k} \sum_{j=1}^{n_i} y_{ij}^2 - \left(\sum_{i=1}^{k} n_i \right) \overline{y}_{..}^2$$

$$= \sum_{i=1}^{k} \sum_{j=1}^{n_i} y_{ij}^2 - \left(\sum_{i=1}^{k} n_i \right) \frac{y_{..}^2}{\left(\sum_{i=1}^{k} n_i \right)^2} = \sum_{i=1}^{k} \sum_{j=1}^{n_i} y_{ij}^2 - \frac{y_{..}^2}{\sum_{i=1}^{k} n_i}$$

$$SS_{\text{E}} = SS_{\text{total}} - SS_{\text{treatment}} = \sum_{i=1}^{k} \sum_{j=1}^{n_i} y_{ij}^2 - \frac{y_{..}^2}{\sum_{i=1}^{k} n_i} - \sum_{i=1}^{k} \frac{y_{i.}^2}{n_i} + \frac{y_{..}^2}{\sum_{i=1}^{k} n_i}$$

$$= \sum_{i=1}^{k} \sum_{j=1}^{n_i} y_{ij}^2 - \sum_{i=1}^{k} \frac{y_{i.}^2}{n_i}$$

Treatment	**Observations**	**Total**
1	64 55 65 46 66 48 58 67 54 58 60	641
2	86 64 75 75 69 87 60 69 71	596
3	89 97 96 68 90 82 80 88 76 64 71 90 78 69	1138
4	57 78 98 76 88 59 77 65	598

$$SS_{\text{total}} = 64^2 + 55^2 + 65^2 + \cdots + 77^2 + 65^2 - \frac{3033^2}{42} = 7391.07$$

$$SS_{\text{treatment}} = \frac{641^2}{11} + \frac{656^2}{9} + \frac{1138^2}{14} + \frac{598^2}{8} - \frac{3033^2}{42} = 3345.644$$

$$SS_{\text{E}} = 7391.07 - 3345.644 = 4045.426$$

The analysis of variance to test the hypotheses:

$$H_0\colon \alpha_1 = \alpha_2 = \alpha_3 = \alpha_4 = 0 \quad \text{vs} \quad H_1\colon \alpha_i \neq 0 \quad \text{for some } i$$

Source	Sum of squares	Degrees of freedom	Mean square	F-ratio	p-Value
Treatment	3345.644	3	1115.2147	10.475573	3.69503E-05
Error	4045.426	38	106.45858		
Total	7391.07	41			

The data are significant. There is sufficient evidence to reject the hypothesis of equality of treatment effect because the p-value $= 0.00003695 \ll 0.01$, the level of significance. □

Next, we present the SAS analysis of the responses of the design in Example 2.2a which draws the same conclusion as we reached in our manual analysis:

SAS PROGRAM FOR EXAMPLE 2.2a

```
data examp2_2a;
input a y@@;
cards;
1 64 1 55 1 65 1 46 1 66 1 48 1 58 1 67 1 54 1 58 1 60
2 86 2 64 2 75 2 75 2 69 2 87 2 60 2 69 2 71
3 89 3 97 3 96 3 68 3 90 3 82 3 80 3 88 3 76 3 64 3 71 3 90 3 78 3 69
4 57 4 78 4 98 4 76 4 88 4 59 4 77 4 65
;
proc glm data=examp2_2a;
class a;
model y=a;
run;
```

SAS OUTPUT FOR EXAMPLE 2.2a

```
                         The GLM Procedure
                      Class Level Information

                     Class      Levels      Values

                       a             4      1 2 3 4

                 Number of Observations Read      42
                 Number of Observations Used      42
                      The GLM Procedure

Dependent Variable: y
                                     Sum of
    Source                 DF        Squares     Mean Square   F Value    Pr > F
    Model                   3    3345.643579     1115.214526     10.48    <.0001
    Error                  38    4045.427850      106.458628
    Corrected Total        41    7391.071429
            R-Square      Coeff Var     Root MSE       y Mean
            0.452660       14.28786     10.31788     72.21429
```

```
Source          DF      Type I SS     Mean Square   F Value   Pr > F
a                3    3345.643579     1115.214526     10.48   <.0001

Source          DF    Type III SS     Mean Square   F Value   Pr > F
a                3    3345.643579     1115.214526     10.48   <.0001
```

2.11 Applications of Orthogonal Contrasts

In this section, we focus on the use of orthogonal contrasts in the study of treatment effects in the CRD. We can use it to test hypothesis about a linear combination of means. It (in the form of orthogonal polynomials) can also be used to fit a response curve for a single quantitative factor set at a number of levels.

2.11.1 Analytical Study of Components of Treatment Effect Using Orthogonal Contrasts

If in the process of carrying out analysis of variance under the fixed effects model of the CRD, we reject the null hypothesis, it means that we have accepted that there is a significant difference between the means of the k treatments in this design. However, this decision does not tell us which means are different from each other. In fact, we will discover that often it is the average of a group of means that differ from the average of another group. Hence we need to further analyze the data to discover the nature of the differences which exist among the k treatment means. To carry out this further analysis, we note that the mean for each treatment, i, is written as

$$\mu_i = \mu + \alpha_i$$

Usually, we estimate μ_i from the data as $\overline{y}_{i.}$, however, we compare treatment means by using the sequence of treatment totals, $\{y_{i.}\}$. We start our further investigation by identifying two treatments whose means appear not to be different from one another and compare them, setting up the appropriate null hypotheses and the alternatives. Sometimes, we may discover that we are interested in finding out whether the average for a set of two treatment means differ from the average for another set of two. To illustrate how this is done, we consider the following example:

Example 2.3

In a pilot study to consider the effect of the concentration of nitrate in a certain fertilizer on the yield of potato, the amount of nitrate added was set at five different concentrations and four observations (yields of four replicate plots) were made for each concentration using a CRD. The responses of the experiment are given in the following table in kilograms per plot.

Nitrate concentration (%)	Potato yields (kg)
0	105, 115, 91, 141
5	135, 131, 145, 175
10	147, 143, 153, 157
15	148, 161, 161, 164
20	132, 149, 155, 164

Analyze the data and compare mean yields for different nitrate concentrations at 1% level of significance.

We see that the total yields for the different levels of nitrate concentrations are

Levels of nitrate concentration (i)	Total yields ($y_{i.}$)
1 (0%)	452
2 (5%)	586
3 (10%)	600
4 (15%)	634
5 (20%)	600

We carry out analysis of variance for the data as follows:

$$SS_{\text{total}} = 105^2 + 115^2 + \cdots + 155^2 + 164^2 - \frac{2872^2}{20} = 8332.8$$

$$SS_{\text{nitrate}} = \frac{452^2 + 586^2 + 600^2 + 634^2 + 600^2}{4} - \frac{2872^2}{20} = 4994.8$$

$$SS_{\text{E}} = 8332.8 - 4994.8 = 3338.0$$

$F(4, 15, 0.01) = 4.89$, clearly indicating that there are significant differences between mean yields for different levels of nitrate concentration (Table 2.7).

The question at this stage is: What is the nature of the differences between the means? As we said earlier, we can answer these questions by testing

TABLE 2.7: ANOVA for Nitrate Concentration Problem

Source	Sum of squares	Degrees of freedom	Mean square	F-ratio	p-Value
Nitrate	4998.4	4	1248.7	5.611	0.005759
Error	3338	15	222.533		
Total	8332.8	19			

hypothesis about the means. For instance, if we think that the mean for 0% and 5% nitrate concentrations may not differ significantly, then we can test this claim by testing the hypothesis

$$H_0 : \mu_1 = \mu_2 \quad \text{vs} \quad H_1 : \mu_1 \neq \mu_2$$

This hypothesis can easily be tested using linear combination of treatment totals

$$y_{1.} - y_{2.} = 0$$

Similarly, if we think that the average yield at 10% and 15% nitrate concentrations may not differ from the average yield for 0% and 5% nitrate concentrations, we test the hypothesis:

$$H_0 : \mu_3 + \mu_4 = \mu_1 + \mu_2 \quad \text{vs} \quad H_1 : \mu_3 + \mu_4 \neq \mu_1 + \mu_2$$

The linear combination of treatment totals for testing this hypothesis is

$$y_{3.} + y_{4.} - y_{1.} - y_{2.} = 0$$

This use of treatment totals in testing the hypotheses implies the use of a linear combination of treatment totals such that

$$C = \sum_{i=1}^{k} c_i y_{i.}$$

provided that $\sum_{i=1}^{k} c_i = 0$. Under such conditions, C is called a contrast and its sum of squares which has one degree of freedom (for the balanced CRD) is written as

$$SS_C = \frac{\left(\sum_{i=1}^{k} c_i y_i\right)^2}{n \sum_{i=1}^{k} c_i^2} \tag{2.25}$$

The formula is slightly adjusted for the unbalanced case and takes the form

$$SS_C = \frac{\left(\sum_{i=1}^{k} c_i y_i\right)^2}{\sum_{i=1}^{k} n_i c_i^2} \tag{2.26}$$

In ANOVA, we divide SS_C by the MS_{E} and compare the resulting statistic with the tabulated F at 1 and $k(n-1)$ degrees of freedom for the balanced case. For the unbalanced case we compare the statistic with tabulated F at 1 and $\sum_{i=1}^{k} n_i - k$ degrees of freedom. There is a special case of contrasts called orthogonal contrasts. This is the case when two contrasts made up of the sequences $\{c_i\}$ and $\{d_i\}$ have the property that

$$\sum_{i=1}^{k} c_i d_i = 0$$

for the balanced CRD, while its form when the design is unbalanced is

$$\sum_{i=1}^{k} n_i c_i d_i = 0$$

We shall use orthogonal contrasts in the next few sections in fitting a polynomial equation to the responses of an experiment. Here, however, we employ this concept on the above design (Example 2.3) on the yield of a chemical experiment at different nitrate concentrations. Using orthogonal contrasts, we can split the treatment sum of squares into $k-1$ independent partitions, each with one degree of freedom. By this method, we study these components of the treatment effect and determine which of them is significant. There are many ways of choosing orthogonal contrasts and some of them depend on the nature of the designed experiment. The nature of the experiment may suggest which treatment means or linear combinations of treatment means should be compared. For instance, in the design of the above experiment (Example 2.3) on nitrate concentrations, we can see that it makes sense to compare the mean yield when there is no nitrate (at 0%) and mean yield at 5% nitrate concentration. It may also suggest the comparison of the average for 0% and 5% nitrate concentrations with that of 10% and 15% nitrate concentrations. Another suggestion is to compare four times the mean at 20% nitrate concentration with the sum of the means at all the lower concentrations (that is at 0%, 5%, 10%, 15%). The above suggested components lead to the following linear combinations of treatment totals, their sums of squares, and the subsequent table of orthogonal contrasts:

$$C_1 = -y_{1.} + y_{2.} = -452 + 586 = 134 \Rightarrow SS_{C_1} = \frac{(134)^2}{4(2)} = 2244.50$$

$$C_2 = y_{3.} + y_{4.} - y_{1.} - y_{2.} = 634 + 600 - 452 - 586 = 196$$

$$\Rightarrow SS_{C_2} = \frac{(196)^2}{4(4)} = 2401.0$$

$$C_3 = y_{3.} - y_{4.} = 600 - 634 = -34 \Rightarrow SS_{C_3} = \frac{(-34)^2}{4(2)} = 144.50$$

$$C_4 = 4y_{5.} - y_{4.} - y_{3.} - y_{2.} - y_{1.} = 4(600) - 634 - 600 - 586 - 452 = 128$$

$$\Rightarrow SS_{C_4} = \frac{(128)^2}{4(20)} = 204.8$$

The table of orthogonal contrasts (Table 2.7a) as suggested by the above choices is as follows:

TABLE 2.7a: Orthogonal Contrasts

Concentration (%)	C_1	C_2	C_3	C_4
0	−1	−1	0	−1
5	1	−1	0	−1
10	0	1	1	−1
15	0	1	−1	−1
20	0	0	0	4

It is very easy to verify that these are orthogonal contrasts. We can see that $SS_{C_1} + SS_{C_2} + SS_{C_3} + SS_{C_4} = 2244.5 + 2401.0 + 144.5 + 204.8 = 4994.8 = SS_{\text{treatment}}$.

We perform ANOVA (Table 2.8) to identify those contrasts that are significant in the design as follows:

TABLE 2.8: ANOVA for the Nitrate Concentration Problem Using Orthogonal Contrasts

Source	**Sum of squares**	**Degrees of freedom**	**Mean square**	***F*-ratio**	***p*-Value**
Nitrate concentration (treatment)	4994.8	4	1248.7	5.611	0.0057587
C_1: $\mu_1 = \mu_2$	2244.5	1	2244.5	10.086	0.0062661
C_2: $\mu_3 + \mu_4 = \mu_1 + \mu_2$	2401.0	1	2401	10.789	0.0050141
C_3: $\mu_3 = \mu_4$	144.5	1	144.5	0.649	0.4330564
C_4: $4\mu_5 = \mu_1 + \mu_2 + \mu_3 + \mu_4$	204.8	1	204.8	0.92	0.3526814
Error	3338.0	15	222.5333		
Total	8332.8	19			

$F(1, 15, 0.01) = 8.68$ and $F(4, 15, 0.01) = 4.89$. Since, C_1 is significant, we reject the null hypothesis H_0: $\mu_1 = \mu_2$ and accept the alternative, H_1: $\mu_1 \neq \mu_2$. Similarly, C_2 is significant, so we reject H_0: $\mu_1 + \mu_2 = \mu_3 + \mu_4$ and accept the alternative, H_1: $\mu_1 + \mu_2 \neq \mu_3 + \mu_4$. □

2.11.1.1 SAS Program for Analysis of Data of Example 2.3

All the analyses carried out in this section regarding data for Example 2.3 can easily be carried out in SAS by invoking the GLM procedure and defining the same contrasts within its syntax to take care of the comparisons of means we carried out by the manual calculations above. We carry this out in the next program:

SAS PROGRAM

```
options nodate ps=65 nonumber;
data conc1;
input conc yield @@;
cards;
1 105 1 115 1 91 1 141 2 135 2 131 2 145 2 175 3 147 3 143
3 153 3 157 4 148 4 161 4 161 4 164 5 132 5 149 5 155 5 164
;
proc print data=conc1;
proc glm data=conc1;
class conc;
model yield=conc;
contrast 'c1' conc 1   −1 0 0   0;
contrast 'c2' conc −1 −1 1 1   0;
contrast 'c3' conc 0   0   1 −1 0;
contrast 'c4' conc 1   1   1 1   −4;
run;
```

SAS OUTPUT

Obs	conc	yield
1	1	105
2	1	115
3	1	91
4	1	141
5	2	135
6	2	131
7	2	145
8	2	175
9	3	147
10	3	143
11	3	153
12	3	157
13	4	148
14	4	161
15	4	161
16	4	164
17	5	132
18	5	149
19	5	155
20	5	164

```
The GLM Procedure

Class Level Information

Class    Levels       Values

conc          5    1 2 3 4 5

Number of Observations Read    20
Number of Observations Used    20
```

```
        The SAS System
       The GLM Procedure
Dependent Variable: yield
                                 Sum of
  Source              DF        Squares    Mean Square    F Value    Pr > F
  Model                4    4994.800000    1248.700000       5.61    0.0058
  Error               15    3338.000000     222.533333
  Corrected Total     19    8332.800000
             R-Square      Coeff Var    Root MSE    yield Mean
             0.599414       10.38827    14.91755      143.6000
  Source              DF    Type I SS      Mean Square    F Value    Pr > F
  conc                 4    4994.800000    1248.700000       5.61    0.0058
  Source              DF    Type III SS    Mean Square    F Value    Pr > F
  conc                 4    4994.800000    1248.700000       5.61    0.0058
  Contrast            DF    Contrast SS    Mean Square    F Value    Pr > F
  c1                   1    2244.500000    2244.500000      10.09    0.0063
  c2                   1    2401.000000    2401.000000      10.79    0.0050
  c3                   1     144.500000     144.500000       0.65    0.4329
  c4                   1     204.800000     204.800000       0.92    0.3526
```

Note that the outputs of SAS agree with results we obtained by manual analysis.

The following table gives another set of orthogonal contrasts which are relevant to this design and can be used in studying the effects of levels of concentration of nitrate in the experiment of Example 2.3:

Concentration (%)	C_1	C_2	C_3	C_4
0	0	1	−1	−1
5	0	0	0	4
10	0	1	1	−1
15	−1	−1	−1	−1
20	1	−1	1	−1

The reader should obtain the linear combinations of treatment totals as well as the sum of squares for the above contrasts and hence perform ANOVA to check which of them is significant (see Exercise 2.5).

2.11.2 Fitting Response Curves in One-Way ANOVA (CRD) Using Orthogonal Polynomial Contrasts

In one-way ANOVA, we investigate a single factor whose levels could either be qualitative or quantitative. When the factor is quantitative, such as time, temperature, pressure, humidity, etc., we can identify the different levels of the factor on a numerical scale. When the factor is qualitative, it is not possible to arrange its levels on a numerical scale. Typical examples include machinists, shifts of workers, batches of raw materials, etc. (as mentioned in this chapter). If the levels of the factor considered is qualitative (e.g., the methods for extracting fruit nutrients in Example 2.2), it does not make sense to think of an intermediate between methods of extraction. However, if the factor is quantitative, such as temperature, there may be a need to develop a response curve from which the estimated responses corresponding to intermediate values of the levels of a factor could be determined. The usual method for obtaining the response curves is regression analysis (in general), which has been discussed in several books. A good detailed treatment can be found in Kleinbaum et al. (1998). Many other books on basic statistics have sections on regression. In this book, we have discussed regression analysis in Chapter 1. Some texts on experimental design have good sections on regression such as Guttman et al. (1982), Chatfield (1983), and Montgomery (2005). As an example of how to use orthogonal contrasts to obtain a polynomial equation for interpolation, we consider another problem.

Example 2.4

A chemist who was studying the effect of reactant concentrations on the reaction rate of a chemical process carried out his experiments using five concentrations of the reactant: 10%, 17.5%, 25%, 32.5%, and 40%. The responses (reaction time in minutes) obtained for four replicates of the experiment at each reactant concentration are shown in the table below. Does reactant concentration have any effect on the reaction time of the process?

10%	**17.5%**	**25%**	**32.5%**	**40%**
64	46	30	20	15
57	42	37	18	13
59	39	33	25	16
62	35	28	22	14

Analysis of variance was carried out for the responses from the experiment in order to test the hypothesis

$$H_0 : \alpha_1 = \alpha_2 = \alpha_3 = \alpha_4 = \alpha_5 = 0 \quad \text{vs} \quad H_1 : \alpha_i \neq 0 \quad \textit{for some } i$$

The sums of squares are

$$\begin{aligned} SS_{\text{total}} &= 64^2 + 57^2 + \cdots + 16^2 + 14^2 - \frac{675^2}{5(4)} = 5335.75 \\ SS_{\text{treatment}} &= \frac{242^2 + 162^2 + 128^2 + 85^2 + 58^2}{4} - \frac{675^2}{5(4)} = 5164.00 \\ SS_{\text{residual}} &= 5335.75 - 5164 = 171.75 \end{aligned}$$

TABLE 2.9: ANOVA for the Study of Effect of Reactant Concentrations on Reaction Time

Source	**Sum of squares**	**Degrees of freedom**	**Mean square**	***F*-ratio**	***p*-Value**
Between treatments	5164.00	4	1291.00	112.75	5.3051E−11
Residual or error or within treatments	171.75	15	11.45		
Total	5335.75	19			

At 5% level of significance, $F(4, 15, 0.05) = 3.06$. From our analysis, the hypothesis of equality of treatments is rejected. The treatment effects are therefore significant. ∎

2.11.2.1 Fitting of Polynomials by the Least Squares Regression Method

Now as we have found that the treatment effects are significant in the above example, after carrying out all the checks to ensure that the model is adequate, we can decide to fit a polynomial equation to the responses in terms of the levels of the factors. Equations obtained this way are very useful for both interpolation as well as extrapolation, making it easy for the estimated responses to be obtained by substituting appropriate values of the factor. Caution is necessary when using equations obtained in this way for extrapolation as there are limits beyond which projections based on them are not reliable. It is not advisable to extrapolate too far from the vicinity of the values of the levels of the factor used. Suppose there are k levels of the factor, it is usually only possible to fit polynomials to such data up to a degree of $k-1$. In practice, it is more advisable to fit the lowest polynomial which adequately represents the data. This has the advantage of reducing the complexity of the fitted model as well as ensuring that there is no overfitting which renders the model ineffective as a tool for the interpolation or projection. In the light of this, it may be useful to fit the lowest possible degree of polynomial which is likely to represent the data adequately and upgrade the degree if lack of fit is detected.

If we decide to fit a first-order polynomial, that is a linear relationship, we can use methods described in Chapter 1 and find the model that fits.

To fit a polynomial of any order b $(1 \leq b \leq k-1)$ to the responses obtained in our experiments, the model is

$$y_i = a_0 + a_1 x_i + a_2 x_i^2 + \cdots + a_b x_i^b + \varepsilon_i \tag{2.27}$$

As usual, $\varepsilon_i \sim \text{NID}(0, \sigma^2)$. x_is represent the levels of the factor. By the theory of least squares regression, there are n normal equations arising from the differentiation of the least squares equation (Equation 2.28)

$$\sum_{i=1}^{n} (y_i - a_0 - a_1 x_i - a_2 x_i^2 - \cdots - a_b x_i^b)^2 = 0 \tag{2.28}$$

with respect to the parameters $a_0, a_1, \ldots, a_b$. These normal equations are

$$\begin{aligned}
na_0 + a_1\sum_{i=1}^{n} x_i + \cdots + a_b\sum_{i=1}^{n} x_i^b &= \sum_{i=1}^{n} y_i \\
a_0\sum_{i=1}^{n} x_i + a_1\sum_{i=1}^{n} x_i^2 + \cdots + a_b\sum_{i=1}^{n} x_i^b &= \sum_{i=1}^{n} x_i y_i \\
\cdot \quad \cdot \quad \cdots \quad \cdot \quad & \cdot \\
\cdot \quad \cdot \quad \cdots \quad \cdot \quad & \cdot \\
\cdot \quad \cdot \quad \cdots \quad \cdot \quad & \cdot \\
a_0\sum_{i=1}^{n} x_i^b + a_1\sum_{i=1}^{n} x^{b+1} + \ldots + a_b\sum_{i=1}^{n} x_i^{2b} &= \sum_{i=1}^{n} x_i^b y_i
\end{aligned}$$

Once these normal equations are known, the parameters of the polynomial equation can be estimated as follows:

$$\begin{bmatrix} \hat{a}_0 \\ \hat{a}_1 \\ \cdot \\ \cdot \\ \cdot \\ \hat{a}_b \end{bmatrix} = \begin{bmatrix} n & \sum_{i=1}^{n} x_i & \cdots & \sum_{i=1}^{n} x_i^b \\ \sum_{i=1}^{n} x_i & \sum_{i=1}^{n} x_i^2 & \cdots & \sum_{i=1}^{n} x_i^{b+1} \\ \cdot & \cdot & \cdots & \cdot \\ \cdot & \cdot & \cdots & \cdot \\ \cdot & \cdot & \cdots & \cdot \\ \sum_{i=1}^{n} x_i^b & \sum_{i=1}^{n} x_i^{b+1} & \cdots & \sum_{i=1}^{n} x_i^{2b} \end{bmatrix}^{-1} \begin{bmatrix} \sum_{i=1}^{n} y_i \\ \sum_{i=1}^{n} x_i y_i \\ \cdot \\ \cdot \\ \cdot \\ \sum_{i=1}^{n} x_i^b y_i \end{bmatrix} \tag{2.29}$$

So that the fitted polynomial equation becomes

$$\hat{y}_i = \hat{a}_0 + \hat{a}_1 x_i + \cdots + \hat{a}_b x_i^b \tag{2.30}$$

where the $\hat{a}_i$ are the same estimates obtained using Equation 2.29. We now apply this theory to the responses of the experiment in Example 2.4. This means that we want to fit a polynomial equation to the responses in that design in terms of reactant concentrations. We shall assume that a second degree polynomial would be adequate for the representation of the responses in terms of the reactant concentration. The second degree polynomial equation to be fitted is

$$y_i = a_0 + a_1 x_i + a_2 x_i^2 + \varepsilon_i \tag{2.31}$$

We prepare a table of the functions found in Equation 2.29 to obtain estimates of the parameters of polynomial equation. However, since we made a number of observations (four of them) at each reactant concentration x_i, we use all of them so that the functions in Equation 2.29 take the forms listed in the following table:

x	y	xy	x^2	x^2y	x^3	x^4
10	64	640	100	6400	1000	10000
10	57	570	100	5700	1000	10000
10	59	590	100	5900	1000	10000
10	62	620	100	6200	1000	10000
17.5	46	805	306.25	14087.5	5359.375	93789.06
17.5	42	735	306.25	12862.5	5359.375	93789.06
17.5	39	682.5	306.25	11943.75	5359.375	93789.06
17.5	35	612.5	306.25	10718.75	5359.375	93789.06
25	30	750	625	18750	15625	390625
25	37	925	625	23125	15625	390625
25	33	825	625	20625	15625	390625
25	28	700	625	17500	15625	390625
32.5	20	650	1056.25	21125	34328.13	1115664
32.5	18	585	1056.25	19012.5	34328.13	1115664
32.5	25	812.5	1056.25	26406.25	34328.13	1115664
32.5	22	715	1056.25	23237.5	34328.13	1115664
40	15	600	1600	24000	64000	2560000
40	13	520	1600	20800	64000	2560000
40	16	640	1600	25600	64000	2560000
40	14	560	1600	22400	64000	2560000
500	675	13537.5	14750	336393.8	481250	16680313

The normal equations are

$$\begin{aligned} 20a_0 + 500a_1 + 14750a_2 &= 675 \\ 500a_0 + 14750a_1 + 481250a_2 &= 13537.5 \\ 14750a_0 + 481250a_1 + 16680313a_2 &= 336393.8 \end{aligned}$$

The problem reduces to the solution of three simultaneous linear equations in a_0, a_1, and a_2, the regression coefficients, so that

$$\begin{bmatrix} a_0 \\ a_1 \\ a_2 \end{bmatrix} = \begin{bmatrix} 20 & 500 & 14750 \\ 500 & 14750 & 481250 \\ 14750 & 481250 & 16680313 \end{bmatrix}^{-1} \begin{bmatrix} 675 \\ 13537.5 \\ 336393.8 \end{bmatrix}$$

$$= \begin{bmatrix} 1.81014 & -0.155732 & 0.0028924 \\ -0.15573 & 0.014554 & -0.0002822 \\ 0.00289 & -0.000282 & 0.0000056 \end{bmatrix} \begin{bmatrix} 675 \\ 13537.5 \\ 336393.8 \end{bmatrix} = \begin{bmatrix} 86.6152 \\ -3.0230 \\ 0.0308 \end{bmatrix}$$

It can be shown on solving that $\hat{a}_0 = 86.6151$, $\hat{a}_1 = -3.023$, and $\hat{a}_2 = 0.0308$. The fitted equation is therefore

$$\hat{y} = 86.6151 - 3.023x + 0.0308x^2 \tag{2.32}$$

There are tests which could be carried out to check that the fitted coefficients are significant. Any reader who is not conversant with these tests should revisit Chapter 1 on regression analysis before proceeding further. Tests such as carried out in Example 1.22 shows that only the quadratic effect is significant. Using the theory of regression analysis, the reader could carry out the test to show that the fitted parameters are significant as an exercise (see Exercise 2.6). We carry out tests for significance of these coefficients in the next section after having discussed the orthogonal polynomial method for fitting polynomial equations.

The SAS program that would fit the quadratic model and test for significance of the parameters of the response curve is as follows:

```
options nonumber nodate ls=74  ps=86;
data react;
input rlevel x @;
xsq=x**2;
xcub=x**3;
Xquart=x**4;
do subject=1 to 4;
input y@@;
output; end;
datalines;
1 10 64 57 59 62
2 17.5 46 42 39 35
3 25 30 37 33 28
4 32.5 20 18 25 22
5 40 15 13 16 14
;
proc reg data=react;
model y=x xsq; run;
```

```
                        The REG Procedure
                         Model: MODEL1
                      Dependent Variable: y

                Number of Observations Read     20
                Number of Observations Used     20

                      Analysis of Variance

                                Sum of          Mean
Source                  DF     Squares        Square    F Value    Pr > F

Model                    2  5118.64286    2559.32143     200.40    <.0001
Error                   17   217.10714      12.77101
Corrected Total         19  5335.75000

             Root MSE           3.57365  R-Square     0.9593
             Dependent Mean    33.75000  Adj R-Sq     0.9545
             Coeff Var         10.58861

                       Parameter Estimates

                      Parameter      Standard
Variable        DF     Estimate         Error    t Value    Pr > |t|

Intercept        1     86.61508       4.80805      18.01      <.0001
x                1     -3.02302       0.43112      -7.01      <.0001
xsq              1      0.03079       0.00849       3.63      0.0021
```

2.11.2.2 Fitting of Response Curves to Data Using Orthogonal Contrasts

When the levels of the factors involved are equally spaced, fitting of polynomial equations are made considerably easier if orthogonal contrast coefficients are used. This method facilitates the fitting of polynomials as well as the testing of hypothesis to determine which order of the polynomial is significant. In this use of orthogonal contrast coefficients, the test is carried out before fitting the polynomial, thereby reducing the workload. As we stated earlier, if a quantitative factor has k levels, it is theoretically possible to fit a polynomial of order $k-1$ to the responses of the experiment in terms of the factor. Contrast coefficients for fitting the orthogonal polynomials have been obtained and their tables appear in Fisher and Yates (1963) and Pearson and Hartley (1966) (see Appendix A10). In general, suppose we wish to fit a kth order polynomial to some responses of an experiment in terms of the controlled variable, x, using orthogonal contrast coefficients; the standard polynomial equation used in regression analysis is given in Equation 2.30 in which k is used to replace $k-1$, but this equation is rewritten in the form

$$y = b_0 + b_1 g_1(x) + \cdots + b_k g_k(x) + \varepsilon_k \tag{2.33}$$

where $g_i(x)$ is a polynomial in x of degree i $(i = 1, 2, 3 \ldots k)$ and the parameter b_i depends on a_i in Equation 2.30. It can be shown that the function g has the following properties:

$$\begin{aligned} &\sum_{x_i} g_i(x_i) g_1(x_i) = 0; \quad i \neq 1 \\ &\sum_{x_i} g_i(x_i) = 0 \end{aligned} \tag{2.34a}$$

where $g_i(x)$ depends on the average of x_i, $\overline{x}$, and the distance between the consecutive values x_i and x_{i+1} is Δ (x_i are the levels of the quantitative factor). Since the contrasts $g_i(x)$ are orthogonal to each other, and Equation 2.34a applies, the normal equations in a multiple regression involving these contrasts are

$$\begin{aligned} &N b_0 = \sum y_i \ (N = kn) \\ &b_1 \sum g_1(x_i) = \sum g_1(x_i) y_i \\ &\vdots \qquad\quad \vdots \qquad\quad \vdots \\ &b_k \sum g_k(x_i) = \sum g_k(x_i) y_i \end{aligned} \tag{2.34b}$$

Other terms of the normal equation become zero on account of the orthogonality properties. The resulting estimates are given as

$$\begin{aligned} b_0 &= \frac{\sum y_i}{N} \quad \text{[because } g_0(x_i) = 1\text{]} \\ b_1 &= \frac{\sum g_1(x_i) y_i}{\sum [g_1(x_i)]^2} \\ \vdots \quad & \quad \vdots \\ b_k &= \frac{\sum g_k(x_i) y_i}{\sum [g_k(x_i)]^2} \end{aligned} \tag{2.34c}$$

It can be shown that the first five orthogonal polynomials are

$$\begin{aligned} g_0(x) &= 1 \\ g_1(x) &= \lambda_1 \left(\frac{x - \overline{x}}{\Delta}\right) \\ g_2(x) &= \lambda_2 \left[\left(\frac{x - \overline{x}}{\Delta}\right)^2 - \frac{c^2 - 1}{12}\right] \\ g_3(x) &= \lambda_3 \left[\left(\frac{x - \overline{x}}{\Delta}\right)^3 - \left(\frac{x - \overline{x}}{\Delta}\right)\left(\frac{3c^2 - 7}{20}\right)\right] \\ g_4(x) &= \lambda_4 \left[\left(\frac{x - \overline{x}}{\Delta}\right)^4 - \left(\frac{x - \overline{x}}{\Delta}\right)^2 \left(\frac{3c^2 - 13}{14}\right) + \frac{3(c^2 - 1)(c^2 - 9)}{560}\right] \end{aligned}$$

where c is the number of levels of the factor being studied or the number of levels of the controlled variable x, which elicits the responses y. $\{\lambda_i\}$ are constants such that the polynomials have integer values. λ_i for $c \leq 10$ can be found in Appendix A10. Since in the Example 2.4, the factor levels are equally spaced, we use the orthogonal contrast coefficients to fit a polynomial to the *responses* y, in terms of the factor x. The orthogonal polynomial contrast coefficients as well as other functions calculated by employing them are shown in Table 2.10.

TABLE 2.10: Orthogonal Polynomial Contrast Coefficients for Example 2.3

		Contrasts (d_i)			
Concentration (x)	**Treatment (total) (y_i)**	**Linear**	**Quadratic**	**Cubic**	**Quartic**
10.00	242	−2	2	−1	1
17.50	162	−1	−1	2	−4
25.00	128	0	−2	0	6
32.50	85	1	−1	−2	−4
40.00	58	2	2	1	1
Effects: $\sum d_i y_i$	675	−445	97	−30	80

Where d_i are the contrasts, the sums of squares are obtained using the formula

$$SS_{\text{contrast}} = \frac{\left(\sum d_i y_i\right)^2}{n \sum d_i^2}$$

$$SS_{\text{linear}} = \frac{(-445)^2}{4[(-2)^2 + (-1)^2 + 0^2 + 1^2 + 2^2]} = 4950.63$$

$$SS_{\text{quadratic}} = \frac{(97)^2}{4[2^2 + (-1)^2 + (-2)^2 + (-1)^2 + 2^2]} = 168.02$$

$$SS_{\text{cubic}} = \frac{(-30)^2}{4[(-1)^2 + 2^2 + 0^2 + (-2)^2 + 1^2]} = 22.50$$

$$SS_{\text{quartic}} = \frac{(80)^2}{4[1^2 + (-4)^2 + 6^2 + (-4)^2 + 1^2]} = 22.86$$

The constants b_i (in Equation 2.34c) are obtained as

$$b_0 = \frac{675}{4(5)} = 33.75$$

$$b_1 = \frac{-445}{4[(-2)^2 + (-1)^2 + 0^2 + 1^2 + 2^2]} = -11.125$$

$$b_2 = \frac{97}{4[2^2 + (-1)^2 + (-2)^2 + (-1)^2 + 2^2]} = 1.732$$

$$b_3 = \frac{-30}{4[(-1)^2 + 2^2 + 0^2 + (-2)^2 + 1^2]} = -0.75$$

$$b_4 = \frac{80}{4[1^2 + (-4)^2 + 6^2 + (-4)^2 + 1^2]} = 0.2857$$

The next step is to test hypothesis to find out which of these estimates of b_i are significant. The analysis of variance are shown in Table 2.11.

TABLE 2.11: ANOVA for Orthogonal Contrast Coefficients

Source	**Sum of squares**	**Degrees of freedom**	**Mean square**	***F*-ratio**	***p*-Value**
Reactant concentration	5164.00	4	1291.00	112.75	5.3051E−11
Linear	4950.63	1	4950.63	432.31	1.796E−12
Quadratic	168.02	1	168.02	14.67	0.00163907
Cubic	22.50	1	22.50	1.96	0.18185608
Quartic	22.86	1	22.86	1.93	0.18503843
Residual	171.75	15	11.45		
Total	5335.75	19			

Only the linear and quadratic effects are significant, indicating the need to fit a second degree polynomial. From the tables in Appendix A10, it can be seen that $\lambda_1 = 1$; $\lambda_2 = 1$; from Table 2.9, $\Delta = 7.50$, so that the polynomial to be fitted by applying the formulae discussed above is

$$\hat{y} = 33.75 - 11.125[(1)(x - 25)/7.5] + 1.732(1)[((x - 25)/7.5)^2 - (5^2 - 1)/12]$$

which on simplification yields

$$\hat{y} = 0.0308x^2 - 3.02x + 86.61$$

which is the same as we obtained by the normal regression analysis (Equation 2.32).

We write a SAS program to carry out the above analysis by orthogonal contrasts.

SAS PROGRAM FOR EXAMPLE 2.3 WITH ORTHOGONAL CONTRASTS:

```
options nonumber nodate ls=74 ps=86;
data react;
input rlevel x @;
do  subject=1 to 4;
input y@@;
output; end;
datalines;
1 10 64 57 59 62
2 17.5 46 42 39 35
3 25 30 37 33 28
4 32.5 20 18 25 22
5 40 15 13 16 14
;
proc glm data=react;
class rlevel;
model  y=rlevel;
contrast  'linear' rlevel -2 -1 0 1 2;
contrast 'quadratic' rlevel  -2 -1 -2 -1 2;
contrast 'cubic' rlevel -1 2 0 -2 1;
contrast  'quartic'  rlevel -1 -4 6 -4 1; run;
```

SAS OUTPUT

```
                    The GLM Procedure
                 Class Level Information
              Class       Levels        Values
              rlevel           5     1 2 3 4 5

             Number of Observations Read     20
             Number of Observations Used     20

                   The SAS System
                         09:12 Thursday, November 1, 2007

                  The GLM Procedure

Dependent Variable: y
```

Source	DF	Sum of Squares	Mean Square	F Value	Pr > F
Model	4	5164.000000	1291.000000	112.75	<.0001
Error	15	171.750000	11.450000		
Corrected Total	19	5335.750000			

R-Square	Coeff Var	Root MSE	y Mean
0.967811	10.02603	3.383785	33.75000

Source	DF	Type I SS	Mean Square	F Value	Pr > F
rlevel	4	5164.000000	1291.000000	112.75	<.0001

Source	DF	Type III SS	Mean Square	F Value	Pr > F
rlevel	4	5164.000000	1291.000000	112.75	<.0001

Contrast	DF	Contrast SS	Mean Square	F Value	Pr > F
linear	1	4950.625000	4950.625000	432.37	<.0001
quadratic	1	168.017857	168.017857	14.67	0.0016
cubic	1	22.500000	22.500000	1.97	0.1813
quartic	1	22.857143	22.857143	2.00	0.1781

2.12 Regression Models for the CRD (One-Way Layout)

In the next few sections, we discuss how we can use regression theory as an alternative way for modeling, analyzing, and interpreting the responses of a typical CRD. The regression models for the one-way experimental layout (CRD) uses dummy variables. There are two ways of coding these variables which lead to different model parameters for the same responses. The first way of defining the dummy or indicator variables in the design is the effects coding method. The other method is called the reference cell method. We will use both methods to model the responses of a typical CRD. Henceforth, when we obtain equivalent regression models for any design, we will illustrate both approaches to the definition of the indicator or dummy variables.

2.12.1 Regression Model for CRD (Effects Coding Method)

The regression model requires us to define a set of $k-1$ dummy variables if there are k groups or k levels for the single factor being investigated in the CRD. Thus, the regression model would be of the form

$$y_{ij} = \mu + \sum_{i=1}^{k-1} \alpha_i X_i + \varepsilon \tag{2.35}$$

where

$$X_i = \begin{cases} -1 & \text{if } y_{ij} \text{ is under group } k \\ 1 & \text{if } y_i \text{ is under group } i \\ 0 & \text{otherwise} \end{cases} \tag{2.35a}$$
$$\text{for } i = 1, 2, \ldots, k-1$$

The parameters of the model can be estimated by using relations which depend on the means of the responses for the groups or levels of the factor in the CRD. They are

$$\mu = \frac{\mu_1 + \mu_2 + \cdots + \mu_k}{k} \tag{2.35b}$$
$$\hat{\alpha}_1 = \mu_1 - \mu$$
$$\hat{\alpha}_2 = \mu_2 - \mu$$
$$\vdots$$
$$\hat{\alpha}_{k-1} = \mu_{k-1} - \mu$$

2.12.2 Regression Model for the Responses of Example 2.2

We will model the responses of the CRD in Example 2.2 using the effects coding method to define the dummy variables, X_i. First, we present the means for the nutrient contents for different methods of nutrient extraction in the following table:

Method 1	**Method 2**	**Method 3**	**Method 4**	**Method 5**
$\hat{\mu}_1$	$\hat{\mu}_2$	$\hat{\mu}_3$	$\hat{\mu}_4$	$\hat{\mu}_5$
10.7	13.4	14.2	17	19.7

Note that $k = 5$, so that the specific model which applies to the responses of the CRD in this example is

$$y_i = \mu + \sum_{i=1}^{4} \alpha_i X_i + \varepsilon \tag{2.35c}$$

where

$$X_i = \begin{cases} -1 & \text{if } y_i \text{ is under group } 5 \\ 1 & \text{if } y_i \text{ is under group } i \\ 0 & \text{otherwise} \end{cases}$$
$$\text{for } i = 1, 2, 3, 4$$

The estimates for the parameters are obtained as

$$\mu = \frac{\mu_1 + \mu_2 + \mu_3 + \mu_4 + \mu_5}{5} = \frac{10.7 + 13.4 + 14.2 + 17 + 19.7}{5} = 15.0$$
$$\hat{\alpha}_1 = 10.7 - 15 = -4.3; \quad \hat{\alpha}_2 = 13.4 - 15 = -1.6$$
$$\hat{\alpha}_3 = 14.2 - 15 = -0.8; \quad \hat{\alpha}_4 = 17 - 15 = 2.0$$

The estimated model is

$$\hat{y} = 15 - 4.3x_1 - 1.6x_2 - 0.8x_3 + 2x_4 \tag{2.36}$$

2.12.2.1 Analysis of Variance to Test the Significance of the Fitted Model

Using the fitted model in Equation 2.36, we estimate each of the responses of the experiment and extract the residuals and perform ANOVA to check for the significance of the model. Recall that by regression theory, the ANOVA tests that the parameters (α_1, α_2, α_3, α_4) of the model are zero against that they are not zero. The significance of the test indicates that the fitted model is significant (Table 2.13).

TABLE 2.12: Estimated Responses and Residuals for Fitted Model 2.36 (Effects Coding Method)

x_1	x_2	x_3	x_4	y	$\hat{y}$ = Estimate	Residual = $y - \hat{y}$	$(y - \hat{y})^2$
1	0	0	0	10	10.7	−0.7	0.49
1	0	0	0	12	10.7	1.3	1.69
1	0	0	0	10	10.7	−0.7	0.49
1	0	0	0	11	10.7	0.3	0.09
1	0	0	0	12	10.7	1.3	1.69
1	0	0	0	9	10.7	−1.7	2.89
1	0	0	0	11	10.7	0.3	0.09
1	0	0	0	13	10.7	2.3	5.29
1	0	0	0	10	10.7	−0.7	0.49
1	0	0	0	9	10.7	−1.7	2.89
0	1	0	0	14	13.4	0.6	0.36
0	1	0	0	14	13.4	0.6	0.36
0	1	0	0	14	13.4	0.6	0.36
0	1	0	0	12	13.4	−1.4	1.96
0	1	0	0	16	13.4	2.6	6.76
0	1	0	0	10	13.4	−3.4	11.56
0	1	0	0	14	13.4	0.6	0.36
0	1	0	0	15	13.4	1.6	2.56
0	1	0	0	10	13.4	−3.4	11.56
0	1	0	0	15	13.4	1.6	2.56
0	0	1	0	17	14.2	2.8	7.84
0	0	1	0	13	14.2	−1.2	1.44
0	0	1	0	15	14.2	0.8	0.64
0	0	1	0	12	14.2	−2.2	4.84
0	0	1	0	16	14.2	1.8	3.24
0	0	1	0	10	14.2	−4.2	17.64
0	0	1	0	15	14.2	0.8	0.64
0	0	1	0	15	14.2	0.8	0.64
0	0	1	0	13	14.2	−1.2	1.44
0	0	1	0	16	14.2	1.8	3.24
0	0	0	1	19	17	2	4
0	0	0	1	15	17	−2	4
0	0	0	1	19	17	2	4

Continued

TABLE 2.12: Continued

x_1	x_2	x_3	x_4	y	$\hat{y}$= Estimate	Residual $= y - \hat{y}$	$(y - \hat{y})^2$
0	0	0	1	14	17	−3	9
0	0	0	1	23	17	6	36
0	0	0	1	12	17	−5	25
0	0	0	1	17	17	0	0
0	0	0	1	17	17	0	0
0	0	0	1	15	17	−2	4
0	0	0	1	19	17	2	4
−1	−1	−1	−1	20	19.7	0.3	0.09
−1	−1	−1	−1	19	19.7	−0.7	0.49
−1	−1	−1	−1	23	19.7	3.3	10.89
−1	−1	−1	−1	16	19.7	−3.7	13.69
−1	−1	−1	−1	20	19.7	0.3	0.09
−1	−1	−1	−1	18	19.7	−1.7	2.89
−1	−1	−1	−1	21	19.7	1.3	1.69
−1	−1	−1	−1	22	19.7	2.3	5.29
−1	−1	−1	−1	18	19.7	−1.7	2.89
−1	−1	−1	−1	20	19.7	0.3	0.09
Total				750	750	0	224.2

We note that if there are $N = kn$ responses which were used to fit a regression model, then

$$S_{yy} = \sum_{i=1}^{N} y_i^2 - \frac{\left(\sum_{i=1}^{N} y_i\right)^2}{N}$$

and

$$SS_{\mathrm{R}} = \sum_{i=1}^{N} (\hat{y}_i - \overline{y})^2$$

and

$$SS_{\mathrm{E}} = \sum_{i=1}^{N} (y_i - \hat{y}_i)^2$$

then $S_{yy} = SS_{\mathrm{R}} + SS_{\mathrm{E}}$, where SS_{E} is the error sum of squares, S_{yy} is the total sum of squares, and SS_{R} is the sum of squares due to regression. From the above table, following extractions of fitted value and residuals, we see that

$$SS_{\mathrm{E}} = 224.2 \quad S_{yy} = 11952 - \frac{750^2}{50} = 702.00 \Rightarrow SS_{\mathrm{R}} = 702 - 224.20 = 477.80$$

We perform ANOVA to test for the significance of fitted model in Table 2.13.

TABLE 2.13: ANOVA for Testing for Significance of Fitted Model 2.36

Source	Sum of squares	Degrees of freedom	Mean square	F-ratio	p-Value
Regression	477.80	4	119.45	23.975	1.14625E−10
Error	244.20	45	4.982222		
Total	702	49			

From the tables of the F-distribution, we see that $F(4{,}45{,}0.01) = 3.77$ indicating that the fitted model is highly significant.

2.12.3 SAS Program for Analysis of Data of Example 2.2 Using Regression Model (Effects Coding Method)

We write the next program to analyze the data of Example 2.2 by using the effects coding method to define the dummy variables in the regression model (Equation 2.35a). In this program, we use PROC REG. We could also have used PROC GLM.

SAS PROGRAM

```
option ps=60 nodate nonumber;
data one_effects;
input x1-x4 y @@;
cards;
1 0 0 0 10 1 0 0 0 12 1 0 0 0 10 1 0 0 0 11 1 0 0 0 12 1 0 0 0 9 1 0 0 0 11 1
0 0 0 13 1 0 0 0 10 1 0 0 0 9 0 1 0 0 14 0 1 0 0 14 0 1 0 0 14 0 1 0 0 12 0 1
0 0 16 0 1 0 0 10 0 1 0 0 14 0 1 0 0 15 0 1 0 0 10 0 1 0 0 15 0 0 1 0 17 0 0
1 0 13 0 0 1 0 15 0 0 1 0 12 0 0 1 0 16 0 0 1 0 10 0 0 1 0 10 0 0 1 0 15 0 0
1 0 15 0 0 1 0 13 0 0 1 0 16 0 0 0 1 19 0 0 0 1 15 0 0 0 1 19 0 0 0 1 14 0 0
0 1 23 0 0 0 1 12 0 0 0 1 17 0 0 0 1 17 0 0 0 1 15 0 0 0 1 19 -1 -1 -1 -1 20
-1 -1 -1 -1 19 -1 -1 -1 -1 23 -1 -1 -1 -1 16 -1 -1 -1 -1 20 -1 -1 -1 -1 18 -1
-1 -1 -1 21 -1 -1 -1 -1 22 -1 -1 -1 -1 18 -1 -1 -1 -1 20
;
proc print data=one_effects; sum x1-x4 y;
proc reg data=one_effects; model y=x1-x4; run;
```

SAS OUTPUT

Obs	x1	x2	x3	x4	y
1	1	0	0	0	10
2	1	0	0	0	12
3	1	0	0	0	10
4	1	0	0	0	11
5	1	0	0	0	12
6	1	0	0	0	9
7	1	0	0	0	11
8	1	0	0	0	13
9	1	0	0	0	10
10	1	0	0	0	9
11	0	1	0	0	14
12	0	1	0	0	14
13	0	1	0	0	14
14	0	1	0	0	12
15	0	1	0	0	16
16	0	1	0	0	10
17	0	1	0	0	14
18	0	1	0	0	15
19	0	1	0	0	10
20	0	1	0	0	15

Obs	x1	x2	x3	x4	y
21	0	0	1	0	17
22	0	0	1	0	13
23	0	0	1	0	15
24	0	0	1	0	12
25	0	0	1	0	16
26	0	0	1	0	10
27	0	0	1	0	15
28	0	0	1	0	15
29	0	0	1	0	13
30	0	0	1	0	16
31	0	0	0	1	19
32	0	0	0	1	15
33	0	0	0	1	19
34	0	0	0	1	14
35	0	0	0	1	23
36	0	0	0	1	12
37	0	0	0	1	17
38	0	0	0	1	17
39	0	0	0	1	15
40	0	0	0	1	19
41	-1	-1	-1	-1	20
42	-1	-1	-1	-1	19
43	-1	-1	-1	-1	23
44	-1	-1	-1	-1	16
45	-1	-1	-1	-1	20
46	-1	-1	-1	-1	18
47	-1	-1	-1	-1	21
48	-1	-1	-1	-1	22
49	-1	-1	-1	-1	18
50	-1	-1	-1	-1	20
	==	==	==	==	===
	0	0	0	0	750

The REG Procedure
Model: MODEL1
Dependent Variable: y

Number of Observations Read	50
Number of Observations Used	50

Analysis of Variance

Source	DF	Sum of Squares	Mean Square	F Value	Pr > F
Model	4	477.80000	119.45000	23.98	<.0001
Error	45	224.20000	4.98222		
Corrected Total	49	702.00000			

Root MSE	2.23209	R-Square	0.6806
Dependent Mean	15.00000	Adj R-Sq	0.6522
Coeff Var	14.88059		

Parameter Estimates

Variable	DF	Parameter Estimate	Standard Error	t Value	Pr > |t|
Intercept	1	15.00000	0.31567	47.52	<.0001
x1	1	-4.30000	0.63133	-6.81	<.0001
x2	1	-1.60000	0.63133	-2.53	0.0148
x3	1	-0.80000	0.63133	-1.27	0.2116
x4	1	2.00000	0.63133	3.17	0.0028

2.12.4 Regression Model for Cell Reference Method

The regression model for the CRD when the cell reference method is used for coding the dummy variables is the same as the model used when effects coding method is used. However, the definition of the dummy variables is different, so also the definition of the intercept of the model, μ. The model along with the dummy variables used for CRD under the reference cell method are as follows:

$$y_i = \mu + \sum_{i=1}^{k-1} \alpha_i X_i + \varepsilon \tag{2.37a}$$

where

$$X_i = \begin{cases} 1 & \text{if } y_i \text{ is in group } i \\ 0 & \text{otherwise} \end{cases}$$
$$\text{for } i = 1, 2, \ldots, k-1$$

In this case, the referenced cell is the kth group or treatment. The estimates of the parameters are obtained as

$$\begin{aligned} \hat{\alpha}_1 &= \mu_1 - \mu_k \\ \hat{\alpha}_1 &= \mu_2 - \mu_k \\ \vdots\ &\quad \vdots \\ \hat{\alpha}_{k-1} &= \mu_{k-1} - \mu_k \\ \hat{\mu} &= \mu_k \end{aligned}$$

The referenced cell for Example 2.2 is the fifth treatment; this means that $k = 5$.

$$y_i = \mu + \sum_{i=1}^{4} \alpha_i X_i + \varepsilon \tag{2.37b}$$

where

$$X_i = \begin{cases} 1 & \text{if } y_i \text{ is in group } i \\ 0 & \text{otherwise} \end{cases}$$
$$\text{for } i = 1, 2, 3, 4$$

We reproduce the table of means of data of Example 2.2.

Means				
Method 1	**Method 2**	**Method 3**	**Method 4**	**Method 5**
$\hat{\mu}_1$	$\hat{\mu}_2$	$\hat{\mu}_3$	$\hat{\mu}_4$	$\hat{\mu}_5$
10.7	13.4	14.2	17	19.7

From the tables and using Equations 2.37a and 2.37b, we see that

$$\mu = \hat{\mu}_{5.} = 19.7; \quad \hat{\alpha}_1 = \hat{\mu}_1 - \hat{\mu}_{5.} = 10.7 - 19.7 = -9.0$$
$$\hat{\alpha}_2 = \hat{\mu}_2 - \hat{\mu}_{5.} = 13.4 - 19.7 = -6.3$$
$$\hat{\alpha}_3 = \hat{\mu}_3 - \hat{\mu}_{5.} = 14.2 - 19.7 = -5.5$$
$$\hat{\alpha}_4 = \hat{\mu}_4 - \hat{\mu}_{5.} = 17 - 19.7 = -2.7$$

The fitted equation is

$$\hat{y} = 19.7 - 9.0x_1 - 6.3x_2 - 5.5x_3 - 2.7x_4 \tag{2.38}$$

Using this fitted equation, we again obtain the estimated responses, extract the residuals, and test for significance of the fitted model, via regression ANOVA (Tables 2.14 and 2.15).

TABLE 2.14: Estimated Responses and Residuals for Fitted Model 2.36 (Reference Cell Coding Method)

x_1	x_2	x_3	x_4	$y=$ **Responses**	$\hat{y}=$ **Estimates**	$y-\hat{y}=$ **Residual**	$(y-\hat{y})^2$
1	0	0	0	10	10.7	−0.7	0.49
1	0	0	0	12	10.7	1.3	1.69
1	0	0	0	10	10.7	−0.7	0.49
1	0	0	0	11	10.7	0.3	0.09
1	0	0	0	12	10.7	1.3	1.69
1	0	0	0	9	10.7	−1.7	2.89
1	0	0	0	11	10.7	0.3	0.09
1	0	0	0	13	10.7	2.3	5.29
1	0	0	0	10	10.7	−0.7	0.49
1	0	0	0	9	10.7	−1.7	2.89
0	1	0	0	14	13.4	0.6	0.36
0	1	0	0	14	13.4	0.6	0.36
0	1	0	0	14	13.4	0.6	0.36
0	1	0	0	12	13.4	−1.4	1.96
0	1	0	0	16	13.4	2.6	6.76
0	1	0	0	10	13.4	−3.4	11.6
0	1	0	0	14	13.4	0.6	0.36
0	1	0	0	15	13.4	1.6	2.56
0	1	0	0	10	13.4	−3.4	11.6
0	1	0	0	15	13.4	1.6	2.56
0	0	1	0	17	14.2	2.8	7.84
0	0	1	0	13	14.2	−1.2	1.44

Continued

TABLE 2.14: Continued

x_1	x_2	x_3	x_4	$y=$ **Responses**	$\hat{y}=$ **Estimates**	$y-\hat{y}=$ **Residual**	$(y-\hat{y})^2$
0	0	1	0	15	14.2	0.8	0.64
0	0	1	0	12	14.2	−2.2	4.84
0	0	1	0	16	14.2	1.8	3.24
0	0	1	0	10	14.2	−4.2	17.6
0	0	1	0	15	14.2	0.8	0.64
0	0	1	0	15	14.2	0.8	0.64
0	0	1	0	13	14.2	−1.2	1.44
0	0	1	0	16	14.2	1.8	3.24
0	0	0	1	19	17	2	4
0	0	0	1	15	17	−2	4
0	0	0	1	19	17	2	4
0	0	0	1	14	17	−3	9
0	0	0	1	23	17	6	36
0	0	0	1	12	17	−5	25
0	0	0	1	17	17	0	0
0	0	0	1	17	17	0	0
0	0	0	1	15	17	−2	4
0	0	0	1	19	17	2	4
0	0	0	0	20	19.7	0.3	0.09
0	0	0	0	19	19.7	−0.7	0.49
0	0	0	0	23	19.7	3.3	10.9
0	0	0	0	16	19.7	−3.7	13.7
0	0	0	0	20	19.7	0.3	0.09
0	0	0	0	18	19.7	−1.7	2.89
0	0	0	0	21	19.7	1.3	1.69
0	0	0	0	22	19.7	2.3	5.29
0	0	0	0	18	19.7	−1.7	2.89
0	0	0	0	20	19.7	0.3	0.09
Total						0	224.20

As stated earlier, if

$$S_{yy} = \sum_{i=1}^{N} y_i^2 - \frac{\left(\sum_{i=1}^{N} y_i\right)^2}{N}$$

and

$$SS_{\text{R}} = \sum_{i=1}^{N} (\hat{y}_i - \overline{y})^2$$

and

$$SS_{\text{E}} = \sum_{i=1}^{N} (y_i - \hat{y}_i)^2$$

then $S_{yy} = SS_{\text{R}} + SS_{\text{E}}$

$$SS_{\text{E}} = 224.20 \quad S_{yy} = 11952 - \frac{750^2}{50} = 702.50 \Rightarrow SS_{\text{R}} = 702 - 224.20 = 477.80$$

TABLE 2.15: ANOVA to Test for Significance of the Fitted Regression Model for Example 2.2

Source	Sum of squares	Degrees of freedom	Mean square	*F*-ratio	*p*-Value
Regression	477.80	4	119.45	23.975	1.14625E−10
Error	224.20	45	4.98222		
Total	702.00	49			

From the tables, we see that $F(0.01,\ 4,\ 45) = 3.77$. The model is significant at 1% level of significance.

2.12.5 SAS Program for Regression Model for Example 2.2 (Reference Cell Coding Method)

Next, we write a SAS program to analyze the data of Example 2.2 when the reference cell coding method is used for defining the dummy variables in the above model Equation 2.37. Again, we use PROC REG in the program.

SAS PROGRAM

```
data regcel;
input X1-X4 y @@;
datalines;
1 0 0 0 10 1 0 0 0  12 1 0 0 0  10 1 0 0 0  11 1 0 0 0  12 1 0 0 0  9 1 0 0
0  11 1 0 0 0  13 1 0 0 0 10 1 0 0 0  9 0 1 0 0  14 0 1 0 0  14 0 1 0 0  14 0
1 0 0  12 0 1 0 0  16 0 1 0 0  10  0 1 0 0 14 0 1 0 0  15 0 1 0 0  10 0 1 0
0  15 0 0 1 0  17 0 0 1 0  13 0 0 1 0  15 0 0 1 0  12 0 0 1 0 16 0 0 1 0  10
0 0 1 0 15 0 0 1 0  15 0 0 1 0  13 0 0 1 0  16 0 0 0 1  19 0 0 0 1  15 0 0 0
1  19 0 0 0 1  14 0 0 0 1  23 0 0 0 1  12 0 0 0 1  17 0 0 0 1  17 0 0 0 1
15 0 0 0 1   19 0 0 0 0 20 0 0 0 0  19 0 0 0 0  23 0 0 0 0   16 0 0 0 0    20
0 0 0 0   18 0 0 0 0  21 0 0 0 0  22 0 0 0 0 18 0 0 0 0  20
;
proc print data=regcel; sum x1-x4;
proc reg data=regcel;
model y=x1-x4; run;
```

SAS OUTPUT

The SAS System

Obs	x1	x2	x3	x4	y
1	1	0	0	0	10
2	1	0	0	0	12
3	1	0	0	0	10
4	1	0	0	0	11
5	1	0	0	0	12
6	1	0	0	0	9
7	1	0	0	0	11
8	1	0	0	0	13
9	1	0	0	0	10
10	1	0	0	0	9
11	0	1	0	0	14

Obs	x1	x2	x3	x4	y
12	0	1	0	0	14
13	0	1	0	0	14
14	0	1	0	0	12
15	0	1	0	0	16
16	0	1	0	0	10
17	0	1	0	0	14
18	0	1	0	0	15
19	0	1	0	0	10
20	0	1	0	0	15
21	0	0	1	0	17
22	0	0	1	0	13
23	0	0	1	0	15
24	0	0	1	0	12
25	0	0	1	0	16
26	0	0	1	0	10
27	0	0	1	0	15
28	0	0	1	0	15
29	0	0	1	0	13
30	0	0	1	0	16
31	0	0	0	1	19
32	0	0	0	1	15
33	0	0	0	1	19
34	0	0	0	1	14
35	0	0	0	1	23
36	0	0	0	1	12
37	0	0	0	1	17
38	0	0	0	1	17
39	0	0	0	1	15
40	0	0	0	1	19
41	0	0	0	0	20
42	0	0	0	0	19
43	0	0	0	0	23
44	0	0	0	0	16
45	0	0	0	0	20
46	0	0	0	0	18
47	0	0	0	0	21
48	0	0	0	0	22
49	0	0	0	0	18
50	0	0	0	0	20
	==	==	==	==	
	10	10	10	10	

The REG Procedure
Model: MODEL1
Dependent Variable: y

Number of Observations Read	50
Number of Observations Used	50

Analysis of Variance

Source	DF	Sum of Squares	Mean Square	F Value	Pr > F
Model	4	477.80000	119.45000	23.98	<.0001
Error	45	224.20000	4.98222		
Corrected Total	49	702.00000			

```
Root MSE               2.23209    R-Square     0.6806
Dependent Mean        15.00000    Adj R-Sq     0.6522
Coeff Var             14.88059

                 Parameter Estimates

                   Parameter    Standard
Variable     DF     Estimate       Error  t Value  Pr > |t|

Intercept     1     19.70000     0.70585    27.91    <.0001
x1            1     -9.00000     0.99822    -9.02    <.0001
x2            1     -6.30000     0.99822    -6.31    <.0001
x3            1     -5.50000     0.99822    -5.51    <.0001
x4            1     -2.70000     0.99822    -2.70    0.0096
```

Note that the parameters of the fitted model obtained by SAS agrees with those obtained manually in Equation 2.38, so does the SAS ANOVA for regression.

2.13 Regression Models for ANOVA for CRD Using Orthogonal Contrasts

Different types of codes exist for coding categorical variables in regression analysis. So far we have used the effects coding and reference cell coding methods to define dummy variables in regression models for responses of CRDs which hitherto were mostly analyzed by traditional ANOVA. Linear multiple regression has the capability to incorporate many of the traditional ANOVA models. This is sometimes accomplished by using orthogonal contrasts rather than dummy variables in the regression models. Some of these orthogonal contrasts are found to produce the same results that are obtained by traditional ANOVA contrast coding. We can use a variety of orthogonal contrasts to obtain regression models for analyzing the outcomes from different designed experiments that would be similar to the traditional ANOVA. Among these contrast coding are simple, forward difference, backward difference, Helmert, reverse Helmert, deviation, and orthogonal polynomial. Orthogonal contrasts for multiple regression have been discussed in Pedhazur (1982), Serlin and Levin (1985), Keppel (1991), and Wendorf (2004). Applications of orthogonal contrast to designed experiments can be found in Tabachnick and Fidell (2007). In developing regression models, we will use orthogonal contrasts because they are most analogous to the user-defined contrasts employed in ANOVA approaches. Among these orthogonal contrasts, we consider those based on Helmert and reverse Helmert coding. The Helmert coding compares the mean of a level of a categorical variable to the mean of subsequent levels. We illustrate this contrast coding with a typical study of five methods used to determine the nutrient contents of a fruit. The table below shows the use of Helmert coding for this study. Typically, if the categorical variable has k

levels, we need to define $k-1$ contrast variables; since there are five methods for extracting nutrients of the fruit, we need four orthogonal contrasts: X_1–X_4.

Extraction method	X_1	X_2	X_3	X_4
1	1	0	0	0
2	−1/4	1	0	0
3	−1/4	−1/3	1	0
4	−1/4	−1/3	−1/2	1
5	−1/4	−1/3	−1/2	−1

In the above table, for contrast variable X_1, we compare mean of method 1 with mean of methods 2–5, while for contrast X_2, we compare mean of method 2 to mean of methods 3–5; for contrast X_3, we compare mean for method 3 to mean of methods 4 and 5 while for contrast X_4 mean of method 4 is compared to mean of method 5. The reverse Helmert coding for the same study is given in the following table; it will produce the same ANOVA as the code in the table above:

Extraction method	X_1	X_2	X_3	X_4
1	−1/4	−1/3	−1/2	−1
2	−1/4	−1/3	−1/2	1
3	−1/4	−1/3	1	0
4	−1/4	1	0	0
5	1	0	0	0

In the above table, for contrast variable X_1, we compare method 5 with methods 1–4, while for contrast X_2 we compare method 4 to methods 3, 2, and 1; for contrast X_3 we compare method 3 to methods 2 and 1, while in X_4 method 4 is compared to method 5. The reverse Helmert coding for the same study given in the table above will produce the same ANOVA as the Helmert coding described previously but different estimates for the parameters of the model. We employ this orthogonal contrast to analyze the responses of the experiment of Example 2.2, which studied five methods for extracting nutrients of a fruit in which the experiment was replicated 10 times.

2.14 Regression Model for Example 2.2 Using Orthogonal Contrasts Coding (Helmert Coding)

There are five levels for extraction method. We need to define four orthogonal contrasts of the regression model for the experiment which is the number of degrees of freedom for the factor extraction method. The regression model for this experiment is

$$y_i = \mu + \alpha_1 X_{1i} + \alpha_2 X_{2i} + \alpha_3 X_{3i} + X_{4i} + \varepsilon_i \quad i = 1, 2, \ldots, 50 \qquad (2.39)$$

Since there are four orthogonal contrasts in the model, using the orthogonal properties discussed and presented in Equations 2.34a through 2.34c, the estimates for the parameters of the model are obtained by the following relations:

$$\hat{\mu} = \frac{\sum y_i}{N} = \overline{y}$$
$$\hat{\alpha}_k = \frac{\sum_{i=1}^{N} X_{ki} y_i}{\sum_{i=1}^{N} X_{ki}^2} \quad k = 1, 2, 3, 4 \tag{2.40}$$

The sums of squares for ANOVA for this model are obtained as (Table 2.15):

$$SS_{X_{ki}} = \frac{\left(\sum X_{ki} Y_i\right)^2}{\sum X_i^2} \quad k = 1, 2, 3, 4 \tag{2.41}$$

To perform ANOVA for this data, we first obtain the sums of squares using Equation 2.41 as follows:

$$SS_{X_1} = \frac{\left(\sum_{i=1}^{50} X_{1i} y_i\right)^2}{\sum_{i=1}^{50} X_{1i}^2} = \frac{(-215)^2}{200} = 231.125$$
$$SS_{X_2} = \frac{\left(\sum_{i=1}^{50} X_{2i} y_i\right)^2}{\sum_{i=1}^{50} X_{2i}^2} = \frac{(-107)^2}{120} = 95.40833$$
$$SS_{X_3} = \frac{\left(\sum_{i=1}^{50} X_{3i} y_i\right)^2}{\sum_{i=1}^{50} X_{3i}^2} = \frac{(-83)^2}{60} = 114.8167$$
$$SS_{X_4} = \frac{\left(\sum_{i=1}^{50} X_{4i} y_i\right)^2}{\sum_{i=1}^{50} X_{4i}^2} = \frac{(-27)^2}{20} = 36.45$$

For the orthogonal contrast regression for Example 2.2, the fitted values and residuals are shown in Table 2.16.

TABLE 2.16: Orthogonal Contrast Regression for Example 2.2: Fitted Values and Residuals

y	x_1	x_2	x_3	x_4	x_1y	x_2y	x_3y	x_4y	x_1^2	x_2^2	x_3^2	x_4^2	$\hat{y}$	$y-\hat{y}$
10	4	0	0	0	40	0	0	0	16	0	0	0	10.7	−0.7
12	4	0	0	0	48	0	0	0	16	0	0	0	10.7	1.3
10	4	0	0	0	40	0	0	0	16	0	0	0	10.7	−0.7
11	4	0	0	0	44	0	0	0	16	0	0	0	10.7	0.3
12	4	0	0	0	48	0	0	0	16	0	0	0	10.7	1.3
9	4	0	0	0	36	0	0	0	16	0	0	0	10.7	−1.7
11	4	0	0	0	44	0	0	0	16	0	0	0	10.7	0.3
13	4	0	0	0	52	0	0	0	16	0	0	0	10.7	2.3
10	4	0	0	0	40	0	0	0	16	0	0	0	10.7	−0.7
9	4	0	0	0	36	0	0	0	16	0	0	0	10.7	−1.7
14	−1	3	0	0	−14	42	0	0	1	9	0	0	13.4	0.6
14	−1	3	0	0	−14	42	0	0	1	9	0	0	13.4	0.6
14	−1	3	0	0	−14	42	0	0	1	9	0	0	13.4	0.6
12	−1	3	0	0	−12	36	0	0	1	9	0	0	13.4	−1.4
16	−1	3	0	0	−16	48	0	0	1	9	0	0	13.4	2.6
10	−1	3	0	0	−10	30	0	0	1	9	0	0	13.4	−3.4
14	−1	3	0	0	−14	42	0	0	1	9	0	0	13.4	0.6
15	−1	3	0	0	−15	45	0	0	1	9	0	0	13.4	1.6
10	−1	3	0	0	−10	30	0	0	1	9	0	0	13.4	−3.4
15	−1	3	0	0	−15	45	0	0	1	9	0	0	13.4	1.6
17	−1	−1	2	0	−17	−17	34	0	1	1	4	0	14.2	2.8
13	−1	−1	2	0	−13	−13	26	0	1	1	4	0	14.2	−1.2
15	−1	−1	2	0	−15	−15	30	0	1	1	4	0	14.2	0.8
12	−1	−1	2	0	−12	−12	24	0	1	1	4	0	14.2	−2.2
16	−1	−1	2	0	−16	−16	32	0	1	1	4	0	14.2	1.8
10	−1	−1	2	0	−10	−10	20	0	1	1	4	0	14.2	−4.2
15	−1	−1	2	0	−15	−15	30	0	1	1	4	0	14.2	0.8
15	−1	−1	2	0	−15	−15	30	0	1	1	4	0	14.2	0.8
13	−1	−1	2	0	−13	−13	26	0	1	1	4	0	14.2	−1.2
16	−1	−1	2	0	−16	−16	32	0	1	1	4	0	14.2	1.8
19	−1	−1	−1	1	−19	−19	−19	19	1	1	1	1	17	2
15	−1	−1	−1	1	−15	−15	−15	15	1	1	1	1	17	−2
19	−1	−1	−1	1	−19	−19	−19	19	1	1	1	1	17	2
14	−1	−1	−1	1	−14	−14	−14	14	1	1	1	1	17	−3
23	−1	−1	−1	1	−23	−23	−23	23	1	1	1	1	17	6
12	−1	−1	−1	1	−12	−12	−12	12	1	1	1	1	17	−5
17	−1	−1	−1	1	−17	−17	−17	17	1	1	1	1	17	0
17	−1	−1	−1	1	−17	−17	−17	17	1	1	1	1	17	0
15	−1	−1	−1	1	−15	−15	−15	15	1	1	1	1	17	−2
19	−1	−1	−1	1	−19	−19	−19	19	1	1	1	1	17	2
20	−1	−1	−1	−1	−20	−20	−20	−20	1	1	1	1	19.7	0.3
19	−1	−1	−1	−1	−19	−19	−19	−19	1	1	1	1	19.7	−0.7
23	−1	−1	−1	−1	−23	−23	−23	−23	1	1	1	1	19.7	3.3
16	−1	−1	−1	−1	−16	−16	−16	−16	1	1	1	1	19.7	−3.7
20	−1	−1	−1	−1	−20	−20	−20	−20	1	1	1	1	19.7	0.3
18	−1	−1	−1	−1	−18	−18	−18	−18	1	1	1	1	19.7	−1.7
21	−1	−1	−1	−1	−21	−21	−21	−21	1	1	1	1	19.7	1.3
22	−1	−1	−1	−1	−22	−22	−22	−22	1	1	1	1	19.7	2.3
18	−1	−1	−1	−1	−18	−18	−18	−18	1	1	1	1	19.7	−1.7
20	−1	−1	−1	−1	−20	−20	−20	−20	1	1	1	1	19.7	0.3
750	0	0	0	0	−215	−107	−83	−27	200	120	60	20	750	0

$$\Rightarrow SS_{\text{methods}} = SS_{X_1} + SS_{X_2} + SS_{X_3} + SS_{X_4}$$
$$= 231.125 + 95.40833 + 114.8167 + 36.45 = 477.8$$
$$SS_{\text{total}} = \sum_{i=1}^{50} y_i^2 - \frac{\left(\sum_{i=1}^{50} y_i\right)^2}{50} = 11952 - \frac{750^2}{50} = 702;$$
$$SS_{\text{E}} = 702 - 477.8 = 224.2$$

where $N = kn$ indicating that there were k groups and n observations. The ANOVA is presented in Table 2.17:

TABLE 2.17: Orthogonal Contrast Regression, ANOVA for Example 2.2

Source	Sum of squares	Degrees of freedom	Mean square	*F*-ratio	*p*-Value
Extraction method	477.8	4	119.45	23.97525	1.14625E−10
Error	224.2	45	4.982222		
Total	702	49			

Using data from Table 2.5, we estimate the parameters for the regression model for Example 2.2, as follows:

$$\hat{\mu} = \frac{\sum_{i=1}^{50} y_i}{50} = \frac{750}{50} = 15 \quad \hat{\alpha}_1 = \frac{\sum_{i=1}^{50} X_{1i} y_i}{\sum_{i=1}^{50} X_{1i}^2} = \frac{-215}{200} = -1.075$$

$$\hat{\alpha}_2 = \frac{\sum_{i=1}^{50} X_{2i} y_i}{\sum_{i=1}^{50} X_{2i}^2} = \frac{-107}{120} = -0.89167$$

$$\hat{\alpha}_3 = \frac{\sum_{i=1}^{50} X_{3i} y_i}{\sum_{i=1}^{50} X_{3i}^2} = \frac{-83}{60} = -1.38333$$

$$\hat{\alpha}_4 = \frac{\sum_{i=1}^{50} X_{4i} y_i}{\sum_{i=1}^{50} X_{4i}^2} = \frac{-27}{20} = -1.35$$

The estimated model is

$$\hat{y} = 15 - 1.075x_1 - 0.89167x_2 - 1.3833x_3 - 1.35x_4 \tag{2.42}$$

With this estimated model, we obtained the predicted responses for each observed response y_i and extracted the associated residuals as shown in the Table 2.15 above. We plot the residuals against the fitted values and the normal probability plot to check for any violation of model assumptions (Figures 2.8 and 2.9).

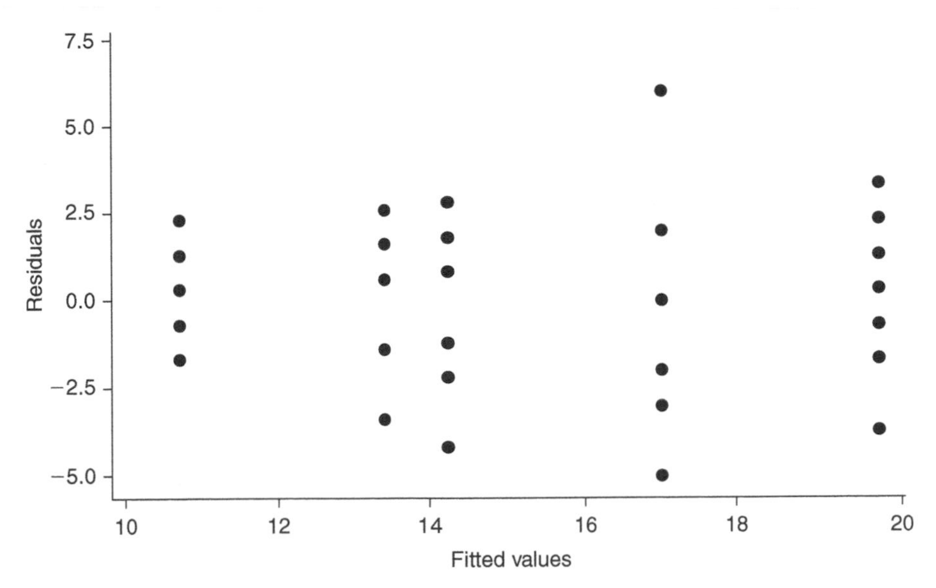

FIGURE 2.8: Scatterplot of residuals against fitted values for the orthogonal contrast regression model of Example 2.2.

The pattern seems random, but there is concern about one with a value of 5.0 which seems to stand out; this is possibly an outlier which may need further investigation. The rest are well within $\pm 2\sqrt{MS_{\mathrm{E}}} = \pm 4.464177$. There is no evidence of unequal variances. The normal probability plot for the residuals is shown in Figure 2.9.

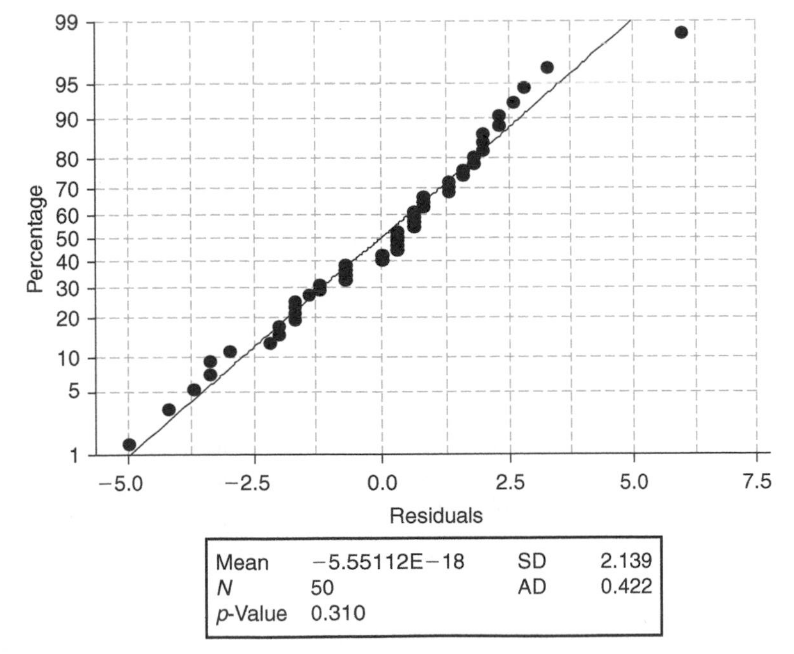

FIGURE 2.9: Normal probability plot for the orthogonal contrast regression model for Example 2.2.

The Anderson–Darling statistic for the residuals is 0.422 which has a p-value of 0.310 indicating that we lack evidence to reject the hypothesis that the residuals are normally distributed.

2.14.1 SAS Analysis of Data of Example 2.2 Using Orthogonal Contrasts (Helmert Coding)

In this section, we present the SAS analysis of the data of Example 2.2 using regression model based on Helmert contrast coding for the regressors.

SAS PROGRAM

```
input y X1-X4 @@;
datalines;
10 4 0 0 0 12 4 0 0 0 10 4 0 0 0 11 4 0 0 0 12 4 0 0 0 9 4 0 0 0 11 4 0 0 0
13 4 0 0 0 10 4 0 0 0 9 4 0 0 0 14 -1 3 0 0 14 -1 3 0 0 14 -1 3 0 0 12 -1 3 0
0 16 -1 3 0 0 10 -1 3 0 0 14 -1 3 0 0 15 -1 3 0 0 10 -1 3 0 0 15 -1 3 0 0 17
-1 -1 2 0 13 -1 -1 2 0 15 -1 -1 2 0 12 -1 -1 2 0 16 -1 -1 2 0 10 -1 -1 2 0 15
-1 -1 2 0 15 -1 -1 2 0 13 -1 -1 2 0 16 -1 -1 2 0 19 -1 -1 -1 1 15 -1 -1 -1 1
19 -1 -1 -1 1 14 -1 -1 -1 1 23 -1 -1 -1 1 12 -1 -1 -1 1 17 -1 -1 -1 1 17 -1 -1
-1 1 15 -1 -1 -1 1 19 -1 -1 -1 1 20 -1 -1 -1 -1 19 -1 -1 -1 -1 23 -1 -1 -1
-1 16 -1 -1 -1 -120 -1 -1 -1 -1 18 -1 -1 -1 -1 21 -1 -1 -1 -1 22 -1 -1 -1 -1
18 -1 -1 -1 -1 20 -1 -1 -1 -1
;
proc print data=nut1;
sum X1 X2 X3 X4 y;
run;
proc glm data=nut1;
model y=x1-x4; run;
```

SAS OUTPUT

Obs	y	X1	X2	X3	X4
1	10	4	0	0	0
2	12	4	0	0	0
3	10	4	0	0	0
4	11	4	0	0	0
5	12	4	0	0	0
6	9	4	0	0	0
7	11	4	0	0	0
8	13	4	0	0	0
9	10	4	0	0	0
10	9	4	0	0	0
11	14	-1	3	0	0
12	14	-1	3	0	0
13	14	-1	3	0	0
14	12	-1	3	0	0
15	16	-1	3	0	0
16	10	-1	3	0	0
17	14	-1	3	0	0

Obs	y	X1	X2	X3	X4
18	15	-1	3	0	0
19	10	-1	3	0	0
20	15	-1	3	0	0
21	17	-1	-1	2	0
22	13	-1	-1	2	0
23	15	-1	-1	2	0
24	12	-1	-1	2	0
25	16	-1	-1	2	0
26	10	-1	-1	2	0
27	15	-1	-1	2	0
28	15	-1	-1	2	0
29	13	-1	-1	2	0
30	16	-1	-1	2	0
31	19	-1	-1	-1	1
32	15	-1	-1	-1	1
33	19	-1	-1	-1	1
34	14	-1	-1	-1	1
35	23	-1	-1	-1	1
36	12	-1	-1	-1	1
37	17	-1	-1	-1	1
38	17	-1	-1	-1	1
39	15	-1	-1	-1	1
40	19	-1	-1	-1	1
41	20	-1	-1	-1	-1
42	19	-1	-1	-1	-1
43	23	-1	-1	-1	-1
44	16	-1	-1	-1	-1
45	20	-1	-1	-1	-1
46	18	-1	-1	-1	-1
47	21	-1	-1	-1	-1
48	22	-1	-1	-1	-1
49	18	-1	-1	-1	-1
50	20	-1	-1	-1	-1
	===	==	==	==	==
	750	0	0	0	0

```
The GLM Procedure

Number of Observations Read    50
Number of Observations Used    50
```

```
The GLM Procedure

Dependent Variable: y
```

Source	DF	Sum of Squares	Mean Square	F Value	Pr > F
Model	4	477.8000000	119.4500000	23.98	<.0001
Error	45	224.2000000	4.9822222		
Corrected Total	49	702.0000000			

R-Square	Coeff Var	Root MSE	y Mean
0.680627	14.88059	2.232089	15.00000

Source	DF	Type I SS	Mean Square	F Value	Pr > F
X1	1	231.1250000	231.1250000	46.39	<.0001
X2	1	95.4083333	95.4083333	19.15	<.0001
X3	1	114.8166667	114.8166667	23.05	<.0001
X4	1	36.4500000	36.4500000	7.32	0.0096

Source	DF	Type III SS	Mean Square	F Value	Pr > F
X1	1	231.1250000	231.1250000	46.39	<.0001
X2	1	95.4083333	95.4083333	19.15	<.0001
X3	1	114.8166667	114.8166667	23.05	<.0001
X4	1	36.4500000	36.4500000	7.32	0.0096

Parameter	Estimate	Standard Error	t Value	Pr > |t|
Intercept	15.00000000	0.31566508	47.52	<.0001
X1	-1.07500000	0.15783254	-6.81	<.0001
X2	-0.89166667	0.20376093	-4.38	<.0001
X3	-1.38333333	0.28816148	-4.80	<.0001
X4	-1.35000000	0.49911032	-2.70	0.0096

Again, we see that the SAS analysis is in accord with the discussions and analysis presented in Section 2.14. In particular, the ANOVA and estimates of parameters of the model are in agreement.

2.15 Regression Model for Example 2.3 Using Orthogonal Contrasts Coding

Again, there are five levels for nitrate concentration. We need to define four orthogonal contrasts of the regression model for the experiment which is the number of degrees of freedom for the factor extraction method. The regression model of Equation 2.39 applies so that

$$y_i = \mu + \alpha_1 X_{1i} + \alpha_2 X_{2i} + \alpha_3 X_{3i} + \alpha_4 X_{4i} + \varepsilon_i \quad i = 1, 2, \ldots, 20 \tag{2.43}$$

Since there are only four observations per nitrate concentration level, then $N = kn = 4(5) = 20$. This implies that in order to estimate the data we use the following relations:

$$\begin{aligned} \hat{\mu} &= \frac{\sum_{i=1}^{20} y_i}{20} = \overline{y} \\ \hat{\alpha}_k &= \frac{\sum_{i=1}^{20} X_{ki} y_i}{\sum_{i=1}^{20} X_{ki}^2} \quad k = 1, 2, 3, 4 \end{aligned} \tag{2.44}$$

The sums of squares for ANOVA for this model are obtained as

$$SS_{X_{ki}} = \frac{\left(\sum_{i=1}^{20} X_{ki} Y_i\right)^2}{\sum_{i=1}^{20} X_i^2} \quad k = 1, 2, 3, 4 \tag{2.45}$$

Notice that in Table 2.19, we used a different set of orthogonal contrasts, the reverse Helmert coding. Using the data, we obtain the estimates for the parameters of the model as follows:

$$\hat{\mu} = \overline{y} = \frac{2872}{20} = 143.6$$

$$\hat{\alpha}_1 = \frac{\sum_{i=1}^{20} X_{1i} y_i}{\sum_{i=1}^{20} X_{1i}^2} = \frac{128}{80} = 1.6 \quad \hat{\alpha}_2 = \frac{\sum_{i=1}^{20} X_{2i} y_i}{\sum_{i=1}^{20} X_{2i}^2} = \frac{264}{48} = 5.5$$

$$\hat{\alpha}_3 = \frac{\sum_{i=1}^{20} X_{3i} y_i}{\sum_{i=1}^{20} X_{3i}^2} = \frac{162}{24} = 6.75 \quad \hat{\alpha}_4 = \frac{\sum_{i=1}^{20} X_{4i} y_i}{\sum_{i=1}^{20} X_{4i}^2} = \frac{134}{8} = 16.75$$

We calculate the sums of squares for the data and carry out ANOVA for the data based on the regression model 2.43 (Table 2.18) as follows:

$$SS_{X_1} = \frac{\left(\sum_{i=1}^{20} X_{1i} y_i\right)^2}{\sum_{i=1}^{20} X_{1i}^2} = \frac{128^2}{80} = 204.8$$

$$SS_{X_2} = \frac{\left(\sum_{i=1}^{20} X_{2i} y_i\right)^2}{\sum_{i=1}^{N} X_{2i}^2} = \frac{264^2}{48} = 1452$$

$$SS_{X_3} = \frac{\left(\sum_{i=1}^{20} X_{3i} y_i\right)^2}{\sum_{i=1}^{N} X_{3i}^2} = \frac{162^2}{24} = 1093.5$$

$$SS_{X_4} = \frac{\left(\sum_{i=1}^{20} X_{4i} y_i\right)^2}{\sum_{i=1}^{20} X_{4i}^2} = \frac{134^2}{8} = 2244.5$$

$$\Rightarrow SS_{\text{nitrate concentration}} = SS_{X_1} + SS_{X_2} + SS_{X_3} + SS_{X_4}$$

$$= 204.8 + 1452 + 1093.5 + 2244.5 = 4994.8$$

$$SS_{\text{total}} = \sum_{i=1}^{20} y_i^2 - \frac{\left(\sum_{i=1}^{20} y_i\right)^2}{20} = 420752 - \frac{2872^2}{20} = 8332.8;$$

$$SS_{\text{E}} = 8332.8 - 4994.8 = 3338$$

TABLE 2.18: ANOVA for Regression Model for Example 2.3 Using Orthogonal Contrasts

Source	**Sum of squares**	**Degrees of freedom**	**Mean square**	***F*-ratio**	***p*-Value**
Nitrate concentration	4994.8	4	1248	5.6113	0.0057587
Error	3338	15	222.53382		
Total	8332.8	19			

Since $F(4, 15, 0.01) = 4.89$, the test is significant for nitrate concentration. The fitted equation for the model is

$$\hat{y} = 143.6 + 1.6x_1 + 5.5x_2 + 6.75x_3 + 16.75x_4 \tag{2.46}$$

As can be seen, we used this equation to obtain the estimated responses. We obtained the residuals (Table 2.19) and plotted them against fitted values (Figure 2.10); we obtained the normal probability plot of the residuals (Figure 2.11).

There seems to be no obvious pattern to the plot. The residuals are all well within $\pm 2\sqrt{MS_{\text{E}}} = \pm 29.8351$, indicating that there are no outliers.

It is not clear from the normal probability plot that the residuals are normally distributed. However, with the Anderson–Darling statistic of 0.406 and associated p-value of 0.319, we lack evidence to reject the null hypothesis that the residuals are normally distributed.

TABLE 2.19: Orthogonal Contrast Regression, Estimates, Residuals for Example 2.3

	y	x_1	x_2	x_3	x_4	x_1y	x_2y	x_3y	x_4y	x_1^2	x_2^2	x_3^2	x_4^2	$\hat{y}$	$y-\hat{y}$
	105	−1	−1	−1	−1	−105	−105	−105	−105	1	1	1	1	113	−8
	115	−1	−1	−1	−1	−115	−115	−115	−115	1	1	1	1	113	2
	91	−1	−1	−1	−1	−91	−91	−91	−91	1	1	1	1	113	−22
	141	−1	−1	−1	−1	−141	−141	−141	−141	1	1	1	1	113	28
	135	−1	−1	−1	1	−135	−135	−135	135	1	1	1	1	146.5	−12
	131	−1	−1	−1	1	−131	−131	−131	131	1	1	1	1	146.5	−16
	145	−1	−1	−1	1	−145	−145	−145	145	1	1	1	1	146.5	−1.5
	175	−1	−1	−1	1	−175	−175	−175	175	1	1	1	1	146.5	28.5
	147	−1	−1	2	0	−147	−147	294	0	1	1	4	0	150	−3
	143	−1	−1	2	0	−143	−143	286	0	1	1	4	0	150	−7
	153	−1	−1	2	0	−153	−153	306	0	1	1	4	0	150	3
	157	−1	−1	2	0	−157	−157	314	0	1	1	4	0	150	7
	148	−1	3	0	0	−148	444	0	0	1	9	0	0	158.5	−11
	161	−1	3	0	0	−161	483	0	0	1	9	0	0	158.5	2.5
	161	−1	3	0	0	−161	483	0	0	1	9	0	0	158.5	2.5
	164	−1	3	0	0	−164	492	0	0	1	9	0	0	158.5	5.5
	132	4	0	0	0	528	0	0	0	16	0	0	0	150	−18
	149	4	0	0	0	596	0	0	0	16	0	0	0	150	−1
	155	4	0	0	0	620	0	0	0	16	0	0	0	150	5
	164	4	0	0	0	656	0	0	0	16	0	0	0	150	14
Total	2872	0	0	0	0	128	264	162	134	80	48	24	8	2872	0

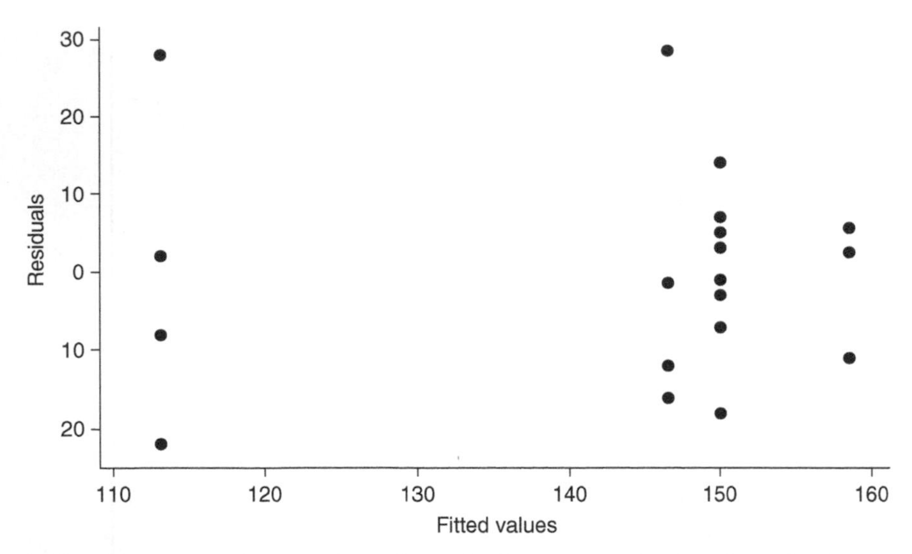

FIGURE 2.10: Scatterplot of residuals against fitted values for Example 2.3—orthogonal contrast regression model.

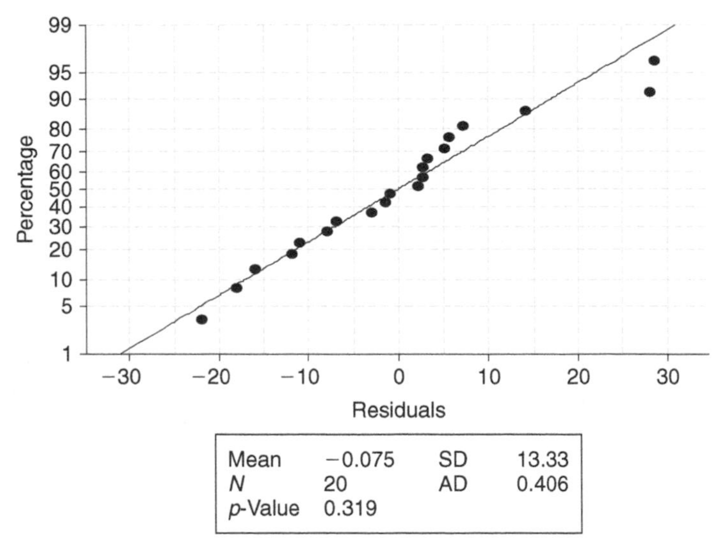

FIGURE 2.11: Normal probability plot of residuals for Example 2.3 studied by an orthogonal contrast regression model.

2.15.1 SAS Analysis of Data of Example 2.3 Using Orthogonal Contrasts (Helmert Coding)

In this section, we present the SAS analysis of the data of Example 2.3 using regression model based on Helmert contrast coding for the regressors.

SAS PROGRAM

```
data cataconc;
input y x1-x4 @@;
datalines;
105 4 0 0 0 115 4 0 0 0 91 4 0 0 0  141 4 0 0 0 135 -1 3 0 0
131 -1 3 0 0 145 -1 3 0 0 175 -1 3 0 0 147 -1 -1 2 0 143 -1 -1 2 0
153 -1 -1 2 0 157 -1 -1 2 0 148 -1 -1 -1 1 161 -1 -1 -1 1 161 -1 -1 -1 1 164
-1 -1 -1 1 132 -1 -1 -1 -1 149 -1 -1 -1 -1 155 -1 -1 -1 -1 164 -1 -1 -1 -1
;
proc print data=cataconc;
sum y x1-x4;
proc glm data=cataconc;
model y=x1-x4;run;
```

SAS OUTPUT

```
 Obs      y     x1     x2     x3     x4

   1    105      4      0      0      0
   2    115      4      0      0      0
   3     91      4      0      0      0
   4    141      4      0      0      0
   5    135     -1      3      0      0
   6    131     -1      3      0      0
   7    145     -1      3      0      0
   8    175     -1      3      0      0
   9    147     -1     -1      2      0
  10    143     -1     -1      2      0
  11    153     -1     -1      2      0
  12    157     -1     -1      2      0
  13    148     -1     -1     -1      1
  14    161     -1     -1     -1      1
  15    161     -1     -1     -1      1
  16    164     -1     -1     -1      1
  17    132     -1     -1     -1     -1
  18    149     -1     -1     -1     -1
  19    155     -1     -1     -1     -1
  20    164     -1     -1     -1     -1
        ====    ==     ==     ==     ==
        2872     0      0      0      0

                The GLM Procedure

     Number of Observations Read        20
     Number of Observations Used        20

The SAS System  11:32 Thursday, January 11, 2007    7

                The GLM Procedure

Dependent Variable: y
                                  Sum of
Source              DF           Squares   Mean Square   F Value   Pr > F

Model                4       4994.800000   1248.700000      5.61   0.0058

Error               15       3338.000000    222.533333

Corrected Total     19       8332.800000

          R-Square   Coeff Var   Root MSE     y Mean
```

```
                 0.599414    10.38827    14.91755    143.6000

Source              DF       Type I SS   Mean Square  F Value    Pr > F

x1                   1     4681.800000   4681.800000    21.04    0.0004
x2                   1      120.333333    120.333333     0.54    0.4735

Source              DF       Type I SS   Mean Square  F Value    Pr > F

x3                   1       48.166667     48.166667     0.22    0.6484
x4                   1      144.500000    144.500000     0.65    0.4329

Source              DF     Type III SS   Mean Square  F Value    Pr > F

x1                   1     4681.800000   4681.800000    21.04    0.0004
x2                   1      120.333333    120.333333     0.54    0.4735
x3                   1       48.166667     48.166667     0.22    0.6484
x4                   1      144.500000    144.500000     0.65    0.4329

                                             Standard
  Parameter            Estimate                 Error      Value    Pr > |t|

  Intercept         143.6000000            3.33566585      43.05      <.0001
  x1                 -7.6500000            1.66783293      -4.59      0.0004
  x2                 -1.5833333            2.15316305      -0.74      0.4735
  x3                 -1.4166667            3.04503238      -0.47      0.6484
  x4                  4.2500000            5.27415080       0.81      0.4329
```

There is an agreement between SAS analysis and previous analysis by manual methods.

2.16 Exercises

Exercise 2.1

Five machines were designed to produce steel pipes whose diameters are to be approximately 4.0 cm. In a pilot study, six pipes produced by each machine as in the table here were chosen at random and their diameters measured. Is there any difference between the diameters of the pipes produced by different machines?

Machine I	Machine II	Machine III	Machine IV	Machine V
3.8	3.9	3.9	4.1	4.1
3.9	3.8	4.2	3.9	4.2
3.9	4.0	4.1	4.0	4.3
3.8	4.0	4.0	4.1	4.1
3.9	4.0	3.8	4.0	4.0
3.8	3.7	4.2	3.9	3.8

Examine the fitted values along with the normal plots of the residuals for the experiment and comment on whether the basic assumptions of the design are satisfied. □

Exercise 2.2
Using regression analysis, use SAS to carry out the analysis of the responses of Example 2.2a containing the one-way experimental layout. There are four treatments with n_1 observations for treatment 1, n_2 observations for treatment 2, ..., and n_4 observations for treatment 4 with the following data:

Treatment	**Observations**
1	64 55 65 46 66 48 58 67 54 58 60
2	86 64 75 75 69 87 60 69 71
3	89 97 96 68 90 82 80 88 76 64 71 90 78 69
4	57 78 98 76 88 59 77 65

□

Exercise 2.3
In an experiment to compare the yields of a process under four different temperatures, the following data were obtained:

15°C	**25°C**	**35°C**	**45°C**
14	35	31	41
12	36	33	31
15	36	30	29
16	28	25	33
17	17	23	6

Is there any significant difference in the yield of the process at different temperatures? Using orthogonal polynomials or otherwise, fit a response curve to the data. □

Exercise 2.4
In a production process, five production lines L1–L5 were chosen at random for study; 11 determinations were made for each production line. The differences between the benchmark number of items expected to be produced per unit time and the actual number produced were recorded; the data are presented as follows:

L1	**L2**	**L3**	**L4**	**L5**	**L1**	**L2**	**L3**	**L4**	**L5**
−16	11	3	6	12	−6	−3	2	−4	8
−8	−5	7	6	3	−3	−4	2	−9	14
−2	−3	2	8	0	1	−2	3	−4	12
−7	1	1	5	3	−9	6	2	3	6
−8	−7	0	15	11	−6	−5	5	−3	5
−8	7	1	6	10					

Assuming a fixed effects model, are the production lines operating on average at the same level? Test at 1% level of significance. Obtain the fitted values, and extract the residuals. Using the residuals, carry out tests to determine whether the normality and equal variance assumptions are satisfied. □

Exercise 2.5
Using the data of Example 2.3, and the following table of contrasts, obtain the linear combinations of treatment totals as well as the sums of squares for the above contrasts and hence perform ANOVA to check which of them is significant.

Concentration (%)	C_1	C_2	C_3	C_4
0	0	1	−1	−1
5	0	0	0	4
10	0	1	1	−1
15	−1	−1	−1	−1
20	1	−1	1	−1

□

Exercise 2.6
Use the theory of regression analysis to carry out the test to show that the fitted parameters in Equation 2.32 for the responses of experiment in Example 2.4 are significant. □

Exercise 2.7
Obtain a regression model for data of Exercise 2.4, using the effects coding method for the dummy variables. Analyze the data, fitting the model and testing for significance of the fitted parameters. Extract residuals for this fit and test for the assumption of normality of the errors. □

Exercise 2.8
Obtain a regression model for data of Exercise 2.4 using the reference cell coding method for dummy variables. Analyze the data, fitting the model and testing for significance of the fitted parameters. Extract residuals for this fit and test for the assumption of normality of the errors. Use L5 as the reference cell. □

Exercise 2.9
Using Bartlett's test, check whether the variances of the data of Exercise 2.1 are homogenous; repeat the same test using Levene's modified test. Do both tests reach the same conclusion? □

Exercise 2.10
The time of reaction to different stimuli was being studied in a CRD. The reaction times measured in seconds are as follows:

S1	*S2*	*S3*	*S4*	*S5*	*S6*
1.7	1.6	2.2	2	1.1	0.9
1.5	1.7	1.9	1.9	1.1	0.8
1.5	1.8	2.3	1.9	1.4	0.8
1.4	1.6	2.1	2.1	1.4	0.9

Test for significant difference between means of stimuli at 1% significance. Extract residuals and test for normality assumption. Using Levene's test, check for homogeneity of the variances. □

Exercise 2.11
Use orthogonal contrasts (Helmert coding) to obtain a regression model for the experiment in Exercise 2.10, estimate its parameters and test for significance of different stimuli. □

Exercise 2.12
Use SAS to analyze the data in Exercise 2.10 using orthogonal contrasts. Extract the residuals and use it to carry out diagnostics (check for model assumptions). □

Exercise 2.13
Write a SAS program to carry out ANOVA and test for significant difference in mean production for the data of Example 2.4. □

References

Anderson, V.L. and McLean, R.A. (1974) *Design of Experiments: A Realistic Approach*, Marcel Dekker, New York.

Barnett, V. and Lewis, T. (1994) *Outliers in Statistical Data*, 3rd ed., Wiley, New York.

Bartlett, M.S. (1934) The problem in statistics of testing several variances. *Proceedings of the Cambridge Philosophical Society*, 30, 164–169.

Bartlett, M.S. (1936) The square root transformation in the analysis of variance. *Journal of Royal Statistical Society*, Suppl. 2, 68–78.

Bartlett, M.S. (1947) The use of transformations. *Biometrics*, 3, 39–52.

Bartlett, M.S. and Kendall, M.G. (1946) The statistical analysis of variance and logarithmic transformations, *Journal of the Royal Statistical Society*, Suppl. 8, 128–138.

Box, G.E.P. and Cox, D.R. (1964) An analysis of transformations. *Journal of the Royal Statistical Society*, Series B, 26, 211–252.

Box, G.E.P., Hunter, J.S., and Hunter, W.G. (2005) *Statistics for Experimenters*, 2nd ed., John Wiley & Sons Inc, Hoboken, NJ.

Brown, M.B. and Forsythe, A.B. (1974) Robust test for equality of variances. *Journal of American Statistical Association*, 69, 364–367.

Burr, I.W. and Foster, L.A. (1972) *A Test of Equality of Variances*, Mimeo Series, 282, University of Purdue, West Lafayette.

Chatfield, C. (1983) *Statistics for Technology*, Chapman and Hall, London.

Cochran, W.G. and Cox, G.M. (1957) *Experimental Designs*, 2nd ed., Wiley, New York.

Conover, W.J., Johnson, M.E., and Johnson, M.M. (1981) A comparative study of tests of homogeneity of variances, with applications to outer continental shelf bidding data, *Technometrics*, 23, 251–261.

Cox, D.R. (1958) *Planning of Experiments*, Wiley, New York.

Duncan, D.B. (1955) Multiple range and multiple F tests, *Biometrics*, 11, 1–42.

Fisher, R.A. and Yates, F. (1963) *Statistical Tables for Biological, Agricultural, and Medical Research*, 4th ed., Oliver and Boyd, Edinburgh.

Guttman, I., Wilks, S.S., and Hunter, J.S. (1982) *Introductory Engineering Statistics*, 3rd ed., Wiley, New York.

John, J.A. and Prescott, P. (1975) Critical values of a test to detect outliers in factorial experiments, *Applied Statistics*, 24, 56–59.

Keppel, G. (1991) *Design and Analysis: A Researcher's Handbook*, 3rd ed., Prentice Hall, Englewood Cliffs NJ.

Keppel, G. and Wickens, T.D. (2004) *Design and Analysis: A Researcher's Handbook*, 4th ed., Prentice Hall, Englewood Cliffs, NJ.

Kleinbaum, D.G., Kupper, L.L., Muller, K.E., and Nizam, A. (1998) *Applied Regression Analysis and Other Multivariable Methods*, Brooks/Cole Publishing Company, Pacific Grove, CA.

Kolmogorov, A.N. (1933) *Grunbegriffe der Wahrscheinlichkeitsrechnung*, Springer, Berlin.

Levene, H. (1960) Robust test for equality of variance. In I. Olkin (Ed.), *Contributions to Probability and Statistics: Essays in Honor of Harold Hotelling*, Stanford University Press, Stanford.

Montgomery, D.C. (2005) *Design and Analysis of Experiments*, 6th ed., Wiley, New York.

Pearson, E.S. and Hartley, H.O. (1966) *Biometrika Tables for Statisticians*, Vol. 1, 3rd ed., Cambridge University Press, Cambridge.

Pedhazur, E.J. (1982) *Multiple Regression in Behavioral Research*, 2nd ed., Holt, Rinehart and Winston, New York.

Serlin, R.C. and Levin, J.R. (1985) Teaching coding schemes for direct interpretation in multiple regression analysis, *Journal of Educational Statistics*, 10, 223–238.

Shapiro, S.S. and Wilks, M.B. (1965) Analysis of variance test for normality, *Biometrika*, 52, 591–611.

Shapiro, S.S., Wilks, M.B., and Chen, H.J. (1968) A comparative study of the various tests of normality, *Journal of the American Statistical Association*, 63, 1343–1372.

Smirnov, V.I. (1939) On the estimation of the discrepancy between empirical curves of distributions for two independent samples. *Bull Math Univ Moscou, Serie Int* 2 (2), 3–14.

Stefansky, W. (1972) Rejecting outliers in factorial designs, *Technometrics*, 14, 469–479.

Tabachnick, B.G. and Fidell, S.F. (2007) *Experimental Design Using ANOVA*, Duxbury Press, Belmont, CA.

Tukey, J.W. (1953) The problem of multiple comparisons, unpublished notes, Princeton University.

Wackerly, D.D., Mendenhall III, W., and Schaefer, R.L. (2002) *Mathematical Statistics with Applications*, 6th ed., Duxbury Press, Belmont, CA.

Welch, B.L. (1951) On the comparison of several mean values: An alternative approach, *Biometrika*, 38, 330–336.

Wendorf, C.A. (2004) Primer on multiple regression coding: Common form and additional case of repeated contrasts, *Understanding Statistics*, 3, 47–57.

Wetherill, G.B. (1981) *Intermediate Statistical Methods*, Chapman and Hall, London.

Chapter 3

Two-Factor Factorial Experiments and Repeated Measures Designs

In sections of Chapter 2, we treated the one-way experimental layout, that is, the completely randomized design (CRD), which involved a single factor set at k levels, or the comparison of means of k groups or k populations. Next, in this chapter, we consider experiments in which the effects of levels of two factors on a response variable are studied. In such studies, resources are usually available so that the designed experiment can be replicated. The data from the experiment are presented in a two-way layout, which involves two factors, each set at a number of levels with replications in every cell, where the cells are determined by the combinations of levels of the factors. Equal or unequal number of measurements can be made in each cell. In the following Section 3.1, we consider the two-factor factorial experiment in its simplest form, with equal number of observations per cell. The two-factor factorial experiment is the simplest form of a crossed design; a designed experiment in which replicates involving all the combinations of levels of all the factors are investigated in at least two replicates is called a full factorial experiment. Factors arranged in a factorial experiment are said to be crossed.

3.1 Full Two-Factor Factorial Experiment (Two-Way ANOVA with Replication)—Fixed Effects Model

Suppose we want to study the effects of two factors, A and B, set at a and b levels, respectively, on a response variable Y. In such situations, we need to make ab observations to obtain at least one response y_{ij} for each combination of levels $i=1, \ldots, a$ and $j=1,2,\ldots,b$. In a full two-factor factorial experiment, for each combination of the levels of the factors A and B, we need to perform experiments to obtain $n(n \geq 2)$ observations. This means that the experiment is replicated n times leading to n observations per cell. For the experiment described above, the appropriate model is

$$y_{ijk} = \mu + \alpha_i + \tau_j + (\alpha\tau)_{ij} + \varepsilon_{ijk} \tag{3.1}$$

where $i = 1, \ldots, a$; $j = 1, \ldots, b$; $k = 1, \ldots, n$. In the model, μ is the overall (grand) mean, γ_i is the effect of the ith level of the factor A, τ_j is the effect of the jth level of the factor B, while $(\alpha\tau)_{ij}$ is the effect of the interaction of the ith level of the factor A and the jth level of the factor B. Since we are considering a fixed effects model, we impose the condition that $\sum_{i=1}^{a}(\alpha\tau)_{ij} = 0$ for all j and $\sum_{j=1}^{b}(\alpha\tau)_{ij} = 0$ for all i. $\varepsilon_{ijk} \sim \text{NID}(0, \sigma)^2$ is the error of the kth replicate for ith level of the factor A and the jth level of the factor B. The basic assumptions mentioned in other fixed effects models, namely, $\sum_{i=1}^{a} \alpha_i = 0$ and $\sum_{j=1}^{b} \tau_j = 0$ form part of the requirement for this model to be considered a fixed effects model. In this design, it is required that all of the ab experiments in each replicate be performed in random order. Here, randomization is carried out across all combinations of levels of the factors in the design. When we have an experiment that conforms to the above descriptions, then the design is considered to be a full two-factor factorial experiment with fixed effects.

The responses obtained in the experiments performed under the assumptions stated above can be arranged in the form shown in Table 3.1.

TABLE 3.1: The Two-Factor Factorial Experiment

	Levels of factor B					
Levels of factor A	B_1	B_2	$\cdots$	B_j	$\cdots$	B_b
A_1	$y_{111}, \ldots, y_{11n}$	$y_{121}, \ldots, y_{12n}$	$\cdots$	$y_{1j1}, \ldots, y_{1jn}$	$\cdots$	$y_{1b1}, \ldots, y_{1bn}$
A_2	$y_{211}, \ldots, y_{21n}$	$y_{221}, \ldots, y_{22n}$	$\cdots$	$y_{2j1}, \ldots, y_{2jn}$	$\cdots$	$y_{2b1}, \ldots, y_{2bn}$
A_i	$y_{i11}, \ldots, y_{i1n}$	$y_{i21}, \ldots, y_{i2n}$	$\cdots$	$y_{ij1}, \ldots, y_{ijn}$	$\cdots$	$y_{ib1}, \ldots, y_{ibn}$
$\vdots$	$\vdots$	$\vdots$	$\vdots$	$\vdots$	$\vdots$	$\vdots$
A_a	$y_{a11}, \ldots, y_{a1n}$	$y_{a21}, \ldots, y_{a2n}$	$\cdots$	$y_{ai1}, \ldots, y_{ajn}$	$\cdots$	$y_{ab1}, \ldots, y_{abn}$

We can obtain the totals of the responses as follows:

$$\begin{aligned} y_{ij.} &= \sum_{k=1}^{n} y_{ijk} \quad & y_{i..} &= \sum_{j=1}^{b}\sum_{k=1}^{n} y_{ijk} \\ y_{.j.} &= \sum_{i=1}^{a}\sum_{k=1}^{n} y_{ijk} \quad & y_{...} &= \sum_{i=1}^{a}\sum_{j=1}^{b}\sum_{k=1}^{n} y_{ijk} \end{aligned} \tag{3.2}$$

The sums of squares can be obtained by the usual partitioning of the total sum of squares into its component parts as follows:

$$SS_{\text{total}} = SS_A + SS_B + SS_{AB} + SS_{\text{E}} \tag{3.3}$$

One of the hypotheses tested under this model is

$$H_0 : (\alpha\tau)_{11} = (\alpha\tau)_{12} = \cdots = (\alpha\tau)_{ab} = 0 \quad \text{vs} \quad H_1 : \text{at least one } (\alpha\tau)_{ij} \neq 0$$

It can be seen that under the null hypothesis (see Mood et al., 1974, p. 245) $\varepsilon_{ijk} \sim \text{NID}(0, \sigma^2)$, SS_{AB}/σ^2 and $SS_{\text{residual}}/\sigma^2$ are independent χ^2 random variables with $(a-1)(b-1)$ and $ab(n-1)$ degrees of freedom. Similar ideas have already been discussed for a similar situation in Chapter 2 (Section 2.7). Consequently, under H_0 the statistic

$$F_0 = \frac{\left[\dfrac{SS_{AB}}{\sigma^2(a-1)(b-1)}\right]}{\left[\dfrac{SS_{\text{residual}}}{\sigma^2(ab(n-1))}\right]} = \frac{ab(n-1)SS_{AB}}{(a-1)(b-1)SS_{\text{residual}}} \tag{3.4}$$

is F distributed with $(a-1)(b-1)$ and $ab(n-1)$ degrees of freedom. Similarly, when the hypothesis to be tested is

$$H_0 : \alpha_1 = \alpha_2 = \cdots = \alpha_a = 0 \quad \text{vs} \quad H_1 : \alpha_i \neq 0 \text{ for } i = 1, 2, \ldots, a$$

then the test statistic

$$F_0 = \frac{\left[\dfrac{SS_{\text{A}}}{\sigma^2(a-1)}\right]}{\left[\dfrac{SS_{\text{residual}}}{\sigma^2(ab(n-1))}\right]} = \frac{ab(n-1)SS_{\text{A}}}{(a-1)SS_{\text{residual}}} \tag{3.5}$$

is F distributed with $a-1$, $ab(n-1)$ degrees of freedom when H_0 is true. The other hypothesis that can be tested under this design is

$$H_0 : \tau_1 = \tau_2 = \cdots = \tau_b = 0 \quad \text{vs} \quad H_1 : \tau_j \neq 0 \text{ for some } j; \;\; j = 1, 2, \ldots, b$$

The statistic

$$F_0 = \frac{\left[\dfrac{SS_{B}}{\sigma^2(b-1)}\right]}{\left[\dfrac{SS_{\text{residual}}}{\sigma^2(ab(n-1))}\right]} = \frac{ab(n-1)SS_{B}}{(b-1)SS_{\text{residual}}} \tag{3.6}$$

to be used for this test also follows the F-distribution with $b-1$ and $ab(n-1)$ degrees of freedom. The ANOVA for the responses of experiments conducted under this design is presented in Table 3.2.

TABLE 3.2: ANOVA for Two-Factor Factorial Experiment

Source	Sum of squares	Degrees of freedom	Mean square	Expected mean square	*F*-ratio
Due to A	SS_A	$a-1$	$\frac{SS_A}{(a-1)}$	$\sigma^2 + \frac{bn}{a-1}\sum_{j=1}^{b}\tau_j^2$	$\frac{MS_A}{MS_{\text{residuals}}}$
Due to B	SS_B	$b-1$	$\frac{SS_B}{(b-1)}$	$\sigma^2 + \frac{an}{b-1}\sum_{i=1}^{a}\alpha_i^2$	$\frac{MS_B}{MS_{\text{residuals}}}$
Due to AB	SS_{AB}	$(a-1)(b-1)$	$\frac{SS_{AB}}{(b-1)(a-1)}$	$\sigma^2 + \frac{n}{(a-1)(b-1)} \times \sum_{i=1}^{a}\sum_{j=1}^{b}v_{ij}^2$	$\frac{MS_{AB}}{MS_{\text{residuals}}}$
Residual	$SS_{\text{residuals}}$	$ab(n-1)$	$\frac{SS_{\text{residuals}}}{(b-1)(a-1)}$	σ^2	
Total	SS_{total}	$abn-1$			

For ease of calculation, the following formulae are useful:

$$
\begin{aligned}
SS_{\text{total}} &= \sum_{i=1}^{a}\sum_{j=1}^{b}\sum_{k=1}^{n} y_{ijk}^2 - \frac{y_{...}^2}{abn} \\
SS_A &= \frac{1}{bn}\sum_{i=1}^{a} y_{i..}^2 - \frac{y_{...}^2}{abn} \\
SS_B &= \frac{1}{an}\sum_{j=1}^{b} y_{.j.}^2 - \frac{y_{...}^2}{abn} \\
SS_{AB} &= \frac{1}{n}\sum_{i=1}^{a}\sum_{j=1}^{b} y_{ij.}^2 - SS_A - SS_B - \frac{y_{...}^2}{abn} \\
SS_{\text{residual}} &= \sum_{i=1}^{a}\sum_{j=1}^{b}\sum_{k=1}^{n} y_{ijk}^2 - \frac{1}{n}\sum_{i=1}^{a}\sum_{j=1}^{b} y_{ij.}^2
\end{aligned}
\tag{3.7}
$$

We consider an example to illustrate this design.

Example 3.1

A company producing Akwette cloth wished to purchase a new set of weaving machines. A dealer presented the company with four types of weaving machines and asked them to make a choice. The company invited five machinists for interview with a view to employing the most skilled. The manager of the company who wishes to make his decision scientifically, engaged a statistician to design an experiment which would help him to test the machines as well as the machinists. As part of the experiment, each candidate was asked to produce three pieces of a specified size of cloth on each of the machines, and the time taken for each of these candidates to produce each of these pieces of cloth was recorded as shown in the following table. The statistician used a two-way experimental design to study the performances of men and machines.

	Machines (M_2)				
Machinists (M_1)	**I**	**II**	**III**	**IV**	**Total**
I	43, 47, 48 (138)	60, 64, 68 (192)	46, 50, 55 (151)	59, 52, 54 (165)	646
II	50, 56, 48 (154)	79, 73, 75 (227)	65, 68, 62 (195)	60, 63, 61 (184)	760
III	72, 64, 75 (211)	80, 88, 84 (252)	62, 70, 73 (205)	70, 71, 76 (217)	885
IV	77, 73, 78 (228)	79, 87, 80 (246)	61, 67, 69 (197)	68, 75, 79 (222)	893
V	67, 73, 66 (206)	84, 80, 80 (244)	69, 83, 74 (226)	62, 60, 57 (179)	855
Total	937	1161	974	967	4039

In order to test the relevant hypotheses and to arrive at a decision, the sums of squares for the factors M_1, M_2, and M_1M_2 were obtained. Using the equations stated earlier in this section, the sums of squares are

$$SS_{\text{total}} = 43^2 + 47^2 + 48^2 + \cdots + 60^2 + 57^2 - \frac{4039^2}{5(4)(3)} = 7406.9833$$

$$SS_{M_1} = \frac{937^2 + 1161^2 + 974^2 + 967^2}{3(4)} - \frac{4039^2}{5(4)(3)} = 2084.9833$$

$$SS_{M_2} = \frac{646^2 + 760^2 + 885^2 + 893^2 + 855^2}{3(5)} - \frac{4039^2}{5(4)(3)} = 3659.2333$$

$$SS_{M_1M_2} = \frac{138^2 + 192^2 + \cdots + 179^2}{3} - 2084.9833 - 3659.2333 - \frac{4039^2}{5(4)(3)}$$

$$= 990.7667$$

$$SS_{\text{residual}} = 7406.9833 - 990.7667 - 2084.9833 - 3659.2333 = 671.999$$

The set of hypotheses to be tested are

$$H_0 : \gamma_1 = \gamma_2 = \cdots = \gamma_5 = 0 \quad \text{vs} \quad H_1 : \text{at least one}\, \gamma_i \neq 0$$

$$H_0' : \tau_1 = \cdots = \tau_4 = 0 \quad \text{vs} \quad H_1' : \text{at least one}\, \tau_j \neq 0$$

$$H_0'' : \nu_{11} = \nu_{12} = \cdots = \nu_{54} = 0 \quad \text{vs} \quad H_1'' : \text{at least one}\, \nu_{ij} \neq 0$$

The ANOVA was carried out, and the results are presented in Table 3.3.

From the tables of the F-distribution, using a level of significance of $\alpha = 1\%$, we obtain $F(4, 40, 0.01) = 3.8283$; $F(3, 40, 0.01) = 4.3126$; and $F(12, 40, 0.01) = 2.6648$. From Table 3.3, we see that the effects of machinists (M_1) and machines (M_2) and the interaction between machines and machinists are all significant at 1% level of significance. This poses some problems in the interpretation of the results. Proper care should, therefore, be taken in interpreting the results of this type of ANOVA. It is worth noting that although the main effects of M_1 and M_2 exceed the critical level of the tabulated F, there is a highly significant interaction, which should not be ignored. We can, therefore, see that the significance of these two main effects has no meaning, since each of them was not acting independently. If we were studying quantitative factors set at different levels, plot of the means of each cell against levels of one of the factors would be the next most appropriate step to be taken in analysis of the responses of the experiment.

This step, it is hoped, would aid the interpretation of the result of the experiment. Any reader interested in this method of interpretation should refer to Montgomery (1984) and Anderson and McLean (1974) for examples in which this method was used. In our case, since our factors are qualitative, it is sufficient to tabulate the cell means of the responses (Table 3.4).

From the tabulation of the means of the different groups according to factors, it is obvious that machinist I turned out the lowest performance time on

TABLE 3.3: Two-Way ANOVA for the Akwette Cloth Company Experiment

Source	Sum of squares	Degrees of freedom	Mean square	F-ratio
M_1	3659.2333	4	914.8083	54.45288
M_2	2084.9833	3	694.9944	41.3687
M_1M_2	990.7667	12	82.5639	4.915
Residual	671.9999	40	16.7999	
Total	7406.9833	59		

TABLE 3.4: Cell Means for the Akwette Cloth Company Experiment

	Machines (M_2)				
Machinists (M_1)	I	II	III	IV	Means
I	46.0	64.00	50.53	55.00	53.83
II	57.33	75.67	65.00	61.33	63.75
III	70.33	84.00	68.33	72.33	73.75
IV	76.00	82.00	65.67	74.00	74.42
V	68.67	81.33	75.33	59.67	71.25
Mean	63.66	77.00	64.07	64.47	67.1

all the four machines. The next machinist who seemed to perform better than the rest (apart from machinist I) is machinist II. On the other hand, apart from machine II whose responses seemed to be the highest, there is apparently no reason to believe that there is any difference between machines I, II, and IV. We will revisit the interaction shortly, but meanwhile, let us show how to analyze the data with a SAS program. □

3.1.1 SAS Analysis of Data of Example 3.1

In the following SAS program, we use the ANOVA procedure to analyze the data for Example 3.1. Here, there is an interaction between the factors—machines and machinists. In stating the model in the program we can either state the effects explicitly as

```
model time = machinist machines machine*machinist;
```

or we can use the following method:

```
model time = machinist|machines;
```

Both are good ways of instructing SAS to carry out the analysis and examine the interaction between machine and machinist as well as the main effects of machine and machinist. The use of the second form definitely allows us to state several main effects and interactions in a model in a more concise way. Its usefulness will be demonstrated if we have three, four, or more factors in the model statement.

SAS PROGRAM FOR RESPONSES OF EXAMPLE 3.1

```
data machines1;
input machinist machines time@@;
  cards;
1 1 43 1 1 47 1 1 48 1 2 60 1 2 64 1 2 68 1 3 46 1 3 50 1 3 55
1 4 59 1 4 52 1 4 54 2 1 50 2 1 56 2 1 48 2 2 79 2 2 73 2 2 75
2 3 65 2 3 68 2 3 62 2 4 60 2 4 63 2 4 61 3 1 72 3 1 64 3 1 75
3 2 80 3 2 88 3 2 84 3 3 62 3 3 70 3 3 73 3 4 70 3 4 71 3 4 76
4 1 77 4 1 73 4 1 78 4 2 79 4 2 87 4 2 80 4 3 61 4 3 67 4 3 69
4 4 68 4 4 75 4 4 79 5 1 67 5 1 73 5 1 66 5 2 84 5 2 80 5 2 80
5 3 69 5 3 83 5 3 74 5 4 62 5 4 60 5 4 57
  ;
proc print data=machines1;proc anova data=machines1;
class machinist machines;
 model time=machinist|machines; run;
```

SAS OUTPUT:

Obs	machinist	machines	time
1	1	1	43
2	1	1	47
3	1	1	48
4	1	2	60
5	1	2	64
6	1	2	68
7	1	3	46
8	1	3	50
9	1	3	55
10	1	4	59
11	1	4	52
12	1	4	54
13	2	1	50
14	2	1	56
15	2	1	48
16	2	2	79
17	2	2	73
18	2	2	75
19	2	3	65
20	2	3	68
21	2	3	62
22	2	4	60
23	2	4	63
24	2	4	61
25	3	1	72
26	3	1	64
27	3	1	75
28	3	2	80
29	3	2	88
30	3	2	84
31	3	3	62
32	3	3	70
33	3	3	73
34	3	4	70
35	3	4	71
36	3	4	76
37	4	1	77
38	4	1	73
39	4	1	78
40	4	2	79
41	4	2	87
42	4	2	80
43	4	3	61
44	4	3	67
45	4	3	69
46	4	4	68
47	4	4	75
48	4	4	79
49	5	1	67

Obs	machinist	machines	time
50	5	1	73
51	5	1	66
52	5	2	84
53	5	2	80
54	5	2	80
55	5	3	69
56	5	3	83
57	5	3	74
58	5	4	62
59	5	4	60
60	5	4	57

The ANOVA Procedure

Dependent Variable: time

Source	DF	Sum of Squares	Mean Square	F Value	Pr > F
Model	19	6734.983333	354.472807	21.10	<.0001
Error	40	672.000000	16.800000		
Corrected Total	59	7406.983333			

R-Square	Coeff Var	Root MSE	time Mean
0.909275	6.088805	4.098780	67.31667

Source	DF	Anova SS	Mean Square	F Value	Pr > F
machinist	4	3659.233333	914.808333	54.45	<.0001
machines	3	2084.983333	694.994444	41.37	<.0001
machinist*machines	12	990.766667	82.563889	4.91	<.0001

The ANOVA Procedure

Level of machinist	Level of machines	N	------------time------------ Mean	Std Dev
1	1	3	46.0000000	2.64575131
1	2	3	64.0000000	4.00000000
1	3	3	50.3333333	4.50924975
1	4	3	55.0000000	3.60555128
2	1	3	51.3333333	4.16333200
2	2	3	75.6666667	3.05505046
2	3	3	65.0000000	3.00000000
2	4	3	61.3333333	1.52752523
3	1	3	70.3333333	5.68624070
3	2	3	84.0000000	4.00000000
3	3	3	68.3333333	5.68624070
3	4	3	72.3333333	3.21455025
4	1	3	76.0000000	2.64575131
4	2	3	82.0000000	4.35889894
4	3	3	65.6666667	4.16333200
4	4	3	74.0000000	5.56776436
5	1	3	68.6666667	3.78593890
5	2	3	81.3333333	2.30940108
5	3	3	75.3333333	7.09459888
5	4	3	59.6666667	2.51661148

3.1.2 Simple Effects of an Independent Variable

Consider the experiment described in Example 3.1 in which the effects of two factors, machines at four levels and machinists at five levels, were investigated. An examination of a factorial experiment reveals that it contains a set of what we could consider as single-factor experiments. For the above experiment, we can subdivide its outcomes into four single-factor experiments based on the four machines used in the experiment as shown in the following tables.

Machinists (M_1)	Machine I
I	
II	
III	
IV	
V	

Machinists (M_1)	Machine II
I	
II	
III	
IV	
V	

Machinists (M_1)	Machine III
I	
II	
III	
IV	
V	

Machinists (M_1)	Machine IV
I	
II	
III	
IV	
V	

In the first table, if the performance times are significant, we would attribute them to machine I. We can make similar attributions to the other three machines II, III, and IV. There are four single-factor experiments, each of which provides information on one of the four machines that are under study. The results from each of these single-factor experiments are called *simple effects*. Thus, the results in all these four tables are the simple effects of machines I through IV, respectively. We can also subdivide the entire experiment into five single-factor experiments based on the five machinists who were involved in the experiment. To carry out the entire study using single-factor experiments as described above, we need to perform nine experiments. Since there should be at least two observations per cell for estimation of error, we need 10 observations in each of these single-factor experiments. Thus, we need 90 observations to carry out all the single-factor experiments. With a factorial design in two factors and only two observations per combination of factor levels (called treatment combinations), the efficiency and the economy of the factorial experiment is made clear by the fact that we could carry out the entire study with only 40 observations. If we design multifactor experiments, we run into the problem of interactions between factors. This problem is not usually observable in single-factor experiments. With significant interaction, we first need to detect its occurrence and interpret it when it occurs.

3.1.3 Absence of Interaction in a Factorial Experiment

In the CRD or single-factor experiments, there are only main effects to be considered. However, when we have multifactor experiments, interactions routinely occur between factors. Occurrence of significance of interactions in a factorial experiment is more important than the significance of main effects. This is because the presence of significant interaction indicates that the factors were not acting alone but together to produce the observed effects. When interaction is not present, the two-factor factorial design is like the study of two separate single-factor experiments. On the other hand, when the effects of one factor changes at the different levels of the second factor, then interaction is present. Suppose in a factorial experiment on two factors, A and B, we observed the following data by making three observations per treatment combination:

	Factor B				
Factor A	**I**	**II**	**III**	**IV**	**Mean**
I	48, 44, 46	62, 66, 64	54, 46, 50	57, 53, 55	53.75
II	53, 47, 50	70, 66, 68	57, 51, 54	63, 55, 59	57.75
III	57, 63, 60	80, 76, 78	66, 62, 64	72, 66, 69	67.75
IV	63, 67, 65	85, 81,83	66, 72, 69	77, 71, 74	72.75
V	69.2, 68, 68.6	86, 87.2, 86.6	75.2, 70, 72.6	76.2, 79, 77.6	76.35
Mean	57.92	75.92	61.92	66.92	65.67

The responses of the above experiment should lead to the following cell and marginal means (Table 3.5). ANOVA for the data indicates the results shown in Table 3.6.

Clearly, there is no interaction effect. A plot of the cell means of A across levels of B is shown in Figure 3.1.

The plots of the means of the responses are parallel to each other showing that there is no interaction. The plot does indicate significant main effect for A. This is signified by the changes in the simple effects from one level

TABLE 3.5: Cell Means for the Hypothetical Two-Factor Factorial Experiment for Levels of A and B

	Factor B				
Factor A	**I**	**II**	**III**	**IV**	**Mean**
I	46	64	50	55	53.75
II	50	68	54	59	57.75
III	60	78	64	69	67.75
IV	65	83	69	74	72.75
V	68.6	86.6	72.6	77.6	76.35
Mean	57.92	75.92	61.92	66.92	65.67

TABLE 3.6: ANOVA for the Hypothetical Two-Factor Factorial Experiment in Factors A and B

Source	Sum of squares	Degrees of freedom	Mean square	F-ratio	p-Value
A	4479.94	4	1119.98	175.77	0.000
B	2711.25	3	903.75	141.83	0.000
$A*B$	0.00	12	0.00	0.00	1.000
Error	254.88	40	6.37		
Total	7446.07	59			

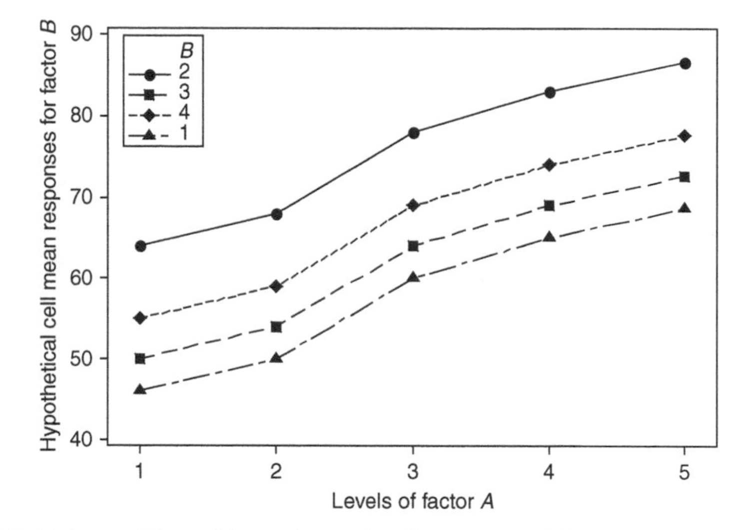

FIGURE 3.1: Plot of hypothetical cell means of factor B across levels of factor A.

of machine to another, which is represented by the marginal means obtained across all levels of B, as is clearly apparent for the plot in Figure 3.2.

A similar set of changes are observable for the plot of the marginal means for the levels of the factor B, indicating the significance of the main effect B. This again would be signified by changes of simple effects from one level of B to another (Figure 3.3). These changes are mirrored by the marginal means obtained across all levels of A (Figure 3.2).

In a typical two-factor factorial experiment, the following could be expected to occur when there is no interaction between the factors:

1. Both main effects could be significant.
2. One main effect could be significant.
3. No main effect could be significant.

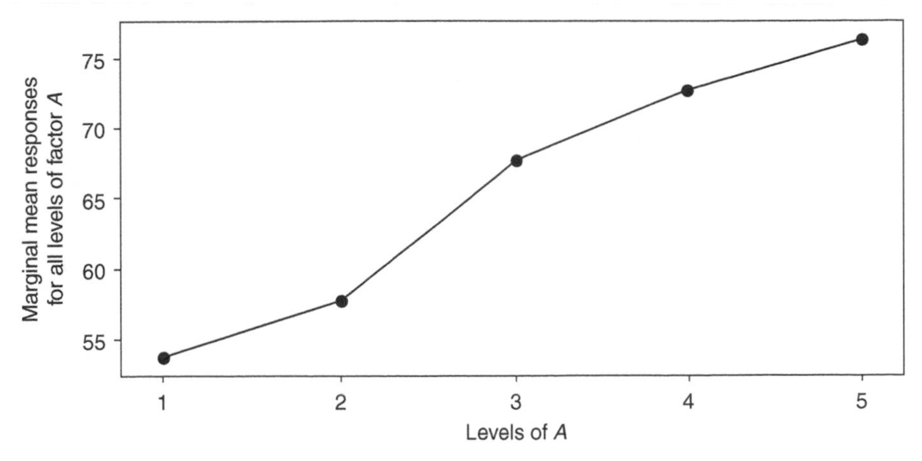

FIGURE 3.2: Plots of marginal means of responses for factor A.

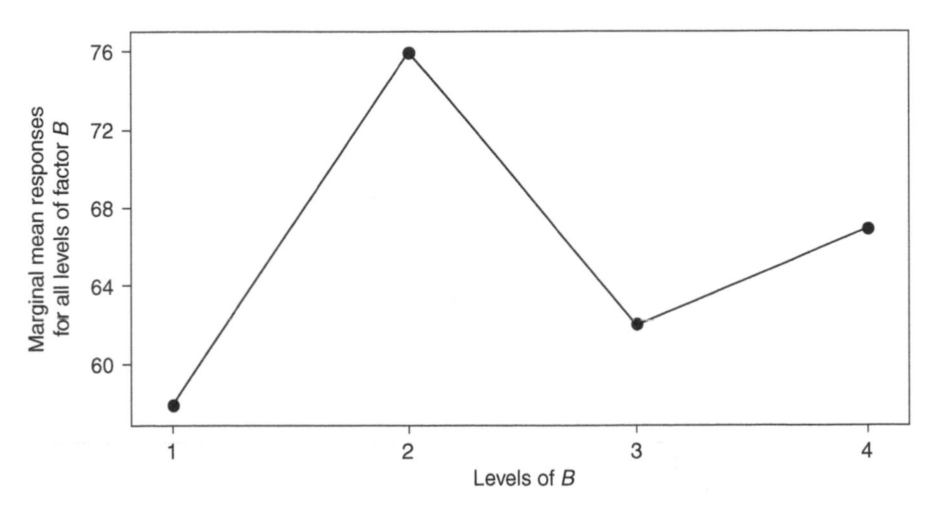

FIGURE 3.3: Plots of marginal means of responses for factor B.

3.1.4 Presence of Interaction in a Factorial Experiment

As mentioned earlier (Section 3.1.3), interactions do routinely occur in multifactor experiments. When they occur, we need to detect them and possibly interpret them. We will illustrate the occurrence of interaction by continuing with the experiment that involves factors A and B as described in the Section 3.1.3. However, this time we will replace the previous data with different responses, which we assume are the outcomes of the experiment. The changes are shown in Tables 3.7 and 3.8.

TABLE 3.7: Data of Hypothetical Two-Factor Factorial Experiment

	Factor *B*			
Factor *A*	**I**	**II**	**III**	**IV**
I	51, 56, 51	63, 57, 62	73, 71, 69	54, 50, 49
II	64, 57, 59	64, 57, 65	73, 73, 70	51, 56, 52
III	73, 76, 73	57, 61, 62	69, 69, 72	57, 63, 57
IV	78, 82, 83	61, 66, 65	72, 76, 74	65, 60, 64
V	91, 86, 90	69, 61, 68	80, 72, 79	69, 61, 62

We analyze the data shown in Table 3.7 as follows:

$$SS_{\text{total}} = 268554 - \frac{3970^2}{4(5)(3)} = 5872.333$$

$$SS_B = \frac{1070^2 + 938^2 + 1092^2 + 870^2}{5(3)} - \frac{3970^2}{4(5)(3)} = 2258.867$$

$$SS_A = \frac{706^2 + 741^2 + 789^2 + 846^2 + 888^2}{5(3)} - \frac{3970^2}{4(5)(3)} = 1843.167$$

$$SS_{A*B} = \frac{158^2 + 182^2 + 213^2 + \cdots + 192^2}{3} - 2258.867 - 1843.167 - \frac{3970^2}{4(5)(3)}$$

$$= 1398.967$$

$$SS_{\text{E}} = 5872.333 - 2258.867 - 1843.167 - 1398.967 = 371.333$$

TABLE 3.8: ANOVA of Hypothetical Data of Two-Factor Factorial Experiment for *A* and *B*

Source	Sum of squares	Degrees of freedom	Mean square	*F*-ratio	*p*-Value
B	2258.867	3	752.9556	81.10839	4.73279 E − 17
A	1843.167	4	460.7917	115.1979	1.01623 E − 19
A∗*B*	1398.967	12	116.5806	9.715046	6.06622 E − 05
Error	371.333	40	9.283325		
Total	5872.333	59			

Clearly, both main effects and interaction are highly significant. As usual, the significance of the main effects is not as important as the significance of the interaction if both occur in a designed experiment. Our next task is to understand the interaction and possibly interpret it. We list the cell and marginal means and plot the cell means to try to understand the nature of interaction.

Cell means

	Factor B				
Factor A	**I**	**II**	**III**	**IV**	**Mean**
I	52.667	60.67	71	51	58.833
II	60	62	72	53	61.75
III	74	60	70	59	65.75
IV	81	64	74	63	70.5
V	89	66	77	64	74
Mean	71.33	62.533	72.8	58	66.167

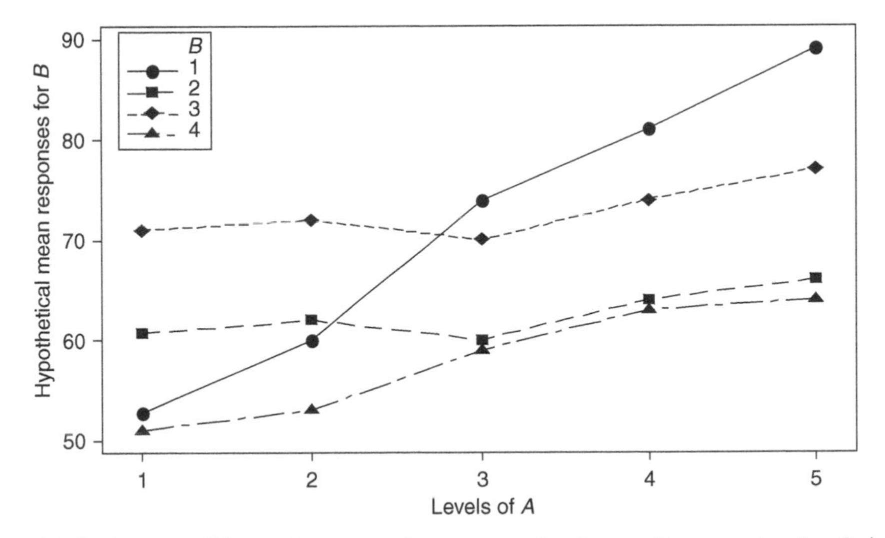

FIGURE 3.4: Plots of means of responses for factor B across levels of A.

In Figure 3.4, we see that the plots of the hypothetical responses are not parallel as in the Section 3.2.1 but are crossed. It is clear that level 1 of B has led to more responses at levels 2–5 of A; levels 2–5 appear to have progressively similarly increased the responses for B but in a less pronounced way. This indicates the presence of interaction. Interaction has occurred because the effects of B on responses have changed for the different levels of the other factor A. Again, we can say that the interaction has occurred because the simple effects of B are not the same for all levels of the factor A.

We may need to determine whether an interaction we observed in our study is *removable* or not. An interaction is removable if we can determine that there is another cognate independent variable that keeps the scores and means unchanged, which if used in the design in place of the one employed would ensure that there is no interaction. To determine whether an interaction is *nonremovable*, we plot the cell means of responses twice, once against each of

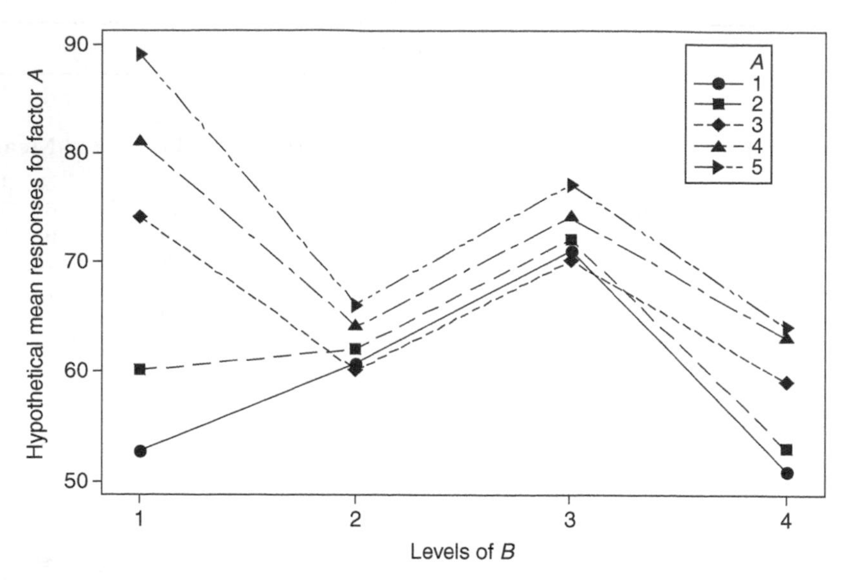

FIGURE 3.5: Plots of hypothetical means of responses for factor A across levels of B.

the independent variables in the design on a line parallel to the x-axis. If either of these plots cross, then we conclude that the interaction is nonremovable. Keppel and Wickens (2004) discuss in more detail the various aspects of the plots that characterize interaction between factors. It is clear from Figures 3.4 and 3.5 that we have nonremovable interaction in the hypothetical data on the two-factor factorial experiment involving A and B.

3.1.5 Interpreting Interaction by Testing Simple Effects

In the analysis of the responses of a two-factor factorial experiment, the best practice when interaction is found to be significant in our ANOVA is to adjust our interpretation of the means by looking separately at the simple effects of a factor at the individual levels of the other factor. Using this method enables us to look at the variability which is attributable to both the interaction and one of the main effects. We shall illustrate this analysis with the above hypothetical responses from a two-factor factorial experiment for A and B. First, we determine which of the two factors should be studied across the levels of the other factor.

3.1.5.1 Choosing the Simple Effects to Test

If interaction has been found significant and we wish to carry out more analysis to understand it, we may have the problem of choosing which simple effect to study for the factors in the two-factor factorial experiment. We need to choose the effect that will best explain the interaction we observed. Keppel and Wickens (2004) list the following heuristic principles that would help us choose the factor whose effect is to be analyzed:

1. Choose to analyze the simple effect of the factor whose main effect sum of squares is larger in ANOVA because it is more likely to provide more information than the factor with the lower sum of squares.
2. Choose the simple effects of the factor with the greater number of levels. It is easier to visualize the effect of the factor with greater number of levels.
3. Choose the treatment factor rather than the blocking or classification factor. Interest in a study is often in the effect of the treatment factor across the levels of the blocking or classification factor.
4. Choose a quantitative factor since its simple effect is easier to interpret.

3.1.5.2 Testing the Simple Effects

Let us study the effect of factor A across individual levels of factor B because it has the higher number of levels. First, let us identify the effects of A at the individual levels of B. The simple effects are shown in Table 3.9.

TABLE 3.9: Simple Effects of A at the Levels of B

Effect of A at level 1 of B	**Effect of A at level 2 of B**	**Effect of A at level 3 of B**	**Effect of A at level 4 of B**
52.67	60.67	71	51
60	62	72	53
74	60	70	59
81	64	74	63
89	66	77	64
Mean → 71.33	62.533	72.8	58

Generally, to carry out the adjusted test, we need to calculate the relevant sums of squares for the effects of the factor at the individual levels of the other factor. The general idea of the adjusted test is to extend the ANOVA table and test for simple effects of one of the factors in the design with a view to gaining insight into the interaction that has been found significant. It is not advisable to interpret interaction by looking at interaction alone, rather looking at it from the simple effects point of view, which incorporates both main effect and interaction, is recommended.

If two factors, A set at a levels and B set at b levels, are studied in a typical two-factor factorial experiment, then the formulae needed to obtain the sums of squares for the desired effects are (using our previous notations)

$$\begin{aligned} SS_{A \text{ at } B_1} &= n\sum_{i=1}^{a}(\overline{Y}_{i1.} - \overline{Y}_{.1.})^2 \\ SS_{A \text{ at } B_2} &= n\sum_{i=1}^{a}(\overline{Y}_{i2.} - \overline{Y}_{.2.})^2 \\ &\vdots \\ SS_{A \text{ at } B_b} &= n\sum_{i=1}^{a}(\overline{Y}_{ib.} - \overline{Y}_{.b.})^2 \end{aligned} \tag{3.8}$$

In order to adjust our analysis of the data in Table 3.9, we calculate the following:

$$\begin{aligned} SS_{A \text{ at } B_1} &= 3[(52.67 - 71.33)^2 + (60 - 71.33)^2 + (74 - 71.33)^2 \\ &\quad + (81 - 71.33)^2 + (89 - 71.33)^2] = 2668.294 \\ SS_{A \text{ at } B_2} &= 3[(60.67 - 62.533)^2 + (62 - 62.533)^2 + (60 - 62.533)^2 \\ &\quad + (64 - 62.533)^2 + (66 - 62.533)^2] \\ &= 73.029 \end{aligned}$$

$$\begin{aligned} SS_{A \text{ at } B_3} &= 3[(71 - 72.8)^2 + (72 - 72.8)^2 + (70 - 72.8)^2 + (74 - 72.8)^2 \\ &\quad + (77 - 72.8)^2] \\ &= 92.4 \end{aligned}$$

$$\begin{aligned} SS_{A \text{ at } B_4} &= 3[(51 - 58)^2 + (53 - 58)^2 + (59 - 58)^2 + (63 - 58)^2 \\ &\quad + (64 - 58)^2] \\ &= 408.0 \end{aligned}$$

It is easy to see that

$$\begin{aligned} &SS_{A \text{ at } B_1} + SS_{A \text{ at } B_2} + SS_{A \text{ at } B_3} + SS_{A \text{ at } B_4} \\ &\quad = 2668.294 + 73.029 + 92.4 + 408.0 \\ &\quad = SS_A + SS_{A*B} \\ &\quad = 1843.167 + 1398.967 \\ &\quad = 3242.148. \end{aligned}$$

We determine which simple effects of A at the levels of the factor B are significant, by carrying out ANOVA, which identifies them and the sums of squares, mean squares, and F-ratios. By this the significant interactions can be identified (Table 3.10).

TABLE 3.10: Extended ANOVA Including Tests for Simple Effects

Source	Sum of squares	Degrees of freedom	Mean square	*F*-ratio	*p*-Value
B	2258.867	3	752.9556	81.10839	4.73279 E − 17
A	1843.167	4	460.7917	115.1979	1.01623 E − 19
$A*B$	1398.967	12	116.5806	9.715046	6.06622 E − 05
A at B_1	2668.294	4	667.0733	71.8577	1.017 E − 17
A at B_2	73.029	4	18.2573	1.0867	0.376072
A at B_3	92.4	4	23.1	1.375	0.259828
A at B_4	408.0	4	102	6.0715	0.000647
Error	371.333	40	9.283325		
Total	5872.333	59			

Only the interactions between A at levels 1 and 4 of B are significant. This agrees with the findings from the plots in Figures 3.4 and 3.5.

3.1.6 Examination of Interaction for Example 3.1 through Analysis of Simple Effects

We return to the data of Example 3.1. Recall that the analysis of the responses of the data of this experiment showed that the interaction effect is significant. We shall attempt to analyze this interaction. We again plot the cell means of responses twice, once against each of the independent variables in the design on the abscissa in Figures 3.6 and 3.7.

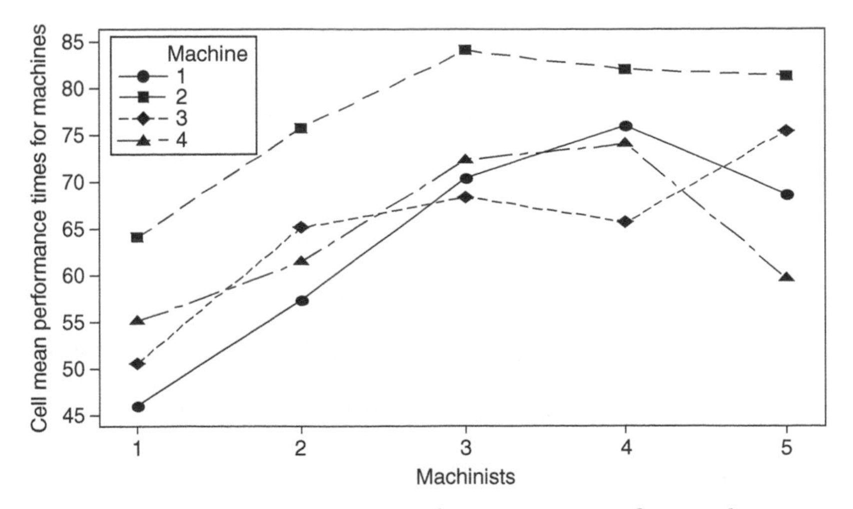

FIGURE 3.6: Plots of mean performance times for machines across different machinists.

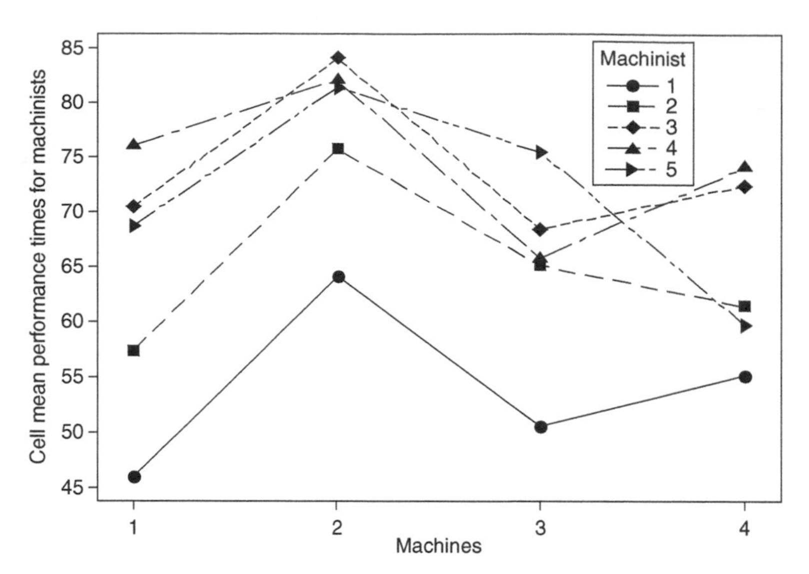

FIGURE 3.7: Plots of cell means of performance times for machinists across the machines.

We can see that the plots are crossed in both figures. This means that the interaction between machine and machinists is nonremovable. We carry out extended ANOVA to determine which simple effects of machinists are significant in order to gain a better understanding of the significant interaction found in previous ANOVA (Table 3.11).

$$SS_{\text{machinists at machine}_1} = 3(46 - 62.466)^2 + 3(81.33 - 62.466)^2$$
$$+ 3(70.33 - 62.466)^2 + 3(76 - 62.466)^2$$
$$+ 3(68.66 - 62.466)^2 = 2035.7334$$

$$SS_{\text{machinists at machine}_2} = 3(64 - 77.4)^2 + 3(75.67 - 77.4)^2 + 3(84 - 77.4)^2$$
$$+ 3(82 - 77.4)^2 + 3(81.33 - 77.4)^2 = 788.1534$$

$$SS_{\text{machinists at machine}_3} = 3(50.33 - 64.9326)^2 + 3(65 - 64.9326)^2$$
$$+ 3(68.33 - 64.9326)^2 + 3(65.67 - 64.9326)^2$$
$$+ 3(75.33 - 64.9326)^2 = 1000.035$$

$$SS_{\text{machinists at machine}_4} = 3(55 - 61.466)^2 + 3(61.33 - 61.466)^2$$
$$+ 3(72.33 - 61.466)^2 + 3(74 - 61.466)^2$$
$$+ 3(59.67 - 61.466)^2 = 825.5428$$

Again, we can see that

$$\begin{aligned} SS_{\text{machinist at machine}_1} + SS_{\text{machinist at machine}_2} + SS_{\text{machinist at machine}_3} \\ + SS_{\text{machinist at machine}_4} &= 2035.733 + 788.1534 + 1000.035 + 825.5428 \\ &= SS_{\text{machinist}} + SS_{\text{machinist*machine}} \\ &= 3659.2333 + 990.7667 = 4649.4643 \end{aligned}$$

TABLE 3.11: Extended ANOVA Testing for Simple Effects for Machinists at Different Levels of Machines

Source	Sum of squares	Degrees of freedom	Mean square	*F*-ratio	*p*-Value
M_1	3659.2333	4	914.8083	54.45288	1.16932 E − 15
M_2	2084.9833	3	694.9944	41.3687	2.46783 E − 12
M_1M_2	990.7667	12	82.5639	4.915	6.37461 E − 05
Machinists at machine$_1$	2035.7334	4	508.9334	30.2938	1.2599 E − 11
Machinists at machine$_2$	788.1534	4	197.0383	11.72854	2.14 E − 06
Machinists at machine$_3$	1000.035	4	250.0087	14.88156	1.57 E − 07
Machinists at machine$_4$	825.5428	4	206.3857	12.28494	1.32 E − 06
Residual	671.9999	40	16.7999		
Total	7406.9833	59			

Clearly, all the simple effects of machinists across the levels of machines are significant.

3.1.7 Testing for Homogeneity of Variances

More analysis could be carried out to confirm these deductions; however, let us test whether the model we have employed is appropriate for the data. First, since we need the variances computed from our model for any further analysis, we carry out the test for the homogeneity of variances and ensure that the variances obtained in our ANOVA are in accordance with the assumed model.

We wish to test the hypothesis of homogeneity of cell variances:

$$H_0 : \sigma_{11}^2 = \sigma_{12}^2 = \cdots = \sigma_{54}^2 \quad \text{vs} \quad H_1 : \text{at least one } \sigma_{ij}^2 \text{ is different}$$

Using the test of Bartlett (1936) described in Chapter 2, we obtain the cell and group standard errors as $s_{\overline{y}(ij)}$ and $s_{\overline{y}}(i..), s_{\overline{y}}(.j.)$ as well as the overall standard error

$$\begin{aligned} s_{\overline{y}(...)} : s_{\overline{y}(ij)} &= \sqrt{(16.62/3)} = 2.253 \quad s_{y(.j.)} = \sqrt{(16.62/12)} = 1.177 \\ s_{y(i..)} &= \sqrt{16.62/15} \quad s_{y(...)} = \sqrt{16.62/60} = 0.526 \end{aligned}$$

so that the standard deviations for observations in different cells are $s_{11} = 2.646$ $s_{12} = 4.163$ $s_{13} = 4.509$ $s_{14} = 3.605$ $s_{21} = 4.163$ $s_{22} = 3.055$

$s_{23} = 2.309$ $s_{24} = 1.528$ $s_{31} = 5.686$ $s_{32} = 4.00$ $s_{33} = 5.686$ $s_{34} = 3.214$
$s_{41} = 2.646$ $s_{42} = 4.359$ $s_{43} = 5.568$ $s_{44} = 3.937$ $s_{51} = 3.786$ $s_{52} = 2.309$
$s_{53} = 7.095$ $s_{54} = 2.517$.

$$A = 40 \log 10(16.62) - 2(21.049565) = 6.7261$$

$$D = 1 + \frac{1}{3(20-1)}\left(\frac{20}{2} - \frac{1}{40}\right) = 1.175$$

$$B = 2.306(6.7261/1.175) = 13.18087$$

Since $\chi^2_{(19,0.05)} = 30.14$, we do not reject the hypothesis of homogeneity of variances. Hence, we do not reject that the error variance (16.7999) used in the above ANOVA is a proper estimate of the common variance of the system.

It is sufficient to say that after an examination of the means in Table 3.11, only $\overline{y}_{11}$ and $\overline{y}_{13}$, and $\overline{y}_{14}$ could be said to belong to the same group with the *lowest* performance time. Since these means are all due to machinist I, we confirm that he turned out the lowest performance time on three of the four machines. It was recommended that machinist I should be chosen. Similarly, our analysis shows that machines I, II, and IV belong to the same group as there is no significant difference between the mean performance times for the tasks on any of these three machines. On the basis of this experiment, it was recommended that any of the machines I, II, and IV should be chosen for purchase by the company.

3.2 Two-Factor Factorial Effects (Random Effects Model)

So far, in most of the designs we have treated, we have assumed the fixed effects model. We have indicated, however, that random effects as well as mixed effects models exist and can be used. Here, we illustrate the random effects model to emphasize that its analysis is different from the fixed effects model. Random effects models are also called components of variance models. Consequently, the equivalent of ANOVA for this design is called analysis of components of variance. The random effects version of the two-factor factorial design is used when we assume that the a levels of A in the design constitute a sample from a population of levels of A and similarly the b levels of B used constitute a sample from a population of levels of B. In practice, therefore, there should be a large number of levels for A from which we select a sample and expect that whatever conclusion we reach about the sample should be applied to all the levels of A. The same principle is adopted with regard to dealing with levels of B. The model for this version of the design is

$$y_{ijk} = \mu + \alpha_i + \tau_j + (\alpha\tau)_{ij} + \varepsilon_{ijk} \qquad (3.9)$$
$$i = 1, 2, \ldots, a \quad j = 1, 2, \ldots, b \quad k = 1, 2, \ldots, n$$

where α_i, τ_j, $(\alpha\tau)_{ij}$, and ε_{ijk} are all random variables. Specifically, we assume that $\alpha_i \sim \text{NID}(0, \sigma_\alpha^2)$; $(\alpha\tau)_{ij} \sim \text{NID}(0, \sigma_{\alpha\tau}^2)$; $\tau_j \sim \text{NID}(0, \sigma_\tau^2)$; $\varepsilon_{ijk} \sim \text{NID}(0, \sigma^2)$. This means that the variance of any observation, y_{ijk} is

$$\text{Var}(y_{ijk}) = \sigma_\alpha^2 + \sigma_\tau^2 + \sigma_{\alpha\tau}^2 + \sigma^2$$

where $\sigma_\alpha^2, \sigma_\tau^2, \sigma_{\alpha\tau}^2$, and σ^2 are called variance components. We are interested in testing the following hypotheses:

$$\begin{aligned} H_0 : \sigma_\alpha^2 = 0 \quad &\text{vs} \quad H_1 : \sigma_\alpha^2 > 0 \\ H_0 : \sigma_\tau^2 = 0 \quad &\text{vs} \quad H_1 : \sigma_\tau^2 > 0 \\ H_0 : \sigma_{\alpha\tau}^2 = 0 \quad &\text{vs} \quad H_1 : \sigma_{\alpha\tau}^2 > 0 \end{aligned} \tag{3.10}$$

In the ANOVA table, the basic form of analysis remains unchanged, save for the way we determine the value of the test statistics. So, we calculate SS_{total}, SS_A, SS_B, SS_{AB} and SS_{E} as we did for the fixed effects model. However, to obtain our test statistics, we need to look at the expected mean squares under the assumptions of the present model (that the effects are random).

$$\begin{aligned} E(MS_A) &= \sigma^2 + n\sigma_{\alpha\tau}^2 + bn\sigma_\alpha^2 \\ E(MS_B) &= \sigma^2 + n\sigma_{\alpha\tau}^2 + an\sigma_\tau^2 \\ E(MS_{AB}) &= \sigma^2 + n\sigma_{\alpha\tau}^2 \\ E(MS_{\text{E}}) &= \sigma^2 \end{aligned} \tag{3.11}$$

From the above expected mean squares, we can easily estimate the variances as

$$\begin{aligned} \hat{\sigma}^2 &= MS_{\text{E}} \\ \hat{\sigma}_\alpha^2 &= \frac{MS_A - MS_{AB}}{bn} \end{aligned} \tag{3.12}$$

$$\begin{aligned} \hat{\sigma}_\tau^2 &= \frac{MS_B - MS_{AB}}{an} \\ \hat{\sigma}_{\alpha\tau}^2 &= \frac{MS_{AB} - MS_{\text{E}}}{n} \end{aligned} \tag{3.13}$$

From the expected mean square, we see that to test

$$H_0 : \sigma_{\alpha\tau}^2 = 0 \quad \text{vs} \quad H_1 : \sigma_{\alpha\tau}^2 > 0$$

(i.e., the hypothesis of no interaction of effects) we need

$$F_0 = \frac{MS_{AB}}{MS_{\text{E}}} \tag{3.14}$$

Since the numerator and the denominator of F_0 have expectation σ^2 only when H_0 is true, then $E(MS_{AB})$ is greater than $E(MS_{\text{E}})$, if and only if H_0

is untrue. Since the numerator and the denominator are independent χ^2 random variables divided by their respective degrees of freedom, then under H_0, F_0 should be distributed as $F[(a-1)(b-1),\ ab(n-1)]$.

Similarly, to test the hypothesis

$$H_0 : \sigma_\alpha^2 = 0 \quad \text{vs} \quad H_1 : \sigma_\alpha^2 > 0$$

we need

$$F_0 = \frac{MS_A}{MS_{AB}} \tag{3.15}$$

which by the same argument we adduced for Equation 3.14 is distributed as $F[a-1,\ (a-1)(b-1)]$ under H_0. And, finally, for testing the hypothesis

$$H_0 : \sigma_\tau^2 = 0 \quad \text{vs} \quad H_1 : \sigma_\tau^2 > 0$$

we need

$$F_0 = \frac{MS_B}{MS_{AB}} \tag{3.16}$$

which by argument similar to those we stated for Equation 3.14 is distributed as $F[b-1,\ (a-1)(b-1)]$ under H_0. In Example 3.2, we apply the results to a two-factor factorial experiment in which the levels of the factors are all assumed to be random.

Example 3.2

Two factors A and B are being studied for their effect on the yield of an industrial process. Three levels of each of the factors A and B are being studied. The data obtained are presented below. Analyze the data assuming that the levels of A and B are random samples from large populations of the levels of A and B, respectively.

Factor *A*	**1**	**2**	**3**	**Total**
1	20, 25, 26, 20 (91)	66, 60, 50, 55 (231)	28, 30, 28, 42 (128)	450
2	20, 38, 30, 29 (117)	74, 50, 50, 59 (233)	45, 30, 42, 55 (172)	522
3	38, 18, 30, 56 (142)	56, 52, 45, 50 (203)	24, 34, 28, 40 (126)	471
Total	350	667	426	1443

Analysis:

We calculate the sums of squares as follows:

$$SS_{\text{total}} = (20^2 + 25^2 + \cdots + 28^2 + 40^2) - \frac{1443^2}{36} = 7514.75$$

$$SS_{\text{A}} = \frac{450^2 + 522^2 + 471^2}{4 \times 3} - \frac{1443^2}{36} = 228.5$$

$$SS_{\text{B}} = \frac{350^2 + 667^2 + 426^2}{4 \times 3} - \frac{1443^2}{36} = 4565.167$$

$$SS_{\text{AB}} = \frac{91^2 + 231^2 + \cdots + 126^2}{4} - 4565.167 - 228.5 - \frac{1443^2}{36} = 575.333$$

$$SS_{\text{E}} = SS_{\text{total}} - SS_A - SS_B - SS_{\text{AB}}$$

$$= 7514.75 - 228.5 - 575.333 - 4565.167 = 2145.75$$

Now, we perform the analysis of components of variance for the random model (Table 3.12).

TABLE 3.12: Analysis of Components of Variance for the Industrial Process Experiment

Source	Sum of squares	Degrees of freedom	Mean square	F-ratio
A	228.5	2	114.25	0.794
B	4565.167	2	2282.5835	15.87
AB	575.333	4	143.8335	1.81
Error	2145.75	27	79.4722	
Total	7514.75	35		

We see that F (2, 4, 0.01) = 18.00 so that only B is significant in this design. So, we can safely say that $H_0 : \sigma^2_\alpha = 0$ in this design for $\alpha = 1\%$. Similarly, $F(4, 27, 0.01) = 4.11$ and $F(4, 27, 0.05) = 2.73$, so we can also say that $H_0 : \sigma^2_{\alpha\tau} = 0$ for this design both at $\alpha = 1\%$ and 5%.

From the above data using Equations 3.12 and 3.13, we can estimate the variances as

$$\hat{\sigma}^2 = 79.4722; \quad \hat{\sigma}^2_\alpha = \frac{114.25 - 143.8335}{3 \times 4} = -2.4653$$

$$\hat{\sigma}^2_\tau = \frac{2282.5835 - 143.8335}{3 \times 4} = 178.23; \quad \hat{\sigma}^2_{\alpha\tau} = \frac{143.8335 - 79.4722}{4} = 16.09$$

The unfortunate thing about estimating the components of variance in this way is that some of the estimates can be negative. The usual practice is to assume that the value of a variance component whose estimate appears as negative is zero. Thus, we set $\hat{\sigma}^2_\alpha = 0$.

Maximum likelihood approaches to the estimation of the variance components do not suffer from the deficiencies of other methods such as negativity. Harville (1975, 1977) and Searle (1968) discuss maximum likelihood approaches to the estimation of components of variance. To carry out variance components estimation, we need to know how to use powerful statistical analysis software. In Section 4.7, we will treat the use of the MIXED procedure in SAS to estimate variance components, which is considered better than the process we just described. □

3.2.1 SAS Analysis of Data of Example 3.2

We write a SAS program to carry out the above analysis. It is best to call the GLM procedure in our program for the analysis. Under this procedure, we have an option within the model statement to declare how the ANOVA tests are to be carried out and to define the associated error to each test. The tests can be performed in SAS if we invoke them in the overall PROC GLM syntax. In the two-factor factorial experiments for A and B, we invoke them as follows:

```
Model y=A|B;
test h=A  e=A*B;
test h=B e=A*B;
```

For both the factors A and B in the design, which are random, we specify in the program that the mean square of the interaction should be used for error as described above. Consequently, SAS provides an extra output (Type III ANOVA), which carries out the specified tests according to instruction and theory.

SAS PROGRAM

```
data twofactrand;
input A B Y@@;
cards;
1 1 20 1 1 25 1 1 26 1 1 20 1 2 66 1 2 60 1 2 50 1 2 55
1 3 28 1 3 30 1 3 28 1 3 42 2 1 20 2 1 38 2 1 30 2 1 29
2 2 74 2 2 50 2 2 50 2 2 59 2 3 45 2 3 30 2 3 42 2 3 55
3 1 38 3 1 18 3 1 30 3 1 56 3 2 56 3 2 52 3 2 45 3 2 50
3 3 24 3 3 34 3 3 28 3 3 40
 ;
 proc glm data=twofactrand;
 class A B;
model y=A B A*B;
test h=a e=a*b;
test h=b e=a*b;run;
```

SAS OUTPUT

Obs	A	B	Y
1	1	1	20
2	1	1	25
3	1	1	26
4	1	1	20
5	1	2	66
6	1	2	60
7	1	2	50
8	1	2	55
9	1	3	28
10	1	3	30
11	1	3	28
12	1	3	42
13	2	1	20
14	2	1	38
15	2	1	30
16	2	1	29
17	2	2	74

Obs	A	B	Y
18	2	2	50
19	2	2	50
20	2	2	59
21	2	3	45
22	2	3	30
23	2	3	42
24	2	3	55
25	3	1	38
26	3	1	18
27	3	1	30
28	3	1	56
29	3	2	56
30	3	2	52
31	3	2	45
32	3	2	50
33	3	3	24
34	3	3	34
35	3	3	28
36	3	3	40

The GLM Procedure

Class Level Information

Class	Levels	Values
A	3	1 2 3
B	3	1 2 3

Number of Observations Read 36
Number of Observations Used 36

The GLM Procedure

Dependent Variable: Y

Source	DF	Sum of Squares	Mean Square	F Value	Pr > F
Model	8	5369.000000	671.125000	8.44	<.0001
Error	27	2145.750000	79.472222		
Corrected Total	35	7514.750000			

R-Square	Coeff Var	Root MSE	Y Mean
0.714462	22.24046	8.914719	40.08333

Source	DF	Type I SS	Mean Square	F Value	Pr > F
A	2	228.500000	114.250000	1.44	0.2551
B	2	4565.166667	2282.583333	28.72	<.0001
A*B	4	575.333333	143.833333	1.81	0.1560

Source	DF	Type III SS	Mean Square	F Value	Pr > F
A	2	228.500000	114.250000	1.44	0.2551
B	2	4565.166667	2282.583333	28.72	<.0001
A*B	4	575.333333	143.833333	1.81	0.1560

Tests of Hypotheses Using the Type III MS for A*B as an Error Term

Source	DF	Type III SS	Mean Square	F Value	Pr > F
A	2	228.500000	114.250000	0.79	0.5123
B	2	4565.166667	2282.583333	15.87	0.0125

3.3 Two-Factor Factorial Experiment (Mixed Effects Model)

It is possible in a two-factor experiment with factors A and B, for one of them to be fixed, while the other is random. In such a situation, we have a *mixed effects model*. In such a model, it is considered that the levels of one of the factors were selected at random from a set of all possible levels that it can assume. The factor with such levels is considered random in the design. Suppose the factor A is random but B is fixed, then for a levels of A chosen at random and b levels of B fixed, the model for the design is

$$y_{ijk} = \mu + \alpha_i + \tau_j + (\alpha\tau)_{ij} + \varepsilon_{ijk} \tag{3.17}$$
$$i = 1, 2, \ldots, a \quad j = 1, 2, \ldots, b \quad k = 1, \ldots, n$$

Then,

$$\sum_{j=1}^{b} \tau_j = 0 \quad \alpha_i \sim \text{NID}(0, \sigma_\alpha^2) \quad (\alpha\tau)_{ij} \sim \text{NID}\left[0, \left(\frac{b-1}{b}\right)\sigma_{\alpha\tau}^2\right]$$
$$\varepsilon_{ijk} \sim \text{NID}(0, \sigma^2)$$

If we assume that summation over all the levels of the fixed factor for the interaction effect $(\alpha\tau)_{ij}$ is zero, that is, $\sum_{j=1}^{b}(\alpha\tau)_{ij} = (\alpha\tau)_{i.} = 0$, and if we define $\text{Var}(\alpha\tau)_{ij}$ as $(b - 1/b)\sigma_{\alpha\tau}^2$ instead of $\sigma_{\alpha\tau}^2$, then we may show that the expected mean squares for this design are

$$\begin{aligned} E(MS_{\text{A}}) &= \sigma^2 + bn\sigma_\alpha^2 \\ E(MS_{\text{B}}) &= \sigma^2 + n\sigma_{\alpha\tau}^2 + \frac{an\sum_{j=1}^{b}\tau_j^2}{b-1} \\ E(MS_{\text{AB}}) &= \sigma^2 + n\sigma_{\alpha\tau}^2 \\ E(MS_{\text{E}}) &= \sigma^2 \end{aligned} \tag{3.18a}$$

The estimates for the variance components are

$$\begin{aligned} \hat{\sigma}_\alpha^2 &= \frac{MS_{\text{A}} - MS_{\text{E}}}{bn} \\ \hat{\sigma}_{\alpha\tau}^2 &= \frac{MS_{\text{AB}} - MS_{\text{E}}}{n} \\ \hat{\sigma}^2 &= MS_{\text{E}} \end{aligned} \tag{3.18b}$$

```
Source            DF      Type I SS    Mean Square    F Value     Pr > F
subject            7    3314.473221     473.496174      38.03     <.0001
treat              6     394.103750      65.683958       5.28     0.0004


Source            DF    Type III SS    Mean Square    F Value     Pr > F
subject            7    3314.473221     473.496174      38.03     <.0001
treat              6     394.103750      65.683958       5.28     0.0004
```

3.5 Mixed RCBD (Involving Two Factors)

In the Section 3.4.2, we described the RMD with replication on one of the factors and the mixed RCBD. Sometimes, however, there is repeated observation on more than one factor. Although this falls into the category of mixed RCBD, its analysis is different from that described in the Section 3.4.2. The design we describe in this section is essentially like a two-way ANOVA with replication on two factors in which the n subjects involved act as the blocks. After thorough randomization of the treatments within subjects across the combinations of the different levels of the factors A and B, the experiments are performed and the responses elicited. For such a designed experiment, the rearranged responses look as presented in Table 3.17.

TABLE 3.17: Observations from Repeated Measures Design (Repeated Observations on Two Factors)

	Factor B (levels)				
Factor A (levels)	**Subjects**	B_1	B_2	...	B_b
A_1	s_1	y_{111}	Y_{121}	...	y_{1b1}
	s_2	y_{112}	y_{122}	...	y_{1b2}
	s_3	y_{113}	y_{123}	...	y_{1b3}
	...	...	...	...	...
	s_n	Y_{11n}	y_{12n}		y_{1bn}
A_2	s_1	y_{211}	y_{221}	...	y_{2b1}
	s_2	y_{212}	y_{222}	...	y_{2b2}
	s_3	y_{213}	y_{223}	...	y_{2b3}
	...	...	...	...	...
	s_n	y_{21n}	y_{22n}		y_{2bn}
...	...	...	...	...	...
A_a	s_1	y_{a11}	y_{a21}	...	y_{ab1}
	s_2	y_{a12}	y_{a22}	...	y_{ab2}
	s_3	y_{a13}	y_{a23}	...	y_{ab3}
	...	...	...	...	...
	s_n	y_{a1n}	y_{a2n}		y_{abn}

In the design presented above, there are two factors A and B that are set at a and b levels, respectively. There are n subjects in the experiment. The appropriate model for this design is

$$y_{ijk} = \mu + \alpha_i + \gamma_j + \beta_k + (\alpha\gamma)_{ij} + \varepsilon_{ijk} \tag{3.24}$$
$$i = 1, \ldots, a \quad j = 1, \ldots, b \quad k = 1, \ldots, n$$

α_i is the effect of ith level of factor A, while γ_j is the effect of the jth level of the factor B. The effect of the kth subject is represented by β_k. $(\alpha\gamma)_{ij}$ represents the effect of interaction between ith level of A and jth level of B. The error ε_{ijk} can be broken down into interactions between the subject effects and the effects of two other independent factors A and B, so that

$$\varepsilon_{ijk} = (\alpha\beta)_{ik} + (\gamma\beta)_{kj} + (\alpha\gamma\beta)_{ijk} \tag{3.25}$$

The mean square for each of the factorial effects that is represented in Equation 3.25 is regarded as an estimate of σ^2 and therefore can be used in ANOVA. For ANOVA, the sums of squares are obtained as follows:

$$SS_{\text{total}} = \sum_{i=1}^{a}\sum_{j=1}^{b}\sum_{k=1}^{n} y_{ijk}^2 - \frac{y_{...}^2}{abn}$$

$$SS_A = \sum_{i=1}^{a} \frac{y_{i.}^2}{bn} - \frac{y_{...}^2}{abn}$$

$$SS_B = \sum_{j=1}^{b} \frac{y_{.j.}^2}{an} - \frac{y_{...}^2}{abn}$$

$$SS_S = \sum_{k=1}^{n} \frac{y_{..k}^2}{ab} - \frac{y_{...}^2}{abn}$$

$$SS_{AB} = \sum_{i=1}^{a}\sum_{j=1}^{b} \frac{y_{ij.}^2}{n} - \sum_{j=1}^{b} \frac{y_{.j.}^2}{an} - \sum_{i=1}^{a} \frac{y_{i.}^2}{bn} + \frac{y_{...}^2}{abn} \tag{3.26}$$

$$SS_{A*S} = \sum_{i=1}^{a}\sum_{k=1}^{n} \frac{y_{i.k}^2}{b} - \sum_{k=1}^{n} \frac{y_{..k}^2}{ab} - \sum_{i=1}^{a} \frac{y_{i.}^2}{bn} + \frac{y_{...}^2}{abn} \tag{3.27}$$

$$SS_{B*S} = \sum_{j=1}^{b}\sum_{k=1}^{n} \frac{y_{.jk}^2}{a} - \sum_{k=1}^{n} \frac{y_{..k}^2}{ab} - \sum_{j=1}^{b} \frac{y_{.j.}^2}{an} + \frac{y_{...}^2}{abn} \tag{3.28}$$

SS_{A*B*S} is obtained by subtraction as

$$SS_{A*B*S} = SS_{\text{total}} - SS_A - SS_{AB} - SS_S - SS_{A*S} - SS_{B*S}$$

TABLE 3.18: Sum of squares, degrees of freedom, and mean square

Sum of squares	Degrees of freedom	Mean square
SS_A	$a-1$	$SS_A/(a-1)$
SS_B	$b-1$	$SS_B/(b-1)$
SS_{AB}	$(a-1)(b-1)$	$SS_{AB}/(a-1)(b-1)$
SS_{A*S}	$(a-1)(n-1)$	$SS_{A*S}/(a-1)(b-1)$
SS_{B*S}	$(b-1)(n-1)$	$SS_{B*S}/(b-1)(n-1)$
SS_{A*B*S}	$(a-1)(b-1)(n-1)$	$SS_{A*B*S}/(a-1)(b-1)(n-1)$

TABLE 3.19: Factorial Effects Tested in ANOVA

Source	*F*-ratio
A	MS_A/MS_{A*S}
B	MS_B/MS_{B*S}
AB	MS_{AB}/MS_{A*B*S}

The degrees of freedom for the sums of squares are as shown in Table 3.18. In the test of hypotheses, the main effects are to be tested using the mean square of the interaction of subjects with that of each factor as estimate of σ^2. The ratio of the mean squares of A and $A \times$ subjects $(A * S)$ is tested for significance. Similarly, the ratio of the mean squares of B and $B \times$ subjects $(B * S)$ is also tested for significance. The ratio of the mean squares of the interaction, AB and $A * B \times$ subjects $(A * B * S)$, is also tested for significance. The F-ratios that are tested in ANOVA are as shown in Table 3.19.

The hypotheses to be tested are presented as follows

$$H_{A0} : \alpha_1 = \alpha_2 = \cdots = \alpha_a = 0 \text{ vs } H_{A1} : \alpha_i \neq 0 \; (\textit{for at least one } i = 1, 2, \ldots, a)$$
$$H_{B0} : \gamma_1 = \gamma_2 = \cdots = \gamma_b = 0 \text{ vs } H_{B1} : \gamma_j \neq 0 \; (\textit{for at least one } j = 1, 2, \ldots, b)$$
$$H_{AB0} : (\alpha\gamma)_{11} = (\alpha\gamma)_{12} = \cdots = (\alpha\gamma)_{ab} = 0 \text{ vs } H_{AB1} : (\alpha\gamma)_{ij} \neq 0$$
$$(\textit{for at least one } (\alpha\gamma)_{ij}).$$

Example 3.4

A study was carried out to find out whether the performances of pupils in reading, comprehension, and arithmetic would improve if they were given vitamin supplements. At the same time, it was thought necessary to find out whether there is any significant difference among the performances of pupils in reading (R), comprehension (C), and arithmetic (A). The sample consisted of 10 pupils; for the first term, they studied without receiving vitamin supplements and were subjected to tests (R, C, and A with the order of taking of the test randomized), which were marked over 30. The pupils received vitamin supplements for the second term, after which they were subjected to the same tests with the numbers changed, again with the order of taking the test randomized. The scores of the pupils are presented as follows:

	V_0				V_1				
Subjects	**R**	**C**	**A**	**Row Subtotal**	**R**	**C**	**A**	**Row Subtotal**	**Row Total**
1	10	13	11	34	17	16	17	50	84
2	12	15	14	41	19	20	20	59	100
3	9	18	13	40	16	22	23	61	101
4	15	17	19	51	18	24	24	66	117
5	16	17	17	50	17	16	17	50	100
6	14	17	16	47	23	16	17	56	103
7	13	18	14	45	22	19	17	58	103
8	10	16	15	41	18	22	20	60	101
9	10	17	18	45	19	24	18	61	106
10	13	22	16	51	17	17	24	58	109
Totals	122	170	153	445	186	196	197	579	1024

V, vitamin levels (no vitamins, vitamins); T, test levels (reading, comprehension, and arithmetic). We calculate the sum of squares for ANOVA is presented in Table 3.20.

$$SS_{\text{total}} = (10^2 + 12^2 + \cdots + 18^2 + 24^2) - \frac{(1024)^2}{60} = 799.7333$$

$$SS_V = \frac{(445^2 + 579^2)}{30} - \frac{(1024)^2}{60} = 299.2666$$

$$SS_T = \frac{(308^2 + 366^2 + 350^2)}{20} - \frac{(1024)^2}{60} = 89.7333$$

$$SS_{VT} = \frac{(122^2 + 170^2 + \cdots + 196^2 + 197^2)}{10} - 17566 - 17775.533$$
$$+ 17476.26667 = 36.13367$$

$$SS_S = \frac{(84^2 + 100^2 + \cdots + 106^2 + 109^2)}{6} - \frac{(1024)^2}{60} = 104.0666$$

$$SS_{VS} = \frac{(34^2 + 41^2 + \cdots + 61^2 + 58^2)}{3} - 17775.533 - 17580.333$$
$$+ 17476.26667 = 61.0703$$

$$SS_{TS} = \frac{(27^2 + 31^2 + \cdots + 36^2 + 40^2)}{2} - 17566 - 17580.333$$

$$+ 17476.26667 = 111.93667$$

$$SS_{VTS} = 18276 - 17901.4 - 17940.667 - 17782 + 17580.333 + 17566$$

$$+ 17775.533 - \frac{(1024)}{60} = 97.5296$$

TABLE 3.20: ANOVA for the Vitamin Experiment

Source	Sum of squares	Degrees of freedom	Mean square	*F*-ratio
S	104.0666	9		
Within subjects				
Vitamins	299.2666	1	299.2666	44.1038
Tests	89.7333	2	44.8666	7.2148
Interaction(*VT*)	36.1336	2	18.0680	3.3346
VS error	61.0703	9	6.7855	
TS error	111.9366	18	6.2187	
VTS error	97.5296	18	5.4183	
Total	799.7333	59		

From the tables, $F(0.01, 1, 9) = 10.6$, $F(0.01, 2, 18) = 6.01$, and $F(0.05, 2, 18) = 3.55$, which means that only vitamins and test effects are significant. Repeated measures and other split-plot designs were discussed in Bruning and Kintz (1987). □

3.5.1 SAS Analysis of Responses of Example 3.4

In this section, we analyze the data of the Example 3.4 by using the GLM procedure in SAS. In this case, we use the REPEATED statement in the syntax of the GLM procedure. We leave off the specification of the second order interaction involving subjects, vitamin, and tests. The sum of squares for this interaction effect is used to estimate error and test for the significance of the interaction between vitamin and tests. The method presented below was the method used for analyzing repeated measures data until the introduction of the PROC MIXED into SAS. In Chapter 4, we will look at the use of PROC MIXED in the analysis of longitudinal data. The SAS program that carries out the desired analysis is presented as follows:

SAS PROGRAM

```
data vitamins;
input subject vitamin test scores @@;
  cards;
  1 1 1 10 1 1 2 13 1 1 3 11 1 2 1 17 1 2 2 16 1 2 3 17
  2 1 1 12 2 1 2 15 2 1 3 14 2 2 1 19 2 2 2 20 2 2 3 20
  3 1 1 9 3 1 2 18 3 1 3 13 3 2 1 16 3 2 2 22 3 2 3 23
  4 1 1 15 4 1 2 17 4 1 3 19 4 2 1 18 4 2 2 24 4 2 3 24
  5 1 1 16 5 1 2 17 5 1 3 17 5 2 1 17 5 2 2 16 5 2 3 17
  6 1 1 14 6 1 2 17 6 1 3 16 6 2 1 23 6 2 2 16 6 2 3 17
  7 1 1 13 7 1 2 18 7 1 3 14 7 2 1 22 7 2 2 19 7 2 3 17
```

```
  8 1 1 10 8 1 2 16 8 1 3 15 8 2 1 18 8 2 2 22 8 2 3 20
  9 1 1 10 9 1 2 17 9 1 3 18 9 2 1 19 9 2 2 24 9 2 3 18
  10 1 1 13 10 1 2 22 10 1 3 16 10 2 1 17 10 2 2 17 10 2 3 24
;
proc glm data=vitamins;
class subject vitamin test;
model scores =vitamin subject vitamin*subject
test*subject test vitamin*test vitamin*test*subject;
test h=vitamin e=vitamin*subject;
test h=test e=test*subject;
test h=vitamin*test e=vitamin*test*subject; run;
```

SAS OUTPUT

```
                              The GLM Procedure

                           Class Level Information

                 Class         Levels          Values
                 subject           10          1 2 3 4 5 6 7 8 9 10
                 vitamin            2          1 2
                 test               3          1 2 3

                   Number of Observations Read           60
                   Number of Observations Used           60

                              The GLM Procedure

Dependent Variable: scores

                                          Sum of
  Source                      DF         Squares    Mean Square    F Value    Pr > F
  Model                       59     799.7333333     13.5548023          .         .
  Error                        0       0.0000000          .
  Corrected Total             59     799.7333333

           R-Square     Coeff Var      Root MSE    scores Mean

           1.000000             .             .       17.06667

  Source                      DF       Type I SS    Mean Square    F Value    Pr > F
  vitamin                      1     299.2666667    299.2666667          .         .
  subject                      9     104.0666667     11.5629630          .         .
  subject*vitamin              9      61.0666667      6.7851852          .         .
  subject*test                20     201.6666667     10.0833333          .         .
  test                         0       0.0000000          .              .         .
  vitamin*test                 2      36.1333333     18.0666667          .         .
  subject*vitamin*test        18      97.5333333      5.4185185          .         .

  Source                      DF     Type III SS    Mean Square    F Value    Pr > F
  vitamin                      1     299.2666667    299.2666667          .         .
  subject                      9     104.0666667     11.5629630          .         .
```

```
Source                 DF    Type III SS    Mean Square    F Value    Pr > F
subject*vitamin         9     61.0666667      6.7851852          .         .
subject*test           18    111.9333333      6.2185185          .         .
test                    2     89.7333333     44.8666667          .         .
vitamin*test            2     36.1333333     18.0666667          .         .
subject*vitamin*test   18     97.5333333      5.4185185          .         .

 Tests of Hypotheses Using the Type III MS for subject*vitamin as an Error Term

Source                 DF    Type III SS    Mean Square    F Value    Pr > F
vitamin                 1    299.2666667    299.2666667      44.11    <.0001

Dependent Variable: scores

  Tests of Hypotheses Using the Type III MS for subject*test as an Error Term

Source                 DF    Type III SS    Mean Square    F Value    Pr > F
test                    2    89.73333333    44.86666667       7.22    0.0050

 Tests of Hypotheses Using the Type III MS for subject*vitamin*test as an Error Term

Source                 DF    Type III SS    Mean Square    F Value    Pr > F
vitamin*test            2    36.13333333    18.06666667       3.33    0.0586
```

Merely invoking repeated in PROC GLM does not ensure that proper analysis is carried out. Thus, if we invoked the GLM procedure with the REPEATED statement as shown below

```
proc glm data=vitamins;
class subject vitamin test;
model scores =vitamin subject vitamin*subject
test*subject test vitamin*test vitamin*test*subject;
repeated subject; run;
```

SAS will output only Types I and III ANOVA presented in the OUTPUT above, leaving out the crucial analysis that expects that the mean square for *vitamin* should be divided by the mean square of the first-order interaction, *subject***vitamin* in order to test for significance; the mean square for *test* should be divided by mean square of the first-order interaction, *subject***test* in order to test for significance of *test*; the mean square of the first-order interaction, *vitamin***test* should be divided by the mean square of the second-order interaction *subject***vitamin***test* in order to test for significance of *vitamin***test* interaction. These tests can be performed in SAS using PROC GLM if we invoke them in test statements after the model statement as follows:

```
test h=vitamin e=vitamin*subject;
test h=test e=test*subject;
test h=vitamin*test e=vitamin*test*subject;
```

Once these tests are invoked, we see that the PROC GLM analysis agrees with the manual analysis we carried out previously. In Chapter 4, we analyze the same data using another procedure in SAS, the PROC MIXED.

3.6 Exercises

Exercise 3.1

In a study of the action potentials of heart membranes in rabbits when subjected to ischemia, membranes from 16 subjects were used. The first eight were randomly chosen and subjected to ischemia with no drug treatment applied, while the next eight were subjected to ischemia and treated with a drug—10^{-4} M adenosine. The action potential characteristics were measured as a percentage decrease in voltage over 120 min, observations being made every 30 min. The responses of the experiment are presented as follows:

No drug + ischemia				
Subjects	**Time (min)**			
	30	**60**	**90**	**120**
1	46.95	53.68	54.61	58.67
2	61.26	60.14	70.58	73.83
3	56.98	61.15	60.06	61.74
4	18.58	33.07	45.35	48.18
5	40.33	46.72	46.55	46.20
6	52.78	53.84	57.78	51.91
7	51.85	61.90	75.43	75.31
8	66.98	69.71	73.65	72.86

10^{-4} M adenosine + ischemia				
Subjects	**Time (min)**			
	30	**60**	**90**	**120**
1	48.95	46.95	46.35	47.50
2	47.72	46.52	44.35	42.73
3	42.27	41.25	60.06	40.41
4	30.08	23.13	45.35	25.84
5	58.06	56.47	46.55	55.68
6	37.40	43.42	57.78	45.73
7	33.78	40.07	75.43	42.78
8	46.52	48.92	73.65	50.12

Analyze the results of this experiment as an RMD and report on the effect of adenosine on the action potentials of the heart membranes when subjected to ischemia. □

Exercise 3.2

Write a SAS program to use PROC GLM to analyze the data of Exercise 3.1. □

Exercise 3.3
In a study of three teaching methods, three teachers were selected to use the three methods to teach students over three terms. At the end of each term, the students were subjected to the same standard test. The data showing the scores of the students over the period of study are shown as follows:

Subjects	Teachers	Term1	Term2	Term3
1	Teacher1	78	78	78
2		82	82	77
3		80	82	86
4		80	80	82
5		79	80	82
6		75	75	83
7		80	83	75
8		76	77	84
9		83	84	81
10		73	74	84
11		75	76	77
12		83	83	75
13		73	79	85
14		75	77	81
1	Teacher2	80	75	78
2		73	71	78
3		72	72	71
4		79	80	78
5		75	78	82
6		73	71	80
7		74	76	74
8		75	82	78
9		72	80	78
10		78	74	76
1	Teacher3	87	84	79
2		85	84	77
3		85	85	77
4		86	84	78
5		85	83	77
6		86	84	78
7		87	85	77
8		80	80	70
9		79	81	69
10		79	80	69
11		78	80	68

□

Exercise 3.4
In pharmacokinetics research, eight mice were subjected to equal doses of a drug for five consecutive days. The purpose of the study was to find out whether their ability to absorb the drug diminishes with use. A blood test was used to measure absorption levels. The following data show the percentages of the drugs absorbed after 1 h following treatment on each of

the 5 days of the experiment:

Mice	**Days of study**				
	1	**2**	**3**	**4**	**5**
1	20	18	17	14	12
2	21	15	14	12	11
3	23	17	15	17	14
4	30	22	18	17	15
5	26	19	16	18	12
6	20	20	21	16	14
7	28	27	22	20	18
8	22	20	20	22	19

Is there any evidence that absorption of this drug decreases with use? Test at 5% level of significance. ▯

Exercise 3.5
Write a SAS program to carry out the analysis of Exercise 3.4 using GLM procedure. ▯

Exercise 3.6
In an experiment to study the effect of two factors, A and B, a two-way layout was used, and the measurements were replicated three times.

Factor B	**Levels of factor** A				
	I	**II**	**III**	**IV**	**V**
1	23, 35, 30	27, 19, 15	28, 36, 39	27, 22, 29	31, 36, 24
2	39, 24, 45	25, 26, 39	29, 46, 32	22, 22, 21	39, 48, 49
3	27, 37, 26	21, 23, 20	42, 28, 39	20, 19, 20	29, 47, 44

Analyze the data and compare the means for factor A using Duncan's multiple range test. ▯

Exercise 3.7
Write a SAS program to carry out the analysis of Exercise 3.7. ▯

Exercise 3.8
A car manufacturer is studying three devices for reducing fuel consumption using three different models of the car to be fitted with the devices. The data obtained are as follows:

	Car models		
Device type	**I**	**II**	**III**
1	44 30 40 55	34 40 35 33	32 30 38 35
2	40 48 49 50	46 49 45 46	40 48 45 45
3	48 40 48 45	34 30 30 39	36 30 34 32

Analyze the data as a full two-factor factorial design and test for significance of all effects at 1% level of significance. □

Exercise 3.9
In the data of Exercise 3.8, obtain the fitted values and extract the residuals for the analysis carried out in Exercise 3.8. Using the residuals, test whether the normality assumption is satisfied. Using the test of Bartlett (1936), determine whether the assumption of homogeneity of variance is satisfied. □

Exercise 3.10
Write a SAS program to analyze the data of Exercise 3.8, assuming the fixed effects model. □

Exercise 3.11
Write a SAS program to analyze the data of Exercise 3.3 using the GLM procedure. □

References

Anderson, V.L. and McLean, R.A. (1974) *Design of Experiments: A Realistic Approach*, Marcel Dekker Inc., New York.

Bartlett, M.S. (1936) The information available in small samples, *Proceedings of Cambridge Philosophical Society*, 32, 560.

Bruning, J.L. and Kintz, B.L. (1987) *Computational Handbook of Statistics*, 3rd ed., Scott, Foresman and Company, Glenview, IL.

Harville, D.A. (1975) Maximum likelihood approaches to variance component estimation and related problems. Technical report No. 75-0175, Aerospace Research Laboratories.

Harville, D.A. (1977) Maximum likelihood approaches to variance component estimation and related problems. *JASA* 72, 320–338.

Keppel, G. and Wickens, T.D. (2004) *Design and Analysis—A Researcher's Handbook*, 4th ed., Prentice Hall, Englewood Cliffs, NJ.

Montgomery, D.C. (1984) *Design and Analysis of Experiments*, 2nd ed., Wiley, New York.

Mood, A.M., Graybill, F.A., and Boes, D.C. (1974) *Introduction to the Theory of Statistics*, 3rd ed., McGraw-Hill, New York.

Searle, S.R. (1968) Another look at Henderson's method of estimating variance components (with discussion). *Biometrics* 24, 749–787.

Chapter 4

Regression Approaches to the Analysis of Responses of Two-Factor Experiments and Repeated Measures Designs

In Chapter 2, we discussed regression approaches to the one-way experimental layout. We defined a set of dummy variables and used it in a regression model to explain the responses in a one-way layout in terms of the factor (the explanatory variable) in the design. The model helped us to analyze the responses of the design. To obtain a similar regression model for a two-way layout, we need to define two sets of dummy variables. The two sets of dummy variables could also be defined using (1) the effects coding or (2) the reference cell methods. We have already demonstrated how both methods work when we developed the regression model for the one-way layout.

4.1 Regression Models for the Two-Way ANOVA (Full Two-Factor Factorial Experiment)

We need to define two sets of dummy variables to set up regression models for the two-way ANOVA with replication, the full two-factor factorial experiment. The typical full two-factor experiment involves two factors A and B set at a and b levels, respectively, and studied in n replicates. In classical fixed effects model, we have to account for the main effects and the interaction of the factors in ANOVA. For this model, n measurements are made for each cell. There are as many cells as the number of combinations of levels of the factors, ab. We need to define two sets of dummy variables corresponding to the main effects and then use their product in a regression model to represent the interaction effect. We can also use both the effects and reference cell coding methods to define the dummy variables to be used in the regression model. Kleinbaum et al. (1998) discuss regression approaches to two-way ANOVA.

4.1.1 Regression Model for Two-Factor Factorial Design with Effects Coding for Dummy Variables

If we use the effects coding to define the dummy variables, then the model that should apply for the full two-factor experiment is

$$y_{ij} = \mu + \sum_{i=1}^{a-1} \alpha_i X_i + \sum_{j=1}^{b-1} \beta_j Z_j + \sum_{i=1}^{a-1} \sum_{j=1}^{b-1} \gamma_{ij} X_i Z_j + \varepsilon \tag{4.1}$$

where

$$X_i = \begin{cases} -1 & \text{if } y_{ij} \text{ is under level } a \text{ of } A \\ 1 & \text{if } y_{ij} \text{ is under level } i \text{ of } A \\ 0 & \text{otherwise} \end{cases}$$

for $i = 1, 2, \ldots, a-1$

and

$$Z_j = \begin{cases} -1 & \text{if } y_{ij} \text{ is under level } b \text{ of } B \\ 1 & \text{if } y_{ij} \text{ is under level } j \text{ of } B \\ 0 & \text{otherwise} \end{cases}$$

for $j = 1, 2, \ldots, b-1$

4.1.2 Estimation of Parameters for the Regression Model for the General Two-Factor Factorial Design with Effect Coding for Dummy Variables

We can estimate the parameters of the regression model when we use effects coding to define the set of dummy variables. To do this, we use the following relations between the parameters and the means for the rows, columns, and cells in the typical two-factor factorial experiment described above:

$$\mu = \mu_{..}$$

$$\alpha_i = \mu_{i.} - \mu_{..} \qquad i = 1, 2, \ldots, a-1$$

$$\beta_j = \mu_{.j} - \mu_{..} \qquad j = 1, 2, \ldots, b-1$$

$$\gamma_{ij} = \mu_{ij} - \mu_{i.} - \mu_{.j} + \mu_{..} \quad i = 1, 2, \ldots, a-1; \quad j = 1, 2, \ldots, b-1$$

$$-\sum_{i=1}^{a-1} \alpha_i = \mu_{a.} - \mu_{..}$$

$$-\sum_{j=1}^{b-1} \beta_j = \mu_{.b} - \mu_{..}$$

$$-\sum_{i=1}^{a-1} \gamma_{ij} = \mu_{aj} - \mu_{a.} - \mu_{.j} + \mu_{..} \quad j = 1, 2, \ldots, b-1$$

$$-\sum_{i=1}^{b-1} \gamma_{ij} = \mu_{ib} - \mu_{i.} - \mu_{.b} + \mu_{..} \quad i = 1, 2, \ldots, a-1$$

4.1.3 Application of the Regression Model for the Two-Factor Factorial Design, with Effect Coding for Dummy Variables to Example 3.1

Let us analyze the responses of the Akwette Cloth Company data, Example 3.1, using the effects coding method. There are two factors in this design, the machinist at 5 levels, and machines at 4 levels. The specific model that applies to the design of Example 3.1 is

$$y_{ij} = \mu + \sum_{i=1}^{4} \alpha_i X_i + \sum_{j=1}^{3} \beta_j Z_j + \sum_{i=1}^{4} \sum_{j=1}^{3} \gamma_{ij} X_i Z_j + \varepsilon \tag{4.2}$$

where

$$X_i = \begin{cases} -1 & \text{if } y_{ij} \text{ is under level 5 of } A \\ 1 & \text{if } y_{ij} \text{ is under level } i \text{ of } A \\ 0 & \text{otherwise} \end{cases}$$
for $i = 1, 2, 3, 4$

$$Z_j = \begin{cases} -1 & \text{if } y_{ij} \text{ is under level 4 of } B \\ 1 & \text{if } y_{ij} \text{ is under level } j \text{ of } B \\ 0 & \text{otherwise} \end{cases}$$
for $j = 1, 2, 3$

4.1.4 Relations for Estimation of Parameters for the Regression Model with Effect Coding for Dummy Variables for Responses of Example 3.1

To estimate the parameters of the regression model Equation 4.2, we have the following relations between the parameters, which are associated with the rows, columns, and cell means:

$$\begin{aligned}
\mu &= \mu_{..} \\
\alpha_i &= \mu_{i.} - \mu_{..} && i = 1, 2, 3, 4 \\
\beta_j &= \mu_{.j} - \mu_{..} && j = 1, 2, 3 \\
\gamma_{ij} &= \mu_{ij} - \mu_{i.} - \mu_{.j} + \mu_{..} && i = 1, 2, 3, 4; \quad j = 1, 2, 3 \\
-\sum_{i=1}^{4} \alpha_i &= \mu_{5.} - \mu_{..} \\
-\sum_{j=1}^{3} \beta_j &= \mu_{.4} - \mu_{..}
\end{aligned}$$

$$-\sum_{i=1}^{4}\gamma_{ij} = \mu_{5j} - \mu_{5.} - \mu_{.j} + \mu_{..} \quad j = 1,2,3$$

$$-\sum_{j=1}^{3}\gamma_{ij} = \mu_{i4} - \mu_{i.} - \mu_{.4} + \mu_{..} \quad i = 1,2,3,4$$

4.1.5 Estimation of Parameters for the Regression Model

The row, column, and cell means for the data of Example 3.1 are presented in Table 4.1. We will use them to obtain the estimated model equation and subsequently use the estimated model to estimate responses, extract residuals, and carry out ANOVA to test for the significance of the fitted model.

TABLE 4.1: The Row, Columns, and Cell Means for Data of Example 3.1

	Machines (M_2)				
Machinists (M_1)	**I**	**II**	**III**	**IV**	**Means**
I	46.0	64.00	50.33	55.00	53.83
II	51.33	75.67	65.00	61.33	63.33
III	70.33	84.00	68.33	72.33	73.75
IV	76.00	82.00	65.67	74.00	74.417
V	68.67	81.33	75.33	59.67	71.25
Mean	62.4666	77.40	64.9333	64.47	67.3166

The means for Example 3.1 presented in the above table indicate that

$$\overline{y}_{..} = 67.3166 \quad \overline{y}_{1.} = 53.8333 \quad \overline{y}_{2.} = 63.3333 \quad \overline{y}_{3.} = 73.75$$
$$\overline{y}_{4.} = 74.4166 \quad \overline{y}_{5.} = 71.25$$
$$\overline{y}_{.1} = 62.466 \quad \overline{y}_{.2} = 77.4 \quad \overline{y}_{.3} = 64.9333 \quad \overline{y}_{.4} = 64.47$$
$$\overline{y}_{11} = 46 \quad \overline{y}_{12} = 64 \quad \overline{y}_{13} = 50.53 \quad \overline{y}_{14} = 55 \quad \overline{y}_{21} = 57.33$$
$$\overline{y}_{22} = 57.33 \quad \overline{y}_{23} = 60.67 \quad \overline{y}_{24} = 61.33 \quad \overline{y}_{31} = 70.33 \quad \overline{y}_{32} = 84$$
$$\overline{y}_{33} = 68.33 \quad \overline{y}_{34} = 72.33 \quad \overline{y}_{41} = 76 \quad \overline{y}_{42} = 82 \quad \overline{y}_{43} = 65.67$$
$$\overline{y}_{44} = 74 \quad \overline{y}_{51} = 68.67 \quad \overline{y}_{52} = 81.33 \quad \overline{y}_{53} = 75.33 \quad \overline{y}_{54} = 59.67$$

In the estimation process, we simply use these sample values to represent their population equivalents in formulae presented earlier (Section 4.1.4) to obtain the estimates of the model. The estimates of the parameters were obtained manually as follows:

$$\hat{\mu}_{..} = 67.3166$$

$$\hat{\alpha}_1 = \hat{\mu}_{1.} - \hat{\mu}_{..} = 53.8333 - 67.3166 = -13.4833$$

$$\hat{\alpha}_2 = \hat{\mu}_{2.} - \hat{\mu}_{..} = 63.3333 - 67.3166 = -3.9833$$

$$\hat{\alpha}_3 = \hat{\mu}_{3.} - \hat{\mu}_{..} = 73.75 - 67.3166 = 6.4334$$

$$\hat{\alpha}_4 = \hat{\mu}_{4.} - \hat{\mu}_{..} = 74.4166 - 67.3166 = 7.10$$

$$\hat{\beta}_1 = \hat{\mu}_{.1} - \hat{\mu}_{..} = 62.4666 - 67.3166 = -4.85$$

$$\hat{\beta}_2 = \hat{\mu}_{.2} - \hat{\mu}_{..} = 77.40 - 67.3166 = 10.0833$$

$$\hat{\beta}_3 = \hat{\mu}_{.} - \hat{\mu}_{..} = 64.9333 - 67.3166 = -2.3833$$

$$\hat{\gamma}_{11} = 46 - 62.4666 - 53.8333 + 67.3166 = -2.9833$$

$$\hat{\gamma}_{12} = 64 - 77.4 - 53.8333 + 67.3166 = 0.083$$

$$\hat{\gamma}_{13} = 50.33 - 64.9333 - 53.8333 + 67.3166 = -1.1167$$

$$\hat{\gamma}_{21} = 51.333 - 62.4666 - 63.333 + 67.3166 = -7.1499$$

$$\hat{\gamma}_{22} = 75.666 - 77.4 - 63.333 + 67.3166 = 2.25$$

$$\hat{\gamma}_{23} = 65 - 64.9333 - 63.333 + 67.3166 = 4.05$$

$$\hat{\gamma}_{31} = 70.333 - 62.466 - 73.75 + 67.3166 = 1.433$$

$$\hat{\gamma}_{32} = 84 - 77.4 - 73.75 + 67.3166 = 0.1667$$

$$\hat{\gamma}_{33} = 68.333 - 64.933 - 73.75 + 67.3166 = -3.0333$$

$$\hat{\gamma}_{41} = 76 - 62.466 - 74.42 + 67.3166 = 6.43$$

$$\hat{\gamma}_{42} = 82 - 77.4 - 74.42 - 67.3166 = -2.5$$

$$\hat{\gamma}_{43} = 65.666 - 64.933 - 74.42 + 67.3166 = -6.366$$

These relations are only true if there are equal numbers of observations per cell, leading to a balanced design. If the numbers of observations differ by cell, the formulae would need to be modified before use. From the above results, the estimated model equation is

$$\begin{aligned}\hat{y} = {} & 67.3166 - 13.4833x_1 - 3.9833x_2 + 6.4334x_3 + 7.10x_4 - 4.85z_1 \\ & + 10.0833z_2 - 2.3833z_3 - 2.9833x_1z_1 + 0.0833x_1z_2 - 1.1167x_1z_3 \\ & - 7.1499x_2z_1 + 2.25x_2z_2 + 4.05x_2z_3 + 1.433x_3z_1 + 0.1667x_3z_2 \\ & - 3.0333x_3z_3 + 6.43x_4z_1 - 2.5x_4z_2 - 6.3666x_4z_3 \end{aligned} \tag{4.3}$$

The estimated responses and the accompanying residuals for the fitted model are shown in Table 4.2:

In Table 4.2:

$$d_1 = x_1z_1 \quad d_2 = x_1z_2 \quad d_3 = x_1z_3 \quad d_4 = x_2z_1 \quad d_5 = x_2z_2 \quad d_6 = x_2z_3$$
$$d_7 = x_3z_1 \quad d_8 = x_3z_2 \quad d_9 = x_3z_3 \quad d_{10} = x_4z_1 \quad d_{11} = x_4z_2 \quad d_{12} = x_4z_3$$

TABLE 4.2: Variables, Estimated Responses, and Residuals for Regression Model for Example 3.1 (Effects Coding Method)

x_1	x_2	x_3	x_4	z_1	z_2	z_3	y	d_1	d_2	d_3	d_4	d_5	d_6	d_7	d_8	d_9	d_{10}	d_{11}	d_{12}	$\hat{y}$	$y_i - \hat{y}_i$	$(y_i - \hat{y}_i)^2$
1	0	0	0	1	0	0	43	1	0	0	0	0	0	0	0	0	0	0	0	46	−3	9
1	0	0	0	1	0	0	47	1	0	0	0	0	0	0	0	0	0	0	0	46	1	1
1	0	0	0	1	0	0	48	1	0	0	0	0	0	0	0	0	0	0	0	46	2	4
1	0	0	0	0	1	0	60	0	1	0	0	0	0	0	0	0	0	0	0	64	−3.9999	15.9992
1	0	0	0	0	1	0	64	0	1	0	0	0	0	0	0	0	0	0	0	64	0.0001	1E−08
1	0	0	0	0	1	0	68	0	1	0	0	0	0	0	0	0	0	0	0	64	4.0001	16.0008
1	0	0	0	0	0	1	46	0	0	1	0	0	0	0	0	0	0	0	0	50.33	−4.3333	18.77749
1	0	0	0	0	0	1	50	0	0	1	0	0	0	0	0	0	0	0	0	50.33	−0.3333	0.111089
1	0	0	0	0	0	1	55	0	0	1	0	0	0	0	0	0	0	0	0	50.33	4.6667	21.77809
1	0	0	0	−1	−1	−1	59	−1	−1	−1	0	0	0	0	0	0	0	0	0	55	4	16
1	0	0	0	−1	−1	−1	52	−1	−1	−1	0	0	0	0	0	0	0	0	0	55	−3	9
1	0	0	0	−1	−1	−1	54	−1	−1	−1	0	0	0	0	0	0	0	0	0	55	−1	1
0	1	0	0	1	0	0	50	0	0	0	1	0	0	0	0	0	0	0	0	51.33	−1.3334	1.777956
0	1	0	0	1	0	0	56	0	0	0	1	0	0	0	0	0	0	0	0	51.33	4.6666	21.77716
0	1	0	0	1	0	0	48	0	0	0	1	0	0	0	0	0	0	0	0	51.33	−3.3334	11.11156
0	1	0	0	0	1	0	79	0	0	0	0	1	0	0	0	0	0	0	0	75.67	3.3334	11.11156
0	1	0	0	0	1	0	73	0	0	0	0	1	0	0	0	0	0	0	0	75.67	−2.6666	7.110756
0	1	0	0	0	1	0	75	0	0	0	0	1	0	0	0	0	0	0	0	75.67	−0.6666	0.444356
0	1	0	0	0	0	1	65	0	0	0	0	0	1	0	0	0	0	0	0	65	0	0
0	1	0	0	0	0	1	68	0	0	0	0	0	1	0	0	0	0	0	0	65	3	9
0	1	0	0	0	0	1	62	0	0	0	0	0	1	0	0	0	0	0	0	65	−3	9
0	1	0	0	−1	−1	−1	60	0	0	0	−1	−1	−1	0	0	0	0	0	0	61.33	−1.3332	1.777422
0	1	0	0	−1	−1	−1	63	0	0	0	−1	−1	−1	0	0	0	0	0	0	61.33	1.6668	2.778222
0	1	0	0	−1	−1	−1	61	0	0	0	−1	−1	−1	0	0	0	0	0	0	61.33	−0.3332	0.111022
0	0	1	0	1	0	0	72	0	0	0	0	0	0	1	0	0	0	0	0	70.33	1.6671	2.779222
0	0	1	0	1	0	0	64	0	0	0	0	0	0	1	0	0	0	0	0	70.33	−6.3329	40.10562
0	0	1	0	1	0	0	75	0	0	0	0	0	0	1	0	0	0	0	0	70.33	4.6671	21.78182
0	0	1	0	0	1	0	80	0	0	0	0	0	0	0	1	0	0	0	0	84	−3.9999	15.9992
0	0	1	0	0	1	0	88	0	0	0	0	0	0	0	1	0	0	0	0	84	4.0001	16.0008
0	0	1	0	0	1	0	84	0	0	0	0	0	0	0	1	0	0	0	0	84	0.0001	1E−08

0	0	1	0	0	0	1	62	0	0	0	0	0	0	0	0	1	0	0	0	68.33	−6.3333	40.11069
0	0	1	0	0	0	1	70	0	0	0	0	0	0	0	0	1	0	0	0	68.33	1.6667	2.777889
0	0	1	0	0	0	1	73	0	0	0	0	0	0	0	0	1	0	0	0	68.33	4.6667	21.77809
0	0	1	0	−1	−1	−1	70	0	0	0	0	0	0	−1	−1	−1	0	0	0	72.33	−2.3335	5.445222
0	0	1	0	−1	−1	−1	71	0	0	0	0	0	0	−1	−1	−1	0	0	0	72.33	−1.3335	1.778222
0	0	1	0	−1	−1	−1	76	0	0	0	0	0	0	−1	−1	−1	0	0	0	72.33	3.6665	13.44322
0	0	0	1	1	0	0	77	0	0	0	0	0	0	0	0	0	1	0	0	76	1.0034	1.006812
0	0	0	1	1	0	0	73	0	0	0	0	0	0	0	0	0	1	0	0	76	−2.9966	8.979612
0	0	0	1	1	0	0	78	0	0	0	0	0	0	0	0	0	1	0	0	76	2.0034	4.013612
0	0	0	1	0	1	0	79	0	0	0	0	0	0	0	0	0	0	1	0	82	−2.9999	8.9994
0	0	0	1	0	1	0	87	0	0	0	0	0	0	0	0	0	0	1	0	82	5.0001	25.001
0	0	0	1	0	1	0	80	0	0	0	0	0	0	0	0	0	0	1	0	82	−1.9999	3.9996
0	0	0	1	0	0	1	61	0	0	0	0	0	0	0	0	0	0	0	1	65.67	−4.6667	21.77809
0	0	0	1	0	0	1	67	0	0	0	0	0	0	0	0	0	0	0	1	65.67	1.3333	1.777689
0	0	0	1	0	0	1	69	0	0	0	0	0	0	0	0	0	0	0	1	65.67	3.3333	11.11089
0	0	0	1	−1	−1	−1	68	0	0	0	0	0	0	0	0	0	−1	−1	−1	74	−6.0032	36.03841
0	0	0	1	−1	−1	−1	75	0	0	0	0	0	0	0	0	0	−1	−1	−1	74	0.9968	0.99361
0	0	0	1	−1	−1	−1	79	0	0	0	0	0	0	0	0	0	−1	−1	−1	74	4.9968	24.96801
−1	−1	−1	−1	1	0	0	67	−1	0	0	−1	0	0	−1	0	0	−1	0	0	68.67	−1.6701	2.789234
−1	−1	−1	−1	1	0	0	73	−1	0	0	−1	0	0	−1	0	0	−1	0	0	68.67	4.3299	18.74803
−1	−1	−1	−1	1	0	0	66	−1	0	0	−1	0	0	−1	0	0	−1	0	0	68.67	−2.6701	7.129434
−1	−1	−1	−1	0	1	0	84	0	−1	0	0	−1	0	0	−1	0	0	−1	0	81.33	2.6668	7.111822
−1	−1	−1	−1	0	1	0	80	0	−1	0	0	−1	0	0	−1	0	0	−1	0	81.33	−1.3332	1.777422
−1	−1	−1	−1	0	1	0	80	0	−1	0	0	−1	0	0	−1	0	0	−1	0	81.33	−1.3332	1.777422
−1	−1	−1	−1	0	0	1	69	0	0	−1	0	0	−1	0	0	−1	0	0	−1	75.33	−6.3332	40.10942
−1	−1	−1	−1	0	0	1	83	0	0	−1	0	0	−1	0	0	−1	0	0	−1	75.33	7.6668	58.77982
−1	−1	−1	−1	0	0	1	74	0	0	−1	0	0	−1	0	0	−1	0	0	−1	75.33	−1.3332	1.777422
−1	−1	−1	−1	−1	−1	−1	62	1	1	1	1	1	1	1	1	1	1	1	1	59.66	2.3369	5.461102
−1	−1	−1	−1	−1	−1	−1	60	1	1	1	1	1	1	1	1	1	1	1	1	59.66	0.3369	0.113502
−1	−1	−1	−1	−1	−1	−1	57	1	1	1	1	1	1	1	1	1	1	1	1	59.66	−2.6631	7.092102
																				4039	0.00	672.0001

We know that $S_{yy} = \sum_{i=1}^{N} y_i^2 - \left(\sum_{i=1}^{N} y_i^2\right)^2 / N$ and $SS_R = \sum_{i=1}^{N} (\hat{y}_i - \overline{y})^2$ and $SS_E = \sum_{i=1}^{N} (y_i - \hat{y}_i)^2$ then $S_{yy} = SS_R + SS_E$. In this design, $N = abn = 5(4)(3) = 60$. From Table 4.2, we see that

$$SS_E = 672.0001 \quad S_{yy} = 279299 - \frac{4039^2}{60} = 7406.9833$$

$$\Rightarrow SS_R = 7406.9833 - 672.0001 = 6734.9832$$

TABLE 4.3: ANOVA to Test the Significance of the Model for Example 3.1 (Effects Coding Method)

Source	Sum of squares	Degrees of freedom	Mean square	F-ratio
Regression	6734.9832	19	354.4728	21.0996
Error	672.0001	40	16.80	
Total	7406.9833	59		

From the table of the F-distribution, we see that $F(19, 40, 0.01) = 2.39$; $F(19, 40, 0.05) = 1.85$, indicating that the fitted model is highly significant. The associated p-value is 3.73226E-15, confirming that the fitted model is highly significant (Table 4.3).

4.1.6 SAS Program for Example 3.1 (Effects Coding for Dummy Variables)

Next, we write a SAS program to carry out the analysis of the data and estimate parameters of the regression model. In the program, we can invoke either the GLM or the REG procedures. Onyiah (2005) discusses the syntax of PROC REG and its uses within SAS. We preferred to use the REG procedure as follows:

SAS PROGRAM

```
data twofact1;
input X1-X4 Z1-Z3 Y@@;
d1 = x1*z1;d2 = x1*z2; d3 = x1*z3;d4 = x2*z1;d5 = x2*z2;d6 = x2*z3;
d7 = x3*z1;d8 = x3*z2; d9 = x3*z3;d10 = x4*z1;d11 = x4*z2;d12 = x4*z3;
datalines;
1 0 0 0 1 0 0 43 1 0 0 0 1 0 0 47 1 0 0 0 1 0 0 48
1 0 0 0 0 1 0 60 1 0 0 0 0 1 0 64 1 0 0 0 0 1 0 68
1 0 0 0 0 0 1 46 1 0 0 0 0 0 1 50 1 0 0 0 0 0 1 55
1 0 0 0 -1 -1 -1 59 1 0 0 0 -1 -1 -1 52 1 0 0 0 -1 -1 -1 54
0 1 0 0 1 0 0 50 0 1 0 0 1 0 0 56 0 1 0 0 1 0 0 48
0 1 0 0 0 1 0 79 0 1 0 0 0 1 0 73 0 1 0 0 0 1 0 75
0 1 0 0 0 0 1 65 0 1 0 0 0 0 1 68 0 1 0 0 0 0 1 62
0 1 0 0 -1 -1 -1 60 0 1 0 0 -1 -1 -1 63 0 1 0 0 -1 -1 -1 61
0 0 1 0 1 0 0 72 0 0 1 0 1 0 0 64 0 0 1 0 1 0 0 75
```

```
0 0 1 0 0 1 0 80 0 0 1 0 0 1 0 88 0 0 1 0 0 1 0 84
0 0 1 0 0 0 1 62 0 0 1 0 0 0 1 70 0 0 1 0 0 0 1 73
0 0 1 0 -1 -1 -1 70 0 0 1 0 -1 -1 -1 71 0 0 1 0 -1 -1 -1 76
0 0 0 1 1 0 0 77 0 0 0 1 1 0 0 73 0 0 0 1 1 0 0 78
0 0 0 1 0 1 0 79 0 0 0 1 0 1 0 87 0 0 0 1 0 1 0 80
0 0 0 1 0 0 1 61 0 0 0 1 0 0 1 67 0 0 0 1 0 0 1 69
0 0 0 1 -1 -1 -1 68 0 0 0 1 -1 -1 -1 75 0 0 0 1 -1 -1 -1 79
-1 -1 -1 -1 1 0 0 67 -1 -1 -1 -1 1 0 0 73 -1 -1 -1 -1 1 0 0 66
-1 -1 -1 -1 0 1 0 84 -1 -1 -1 -1 0 1 0 80 -1 -1 -1 -1 0 1 0 80
-1 -1 -1 -1 0 0 1 69 -1 -1 -1 -1 0 0 1 83 -1 -1 -1 -1 0 0 1 74
-1 -1 -1 -1 -1 -1 -1 62 -1 -1 -1 -1 -1 -1 -1 60 -1 -1 -1 -1 -1 -1 -1 57
;
proc print data=twofact1;
proc reg data=twofact1;
model y=x1-x4 z1-z3 d1-d12; run;
```

SAS OUTPUT

Obs	X1	X2	X3	X4	Z1	Z2	Z3	Y	d1	d2	d3	d4	d5	d6	d7	d8	d9	d10	d11	d12
1	1	0	0	0	1	0	0	43	1	0	0	0	0	0	0	0	0	0	0	0
2	1	0	0	0	1	0	0	47	1	0	0	0	0	0	0	0	0	0	0	0
3	1	0	0	0	1	0	0	48	1	0	0	0	0	0	0	0	0	0	0	0
4	1	0	0	0	0	1	0	60	0	1	0	0	0	0	0	0	0	0	0	0
5	1	0	0	0	0	1	0	64	0	1	0	0	0	0	0	0	0	0	0	0
6	1	0	0	0	0	1	0	68	0	1	0	0	0	0	0	0	0	0	0	0
7	1	0	0	0	0	0	1	46	0	0	1	0	0	0	0	0	0	0	0	0
8	1	0	0	0	0	0	1	50	0	0	1	0	0	0	0	0	0	0	0	0
9	1	0	0	0	0	0	1	55	0	0	1	0	0	0	0	0	0	0	0	0
10	1	0	0	0	-1	-1	-1	59	-1	-1	-1	0	0	0	0	0	0	0	0	0
11	1	0	0	0	-1	-1	-1	52	-1	-1	-1	0	0	0	0	0	0	0	0	0
12	1	0	0	0	-1	-1	-1	54	-1	-1	-1	0	0	0	0	0	0	0	0	0
13	0	1	0	0	1	0	0	50	0	0	0	1	0	0	0	0	0	0	0	0
14	0	1	0	0	1	0	0	56	0	0	0	1	0	0	0	0	0	0	0	0
15	0	1	0	0	1	0	0	48	0	0	0	1	0	0	0	0	0	0	0	0
16	0	1	0	0	0	1	0	79	0	0	0	0	1	0	0	0	0	0	0	0
17	0	1	0	0	0	1	0	73	0	0	0	0	1	0	0	0	0	0	0	0
18	0	1	0	0	0	1	0	75	0	0	0	0	1	0	0	0	0	0	0	0
19	0	1	0	0	0	0	1	65	0	0	0	0	0	1	0	0	0	0	0	0
20	0	1	0	0	0	0	1	68	0	0	0	0	0	1	0	0	0	0	0	0
21	0	1	0	0	0	0	1	62	0	0	0	0	0	1	0	0	0	0	0	0
22	0	1	0	0	-1	-1	-1	60	0	0	0	-1	-1	-1	0	0	0	0	0	0
23	0	1	0	0	-1	-1	-1	63	0	0	0	-1	-1	-1	0	0	0	0	0	0
24	0	1	0	0	-1	-1	-1	61	0	0	0	-1	-1	-1	0	0	0	0	0	0
25	0	0	1	0	1	0	0	72	0	0	0	0	0	0	1	0	0	0	0	0
26	0	0	1	0	1	0	0	64	0	0	0	0	0	0	1	0	0	0	0	0
27	0	0	1	0	1	0	0	75	0	0	0	0	0	0	1	0	0	0	0	0
28	0	0	1	0	0	1	0	80	0	0	0	0	0	0	0	1	0	0	0	0
29	0	0	1	0	0	1	0	88	0	0	0	0	0	0	0	1	0	0	0	0
30	0	0	1	0	0	1	0	84	0	0	0	0	0	0	0	1	0	0	0	0
31	0	0	1	0	0	0	1	62	0	0	0	0	0	0	0	0	1	0	0	0
32	0	0	1	0	0	0	1	70	0	0	0	0	0	0	0	0	1	0	0	0
33	0	0	1	0	0	0	1	73	0	0	0	0	0	0	0	0	1	0	0	0
34	0	0	1	0	-1	-1	-1	70	0	0	0	0	0	0	-1	-1	-1	0	0	0
35	0	0	1	0	-1	-1	-1	71	0	0	0	0	0	0	-1	-1	-1	0	0	0

Obs	X1	X2	X3	X4	Z1	Z2	Z3	Y	d1	d2	d3	d4	d5	d6	d7	d8	d9	d10	d11	d12
36	0	0	1	0	-1	-1	-1	76	0	0	0	0	0	0	-1	-1	-1	0	0	0
37	0	0	0	1	1	0	0	77	0	0	0	0	0	0	0	0	0	1	0	0
38	0	0	0	1	1	0	0	73	0	0	0	0	0	0	0	0	0	1	0	0
39	0	0	0	1	1	0	0	78	0	0	0	0	0	0	0	0	0	1	0	0
40	0	0	0	1	0	1	0	79	0	0	0	0	0	0	0	0	0	0	1	0
41	0	0	0	1	0	1	0	87	0	0	0	0	0	0	0	0	0	0	1	0
42	0	0	0	1	0	1	0	80	0	0	0	0	0	0	0	0	0	0	1	0
43	0	0	0	1	0	0	1	61	0	0	0	0	0	0	0	0	0	0	0	1
44	0	0	0	1	0	0	1	67	0	0	0	0	0	0	0	0	0	0	0	1
45	0	0	0	1	0	0	1	69	0	0	0	0	0	0	0	0	0	0	0	1
46	0	0	0	1	-1	-1	-1	68	0	0	0	0	0	0	0	0	0	-1	-1	-1
47	0	0	0	1	-1	-1	-1	75	0	0	0	0	0	0	0	0	0	-1	-1	-1
48	0	0	0	1	-1	-1	-1	79	0	0	0	0	0	0	0	0	0	-1	-1	-1
49	-1	-1	-1	-1	1	0	0	67	-1	0	0	-1	0	0	-1	0	0	-1	0	0
50	-1	-1	-1	-1	1	0	0	73	-1	0	0	-1	0	0	-1	0	0	-1	0	0
51	-1	-1	-1	-1	1	0	0	66	-1	0	0	-1	0	0	-1	0	0	-1	0	0
52	-1	-1	-1	-1	0	1	0	84	0	-1	0	0	-1	0	0	-1	0	0	-1	0
53	-1	-1	-1	-1	0	1	0	80	0	-1	0	0	-1	0	0	-1	0	0	-1	0
54	-1	-1	-1	-1	0	1	0	80	0	-1	0	0	-1	0	0	-1	0	0	-1	0
55	-1	-1	-1	-1	0	0	1	69	0	0	-1	0	0	-1	0	0	-1	0	0	-1
56	-1	-1	-1	-1	0	0	1	83	0	0	-1	0	0	-1	0	0	-1	0	0	-1
57	-1	-1	-1	-1	0	0	1	74	0	0	-1	0	0	-1	0	0	-1	0	0	-1
58	-1	-1	-1	-1	-1	-1	-1	62	1	1	1	1	1	1	1	1	1	1	1	1
59	-1	-1	-1	-1	-1	-1	-1	60	1	1	1	1	1	1	1	1	1	1	1	1
60	-1	-1	-1	-1	-1	-1	-1	57	1	1	1	1	1	1	1	1	1	1	1	1

Analysis of Variance

Source	DF	Sum of Squares	Mean Square	F Value	Pr > F
Model	19	6734.98333	354.47281	21.10	<.0001
Error	40	672.00000	16.80000		
Corrected Total	59	7406.98333			

Root MSE	4.09878	R-Square	0.9093
Dependent Mean	67.31667	Adj R-Sq	0.8662
Coeff Var	6.08880		

Parameter Estimates

Variable	DF	Parameter Estimate	Standard Error	t Value	Pr > |t|
Intercept	1	67.31667	0.52915	127.22	<.0001
X1	1	-13.48333	1.05830	-12.74	<.0001
X2	1	-3.98333	1.05830	-3.76	0.0005
X3	1	6.43333	1.05830	6.08	<.0001
X4	1	7.10000	1.05830	6.71	<.0001
Z1	1	-4.85000	0.91652	-5.29	<.0001
Z2	1	10.08333	0.91652	11.00	<.0001
Z3	1	-2.38333	0.91652	-2.60	0.0130
d1	1	-2.98333	1.83303	-1.63	0.1115
d2	1	0.08333	1.83303	0.05	0.9640
d3	1	-1.11667	1.83303	-0.61	0.5458

Variable	DF	Parameter Estimate	Standard Error	t Value	Pr > |t|
d4	1	-7.15000	1.83303	-3.90	0.0004
d5	1	2.25000	1.83303	1.23	0.2268
d6	1	4.05000	1.83303	2.21	0.0329
d7	1	1.43333	1.83303	0.78	0.4388
d8	1	0.16667	1.83303	0.09	0.9280
d9	1	-3.03333	1.83303	-1.65	0.1058
d10	1	6.43333	1.83303	3.51	0.0011
d11	1	-2.50000	1.83303	-1.36	0.1802
d12	1	-6.36667	1.83303	-3.47	0.0012

The SAS analysis, its estimates, and its ANOVA for the model agree with the manual analysis, estimates, and the ANOVA, which were previously obtained elsewhere in the text.

4.1.7 Splitting Regression Sum of Squares according to Factorial Effect and Performing ANOVA

Sometimes, we may need to test for significance of each main factor and the interaction. This is carried out by using multiple model statements in the PROC REG syntax. Upon execution of the program, we use the outputs to obtain the sums of squares that pertain to each main effect and the interaction and carry out ANOVA. The typical statements that would accomplish this objective for Example 3.1 are

```
proc print data=twofact1;
proc reg data=twofact1;
model y=x1-x4;
model y=x1-x4 z1-z3;
model y=x1-x4 z1-z3 d1-d12;
run;
```

When the program is executed, we obtain the following sets of outputs:

SAS OUTPUT

```
The REG Procedure
Model: MODEL1
Dependent Variable: Y

Number of Observations Read    60
Number of Observations Used    60
```

Analysis of Variance

Source	DF	Sum of Squares	Mean Square	F Value	Pr > F
Model	4	3659.23333	914.80833	13.43	<.0001
Error	55	3747.75000	68.14091		
Corrected Total	59	7406.98333			

```
                    The REG Procedure
                     Model: MODEL2
                 Dependent Variable: Y

         Number of Observations Read     60
         Number of Observations Used     60

                  Analysis of Variance

                            Sum of          Mean
Source              DF     Squares        Square   F Value   Pr > F

Model                7   5744.21667    820.60238     25.66   <.0001
Error               52   1662.76667     31.97628
Corrected Total     59   7406.98333

                    The REG Procedure
                     Model: MODEL3
                 Dependent Variable: Y

         Number of Observations Read     60
         Number of Observations Used     60

                  Analysis of Variance

                            Sum of          Mean
Source              DF     Squares        Square   F Value   Pr > F

Model               19   6734.98333    354.47281     21.10   <.0001
Error               40    672.00000     16.80000
Corrected Total     59   7406.98333
```

From the outputs for models 1–3 above we find that

$$SS_{\text{machinist}} = SS_{x_1-x_4} = 3659.2333$$

$$SS_{\text{machines}} = SS_{z_1-z_3} = 5744.21667 - 3859.2333 = 2084.893$$

$$SS_{\text{machinist*machines}} = SS_{x_1z_1-x_4z_3} = 6734.9833 - 5744.21667 = 990.7667$$

The ANOVA is shown in Table 4.4.

TABLE 4.4: ANOVA Using Results from Fitting Regression Model for Example 3.1

Source	**Sum of squares**	**Degrees of freedom**	**Mean square**	***F*-ratio**	***p*-Value**
Machinist	3859.233	4	964.8083	57.42907	4.77691E−16
Machines	2084.983	3	694.9944	41.36872	2.46782E−12
Machinist∗machine	990.7667	12	82.56389	4.914517	6.38092E−05
Error	672	40	16.8		
Total	7406.983	59			

The above method can be used for any regression model developed for any experimental design to obtain relevant sums of squares and carry out ANOVA.

4.2 Regression Model for Two-Factor Factorial Experiment Using Reference Cell Coding for Dummy Variables

If the reference cell is cell (5,4) when we use the cell reference coding to define the dummy variables in the model for Example 3.1, the regression model is the same as was stated in Equation 4.1. However, the definitions of the dummy variables are different as shown below:

$$y_{ij} = \mu + \sum_{i=1}^{4} \alpha_i X_i + \sum_{j=1}^{3} \beta_j Z_j + \sum_{i=1}^{4}\sum_{j=1}^{3} \gamma_{ij} X_i Z_j + \varepsilon \qquad (4.4)$$

where

$$X_i = \begin{cases} 1 & \text{if } y_{ij} \text{ is under level } i \text{ of machinist} \\ 0 & \text{otherwise} \end{cases}$$
for $i = 1, 2, 3, 4$

$$Z_j = \begin{cases} 1 & \text{if } y_{ij} \text{ is under level } j \text{ of machines} \\ 0 & \text{otherwise} \end{cases}$$
for $j = 1, 2, 3$

To estimate the parameters of the regression model when we use reference cell coding to define the set of dummy variables, we have the following relations by using cell means (if cell (5,4) is the reference cell):

$$\begin{aligned} \mu_{ij} &= \mu + \alpha_i + \beta_j + \gamma_{ij} && i = 1, 2, 3, 4;\ j = 1, 2, 3 \\ \mu_{5j} &= \mu + \beta_j && j = 1, 2, 3 \\ \mu_{i4} &= \mu + \alpha_i && i = 1, 2, 3, 4 \\ \mu_{54} &= \mu \end{aligned}$$

Using the row and column marginal means, we obtain the following relations:

$$\mu_{i.} = \mu + \alpha_i + \sum_{j=1}^{4-1} \frac{\beta_j + \gamma_{ij}}{4} \qquad i = 1, 2, 3, 4$$

$$\mu_{5.} = \mu + \sum_{j=1}^{4-1} \frac{\beta_j}{4}$$

$$\mu_{.j} = \mu + \beta_j + \sum_{i=1}^{5-1} \frac{\alpha_i + \gamma_{ij}}{5} \quad j = 1, 2, 3$$

$$\mu_{.4} = \mu + \sum_{i=1}^{5-1} \frac{\alpha_i}{5}$$

These relations can only be used when equal numbers of observations are made per cell. Otherwise, we need to amend the relations.

4.2.1 Estimation of Parameters for the Regression Model with Reference Cell Coding for Dummy Variables (Example 3.1)

In this section, we use the formulae stated above to estimate the parameters of the model we used for Example 3.1 which was stated in Equation 4.4. In the following analysis, we define the dummy variables by using cell (5,4) as the reference cell.

Here, we obtain the estimates of the parameters related to the column factor as follows:

$$\mu = \mu_{54} = 59.6667$$
$$\mu_{51} = 59.6667 + \beta_1 \Rightarrow \hat{\beta}_1 = 68.6667 - 59.6667 = 9.0$$
$$\mu_{52} = 59.6667 + \beta_2 \Rightarrow \hat{\beta}_2 = 81.3333 - 59.6667 = 21.6667$$
$$\mu_{53} = 59.6667 + \beta_3 \Rightarrow \hat{\beta}_3 = 75.3333 - 59.6667 = 15.6667$$

Next, we obtain estimates of the parameters for the row factor as follows:

$$\mu_{i4} = \mu + \alpha_i \quad i = 1, 2, 3, 4$$
$$\mu_{14} = 55 = 59.6667 + \alpha_1 \Rightarrow \hat{\alpha}_1 = 55 - 59.6667 = -4.6667$$
$$\mu_{24} = 61.3333 = 59.6667 + \alpha_2 \Rightarrow \hat{\alpha}_2 = 61.3333 - 59.6667 = 1.6667$$
$$\mu_{34} = 72.3333 = 59.6667 + \alpha_3 \Rightarrow \hat{\alpha}_3 = 72.3333 - 59.6667 = 12.6667$$
$$\mu_{44} = 74 = 59.6667 + \alpha_4 \Rightarrow \hat{\alpha}_4 = 74 - 59.6667 = 14.3333$$

To obtain the parameters for the interaction, we note that

$$\overline{y}_{ij} = \hat{\mu} + \hat{\alpha}_i + \hat{\beta}_j + \gamma_{ij} \quad i = 1, 2, 3, 4; \; j = 1, 2, 3$$
$$\Rightarrow \hat{\gamma}_{ij} = \overline{y}_{ij} - \hat{\mu} - \hat{\alpha}_i - \hat{\beta}_j$$
$$\hat{\gamma}_{11} = \overline{y}_{11} - \overline{y}_{..} - \hat{\alpha}_1 - \hat{\beta}_1 \Rightarrow \hat{\gamma}_{11} = 46 - 59.6667 + 4.6667 - 9 = 18$$

$$\hat{\gamma}_{12} = \overline{y}_{12} - \overline{y}_{..} - \hat{\alpha}_1 - \hat{\beta}_2 \Rightarrow \hat{\gamma}_{12} = 64 - 59.6667 + 4.6667 - 21.6667 = -12.6667$$

$$\hat{\gamma}_{13} = \overline{y}_{13} - \overline{y}_{..} - \hat{\alpha}_1 - \hat{\beta}_3 \Rightarrow \hat{\gamma}_{13} = 50.3333 - 59.6667 + 4.6667 - 15.6667 = -20.3333$$

$$\hat{\gamma}_{21} = \overline{y}_{21} - \overline{y}_{..} - \hat{\alpha}_2 - \hat{\beta}_1 \Rightarrow \hat{\gamma}_{21} = 51.3333 - 59.6667 - 1.6667 - 9 = -19.00$$

$$\hat{\gamma}_{22} = \overline{y}_{22} - \overline{y}_{..} - \hat{\alpha}_2 - \hat{\beta}_2 \Rightarrow \hat{\gamma}_{22} = 75.6667 - 59.6667 - 1.6667 - 21.6667 = -7.3333$$

$$\hat{\gamma}_{23} = \overline{y}_{23} - \overline{y}_{..} - \hat{\alpha}_2 - \hat{\beta}_3 \Rightarrow \hat{\gamma}_{23} = 65.00 - 59.6667 - 1.6667 - 15.6667 = -12.00$$

$$\hat{\gamma}_{31} = \overline{y}_{31} - \overline{y}_{..} - \hat{\alpha}_3 - \hat{\beta}_1 \Rightarrow \hat{\gamma}_{31} = 70.3333 - 59.6667 - 12.6667 - 9 = -11.00$$

$$\hat{\gamma}_{32} = \overline{y}_{32} - \overline{y}_{..} - \hat{\alpha}_3 - \hat{\beta}_2 \Rightarrow \hat{\gamma}_{32} = 84 - 59.6667 - 12.6667 - 21.6667 = -10.00$$

$$\hat{\gamma}_{33} = \overline{y}_{33} - \overline{y}_{..} - \hat{\alpha}_3 - \hat{\beta}_3 \Rightarrow \hat{\gamma}_{33} = 68.3333 - 59.6667 - 12.6667 - 15.6667 = -19.6667$$

$$\hat{\gamma}_{41} = \overline{y}_{41} - \overline{y}_{..} - \hat{\alpha}_4 - \hat{\beta}_1 \Rightarrow \hat{\gamma}_{41} = 76 - 59.6667 - 14.3333 - 9 = -7.00$$

$$\hat{\gamma}_{42} = \overline{y}_{42} - \overline{y}_{..} - \hat{\alpha}_4 - \hat{\beta}_2 \Rightarrow \hat{\gamma}_{42} = 82 - 59.6667 - 14.3333 - 21.6667 = -13.6667$$

$$\hat{\gamma}_{43} = \overline{y}_{43} - \overline{y}_{..} - \hat{\alpha}_4 - \hat{\beta}_3 \Rightarrow \hat{\gamma}_{43} = 65.6667 - 59.6667 - 14.3333 - 15.6667 = -24.00$$

The fitted regression model is

$$\begin{aligned}\hat{y} = {} & 59.6667 - 4.6667x_1 + 1.6667x_2 + 12.6667x_3 + 14.3333x_4 + 9z_1 \\ & + 21.6667z_2 + 15.6667z_3 + 18x_1z_1 - 12.6667x_1z_2 - 20.3333x_1z_3 \\ & - 19x_2z_1 - 7.3333x_2z_2 - 12x_2z_3 - 11x_3z_1 - 10x_3z_2 \\ & - 19.6667x_3z_3 - 7x_4z_1 - 13.6667x_4z_2 - 24x_4z_3 \end{aligned} \tag{4.5}$$

The fitted values and extracted residuals are shown in the following table in which the d_i are as previously described. ANOVA to test for significance of the model is shown in Table 4.5.

TABLE 4.5: Variables, Estimated Responses, and Residuals for Regression Model for Example 3.1 (Reference Cell Coding Method)

x_1	x_2	x_3	x_4	z_1	z_2	z_3	y	d_1	d_2	d_3	d_4	d_5	d_6	d_7	d_8	d_9	d_{10}	d_{11}	d_{12}	$\hat{y}$	$y_i - \hat{y}_i$	$(y_i - \hat{y}_i)^2$
1	0	0	0	1	0	0	43	1	0	0	0	0	0	0	0	0	0	0	0	46	−3	9
1	0	0	0	1	0	0	47	1	0	0	0	0	0	0	0	0	0	0	0	46	1	1
1	0	0	0	1	0	0	48	1	0	0	0	0	0	0	0	0	0	0	0	46	2	4
1	0	0	0	0	1	0	60	0	1	0	0	0	0	0	0	0	0	0	0	64	−4	16
1	0	0	0	0	1	0	64	0	1	0	0	0	0	0	0	0	0	0	0	64	0	0
1	0	0	0	0	1	0	68	0	1	0	0	0	0	0	0	0	0	0	0	64	4	16
1	0	0	0	0	0	1	46	0	0	1	0	0	0	0	0	0	0	0	0	50.33	−4.333	18.778
1	0	0	0	0	0	1	50	0	0	1	0	0	0	0	0	0	0	0	0	50.33	−0.333	0.1112
1	0	0	0	0	0	1	55	0	0	1	0	0	0	0	0	0	0	0	0	50.33	4.6666	21.777
1	0	0	0	0	0	0	59	0	0	0	0	0	0	0	0	0	0	0	0	55	4	16
1	0	0	0	0	0	0	52	0	0	0	0	0	0	0	0	0	0	0	0	55	−3	9
1	0	0	0	0	0	0	54	0	0	0	0	0	0	0	0	0	0	0	0	55	−1	1
0	1	0	0	1	0	0	50	0	0	0	1	0	0	0	0	0	0	0	0	51.33	−1.333	1.778
0	1	0	0	1	0	0	56	0	0	0	1	0	0	0	0	0	0	0	0	51.33	4.6666	21.777
0	1	0	0	1	0	0	48	0	0	0	1	0	0	0	0	0	0	0	0	51.33	−3.333	11.112
0	1	0	0	0	1	0	79	0	0	0	0	1	0	0	0	0	0	0	0	75.67	3.3332	11.11
0	1	0	0	0	1	0	73	0	0	0	0	1	0	0	0	0	0	0	0	75.67	−2.667	7.1118
0	1	0	0	0	1	0	75	0	0	0	0	1	0	0	0	0	0	0	0	75.67	−0.667	0.4446
0	1	0	0	0	0	1	65	0	0	0	0	0	1	0	0	0	0	0	0	65	−1E−04	1E−08
0	1	0	0	0	0	1	68	0	0	0	0	0	1	0	0	0	0	0	0	65	2.9999	8.9994
0	1	0	0	0	0	1	62	0	0	0	0	0	1	0	0	0	0	0	0	65	−3	9.0006
0	1	0	0	0	0	0	60	0	0	0	0	0	0	0	0	0	0	0	0	61.33	−1.333	1.778
0	1	0	0	0	0	0	63	0	0	0	0	0	0	0	0	0	0	0	0	61.33	1.6666	2.7776
0	1	0	0	0	0	0	61	0	0	0	0	0	0	0	0	0	0	0	0	61.33	−0.333	0.1112
0	0	1	0	1	0	0	72	0	0	0	0	0	0	1	0	0	0	0	0	70.33	1.6666	2.7776
0	0	1	0	1	0	0	64	0	0	0	0	0	0	1	0	0	0	0	0	70.33	−6.333	40.112
0	0	1	0	1	0	0	75	0	0	0	0	0	0	1	0	0	0	0	0	70.33	4.6666	21.777
0	0	1	0	0	1	0	80	0	0	0	0	0	0	0	1	0	0	0	0	84	−4	16.001
0	0	1	0	0	1	0	88	0	0	0	0	0	0	0	1	0	0	0	0	84	3.9999	15.999
0	0	1	0	0	1	0	84	0	0	0	0	0	0	0	1	0	0	0	0	84	−1E−04	1E−08

0	0	1	0	0	0	1	62	0	0	0	0	0	0	0	0	1	0	0	0	68.33	−6.333	40.112
0	0	1	0	0	0	1	70	0	0	0	0	0	0	0	0	1	0	0	0	68.33	1.6666	2.7776
0	0	1	0	0	0	1	73	0	0	0	0	0	0	0	0	1	0	0	0	68.33	4.6666	21.777
0	0	1	0	0	0	0	70	0	0	0	0	0	0	0	0	0	0	0	0	72.33	−2.333	5.4448
0	0	1	0	0	0	0	71	0	0	0	0	0	0	0	0	0	0	0	0	72.33	−1.333	1.778
0	0	1	0	0	0	0	76	0	0	0	0	0	0	0	0	0	0	0	0	72.33	3.6666	13.444
0	0	0	1	1	0	0	77	0	0	0	0	0	0	0	0	0	1	0	0	76	1	1
0	0	0	1	1	0	0	73	0	0	0	0	0	0	0	0	0	1	0	0	76	−3	9
0	0	0	1	1	0	0	78	0	0	0	0	0	0	0	0	0	1	0	0	76	2	4
0	0	0	1	0	1	0	79	0	0	0	0	0	0	0	0	0	0	1	0	82	−3	9
0	0	0	1	0	1	0	87	0	0	0	0	0	0	0	0	0	0	1	0	82	5	25
0	0	0	1	0	1	0	80	0	0	0	0	0	0	0	0	0	0	1	0	82	−2	4
0	0	0	1	0	0	1	61	0	0	0	0	0	0	0	0	0	0	0	1	65.67	−4.667	21.778
0	0	0	1	0	0	1	67	0	0	0	0	0	0	0	0	0	0	0	1	65.67	1.3333	1.7777
0	0	0	1	0	0	1	69	0	0	0	0	0	0	0	0	0	0	0	1	65.67	3.3333	11.111
0	0	0	1	0	0	0	68	0	0	0	0	0	0	0	0	0	0	0	0	74	−6	36
0	0	0	1	0	0	0	75	0	0	0	0	0	0	0	0	0	0	0	0	74	1	1
0	0	0	1	0	0	0	79	0	0	0	0	0	0	0	0	0	0	0	0	74	5	25
0	0	0	0	1	0	0	67	0	0	0	0	0	0	0	0	0	0	0	0	68.67	−1.667	2.7779
0	0	0	0	1	0	0	73	0	0	0	0	0	0	0	0	0	0	0	0	68.67	4.3333	18.777
0	0	0	0	1	0	0	66	0	0	0	0	0	0	0	0	0	0	0	0	68.67	−2.667	7.1113
0	0	0	0	0	1	0	84	0	0	0	0	0	0	0	0	0	0	0	0	81.33	2.6666	7.1108
0	0	0	0	0	1	0	80	0	0	0	0	0	0	0	0	0	0	0	0	81.33	−1.333	1.778
0	0	0	0	0	1	0	80	0	0	0	0	0	0	0	0	0	0	0	0	81.33	−1.333	1.778
0	0	0	0	0	0	1	69	0	0	0	0	0	0	0	0	0	0	0	0	75.33	−6.333	40.112
0	0	0	0	0	0	1	83	0	0	0	0	0	0	0	0	0	0	0	0	75.33	7.6666	58.777
0	0	0	0	0	0	1	74	0	0	0	0	0	0	0	0	0	0	0	0	75.33	−1.333	1.778
0	0	0	0	0	0	0	62	0	0	0	0	0	0	0	0	0	0	0	0	59.67	2.3333	5.4443
0	0	0	0	0	0	0	60	0	0	0	0	0	0	0	0	0	0	0	0	59.67	0.3333	0.1111
0	0	0	0	0	0	0	57	0	0	0	0	0	0	0	0	0	0	0	0	59.67	−2.667	7.1113
							4039													4039	0	672

The values at the foot of the table given in pp. 234–235 are the totals for the variables.

The ANOVA to test for significance of the model is carried out in Table 4.6.

$$SS_{\mathrm{E}} = 672.0$$

$$S_{yy} = 279299 - \frac{4039^2}{60} = 7406.9833$$

$$\Rightarrow SS_{\mathrm{R}} = 7406.9833 - 672.0 = 6734.9833$$

TABLE 4.6: ANOVA to Test the Significance of the Model (Reference Cell Coding)

Source	Sum of squares	Degrees of freedom	Mean square	*F*-ratio
Regression	6734.9833	19	354.4728	21.0996
Error	672.0001	40	16.80	
Total	7406.9833	59		

From the tables of F-distribution, we see that $F(19, 40, 0.01) = 2.39$ and $F(19, 40, 0.05) = 1.85$, indicating that the fitted model is significant.

4.2.1.1 SAS Program

The SAS program for analyzing the data of Example 3.1 using reference cell coding for defining dummy variables in the model is presented below. For this analysis, we use cell (5,4) for reference.

SAS PROGRAM

```
data twofact1a;
input X1-X4 Z1-Z3 Y@@;
d1=x1*z1;d2=x1*z2; d3=x1*z3;d4=x2*z1;d5=x2*z2;d6=x2*z3;
d7=x3*z1;d8=x3*z2; d9=x3*z3;d10=x4*z1;d11=x4*z2;d12=x4*z3;
datalines;
1 0 0 0 1 0 0 43 1 0 0 0 1 0 0 47 1 0 0 0 1 0 0 48
1 0 0 0 0 1 0 60 1 0 0 0 0 1 0 64 1 0 0 0 0 1 0 68
1 0 0 0 0 0 1 46 1 0 0 0 0 0 1 50 1 0 0 0 0 0 1 55
1 0 0 0 0 0 0 59 1 0 0 0 0 0 0 52 1 0 0 0 0 0 0 54
0 1 0 0 1 0 0 50 0 1 0 0 1 0 0 56 0 1 0 0 1 0 0 48
0 1 0 0 0 1 0 79 0 1 0 0 0 1 0 73 0 1 0 0 0 1 0 75
0 1 0 0 0 0 1 65 0 1 0 0 0 0 1 68 0 1 0 0 0 0 1 62
0 1 0 0 0 0 0 60 0 1 0 0 0 0 0 63 0 1 0 0 0 0 0 61
0 0 1 0 1 0 0 72 0 0 1 0 1 0 0 64 0 0 1 0 1 0 0 75
0 0 1 0 0 1 0 80 0 0 1 0 0 1 0 88 0 0 1 0 0 1 0 84
0 0 1 0 0 0 1 62 0 0 1 0 0 0 1 70 0 0 1 0 0 0 1 73
0 0 1 0 0 0 0 70 0 0 1 0 0 0 0 71 0 0 1 0 0 0 0 76
0 0 0 1 1 0 0 77 0 0 0 1 1 0 0 73 0 0 0 1 1 0 0 78
0 0 0 1 0 1 0 79 0 0 0 1 0 1 0 87 0 0 0 1 0 1 0 80
0 0 0 1 0 0 1 61 0 0 0 1 0 0 1 67 0 0 0 1 0 0 1 69
0 0 0 1 0 0 0 68 0 0 0 1 0 0 0 75 0 0 0 1 0 0 0 79
0 0 0 0 1 0 0 67 0 0 0 0 1 0 0 73 0 0 0 0 1 0 0 66
```

```
0 0 0 0 0 1 0 84 0 0 0 0 0 1 0 80 0 0 0 0 0 1 0 80
0 0 0 0 0 0 1 69 0 0 0 0 0 0 1 83 0 0 0 0 0 0 1 74
0 0 0 0 0 0 0 62 0 0 0 0 0 0 0 60 0 0 0 0 0 0 0 57
;
proc print data=twofact1a; sum x1-x4 z1-z3 d1-d12 y;
proc reg data=twofact1a;
model y=x1-x4 z1-z3 d1-d12; run;
```

SAS OUTPUT:

Obs	X1	X2	X3	X4	Z1	Z2	Z3	Y	d1	d2	d3	d4	d5	d6	d7	d8	d9	d10	d11	d12
1	1	0	0	0	1	0	0	43	1	0	0	0	0	0	0	0	0	0	0	0
2	1	0	0	0	1	0	0	47	1	0	0	0	0	0	0	0	0	0	0	0
3	1	0	0	0	1	0	0	48	1	0	0	0	0	0	0	0	0	0	0	0
4	1	0	0	0	0	1	0	60	0	1	0	0	0	0	0	0	0	0	0	0
5	1	0	0	0	0	1	0	64	0	1	0	0	0	0	0	0	0	0	0	0
6	1	0	0	0	0	1	0	68	0	1	0	0	0	0	0	0	0	0	0	0
7	1	0	0	0	0	0	1	46	0	0	1	0	0	0	0	0	0	0	0	0
8	1	0	0	0	0	0	1	50	0	0	1	0	0	0	0	0	0	0	0	0
9	1	0	0	0	0	0	1	55	0	0	1	0	0	0	0	0	0	0	0	0
10	1	0	0	0	0	0	0	59	0	0	0	0	0	0	0	0	0	0	0	0
11	1	0	0	0	0	0	0	52	0	0	0	0	0	0	0	0	0	0	0	0
12	1	0	0	0	0	0	0	54	0	0	0	0	0	0	0	0	0	0	0	0
13	0	1	0	0	1	0	0	50	0	0	0	1	0	0	0	0	0	0	0	0
14	0	1	0	0	1	0	0	56	0	0	0	1	0	0	0	0	0	0	0	0
15	0	1	0	0	1	0	0	48	0	0	0	1	0	0	0	0	0	0	0	0
16	0	1	0	0	0	1	0	79	0	0	0	0	1	0	0	0	0	0	0	0
17	0	1	0	0	0	1	0	73	0	0	0	0	1	0	0	0	0	0	0	0
18	0	1	0	0	0	1	0	75	0	0	0	0	1	0	0	0	0	0	0	0
19	0	1	0	0	0	0	1	65	0	0	0	0	0	1	0	0	0	0	0	0
20	0	1	0	0	0	0	1	68	0	0	0	0	0	1	0	0	0	0	0	0
21	0	1	0	0	0	0	1	62	0	0	0	0	0	1	0	0	0	0	0	0
22	0	1	0	0	0	0	0	60	0	0	0	0	0	0	0	0	0	0	0	0
23	0	1	0	0	0	0	0	63	0	0	0	0	0	0	0	0	0	0	0	0
24	0	1	0	0	0	0	0	61	0	0	0	0	0	0	0	0	0	0	0	0
25	0	0	1	0	1	0	0	72	0	0	0	0	0	0	1	0	0	0	0	0
26	0	0	1	0	1	0	0	64	0	0	0	0	0	0	1	0	0	0	0	0
27	0	0	1	0	1	0	0	75	0	0	0	0	0	0	1	0	0	0	0	0
28	0	0	1	0	0	1	0	80	0	0	0	0	0	0	0	1	0	0	0	0
29	0	0	1	0	0	1	0	88	0	0	0	0	0	0	0	1	0	0	0	0
30	0	0	1	0	0	1	0	84	0	0	0	0	0	0	0	1	0	0	0	0
31	0	0	1	0	0	0	1	62	0	0	0	0	0	0	0	0	1	0	0	0
32	0	0	1	0	0	0	1	70	0	0	0	0	0	0	0	0	1	0	0	0
33	0	0	1	0	0	0	1	73	0	0	0	0	0	0	0	0	1	0	0	0
34	0	0	1	0	0	0	0	70	0	0	0	0	0	0	0	0	0	0	0	0
35	0	0	1	0	0	0	0	71	0	0	0	0	0	0	0	0	0	0	0	0
36	0	0	1	0	0	0	0	76	0	0	0	0	0	0	0	0	0	0	0	0
37	0	0	0	1	1	0	0	77	0	0	0	0	0	0	0	0	0	1	0	0
38	0	0	0	1	1	0	0	73	0	0	0	0	0	0	0	0	0	1	0	0
39	0	0	0	1	1	0	0	78	0	0	0	0	0	0	0	0	0	1	0	0
40	0	0	0	1	0	1	0	79	0	0	0	0	0	0	0	0	0	0	1	0
41	0	0	0	1	0	1	0	87	0	0	0	0	0	0	0	0	0	0	1	0
42	0	0	0	1	0	1	0	80	0	0	0	0	0	0	0	0	0	0	1	0
43	0	0	0	1	0	0	1	61	0	0	0	0	0	0	0	0	0	0	0	1
44	0	0	0	1	0	0	1	67	0	0	0	0	0	0	0	0	0	0	0	1

Obs	X1	X2	X3	X4	Z1	Z2	Z3	Y	d1	d2	d3	d4	d5	d6	d7	d8	d9	d10	d11	d12
45	0	0	0	1	0	0	1	69	0	0	0	0	0	0	0	0	0	0	0	1
46	0	0	0	1	0	0	0	68	0	0	0	0	0	0	0	0	0	0	0	0
47	0	0	0	1	0	0	0	75	0	0	0	0	0	0	0	0	0	0	0	0
48	0	0	0	1	0	0	0	79	0	0	0	0	0	0	0	0	0	0	0	0
49	0	0	0	0	1	0	0	67	0	0	0	0	0	0	0	0	0	0	0	0
50	0	0	0	0	1	0	0	73	0	0	0	0	0	0	0	0	0	0	0	0
51	0	0	0	0	1	0	0	66	0	0	0	0	0	0	0	0	0	0	0	0
52	0	0	0	0	0	1	0	84	0	0	0	0	0	0	0	0	0	0	0	0
53	0	0	0	0	0	1	0	80	0	0	0	0	0	0	0	0	0	0	0	0
54	0	0	0	0	0	1	0	80	0	0	0	0	0	0	0	0	0	0	0	0
55	0	0	0	0	0	0	1	69	0	0	0	0	0	0	0	0	0	0	0	0
56	0	0	0	0	0	0	1	83	0	0	0	0	0	0	0	0	0	0	0	0
57	0	0	0	0	0	0	1	74	0	0	0	0	0	0	0	0	0	0	0	0
58	0	0	0	0	0	0	0	62	0	0	0	0	0	0	0	0	0	0	0	0
59	0	0	0	0	0	0	0	60	0	0	0	0	0	0	0	0	0	0	0	0
60	0	0	0	0	0	0	0	57	0	0	0	0	0	0	0	0	0	0	0	0
	==	==	==	==	==	==	==	====	==	==	==	==	==	==	==	==	==	===	===	===
	12	12	12	12	15	15	15	4039	3	3	3	3	3	3	3	3	3	3	3	3

The REG Procedure
Model: MODEL1
Dependent Variable: Y

Number of Observations Read	60
Number of Observations Used	60

Analysis of Variance

Source	DF	Sum of Squares	Mean Square	F Value	Pr > F
Model	19	6734.98333	354.47281	21.10	<.0001
Error	40	672.00000	16.80000		
Corrected Total	59	7406.98333			

Root MSE	4.09878	R-Square	0.9093
Dependent Mean	67.31667	Adj R-Sq	0.8662
Coeff Var	6.08880		

Parameter Estimates

Variable	DF	Parameter Estimate	Standard Error	t Value	Pr > |t|
Intercept	1	59.66667	2.36643	25.21	<.0001
X1	1	-4.66667	3.34664	-1.39	0.1709
X2	1	1.66667	3.34664	0.50	0.6212
X3	1	12.66667	3.34664	3.78	0.0005
X4	1	14.33333	3.34664	4.28	0.0001
Z1	1	9.00000	3.34664	2.69	0.0104
Z2	1	21.66667	3.34664	6.47	<.0001
Z3	1	15.66667	3.34664	4.68	<.0001
d1	1	-18.00000	4.73286	-3.80	0.0005
d2	1	-12.66667	4.73286	-2.68	0.0107
d3	1	-20.33333	4.73286	-4.30	0.0001
d4	1	-19.00000	4.73286	-4.01	0.0003
d5	1	-7.33333	4.73286	-1.55	0.1292

Variable	DF	Parameter Estimate	Standard Error	t Value	Pr > |t|
d6	1	-12.00000	4.73286	-2.54	0.0152
d7	1	-11.00000	4.73286	-2.32	0.0253
d8	1	-10.00000	4.73286	-2.11	0.0409
d9	1	-19.66667	4.73286	-4.16	0.0002
d10	1	-7.00000	4.73286	-1.48	0.1470
d11	1	-13.66667	4.73286	-2.89	0.0062
d12	1	-24.00000	4.73286	-5.07	<.0001

Looking at the estimates of the parameters carried out by SAS as it fitted the model and performed ANOVA to test its significance, we see that they agree with the estimates previously obtained by manual calculation.

4.3 Use of SAS for the Analysis of Responses of Mixed Models

In this section, our data analyses will utilize the MIXED procedure or PROC MIXED as implemented in SAS software. The MIXED procedure was devised to extend the capability of SAS to analyze data, which hitherto could not be handled properly by the GLM procedure. The procedure is designed in the same style as PROC GLM. It features the ability to specify various covariance structures with similar language as was used in PROC GLM, including the MODEL, REPEATED, RANDOM, CONTRAST, ESTIMATE, and LSMEANS statements. Restricted/residual maximum likelihood (REML) and maximum likelihood (ML) estimations were implemented by using a form of the Newton–Raphson algorithm. It has the ability to handle unbalanced data and the ability to create SAS data sets, which correspond to virtually any type of table. It features subject group effects, which enable the modeling of blocking and heterogeneity. It provides appropriate standard errors and corresponding F- and t-tests for all specified and estimable linear combinations of fixed and random effects. PROC MIXED can utilize the SAS ODS system that provides versatility in the display of SAS outputs. In general, the matrix form of the standard linear regression model can be stated as

$$\mathbf{Y} = \mathbf{X}\mathbf{b} + \mathbf{E} \tag{4.6}$$

where

- $\mathbf{X}$ is the matrix of controllable or independent variables
- $\mathbf{Y}$ is the vector of the response or outcome variables
- $\mathbf{b}$ is the vector of the parameters of the regression model
- $\mathbf{E}$ is the vector of random errors associated with the response vectors

It is well known under the ordinary least squares analysis of this model that the normal equations are obtained as

$$\mathbf{X}'\mathbf{X}\mathbf{b} = \mathbf{X}'\mathbf{Y} \tag{4.7}$$

Here, without making any assumptions about the actual distribution of $\mathbf{E}$, the error vector, if we represent the generalized inverse of $(\mathbf{X'X})$ as $(\mathbf{X'X})^{-}$, then explicit solutions for the parameters of the model are obtained as

$$\hat{\mathbf{b}} = (\mathbf{X'X})^{-}(\mathbf{X'Y}) \tag{4.8}$$

$\hat{\mathbf{b}}$ in Equation 4.8 is the ordinary least squares estimator of the parameter vector $\mathbf{b}$ no matter what distribution we assume for $\mathbf{E}$. Also by classical statistical theory, if the errors are normally and independently distributed with a common variance σ^2, then we see that $\hat{\mathbf{b}}$ is a minimum variance unbiased estimator (MVUE) of $\mathbf{b}$, and the covariance matrix for the vector of estimators $\hat{\mathbf{b}}$ for the vector of parameters $\mathbf{b}$ is $\sigma^2(\mathbf{X'X})^{-}$. Also if $\mathbf{C}$ constitutes a vector of coefficients so that a linear combination $\mathbf{C'b}$ is estimable, then $\mathbf{C'}\hat{\mathbf{b}}$ is the best linear unbiased estimator (BLUE) of $\mathbf{C'b}$. If there are t treatments and b subjects in the design, a scalar statistical model that describes the general mixed model for a typical RMD is

$$y_{ijk} = \mu + \tau_i + s_{ij} + \gamma_k + \theta_{ik} + E_{ijk} \tag{4.9}$$

where $i = 1, 2, \ldots, t$; $j = 1, 2, \ldots, b$; and μ, α_i, τ_k, and θ_{ik} are fixed parameters, so that the mean of the ith treatment at time k is $\mu_{ik} = \mu + \tau_i + \gamma_k + \theta_{ik}$.

In its matrix-form the general linear mixed model is written as an extension of the model in Equation 4.6 so that

$$\mathbf{Y} = \mathbf{Xb} + \mathbf{Zw} + \mathbf{E} \tag{4.10}$$

where $\mathbf{w}$ and $\mathbf{E}$ are both multivariate with mean vector $\mathbf{0}$, and covariance matrices $\mathbf{G}$ and $\mathbf{R}$, respectively. Thus, we can say that $\mathbf{w} \sim \text{MVN}(\mathbf{0}, \mathbf{G})$ and $\mathbf{E} \sim \text{MVN}(\mathbf{0}, \mathbf{R})$ with no other constraint on the matrices $\mathbf{G}$ and $\mathbf{R}$, except that they be positive definite. In Equation 4.9, s_{ij} constitutes the $\mathbf{w}$ term in Equation 4.10, while E_{ijk} make up $\mathbf{E}$ term in Equation 4.10. Thus, $s_{ij} \sim \text{MVN}(\mathbf{0}, \sigma_s^2\mathbf{I})$ while $E_{ijk} \sim \text{MVN}(\mathbf{0}, \sigma^2\mathbf{I})$. With these, the variance–covariance matrix for $\mathbf{Y}$ is obtained as $\mathbf{\Sigma}(\mathbf{Y}) = \sigma_s^2\mathbf{ZZ'} + \sigma^2\mathbf{I}$, a block diagonal matrix with each block corresponding to each subject in the design. From the model (Equation 4.10), if we denote the covariance matrix of the response vector $\mathbf{Y}$ as $\mathbf{V}$ we can easily show that

$$\mathbf{V} = \mathbf{ZGZ'} + \mathbf{R} \tag{4.11}$$

If we set $\mathbf{R} = \sigma^2\mathbf{I}$ and $\mathbf{Z} = 0$, then Equation 4.10 reduces to Equation 4.6, that is, the mixed model reduces to the standard general linear model.

Taking expectation of $\mathbf{Y}$ in Equation 4.10, then

$$\mathbf{E}(\mathbf{Y}) = \mathbf{E}(\mathbf{Xb} + \mathbf{Zw} + \mathbf{E}) = \mathbf{Xb} \tag{4.12}$$

Thus, it can be seen that $\mathbf{Y} \sim \text{MVN}(\mathbf{Xb}, \mathbf{ZGZ'} + \mathbf{R})$. Therefore, under the general linear mixed model, the estimates of the parameters of the regression model is obtained as

$$\hat{\mathbf{b}} = (\mathbf{XV}^{-1}\mathbf{X'})^{-}\mathbf{XV}^{-1}\mathbf{Y} \tag{4.13}$$

Following from the foregoing, the variance of the linear combination of the vector of parameters, $\mathbf{C}'\hat{\mathbf{b}}$ is obtained as

$$\mathbf{Var}(\mathbf{C}'\hat{\mathbf{b}}) = \mathbf{C}'(\mathbf{X}\mathbf{V}^{-1}\mathbf{X}')\mathbf{C} \tag{4.14}$$

With these results, we can make inferences about any linear combination of the vector of parameters $\mathbf{b}$.

4.3.1 Implementation of the General Linear Mixed Model in PROC MIXED in SAS

The theory we discussed above about the parameter vector $\mathbf{b}$ has been implemented in the MIXED procedure in SAS, making it a more versatile procedure to use in analyzing responses of some complex designs than the GLM procedure. It should be noted that the covariance matrices $\mathbf{G}$ and $\mathbf{R}$ in the above theory are functions of unknown parameters. Under the MIXED model, estimation is more difficult than in the general linear model, making least squares method inadequate for the purpose. This is because under this model, we do not only have $\mathbf{b}$, but we have unknown parameters in $\mathbf{w}$, $\mathbf{G}$, and $\mathbf{R}$. The best method is to use the general least squares (GLS) to minimize $(\mathbf{Y}-\mathbf{Xb})'\mathbf{V}^{-1}(\mathbf{Y}-\mathbf{Xb})$, but the GLS requires the knowledge of $\mathbf{V}$ and often it is unknown. There are two ways of dealing with this difficulty (1) to decide reasonable estimate for $\mathbf{V}$ and use it in the minimization, reducing the problem to simply obtaining a reasonable estimate of $\mathbf{G}$ and $\mathbf{R}$ or (2) to use likelihood-related methods. This enables us to take advantage of the assumption that both $\mathbf{E}$ and $\mathbf{w}$ are normally distributed and use it in the process. This approach is suggested by Hartley and Rao (1967), Laird and Ware (1982), Jennrich and Schluchter (1986), among others. The MIXED procedure uses two of such likelihood-based methods, ML and REML to estimate $\mathbf{G}$ and $\mathbf{R}$. Implementation of this approach in SAS was carried out in the MIXED procedure, which constructs an objective function associated with either the ML or the REML and maximizes it over all unknown parameters. The MIXED procedure also uses a third approach that is based on a form of method of moments called **MIVQUE(0)**. The natural logarithm of the ML and the REML that were implemented in the MIXED procedure are as follows: If $\mathbf{k}=\mathbf{Y}-\mathbf{X}(\mathbf{X}'\mathbf{V}^{-1}\mathbf{X})\mathbf{X}\mathbf{V}^{-1}\mathbf{Y}$, then

$$\ln_{\mathrm{ML}}(\mathbf{G},\mathbf{R}) = -\frac{1}{2}\ln|\mathbf{V}| - \frac{1}{2}\mathbf{k}'\mathbf{V}^{-1}\mathbf{k} - \frac{n}{2}\ln(2\pi) \tag{4.15}$$

$$\ln_{\mathrm{REML}}(\mathbf{G},\mathbf{R}) = -\frac{1}{2}\ln|\mathbf{V}| - \frac{1}{2}\ln|\mathbf{X}'\mathbf{V}^{-1}\mathbf{X}| - \frac{1}{2}\mathbf{k}'\mathbf{V}^{-1}\mathbf{k} - \frac{n-r}{2}\ln(2\pi) \tag{4.16}$$

Under the ML or the REML, we can obtain estimates of $\mathbf{G}$ and $\mathbf{R}$ as $\hat{\mathbf{G}}$ and $\hat{\mathbf{R}}$. Using these results, we can find the estimates of $\mathbf{b}$ and $\mathbf{w}$ by solving the mixed model equations

$$\begin{bmatrix} \mathbf{X}'\hat{\mathbf{R}}^{-1}\mathbf{X} & \mathbf{X}'\hat{\mathbf{R}}^{-1}\mathbf{Z} \\ \mathbf{Z}'\hat{\mathbf{R}}^{-1}\mathbf{X} & \mathbf{Z}'\hat{\mathbf{R}}^{-1}\mathbf{Z} + \hat{\mathbf{G}}^{-1} \end{bmatrix} \begin{bmatrix} \hat{\mathbf{b}} \\ \hat{\mathbf{w}} \end{bmatrix} = \begin{bmatrix} \mathbf{X}'\hat{\mathbf{R}}^{-1}\mathbf{Y} \\ \mathbf{Z}'\hat{\mathbf{R}}^{-1}\mathbf{X} \end{bmatrix} \tag{4.17}$$

from which we obtain

$$\hat{\mathbf{b}} = (\mathbf{X}'\hat{\mathbf{V}}^{-1}\mathbf{X})\mathbf{X}'\hat{\mathbf{V}}^{-1}\mathbf{Y} \tag{4.18}$$

$$\hat{\mathbf{w}} = \mathbf{G}\mathbf{Z}\hat{\mathbf{V}}^{-1}(\mathbf{Y} - \mathbf{X}\hat{\beta}) \tag{4.19}$$

4.4 Use of PROC Mixed in the Analysis of Responses of RMD in SAS

A typical RMD involves experiments with treatments repeated over time on human or animal subjects. Because of the ability of the subjects to retain or carry over the effects of earlier treatments to subsequent treatments, the responses are believed to be correlated, especially those that are closer to each other. One of the methods used for the analysis of responses of the RMD is to treat them as if they are coming from a typical split-plot design. In this case, methods similar to those used for analyzing the factorial experiment with random factors are used to analyze the responses. In SAS, this type of analysis is carried out as ANOVA, using GLM procedure with the RANDOM statement. Another method allows for both univariate and multivariate analysis methods to be applied to linear transformations of responses of the repeated measures experiment. Such transformations may be means, differences between responses at different times, points in the design, or some other regression parameters. The analysis is carried out in SAS by invoking the REPEATED statement. A third method deals with the analysis of responses of the RMD by using a mixed model approach with special parametric structures on the covariance matrices. This is implemented in SAS under the MIXED procedure using the REPEATED statement. The MIXED procedure is also used for analyzing data from the mixed RCBD design in which the subjects are treated as blocks and the order of treatment is randomized within blocks as described in Sections 3.5.1 and 3.5.2. As we stated in Section 4.3, the covariance structure needed in carrying out the analysis of the responses of a repeated measures experiment are often unknown. We have to use reasonable estimated covariances to analyze the data. Many structures can be used for covariance matrices. They can be found in SAS/STAT User's Guide (1999, Version 8, pp. 2138–2139). Here, we consider four of these structures for the covariance matrices used in the analysis (a) variance components (VC), (b) compound symmetry (CS), (c) autoregressive one [AR(1)], and (d) unstructured (UN). The general mixed model is shown in Equation 4.22 and reproduced here as

$$\mathbf{Y} = \mathbf{X}\mathbf{b} + \mathbf{Z}\mathbf{w} + \mathbf{E}$$

Under this model, the REPEATED statement is used in the MIXED procedure to define covariance structure on the vector $\mathbf{w}$, that is, to define the matrix $\mathbf{R}$. The matrix $\mathbf{R}$ used for the RMD can also be defined as a block diagonal, each block corresponding to each subject in the experiment. To illustrate the different forms of the covariance structures implemented in MIXED procedure in SAS, we find the covariance structure for a subject on which measurement was made six times. We refer to Tables 41.3 through 41.5 on pages 2138–2139 of SAS/STAT User's Guide (1999), Version 8. In the discussion of VC below, we illustrate with a 6×6 variance–covariance matrix. Equivalents for other dimensions for the variance–covariance matrix can be similarly defined.

a. *Variance components.* When we use the VC structure to the RMD described above, we obtain a covariance structure of the form

$$\mathbf{R(6,6)} = \begin{bmatrix} \sigma_B^2 & 0 & 0 & 0 & 0 & 0 \\ 0 & \sigma_B^2 & 0 & 0 & 0 & 0 \\ 0 & 0 & \sigma_B^2 & 0 & 0 & 0 \\ 0 & 0 & 0 & \sigma_{AB}^2 & 0 & 0 \\ 0 & 0 & 0 & 0 & \sigma_{AB}^2 & 0 \\ 0 & 0 & 0 & 0 & 0 & \sigma_{AB}^2 \end{bmatrix}$$

The Bs refer to the subject, so that σ_B^2 is the variance of B and σ_{AB}^2 refers to the variance of the interaction between subject and treatment. This is the default setting for the covariance structure under the MIXED procedure in SAS.

b. *Compound symmetry structure.* If we consider the CS structure, then the covariance for an individual on whom measurement was repeated six times, then the 6×6 covariance matrix ($\mathbf{R}$) would be of the form

$$\mathbf{R(6,6)} = \begin{bmatrix} \sigma^2+\sigma_1 & \sigma_1 & \sigma_1 & \sigma_1 & \sigma_1 & \sigma_1 \\ \sigma_1 & \sigma^2+\sigma_1 & \sigma_1 & \sigma_1 & \sigma_1 & \sigma_1 \\ \sigma_1 & \sigma_1 & \sigma^2+\sigma_1 & \sigma_1 & \sigma_1 & \sigma_1 \\ \sigma_1 & \sigma_1 & \sigma_1 & \sigma^2+\sigma_1 & \sigma_1 & \sigma_1 \\ \sigma_1 & \sigma_1 & \sigma_1 & \sigma_1 & \sigma^2+\sigma_1 & \sigma_1 \\ \sigma_1 & \sigma_1 & \sigma_1 & \sigma_1 & \sigma_1 & \sigma^2+\sigma_1 \end{bmatrix}$$

where σ^2 and σ_1 are the constant variance and constant covariance respectively, which are the properties of the CS structure.

c. *Autoregressive one covariance structure.* Application of the autoregressive one [AR(1)] structure to the covariance of a subject in a RMD with six measurements per individuals leads to a 6×6 variance–covariance matrix of the form

$$\mathbf{R(6,6)} = \begin{bmatrix} \sigma^2 & \sigma^2\rho & \sigma^2\rho^2 & \sigma^2\rho^3 & \sigma^2\rho^4 & \sigma^2\rho^5 \\ \sigma^2\rho & \sigma^2 & \sigma^2\rho & \sigma^2\rho^2 & \sigma^2\rho^3 & \sigma^2\rho^4 \\ \sigma^2\rho^2 & \sigma^2\rho & \sigma^2 & \sigma^2\rho & \sigma^2\rho^2 & \sigma^2\rho^3 \\ \sigma^2\rho^3 & \sigma^2\rho^2 & \sigma^2\rho & \sigma^2 & \sigma^2\rho & \sigma^2\rho^2 \\ \sigma^2\rho^4 & \sigma^2\rho^3 & \sigma^2\rho^2 & \sigma^2\rho & \sigma^2 & \sigma^2\rho \\ \sigma^2\rho^5 & \sigma^2\rho^4 & \sigma^2\rho^3 & \sigma^2\rho^2 & \sigma^2\rho & \sigma^2 \end{bmatrix}$$

Under this structure, if two observations are m units apart, then the correlation between them is considered to be ρ^m, and the variances σ^2 and $\rho\sigma^2$ are not confounded.

d. *Unstructured covariance.* When we consider the covariance for **R** to be UN, we have the following 6×6 variance–covariance matrix:

$$\mathbf{R(6,6)} = \begin{bmatrix} \sigma_1^2 & \sigma_{21} & \sigma_{31} & \sigma_{41} & \sigma_{51} & \sigma_{61} \\ \sigma_{21} & \sigma_2^2 & \sigma_{32} & \sigma_{42} & \sigma_{52} & \sigma_{62} \\ \sigma_{31} & \sigma_{32} & \sigma_3^2 & \sigma_{43} & \sigma_{53} & \sigma_{63} \\ \sigma_{41} & \sigma_{42} & \sigma_{43} & \sigma_4^2 & \sigma_{54} & \sigma_{64} \\ \sigma_{51} & \sigma_{52} & \sigma_{53} & \sigma_{54} & \sigma_5^2 & \sigma_{65} \\ \sigma_{61} & \sigma_{62} & \sigma_{63} & \sigma_{64} & \sigma_{65} & \sigma_6^2 \end{bmatrix}$$

In this case, no mathematical structure is imposed on the covariance matrix. The UN covariance structure is the most general of all the covariance structures employed in the mixed model, when it is analyzed by the MIXED procedure in SAS.

4.4.1 Choosing the Covariance Structure to Use

To analyze the data from a RMD using the MIXED procedure, we need to determine which of the methods we should use to choose a reasonable covariance structure for **R.** The method to use depends on looking at two measures, which are given as part of the output, namely, AIC, AICC, and BIC. AIC is the Akaike's Information Criterion, Akaike (1974), while AICC is the finite-sample corrected version of AIC (Hurvich and Tsai, 1989; Burnham and Anderson, 1998). BIC is the Bayesian Information Criterion (Schwarz, 1978). Weiss (2005) describes general covariance structures that can be used. Basically, the smaller the value of AIC, AICC, or BIC for any covariance structure, the better is that structure for estimating the covariance structure to use in the MIXED procedure for analyzing the responses of a RMD. We will illustrate this method in the analysis of the responses of a RMD in Example 3.3.

4.4.2 Analysis of Responses of the Experiment on Guinea-Pig Ventricular Mycotes (Example 3.3) with PROC MIXED in SAS

For the PROC MIXED to analyze the data of this experiment, it uses reference cell coding method to define the dummy variables to be utilized in the analysis. Cell (8,7) was used to define the dummy variables. Thus, for the model we employed if X_i and Z_j are the dummy variables for subjects and time, then

$$y_{ij} = \mu + \sum_{i=1}^{7} \beta_i X_i + \sum_{j=1}^{6} \gamma_j Z_j + \varepsilon \tag{4.20}$$

where

$$X_i = \begin{cases} 1 & \text{if } y_{ij} \text{ is under subject } i \\ 0 & \text{otherwise} \end{cases}$$
for $i = 1, 2, \ldots, 7$

$$Z_j = \begin{cases} 1 & \text{if } y_{ij} \text{ is under subject } j \\ 0 & \text{otherwise} \end{cases}$$
for $j = 1, 2, \ldots, 6$

To carry out the analysis of the data, we need to choose which covariance structure to use. It turns out that for our data, under the REML method, convergence with a positive definite last Hessian occurs only for the VC structure. We choose to use VC structure in our program, which is presented below.

Hessian: This is a matrix of second-order derivatives for a scalar-valued function. If a function f has n variables, the Hessian matrix is the $n \times n$ matrix whose components are equal to the partial second-order derivatives for the function f. Newton's methods for optimization are generally used to compute the Hessian.

SAS PROGRAM

```
data repeatd2;
input person time y @@;
y=-y;
datalines;
1 1 94.1  1 2 92.40 1 3 92.09 1 4 91.42 1 5 91.06 1 6 90.38 1 7 90.22
2 1 84.93 2 2 81.59 2 3 74.87 2 4 83.90 2 5 73.22 2 6 86.59 2 7 90.03
3 1 75.99 3 2 77.01 3 3 67.97 3 4 69.95 3 5 70.60 3 6 69.80 3 7 69.29
4 1 80.01 4 2 71.96 4 3 62.82 4 4 64.10 4 5 69.01 4 6 65.28 4 7 59.55
5 1 85.47 5 2 83.00 5 3 84.58 5 4 83.96 5 5 84.16 5 6 83.46 5 7 85.15
6 1 78.99 6 2 73.96 6 3 73.07 6 4 72.63 6 5 73.02 6 6 70.74 6 7 73.00
7 1 83.86 7 2 77.59 7 3 75.76 7 4 75.27 7 5 74.59 7 6 73.31 7 7 71.66
8 1 79.36 8 2 75.36 8 3 71.36 8 4 63.95 8 5 64.82 8 6 67.71 8 7 68.37
;
proc print data=repeatd2; sum y;
```

```
proc mixed data=repeatd2 covtest;
class person time;
model y= time person /s;
repeated/type=vc subject= person;
```

SAS OUTPUT

Obs	person	time	y
1	1	1	-94.10
2	1	2	-92.40
3	1	3	-92.09
4	1	4	-91.42
5	1	5	-91.06
6	1	6	-90.38
7	1	7	-90.22
8	2	1	-84.93
9	2	2	-81.59
10	2	3	-74.87
11	2	4	-83.90
12	2	5	-73.22
13	2	6	-86.59
14	2	7	-90.03
15	3	1	-75.99
16	3	2	-77.01
17	3	3	-67.97
18	3	4	-69.95
19	3	5	-70.60
20	3	6	-69.80
21	3	7	-69.29
22	4	1	-80.01
23	4	2	-71.96
24	4	3	-62.82
25	4	4	-64.10
26	4	5	-69.01
27	4	6	-65.28
28	4	7	-59.55
29	5	1	-85.47
30	5	2	-83.00
31	5	3	-84.58
32	5	4	-83.96
33	5	5	-84.16
34	5	6	-83.46
35	5	7	-85.15
36	6	1	-78.99
37	6	2	-73.96
38	6	3	-73.07
39	6	4	-72.63
40	6	5	-73.02
41	6	6	-70.74
42	6	7	-73.00
43	7	1	-83.86
44	7	2	-77.59
45	7	3	-75.76
46	7	4	-75.27
47	7	5	-74.59

```
Obs      person    time        y
 48         7        6      -73.31
 49         7        7      -71.66
 50         8        1      -79.36
 51         8        2      -75.36
 52         8        3      -71.36
 53         8        4      -63.95
 54         8        5      -64.82
 55         8        6      -67.71
 56         8        7      -68.37
                            ========
                            -4318.30
                The Mixed Procedure

                 Model Information

Data Set                       WORK.REPEATD2
Dependent Variable             y
Covariance Structure           Variance Components
Estimation Method              REML
Residual Variance Method       Parameter
Fixed Effects SE Method        Model-Based
Degrees of Freedom Method      Between-Within

              Class Level Information

 Class       Levels    Values

 person           8    1 2 3 4 5 6 7 8
 time             7    1 2 3 4 5 6 7

                    Dimensions

       Covariance Parameters          1
       Columns in X                  16
       Columns in Z                   0
       Subjects                      56
       Max Obs Per Subject            1

              Number of Observations

   Number of Observations Read             56
   Number of Observations Used             56
   Number of Observations Not Used          0

                Iteration History

Iteration    Evaluations   -2 Res Log Like   Criterion

        0              1      251.20177212
        1              1      251.20177212  0.00000000

             Convergence criteria met.

               The Mixed Procedure

           Covariance Parameter Estimates

 Cov                      Standard        Z
 Parm        Estimate        Error    value       Pr Z

 person       12.4502       2.7169     4.58     <.0001
```

Fit Statistics

-2 Res Log Likelihood	251.2
AIC (smaller is better)	253.2
AICC (smaller is better)	253.3
BIC (smaller is better)	255.2

Null Model Likelihood Ratio Test

DF	Chi-Square	Pr > ChiSq
0	0.00	1.0000

Solution for Fixed Effects

Effect	person	time	Estimate	Standard Error	DF	t Value	Pr > |t|
Intercept			-68.9291	1.7642	42	-39.07	<.0001
time		1	-6.9300	1.7642	42	-3.93	0.0003
time		2	-3.2000	1.7642	42	-1.81	0.0769
time		3	0.5938	1.7642	42	0.34	0.7381
time		4	0.2613	1.7642	42	0.15	0.8830
time		5	0.8488	1.7642	42	0.48	0.6330
time		6	2.03E-13	1.7642	42	0.00	1.0000
time		7	0	.	.	.	.
person	1		-21.5343	1.8861	42	-11.42	<.0001
person	2		-12.0286	1.8861	42	-6.38	<.0001
person	3		-1.3829	1.8861	42	-0.73	0.4675
person	4		2.6000	1.8861	42	1.38	0.1753
person	5		-14.1214	1.8861	42	-7.49	<.0001
person	6		-3.4971	1.8861	42	-1.85	0.0707
person	7		-5.8729	1.8861	42	-3.11	0.0033
person	8		0	.	.	.	.

The Mixed Procedure

Type 3 Tests of Fixed Effects

Effect	Num DF	Den DF	F Value	Pr > F
time	6	42	5.28	0.0004
person	7	42	38.03	<.0001

From the output of the above analysis, it is clear that both time and person effects are highly significant.

4.4.3 Residual Analysis for the Guinea-Pig Mycotes Experiment (Example 3.3)

Using the fitted regression model, we obtain the estimates of the responses for the experiment and their attendant residuals (Table 4.7).

In the next section, we use graphical methods to examine some of the model assumptions. Generally, we assumed that the errors are normally independently distributed with zero mean and equal variances. We use different graphical plots of the residuals to test that these assumptions are satisfied.

TABLE 4.7: Fitted Values and Residuals for Example 3.3

Person	Time	y	$\hat{y}$	$(y-\hat{y})$ = residuals
1	1	−94.1	−97.3934	3.2934
1	2	−92.4	−93.6634	1.2634
1	3	−92.09	−89.8696	−2.2204
1	4	−91.42	−90.2021	−1.2179
1	5	−91.06	−89.6146	−1.4454
1	6	−90.38	−90.4634	0.0834
1	7	−90.22	−90.4634	0.2434
2	1	−84.93	−87.8877	2.9577
2	2	−81.59	−84.1577	2.5677
2	3	−74.87	−80.3639	5.4939
2	4	−83.9	−80.6964	−3.2036
2	5	−73.22	−80.1089	6.8889
2	6	−86.59	−80.9577	−5.6323
2	7	−90.03	−80.9577	−9.0723
3	1	−75.99	−77.242	1.252
3	2	−77.01	−73.512	−3.498
3	3	−67.97	−69.7182	1.7482
3	4	−69.95	−70.0507	0.1007
3	5	−70.6	−69.4632	−1.1368
3	6	−69.8	−70.312	0.512
3	7	−69.29	−70.312	1.022
4	1	−80.01	−73.2591	−6.7509
4	2	−71.96	−69.5291	−2.4309
4	3	−62.82	−65.7353	2.9153
4	4	−64.1	−66.0678	1.9678
4	5	−69.01	−65.4803	−3.5297
4	6	−65.28	−66.3291	1.0491
4	7	−59.55	−66.3291	6.7791
5	1	−85.47	−89.9805	4.5105
5	2	−83	−86.2505	3.2505
5	3	−84.58	−82.4567	−2.1233
5	4	−83.96	−82.7892	−1.1708
5	5	−84.16	−82.2017	−1.9583
5	6	−83.46	−83.0505	−0.4095
5	7	−85.15	−83.0505	−2.0995
6	1	−78.99	−79.3562	0.3662
6	2	−73.96	−75.6262	1.6662
6	3	−73.07	−71.8324	−1.2376
6	4	−72.63	−72.1649	−0.4651
6	5	−73.02	−71.5774	−1.4426
6	6	−70.74	−72.4262	1.6862
6	7	−73	−72.4262	−0.5738
7	1	−83.86	−81.732	−2.128
7	2	−77.59	−78.002	0.412
7	3	−75.76	−74.2082	−1.5518
7	4	−75.27	−74.5407	−0.7293
7	5	−74.59	−73.9532	−0.6368
7	6	−73.31	−74.802	1.492
7	7	−71.66	−74.802	3.142

Continued

TABLE 4.7: Continued

Person	Time	y	$\hat{y}$	$(y - \hat{y})$ = residuals
8	1	−79.36	−75.8591	−3.5009
8	2	−75.36	−72.1291	−3.2309
8	3	−71.36	−68.3353	−3.0247
8	4	−63.95	−68.6678	4.7178
8	5	−64.82	−68.0803	3.2603
8	6	−67.71	−68.9291	1.2191
8	7	−68.37	−68.9291	0.5591
Total		−4318.3	−4318.3	0.00

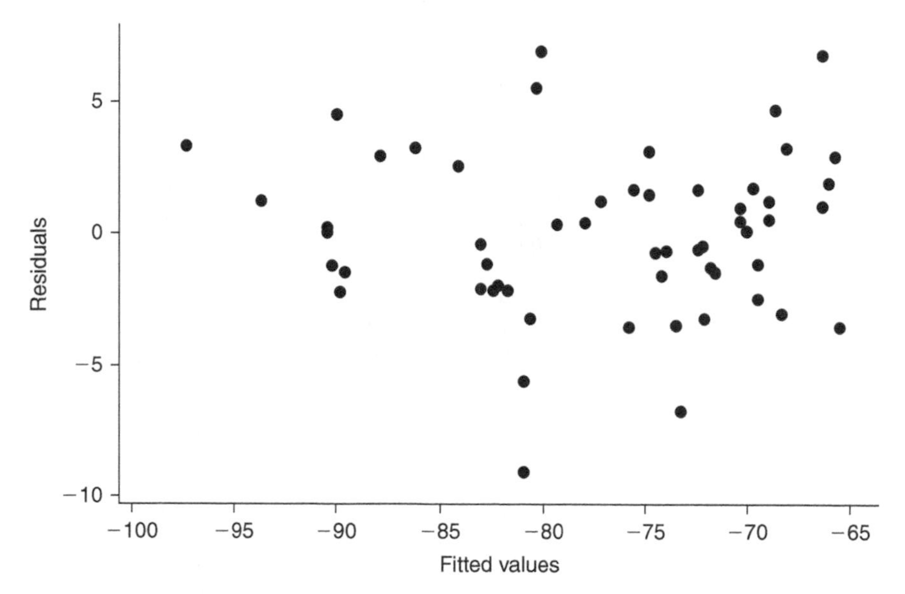

FIGURE 4.1: Scatterplot of residuals against fitted values for the guinea-pig mycotes experiment.

Here, we plot three graphs: a scatterplot of residuals vs fitted values, a histogram of the residuals, and a probability plot of the residuals. A plot of the residuals against fitted values is presented in Figure 4.1.

The scatterplot of the residuals against the fitted values (Figure 4.1) does not indicate any obvious departures from model assumptions. The plot seems patternless, and therefore random. Also, only very few of the residuals (three of them) lie outside the $\pm 2\sqrt{MSE}$ whose value is ± 7.057.

The histogram of the residuals (Figure 4.2) appears reasonably normal. It, therefore, lends credence to the normality assumption for the residuals of the model.

It is not very clear that the assumption of normality is satisfied from the above plot in Figure 4.3, because all the cumulative probability plots for the residuals are not on a straight line. The Anderson–Darling statistic for testing

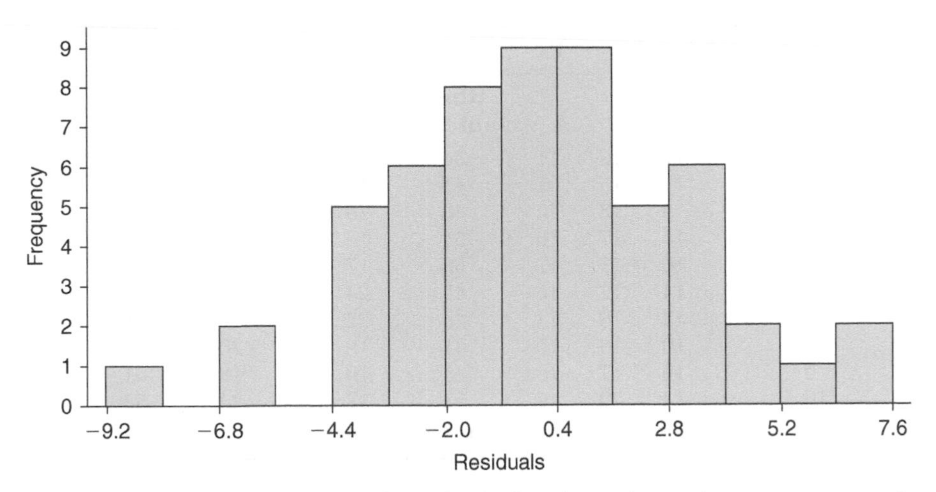

FIGURE 4.2: Histogram of residuals for the guinea-pig mycotes experiment.

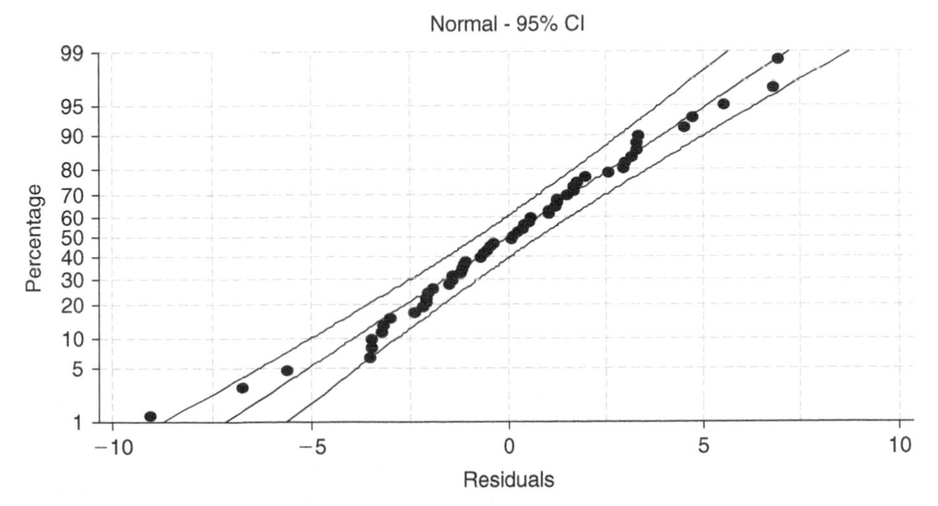

FIGURE 4.3: Probability plot of the residuals for the guinea-pig mycotes experiment.

normality is 0.268, which is small, with a high p-value of 0.671, indicating that we could not reject the hypothesis of normality of the residuals.

4.4.4 Analysis of Responses of Example 3.4 Using PROC MIXED in SAS

We reproduce the data of Example 3.4 here to facilitate the understanding of the regression model, which will be fitted in our SAS program using the PROC MIXED.

	V_0				V_1			
Subjects	**R**	**C**	**A**	**Row subtotal**	**R**	**C**	**A**	**Row subtotal**
1	10	13	11	34	17	16	17	50
2	12	15	4	41	19	20	20	59
3	9	18	13	40	16	22	23	61
4	15	17	19	51	18	24	24	66
5	16	17	17	50	17	16	17	50
6	14	17	16	47	23	16	17	56
7	13	18	14	45	22	19	17	58
8	10	16	15	41	18	22	20	60
9	10	17	18	45	19	24	18	61
10	13	22	16	51	17	17	24	58
Total	122	170	153	445	186	196	197	579

4.4.4.1 Choosing Covariance Structure for Example 3.4

Because the variance structure for the data in Example 3.4 is unknown, the fitted models with the four covariance structures VC, CS, UN, and AR(1) were used to analyze the data to determine which of them is best suited to use in analyzing and interpreting the data. The values obtained for AIC, AICC, and BIC using the ML method [based on the solution of (4.15)] are as follows:

Criterion	**VC**	**CS**	**UN**	**AR(1)**
AIC	294.2	294.1	301.5	293.6
AICC	296.3	296.9	348.8	296.5
BIC	296.3	296.5	309.7	296.1

Based on the above table, because smaller is better, it is clear that we cannot choose the UN covariance structure as it turns out the highest value for all criteria. If we discount AICC, we see that the smallest values were obtained for the structure AR(1), the autoregressive one. There is not much between the AR(1) and VC or CS which for AIC have the values 293.6, 294.2 and 294.1 respectively. For BIC the values are 296.1, 296.3 and 296.5 respectively. So, we will use the AR(1) structure.

A similar table under the REML method [based on the solution of (4.16)], which is the default in the MIXED procedure syntax in SAS is as follows:

Criterion	**VC**	**CS**	**UN**	**AR(1)**
AIC	273.7	273.8	284.3	273.4
AICC	273.7	274.0	313.1	273.6
BIC	274.0	274.4	290.6	274.0

We have come to similar conclusions as we reached by considering the table produced under the ML method.

4.4.4.2 Specifying the Model and Analyzing Responses of Example 3.4

As we stated earlier, the MIXED procedure uses the reference cell coding for dummy variables, so that if X_i and Z_j represent the dummies related to Vitamin (V_0,V_1) and Tests (R = *read*, C = *comp*, and A = *arit*) in the model, then the fitted model is

$$y_{ij} = \mu + \alpha X_1 + \sum_{j=1}^{2} \beta_j Z_j + \sum_{j=1}^{6} \gamma_{1j} X_1 Z_j + \varepsilon \qquad (4.21)$$

where

$$X_i = \begin{cases} 1 & \text{if } y_{ij} \text{ is under vitamin } i \\ 0 & \text{otherwise} \end{cases}$$

for $i = V_0$

so we have only X_1 in the model. Similarly,

$$Z_j = \begin{cases} 1 & \text{if } y_{ij} \text{ is under test } j \\ 0 & \text{otherwise} \end{cases}$$

for $j = arit,\ comp$

In the above model, the terms relating to (1) subjects, (2) first-order interaction between subjects and tests, (3) first-order interaction between subjects and vitamin, and (4) second-order interaction between subject, vitamin, and tests have been omitted because we are not testing for them. However, it is clear (see the output of the program) that the mean square for interaction between subjects and vitamin is used as the estimate of error variance in testing for significance of vitamin. Similarly, the mean square for interaction between subjects and tests is used as the estimate of error variance in testing for significance of tests. Also, the mean square for interaction between subjects, tests, and vitamin is used as the estimate of error variance in testing for significance of the interaction between vitamin and test. We write a program using the MIXED procedure to analyze the data of that experiment in terms of the parameterization of Equation 4.1. The SAS program for the analysis of the responses is given with its output as follows:

SAS PROGRAM

```
options nodate nonumber ps=65 ls=80;
data nutrit1;
      input Person vitamin$ y1 y2 y3 @@;
      y=y1; Test='read'; output;
      y=y2; test='comp'; output;
      y=y3; test='arit'; output;
       drop y1-y3;
      datalines;
```

```
1  v0 10 13 11 2 v0 12 15 14 3 v0 9 18 13 4 v0 15 17 19 5 v0 16 17 17
6  v0 14 17 16 7 v0 13 18 14 8 v0 10 16 15 9 v0 10 17 18 10 v0 13 22 16
1  v1 17 16 17 2 v1 19 20 20 3 v1 16 22 23 4 v1 18 24 24 5 v1 17 16 17
6  v1 23 16 17 7 v1 22 19 17 8 v1 18 22 20 9 v1 19 24 18 10 v1 17 17 24
   ;
 proc print data=nutrit1;
   proc mixed data=nutrit1 method=reml covtest;
      class Person vitamin test;
      model y = vitamin test vitamin*test/s;
     repeated / type=ar(1) subject=Person r;
   run;
```

OUTPUT:

```
                          The Mixed Procedure

                         Model Information

         Data Set                     WORK.NUTRIT1
         Dependent Variable           y
         Covariance Structure         Autoregressive
         Subject Effect               Person
         Estimation Method            REML
         Residual Variance Method     Profile
         Fixed Effects SE Method      Model-Based
         Degrees of Freedom Method    Between-Within

             Class Level Information

          Class       Levels    Values

          Person          10    1 2 3 4 5 6 7 8 9 10
          vitamin          2    v0 v1
          Test             3    arit comp read

                          Dimensions

             Covariance Parameters          2
             Columns in X                  12
             Columns in Z                   0
             Subjects                      10
             Max Obs Per Subject            6

                  Number of Observations

        Number of Observations Read             60
        Number of Observations Used             60
        Number of Observations Not Used          0

                     Iteration History

 Iteration    Evaluations    -2 Res Log Like    Criterion

         0              1       271.65210829
         1              2       269.38722197   0.00000000
```

Convergence criteria met.

Estimated R Matrix for Person 1

Row	Col1	Col2	Col3	Col4	Col5	Col6
1	6.9436	1.5552	0.3483	0.07802	0.01748	0.003914
2	1.5552	6.9436	1.5552	0.3483	0.07802	0.01748
3	0.3483	1.5552	6.9436	1.5552	0.3483	0.07802
4	0.07802	0.3483	1.5552	6.9436	1.5552	0.3483
5	0.01748	0.07802	0.3483	1.5552	6.9436	1.5552
6	0.003914	0.01748	0.07802	0.3483	1.5552	6.9436

The Mixed Procedure

Covariance Parameter Estimates

Cov Parm	Subject	Estimate	Standard Error	Z Value	Pr Z
AR(1)	Person	0.2240	0.1456	1.54	0.1241
Residual		6.9436	1.3942	4.98	<.0001

Fit Statistics

-2 Res Log Likelihood	269.4
AIC (smaller is better)	273.4
AICC (smaller is better)	273.6
BIC (smaller is better)	274.0

Null Model Likelihood Ratio Test

DF	Chi-Square	Pr > ChiSq
1	2.26	0.1323

Solution for Fixed Effects

Effect	vitamin	Test	Estimate	Standard Error	DF	t Value	Pr > |t|
Intercept			18.6000	0.8333	9	22.32	<.0001
vitamin	v0		-6.4000	1.1718	9	-5.46	0.0004
vitamin	v1		0	.	.	.	.
Test		arit	1.1000	1.1485	18	0.96	0.3509
Test		comp	1.0000	1.0381	18	0.96	0.3482
Test		read	0	.	.	.	.
vitamin*Test	v0	arit	2.0000	1.7084	18	1.17	0.2570
vitamin*Test	v0	comp	3.8000	1.4823	18	2.56	0.0195
vitamin*Test	v0	read	0	.	.	.	.
vitamin*Test	v1	arit	0	.	.	.	.
vitamin*Test	v1	comp	0	.	.	.	.
vitamin*Test	v1	read	0	.	.	.	.

Type 3 Tests of Fixed Effects

Effect	Num DF	Den DF	F Value	Pr > F
vitamin	1	9	35.59	0.0002
Test	2	18	8.23	0.0029
vitamin*Test	2	18	3.36	0.0577

4.4.4.3 Discussion of SAS Output for Example 3.4

Consider the tests carried out under the "Type 3 Tests for Fixed Effects." The effect of vitamin whose mean square has χ^2 with 1 degree of freedom is tested by dividing its mean square by the mean square of the interaction between vitamin and subjects. Subjects have 9 degrees of freedom, so that the ratio under the null hypothesis that the effects for levels of vitamin are all zero is distributed as F with 1 and 9 degrees of freedom. This high significance indicates that taking vitamin has significant effect on the performance of the students. Similarly for the test effect, its mean square, which is distributed as χ^2 with 2 degrees of freedom, is tested by dividing it with the mean square of the interaction between subject and tests, which has 18 degrees of freedom. Under the null hypothesis that the effects of the different levels of test is zero, the resulting statistic is distributed as F with 2 and 18 degrees of freedom. The test effect is also highly significant, indicating differential performance in Reading, Comprehension, and Arithmetic. The interaction of test and vitamin, whose mean square is χ^2 distributed with 2 degrees of freedom, is divided by the mean square of the second-order interaction of subjects, tests, and vitamin. This is χ^2 distributed with 18 degrees of freedom, yielding a statistic that is F-distributed with 2 and 18 degrees of freedom with a p-value of 0.0577. If we were to allow Type I error of size 6%, then this effect would be significant. It would then mean that vitamin and tests are jointly contributing to the observed changes in the responses of the study.

From the above output, the estimated regression equations based on the above analysis with MIXED procedure are

$$
\begin{aligned}
&\text{Under } V_0: \\
&\quad \text{arit}: \hat{y} = 12.2 + 1.1 * arit + 2arit * V_0 \\
&\quad \text{comp}: \hat{y} = 12.2 + 1.0 * comp + 3.8 * comp * V_0 \\
&\quad \text{read}: \hat{y} = 12.2 \\
&\text{Under } V_1: \\
&\quad \text{arit}: \hat{y} = 18.6 + 1.1 * arit \\
&\quad \text{comp}: \hat{y} = 18.6 + 1.0 * comp \\
&\quad \text{read}: \hat{y} = 18.6
\end{aligned}
$$

To obtain the estimates for the responses, we replace each of *arit, comp*, and V_0 with 1 if they are specified for a subject. Other values such as *read* and V_1 will be replaced by 0. Using them, we obtain the estimated responses and residuals for each observation as shown in Table 4.8.

4.5 Residual Analysis for the Vitamin Experiment (Example 3.4)

As we did in the Section 4.4, we plot three graphs to find out whether the model assumptions are validated. In Figure 4.4, we present the scatterplot of

TABLE 4.8: Fitted Values and Residuals for Example 3.4

Person	Vitamin	y	Test	$\hat{y}$	$y-\hat{y}$
1	V_0	10	*read*	12.2	−2.2
1	V_0	13	*comp*	17	−4
1	V_0	11	*arit*	15.3	−4.3
2	V_0	12	*read*	12.2	−0.2
2	V_0	15	*comp*	17	−2
2	V_0	14	*arit*	15.3	−1.3
3	V_0	9	*read*	12.2	−3.2
3	V_0	18	*comp*	17	1
3	V_0	13	*arit*	15.3	−2.3
4	V_0	15	*read*	12.2	2.8
4	V_0	17	*comp*	17	0
4	V_0	19	*arit*	15.3	3.7
5	V_0	16	*read*	12.2	3.8
5	V_0	17	*comp*	17	0
5	V_0	17	*arit*	15.3	1.7
6	V_0	14	*read*	12.2	1.8
6	V_0	17	*comp*	17	0
6	V_0	16	*arit*	15.3	0.7
7	V_0	13	*read*	12.2	0.8
7	V_0	18	*comp*	17	1
7	V_0	14	*arit*	15.3	−1.3
8	V_0	10	*read*	12.2	−2.2
8	V_0	16	*comp*	17	−1
8	V_0	15	*arit*	15.3	−0.3
9	V_0	10	*read*	12.2	−2.2
9	V_0	17	*comp*	17	0
9	V_0	18	*arit*	15.3	2.7
10	V_0	13	*read*	12.2	0.8
10	V_0	22	*comp*	17	5
10	V_0	16	*arit*	15.3	0.7
1	V_1	17	*read*	18.6	−1.6
1	V_1	16	*comp*	19.6	−3.6
1	V_1	17	*arit*	19.7	−2.7
2	V_1	19	*read*	18.6	0.4
2	V_1	20	*comp*	19.6	0.4
2	V_1	20	*arit*	19.7	0.3
3	V_1	16	*read*	19.7	−3.7
3	V_1	22	*comp*	19.6	2.4
3	V_1	23	*arit*	18.6	4.4
4	V_1	18	*read*	19.7	−1.7
4	V_1	24	*comp*	19.6	4.4
4	V_1	24	*arit*	18.6	5.4
5	V_1	17	*read*	19.7	−2.7
5	V_1	16	*comp*	19.6	−3.6
5	V_1	17	*arit*	18.6	−1.6
6	V_1	23	*read*	19.7	3.3
6	V_1	16	*comp*	19.6	−3.6
6	V_1	17	*arit*	18.6	−1.6
7	V_1	22	*read*	19.7	2.3
7	V_1	19	*comp*	19.6	−0.6

Continued

TABLE 4.8: Continued

Person	Vitamin	y	Test	$\hat{y}$	$y-\hat{y}$
7	V_1	17	*arit*	18.6	−1.6
8	V_1	18	*read*	19.7	−1.7
8	V_1	22	*comp*	19.6	2.4
8	V_1	20	*arit*	18.6	1.4
9	V_1	19	*read*	19.7	−0.7
9	V_1	24	*comp*	19.6	4.4
9	V_1	18	*arit*	18.6	−0.6
10	V_1	17	*read*	19.7	−2.7
10	V_1	17	*comp*	19.6	−2.6
10	V_1	24	*arit*	18.6	5.4
Total					0

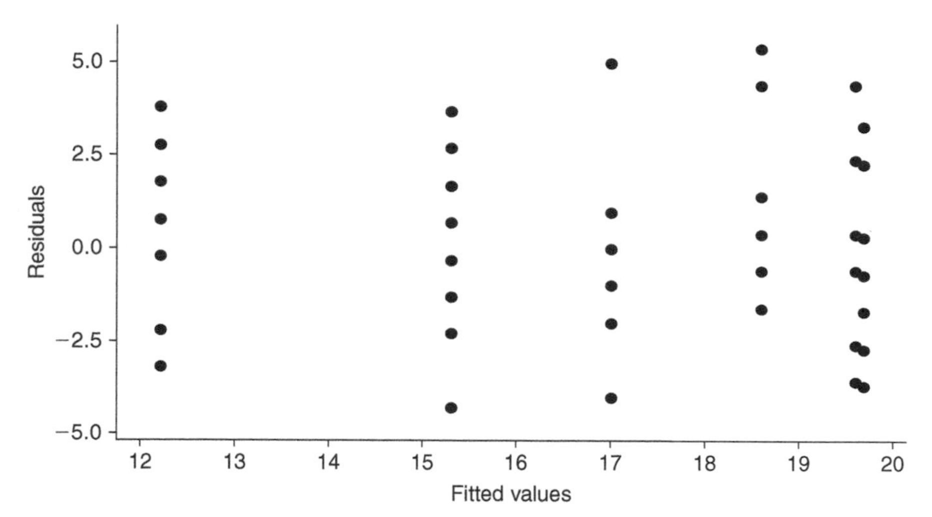

FIGURE 4.4: Scatterplot of residuals against fitted values for the vitamin experiment.

residuals against fitted values, while the histogram is shown in Figure 4.5 and the normal probability plot in Figure 4.6.

No obvious pattern is indicated by the above plot (Figure 4.4), indicating that the randomness assumption is satisfied. The residuals are all within the general rule that it should lie in the range $\pm 2\sqrt{MSE}$, which in this case is ± 5.19. There is no reason from the graph to suspect heteroscedasticity for the errors.

Figure 4.5 can be construed to broadly indicate a pattern that is reasonably close to normality. Although the data is not as close to normality as we were able to observe for the residuals of the analysis for Example 3.7, nevertheless it is normal.

Looking at the normal probability plot of residuals in Figure 4.6, it is not clear that the assumption of normality for residuals is met. However, the

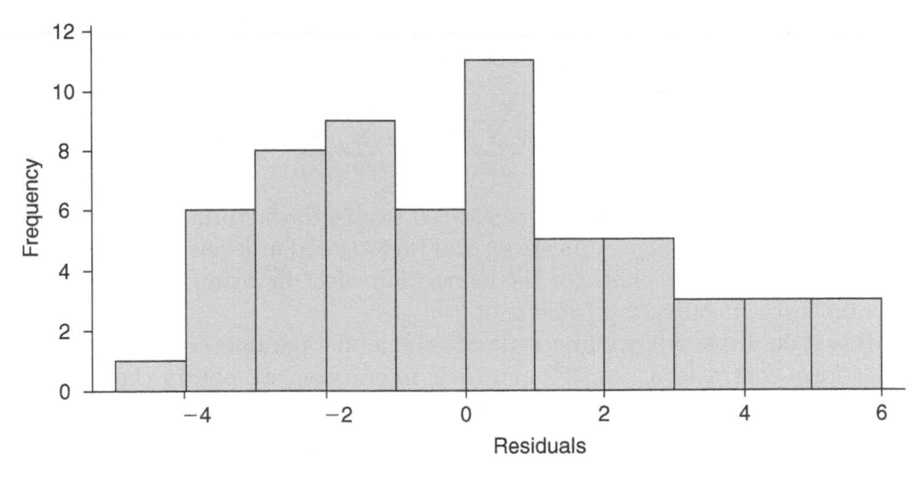

FIGURE 4.5: Histogram of the residuals for the vitamin experiment.

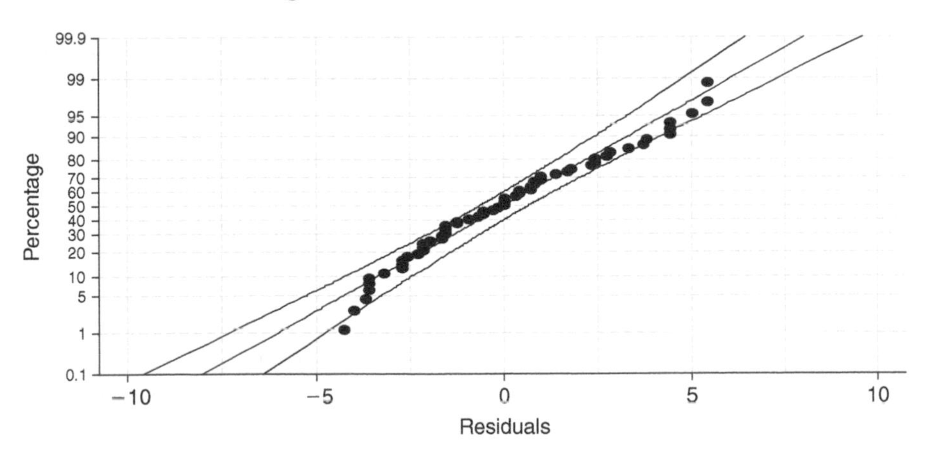

FIGURE 4.6: Normal probability plot of residuals for the vitamin experiment.

Anderson–Darling statistic is 0.603, with a p-value of 0.112 (compare with a typical level of significance of 0.05), which is large, enabling us not to reject the null hypothesis that the residuals are normally distributed.

4.6 Regression Model of the Two-Factor Factorial Design Using Orthogonal Contrasts (Example 3.1)

We can also obtain the regression model for the two-factor factorial design when we use orthogonal contrasts as weighting coefficients. The weighting coefficients may be used to leave out some levels of a variable or separate some levels of a variable for comparison, or to divide the levels of a factor into groups to compare their means. The required model is of the form

$$y_{ij} = \mu + \sum_{i=1}^{a-1} \alpha_i X_i + \sum_{j=1}^{b-1} \beta_j Z_j + \sum_{i=1}^{a-1}\sum_{j=1}^{b-1} \gamma_{ij} X_i Z_j + \varepsilon \tag{4.22}$$

The above model represents a replicated two-factor factorial experiment, in which the effects of the factors are set and investigated at levels a and b, respectively. The model accounts for the interaction effect by using the products of the orthogonal contrasts (Table 4.9).

Based on Table 4.9, we first estimate the model parameters using the fact that for a contrasts X_i and Z_j and their interaction, we obtain the estimate of its coefficient in the regression model as

$$\hat{\alpha}_i = \frac{\sum X_i Y_i}{\sum X_i^2} \quad \hat{\beta}_j = \frac{\sum Z_j Y_i}{\sum Z_j^2} \quad \hat{\gamma}_{ij} = \frac{\sum\sum X_i Z_j Y_i}{\sum (X_i Z_j)^2}$$

Below we present a table containing the sums of squares and cross products stated in the above formulae, which are needed for estimating parameters of the regression model (Table 4.10).

In this and subsequent tables, we obtain the values of some functions in the analysis of the responses of Example 3.5

$$d_1 = x_1 z_1 \quad d_2 = x_1 z_2 \quad d_3 = x_1 z_3 \quad d_4 = x_2 z_1 \quad d_5 = x_2 z_2$$
$$d_6 = x_2 z_3 \quad d_7 = x_3 z_1 \quad d_8 = x_3 z_2 \quad d_9 = x_3 z_3 \quad d_{10} = x_4 z_1$$
$$d_{11} = x_4 z_2 \quad d_{12} = x_4 z_3$$

Using these relations and Table 4.9 above we see that

$$\hat{\mu} = \frac{4039}{60} = 67.31667 \quad \hat{\alpha}_1 = \frac{-809}{240} = -3.3708 \quad \hat{\alpha}_2 = \frac{-353}{144} = -2.4513$$

$$\hat{\alpha}_3 = \frac{22}{72} = 0.3056 \quad \hat{\alpha}_4 = \frac{38}{24} = 1.583 \quad \hat{\beta}_1 = \frac{-291}{180} = -1.6167$$

$$\hat{\beta}_2 = \frac{381}{90} = 4.2333 \quad \hat{\beta}_3 = \frac{7}{30} = 0.2333$$

$$\hat{\gamma}_{11} = \frac{-179}{720} = -0.2486 \quad \hat{\gamma}_{12} = \frac{-41}{360} = -0.1139 \quad \hat{\gamma}_{13} = \frac{-77}{120} = -0.641667$$

$$\hat{\gamma}_{21} = \frac{-379}{432} = -0.8773 \quad \hat{\gamma}_{22} = \frac{-13}{216} = -0.0602 \quad \hat{\gamma}_{23} = \frac{23}{72} = 0.31944$$

$$\hat{\gamma}_{31} = \frac{-70}{216} = -0.32407 \quad \hat{\gamma}_{32} = \frac{8}{108} = 0.0741 \quad \hat{\gamma}_{33} = \frac{-46}{36} = -1.2778$$

$$\hat{\gamma}_{41} = \frac{50}{72} = 0.6944 \quad \hat{\gamma}_{42} = \frac{-10}{36} = -0.2778 \quad \hat{\gamma}_{43} = \frac{-72}{12} = -6.0$$

TABLE 4.9: Orthogonal Contrasts for Responses of Example 3.1

Combinations of levels of factors	x_1	x_2	x_3	x_4	z_1	z_2	z_3	y	d_1	d_2	d_3	d_4	d_5	d_6	d_7	d_8	d_9	d_{10}	d_{11}	d_{12}
	4	0	0	0	3	0	0	43	12	0	0	0	0	0	0	0	0	0	0	0
a₁b₁	4	0	0	0	3	0	0	47	12	0	0	0	0	0	0	0	0	0	0	0
	4	0	0	0	3	0	0	48	12	0	0	0	0	0	0	0	0	0	0	0
	4	0	0	0	–1	2	0	60	–4	8	0	–0	0	0	–0	0	0	–0	0	0
a₁b₂	4	0	0	0	–1	2	0	64	–4	8	0	–0	0	0	–0	0	0	–0	0	0
	4	0	0	0	–1	2	0	68	–4	8	0	–0	0	0	–0	0	0	–0	0	0
	4	0	0	0	–1	–1	1	46	–4	–4	4	–0	–0	0	–0	–0	0	–0	–0	0
a₁b₃	4	0	0	0	–1	–1	1	50	–4	–4	4	–0	–0	0	–0	–0	0	–0	–0	0
	4	0	0	0	–1	–1	1	55	–4	–4	4	–0	–0	0	–0	–0	0	–0	–0	0
	4	0	0	0	–1	–1	–1	59	–4	–4	–4	–0	–0	–0	–0	–0	–0	–0	–0	–0
a₁b₄	4	0	0	0	–1	–1	–1	52	–4	–4	–4	–0	–0	–0	–0	–0	–0	–0	–0	–0
	4	0	0	0	–1	–1	–1	54	–4	–4	–4	–0	–0	–0	–0	–0	–0	–0	–0	–0
	–1	3	0	0	3	0	0	50	–3	–0	–0	9	0	0	0	0	0	0	0	0
a₂b₁	–1	3	0	0	3	0	0	56	–3	–0	–0	9	0	0	0	0	0	0	0	0
	–1	3	0	0	3	0	0	48	–3	–0	–0	9	0	0	0	0	0	0	0	0
	–1	3	0	0	–1	2	0	79	1	–2	–0	–3	6	0	–0	0	0	–0	0	0
a₂b₂	–1	3	0	0	–1	2	0	73	1	–2	–0	–3	6	0	–0	0	0	–0	0	0
	–1	3	0	0	–1	2	0	75	1	–2	–0	–3	6	0	–0	0	0	–0	0	0
	–1	3	0	0	–1	–1	1	65	1	1	–1	–3	–3	3	–0	–0	0	–0	–0	0
a₂b₃	–1	3	0	0	–1	–1	1	68	1	1	–1	–3	–3	3	–0	–0	0	–0	–0	0
	–1	3	0	0	–1	–1	1	62	1	1	–1	–3	–3	3	–0	–0	0	–0	–0	0
	–1	3	0	0	–1	–1	–1	60	1	1	1	–3	–3	–3	–0	–0	–0	–0	–0	–0
a₂b₄	–1	3	0	0	–1	–1	–1	63	1	1	1	–3	–3	–3	–0	–0	–0	–0	–0	–0
	–1	3	0	0	–1	–1	–1	61	1	1	1	–3	–3	–3	–0	–0	–0	–0	–0	–0
	–1	–1	2	0	3	0	0	72	–3	–0	–0	–3	–0	–0	6	0	0	0	0	0
a₃b₁	–1	–1	2	0	3	0	0	64	–3	–0	–0	–3	–0	–0	6	0	0	0	0	0
	–1	–1	2	0	3	0	0	75	–3	–0	–0	–3	–0	–0	6	0	0	0	0	0
	–1	–1	2	0	–1	2	0	80	1	–2	–0	1	–2	–0	–2	4	0	–0	0	0

Continued

TABLE 4.9: Continued

Combinations of levels of factors	x_1	x_2	x_3	x_4	z_1	z_2	z_3	y	d_1	d_2	d_3	d_4	d_5	d_6	d_7	d_8	d_9	d_{10}	d_{11}	d_{12}
a_3b_2	−1	−1	2	0	−1	2	0	88	1	−2	−0	1	−2	−0	−2	4	0	−0	0	0
	−1	−1	2	0	−1	2	0	84	1	−2	−0	1	−2	−0	−2	4	0	−0	0	0
	−1	−1	2	0	−1	−1	1	62	1	1	−1	1	1	−1	−2	−2	2	−0	−0	0
a_3b_3	−1	−1	2	0	−1	−1	1	70	1	1	−1	1	1	−1	−2	−2	2	−0	−0	0
	−1	−1	2	0	−1	−1	1	73	1	1	−1	1	1	−1	−2	−2	2	−0	−0	0
	−1	−1	2	0	−1	−1	−1	70	1	1	1	1	1	1	−2	−2	−2	−0	−0	−0
a_3b_4	−1	−1	2	0	−1	−1	−1	71	1	1	1	1	1	1	−2	−2	−2	−0	−0	−0
	−1	−1	2	0	−1	−1	−1	76	1	1	1	1	1	1	−2	−2	−2	−0	−0	−0
	−1	−1	−1	1	3	0	0	77	−3	−0	−0	−3	−0	−0	−3	−0	−0	3	0	0
a_4b_1	−1	−1	−1	1	3	0	0	73	−3	−0	−0	−3	−0	−0	−3	−0	−0	3	0	0
	−1	−1	−1	1	3	0	0	78	−3	−0	−0	−3	−0	−0	−3	−0	−0	3	0	0
	−1	−1	−1	1	−1	2	0	79	1	−2	−0	1	−2	−0	1	−2	−0	−1	2	0
a_4b_2	−1	−1	−1	1	−1	2	0	87	1	−2	−0	1	−2	−0	1	−2	−0	−1	2	0
	−1	−1	−1	1	−1	2	0	80	1	−2	−0	1	−2	−0	1	−2	−0	−1	2	0
	−1	−1	−1	1	−1	−1	1	61	1	1	−1	1	1	−1	1	1	−1	−1	−1	1
a_4b_3	−1	−1	−1	1	−1	−1	1	67	1	1	−1	1	1	−1	1	1	−1	−1	−1	1
	−1	−1	−1	1	−1	−1	1	69	1	1	−1	1	1	−1	1	1	−1	−1	−1	1
	−1	−1	−1	1	−1	−1	−1	68	1	1	1	1	1	1	1	1	1	−1	−1	−1
a_4b_4	−1	−1	−1	1	−1	−1	−1	75	1	1	1	1	1	1	1	1	1	−1	−1	−1
	−1	−1	−1	1	−1	−1	−1	79	1	1	1	1	1	1	1	1	1	−1	−1	−1
	−1	−1	−1	−1	3	0	0	67	−3	−0	−0	−3	−0	−0	−3	−0	−0	−3	−0	−0
a_5b_1	−1	−1	−1	−1	3	0	0	73	−3	−0	−0	−3	−0	−0	−3	−0	−0	−3	−0	−0
	−1	−1	−1	−1	3	0	0	66	−3	−0	−0	−3	−0	−0	−3	−0	−0	−3	−0	−0
	−1	−1	−1	−1	−1	2	0	84	1	−2	−0	1	−2	−0	1	−2	−0	1	−2	−0
a_5b_2	−1	−1	−1	−1	−1	2	0	80	1	−2	−0	1	−2	−0	1	−2	−0	1	−2	−0
	−1	−1	−1	−1	−1	2	0	80	1	−2	−0	1	−2	−0	1	−2	−0	1	−2	−0
	−1	−1	−1	−1	−1	−1	1	69	1	1	−1	1	1	−1	1	1	−1	1	1	−1
a_5b_3	−1	−1	−1	−1	−1	−1	1	83	1	1	−1	1	1	−1	1	1	−1	1	1	−1
	−1	−1	−1	−1	−1	−1	1	74	1	1	−1	1	1	−1	1	1	−1	1	1	−1
	−1	−1	−1	−1	−1	−1	−1	62	1	1	1	1	1	1	1	1	1	1	1	1
a_5b_4	−1	−1	−1	−1	−1	−1	−1	60	1	1	1	1	1	1	1	1	1	1	1	1
	−1	−1	−1	−1	−1	−1	−1	57	1	1	1	1	1	1	1	1	1	1	1	1
Total	0	0	0	0	0	0	0	4039	0	0	0	0	0	0	0	0	0	0	0	0

TABLE 4.10: Sum of Squares and Cross Products for Orthogonal Contrasts Regression Model for Example 3.1

Sum of squares and cross products	x_1	x_2	x_3	x_4	z_1	z_2	z_3	d_1	d_2	d_3
$\sum X_i$ or $\sum Z_j$	0	0	0	0	0	0	0	0	0	0
$\sum\sum X_iZ_jY_i$ or $\sum X_iY_i$	−809	−353	22	38	−291	381	7	−179	−41	−77
$\sum X_i^2$ or $\sum Z_j^2$	240	144	72	24	180	90	30	720	360	120

	d_4	d_5	d_6	d_7	d_8	d_9	d_{10}	d_{11}	d_{12}
$\sum X_i$ or $\sum Z_j$	0	0	0	0	0	0	0	0	0
$\sum\sum X_iZ_jY_i$ or $\sum X_iY_i$	−379	−13	23	−70	8	−46	50	−10	−72
$\sum X_i^2$ or $\sum Z_j^2$	432	216	72	216	108	36	72	36	12

The above estimates for the parameters lead to the following estimated regression equation, which relates the contrasts with the responses of this experiment

$$\begin{aligned}\hat{y} = {} & 67.31667 - 3.3708x_1 - 2.4513x_2 + 0.3056x_3 + 1.583x_4 - 1.6167z_1 \\ & + 4.2333z_2 + 0.2333z_3 - 0.2486x_1z_1 - 0.1139x_1z_2 - 0.641667x_1z_3 \\ & - 0.8773x_2z_1 - 0.0602x_2z_2 + 0.31944x_2z_3 - 0.32407x_3z_1 + 0.0741x_3z_2 \\ & - 1.2778x_3z_3 + 0.6944x_4z_1 - 0.2778x_4z_2 - 6.0x_4z_3 \qquad (4.23)\end{aligned}$$

With this, we obtain the estimated responses for the experiment and extract the associated residuals that are presented in the Table 4.11. We plot the residuals against fitted values; we obtain a normal probability plot for the residuals.

Next, we present the scatterplot of the residuals against fitted values (Figure 4.7).

There is no systematic pattern to the residuals indicating randomness; each of the residuals is well within the $\pm 2\sqrt{MSE}$, which has a value of ± 8.1976. There are therefore no outliers. The normal probability plot for the residuals is presented in Figure 4.8.

The plot does not seem linear as some of the cumulative probabilities are not on a straight line. The Anderson–Darling statistic for the residuals is 0.484 with a p-value of 0.221, so we could not reject the hypothesis that the residuals are normally distributed. With the fitting of the regression model, we can calculate the sums of squares and perform ANOVA that is similar to the ANOVA under the traditional fixed effects model for a two-factor factorial design, which we discussed in Chapter 3. The sums of squares for ANOVA are obtained by using properties of the contrasts x_1–x_4 and z_1–z_3 and their interactions as shown in Table 4.11a. Thereafter, they are combined to obtain the sums of squares of the main effects and interactions to perform ANOVA. To obtain the sums

TABLE 4.11: Fitted Values and Residuals for Orthogonal Contrasts Regression Model for Example 3.1

x_1	x_2	x_3	x_4	z_1	z_2	z_3	y	d_1	d_2	d_3	d_4	d_5	d_6	d_7	d_8	d_9	d_{10}	d_{11}	d_{12}	$\hat{y}$	$(y-\hat{y})$	$(y_i-\hat{y}_i)^2$
4	0	0	0	3	0	0	43	12	0	0	0	0	0	0	0	0	0	0	0	46	−3	9
4	0	0	0	3	0	0	47	12	0	0	0	0	0	0	0	0	0	0	0	46	1	1
4	0	0	0	3	0	0	48	12	0	0	0	0	0	0	0	0	0	0	0	46	2	4
4	0	0	0	−1	2	0	60	−4	8	0	−0	0	0	−0	0	0	−0	0	0	64	−4	16
4	0	0	0	−1	2	0	64	−4	8	0	−0	0	0	−0	0	0	−0	0	0	64	0	0
4	0	0	0	−1	2	0	68	−4	8	0	−0	0	0	−0	0	0	−0	0	0	64	4	16
4	0	0	0	−1	−1	1	46	−4	−4	4	−0	−0	0	−0	−0	0	−0	−0	0	50.33	−4.33	18.7489
4	0	0	0	−1	−1	1	50	−4	−4	4	−0	−0	0	−0	−0	0	−0	−0	0	50.33	−0.33	0.1089
4	0	0	0	−1	−1	1	55	−4	−4	4	−0	−0	0	−0	−0	0	−0	−0	0	50.33	4.67	21.8089
4	0	0	0	−1	−1	−1	59	−4	−4	−4	−0	−0	−0	−0	−0	−0	−0	−0	−0	55	4	16
4	0	0	0	−1	−1	−1	52	−4	−4	−4	−0	−0	−0	−0	−0	−0	−0	−0	−0	55	−3	9
4	0	0	0	−1	−1	−1	54	−4	−4	−4	−0	−0	−0	−0	−0	−0	−0	−0	−0	55	−1	1
−1	3	0	0	3	0	0	50	−3	−0	−0	9	0	0	0	0	0	0	0	0	51.33	−1.33	1.7689
−1	3	0	0	3	0	0	56	−3	−0	−0	9	0	0	0	0	0	0	0	0	51.33	4.67	21.8089
−1	3	0	0	3	0	0	48	−3	−0	−0	9	0	0	0	0	0	0	0	0	51.33	−3.33	11.0889
−1	3	0	0	−1	2	0	79	1	−2	−0	−3	6	0	−0	0	0	−0	0	0	75.67	3.33	11.0889
−1	3	0	0	−1	2	0	73	1	−2	−0	−3	6	0	−0	0	0	−0	0	0	75.67	−2.67	7.1289
−1	3	0	0	−1	2	0	75	1	−2	−0	−3	6	0	−0	0	0	−0	0	0	75.67	−0.67	0.4489
−1	3	0	0	−1	−1	1	65	1	1	−1	−3	−3	3	−0	−0	0	−0	−0	0	65	0	0
−1	3	0	0	−1	−1	1	68	1	1	−1	−3	−3	3	−0	−0	0	−0	−0	0	65	3	9
−1	3	0	0	−1	−1	1	62	1	1	−1	−3	−3	3	−0	−0	0	−0	−0	0	65	−3	9
−1	3	0	0	−1	−1	−1	60	1	1	1	−3	−3	−3	−0	−0	−0	−0	−0	−0	61.33	−1.33	1.7689
−1	3	0	0	−1	−1	−1	63	1	1	1	−3	−3	−3	−0	−0	−0	−0	−0	−0	61.33	1.67	2.7889
−1	3	0	0	−1	−1	−1	61	1	1	1	−3	−3	−3	−0	−0	−0	−0	−0	−0	61.33	−0.33	0.1089
−1	−1	2	0	3	0	0	72	−3	−0	−0	−3	−0	−0	6	0	0	0	0	0	70.33	1.67	2.7889
−1	−1	2	0	3	0	0	64	−3	−0	−0	−3	−0	−0	6	0	0	0	0	0	70.33	−6.33	40.0689
−1	−1	2	0	3	0	0	75	−3	−0	−0	−3	−0	−0	6	0	0	0	0	0	70.33	4.67	21.8089
−1	−1	2	0	−1	2	0	80	1	−2	−0	1	−2	−0	−2	4	0	−0	0	0	84	−4	16
−1	−1	2	0	−1	2	0	88	1	−2	−0	1	−2	−0	−2	4	0	−0	0	0	84	4	16
−1	−1	2	0	−1	2	0	84	1	−2	−0	1	−2	−0	−2	4	0	−0	0	0	84	0	0

−1	−1	2	0	−1	−1	1	62	1	1	−1	1	1	−1	−2	−2	2	−0	−0	0	68.33	−6.33	40.0689
−1	−1	2	0	−1	−1	1	70	1	1	−1	1	1	−1	−2	−2	2	−0	−0	0	68.33	1.67	2.7889
−1	−1	2	0	−1	−1	1	73	1	1	−1	1	1	−1	−2	−2	2	−0	−0	0	68.33	4.67	21.8089
−1	−1	2	0	−1	−1	−1	70	1	1	1	1	1	1	−2	−2	−2	−0	−0	−0	72.33	−2.33	5.4289
−1	−1	2	0	−1	−1	−1	71	1	1	1	1	1	1	−2	−2	−2	−0	−0	−0	72.33	−1.33	1.7689
−1	−1	2	0	−1	−1	−1	76	1	1	1	1	1	1	−2	−2	−2	−0	−0	−0	72.33	3.67	13.4689
−1	−1	−1	1	3	0	0	77	−3	−0	−0	−3	−0	−0	−3	−0	−0	3	0	0	76	1	1
−1	−1	−1	1	3	0	0	73	−3	−0	−0	−3	−0	−0	−3	−0	−0	3	0	0	76	−3	9
−1	−1	−1	1	3	0	0	78	−3	−0	−0	−3	−0	−0	−3	−0	−0	3	0	0	76	2	4
−1	−1	−1	1	−1	2	0	79	1	−2	−0	1	−2	−0	1	−2	−0	−1	2	0	82	−3	9
−1	−1	−1	1	−1	2	0	87	1	−2	−0	1	−2	−0	1	−2	−0	−1	2	0	82	5	25
−1	−1	−1	1	−1	2	0	80	1	−2	−0	1	−2	−0	1	−2	−0	−1	2	0	82	−2	4
−1	−1	−1	1	−1	−1	1	61	1	1	−1	1	1	−1	1	1	−1	−1	−1	1	65.67	−4.67	21.8089
−1	−1	−1	1	−1	−1	1	67	1	1	−1	1	1	−1	1	1	−1	−1	−1	1	65.67	1.33	1.7689
−1	−1	−1	1	−1	−1	1	69	1	1	−1	1	1	−1	1	1	−1	−1	−1	1	65.67	3.33	11.0889
−1	−1	−1	1	−1	−1	−1	68	1	1	1	1	1	1	1	1	1	−1	−1	−1	74	−6	36
−1	−1	−1	1	−1	−1	−1	75	1	1	1	1	1	1	1	1	1	−1	−1	−1	74	1	1
−1	−1	−1	1	−1	−1	−1	79	1	1	1	1	1	1	1	1	1	−1	−1	−1	74	5	25
−1	−1	−1	−1	3	0	0	67	−3	−0	−0	−3	−0	−0	−3	−0	−0	−3	−0	−0	68.67	−1.67	2.7889
−1	−1	−1	−1	3	0	0	73	−3	−0	−0	−3	−0	−0	−3	−0	−0	−3	−0	−0	68.67	4.33	18.7489
−1	−1	−1	−1	3	0	0	66	−3	−0	−0	−3	−0	−0	−3	−0	−0	−3	−0	−0	68.67	−2.67	7.1289
−1	−1	−1	−1	−1	2	0	84	1	−2	−0	1	−2	−0	1	−2	−0	1	−2	−0	81.33	2.67	7.1289
−1	−1	−1	−1	−1	2	0	80	1	−2	−0	1	−2	−0	1	−2	−0	1	−2	−0	81.33	−1.33	1.7689
−1	−1	−1	−1	−1	2	0	80	1	−2	−0	1	−2	−0	1	−2	−0	1	−2	−0	81.33	−1.33	1.7689
−1	−1	−1	−1	−1	−1	1	69	1	1	−1	1	1	−1	1	1	−1	1	1	−1	75.33	−6.33	40.0689
−1	−1	−1	−1	−1	−1	1	83	1	1	−1	1	1	−1	1	1	−1	1	1	−1	75.33	7.67	58.8289
−1	−1	−1	−1	−1	−1	1	74	1	1	−1	1	1	−1	1	1	−1	1	1	−1	75.33	−1.33	1.7689
−1	−1	−1	−1	−1	−1	−1	62	1	1	1	1	1	1	1	1	1	1	1	1	59.66	2.34	5.4756
−1	−1	−1	−1	−1	−1	−1	60	1	1	1	1	1	1	1	1	1	1	1	1	59.66	0.34	0.1156
−1	−1	−1	−1	−1	−1	−1	57	1	1	1	1	1	1	1	1	1	1	1	1	59.66	−2.66	7.0756
Total							4039													4039	0	672.0005

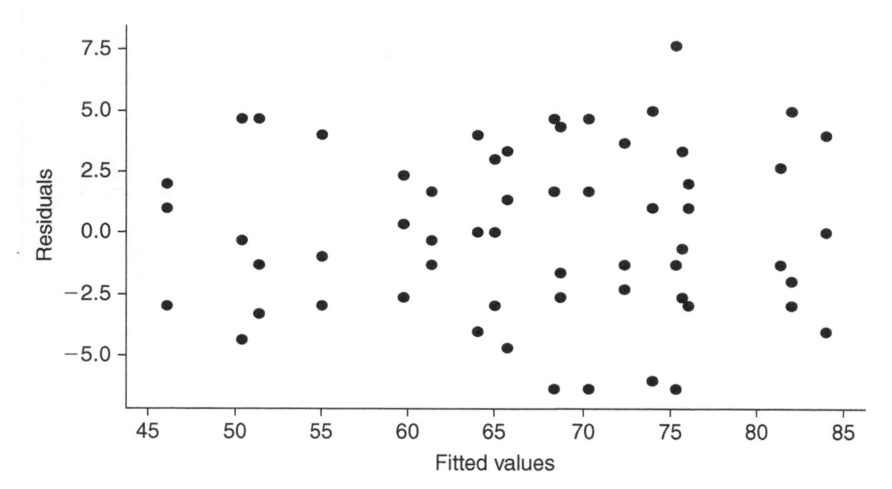

FIGURE 4.7: Scatterplot of residuals against fitted values for the Akwette-cloth experiment.

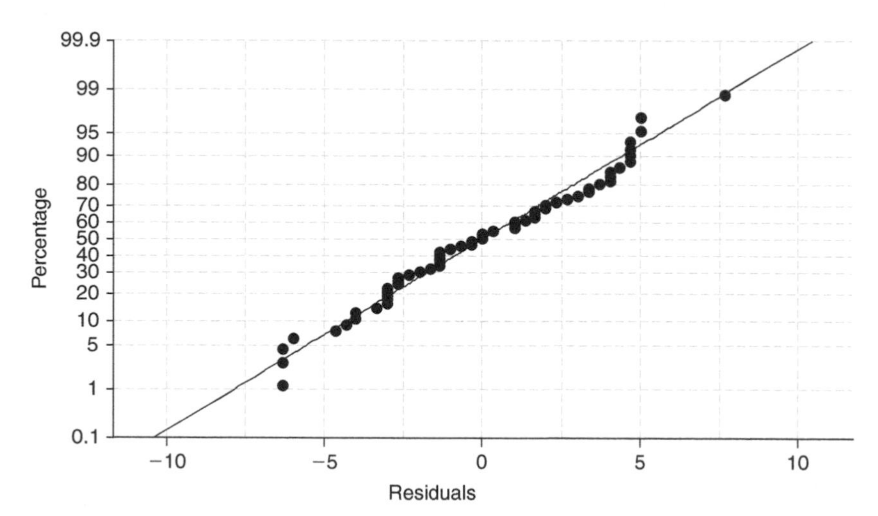

FIGURE 4.8: Normal probability plot of the residuals for the Akwette-cloth experiment.

of squares for these contrasts and their interactions, we use the following relations:

$$SS_{X_i} = \frac{(\sum X_i Y_i)^2}{\sum X_i^2}$$

$$SS_{Z_j} = \frac{(\sum Z_j Y_i)^2}{\sum Z_j^2}$$

$$SS_{X_i Z_j} = \frac{(\sum \sum X_i Z_j Y_i)^2}{\sum (X_i Z_j)^2}$$

Thus we calculate the sum of squares in Table 4.11a below:

TABLE 4.11a: Sums of Squares for Orthogonal Contrasts Regression Model for Example 3.1

Contrast	Sum of squares	Contrast	Sum of squares
x_1	$\frac{(-809)^2}{240} = 2727.0004$	d_4	$\frac{(-379)^2}{432} = 332.5032$
x_2	$\frac{(-353)^2}{144} = 865.34028$	d_5	$\frac{(-13)^2}{216} = 0.7824$
x_3	$\frac{22^2}{72} = 6.7222$	d_6	$\frac{23^2}{72} = 7.3472$
x_4	$\frac{38^2}{24} = 60.16667$	d_7	$\frac{(-70)^2}{216} = 22.6852$
z_1	$\frac{(-291)^2}{180} = 470.45$	d_8	$\frac{8^2}{108} = 0.5926$
z_2	$\frac{381^2}{90} = 1612.90$	d_9	$\frac{(-46)^2}{36} = 58.7778$
z_3	$\frac{7^2}{30} = 1.6333$	d_{10}	$\frac{50^2}{72} = 34.7222$
d_1	$\frac{(-179)^2}{720} = 44.5014$	d_{11}	$\frac{(-10)^2}{36} = 2.7778$
d_2	$\frac{(-41)^2}{360} = 4.6694$	d_{12}	$\frac{(-72)^2}{12} = 432$
d_3	$\frac{(-77)^2}{120} = 49.408$	$SS_{\text{total}} = \sum Y^2 - \frac{(\sum Y)^2}{N}$	$27929 - \frac{4039^2}{60} = 7406.9833$

From Table 4.11a we see that (Table 4.12):

$$
\begin{aligned}
SS_{\text{machinists}(\text{M}_1)} &= SS_{X_1} + SS_{X_2} + SS_{X_3} + SS_{X_4} \\
&= 2727.0004 + 865.34028 + 6.7222 + 60.16667 \\
&= 3659.22957 \\
SS_{\text{machines}(\text{M}_2)} &= SS_{Z_1} + SS_{Z_2} + SS_{Z_3} \\
&= 470.45 + 1612.90 + 1.6333 \\
&= 2084.9833 \\
SS_{\text{machinists}|\text{machines}(\text{M}_1\text{M}_2)} &= SS_{d_1} + SS_{d_2} + \cdots + SS_{d_{12}} \\
&= 44.5014 + 4.6694 + 49.408 + 332.5032 + 0.7824 \\
&\quad + 7.3472 + 22.6852 + 0.5926 + 58.7778 \\
&\quad + 34.7222 + 2.7778 + 432 \\
&= 990.7671
\end{aligned}
$$

$$SS_{\text{E}} = SS_{\text{total}} - SS_{\text{machinists}(\text{M}_1)} - SS_{\text{machines}(\text{M}_2)} - SS_{\text{machinists}|\text{machines}(\text{M}_1\text{M}_2)}$$
$$= 7406.9833 - 3659.22957 - 2084.9833 - 990.7671$$
$$= 672.00333$$

TABLE 4.12: ANOVA for Example 3.1 Based on the Above Orthogonal Contrast Regression Approach

Source	**Sum of squares**	**Degrees of freedom**	**Mean square**	***F*-ratio**
Machinists	3659.22957	4	914.8074	54.4528
Machines	2084.9833	3	694.9944	41.3687
Machinists*Machines	990.7672	12	82.5639	4.9145
Error	672.00323	40	16.8	
Total	7406.9833	59		

The above analysis confirms the results we obtained for the traditional analysis we carried out for the data in Example 3.1.

4.7 Use of PROC Mixed in SAS to Estimate Variance Components When Levels of Factors Are Random

In Chapter 3, we attempted to estimate components of variance (VC) when factors A and B were random in the factorial design of Example 3.2. We found that one of the components was negative, which should not be because variances are by definition nonnegative. There is some flaw in the method of estimation we adopted because it frequently leads to negative estimates for VC. It was suggested (Eliason, 1993) that use be made of method of ML to obtain the estimates for the VC to get better results. ML estimation of VC requires the use of powerful statistical software. In Sections 4.3 through 4.5, we discussed one tool designed for such estimation—the MIXED procedure in SAS software. We had discussed its use in the analysis of data from RMDs.

Here, we discuss how to use the MIXED procedure in SAS software to carry out the ML estimation of VC. The method we discuss would be similar to those suggested by Littell et al. (2002). In the SAS environment, the procedures that are suitable for ML estimation are VARCOMP and MIXED.

The MIXED procedure produces three types of estimates for the VC in accordance with its implementation which we described in Section 4.3. These are **REML**, **ML**, and **MIVQUE(0)**, relating to restricted maximum likelihood, maximum likelihood, and method of moments, respectively. In a typical full two-factor factorial experiment with random effects, we would state the model and the associated assumptions as follows:

$$\begin{gathered} y_{ijk} = \mu + \alpha_i + \tau_j + (\alpha\tau)_{ij} + \varepsilon_{ijk} \\ i = 1, 2, \ldots, a \;\; j = 1, 2, \ldots, b \;\; k = 1, 2, \ldots, n \end{gathered} \tag{4.24}$$

where α_i, τ_j, $(\alpha\tau)_{\text{ij}}$, and ε_{ijk} are all random variables. Specifically, we assume that $\alpha_i \sim \text{NID}(0, \sigma^2_\alpha)$; $(\alpha\tau)_{ij} \sim \text{NID}(0, \sigma^2_{\alpha\tau})$; $\tau_j \sim \text{NID}(0, \sigma^2_\tau)$; $\varepsilon_{ijk} \sim \text{NID}(0, \sigma^2)$. This means that the variance of any observation, y_{ijk} is

$$\text{Var}(y_{ijk}) = \sigma^2_\alpha + \sigma^2_\tau + \sigma^2_{\alpha\tau} + \sigma^2$$

where $\sigma^2_\alpha, \sigma^2_\tau, \sigma^2_{\alpha\tau}$, and σ^2 are called VC.

In the MIXED procedure, REML and ML estimates of components of variance are based on the assumption of normality stated in the model described in Equation 4.24. To describe REML and ML, we earlier presented the natural log of the ML and REML in Equations 4.15 and 4.16, which were implemented in the MIXED procedure of SAS, under which we planned to estimate the components of variance described in Equation 4.24. The MIXED procedure is capable of dealing with other more complicated estimation of VC. It is very powerful for that purpose and other analyses, which come under the mixed effects model.

The MIXED procedure also allows us to carry out Z-tests (based on Wald's Z, where each estimate is divided by its standard error to obtain Z) of the following sets of hypotheses connected with its estimation of the VC:

$$\begin{aligned} H_0 &: \sigma^2_\alpha = 0 \quad \text{vs} \quad H_1 : \sigma^2_\alpha > 0 \\ H_0 &: \sigma^2_\tau = 0 \quad \text{vs} \quad H_1 : \sigma^2_\tau > 0 \\ H_0 &: \sigma^2_{\alpha\tau} = 0 \quad \text{vs} \quad H_1 : \sigma^2_{\alpha\tau} > 0 \end{aligned} \tag{4.25}$$

The use of Z-scores to test the above hypotheses is valid when the sampling distributions of each of these estimators $\hat{\sigma}^2_\alpha, \hat{\sigma}^2_\tau$, and $\hat{\sigma}^2_{\alpha\tau}$ can be approximated by the normal distribution. This approximation is inappropriate when the number of levels of the random factors is small; however, if the number of levels of the random factor is large, we can use the Z-test for all the hypotheses in Equation 4.25. For this reason, we may not be able to use the MIXED procedure to test the hypothesis in the context of Example 3.2 where both A and B are random, and each has only three levels, as it may be inappropriate. Here we present another example in which the levels of the factor may be adequate as it has 24 levels for factor A and three levels for factor B.

Example 4.1

In the following experiment, samples of times it took for an item to be manufactured in a sample of 24 different operating days out of many were recorded, using a sample of three methods out of many. Here, days and methods are considered random. Perform ANOVA and obtain estimates for components of variance for the responses of this experiment.

	Methods (B)		
Days (A)	**I**	**II**	**III**
1	20,25,28	56,66,60	28,30,22
2	26,20,27	50,55,59	38,42,41
3	30,38,28	74,50,62	25,20,22
4	30,29,33	40,39,41	22,25,29
5	38,18,21	56,52,42	24,24,20
6	30,36,40	45,50,55	28,20,28
7	40,30,42	59,30,55	39,34,38
8	24,29,32	50,54,49	25,22,22
9	32,38,30	66,61,68	25,23,25
10	38,34,34	64,69,65	25,29,24
11	30,30,30	74,65,70	33,38,40
12	32,37,35	41,37,42	21,25,28
13	40,30,32	69,67,69	26,26,22
14	31,38,43	68,65,68	20,22,26
15	31,34,34	61,65,68	26,28,23
16	35,30,34	66,67,65	21,23,24
17	24,28,29	56,62,63	30,30,25
18	29,24,29	50,57,54	38,42,43
19	32,39,30	64,50,65	27,26,32
20	30,29,35	44,37,40	22,27,33
21	38,18,21	52,54,44	24,24,20
22	31,34,41	45,53,53	29,21,29
23	42,32,44	59,30,57	39,34,38
24	24,24,33	54,56,50	27,25,27

□

4.7.1 SAS Program for Analysis of Data of Example 4.1

In using the MIXED procedure to estimate the components of variance in the data for the design of Example 4.1, every effect in Equation 4.24 is random, while only the intercept μ is fixed. Thus, the model is mixed. Therefore, in the model statement with the MIXED procedure syntax, y (the response variable) is equated to nothing, while the rest of the effects are declared in the subsequent random statement.

In our program, we first carry out the traditional analysis of variance for random effects using the GLM procedure in SAS. The outcome of the analysis is more properly displayed in the Type III ANOVA. Thereafter, we invoke the MIXED procedure for the estimation of the VC and tests of hypotheses of Equation 4.25 using Wald's Z.

SAS PROGRAM

```
data newrand1;
input a b y@@;
cards;
```

```
1 1 20 1 1 25 1 1 28 1 2 56 1 2 66 1 2 60 1 3 28 1 3 30 1 3 22
2 1 26 2 1 20 2 1 27 2 2 50 2 2 55 2 2 59 2 3 38 2 3 42 2 3 41
3 1 30 3 1 38 3 1 28 4 1 30 4 1 29 4 1 33 3 2 74 3 2 50 3 2 62 4 2 40 4 2 39
4 2 41 3 3 25 3 3 20 3 3 22 4 3 22 4 3 25 4 3 29 5 1 38 5 1 18 5 1 21 6 1 30
6 1 36 6 1 40 5 2 56 5 2 52 5 2 42 6 2 45 6 2 50 6 2 55 5 3 24 5 3 24 5 3 20
6 3 28 6 3 20 6 3 28 7 1 40 7 1 30 7 1 42 7 2 59 7 2 30 7 2 55 7 3 39 7 3 34
7 3 38 8 1 24 8 1 29 8 1 32 8 2 50 8 2 54 8 2 49 8 3 25 8 3 22 8 3 22 9 1 32
9 1 38 9 1 30 9 2 66 9 2 61 9 2 68 9 3 25 9 3 23 9 3 25 10 1 38 10 1 34 10 1
34 10 2 64 10 2 69 10 2 65 10 3 25 10 3 29 10 3 24 11 1 30 11 1 30 11
1 30 12 1 32 12 1 37 12 1 35 11 2 74 11 2 65 11 2 70 12 2 41 12 2 37 12 2 42
11 3 33 11 3 38 11 3 40 12 3 21 12 3 25 12 3 28 13 1 40 13 1 30 13 1 32 14 1
31 14 1 38 14 1 43 13 2 69 13 2 67 13 2 69 14 2 68 14 2 65 14 2 68 13 3 26
13 3 26 13 3 22 14 3 20 14 3 22 14 3 26 15 1 32 15 1 34 15 1 34 15 2 61 15 2
65 15 2 68 15 3 26 15 3 28 15 3 23 16 1 35 16 1 30 16 1 34 16 2 66 16 2 67 16
2 65 16 3 21 16 3 23 16 3 24 17 1 24 17 1 28 17 1 29 17 2 56 17 2 62 17 2 63
17 3 30 17 3 30 17 3 25 18 1 29 18 1 24 18 1 29 18 2 50 18 2 57 18 2 54 18 3
38 18 3 42 18 3 43 19 1 32 19 1 39 19 1 30 20 1 30 20 1 29 20 1 35 19 2 64 19
2 50 19 2 65 20 2 44 20 2 37 20 2 40 19 3 27 19 3 26 19 3 32 20 3 22 20 3 27
20 3 33 21 1 38 21 1 18 21 1 21 22 1 31 22 1 34 22 1 41 21 2 52 21 2 54 21 2
44 22 2 45 22 2 53 22 2 53 21 3 24 21 3 24 21 3 20 22 3 29 22 3 21 22 3 29
23 1 42 23 1 32 23 1 44 23 2 59 23 2 30 23 2 57 23 3 39 23 3 34 23 3 38 24 1
24 24 1 29 24 1 33 24 2 54 24 2 56 24 2 50 24 3 27 24 3 25 24 3 27
;
proc print data=newrand1;
PROC GLM DATA=NEWRAND1;
 class A B;
model y=A B A*B;
test h=a e=a*b;
test h=b e=a*b;
proc mixed data=newrand1 METHOD=REML COVTEST;
CLASS A B;
MODEL Y=;
RANDOM A B A*B; RUN;
```

SAS OUTPUT

```
                              The GLM Procedure

                           Class Level Information

  Class   Levels   Values
  a           24   1 2 3 4 5 6 7 8 9 10 11 12 13 14 15 16 17 18 19 20 21 22 23 24
  b            3   1 2 3

                  Number of Observations Read       216
                   Number of Observations Used      216

                          The GLM Procedure

Dependent Variable: y
                                            Sum of
      Source                    DF         Squares     Mean Square    F Value    Pr > F

      Model                     71     42828.88426       603.22372      23.37    <.0001

      Error                    144      3716.66667        25.81019

      Corrected Total          215     46545.55093
```

R-Square	Coeff Var	Root MSE	y Mean
0.920150	13.23237	5.080373	38.39352

Source	DF	Type I SS	Mean Square	F Value	Pr > F
a	23	3017.10648	131.17854	5.08	<.0001
b	2	33245.84259	16622.92130	644.05	<.0001
a*b	46	6565.93519	142.73772	5.53	<.0001

Source	DF	Type III SS	Mean Square	F Value	Pr > F
a	23	3017.10648	131.17854	5.08	<.0001
b	2	33245.84259	16622.92130	644.05	<.0001
a*b	46	6565.93519	142.73772	5.53	<.0001

Tests of Hypotheses Using the Type III MS for a*b as an Error Term

Source	DF	Type III SS	Mean Square	F Value	Pr > F
a	23	3017.10648	131.17854	0.92	0.5756
b	2	33245.84259	16622.92130	116.46	<.0001

The Mixed Procedure

Model Information

Data Set	WORK.NEWRAND1
Dependent Variable	y
Covariance Structure	Variance Components
Estimation Method	REML
Residual Variance Method	Profile
Fixed Effects SE Method	Model-Based
Degrees of Freedom Method	Containment

Class Level Information

Class	Levels	Values
a	24	1 2 3 4 5 6 7 8 9 10 11 12 13 14 15 16 17 18 19 20 21 22 23 24
b	3	1 2 3

Dimensions

Covariance Parameters	4
Columns in X	1
Columns in Z	99
Subjects	1
Max Obs Per Subject	216

Number of Observations

Number of Observations Read	216
Number of Observations Used	216
Number of Observations Not Used	0

```
                         Iteration History

Iteration    Evaluations    -2 Res Log Like       Criterion

        0              1      1771.69181255
        1              2      1443.48815100      0.00000008
        2              1      1443.48810992      0.00000000

                      Convergence criteria met.

                        The Mixed Procedure

                   Covariance Parameter Estimates

                               Standard        Z
Cov Parm      Estimate            Error    Value        Pr Z

a                    0                .        .           .
b               228.95           230.87     0.99      0.1607
a*b            37.6914           7.9467     4.74      <.0001
Residual       25.8102           3.0418     8.49      <.0001
                          Fit Statistics

              -2 Res Log Likelihood           1443.5
              AIC (smaller is better)         1449.5
              AICC (smaller is better)        1449.6
              BIC (smaller is better)         1453.0
```

Using the output of the GLM procedure for Type III SAS, we obtain the estimate of the components of variance by the method described in Equations 3.12 and 3.13 (applied to Example 3.2). The estimates are

$$\hat{\sigma}^2 = 25.81019$$

$$\hat{\sigma}_\alpha^2 = \frac{131.17854 - 142.73772}{3 \times 3} = -1.28435$$

$$\hat{\sigma}_\beta^2 = \frac{16622.9213 - 142.73772}{3 \times 24} = 228.8914$$

$$\hat{\sigma}_{\alpha\beta}^2 = \frac{142.73772 - 25.81019}{3} = 38.97584$$

Manual analysis shows that $\hat{\sigma}_\alpha^2$ is negative. However, by using the REML method in MIXED procedure in SAS, we see that SAS constrains this estimate to be zero. Note in the table below that the invocation of the restricted maximum likelihood (REML) leads to similar estimates as those obtained by using the random statements in the GLM procedure

Proc GLM with random statement or manual analysis for random effects	**Proc Mixed with REML**
$\hat{\sigma}^2 = 25.81019$	$\hat{\sigma}^2 = 25.81019$
$\hat{\sigma}_\alpha^2 = -1.28435$	$\hat{\sigma}_\alpha^2 = 0$
$\hat{\sigma}_\beta^2 = 228.8914$	$\hat{\sigma}_\beta^2 = 228.95$
$\hat{\sigma}_{\alpha\beta}^2 = 38.97584$	$\hat{\sigma}_{\alpha\beta}^2 = 37.6914$

4.8 Exercises

Exercise 4.1
Define the dummy variables by reference cell method, using cell (4,3) as the reference cell, and obtain a regression model for the Akwette cloth design of Example 3.1. Estimate the model parameters, obtain the estimated responses, extract the residuals for the fitted model by manual methods, and test for significance of the fitted model. □

Exercise 4.2
Write a SAS program using the model of Exercise 4.1 to carry out the analysis of the data described in Exercise 4.1. □

Exercise 4.3
Using cell reference method to define the dummy variables, with cell (3,3) as the reference cell obtain the regression model for the data of Exercise 3.8. Obtain the model parameters and test for significance of the model. Using the residuals of this model, check that the normality assumption is satisfied. □

Exercise 4.4
Using effects coding method to define the dummy variables, obtain the regression model for the data of Exercise 3.8. Estimate the model parameters and test for significance of the model. □

Exercise 4.5
Write a SAS program to analyze the data under the model of Exercise 4.3. □

Exercise 4.6
Repeat Exercise 4.5 for the model defined under Exercise 4.4. □

Exercise 4.7
Fit a regression model to the data of Exercise 3.8 using orthogonal contrasts. Perform ANOVA to test for significance of the model. Obtain the fitted values and residuals for this design. Obtain a plot of residuals against fitted values and comment on its pattern. Obtain a normal probability plot of the residuals for this model. □

Exercise 4.8
Write a SAS program to analyze the repeated measures data of Exercise 3.1 using the MIXED procedure. □

Exercise 4.9
Write a SAS program to analyze the repeated measures data of Exercise 3.5 using the MIXED procedure. □

Exercise 4.10
Write a SAS program to analyze the repeated measures data of Exercise 3.3 using the MIXED procedure. □

Exercise 4.11
Write a SAS program to analyze the data of Example 3.1 using orthogonal contrasts and the model (4.22). □

Exercise 4.12
Write a SAS program to analyze the data of Exercise 3.8 using orthogonal contrasts and the model developed for the data in Exercise 4.7. □

Exercise 4.13
For the model fitted for Exercise 4.7, obtain the fitted responses and the residuals. Check that the normality assumption is satisfied. □

References

Akaike, H. (1974) A new look at statistical model indentification, *IEEE Transaction on Automatic Control*, AC-19, 716–723.

Burnham, K.P. and Anderson, D.R. (1998) *Model Selection and Inference: A Practical Information—Theoretic Approach,* Springer-Verlag, New York.

Eliason, S.R. (1993) *Maximum Likelihood Estimation: Logic and Practice*, Sage, Newbury Park, CA.

Hartley, H.O. and Rao, J.N.K. (1967) Maximum likelihood estimation for the mixed analysis of variance model, *Biometrika*, 54, 93–108.

Hurvich, C.M. and Tsai, C.L. (1989) Regression and time series model selection in small samples, *Biometrika*, 76, 297–307.

Jennrich, R.I. and Schluchter, M.D. (1986) Unbalanced repeated measures models with structured covariance matrices, *Biometrics*, 42, 805–820.

Kleinbaum, D.G., Kupper, L.L., Muller, K.E., and Nizam, A. (1998) *Applied Regression Analysis and Other Multivariable Methods*, Brookes/Cole Publishing Company, Pacific Grove, CA.

Laird, N.M. and Ware, J.H. (1982) Random-effects model for longitudinal data, *Biometrics*, 38, 963–974.

Littell, R.C., Milliken, G.A., Stroup, W.W., and Wolfinger, R.D. (2002) *SAS System for Mixed Models*, SAS Institute Inc., Cary, NC.

Onyiah, L.C. (2005) *SAS Programming and Data Analysis: A Theory and Program-Driven Approach.* University Press of America, Lanham, MD.

SAS/STAT User's Guide (1999) 2, SAS Institute Inc., Cary, NC.

Schwarz, G. (1978) Estimating the dimension of a model, *Annals of Statistics*, 6, 461–464.

Weiss, R.E. (2005) *Modelling Longitudinal Data*, Springer, New York.

Chapter 5

Designs with Randomization Restriction—Randomized Complete Block, Latin Squares, and Related Designs

5.1 Randomized Complete Block Design

The randomized complete block design (RCBD) is used for studying the effect of treatment in the presence of a nuisance factor, which is not of primary interest in the study. Ordinarily, the study that is modeled as RCBD would have been treated as a completely randomized design (CRD). The nuisance factor (usually called the block), is designed into the experiment to ensure that if its effect is large, it would not inflate the size of the error sum of squares, carry through to the ANOVA, and thus depress the sensitivity of the ANOVA to detect the significance of the treatment effect. For instance, an investigation may be such that it stretches over different shifts of workers of differing skills. The shifts in such a study could be treated as blocks. Alternatively, treatments may be such that they are carried out by different groups of workers with differing expertise and experience. The groups could be treated as blocks and the investigation carried out under RCBD. We discuss the RCBD in the next few sections. The RCBD is important because after initially being applied to selected agricultural experiments, it is now used extensively in agricultural, medical and industrial experiments. For example, in agricultural experiments, there are often different fertilizers tested on various types of soils to determine which one gives the best yield for a crop. The types of soils may be clay-like, loamy, humus, or sandy. To design an experiment to study the yields of soybean under five fertilizer treatments on soils that are clay-like, loamy, humus, or sandy, the soil type is treated as the nuisance factor in the experiment and would be designed into the experiment as the block.

Suppose that five different fertilizers, T_1, T_2, T_3, T_4, and T_5, are to be studied under the soil types stated above. Randomization of the order of application of each fertilizer is carried out at each level of the block (clay-like, loamy, humus, sandy). Thus, blocking represents a restriction on randomization since we do not carry out randomization over the entire experimental material as in CRD, but only within the confines of each level of the block. The result of this randomization could be laid out as in the table below in the actual planting.

Soil type (block)			
Clay-like	**Loamy**	**Humus**	**Sandy**
T_2	T_4	T_2	T_5
T_1	T_3	T_5	T_1
T_5	T_5	T_1	T_4
T_3	T_2	T_4	T_2
T_4	T_1	T_3	T_4

In Chapter 2, while studying materials for making soles of shoes, we came across a special case of the RCBD (Idea III, paired comparison). We employed randomization as a way of converting uncontrolled variation (in this case, the effects of the left legs of the tennis players) into random error. We achieved randomization by assigning shoes made from the two materials at random to each leg of each tennis player. By doing so, we had a better design than those proposed under Ideas I and II. There is, however, another way of improving the efficiency of our models—by blocking. Although blocking was present in Idea III, it was only subtly involved in our analysis. Later, we shall rearrange the results of that experiment and analyze it with the general theory of the RCBD, to make the blocking in this design more explicit. In Chapter 2, the pairing of the units in the paired comparison experiment, and the subsequent subtraction of one of the observations in each block from the other, eliminated the block effect. All that was left behind was the difference in treatment effects and the differences in errors.

To make this fact clear, consider an observation in the paired comparison experiment. We shall represent the treatment effect by γ_i (for treatment i; for our experiment and indeed any other paired comparison experiment, $i=1,\ 2$), and the block effect, which we shall represent by β_j (for the jth block effect; $j=1,2,\ldots,n$; $n=12$ for the design represented by Idea III in the shoe materials experiment discussed in Chapter 2). The overall mean is μ and with random error $\varepsilon_{ij} \sim \text{NID}\,(0,\ \sigma^2)$, the appropriate representation of each response in our experiment is given in the following equation:

$$y_{ij} = \mu + \gamma_i + \beta_j + \varepsilon_{ij} \quad i = 1, 2, \ldots, k; \quad j = 1, 2, \ldots, b \tag{5.1}$$

Consider two observations in the same block j, each of which received one of the two treatments, A or B. These responses could be written as

$$y_{1j} = \mu + \gamma_1 + \beta_j + \varepsilon_{1j} \quad y_{2j} = \mu + \gamma_2 + \beta_j + \varepsilon_{2j}$$

so that the difference between them is

$$y_{1j} - y_{2j} = (\gamma_1 - \gamma_2) + (\varepsilon_{1j} - \varepsilon_{2j}) \tag{5.2}$$

This clearly indicates that in a very simple manner, paired comparison eliminated the effect of a factor in which we had little interest, namely, the blocks, thus enabling us to study the differences in treatments (sole materials). In the design of experiments, one strategy used to advantage when the experimental material to be employed in our study is the concern, is to group our experimental units into homogeneous units (called blocks) and compare several treatments within the blocks. By doing so, we eliminate the differences between the blocks from our experimental error and increase the precision of our result. The other advantages of this design are that there are no restrictions on the number of blocks that could be employed, or in the number of replications to be used in an experiment, except those imposed by resource constraints. Moreover, the method of analysis is relatively simple. Soil types and fertilizers are good examples of agricultural experiments that benefit from blocking. In manufacturing, experimental materials may come from different batches of raw materials, each of which could be treated as a block to remove differences in batches from the experimental error. Further, when the performance of each full replicate of an experiment is carried out on a different day or carried out by a different group of workers, each replicate could be regarded as a block to eliminate the effect of days or groups on the responses. In experiments involving humans and animals, some peculiarities of the units used in the experiment may dictate that membership of a block be based on age, sex, weight, or some other characteristics of the individuals. As already mentioned, the overall aim is to choose a block so that the units in each of the blocks are more homogeneous than the whole aggregate of the units to which the treatments are to be applied.

To obtain the RCBD, when k treatments are to be applied, which are replicated b times, we simply divide the entire experimental materials into b blocks of k units each; the total number of units required for the experiment is kb. In Chapter 2, we used 12 tennis players (each with right and left legs) to study sole materials of types A and B. So there were two treatments with 12 blocks and the number of units required for the experiments was 24. To summarize our discussion of the RCBD, we note that this design involves the grouping of all experimental units in blocks. Randomization is then applied separately within each block to assign treatments. Suppose we have k treatments with b replications, we could arrange each replicate in a block. The arrangements of experimental units in a typical RCBD are shown in Table 5.1. Although the responses have been presented in order in Table 5.1, it has to be emphasized that the choice of units in each block, receiving any particular treatment is completely random. The randomization and assignment of treatments within blocks are usually accomplished in the design stages. Here, however, we present the data in the following systematic order, which is more convenient for the discussion of the analysis of the responses of the experiment:

$$\overline{y}_{.j} = \frac{1}{k}\sum_{i=1}^{k} y_{ij} \quad \overline{y}_{i.} = \frac{1}{b}\sum_{j=1}^{b} y_{ij} \quad \overline{y}_{..} = \frac{1}{bk}\sum_{i=1}^{k}\sum_{j=1}^{b} y_{ij} \tag{5.3}$$

TABLE 5.1: Arrangement of Units in a Typical Randomized Complete Block

	Blocks						
Treatments	**1**	**2**	...	$b-1$	b	**Total**	**Mean**
1	y_{11}	y_{12}	...	y_{1b-1}	y_{1b}	$y_{1.}$	$\overline{y}_{1.}$
2	y_{21}	y_{21}	...	y_{2b-1}	y_{2b}	$y_{2.}$	$\overline{y}_{2.}$
$\vdots$	$\vdots$	$\vdots$	$\vdots$	$\vdots$	$\vdots$	$\vdots$	$\vdots$
k	y_{k1}	y_{k2}	...	y_{kb-1}	y_{kb}	$y_{k\cdot}$	$\overline{y}_{k\cdot}$
Total			...		$y_{.b}$	$y_{\cdot\cdot}$	$\overline{y}_{\cdot\cdot}$
Mean	$\overline{y}_{.1}$	$\overline{y}_{.2}$	...	$\overline{y}_{.b-1}$	$\overline{y}_{.b}$	$\overline{y}_{..}$	

The model in Equation 5.1 represents the RCBD. By adopting this model, we are assuming that the effects of treatments and blocks are additive. The model is satisfactory because even when the effects are considered to be multiplicative, a logarithmic transformation of the responses would result in an additive model. Further, we assume that

$$\sum_{i=1}^{k} \gamma_i = \sum_{j=1}^{b} \beta_j = 0 \tag{5.4}$$

This is the fixed effects model for RCBD. It is, of course, possible to use a random effects model for RCBD. In such cases, the levels of block and treatment in the design would be considered as random samples of levels selected from a population of levels available for treatment and block. Just as in the one-way experimental layout (CRD), the randomization of treatment applications to units within each block or the levels of the block factor is considered important, and is meticulously carried out. This ensures that any systematic error within each level of block is eliminated or converted into random error.

The total sum of squares from the least squares analysis for this model (RCBD), is similar in some respects to that of the one-way ANOVA, except that here the summation is over the levels of two factors (treatments and block)

$$SS_{\text{total}} = \sum_{i=1}^{k}\sum_{j=1}^{b}(y_{ij} - \overline{y}_{..})^2$$

which can be rewritten as

$$\begin{aligned} SS_{\text{total}} &= \sum_{i=1}^{k}\sum_{j=1}^{b}(y_{ij} - \overline{y}_{i.} + \overline{y}_{i.} - \overline{y}_{.j} + \overline{y}_{.j} - \overline{y}_{..} + \overline{y}_{..} - \overline{y}_{..})^2 \\ &= \sum_{i=1}^{k}\sum_{j=1}^{b}[(y_{ij} - \overline{y}_{i.} - \overline{y}_{.j} + \overline{y}_{..}) + (\overline{y}_{i.} - \overline{y}_{..}) + (\overline{y}_{.j} - \overline{y}_{..})]^2 \end{aligned}$$

$$= \sum_{i=1}^{k}\sum_{j=1}^{b}(y_{ij} - \overline{y}_{i.} - \overline{y}_{.j} + \overline{y}_{..})^2 + \sum_{i=1}^{k}\sum_{j=1}^{b}(\overline{y}_{i.} - \overline{y}_{..})^2$$

$$+ \sum_{i=1}^{k}\sum_{j=1}^{b}(\overline{y}_{.j} - \overline{y}_{..})^2 + \text{cross-product terms (CPT)}$$

The cross-product terms sum to zero leading to Equation 5.5:

$$SS_{\text{total}} = SS_{\text{residual}} + SS_{\text{treatment}} + SS_{\text{block}} \tag{5.5}$$

It can also be shown that

$$ESS_{\text{residual}} = (k-1)(b-1)\sigma^2 \tag{5.6a}$$

$$ESS_{\text{treatment}} = (k-1)\sigma^2 + b\sum_{i=1}^{k}\gamma_i^2 \tag{5.6b}$$

$$ESS_{\text{block}} = (b-1)\sigma^2 + k\sum_{j=1}^{b}\beta_j \tag{5.6c}$$

The ANOVA that follows is based on the assumption that $\gamma_i = 0$, for all i. It can be shown (Mood et al., 1974, p. 245) that under this assumption $SS_{\text{residual}}/\sigma^2$ and $SS_{\text{treatment}}/\sigma^2$ are distributed as χ^2 random variables with $(k-1)(b-1)$ and $k-1$ degrees of freedom, respectively. Regardless of the value of each β_j, as long as γ_is are all zero, then

$$F_0 = \frac{\left[\dfrac{SS_{\text{treatment}}}{\sigma^2(k-1)}\right]}{\left[\dfrac{SS_{\text{residual}}}{\sigma^2(k-1)(b-1)}\right]} = \frac{(b-1)SS_{\text{treatment}}}{SS_{\text{residual}}} \tag{5.7}$$

where F_0 is a random variable distributed as F with $(k-1)$ and $(k-1)(b-1)$ degrees of freedom. Similarly, some statisticians believe that when $\beta_j = 0$, for all j, then $SS_{\text{blocks}}/\sigma^2$ is also χ^2 distributed with $b-1$ degrees of freedom, so that

$$F_0 = \frac{\left[\dfrac{SS_{\text{block}}}{\sigma^2(b-1)}\right]}{\left[\dfrac{SS_{\text{residual}}}{\sigma^2(k-1)(b-1)}\right]} = \frac{(k-1)SS_{\text{block}}}{SS_{\text{residual}}} \tag{5.8}$$

where F is distributed with $b-1$, and $(k-1)(b-1)$ degrees of freedom, irrespective of the values γ_i takes. Some statisticians argue that the other set of hypotheses related to the blocks could not be tested using F-ratio. Testing for the block differences in RCBD with $F = MS_{\text{block}}/MS_{\text{E}}$ is not considered to be a good idea. There is no universal agreement on this issue. A number of standard statistics books use this test and recommend it without reservations. Among such books are Wackerly et al. (2002, pp. 653–658); and Bowerman and O'Connell (1990, pp. 923–942). Anderson and McLean (1974) objected to the use of F-statistic as a test for comparing the block means. They state that blocking is a restriction on randomization and argue that since only treatments were randomized within each block, the blocking does not make the F-statistic a meaningful test for comparing block means. The use of F-test for blocking, they posit, represents a test of block means and the randomization restriction. However, Box et al. (1978) argue that the usual ANOVA F-test can be justified on the basis of randomization alone without utilizing the normality assumption. They believe that $F = MS_{\text{block}}/MS_{\text{E}}$ is F-distributed with $b-1$ and $(k-1)(b-1)$ degrees of freedom. Others disagree. Montgomery (2005) states that in general, the use of $F = MS_{\text{block}}/MS_{\text{E}}$ to compare block means is not a good idea, but in practice its use to investigate the effect of the blocking variable is reasonable as an approximate way of testing whether blocking was effective. Thus the set of hypotheses, which could be tested on the bases of the foregoing deductions for the RCBD are as follows:

$$H_0\colon \gamma_1 = \gamma_2 = \cdots = \gamma_k = 0 \quad \text{vs} \quad H_1\colon \gamma_i \neq 0 \text{ (for some } i)$$

For the above reasons, the block F-statistic is omitted in the following ANOVA table for RCBD.

TABLE 5.2: ANOVA for a Randomized Block Experiment

Source	**Sum of squares**	**Degrees of freedom**	**Mean square**	***F*-ratio**
Due to treatment	$SS_{\text{treatment}}$	$k-1$	$SS_{\text{treatment}}/(k-1)$	$MS_{\text{treatment}}/MS_{\text{E}}$
Due to block	SS_{block}	$b-1$	$SS_{\text{block}}/(b-1)$	
Residual	SS_{E}	$(k-1)(b-1)$	$SS_{\text{E}}/(b-1)(k-1)$	
Total	SS_{total}	$bk-1$		

Table 5.2 shows the ANOVA resulting from the model, which should be used for testing these hypotheses.

From the above, we can easily obtain the expected mean squares from Equations 5.9a through 5.9c as follows:

$$EMS_{\text{treatment}} = \sigma^2 + \frac{b\sum_{i=1}^{k} \gamma_i^2}{k-1} \tag{5.9a}$$

$$EMS_{\text{block}} = \sigma^2 + \frac{k\sum_{j=1}^{l} \beta_j^2}{b-1} \tag{5.9b}$$

$$EMS_{\text{E}} = \sigma^2 \tag{5.9c}$$

For easier calculation, it is recommended that the following forms of the sums of squares as in Equations 5.10 through 5.13 (which can easily be obtained by algebraic manipulation of the formulae given earlier) be used:

$$SS_{\text{total}} = \sum_{i=1}^{k}\sum_{j=1}^{b} y_{ij}^2 - \frac{y_{..}^2}{bk} \tag{5.10}$$

$$SS_{\text{treatment}} = \frac{1}{b}\sum_{i=1}^{k} y_{i.}^2 - \frac{y_{..}^2}{bk} \tag{5.11}$$

$$SS_{\text{block}} = \frac{1}{k}\sum_{j=1}^{b} y_{.j}^2 - \frac{y_{..}^2}{bk} \tag{5.12}$$

$$SS_{\text{E}} = \sum_{i=1}^{k}\sum_{j=1}^{b} y_{ij}^2 - \frac{1}{k}\sum_{j=1}^{b} y_{.j}^2 - \frac{1}{b}\sum_{i=1}^{k} y_{i.}^2 + \frac{y_{..}^2}{bk}$$

or

$$SS_{\text{E}} = SS_{\text{total}} - SS_{\text{treatment}} - SS_{\text{block}} \tag{5.13a}$$

5.2 Testing for Differences in Block Means

The RCBD is a study of treatments, in circumstances in which blocking is necessary. So the significance of the treatment effect is tested in ANOVA, using the ratio of the $MS_{\text{treatment}}$ and MS_{E}. Sometimes, we may be interested in comparing block means. There are disagreements regarding what should be done. We discussed the arguments in the Section 5.1. To illustrate, we analyze the responses of a designed experiment as an example.

Example 5.1

A microbiologist decided to study the effect of temperature on the growth of algae in a medium. Since the algae were obtained from a random sample of locations of size three from several different locations in the seas, it was suspected that their sources could lead to differential growth of the algae in the medium. Consequently, it was decided to test whether varying temperature-levels are responsible for the observed differences in the growth of the algae. The algae source is designed into the experiment as a blocking factor since interest is in the effect of temperature on the growth of algae. The growth of the algae at five settings of temperature were studied and the following responses were obtained. Note that the responses refer to the number of molds of each algae found in 1 mL of the growth suspension, 3 days after the growth began and after the growth suspension had been diluted 100 times. The molds were usually counted under microscope.

	Algae source (block)		
Temperature (treatment)	**I**	**II**	**III**
I	23	34	19
II	31	37	22
III	52	69	16
IV	35	37	30
V	39	47	31

To analyze the data, we can expand the table further by adding the totals for the rows and columns as follows:

	Algae source (block)			
Temperature	**I**	**II**	**III**	**Total**
I	23	34	19	76
II	31	37	22	90
III	52	69	16	137
IV	35	37	30	102
V	39	47	31	117
Total	180	224	118	522

$$SS_{\text{total}} = 23^2 + 34^2 + 19^2 + \cdots + 31^2 - \frac{522^2}{3(5)} = 2600.40$$

$$SS_{\text{algae type}} = \frac{180^2 + 224^2 + 118^2}{5} - \frac{522^2}{3(5)} = 1134.40$$

$$SS_{\text{temperature}} = \frac{76^2 + 90^2 + 137^2 + 102^2 + 117^2}{3} - \frac{522^2}{3(5)} = 747.07$$

$$SS_{\text{E}} = 2600.40 - 1134.40 - 747.07 = 718.93$$

TABLE 5.3: ANOVA for the Microbiologist's Algae Experiment

Source	Sum of squares	Degrees of freedom	Mean square	F-ratio
Temperature	747.07	4	186.768	2.08
Source of algae	1134.40	2	567.200	
Residual	718.93	8	89.867	
Total	2600.40	14		

From the tables, $F(4, 8, 0.05) = 3.84$ and $F(2, 8, 0.05) = 4.46$. We therefore reject the hypothesis that varying temperature exerts significant effect on the growth of algae. On the other hand, there is a large mean square for the source of algae; this leads us to say tentatively that there is differential growth of algae due to source. Further analysis to know which algae grows best in this medium would have been carried out using any of the methods described in the Sections 5.1 (e.g., Duncan's multiple range test), if temperature was found significant (Table 5.3). (Any interested reader is welcome to carry out this analysis as an exercise.) □

5.2.1 Relative Efficiency of the RCBD to CRD

We know that if we analyzed the responses of a typical RCBD experiment as if they came from a CRD when the block effects are large, the mean square error would have been larger, making the test on treatment less sensitive. Possibly, it could have affected the conclusions we would reach about the effects of treatments in ANOVA. To consider the efficiency of the RCBD relative to the CRD, we define a quantity R as in Equation 5.13b, where

$$R = \frac{(df_{\text{RCBD}} + 1)(df_{\text{CRD}} + 3)}{(df_{\text{RCBD}} + 3)(df_{\text{CRD}} + 1)} \cdot \frac{\sigma^2_{\text{CRD}}}{\sigma^2_{\text{RCBD}}} \tag{5.13b}$$

where df_{RCBD} and df_{CRD} are the degrees of freedom of error in the RCBD and CRD, while σ^2_{CRD} and σ^2_{RCBD} are the error variances for CRD and RCBD, respectively. We use the mean square errors for blocks and error in the RCBD to estimate the error variances for different designs (Cochran and Cox, 1957, pp. 112–114) as in Equations 5.13c and 5.13d, so that

$$\hat{\sigma}^2_{\text{CRD}} = \frac{(b-1)MS_{\text{block}} + b(k-1)MS_{\text{E}}}{bk-1} \tag{5.13c}$$

$$\hat{\sigma}^2_{\text{RCBD}} = MS_{\text{E}} \tag{5.13d}$$

5.2.1.1 Application to Example 5.1

We illustrate the use of the above formulae to find the efficiency of RCBD relative to CRD, applying it to the RCBD in Example 5.1. First, we estimate the error variances as follows:

$$\hat{\sigma}^2_{\text{CRD}} = \frac{(b-1)MS_{\text{block}} + b(k-1)MS_{\text{E}}}{bk-1}$$
$$= \frac{(3-1)(567.2) + 3(5-1)(89.867)}{5(3)-1} = 158.0574$$
$$\hat{\sigma}^2_{\text{RCBD}} = MS_{\text{E}} = 89.867$$

$df_{\text{RCBD}} = 8$; $df_{\text{CRD}} = 10$. We now calculate R, the measure of relative efficiency as

$$R = \frac{(8+1)(10+3)}{(8+3)(10+1)} \times \frac{158.0574}{89.867} = 1.7007$$

This result indicates that we should have used about two times as many replicates for a CRD to achieve the same level of sensitivity in the algae experiment as we obtained by blocking. We note that if blocking was not needed, we would have made the test on treatment less sensitive because we lost $b-1$ degrees of freedom from the degrees of freedom for error.

5.2.1.2 Residuals and Parameters Estimates in the RCBD

We can obtain the estimates of the parameters of the model 5.1 as follows:

$$\hat{\gamma}_i = \overline{y}_{i.} - \overline{y}_{..}$$
$$\hat{\beta}_j = \overline{y}_{.j} - \overline{y}_{..} \tag{5.13e}$$

The residuals are obtained for the RCBD as

$$\hat{\varepsilon}_{ij} = y_{ij} - \overline{y}_{i.} - \overline{y}_{.j} + \overline{y}_{..} \tag{5.13f}$$

The estimates of the effects for the data of Example 5.1 are

$$\overline{y}_{1.} = \frac{76}{3} = 25.3333 \quad \overline{y}_{2.} = \frac{90}{3} = 30 \quad \overline{y}_{3.} = \frac{137}{3} = 45.6667$$

$$\overline{y}_{4.} = \frac{102}{3} = 34 \quad \overline{y}_{5.} = \frac{117}{3} = 39 \quad \overline{y}_{.1} = \frac{180}{5} = 36 \quad \overline{y}_{.2} = \frac{224}{5} = 44.8$$

$$\overline{y}_{.3} = \frac{76}{3} = 23.6 \quad \overline{y}_{..} = \frac{522}{15} = 34.8$$

$\hat{\alpha}_1 = 25.3333 - 34.8 = -9.46667 \quad \hat{\alpha}_2 = 30 - 34.8 = -4.8$

$\hat{\alpha}_3 = 45.6667 - 34.8 = 10.8667 \quad \hat{\alpha}_4 = 34 - 34.8 = -0.8$

$\hat{\alpha}_5 = 39 - 34.8 = 4.2 \quad \hat{\beta}_1 = 36 - 34.8 = 1.2 \quad \hat{\beta}_2 = 44.8 - 34.8 = 10$

$\hat{\beta}_3 = 23.6 - 34.8 = -11.2 \quad \hat{\mu} = 34.8$

We obtained the fitted values and residuals as shown in Table 5.4.

Next, we present four plots involving the residuals. The first one is a plot of residuals against fitted values (Figure 5.1). The other is the normal probability plot of residuals for Example 5.1 (Figure 5.2). The others are

TABLE 5.4: Fitted Values and Residuals for Example 5.1

Temperature	Algae	y	Estimates	Residuals
1	1	23	26.5333	−3.5333
1	2	34	35.3333	−1.3333
1	3	19	14.1333	4.8667
2	1	31	31.2	−0.2
2	2	37	40	−3
2	3	22	18.8	3.2
3	1	52	46.8667	5.1333
3	2	69	55.6667	13.3333
3	3	16	34.4667	−18.4667
4	1	35	35.2	−0.2
4	2	37	44	−7
4	3	30	22.8	7.2
5	1	39	40.2	−1.2
5	2	47	49	−2
5	3	31	27.8	3.2

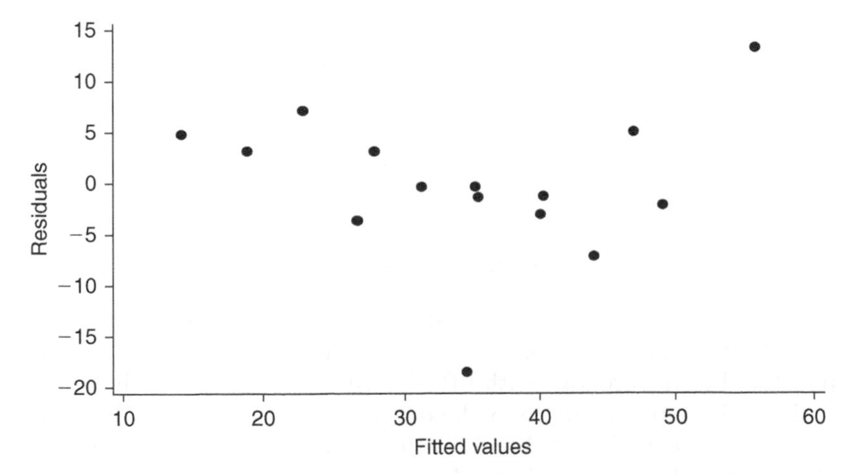

FIGURE 5.1: Scatterplot of residuals against fitted values for Example 5.1.

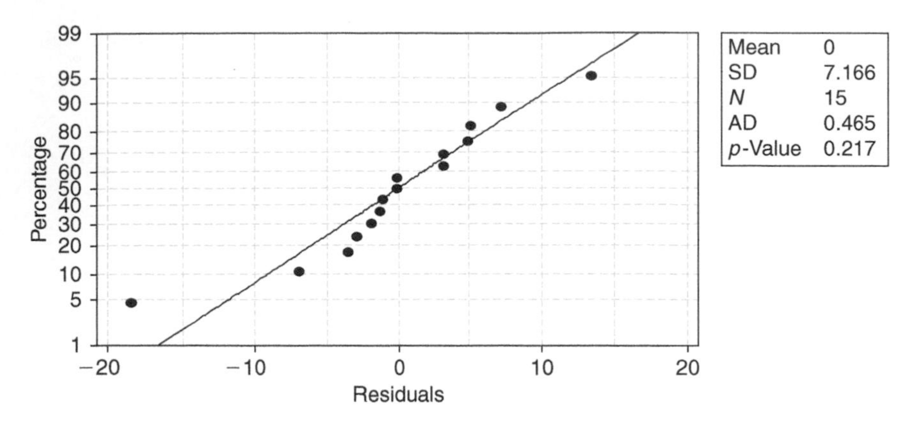

FIGURE 5.2: Normal probability plot of residuals for Example 5.1.

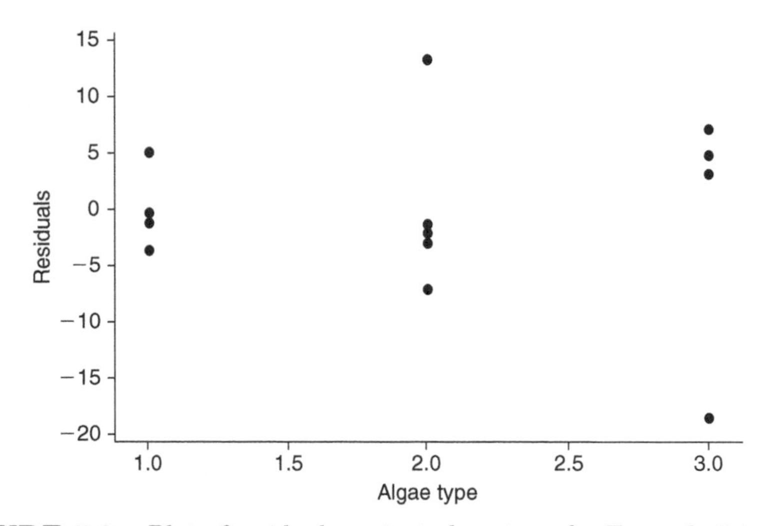

FIGURE 5.3: Plot of residuals against algae type for Example 5.1.

the plots of the residuals against algae type and against temperature levels (Figures 5.3 and 5.4).

5.2.2 SAS Analysis of Responses of Example 5.1

We present the SAS analysis of data of Example 5.1 using the GLM procedure. In SAS programs for analysis of RCBD data, we can also replace PROC GLM in the program with PROC ANOVA to carry out the same analysis. There is some difference in the outputs because PROC GLM provides information on both TYPE I and TYPE III sums of squares for the factors in the design, which PROC ANOVA does not provide.

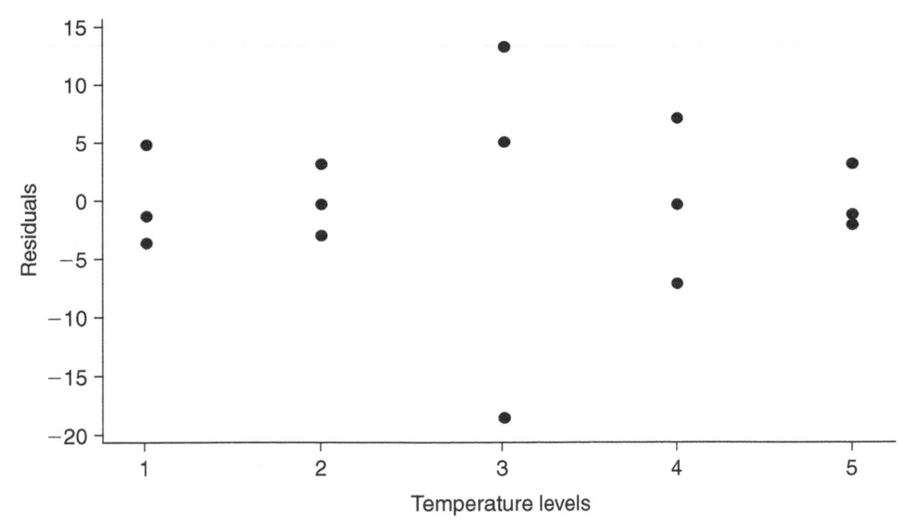

FIGURE 5.4: Plot of residuals against temperature levels for Example 5.1.

PROGRAM FOR EXAMPLE 5.1

```
data algae;
input temp algae y@@;
datalines;
1  1  23  1  2  34  1  3  19  2  1  31  2  2  37  2  3  22
3  1  52  3  2  69  3  3  16  4  1  35  4  2  37  4  3  30
5  1  39  5  2  47  5  3  31
;
proc print data=algae;
proc glm data=algae;
class temp algae;
model y=temp algae; run;
```

SAS OUTPUT

Obs	temp	algae	y
1	1	1	23
2	1	2	34
3	1	3	19
4	2	1	31
5	2	2	37
6	2	3	22
7	3	1	52
8	3	2	69
9	3	3	16
10	4	1	35
11	4	2	37
12	4	3	30
13	5	1	39
14	5	2	47
15	5	3	31

```
                          The GLM Procedure

Dependent Variable: y

                                   Sum of
Source                  DF        Squares    Mean Square    F Value    Pr > F

Model                    6    1881.466667     313.577778       3.49    0.0533

Error                    8     718.933333      89.866667

Corrected Total         14    2600.400000


          R-Square     Coeff Var      Root MSE        y Mean

          0.723530      27.24081      9.479803      34.80000

Source                  DF      Type I SS    Mean Square    F Value    Pr > F
temp                     4     747.066667     186.766667       2.08    0.1757
algae                    2    1134.400000     567.200000       6.31    0.0226

Source                  DF    Type III SS    Mean Square    F Value    Pr > F

temp                     4     747.066667     186.766667       2.08    0.1757
algae                    2    1134.400000     567.200000       6.31    0.0226
```

Next, we consider the following example to illustrate the RCBD:

Example 5.2

In a small experiment, at Uzo-Uwani Farm, to test the performances of four different fertilizers on yields of cassava, the plot to be used was divided into eight subplots and treated as blocks. Treatments were applied at random in each block, with each treatment appearing once in each block. The yield of cassava (in kilograms) harvested at the end of 6 months and the random assignments in each block used in the actual experiments are presented in Table 5.5.

The same data are rearranged in Table 5.6 to neatly conform to the formulae we use for manual analysis of the responses.

The calculation of the sums of squares is greatly simplified by using the formulae in Equations 5.10 through 5.13a. Using the responses for the experiment on fertilizer treatments on cassava, we obtain the sums of squares as follows:

TABLE 5.5: Cassava Yields under Different Fertilizer Treatments at Uzo-Uwani Showing Actual Treatments in the Field

Cassava Plots (Blocks)							
I	**II**	**III**	**IV**	**V**	**VI**	**VII**	**VIII**
8 A	2 C	10 D	7 A	5 D	3 D	2 A	8 C
10 C	3 B	9 C	6 D	2 C	10 B	8 D	3 B
8 B	6 A	13 B	10 C	3 D	13 A	10 C	9 A
8 D	3 D	13 A	8 B	4 A	3 C	7 B	6 D

TABLE 5.6: Rearranged Data on Fertilizer Experiment at Uzo-Uwani

	Blocks								
Treatment	**I**	**II**	**III**	**IV**	**V**	**VI**	**VII**	**VIII**	$y_{i.}$
A	8	6	13	7	4	13	2	9	62
B	8	3	10	8	5	10	7	3	54
C	10	2	9	10	2	3	10	8	54
D	8	3	10	6	3	3	8	6	47
$y_{.j}$	34	14	42	31	14	29	27	26	217

$$SS_{\text{total}} = 8^2 + 6^2 + 13^2 + \cdots + 6^2 - \frac{217^2}{8(4)} = 325.469$$

$$SS_{\text{block}} = \frac{34^2 + 14^2 + 42^2 + \cdots + 26^2}{4} - \frac{217^2}{8(4)} = 158.219$$

$$SS_{\text{treatment}} = \frac{62^2 + 54^2 + 54^2 + \cdots + 47^2}{4} - \frac{217^2}{8(4)} = 14.094$$

$$SS_{\text{residual}} = 325.469 - 158.219 - 14.094 = 153.156$$

As usual, the hypothesis we are interested in is as follows:

$$H_0\colon \gamma_1 = \gamma_2 = \gamma_3 = \gamma_4 = 0 \quad \text{vs} \quad H_1\colon \gamma_i \neq 0 \quad \text{for } i = 1, 2, 3, 4$$

The block effects are not of interest here; they are sometimes tested to find out whether blocking has been effective.

TABLE 5.7: ANOVA for the Fertilizer Experiment

Source	Sum of squares	Degrees of freedom	Mean square	F-ratio
Treatment	14.094	3	4.698	0.644
Block	158.219	7	22.603	
Residual	153.156	21	7.293	
Total	325.469	31		

From Tables A5–A7, $F(3, 21, 0.05) = 3.07$, so we accept the null hypothesis H_0. Interest in this particular study was not on the block effects. However, we note that the block mean square is large which may indicate that blocking has been effective. ∎

5.2.3 SAS Analysis of Data of Example 5.2

We write a program to carry out the analysis of data of Example 5.2. In this analysis, we again use the GLM procedure in SAS. It can also be analyzed by using the ANOVA procedure for the same problem as we stated previously. We will assume that the plots constitute a random sample selected from a population of other plots.

SAS PROGRAM

```
data cassava;
input fert plot y@@;
datalines;
1 1 8 1 2 6 1 3 13 1 4 7 1 5 4 1 6 13 1 7 2 1 8 9
2 1 8 2 2 3 2 3 10 2 4 8 2 5 5 2 6 10 2 7 7 2 8 3
3 1 10 3 2 2 3 3 9 3 4 10 3 5 2 3 6 3 3 7 10 3 8 8
4 1 8 4 2 3 4 3 10 4 4 6 4 5 3 4 6 3 4 7 8 4 8 6
;
proc print data=cassava;
proc glm data=cassava;
class fert plot;
model y=fert plot; random plot; run;
```

SAS OUTPUT

Obs	fert	plot	y
1	1	1	8
2	1	2	6
3	1	3	13
4	1	4	7
5	1	5	4
6	1	6	13
7	1	7	2
8	1	8	9
9	2	1	8
10	2	2	3
11	2	3	10
12	2	4	8
13	2	5	5
14	2	6	10
15	2	7	7
16	2	8	3
17	3	1	10

Obs	fert	plot	y
18	3	2	2
19	3	3	9
20	3	4	10
21	3	5	2
22	3	6	3
23	3	7	10
24	3	8	8
25	4	1	8
26	4	2	3
27	4	3	10
28	4	4	6
29	4	5	3
30	4	6	3
31	4	7	8
32	4	8	6

```
              The GLM Procedure
           Class Level Information

   Class          Levels    Values

   fert                4    1 2 3 4

   plot                8    1 2 3 4 5 6 7 8

   Number of Observations Read         32
   Number of Observations Used         32

Dependent Variable: y
```

Source	DF	Sum of Squares	Mean Square	F Value	Pr > F
Model	10	172.3125000	17.2312500	2.36	0.0466
Error	21	153.1562500	7.2931548		
Corrected Total	31	325.4687500			

R-Square	Coeff Var	Root MSE	y Mean
0.529429	39.82428	2.700584	6.781250

Source	DF	Type I SS	Mean Square	F Value	Pr > F
fert	3	14.0937500	4.6979167	0.64	0.5952
plot	7	158.2187500	22.6026786	3.10	0.0208

Source	DF	Type III SS	Mean Square	F Value	Pr > F
fert	3	14.0937500	4.6979167	0.64	0.5952
plot	7	158.2187500	22.6026786	3.10	0.0208

Example 5.3

For the sake of curiosity, we wish to perform another analysis of data obtained from the experiment performed to test materials A and B for making soles of tennis shoes (Chapter 2, Example 2.1). This time, we arrange the data in blocks. There are 12 blocks with two treatments (12 shoes with sole materials A and B), which were assigned at random to the two units in a block (player's right and left feet).

	Blocks											
Treatments	**1**	**2**	**3**	**4**	**5**	**6**	**7**	**8**	**9**	**10**	**11**	**12**
A	13.0	8.0	10.7	14.1	10.5	6.4	9.3	10.6	8.6	13.1	12.2	11.4
B	13.8	8.6	11.0	14.0	11.6	6.2	9.6	11.1	9.1	13.4	11.8	10.8
Total	26.8	16.6	21.7	28.1	22.1	12.6	18.9	21.7	17.7	26.5	24	22.2

Grand total: 258.9

Treatment: A B

Total: 127.9 131

$$SS_{\text{total}} = [(13.0)^2 + (8.0)^2 + \cdots + (10.8)^2] - \frac{(258.9)^2}{2(12)} = 115.8663$$

$$SS_{\text{block}} = \frac{[(26.8)^2 + (16.6)^2 + \cdots + (22.2)^2]}{2} - \frac{(258.9)^2}{2(12)} = 114.0913$$

$$SS_{\text{treatment}} = \frac{[(127.9)^2 + (131)^2]}{12} - \frac{(258.9)^2}{2(12)} = 0.4404$$

$$SS_{\text{residual}} = 115.8663 - 114.0913 - 0.4404 = 1.3746$$

TABLE 5.8: ANOVA Table for the Shoe Material Experiment

Source	**Sum of squares**	**Degrees of freedom**	**Mean square**	***F*-ratio**
Treatment	0.4404	1	0.4404	3.2043
Block	114.0913	11	10.3719	
Residual	1.3746	11	0.12496	
Total	112.0055	23		

Because $F(1,11,0.05) = 4.84$, we conclude that there is no significant difference between the shoe materials A and B (this is the same result obtained by using paired comparison). On the other hand, the block mean square is large and could have been significant. This large mean square for block was eliminated from the error which could have made it impossible for us to detect any significant difference between shoe materials A and B. In fact if sum of squares due to block had been included in the residual mean square; that is, if we did not use RCBD, and instead used the CRD, we would have concluded that the treatment effect is not significant in the ANOVA of Table 5.8. Any interested reader should check that under the CRD, although the $MS_{\text{treatment}} = 0.4405$ (same as in this design), the MS_{residual}, if the data is analyzed in a one-way experimental layout, is 5.24845 (more than 42-fold increase in the estimate of the variance using the MS_{residual}), which would have distorted any conclusions. It would have led to the conclusion that there was no significant difference between the wears of the shoe materials A and B; similar to the conclusions we reached by pooled variance t-test (Idea I), which did not recognize blocking in the designed experiment. □

5.2.4 SAS Analysis of Responses of Example 5.3

Next, we write a SAS program to analyze the responses of the study of soles for making soles of gym shoes (Example 5.3). In this analysis, the two legs of each boy make up the block of treatments. There are two levels of treatment (two types of sole material). Thus, there are ten blocks and two treatments. The responses are the wears on the soles after use by the boys for a specified period.

SAS PROGRAM

```
data tennis;
input soletype Player wear @@;
datalines;
1 1 13.0 1 2 8.0 1 3 10.7 1 4 14.1 1 5 10.5
1 6 6.4 1 7 9.3 1 8 10.6 1 9 8.6 1 10 13.1 1 11 12.2 1 12 11.4
2 1 13.8 2 2 8.6 2 3 11.0 2 4 14.0 2 5 11.6
2 6 6.2 2 7 9.6 2 8 11.1 2 9 9.1 2 10 13.4 2 11 11.8 2 12 10.8
;
proc anova data=tennis;
class player soletype;
model wear=soletype player; run;
```

SAS OUTPUT

```
                         The ANOVA Procedure

                       Class Level Information

             Class         Levels    Values

             player            12    1 2 3 4 5 6 7 8 9 10 11 12

             soletype           2    1 2

                Number of Observations Read          24
                Number of Observations Used          24

                         The ANOVA Procedure

Dependent Variable: wear

                                       Sum of
 Source                      DF       Squares     Mean Square    F Value    Pr > F
 Model                       12   114.4916667       9.5409722      76.35    <.0001
 Error                       11     1.3745833       0.1249621
 Corrected Total             23   115.8662500
```

R-Square	Coeff Var	Root MSE	wear Mean
0.988136	3.276939	0.353500	10.78750

Source	DF	Anova SS	Mean Square	F Value	Pr > F
soletype	1	0.4004167	0.4004167	3.20	0.1010
player	11	114.0912500	10.3719318	83.00	<.0001

Source	DF	Anova SS	Mean Square	F Value	Pr > F
soletype	1	0.8405000	0.8405000	11.21	0.0085
player	9	110.4905000	12.2767222	163.81	<.0001

5.3 Estimation of a Missing Value in the RCBD

Sometimes due to error, lack of due care in handling data, or some other problems unforeseen at the start of an experiment, one of the observations in RCBD could be missing. The model of the RCBD is usually of the form

$$y_{ij} = \mu + \gamma_i + \beta_j + \varepsilon_{ij} \quad i = 1, 2, \ldots, k; \quad j = 1, 2, \ldots, b$$

To fully carry out the analysis of the designed experiment, we need to estimate this value. Let us assume that the effects of both blocks and treatment are fixed. From the least squares principle, it is desirable to choose an estimate for the missing value so as to minimize the sum of squares for error. To estimate the missing value, we need to minimize the sum of squares error in this design with respect to the missing value.

Now SS_{E} can be written as

$$\sum_{i=1}^{a}\sum_{j=1}^{b}(y_{ij} - \bar{y}_{i.} - \bar{y}_{.j} + \bar{y}_{..})^2 = \sum_{i=1}^{a}\sum_{j=1}^{b} y_{ij}^2 - \frac{1}{b}\sum_{i=1}^{a}\left(\sum_{j=1}^{b} y_{ij}\right)^2 - \frac{1}{a}\sum_{j=1}^{b}\left(\sum_{i=1}^{a} y_{ij}\right)^2 + \frac{1}{ab}\left(\sum_{i=1}^{a}\sum_{j=1}^{b} y_{ij}\right)^2$$

If we denote the missing value by x, we note that we can now rewrite SS_{E} as

$$SS_{\mathrm{E}} = x^2 - \frac{1}{b}(y'_{i.} + x)^2 + \frac{1}{a}(y'_{.j} + x)^2 + \frac{1}{ab}(y'_{..} + x)^2 + R$$

R is a remainder quantity from SS_{E}, which does not contain the quantity, x. So that the total value y' is exclusive of the missing value as in the following equation:

$$\frac{\partial SS_{\mathrm{E}}}{\partial x} = 2x - \frac{2}{b}(y'_{i.} + x) - \frac{2}{a}(y'_{.j} + x) + \frac{2}{ab}(y'_{..} + x) = 0$$

(at the minimum point)

$$\Rightarrow abx - a(y'_{i.} + x) - b(y'_{.j} + x) + (y'_{..} + x) = 0$$

$$x(ab - a - b + 1) - ay'_{i.} - by'_{.j} + y'_{..} = 0$$

$$\hat{x} = \frac{(ay'_{i.} + by'_{.j} - y'_{..})}{ab - a - b - 1} = \frac{ay'_{i.} + by'_{.j} - y'_{..}}{(a-1)(b-1)} \tag{5.14}$$

After estimation in carrying out ANOVA, the usual degrees of freedom for error and therefore, total for the RCBD are both reduced by 1 to reflect the actual number of observations in the experiment. The estimated missing data point is simply used as a placeholder, which guarantees that analyses could proceed and does not change the degrees of freedom available from observed data.

Now we utilize this result in a designed experiment with a missing value in the following example:

Example 5.4

An experiment was performed to study the effects of five fertilizer treatments on a crop in six plots. At the end of the experiment, it was discovered that one of the observations (in the fifth plot and which was treated with the third fertilizer) was missing. Estimate this missing observation and carry out ANOVA for the data.

	Plots						
Treatments	**1**	**2**	**3**	**4**	**5**	**6**	**Total**
1	34	36	61	24	43	42	240
2	30	30	30	20	20	30	160
3	44	44	51	26	X	27	177
4	40	40	50	41	41	42	234
5	39	39	49	46	44	39	241
Total	137	189	241	157	148	180	1052

So that the missing value is estimated as

$$\hat{x} = \frac{5y'_{3.} + 6y'_{.5} - y'_{..}}{(5-1)(6-1)} = \frac{5(177) + 6(148) - 1052}{4(5)} = 36.05$$

	Blocks						
Treatments	**1**	**2**	**3**	**4**	**5**	**6**	**Total**
1	34	36	61	24	43	42	240
2	30	30	30	20	20	30	160
3	29	44	51	26	36.05	27	213.05
4	20	40	50	41	41	42	234
5	24	39	49	46	44	39	241
Total	137	189	241	157	184.05	180	1088.05

Once we replace $\hat{x}$ by 36.05, we obtain the sums of squares corresponding to different levels of block and treatment as follows:

$$y_{..} = 1088.05 \quad y_{1.} = 240 \quad y_{2.} = 160 \quad y_{3.} = 213.05 \quad y_{4.} = 234 \quad y_{5.} = 241$$
$$y_{.1} = 137 \quad y_{.2} = 189 \quad y_{.3} = 241 \quad y_{.4} = 157 \quad y_{.5} = 184.05 \quad y_{.6} = 180$$

$$SS_{\text{total}} = \frac{34^2 + 30^2 + \cdots + 42^2 + 39^2}{1} - \frac{(1088.05)^2}{30} = 3079.84$$
$$SS_{\text{block}} = \frac{137^2 + \cdots + 180^2}{5} - \frac{(1088.05)^2}{30} = 1237.12$$
$$SS_{\text{treatment}} = \frac{240^2 + 160^2 + \cdots + 241^2}{6} - \frac{(1088.05)^2}{30} = 776.12$$
$$SS_{\text{E}} = 3079.84 - 1237.12 - 776.12 = 1066.60$$

TABLE 5.9: Analysis of Variance for the Missing Value Problem

Source	Sum of squares	Degrees of freedom	Mean square	F-ratio
Treatment	776.12	4	194.03	3.64
Block	1237.12	5	247.42	
Error	1066.60	19	53.33	
Total	3079.84	28		

We find that $F(4,\ 20,\ 0.05) = 2.87$; $F(4,\ 20,\ 0.01) = 4.45$. Treatments are significant for $\alpha = 2.5\%$ because the p-value for the above test is 0.021961. Notice the large mean square for the block; once again a large mean square for block was not allowed to increase the mean square error, and mask the significance of the treatment effect (Table 5.9). This was good as it made our estimate for the error variance smaller, making our test for significance of treatment effect more precise. □

5.4 Latin Squares

In the CRDs described in Chapters 2 and 3 there were no controls placed on sources of variation. In the RCBD, which we described in the Section 5.3, we were able to control one source of variation by blocking. The result is that we achieved a more efficient design. Sometimes, however, the need may arise for more than one source of variation to be controlled. One design, which allows us to control two sources of variation, is the Latin square.

As the name suggests, the Latin square design requires $r \times r$ or r^2 experimental units, if r treatments or factors are to be studied. These units are first assigned to r groups of r treatments based on one source of variation. This is usually called the row classification. The units are further classified into the r groups of r units each; this second assignment is based on the second source of variation, which is called the column classification. Treatments are allowed to occur only once in each column or row. In effect, blocking applied simultaneously in two different directions is achieved by the Latin square, thereby eliminating two sources of extraneous variation. If we had a Latin square design with five units in each row and column, there will be 5×5 units in the design. The nonrandomized units will look somewhat as follows:

A B C D E
B C D E A
C D E A B
D E A B C
E A B C D

Essentially, the Latin squares design represents two restrictions on randomization. In actual field arrangement during experimentation, row-wise randomization could be carried out first, followed by column-wise randomization. Any scheme that ensures proper randomizations could be used. For instance, one may decide to write each of the letters representing each of the treatments to be used in the experiment r times (for a $r \times r$ Latin square design), experiment and put the r letters in a bag, shuffle properly before proceeding to choose one at a time to obtain the order in which the treatments should appear in the row. This should be carried out for all the rows. The same should be repeated for all the columns. In this way, row-wise and column-wise randomizations could be accomplished. A full discussion of randomization methods in Latin squares can be found in Fisher and Yates (1953). Consider the 5×5 Latin square design. We shall use a table of random digits to obtain an arrangement, which will lead to the randomization of units in both the rows and columns while seeking to conform to the randomization restrictions on the design. When randomization restrictions are achieved, the possible order in which the treatments could be applied is shown in Table 5.10. From the table of random digits in Appendices A11 and A12, we obtain the following five-digit random numbers, rank them, and use them to assign units to the rows. The same is repeated for assigning units to each column.

TABLE 5.10: Random Numbers for Row-Wise Randomization

Random number	Rank	Position in row
07514	1	5
87024	4	4
49326	3	3
27757	2	2
99721	5	1

The above outcome means that the letter currently in position 5 in each row should be put in position 1 and vice versa. All other positions of other letters remain unchanged. With this outcome, we rearrange the units in the 5×5 Latin square design above to obtain the following rows:

E B C D A
A C D E B
B D E A C
C E A B D
D A B C E

Repeating the same process of randomization yielded Table 5.11 of random numbers.

TABLE 5.11: Random Numbers for Column-Wise Randomization

Random numbers	Rank	Position in column
48307	5	5
31721	2	4
35567	3	3
41460	4	2
15369	1	1

The resulting arrangements of units in the column are presented as follows:

E B C D A
C E A B D
B D E A C
A C D E B
D A B C E

Having completed the randomization, the application of treatments is carried out in this order in the field, or the set of experiments are performed in the above order. It should be noted that another experimenter carrying out the above randomization could have come out with an arrangement different from the above discussed, which would be correct as long as the process was strictly adhered to.

The appropriate linear model for representing the Latin square design as in Equation 5.15 is

$$y_{ijk} = \mu + \beta_i + \tau_j + \gamma_k + \varepsilon_{ijk} \quad i = 1, \ldots, r \;\; j = 1, \ldots, r \;\; k = 1, \ldots, r \qquad (5.15a)$$

where μ is the grand mean, γ_k refers to the effect of treatment k or kth level of a factor, τ_j is the effect of jth level of column-wise blocking while β_i is the effect of ith level of row-wise blocking. The rows and columns of Latin square designs are used to control error. Consequently, it is only reasonable that the rows and columns should have random effects. The following assumptions are made about the model: $\beta_i \sim \text{NID}(0, \sigma_\beta^2)$; $\tau_j \sim \text{NID}(0, \sigma_\tau^2)$. The treatment effects can either be fixed (which is more often used) or random. If they are fixed, then $\sum_{k=1}^{r} \gamma_k = 0$. If they are random, then $\gamma_k \sim \text{NID}(0, \sigma_\gamma^2)$ while

$\varepsilon_{ijk} \sim \text{NID}(0, \sigma^2)$. In this model, only one of the positions of the units in the row (i) and column (j) is required to fully identify any treatment k.

As usual, the total sum of squares resulting from the responses of an experiment, performed according to this model, could be partitioned for ANOVA as follows (Table 5.12):

$$\begin{aligned}
SS_{\text{total}} &= \sum_{i=1}^{r}\sum_{j=1}^{r}\sum_{k=1}^{r}(y_{ijk} - \overline{y}_{...}) \\
&= \sum_{i=1}^{r}\sum_{j=1}^{r}\sum_{k=1}^{r}\left[(y_{ijk} - \overline{y}_{i..} - \overline{y}_{.j.} - \overline{y}_{..k} + 2\overline{y}_{...})\right]^2 + r\sum_{i=1}^{r}(\overline{y}_{i..} - \overline{y}_{...})^2 \\
&\quad + r\sum_{j=1}^{r}(\overline{y}_{.j.} - \overline{y}_{...})^2 + r\sum_{k=1}^{r}(\overline{y}_{..k} - \overline{y}_{...})^2 \\
&\quad + \text{cross-product terms whose sum is zero.}
\end{aligned}$$

From the above we can see as in Equation 5.15a that

$$SS_{\text{total}} = SS_{\text{residual}} + SS_{\text{rows}} + SS_{\text{columns}} + SS_{\text{treatment}} \tag{5.15b}$$

To obtain the sums of squares, the calculations are greatly reduced when the following formulae are used:

$$\begin{aligned}
SS_{\text{total}} &= \sum_{i=1}^{r}\sum_{j=1}^{r}\sum_{k=1}^{r} y_{ijk}^2 - \frac{y_{...}^2}{r^2} \\
SS_{\text{rows}} &= \frac{1}{r}\sum_{i=1}^{r} y_{i..}^2 - \frac{y_{...}^2}{r^2} \\
SS_{\text{columns}} &= \frac{1}{r}\sum_{j=1}^{r} y_{.j.}^2 - \frac{y_{...}^2}{r^2} \\
SS_{\text{treatment}} &= \frac{1}{r}\sum_{k=1}^{r} y_{..k}^2 - \frac{y_{...}^2}{r^2} \\
SS_{\text{residual}} &= SS_{\text{total}} - SS_{\text{rows}} - SS_{\text{columns}} - SS_{\text{treatment}}
\end{aligned}$$

TABLE 5.12: ANOVA for the Latin Square Design

Source	Sum of squares	Degrees of freedom	Mean square	*F*-ratio
Treatment	$SS_{\text{treatment}}$	$r-1$	$\frac{SS_{\text{treatment}}}{r-1}$	$\frac{MS_{\text{treatment}}}{MS_{\text{residual}}}$
Rows	SS_{rows}	$r-1$	$\frac{SS_{\text{rows}}}{r-1}$	$\frac{MS_{\text{rows}}}{MS_{\text{residual}}}$
Columns	SS_{columns}	$r-1$	$\frac{SS_{\text{columns}}}{r-1}$	$\frac{MS_{\text{columns}}}{MS_{\text{residual}}}$
Residual	SS_{residual}	$(r-1)(r-2)$	$\frac{SS_{\text{residual}}}{(r-1)(r-2)}$	
Total	SS_{total}	r^2-1		

5.4.1 Use of Latin Squares for Factorial Experiments

Although the major use of the Latin square design is for blocking out and isolating factors in which the experimenter is not primarily interested, it is possible to use the Latin square to study factors for which there are no randomization restrictions. In this case, it will be possible to study three factors assigned under row, columns, and letters with only r^2 observations. In such designs, the experimenter assumes that there is no interaction between the factors. The experiment is performed as usual; formulae and results that were previously described apply. If it is suspected that interaction between some of the factors might occur, then Latin square would be inappropriate for studying such factors.

5.5 Some Expected Mean Squares

Under the assumptions mentioned earlier, the expected mean squares for the Latin squares model when treatment effects are assumed to be fixed are as follows:

$$\begin{aligned} EMS_{\text{treatment}} &= \sigma^2 + \frac{r}{r-1}\sum_{k=1}^{r} \gamma_k^2 \\ EMS_{\text{columns}} &= \sigma^2 + r\sigma_\tau^2 \\ EMS_{\text{rows}} &= \sigma^2 + r\sigma_\beta^2 \\ EMS_{\text{residual}} &= \sigma^2 \end{aligned}$$

The hypothesis, which is usually tested in this design, investigates whether treatment effects are significant. Of course, it is of interest to the experimenter to find out whether row-wise or column-wise blocking has been effective. In this case, a significant row or column effect indicates effectiveness. Even when they are not significant, they still serve to make our ANOVA more precise and the conclusions we reach on the treatments more accurate. If the Latin square is being used for studying three factors, then the rows and column effects are tested for significance of the factorial effects which they represent. The primary hypothesis to be tested for the Latin square design (i.e., for treatment effect) can be written as

$$H_0 : \gamma_1 = \gamma_2 = \cdots = \gamma_k = 0 \quad \text{vs} \quad H_1 : \gamma_k \neq 0 \text{ for some } k$$

5.5.1 Estimation of Treatment Effects and Confidence Intervals for Differences between Two Treatments

If our aim is to obtain point estimates of treatment effects in Latin square design, we can represent the treatment mean for treatment k as μ_k, where

$$\hat{\mu}_k = \overline{y}_{..k} = \frac{y_{..k}}{r} \tag{5.15c}$$

The variance of $\overline{y}_{..k}$ is obtained as follows:

$$\text{Var}(\overline{y}_{..k}) = \frac{s^2}{r} = \frac{MS_{\text{residual}}}{r} \tag{5.16a}$$

Since $\varepsilon_{ijk} \sim \text{NID}(0, \sigma^2)$, the $100(1-\alpha)\%$ confidence interval estimate of the mean effect

$$\overline{y}_{..k} \pm t[\alpha, (r-1)(r-2)]\sqrt{\frac{MS_{\text{residual}}}{r}} \tag{5.16b}$$

If the null hypothesis is rejected, the $100(1-\alpha)\%$ confidence interval for the difference between two means $(\mu_k - \mu_m)$ is obtained as

$$(\overline{y}_{..k} - \overline{y}_{..m}) \pm t[\alpha, (r-1)(r-2)]\sqrt{\frac{2MS_{\text{residual}}}{r}} \tag{5.16c}$$

because

$$\hat{\sigma}(\overline{y}_{..k} - \overline{y}_{..m}) = \sqrt{\frac{2MS_{\text{residual}}}{r}} \tag{5.16d}$$

We now illustrate the design with the following example:

Example 5.5

A car manufacturer was concerned about two issues: rising cost of fuel and pollution from car emissions. The manufacturer decided to produce more fuel-efficient cars by carrying out an experiment to test a set of five prototypes of burners, designed to be fitted to the new generation of cars to be produced. These burners were designated (A, B, C, and D). The old burner (E), which had been in use, was then added to compare them to see whether the new burners had improved fuel efficiency. The car manufacturer planned to use five different drivers and five different cars in the experiments. Realizing that the skills of the drivers could be different and suspecting that cars could have an effect on fuel consumption, it was decided to use a 5×5 Latin square design to study the efficiency of the burners. Randomization led to the following layout. Experiments were performed, changing the burners in each car in the order shown. Cars were driven on the same road, at the same speed, for the same distance before fuel consumption was measured in miles per gallon.

	Drivers					
Cars	**1**	**2**	**3**	**4**	**5**	**Total**
1	E 54	A 76	C 60	D 59	B 60	309
2	D 57	E 50	B 69	C 54	A 70	300
3	C 53	D 66	A 75	B 59	E 50	303
4	B 60	C 60	E 53	A 64	D 60	297
5	A 64	B 70	D 69	E 51	C 56	310
Total	288	322	326	287	296	1519

Analysis: Treatment totals

A	**B**	**C**	**D**	**E**
349	318	283	311	258

The null hypothesis of major interest is as follows:

$$H_0 : \gamma_1 = \gamma_2 = \cdots = \gamma_5 = 0 \quad \text{vs} \quad H_1 : \gamma_k \neq 0 \quad \text{for } k = 1, 2, 3, 4, 5$$

The sums of squares are obtained as follows:

$$\begin{aligned} SS_{\text{total}} &= (54^2 + 76^2 + \cdots + 51^2 + 56^2) - (1519)^2/25 \\ &= 93669 - 92294.44 = 1374.56 \\ SS_{\text{treatment}} &= (349^2 + 318^2 + 283^2 + 311^2 + 258^2)/5 - (1519)^2/25 \\ &= 466299/5 - 92294.44 = 965.36 \\ SS_{\text{rows}} &= (309^2 + 300^2 + 303^2 + 297^2 + 310^2)/5 - (1519)^2/25 \\ &= 461599/5 - 92294.44 = 25.36 \\ SS_{\text{columns}} &= (288^2 + 322^2 + 326^2 + 287^2 + 296^2)/5 - (1519)^2/25 \\ &= 462889/5 - 92294.44 = 283.36 \\ SS_{\text{residual}} &= 1374.56 - 965.36 - 25.36 - 283.6 = 100.48 \end{aligned}$$

By carrying out the ANOVA, we obtain the data shown in Table 5.13.

TABLE 5.13: ANOVA for the Car Burner Experiment

Source	**Sum of squares**	**Degrees of freedom**	**Mean square**	***F*-ratio**
Treatment (burner)	965.36	4	241.34	28.82
Rows	25.36	4	6.34	0.76
Columns	283.36	4	70.84	8.46
Residual	100.48	12	8.37	
Total	1374.56	24		

From the tables, $F(4, 12, 0.01) = 5.412$, which indicates that only the driver (columns) and treatment effects are significant. The rows (cars) effect is not significant. The design used was effective because the sources of extraneous errors were blocked off, thereby making the test more sensitive in detecting the effects of the treatments. Besides the necessary checks that should be carried out on model adequacy, comparison of means could be carried out using Duncan's multiple range test, which was discussed in Chapter 2. The comparison is left as an exercise for the reader. Since the treatment effects are significant, we may wish to obtain the $100(1-\alpha)\%$ confidence intervals for the differences between any two of the means of A, B, C, D, and E. In particular, we may wish to find the confidence intervals for the differences between the means for the new burners and the old model, E. Here, we carry out the test. First, we note that

$$\overline{Y}_{\mathrm{A}} = 69.8 \quad \overline{Y}_{\mathrm{B}} = 79.5 \quad \overline{Y}_{\mathrm{C}} = 56.6 \quad \overline{Y}_{\mathrm{D}} = 62.2 \quad \overline{Y}_{\mathrm{E}} = 51.60$$

so that 99% confidence interval for the difference in the mean performances of burners A and E is

$$\overline{Y}_{\mathrm{A}} - \overline{Y}_{\mathrm{E}} \pm t[\alpha/2, (r-1)(r-2)]\sqrt{\frac{2MS_{\mathrm{residual}}}{r}}$$

$$\text{i.e., } 18.2 \pm 3.055\sqrt{[2(8.373/5)]}$$

$$18.2 \pm 5.49$$

$$(12.710, 23.690)$$

The 99% confidence interval for the difference between B and E is

$$63.6 - 51.6 \pm 5.49$$

$$12.00 \pm 5.49$$

$$(6.51, 17.49)$$

The 99% confidence interval for the difference between C and E is

$$56.6 - 51.6 \pm 5.49$$

$$5.0 \pm 5.49$$

$$(-0.49, 10.49)$$

The 99% confidence interval for difference between D and E is

$$62.2 - 51.6 \pm 5.49$$

$$10.6 \pm 5.49$$

$$(5.11, 16.09)$$

The $100(1-\alpha)\%$ confidence intervals for differences between other means can similarly be found. These confidence intervals indicate differences between E and A, B, and D, but not C.

Next, we carry out the same analysis using SAS. We write a SAS program and present its output. □

SAS PROGRAM

```
data latin4_1;
input cars drivers burners y@@;
datalines;
1 1 5 54 1 2 1 76 1 3 3 60 1 4 4 59 1 5 2 60
2 1 4 57 2 2 5 50 2 3 2 69 2 4 3 54 2 5 1 70
3 1 3 53 3 2 4 66 3 3 1 75 3 4 2 59 3 5 5 50
4 1 2 60 4 2 3 60 4 3 5 53 4 4 1 64 4 5 4 60
5 1 1 64 5 2 2 70 5 3 4 69 5 4 5 51 5 5 3 56
;
proc anova data=latin4_1;
class cars drivers burners;
model y=cars drivers burners;
means burners/lsd;
run;
```

SAS OUTPUT

```
                     The ANOVA Procedure

                   Class Level Information

                 Class      Levels        Values

                 cars            5        1 2 3 4 5

                 drivers         5        1 2 3 4 5

                 burners         5        1 2 3 4 5

Dependent Variable: y

                                     Sum of
    Source                 DF       Squares    Mean Square    F Value    Pr > F

    Model                  12   1274.080000     106.173333      12.68    <.0001

    Error                  12    100.480000       8.373333

    Corrected Total        24   1374.560000

                  R-Square     Coeff Var     Root MSE       y Mean

                  0.926900      4.762461     2.893671     60.76000
```

```
Source                   DF       Anova SS     Mean Square    F Value  Pr > F

cars                      4     25.3600000       6.3400000       0.76  0.5725
drivers                   4    283.3600000      70.8400000       8.46  0.0017
burners                   4    965.3600000     241.3400000      28.82  <.0001

                          t Tests (LSD) for y

NOTE: This test controls the Type I comparisonwise error rate, not the
experimentwise error rate.

                          Alpha                             0.05
                          Error Degrees of Freedom            12
                          Error Mean Square             8.373333
                          Critical Value of t            2.17881
                          Least Significant Difference 3.9875

        Means with the same letter are not significantly different.

            t Grouping          Mean       N    burners

                     A        69.800       5    1
                     B        63.600       5    2
                     B
                     B        62.200       5    4
                     C        56.600       5    3
                     D        51.600       5    5
```

The analysis of the data using PROC ANOVA in SAS confirms the results of the manual analysis we previously carried out. In the comparison of means, the use of LSD, which is akin to confidence interval, confirms that only treatments C and E have means, which are not significantly different from each other.

5.6 Replications in Latin Square Design

One disadvantage of a small Latin square experiment is that it provides very small number of error degrees of freedom. The consequence of this is that the estimated σ^2 might become very large and therefore restrict the experimenter's ability to detect any significant effects in the design. For instance a 3×3 Latin square design provides only 2 degrees of freedom for error. The 4×4 Latin square has only 4 error degrees of freedom. When small Latin squares are used, it is necessary to replicate them to increase the error degrees of freedom. Replications in this, like in every other design, help to improve the efficiency of the design and strengthen the conclusions reached. There are several ways of replicating a Latin square design. Here, we have discussed only four. We will see how they apply without loss of generality to the design in Example 5.5. Suppose that the Latin square with 5×5 units, designed for the test of the car engine burners used in Example 5.5 was to be replicated n times, then the four possible designs are as follows:

Case I
Design the experiments using the same drivers and the same cars for all the replicates of the design. □

Case II
Use the same cars for all the replicates of the experiments and use different drivers for each replicate. □

Case III
Change the cars used for each replicate but retain the same drivers throughout the experiments. □

Case IV
The next design is to change all the cars and all the drivers for all the experiments. □

From the point of view of economy, all the designs mentioned may not have equal appeal. However, there are possible ways of designing the experiment to study the effects of the burners, and their different analyses. Each of the designs leads to a different way of calculating the sums of squares for use in ANOVA. The calculations are discussed as follows:

CASE I
A Latin square design with $r \times r$ elements and n replicates.

Let each response be y_{ijkl}, where $l = 1, \ldots, n$ and $i, j, k = 1, \ldots, r$. For such a design,

$$\begin{aligned}
SS_{\text{total}} &= \sum_{i=1}^{r}\sum_{j=1}^{r}\sum_{k=1}^{r}\sum_{l=1}^{n} y_{ijkl}^2 - \frac{y_{....}^2}{nr^2} \\
SS_{\text{rows}} &= \frac{1}{nr}\sum_{i=1}^{r} y_{i...}^2 - \frac{y_{....}^2}{nr^2} \\
SS_{\text{columns}} &= \frac{1}{nr}\sum_{j=1}^{r} y_{.j..}^2 - \frac{y_{....}^2}{nr^2} \\
SS_{\text{treatments}} &= \frac{1}{nr}\sum_{i=1}^{r} y_{..k.}^2 - \frac{y_{....}^2}{nr^2} \\
SS_{\text{replicates}} &= \frac{1}{r^2}\sum_{l=1}^{r} y_{...l}^2 - \frac{y_{....}^2}{nr^2} \\
SS_{\text{residual}} &= SS_{\text{total}} - SS_{\text{rows}} - SS_{\text{columns}} - SS_{\text{treats}} - SS_{\text{replicates}}
\end{aligned} \tag{5.16e}$$

Table 5.14 presents the ANOVA table for this design. □

TABLE 5.14: ANOVA for Replicated Latin Square Design (Case I)

Source	Sum of squares	Degrees of freedom	Mean square	*F*-ratio
Treatment	$SS_{\text{treatment}}$	$r-1$	$\frac{SS_{\text{treatment}}}{r-1}$	$\frac{MS_{\text{treatment}}}{MS_{\text{residual}}}$
Columns	SS_{columns}	$r-1$	$\frac{SS_{\text{columns}}}{r-1}$	$\frac{MS_{\text{columns}}}{MS_{\text{residual}}}$
Rows	SS_{rows}	$r-1$	$\frac{SS_{\text{rows}}}{r-1}$	$\frac{MS_{\text{rows}}}{MS_{\text{residual}}}$
Replicates	$SS_{\text{replicates}}$	$n-1$	$\frac{SS_{\text{replicates}}}{n-1}$	$\frac{MS_{\text{replicates}}}{MS_{\text{residual}}}$
Residual	SS_{residual}	$(r-1)[n(r+1)-3]$	$\frac{SS_{\text{residual}}}{(r-1)[n(r+1)-3]}$	
Total	SS_{total}	nr^2-1		

CASE II

A $r \times r$ Latin square design with n replicates.

As in Case I, let each response be y_{ijkl}, $l=1,\ldots,n; i,j,k=1,\ldots,r$;

$$SS_{\text{total}} = \sum_{i=1}^{r}\sum_{j=1}^{r}\sum_{k=1}^{r}\sum_{l=1}^{n} y_{ijkl}^2 - \frac{y_{....}^2}{nr^2}$$

$$SS_{\text{treatments}} = \frac{1}{nr}\sum_{k=1}^{r} y_{..k.}^2 - \frac{y_{....}^2}{nr^2}$$

$$SS_{\text{columns}} = \frac{1}{r}\sum_{i=1}^{r}\sum_{l=1}^{n} y_{i..l.}^2 - \frac{1}{r^2}\sum_{l=1}^{n} y_{...l}^2$$

$$SS_{\text{columns}} = \frac{1}{nr}\sum_{j=1}^{r} y_{.j..}^2 - \frac{y_{....}^2}{nr^2}$$

$$SS_{\text{replicates}} = \frac{1}{r^2}\sum_{l=1}^{n} y_{...l}^2 - \frac{y_{....}^2}{nr^2}$$

$$SS_{\text{E}} = SS_{\text{total}} - SS_{\text{rows}} - SS_{\text{columns}} - SS_{\text{replicates}} \tag{5.17}$$

The ANOVA table for this design is shown in Table 5.15. □

CASE III

A $r \times r$ Latin square design with n replicates:

The design and its ANOVA are virtually the same for Case II but for a few changes, namely, those related to SS_{rows} (Table 5.16). For this design, the sum of squares for the rows reverts to SS_{rows} for Case I, with $r-1$ degrees of freedom; SS_{columns} changes to

$$SS_{\text{columns}} = \frac{1}{r}\sum_{j=1}^{r}\sum_{l=1}^{n} y_{.j.l}^2 - \frac{1}{r^2}\sum_{l=1}^{n} y_{...l}^2$$

with $n(r-1)$ degrees of freedom. □

CASE IV

A $r \times r$ Latin square design with n replicates.

The sums of squares are calculated as

$$SS_{\text{rows}} = \frac{1}{r}\sum_{i=1}^{r}\sum_{l=1}^{n} y_{i..l}^2 - \frac{1}{r^2}\sum_{l=1}^{n} y_{...l}^2$$

$$SS_{\text{columns}} = \frac{1}{r}\sum_{j=1}^{r}\sum_{l=1}^{n} y_{.j.l}^2 - \frac{1}{r^2}\sum_{l=1}^{n} y_{...l}^2 \tag{5.18}$$

Each of the rows and columns sums of squares, has $n(r-1)$ degrees of freedom. The rest of the sums of squares are calculated as in Case I above. The ANOVA for this design is presented in Table 5.17. □

TABLE 5.15: ANOVA for Replicated Latin Square Design (Case II)

Source	Sum of squares	Degrees of freedom	Mean square	F-ratio
Treatment	$SS_{\text{treatment}}$	$r-1$	$\frac{SS_{\text{treatment}}}{r-1}$	$\frac{MS_{\text{treatment}}}{MS_{\text{residual}}}$
Columns	SS_{columns}	$r-1$	$\frac{SS_{\text{columns}}}{r-1}$	$\frac{MS_{\text{columns}}}{MS_{\text{residual}}}$
Rows	SS_{rows}	$n(r-1)$	$\frac{SS_{\text{rows}}}{r-1}$	$\frac{MS_{\text{rows}}}{MS_{\text{residual}}}$
Replicates	$SS_{\text{replicates}}$	$n-1$	$\frac{SS_{\text{replicates}}}{n-1}$	$\frac{MS_{\text{replicates}}}{MS_{\text{residual}}}$
Residual	SS_{residual}	$(r-1)(nr-2)$	$\frac{SS_{\text{residual}}}{(r-1)(nr-2)}$	
Total	SS_{total}	nr^2-1		

TABLE 5.16: ANOVA for Replicated Latin Square Design (Case III)

Source	Sum of squares	Degrees of freedom	Mean square	F-ratio
Treatment	$SS_{\text{treatment}}$	$r-1$	$\frac{SS_{\text{treatment}}}{r-1}$	$\frac{MS_{\text{treatment}}}{MS_{\text{residual}}}$
Columns	SS_{columns}	$n(r-1)$	$\frac{SS_{\text{columns}}}{n(r-1)}$	$\frac{M_{\text{columns}}}{MS_{\text{residual}}}$
Rows	SS_{rows}	$r-1$	$\frac{SS_{\text{rows}}}{r-1}$	$\frac{MS_{\text{rows}}}{MS_{\text{residual}}}$
Replicates	$SS_{\text{replicates}}$	$n-1$	$\frac{SS_{\text{replicates}}}{n-1}$	$\frac{MS_{\text{replicates}}}{MS_{\text{residual}}}$
Residual	SS_{residual}	$(r-1)(nr-2)$	$\frac{SS_{\text{residual}}}{(r-1)(nr-2)}$	
Total	SS_{total}	nr^2-1		

TABLE 5.17: ANOVA for Replicated Latin Square Design (Case IV)

Source	Sum of squares	Degrees of freedom	Mean square	F-ratio
Treatment	$SS_{\text{treatment}}$	$r-1$	$\frac{SS_{\text{treatment}}}{r-1}$	$\frac{MS_{\text{treatment}}}{MS_{\text{residual}}}$
Columns	SS_{columns}	$n(r-1)$	$\frac{SS_{\text{columns}}}{n(r-1)}$	$\frac{M_{\text{columns}}}{MS_{\text{residual}}}$
Rows	SS_{rows}	$n(r-1)$	$\frac{SS_{\text{rows}}}{n(r-1)}$	$\frac{MS_{\text{rows}}}{MS_{\text{residual}}}$
Replicates	$SS_{\text{replicates}}$	$n-1$	$\frac{SS_{\text{replicates}}}{n-1}$	$\frac{MS_{\text{replicates}}}{MS_{\text{residual}}}$
Residual	SS_{residual}	$(r-1)[n(r-1)-1]$	$\frac{SS_{\text{residual}}}{(r-1)[n(r-1)-1]}$	
Total	SS_{total}	nr^2-1		

Below we present an example of a replicated $r \times r$ Latin square design. This is an example in the mold of Case III above in which the columns are kept constant throughout the replication.

Example 5.6

A company, which manufactures high-quality paper fiber from cotton, is carrying out research that should lead to improved production methods. They have just devised three new manufacturing processes, which they believe would improve the quantity of paper they could produce from a given weight of raw material (cotton). They are aware that the quantity produced may depend on the skills of the operators. The cotton being used for the manufacture of these products is obtained from four sources: Nigeria, United States of America, India, and Uzbekistan. It is believed that the sources of raw material may affect the output of the processes; consequently, they decided to use 12 operators for the experiment. Since it is also the intention to compare the new methods with the old method, which has been in use, they decided to use a 4×4 Latin square design in three replicates, with 12 different operators being used in each replicate. The sources of raw material for the different manufacturing processes are represented in the design by the columns. The rows represent the operators while the old manufacturing process is represented by the letter A, the rest (i.e., B, C, and D) are the new methods. The arrangements of the design after randomization are shown below. The responses are the number of standard A4 sheets produced from 1 kg of raw material. ▯

	Source of cotton				
			Replicate I		
Operators ↓	**I**	**II**	**III**	**IV**	**Total**
I	C 285	B 309	D 287	A 211	1092
II	A 208	C 239	B 317	D 249	1013
III	B 285	D 286	A 212	C 295	1078
IV	D 265	A 209	C 324	B 301	1099
			Replicate II		
V	B 284	A 215	D 278	C 278	1055
VI	A 208	D 279	C 314	B 311	1112
VII	C 264	B 312	A 213	D 290	1079
VIII	D 290	C 301	B 291	A 201	1083
			Replicate III		
IX	D 280	A 204	C 296	B 285	1065
X	B 265	C 243	D 293	A 211	1012
XI	C 285	B 275	A 210	D 249	1019
XII	A 218	D 290	B 260	C 275	1043
Total	3137	3162	3295	3156	12,750

We analyze the data

Treatments	A	B	C	D
Means	2520	3495	3399	3336

$$\begin{aligned}
SS_{\text{total}} &= (285^2 + 309^2 + \cdots + 260^2 + 275^2) - (12750)^2/48 \\
&= 3451408 - 3386718.75 = 64689.25 \\
SS_{\text{treatment}} &= (2520^2 + 3495^2 + 3399^2)/12 - (12750)^2/48 \\
&= 3437293.5 - 3386718.75 = 50574.75 \\
SS_{\text{rows}} &= (1092^2 + 1013^2 + \cdots + 1019^2 + 1043^2)/4 \\
&\quad -(4139^2 + 4329^2 + 4282^2)/16 \\
&= (13559796)/4 - 54207086/16 = 2006.13 \\
SS_{\text{columns}} &= (3137^2 + 3162^2 + 3295^2 + 3156^2)/12 - (12750)^2/48 \\
&= 3388031.17 - 3386718.75 = 1312.42
\end{aligned}$$

TABLE 5.18: ANOVA for the Paper Company Experiment

Source	Sum of squares	Degrees of freedom	Mean square	F-ratio
Treatments	50574.75	3	16858.25	52.84
Replicates	1224.13	2	612.06	1.91
Operators	2006.13	9	222.89	<1.0
Raw material	1312.42	3	437.33	1.37
Residuals	9571.82	30	319.10	
Total	64689.25	47		

$$SS_{\text{replicates}} = (4139^2 + 4329^2 + 4282^2)/16 - (12750)^2/48$$
$$= 54207086/16 - 3386718.75 = 1224.13$$

$$SS_{\text{residual}} = 64689.25 - 2006.13 - 1224.13 - 50574.75 = 9571.82$$

The hypothesis to be tested is shown in Table 5.18.

$$H_0 : \gamma_1 = \gamma_2 = \gamma_3 = \gamma_4 = 0 \quad \text{vs} \quad H_1 : \gamma_k \neq 0 \quad \text{for some } k = 1, 2, 3, 4$$

Since F (3, 30, 0.01) = 7.56, we reject the hypothesis of no treatment effect. This means that there is a significant difference between the means for the manufacturing processes. We can use the least significant difference (LSD) described in Chapter 2 to compare the means of the production methods. The means of different methods are given as follows:

Method	A	B	C	D
Mean	210	291.25	283.25	278

Standard error = 7.98

Since we rejected the hypothesis at 1% level of significance, we calculate the LSD to study how the means are related to each other. Since $s^2 = 319.10$ and $t[\alpha/2, (n-1)(nr-2)] = t(0.005, 30) = 2.750$,

$$\text{LSD} = t[\alpha/2, (n-1)(nr-2)]\sqrt{\frac{2MS_{\text{residual}}}{nr}}$$

i.e.,

$$\text{LSD} = 2.750 \times 7.2927 = 20.05$$

To compare the means, we rank them in ascending order of their magnitudes

$$Y_{\text{A}} < Y_{\text{D}} < Y_{\text{C}} < Y_{\text{B}}$$

The differences between these means are as follows:

$$\overline{Y}_{\text{B}} - \overline{Y}_{\text{C}} = 8.0$$
$$\overline{Y}_{\text{B}} - \overline{Y}_{\text{D}} = 13.25$$
$$\overline{Y}_{\text{B}} - \overline{Y}_{\text{A}} = 81.25$$
$$\overline{Y}_{\text{C}} - \overline{Y}_{\text{D}} = 5.25$$
$$\overline{Y}_{\text{C}} - \overline{Y}_{\text{A}} = 73.25$$
$$\overline{Y}_{\text{D}} - \overline{Y}_{\text{A}} = 68.0$$

From our analysis, we can conclude that the new methods, B, C, and D are not significantly different from each other because the differences between their means are all below the LSD. They are all significantly different from the old method A, indicating clearly that the new methods have improved the efficiency of paper fiber production. The next SAS program carries out the same analysis for Example 5.2. We present the output, which agrees with the findings we obtained by manual analysis.

SAS PROGRAM

```
data latin4_2;

input operator source method y@@;
datalines;
1 1 3 285 1 2 2 309 1 3 4 287 1 4 1 211 2 1 1 208 2 2 3 239 2 3 2 317 2 4 4
249 3 1 2 285 3 2 4 286 3 3 1 212 3 4 3 295 4 1 4 265 4 2 1 209 4 3 3 324 4 4
2 301 5 1 2 284 5 2 1 215 5 3 4 278 5 4 3 278 6 1 1 208 6 2 4 279 6 3 3 314 6
4 2 311 7 1 3 264 7 2 2 312 7 3 1 213 7 4 4 290 8 1 4 290 8 2 3 301 8 3 2 291
8 4 1 201 9 1 4 280 9 2 1 204 9 3 3 296 9 4 2 285 10 1 2 265 10 2 3 243 10 3
4 293 10 4 1 211 11 1 3 285 11 2 2 275 11 3 1 210 11 4 4 249 12 1 1 218 12 2
4 290 12 3 2 260 12 4 3 275
;
proc print data=latin4_2; sum y;
proc anova data=latin4_2;
class operator source method;
model y=operator source method;
means method/lsd; run;
```

SAS OUTPUT

Obs	operator	source	method	y
1	1	1	3	285
2	1	2	2	309
3	1	3	4	287
4	1	4	1	211
5	2	1	1	208
6	2	2	3	239
7	2	3	2	317
8	2	4	4	249
9	3	1	2	285
10	3	2	4	286
11	3	3	1	212
12	3	4	3	295
13	4	1	4	265
14	4	2	1	209
15	4	3	3	324
16	4	4	2	301
17	5	1	2	284
18	5	2	1	215
19	5	3	4	278
20	5	4	3	278
21	6	1	1	208
22	6	2	4	279

```
 Obs    operator    source    method      y

  23        6          3         3       314
  24        6          4         2       311
  25        7          1         3       264
  26        7          2         2       312
  27        7          3         1       213
  28        7          4         4       290
  29        8          1         4       290
  30        8          2         3       301
  31        8          3         2       291
  32        8          4         1       201
  33        9          1         4       280
  34        9          2         1       204
  35        9          3         3       296
  36        9          4         2       285
  37       10          1         2       265
  38       10          2         3       243
  39       10          3         4       293
  40       10          4         1       211
  41       11          1         3       285
  42       11          2         2       275
  43       11          3         1       210
  44       11          4         4       249
  45       12          1         1       218
  46       12          2         4       290
  47       12          3         2       260
  48       12          4         3       275
                                        =====
                                        12750
                The ANOVA Procedure

             Class Level Information

 Class          Levels    Values

 operator           12    1 2 3 4 5 6 7 8 9 10 11 12

 source              4    1 2 3 4

 method              4    1 2 3 4

        Number of Observations Read          48
        Number of Observations Used          48

                  The ANOVA Procedure

Dependent Variable: y
                                Sum of
  Source               DF      Squares     Mean Square    F Value    Pr > F

  Model                17  55117.41667      3242.20098      10.16    <.0001

  Error                30   9571.83333       319.06111

  Corrected Total      47  64689.25000
```

```
            R-Square     Coeff Var      Root MSE        y Mean

            0.852034      6.724624      17.86228      265.6250

Source              DF        Anova SS    Mean Square    F Value    Pr > F

operator            11      3230.25000      293.65909       0.92    0.5341
source               3      1312.41667      437.47222       1.37    0.2705
method               3     50574.75000    16858.25000      52.84    <.0001


                        The ANOVA Procedure

                        t Tests (LSD) for y

 NOTE: This test controls the Type I comparisonwise error rate, not the
experimentwise error rate.

                       Alpha                              0.05
                       Error Degrees of Freedom             30
                       Error Mean Square              319.0611
                       Critical Value of t             2.04227
                       Least Significant Difference     14.893

              Means with the same letter are not significantly different.

                t Grouping          Mean       N    method

                         A       291.250      12    2
                         A
                         A       283.250      12    3
                         A
                         A       278.000      12    4

                         B       210.000      12    1
```

We again illustrate with another example, using the model represented by Case I above.

Example 5.7

In an industrial experiment, the effects of five catalysts on the reaction time of a chemical process were being studied. Five laboratories were being used for the study along with raw material obtained from five different mines. To isolate and eliminate laboratory differences and raw material variations due to source, a 5×5 Latin square design was used. The coded form of the responses (reaction times) obtained is presented below. Suppose the above design for testing catalysts was replicated in accordance with Case I above. The three replicates were run and the outcome of the experiments were as follows:

	Laboratories				
Mines	**1**	**2**	**3**	**4**	**5**
1	A = 3	B = 1	C = 2	D = 11	E = 6
2	B = −4	C = 3	D = 11	E = 8	A = 8
3	C = −3	D = 7	E = 5	A = 9	B = 3
4	D = 5	E = 12	A = 7	B = 5	C = 4
5	E = 1	A = 11	B = 1	C = 11	D = 13

	Laboratories				
Mines	**1**	**2**	**3**	**4**	**5**
1	A = 2	B = 4	C = 2	D = 5	E = 3
2	B = −2	C = 3	D = 10	E = 5	A = 4
3	C = −5	D = 4	E = 3	A = 8	B = 3
4	D = 4	E = 9	A = 6	B = 3	C = 4
5	E = 2	A = 8	B = 3	C = 10	D = 8

	Laboratories				
Mines	**1**	**2**	**3**	**4**	**5**
1	A = 3	B = 2	C = 2	D = 5	E = 4
2	B = −4	C = 3	D = 11	E = 8	A = 5
3	C = −4	D = 7	E = 4	A = 9	B = 3
4	D = 5	E = 7	A = 4	B = 5	C = 3
5	E = 3	A = 10	B = 3	C = 9	D = 10

In this case, raw materials from five different mines are used in the entire design.

Total	**1**	**2**	**3**	**4**	**5**
Laboratories	6	91	74	105	81
Mines	49	69	53	83	103
Catalysts	97	26	44	110	80
Replicates	134	106	117		
Overall	357				

$$
\begin{aligned}
SS_{\text{total}} &= \left[3^2 + (-4)^2 + \cdots + 3^2 + 10^2\right] - 357^2/3(25) = 1115.68 \\
SS_{\text{catalysts}} &= (1/15)(97^2 + 26^2 + \cdots + 80^2) - 357^2/(3)(25) = 335.4133 \\
SS_{\text{mines}} &= (1/15)(49^2 + 69^2 + \cdots + 103^2) - 357^2/(3)(25) = 131.9467 \\
SS_{\text{laboratories}} &= (1/15)(6^2 + 91^2 + 74^2 + 105^2 + 81^2) - 357^2/(3)(25) = 392.6163 \\
SS_{\text{replicates}} &= (1/25)(134^2 + 117^2 + 106^2) - 357^2/(3)(25) = 15.92 \\
SS_{\text{E}} &= 1115.68 - 335.4133 - 131.9467 - 388.6163 - 15.92 = 239.7837
\end{aligned}
$$

TABLE 5.19: ANOVA for the Replicated Catalyst Experiment

Source	Sum of squares	Degrees of freedom	Mean square	F-ratio
Treatment (catalysts)	335.4133	4	83.8533	20.9822
Mines	131.9467	4	32.9867	8.2541
Laboratories	392.61	4	98.1525	24.56
Replicates	15.92	2	7.96	1.992
Error	239.7837	60	3.9964	
Total	1115.680	74		

$F(4, 60, 0.01) = 3.65$; $F(2, 60, 0.01) = 4.98$; $F(2, 60, 0.05) = 3.15$. All effects are significant at 1% level of significance except the replicate effect, which is not even significant at 10% level of significance (Table 5.19).

Next, we present the SAS analysis of the responses of this experiment. □

SAS PROGRAM

```
data latinsq2;
input mines lab catalyst replicate yield@@;
datalines;
1 1 1 1 3 1 2 2 1 1 1 3 3 1 2 1 4 4 1 5 1 5 5 1 6 2 1 2 1 -4
2 2 3 1 3 2 3 4 1 11 2 4 5 1 8 2 5 1 1 8
3 1 3 1 -3 3 2 4 1 7 3 3 5 1 5 3 4 1 1 9 3 5 2 1 3
4 1 4 1 5 4 2 5 1 12 4 3 1 1 7 4 4 2 1 5 4 5 3 1 4
5 1 5 1 1 5 2 1 1 11 5 3 2 1 1 5 4 3 1 11 5 5 4 1 13
1 1 1 2 2 1 2 2 2 4 1 3 3 2 2 1 4 4 2 5 1 5 5 2 3
2 1 2 2 -2 2 2 3 2 3 2 3 4 2 10 2 4 5 2 5 2 5 1 2 4
3 1 3 2 -5 3 2 4 2 4 3 3 5 2 3 3 4 1 2 8 3 5 2 2 3
4 1 4 2 4 4 2 5 2 9 4 3 1 2 6 4 4 2 2 3 4 5 3 2 4 5 1 5 2 2
5 2 1 2 8 5 3 2 2 3 5 4 3 2 10 5 5 4 2 8 1 1 1 3 3 1 2 2 3
2 1 3 3 3 2 1 4 4 3 5 1 5 5 3 4 2 1 2 3 -4 2 2 3 3 3 2 3
4 3 11 2 4 5 3 8 2 5 1 3 5 3 1 3 3 -4 3 2 4 3 7 3 3 5 3 4
3 4 1 3 9 3 5 2 3 3 4 1 4 3 5 4 2 5 3 7 4 3 1 3 4 4 4 2 3 5
4 5 3 3 3 5 1 5 3 3 5 2 1 3 10 5 3 2 3 3 5 4 3 3 9 5 5 4 3 10
;
proc print data=latinsq2; run;
proc anova  data=latinsq2;class mines lab catalyst replicate;
model yield=mines lab catalyst replicate;run;
```

OUTPUT

Obs	mines	lab	catalyst	replicate	yield
1	1	1	1	1	3
2	1	2	2	1	1
3	1	3	3	1	2
4	1	4	4	1	5
5	1	5	5	1	6
6	2	1	2	1	-4
7	2	2	3	1	3
8	2	3	4	1	11
9	2	4	5	1	8

Obs	mines	lab	catalyst	replicate	yield
10	2	5	1	1	8
11	3	1	3	1	-3
12	3	2	4	1	7
13	3	3	5	1	5
14	3	4	1	1	9
15	3	5	2	1	3
16	4	1	4	1	5
17	4	2	5	1	12
18	4	3	1	1	7
19	4	4	2	1	5
20	4	5	3	1	4
21	5	1	5	1	1
22	5	2	1	1	11
23	5	3	2	1	1
24	5	4	3	1	11
25	5	5	4	1	13
26	1	1	1	2	2
27	1	2	2	2	4
28	1	3	3	2	2
29	1	4	4	2	5
30	1	5	5	2	3
31	2	1	2	2	-2
32	2	2	3	2	3
33	2	3	4	2	10
34	2	4	5	2	5
35	2	5	1	2	4
36	3	1	3	2	-5
37	3	2	4	2	4
38	3	3	5	2	3
39	3	4	1	2	8
40	3	5	2	2	3
41	4	1	4	2	4
42	4	2	5	2	9
43	4	3	1	2	6
44	4	4	2	2	3
45	4	5	3	2	4
46	5	1	5	2	2
47	5	2	1	2	8
48	5	3	2	2	3
49	5	4	3	2	10
50	5	5	4	2	8
51	1	1	1	3	3
52	1	2	2	3	2
53	1	3	3	3	2
54	1	4	4	3	5
55	1	5	5	3	4
56	2	1	2	3	-4
57	2	2	3	3	3
58	2	3	4	3	11
59	2	4	5	3	8
60	2	5	1	3	5
61	3	1	3	3	-4

Obs	mines	lab	catalyst	replicate	yield
62	3	2	4	3	7
63	3	3	5	3	4
64	3	4	1	3	9
65	3	5	2	3	3
66	4	1	4	3	5
67	4	2	5	3	7
68	4	3	1	3	4
69	4	4	2	3	5
70	4	5	3	3	3
71	5	1	5	3	3
72	5	2	1	3	10
73	5	3	2	3	3
74	5	4	3	3	9
75	5	5	4	3	10

The ANOVA Procedure

Class Level Information

Class	Levels	Values
mines	5	1 2 3 4 5
lab	5	1 2 3 4 5
catalyst	5	1 2 3 4 5
replicate	3	1 2 3

Number of Observations Read 75
Number of Observations Used 75

The ANOVA Procedure

Dependent Variable: yield

Source	DF	Sum of Squares	Mean Square	F Value	Pr > F
Model	14	875.893333	62.563810	15.65	<.0001
Error	60	239.786667	3.996444		
Corrected Total	74	1115.680000			

R-Square	Coeff Var	Root MSE	yield Mean
0.785076	41.99813	1.999111	4.760000

Source	DF	Anova SS	Mean Square	F Value	Pr > F
mines	4	131.9466667	32.9866667	8.25	<.0001
lab	4	392.6133333	98.1533333	24.56	<.0001
catalyst	4	335.4133333	83.8533333	20.98	<.0001
replicate	2	15.9200000	7.9600000	1.99	0.1454

5.6.1 Treatment of Residuals in Latin Squares

As we discussed in Chapter 2, as part of the analysis, it is always necessary after each investigation to obtain and plots of the residuals to check the adequacy of the model. In the case of the Latin squares, the residuals are calculated as

$$\varepsilon_{ijk} = y_{ijk} - \hat{y}_{ijk} = y_{ijk} - \overline{y}_{i..} - \overline{y}_{.j.} - \overline{y}_{..k} + 2\overline{y}$$

The residuals should be plotted against the rows, columns, and treatments to check that basic assumptions about the model are not violated. Normal probability plots of residuals should indicate whether there is a marked departure from the normality assumption for the errors. We can also investigate homogeneity of variances.

5.6.2 Estimation of Missing Value in Unreplicated Latin Square Designs

Sometimes in a Latin square design, one observation may be missing. The missing value may be estimated for the $r \times r$ Latin square by least squares analysis. The method employed is discussed below. Here, estimation for the unreplicated Latin square design is treated. The procedure can be used to estimate the missing values when the Latin square design is replicated. It can also be used for the estimation of more than one missing value by adopting a sequential approach.

The model for the unreplicated Latin square design with fixed effects is

$$y_{ijk} = \mu + \alpha_i + \beta_j + \gamma_k + \varepsilon_{ijk} \quad i, j, k = 1, \ldots, r$$

$$\varepsilon_{ijk} \sim \text{NID}(0, \sigma^2) \quad \sum_{i=1}^{r} \alpha_i = \sum_{j=1}^{r} \beta_j = \sum_{k=1}^{r} \gamma_k = 0$$

To estimate the missing value, we need to choose a value which minimizes the variance, that is, the error sum of squares or the residual sum of squares. From this model, the sum of squares for error is obtained as

$$SS_{\text{residual}} = SS_{\text{total}} - SS_{\text{rows}} - SS_{\text{columns}} - SS_{\text{Latin (or treatment)}}$$

This can be written as

$$SS_{\text{E}} = \sum_{i=1}^{r}\sum_{j=1}^{r}\sum_{k=1}^{r} y_{ijk}^2 - \frac{y_{...}^2}{r^2} - \frac{\sum_{j=1}^{r} y_{.j.}^2}{r} + \frac{y_{...}^2}{r^2} - \frac{\sum_{k=1}^{r} y_{..k}^2}{r} + \frac{y_{...}^2}{r^2} - \frac{\sum_{i=1}^{r} y_{i..}^2}{r} + \frac{y_{...}^2}{r^2}$$

$$= \sum_{i=1}^{r}\sum_{j=1}^{r}\sum_{k=1}^{r} y_{ijk}^2 - \frac{y_{...}^2}{r^2} - \frac{\sum_{i=1}^{r} y_{i..}^2}{r} + \frac{y_{...}^2}{r^2} - \frac{\sum_{j=1}^{r} y_{.j.}^2}{r} + \frac{y_{...}^2}{r^2} - \frac{\sum_{k=1}^{r} y_{..k}^2}{r} + \frac{y_{...}^2}{r^2}$$

$$= \sum_{i=1}^{r}\sum_{j=1}^{r}\sum_{k=1}^{r} y_{ijk}^2 - \frac{\sum_{i=1}^{r} y_{i..}^2}{r} - \frac{\sum_{j=1}^{r} y_{.j.}^2}{r} - \frac{\sum_{k=1}^{r} y_{..k}^2}{r} + \frac{2y_{...}^2}{r^2}$$

Minimizing SS_{E} with respect to the missing value (which we shall denote by x) is equivalent to minimizing the following:

$$L = x^2 - \frac{1}{r}(y'_{i..} + x)^2 - \frac{1}{r}(y'_{.j.} + x)^2 - \frac{1}{r}(y'_{..k} + x)^2 + \frac{2}{r}(y'_{...} + x)^2 + R$$

R is a remainder quantity which does not contain x (where $y'_{...}, y'_{i..}, y'_{.j.}, y'_{..k}$ represent the totals for all rows, columns, and treatments, without the missing value x). Minimizing,

$$\begin{aligned}
\frac{\partial L}{\partial x} &= 2x - \frac{2}{r}(y'_{i..} + x) - \frac{2}{r}(y'_{.j.} + x) - \frac{2}{r}(y'_{..k} + x) + \frac{4}{r^2}(y'_{...} + x) = 0 \\
&= 2x\left[1 - \frac{1}{r} - \frac{1}{r} - \frac{1}{r} + \frac{2}{r^2}\right] = \frac{2}{r^2}\left[r(y'_{i..} + y'_{.j.} + y'_{..k}) - 2y'_{...}\right] \\
&= x\left(r^2 - 3r + 2\right) = r(y'_{i..} + y'_{.j.} + y'_{..k}) - 2y'_{...} \\
\hat{x} &= \frac{r(y'_{i..} + y'_{.j.} + y'_{..k}) - 2y'_{...}}{(r-1)(r-2)} \qquad (5.19)
\end{aligned}$$

Once the missing value has been estimated, it is used to replace the missing value in the Latin square and the analysis of the design is carried out as discussed in the Section 5.4. Again, in the ensuing ANOVA, the usual degrees of freedom for error and, therefore, for total for the $r \times r$ Latin square design are both reduced by 1 to reflect the actual number of observations in the experiment for the same reasons we adduced in the treatment of missing values under RCBD.

Next, we illustrate the method by treating an example as follows:

Example 5.8

As a pilot study, only one replicate of the experiment in Example 5.3 was run. The coded form of the responses (reaction times) obtained is presented below. The observation marked x was missing. Estimate x and hence carry out ANOVA and determine which catalyst is to be preferred.

	Laboratories				
Mines	**1**	**2**	**3**	**4**	**5**
1	A = 3	B = 1	C = 2	D = 5	E = 6
2	B = −4	C = 3	D = 11	E = 8	A = 8
3	C = −3	D = 7	E = x	A = 9	B = 3
4	D = 5	E = 12	A = 7	B = 5	C = 4
5	E = 1	A = 11	B = 1	C = 11	D = 13

Total					
Laboratories	1	2	3	4	5
	2	34	21	37	34
Mines	1	2	3	4	5
	17	26	16	33	37
Catalysts	1	2	3	4	5
	38	6	17	41	27
Overall total = 129					

$$\hat{x} = \frac{5(y'_{3..} + y'_{.3.} + y'_{..5} - 2y'_{...})}{12} = [5(27{+}16{+}21){-}2(129)]/12 = 65/12 = 5.1667$$

Total					
Laboratories	1	2	3	4	5
	2	34	26.1667	37	34
Mines	1	2	3	4	2
	17	26	21.1667	33	37
Catalysts	1	2	3	4	5
	38	6	17	41	32.1667

Now we calculate the sums of squares and carry out analysis of variance for this design as follows:

$$SS_{\text{total}} = (3^2 + 3^2 + \cdots + 13^2) - (134.1667)^2/25 = 475.6666$$

$$SS_{\text{mines}} = 1/5(17^2 + \cdots + 37^2) - (134.1667)^2/25 = 54.1777$$

$$SS_{\text{laboratories}} = 1/5(2^2 + \cdots + 34^2) - (134.1667)^2/25 = 168.9111$$

$$SS_{\text{Latin(catalysts)}} = 1/5(38^2 + \cdots + 32.1667^2) - (134.1667)^2/25 = 176.9112$$

$$SS_{\text{E}} = 475.6666 - 168.9111 - 54.1777 - 176.9112 = 75.6666$$

We carry out ANOVA as shown in Table 5.20.

TABLE 5.20: ANOVA for the Missing Value Problem

Source	Sum of squares	Degrees of freedom	Mean square	*F*-ratio
Mines	54.1777	4	13.5444	1.9690
Laboratories	168.9111	4	42.2278	6.1388
Catalysts	176.9112	4	44.2278	6.4296
Residual	75.6666	11	6.87878	
Total	475.6666	23		

$F(4,\ 24,\ 0.01) = 4.22$, so that laboratories and catalysts have significant effects. But our interest is restricted to catalysts alone. However, we are satisfied that sources of significant error have been effectively designed into the experiment to enable us to detect significant catalyst effects. The reader can carry out the rest of the analysis (such as Duncan's test) to identify the best catalyst and can also check the adequacy of the model using methods described elsewhere in the book. □

5.7 Graeco-Latin Square Design

Another design, which is related to the Latin square is the Graeco-Latin square. The design is achieved by superimposing a second $r \times r$ Latin square design on another $r \times r$ Latin square design, and representing the superimposed design by Greek letters, provided that each Greek letter appears only once with any Latin letter in the entire design. Two Latin squares arranged in this manner are said to be orthogonal.

The Graeco-Latin square design enables us to investigate up to four factors in a single design. Two of these are set in the rows and columns, while the other two factors may be represented by the Greek and Latin letters. With this design, some variations from sources considered extraneous could be isolated and removed. The design exists in theory for all $r \geq 3$ except for $r = 6$ and 10. It is necessary to ensure that there is no interaction between the factors if the Graeco-Latin square design is to be used for the study of a number of factors.

One disadvantage of the Graeco-Latin square design, is that for small r, only a few degrees of freedom is available for experimental error. For instance, for $r = 3$, there is no degree of freedom for error; for $r = 4$, the degree of freedom is only 3. Another disadvantage is that the number of units required in an experiment increases rapidly as r increases.

In the Latin square design, we achieved blocking in two directions and we were able to eliminate variations from two extraneous sources. The Graeco-Latin square design further extends this principle since it enables us to isolate three sources of uncontrolled error and deal with them separately. Thus, the Graeco-Latin square design allows us to investigate four factors set at r levels by using rows, columns, Latin letters, and Greek letters in a $r \times r$ Graeco-Latin square design. A typical $r \times r$ Graeco-Latin square design has the linear model

$$y_{ijkl} = \mu + \beta_i + \tau_j + \gamma_k + \delta_l + \varepsilon_{ijkl} \quad i, j, k, l = 1, 2, \ldots, r \tag{5.20}$$

where y_{ijkl} is the observation in the ith row, which is represented by the jth Latin letter but which has the kth Greek letter attached to it and which also appears at the lth column. The symbol τ_j refers to the effect of treatment

TABLE 5.21: 5 × 5 Graeco-Latin Square Design

	Columns				
Rows	**1**	**2**	**3**	**4**	**5**
1	Cϕ	Dω	Eρ	Bφ	Aθ
2	Bω	Cρ	Dφ	Aθ	Eϕ
3	Eρ	Aφ	Bθ	Dϕ	Cω
4	Aφ	Bθ	Cϕ	Eω	Dρ
5	Dθ	Eϕ	Aω	Cρ	Bφ

designated by the jth Latin letter and γ_k refers to the effect of the factor represented by kth Greek letter (Table 5.21). In the entire experiment, there are r^2 observations and only two subscripts of the four (i, j, k, and l) are sufficient to fully identify any observation. $\varepsilon_{ijkl} \sim \text{NID}(0, \sigma^2)$.

As usual, it is possible to partition the total sum of squares of all variations into different components for the purposes of ANOVA. Essentially, the basic assumptions for the Graeco-Latin square design are the same as for the Latin square design.

The sum of squares for Graeco-Latin square design can be split into its component parts as follows:

$$\begin{aligned}
SS_{\text{total}} &= \sum_{i=1}^{r}\sum_{j=1}^{r}\sum_{k=1}^{r}\sum_{l=1}^{r}(y_{ijkl} - \overline{y}_{....})^2 \\
&= \sum_{i=1}^{r}\sum_{j=1}^{r}\sum_{k=1}^{r}\sum_{l=1}^{r}(y_{ijkl} - +\overline{y}_{i...} - \overline{y}_{i...} + \overline{y}_{.j..} - \overline{y}_{.j..} + \overline{y}_{..k.} - \overline{y}_{..k.} + \overline{y}_{...l} \\
&\qquad - \overline{y}_{...l} + \overline{y}_{....} - \overline{y}_{....} + \overline{y}_{....} - \overline{y}_{....} + \overline{y}_{....} - \overline{y}_{....} - \overline{y}_{....})^2 \\
&= \sum_{i=1}^{r}\sum_{j=1}^{r}\sum_{k=1}^{r}\sum_{l=1}^{r}[(y_{ijkl} - \overline{y}_{i...} - \overline{y}_{.j..} - \overline{y}_{..k.} - \overline{y}_{...l} + 3\overline{y}_{....}) + (\overline{y}_{i...} - \overline{y}_{....}) \\
&\qquad + (\overline{y}_{.j..} - \overline{y}_{....}) + (\overline{y}_{..k.} - \overline{y}_{....}) + (\overline{y}_{...l} - \overline{y}_{....})]^2 \\
&= \sum_{i=1}^{r}\sum_{j=1}^{r}\sum_{k=1}^{r}\sum_{l=1}^{r}(y_{ijkl} - \overline{y}_{i...} - \overline{y}_{.j..} - \overline{y}_{..k.} - \overline{y}_{...l} + 3\overline{y}_{....})^2 \\
&\qquad + r\sum_{i=1}^{r}(\overline{y}_{i...} - \overline{y}_{....})^2 + r\sum_{j=1}^{r}(\overline{y}_{.j..} - \overline{y}_{....})^2 \\
&\qquad + r\sum_{k=1}^{r}(\overline{y}_{..k.} - \overline{y}_{....})^2 + r\sum_{l=1}^{r}(\overline{y}_{...l} - \overline{y}_{....})^2
\end{aligned}$$

$$\Rightarrow SS_{\text{total}} = SS_{\text{E}} + SS_{\text{row-factor}} + SS_{\text{column-factor}} + SS_{\text{Latin-factor}} + SS_{\text{Greek-factor}}$$

For calculation purposes, the sums of squares arising from different sources could be obtained with the following formulae:

$$
\begin{aligned}
SS_{\text{total}} &= \sum_{i=1}^{r}\sum_{j=1}^{r}\sum_{k=1}^{r}\sum_{l=1}^{r} y_{ijkl}^2 - \frac{y_{....}^2}{r^2} \\
SS_{\text{rows}} &= \sum_{i=1}^{r} \frac{y_{i...}^2}{r} - \frac{y_{....}^2}{r^2} \\
SS_{\text{Latin}} &= \sum_{j=1}^{r} \frac{y_{.j..}^2}{r} - \frac{y_{....}^2}{r^2} \qquad (5.21) \\
SS_{\text{Greek}} &= \sum_{k=1}^{r} \frac{y_{..k.}^2}{r} - \frac{y_{....}^2}{r^2} \\
SS_{\text{columns}} &= \sum_{l=1}^{r} \frac{y_{...l}^2}{r} - \frac{y_{....}^2}{r^2} \\
SS_{\text{residual}} &= SS_{\text{total}} - SS_{\text{rows}} - SS_{\text{columns}} - SS_{\text{Greek}} - SS_{\text{Latin}}
\end{aligned}
$$

The hypothesis to be tested for this design is

$$H_0 : \gamma_1 = \gamma_2 = ... = \gamma_k = 0 \quad \text{vs} \quad H_1 : \gamma_k \neq 0 \quad \text{for some } k$$

If the use of Graeco-Latin square design is solely for the isolation and removal of variations from extraneous sources, then those extraneous factors are only tested in ANOVA to ensure that the design is effective for this purpose. The general ANOVA for this design is presented in Table 5.22.

TABLE 5.22: The General ANOVA for the Graeco-Latin Square Design

Source	**Sum of squares**	**Degrees of freedom**	**Mean square**	***F*-ratio**
Latin (treatments)	SS_{Latin}	$r-1$	$SS_{\text{Latin}}/(r-1)$	$\frac{MS_{\text{Latin}}}{MS_{\text{residual}}}$
Greek	SS_{Greek}	$r-1$	$SS_{\text{Greek}}/(r-1)$	$\frac{MS_{\text{Greek}}}{MS_{\text{residual}}}$
Rows	SS_{rows}	$r-1$	$SS_{\text{rows}}/(r-1)$	$\frac{MS_{\text{rows}}}{MS_{\text{residual}}}$
Columns	SS_{columns}	$r-1$	$SS_{\text{columns}}/(r-1)$	$\frac{MS_{\text{columns}}}{MS_{\text{residual}}}$
Residual	SS_{residual}	$(r-1)(r-3)$	$SS_{\text{residual}}/(r-1)(r-3)$	
Total	SS_{total}	r^2-1		

We present an example of Graeco-Latin square design.

Example 5.9

A microbiologist, who was mainly interested in the differential growth of some algae in acetic acid, suspects that the pH level, temperature, and size of vessel in which the algae is grown could exert influence on the growth of algae. Although her aim was to study the differences in the growth of these algae, she decided to use a 5×5 Graeco-Latin square design to carry out her investigation to isolate the variations due to other sources, which are not of primary interest. Five different algae were being studied in five different types of vessels with five settings of the pH level. The Greek letters represent temperature at five levels, the columns represent five settings of pH, rows represent the vessels, while the algae types are represented by the Latin letters. The responses as well as the design are shown in Table 5.23. The observations are the number of molds of the algae obtained after 3 days of growth from 1 mL of the suspension, after it had been diluted to 1 part growth suspension to 100 parts of water. □

TABLE 5.23: The 5×5 Graeco-Latin Square Experiment to Study Algae

	Columns					
Rows	**1**	**2**	**3**	**4**	**5**	**Total**
1	Cω 57	Dψ 47	Eθ 60	Bρ 40	Aϕ 19	223
2	Bϕ 55	Cρ 43	Dω 38	Aθ 23	Eψ 71	230
3	Eρ 74	Aω 14	Bψ 56	Dϕ 26	Cθ 54	224
4	Aψ 17	Bθ 31	Cϕ 47	Eω 61	Dρ 23	179
5	Dθ 30	Eϕ 70	Aρ 17	Cψ 70	Bω 44	231
Total	233	205	218	220	211	1087

The total for the Greek and Latin letters are

Letters	A	B	C	D	E	θ	ϕ	ω	ρ	ψ
Total	90	226	271	164	336	198	217	214	197	261

$$\begin{aligned}
SS_{\text{total}} &= (57^2 + 55^2 + \cdots + 23^2 + 44^2) - 1087^2/25 = 55941 - 47262.76 \\
&= 8678.24 \\
SS_{\text{Latin}} &= (90^2 + 226^2 + 271^2 + 164^2 + 211^2)/5 - 1087^2/25 \\
&= 54481.8 - 47262.76 = 7219.04 \\
SS_{\text{Greek}} &= (198^2 + 217^2 + 214^2 + 197^2 + 261^2)/5 - 1087^2/25 \\
&= 47803.8 - 47262.76 = 541.04 \\
SS_{\text{rows}} &= (223^2 + 230^2 + 224^2 + 179^2 + 231^2)/5 - 1087^2/25 \\
&= 47641.8 - 47262.76 = 378.64 \\
SS_{\text{columns}} &= (233^2 + 205^2 + 218^2 + 220^2 + 211^2)/5 - 1087^2/25 \\
&= 47351.8 - 47262.76 = 89.04 \\
SS_{\text{residual}} &= 8678.24 - 7219.04 - 541.04 - 378.64 - 89.04 = 450.48
\end{aligned}$$

TABLE 5.24: ANOVA for Growth of Algae in Acetic Acid

Source	Sum of squares	Degrees of freedom	Mean square	F-ratio
Latin(algae)	7219.04	4	1804.76	32.05
Rows(vessels)	378.64	4	94.66	1.681
Greek(temp.)	541.04	4	135.26	2.402
Columns (pH)	89.04	4	22.26	0.396
Residuals	450.48	8	56.31	
Total	8678.24	24		

From the tables, $F(4, 8, 0.05) = 3.84$. The analysis of the responses of the experiment (Table 5.24) indicates that type of algae has a highly significant effect. The analysis also shows that temperature, size of vessels, and pH levels have not contributed significantly to the growth of the algae.

The means for the growth of the five types of algae are as follows:

Algae	A	B	C	D	E
Mean	18.0	45.2	54.2	32.8	67.2

Standard error = 3.36

It is possible to use Duncan's multiple range test to compare these means—interested readers can carry this out as an exercise. We compare these means using the LSD. Since $s^2 = 56.31$ and we used 5% level of significance for our test, $t[\alpha/2, (r-1)(r-3)] = t(0.025, 8) = 2.306$,

$$\text{LSD} = 2.306 \times 4.7459 = 10.944$$

Arranging the means in ascending order of magnitude leads to $\overline{Y}_{\text{A}} < \overline{Y}_{\text{D}} < \overline{Y}_{\text{B}} < \overline{Y}_{\text{C}} < \overline{Y}_{\text{E}}$

$$\begin{aligned}
\overline{Y}_{\text{E}} - \overline{Y}_{\text{C}} &= 13.0 \\
\overline{Y}_{\text{E}} - \overline{Y}_{\text{B}} &= 22.0 \\
\overline{Y}_{\text{E}} - \overline{Y}_{\text{A}} &= 49.2 \\
\overline{Y}_{\text{C}} - \overline{Y}_{\text{B}} &= 9.0 \\
\overline{Y}_{\text{C}} - \overline{Y}_{\text{A}} &= 36.2 \\
\overline{Y}_{\text{B}} - \overline{Y}_{\text{D}} &= 12.40 \\
\overline{Y}_{\text{B}} - \overline{Y}_{\text{A}} &= 27.2 \\
\overline{Y}_{\text{D}} - \overline{Y}_{\text{E}} &= 14.8
\end{aligned}$$

On the basis of the analysis, it can be said that there is no significant difference between C and B. The rest (A, D, E) are significantly different from one another. The experiment concludes that the differential growth of algae in acetic acid is mainly influenced by the type of algae. Although there is not much difference between the mean growths of C and B, there are differences among the rest. Moreover, acetic acid appears to favor the growth of E, C, and B more than the rest.

5.7.1 SAS Analysis of Data of Example 5.9

Next, we write a program to analyze the data of Example 5.9. For this Graeco-Latin design, we use PROC GLM for the analysis. We can also use PROC ANOVA to carry out this analysis.

SAS PROGRAM

```
options nodate nonumber ls=70  ps=80;
data grklat;
input vessel pH algae temp y@@;
cards;
1 1 3 3 57 1 2 4 5 47 1 3 5 1 60 1 4 2 4 40 1 5 1 2 19 2 1 2 2 55 2 2 3 4 43
2 3 4 3 38 2 4 1 1 23 2 5 5 5 71 3 1 5 4 74 3 2 1 3 14 3 3 2 5 56 3 4 4 2 26
3 5 3 1 54
4 1 1 5 17 4 2 2 1 31 4 3 3 2 47 4 4 5 3 61 4 5 4 4 23 5 1 4 1 30 5 2 5 2 70
5 3 1 4 17
5 4 3 5 70 5 5 2 3 44
;
proc print data=grklat; run;
title 'Graeco-Latin squares problem:example 5.9';
proc glm data=grklat;
class vessel pH algae temp;
model y=vessel pH algae temp;
means algae/duncan;run;
```

SAS OUTPUT

Graeco-Latin squares problem:example 5.9

Obs	vessel	pH	algae	temp	y
1	1	1	3	3	57
2	1	2	4	5	47
3	1	3	5	1	60
4	1	4	2	4	40
5	1	5	1	2	19
6	2	1	2	2	55
7	2	2	3	4	43
8	2	3	4	3	38
9	2	4	1	1	23
10	2	5	5	5	71
11	3	1	5	4	74
12	3	2	1	3	14
13	3	3	2	5	56
14	3	4	4	2	26
15	3	5	3	1	54

Obs	vessel	pH	algae	temp	y
16	4	1	1	5	17
17	4	2	2	1	31
18	4	3	3	2	47
19	4	4	5	3	61
20	4	5	4	4	23
21	5	1	4	1	30
22	5	2	5	2	70
23	5	3	1	4	17
24	5	4	3	5	70
25	5	5	2	3	44

Class Level Information

Class	Levels	Values
vessel	5	1 2 3 4 5
pH	5	1 2 3 4 5
algae	5	1 2 3 4 5
temp	5	1 2 3 4 5

Number of Observations Read	25
Number of Observations Used	25

Dependent Variable: y

Source	DF	Sum of Squares	Mean Square	F Value	Pr > F
Model	16	8227.760000	514.235000	9.13	0.0018
Error	8	450.480000	56.310000		
Corrected Total	24	8678.240000			

R-Square	Coeff Var	Root MSE	y Mean
0.948091	17.25851	7.503999	43.48000

Source	DF	Anova SS	Mean Square	F Value	Pr > F
vessel	4	378.640000	94.660000	1.68	0.2463
pH	4	89.040000	22.260000	0.40	0.8069
algae	4	7219.040000	1804.760000	32.05	<.0001
temp	4	541.040000	135.260000	2.40	0.1356

Duncan's Multiple Range Test for y

NOTE: This test controls the Type I comparisonwise error rate, not the experimentwise error rate.

```
        Alpha                              0.05
        Error Degrees of Freedom              8
        Error Mean Square                 56.31

Number of Means          2          3          4          5
Critical Range       10.94      11.40      11.66      11.82

Means with the same letter are not significantly different.

Duncan Grouping          Mean       N    algae

              A        67.200       5    5

              B        54.200       5    3
              B
              B        45.200       5    2

              C        32.800       5    4

              D        18.000       5    1
```

5.8 Estimation of Parameters of the Model and Extraction of Residuals

The estimates for the effects in this design are obtained as

$$\begin{aligned} \hat{\mu} &= \overline{y}_{....} \\ \hat{\alpha}_i &= \overline{y}_{i...} - \overline{y}_{....} \\ \hat{\beta}_j &= \overline{y}_{.j..} - \overline{y}_{....} \\ \hat{\gamma}_k &= \overline{y}_{..k.} - \overline{y}_{....} \\ \hat{\delta}_l &= \overline{y}_{...l} - \overline{y}_{....} \end{aligned} \tag{5.22}$$

The fitted values or estimates of each response in the design is

$$\hat{y}_{ijkl} = \overline{y}_{i...} + \overline{y}_{.j..} + \overline{y}_{..k.} + \overline{y}_{...l} - 3\overline{y}_{....} \tag{5.23}$$

so that the residuals are obtained as

$$\hat{\varepsilon}_{ijkl} = y_{ijkl} - \overline{y}_{i...} - \overline{y}_{.j..} - \overline{y}_{..k.} - \overline{y}_{...l} + 3\overline{y}_{....} \tag{5.24}$$

5.8.1 Application to Data of Example 5.9

Using Equation 5.22, we obtain estimates of parameters of the model for data of the experiment in Example 5.9 as follows:

$\hat{\mu} = 43.48$

$\hat{\alpha}_1 = 44.6 - 43.48 = 1.12$ $\hat{\alpha}_2 = 46 - 43.48 = 2.52$ $\hat{\alpha}_3 = 44.8 - 43.48 = 1.32$

$\hat{\alpha}_4 = 35.8 - 43.48 = 7.68$ $\hat{\alpha}_5 = 46.2 - 43.48 = 2.72$

$\hat{\beta}_1 = 46.6 - 43.48 = 3.12$ $\hat{\beta}_2 = 41 - 43.48 = -2.48$ $\hat{\beta}_3 = 43.6 - 43.48 = 0.12$

$\hat{\beta}_4 = 44 - 43.48 = 0.52$ $\hat{\beta}_5 = 42.2 - 43.48 = -1.28$

$\hat{\gamma}_1 = 18 - 43.48 = -25.48$ $\hat{\gamma}_2 = 45.2 - 43.48 - 1.72$ $\hat{\gamma}_3 = 54.2 - 43.48 = 10.72$

$\hat{\gamma}_4 = 32.8 - 43.48 = -10.68$ $\hat{\gamma}_5 = 67.2 - 43.48 = 23.72$

$\hat{\delta}_1 = 39.6 - 43.48 = -3.88$ $\hat{\delta}_2 = 43.4 - 43.48 = -0.08$

$\hat{\delta}_3 = 42.8 - 43.48 = -0.68$ $\hat{\delta}_4 = 39.4 - 43.48 = -4.08$

$\hat{\delta}_5 = 52.2 - 43.48 = 8.72$

Here, using Equations 5.23 and 5.24, we obtain the estimates for the responses and residuals for the experiment in Example 5.9. These are presented in Table 5.25.

Plotting the residuals against fitted values for Example 5.9, we obtain Figure 5.5.

The plot indicates nothing unusual. It has no obvious pattern, and the values of the residuals lie well between $\pm 2\sqrt{MS_E}$. Next, we obtain the normal probability plot for the residuals as shown in Figure 5.6.

The plot indicates that the normality assumption is satisfied.

TABLE 5.25: Estimates of Responses and Residuals for Example 5.9

y_{ijkl}	Observed values	Fitted values	Residuals
y_{1133}	57	57.76	−0.76
y_{1245}	47	40.16	6.84
y_{1351}	60	64.56	−4.56
y_{1424}	40	42.76	−2.76
y_{1512}	19	17.76	1.24
y_{2122}	55	50.76	4.24
y_{2234}	43	50.16	−7.16
y_{2343}	38	34.76	3.24
y_{2411}	23	17.16	5.84
y_{2555}	71	77.16	−6.16
y_{3154}	74	67.56	6.44
y_{3213}	14	16.16	−2.16
y_{3325}	56	55.36	0.64
y_{3442}	26	34.56	−8.56
y_{3531}	54	50.36	3.64
y_{4115}	17	22.16	−5.16
y_{4221}	31	31.16	−0.16
y_{4332}	47	46.56	0.44
y_{4453}	61	59.36	1.64
y_{4544}	23	19.76	3.24
y_{5141}	30	34.76	−4.76
y_{5252}	70	67.36	2.64
y_{5314}	17	16.76	0.24
y_{5435}	70	66.16	3.84
y_{5523}	44	45.96	−1.96

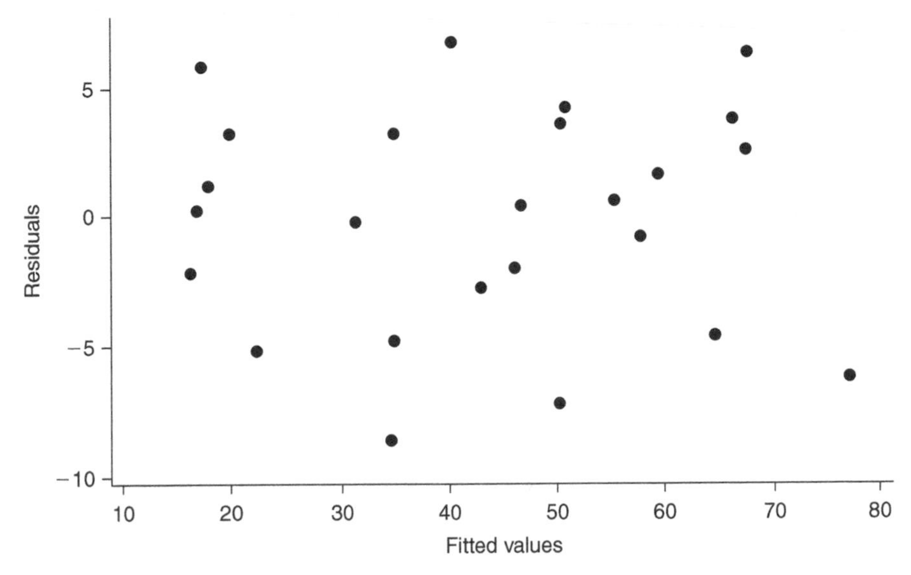

FIGURE 5.5: Plot of residuals against fitted values for Example 5.9.

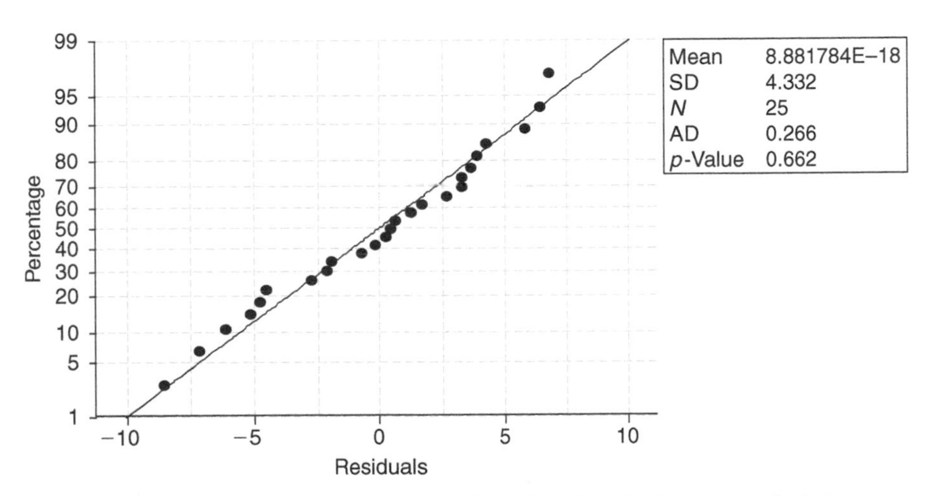

FIGURE 5.6: Normal probability plot of residuals for Example 5.9.

5.9 Exercises

Exercise 5.1

In a farm settlement program on growth of cassava, five varieties of fertilizers were studied. Since the farm settlements were located in different parts of the country, five locations were chosen for the study and five varieties of cassava were also studied. The 5 × 5 Latin square design was chosen for the study and the data below represent the yield in kilograms after 6 months:

	Location				
Variety	**I**	**II**	**III**	**IV**	**V**
I	B 2.25	A 5.6	D 6.75	E 3.1	C 4.25
II	E 2.97	D 7.3	C 3.98	A 4.9	B 3.20
III	A 4.1	C 5.2	B 4.2	D 6.2	E 3.6
IV	D 5.9	B 4.7	E 4.9	C 4.4	A 6.21
V	C 3.8	E 3.8	A 7.4	B 3.9	D 5.88

Are any of the factors like location, variety, and fertilizer type significant? (fertilizer types are represented by the Latin letters, A, B, C, D, and E). For any of the factors that are significant, compare their means. Check that basic assumptions such as randomness, normality of errors, are not violated. Is there any outlier? ▯

Exercise 5.2

In a similar experiment involving maize of different varieties, it was considered that the spacing of the maize could influence yield, so five spacing methods were studied in addition to five varieties of maize and five locations as well as five fertilizer treatments. A Graeco-Latin square design was chosen for the study and the yields of maize per hectare are presented below

	Location				
Variety	**I**	**II**	**III**	**IV**	**V**
I	βC 5.65	δD 7.68	θE 8.75	νB 4.32	φA 5.27
II	φB 3.79	νC 8.35	βD 4.98	θA 5.94	δE 7.50
III	νE 8.12	βA 6.27	δB 4.22	φD 7.29	θC 4.71
IV	δA 7.93	θB 4.77	φC 6.92	βE 8.48	νD 6.5
V	θD 4.85	φE 8.88	νA 8.45	δC 4.49	βB 4.88

In the design, rows = variety; columns = location; Latin = fertilizer type, and Greek = spacing. Is there any significant difference between the yields under different fertilizer treatments? Carry out the plots of the residuals against the following: (1) fitted values, (2) location, (3) variety, (4) fertilizer treatment, and (5) spacing. How would you interpret the normal probability plot of residuals in relation to basic assumptions in this design? ▯

Exercise 5.3
At the pilot stages of the steel production program by a National Steel Company, a 3 × 3 Graeco-Latin square experiment was carried out to compare the tensile strengths of steel rods produced under different carbon treatments. The company sourced its iron ore from three mines at I, O, and M. It has steel plants in three cities, A, W, and J. Since the operators in each of these plants have been trained by different nations, A, I, and S, it was thought that their skills might vary. There were also three levels of carbon in the steel produced. Three replicates of the 3 × 3 Graeco-Latin square design were obtained as follows:

	Mines		
Steel plants	**I**	**O**	**M**
A	Aγ 181	Bβ 124	Cα 217
W	Bα 200	Cγ 153	Aβ 199
J	Cβ 179	Aα 169	Bγ 210
A	Aγ 184	Bβ 130	Cα 220
W	Bα 193	Cγ 146	Aβ 194
J	Cβ 168	Aα 157	Bγ 206
A	Aγ 177	Bβ 126	Cα 214
W	Bα 197	Cγ 149	Aβ 191
J	Cβ 173	Aα 163	Bγ 213

Rows = steel plants; Columns = mines of iron ore; Latin = operators; Greek = Carbon content. Analyze the data fully and write a short report to the steel company. ▯

Exercise 5.4
In an industrial research on extraction of flavoring oils, raw materials were sourced from five different countries and each of was subjected to extraction using vessels of five different sizes, each of which was set at five different extraction temperatures, while five types of chemicals were employed for the extraction. A Graeco-Latin square design was used to study the quantity of oil recovered for the same weight of raw material and the yields are presented below:

	Type of chemical used				
Country	**I**	**II**	**III**	**IV**	**V**
1	Aα 28	Bε 18	Cγ 19	Dδ 16	Eβ 14
2	Bγ 20	Cδ 23	Dβ 18	Eα 12	Aε 23
3	Cβ 20	Dα 13	Eε 17	Aγ 26	Bδ 13
4	Dε 15	Eγ 16	Aδ 22	Bβ 14	Cα 17
5	Eδ 11	Aβ 26	Bα 17	Cε 18	Dγ 14

Latin letters = vessel sizes; Greek letters = extraction temperature. Analyze the responses of this experiment and suggest the best extraction conditions. ▯

Exercise 5.5

In the training of machinists, 16 apprentices were tested on four tasks, which require machine skills. The different skills were tested on different days. A 4×4 Latin square design replicated four times was employed for this study. The Latin letters represent the tests; the rows, the subjects; while the columns represent the days. Their scores in the tests are presented as follows:

	Days			
Apprentices	**I**	**II**	**III**	**IV**
1	A 30	C 49	D 56	B 49
2	C 44	D 57	B 46	A 45
3	D 55	B 50	A 49	C 62
4	B 51	A 52	C 61	D 68
5	B 41	A 42	C 55	D 62
6	D 48	C 53	A 42	B 49
7	C 46	B 47	D 64	A 47
8	A 41	D 66	B 55	C 62
9	B 40	A 45	D 60	C 52
10	A 30	B 47	C 50	D 61
11	D 46	C 47	A 36	B 43
12	C 48	D 61	B 50	A 45
13	A 39	B 48	C 41	D 59
14	B 55	C 58	D 52	A 34
15	C 49	D 66	A 44	B 41
16	D 64	A 33	B 59	C 42

Analyze the responses and check the usual assumptions. Are there any significant differences between the mean performances of apprentices? □

Exercise 5.6

An engineer studying the occurrence of an event in a certain process, which depends on the time of the day, and day of the week, had to use a number of observers to record the incidents. Since the occurrences of the event in 5 days of the week were being studied, five observers and five time periods were used to design a 5×5 Latin square experiment. A new set of observers was used each week and the study took 3 weeks. The results obtained are presented as follows:

	Observers				
Days of Week	**I**	**II**	**III**	**IV**	**V**
Monday	20A	21B	21C	28D	12E
Tuesday	79B	60C	62D	45E	59A
Wednesday	48C	27D	21E	31A	51B
Thursday	94D	17E	71A	72B	96C
Friday	20E	18A	24B	10C	33D

Days of Week	Observers				
	VI	**VII**	**VIII**	**IX**	**X**
Monday	16A	19B	21E	28C	12D
Tuesday	49B	48C	62A	45D	59E
Wednesday	38C	24D	21B	31E	41A
Thursday	82D	27E	71C	72A	86B
Friday	27E	19A	24D	10B	36C

Days of Week	XI	XII	XIII	XIV	XV
Monday	21D	17C	14E	22B	6A
Tuesday	81E	67D	77A	60C	69B
Wednesday	50A	30E	25B	37D	48C
Thursday	91B	20A	73C	68E	79D
Friday	30C	28B	29D	16A	29E

Latin letters refer to periods of the day. Analyze the data using Case III as the design. □

Exercise 5.7

In phytochemical research, the percentage content of a plant chemical extract used for drug production was being studied. It was believed that the contents could vary according to the country of origin of the plant and the plant type from which it is extracted, as well as the method of extraction. A 5×5 Latin-square design was employed for the study and the responses in percentages are presented as follows:

Country	Type of plant				
	I	**II**	**III**	**IV**	**V**
I	2.67E	3.15D	4.29A	4.95C	5.62B
II	1.40B	1.77E	1.40D	3.54A	2.93C
III	3.32C	4.53B	4.50E	3.99D	3.68A
IV	1.92A	1.07C	2.29B	2.85E	3.08D
V	1.88D	2.16A	2.83C	2.38B	3.51E

Rows = country of origin, Columns = type of plant, and Latin letters = extraction method. Analyze the data fully and report your findings. □

Exercise 5.8

Location	Chemicals			
	A	**B**	**C**	**D**
1	10.1	11.4	9.9	12.1
2	12.2	12.9	12.3	13.4
3	11.9	12.7	11.4	12.9

An experimenter wanted to compare the effects of four chemicals A, B, C, and D on water hardness. Hard water from three different locations was used in the study, but the location of hard water was considered a nuisance in the study. RCBD would be used to analyze the data. Analyze the data and test for differences between the chemicals. Test at 5% level of significance. ▯

Exercise 5.9

	Extraction methods			
Source	**I**	**II**	**III**	**IV**
I	80	85	90	95
II	85	89	94	96
III	89	91	99	100
IV	92	100	105	110

A company is studying four methods for extracting an important bio-fuel from a plant species. It obtained the plants from four sources (countries) and thought that the source might affect the amount of bio-fuel extracted. It decided to treat the source as block in the study of the extraction method. It arranged to study the four methods and the amount of bio-fuel that can be extracted from one unit of biomaterial from each source. The data represents the amount of bio-fuel in liters extracted from one unit of biomaterial. Analyze the data stating your hypotheses and conclusions clearly. Test at 2.5% level of significance. ▯

Exercise 5.10

To compare five varieties of wheat, a RCBD with four blocks (plots) was used. Unfortunately, one unit was damaged very badly and the yield from that unit had to be ignored. Estimate the yield from the unit, and hence completely analyze the data stating your conclusions fully.

	Wheat varieties				
Plots	**A**	**B**	**C**	**D**	**E**
1	2	2	4	2	4
2	6	3	x	4	2
3	8	3	5	3	2
4	5	2	4	2	1

▯

Exercise 5.11

A randomized block experiment was performed to study six corn varieties, which were grown in four experimental parcels of land in different locations, treated as blocks. The yields are recorded below in hundreds of bushels per acre

	Corn varieties					
Parcel	**I**	**II**	**III**	**IV**	**V**	**VI**
1	48.8	50.6	47.8	37.2	37.2	45.9
2	48.3	58.8	42.8	35.1	38.0	43.5
3	49.5	51.2	55.9	34.2	38.8	43.8
4	59.5	59.5	57.8	39.7	36.4	47.3

Analyze the responses fully and report your findings. Carry out the tests at level of significance, $\alpha = 5\%$. □

Exercise 5.12

The data below were obtained from a randomized block experiment in which five treatments were being compared in four blocks

	Treatments				
Blocks	**I**	**II**	**III**	**IV**	**V**
1	18	10	8	13	9
2	14	13	10	9	12
3	17	11	10	10	13
4	12	13	12	9	9

Carry out a full analysis to compare the treatments. Extract fitted values and residuals and check for normality assumption. □

Exercise 5.13

An experiment was performed to compare the quantity of oleoresin derivable from a unit weight of ginger obtained from five countries. Five students with differing skills were given a unit weight of ginger from each of the five countries to determine the oleoresin contents in milligrams.

	Students				
Ginger type	**I**	**II**	**III**	**IV**	**V**
Ginger I	7	7	15	11	9
Ginger II	12	17	12	18	18
Ginger III	14	18	18	19	23
Ginger IV	19	25	22	19	23
Ginger V	7	10	11	15	11

Analyze the data and compare the mean oleoresin contents of the five gingers at 1% level of significance. Use Duncan's multiple range test in a post ANOVA analysis to group the ginger means. □

Exercise 5.14
Use SAS to carry out the analysis required in Exercise 5.13, assuming fixed effects for both factors. □

Exercise 5.15
Use SAS to carry out the analysis of the responses in the experiment in Exercise 5.4. □

Exercise 5.16
Use SAS to carry out the analysis of the responses in the experiment in Exercise 5.3. □

Exercise 5.17
Use SAS to carry out the analysis of the responses in the experiment in Exercise 5.2. □

References

Anderson, V.L. and McLean, R.A. (1974) *Design of Experiments: A Realistic Approach,* Marcel Dekker, New York.

Bowerman, B.L. and O'Connell, R.T. (1990) *Linear Statistical Models: An Applied Approach*, 2nd ed., PWS-Kent Publishing Company, Boston.

Box, G.E., Hunter, J.S., and Hunter, W.G. (1978) *Statistics for Experimenters*, Wiley, New York.

Cochran, W.G. and Cox, G.M. (1957). *Experimental Designs*, 2nd ed., Wiley, New York.

Fisher, R.A. and Yates, F. (1953) *Statistical Tables for Biological, Agricultural, and Medical Research*, 4th ed., Oliver and Boyd, Edinburgh.

Montgomery, D.C. (2005) *Design and Analysis of Experiments*, 6th ed., Wiley, New York.

Mood, A.M., Graybill, F.A., and Boes, D.C. (1974) *Introduction to the Theory of Statistics*, 3rd ed., McGraw-Hill, New York.

Wackerly, D.D., Mendenhall III, W., and Scheaffer, R.L. (2002) *Mathematical Statistics with Applications*, 6th ed., Duxbury Press, Belmont, CA.

Chapter 6

Regression Models for Randomized Complete Block, Latin Squares, and Graeco-Latin Square Designs

6.1 Regression Models for the Randomized Complete Block Design

In the first section of Chapter 5, we discussed the design and analysis of the responses of the randomized complete block design (RCBD). We analyzed the responses of this design using classical models; we also wrote SAS programs for the analysis of the responses of RCBD experiments based on that model. In this section, we develop a regression model for the RCBD. The use of regression models to represent the responses of designed experiments was briefly discussed in the introductory parts of this book. We had used regression to model the responses of the typical completely randomized design (CRD). For the RCBD, we intend to use regression models based on both effects and reference cell coding methods.

6.1.1 Dummy Variables Regression Model for RCBD with the Effects Coding Method

Two sets of dummy variables need to be defined to obtain a regression model that represents RCBD. The dummy variables would represent the treatment and block. We use the effects coding method to define the two sets of dummy variables in the model, namely, X_i which contains $k-1$ dummy variables, and Z_j which contains $b-1$ dummy variables. The number of dummy variables in each of the two sets mirrors the levels of the two explanatory variables in the design, treatment, and block. We can see that the number of dummy variables in a set is obtained as the number of levels of each factor minus one. Based on the foregoing discussion, we define the dummy variables and obtain the model for the RCBD as follows:

$$X_i = \begin{cases} -1 & \text{if } y_{ij} \text{ is under the } k\text{th treatment} \\ 1 & \text{if } y_{ij} \text{ is under the } i\text{th treatment} \\ 0 & \text{otherwise} \end{cases}$$

for $i = 1, 2, \ldots, k-1$

$$Z_j = \begin{cases} -1 & \text{if } y_{ij} \text{ is under the } b\text{th block} \\ 1 & \text{if } y_{ij} \text{ is under the } j\text{th block} \\ 0 & \text{otherwise} \end{cases}$$

for $j = 1, 2, \ldots, b-1$

Thus, the regression model that emerges for the analysis of the responses of the RCBD when we use the above dummy variables is

$$y_{ij} = \mu + \sum_{i=1}^{k-1} \gamma_i X_i + \sum_{j=1}^{b-1} \beta_j Z_j + \varepsilon \tag{6.1}$$

This is the typical regression model for any RCBD with b levels for block and k levels for treatment in which the dummy variable is defined according to the effects coding method.

6.1.2 Estimation of Parameters of the Regression Model for RCBD (Effects Coding Method)

We represented the RCBD as a two-way layout in which there is only one observation in any cell (i, j), $(i = 1, 2, \ldots, k;\ j = 1, 2, \ldots, b)$ of the layout. However, there is nothing that restricts us from replicating a RCBD. If that happens, we would have at least two observations per cell. In such situations, cell means are estimated by averaging the values in each cell. If the RCBD is not replicated, then the single cell value represents an estimate of the equivalent cell mean. In Table 6.1, we present means for the typical RCBD.

TABLE 6.1: Means for a Typical RCBD

	Blocks					
Treatments	**1**	**2**	...	**$b-1$**	**b**	**Means**
1	μ_{11}	μ_{12}	...	μ_{1b-1}	μ_{1b}	$\mu_{1.}$
2	μ_{21}	μ_{22}	...	μ_{2b-1}	μ_{2b}	$\mu_{2.}$
⋮	⋮	⋮	...	⋮	⋮	⋮
k	μ_{k1}	μ_{k2}	...	μ_{kb-1}	μ_{kb}	$\mu_{k.}$
Means	$\mu_{.1}$	$\mu_{.2}$	...	$\mu_{.b-1}$	$\mu_{.b}$	$\mu_{..}$

Under the model 6.1, with the various means described in Table 6.1, we have

$$\mu = \mu_{..}$$
$$\gamma_i = \mu_{i.} - \mu_{..} \quad i = 1, 2, \ldots, k-1$$
$$\beta_j = \mu_{.j} - \mu_{..} \quad j = 1, 2, \ldots, b-1$$
$$-\sum_{i=1}^{k-1} \gamma_i = \mu_{k.} - \mu_{..}$$
$$-\sum_{j=1}^{b-1} \beta_j = \mu_{.b} - \mu_{..}$$

The cell means defined in Table 6.1 have the following relationships with the parameters of the regression model:

$$\mu_{ij} = \mu_{..} + \gamma_i + \beta_j \quad i = 1, 2, \ldots, k-1 \; j = 1, 2, \ldots, b-1$$
$$\mu_{kj} = \mu_{..} - \sum_{i=1}^{k-1} \gamma_i + \beta_j$$
$$\mu_{ib} = \mu_{..} - \sum_{j=1}^{b-1} \beta_j + \gamma_i$$
$$\mu_{kb} = \mu_{..} - \sum_{i=1}^{k-1} \gamma_i - \sum_{j=1}^{b-1} \beta_j$$
$$\mu_i = \mu_{..} + \gamma_i$$
$$\mu_{.j} = \mu + \beta_j$$

We will estimate the parameters of the regression model by using these relationships. As the actual cell means $\mu_{i.}, \mu_{.j}, \mu_{.b}, \mu_{k.}$ are unknown, we replace them in the above equations with their sample data equivalents, namely, $\overline{y}_{i.}, \overline{y}_{.j}, \overline{y}_{.b}, \overline{y}_{k.}$. With these sample values, we estimate the parameters as follows:

$$\hat{\gamma}_i = \overline{y}_{i.} - \overline{y}_{..} \quad i = 1, 2, \ldots, k-1$$
$$\hat{\beta}_j = \overline{y}_{.j} - \overline{y}_{..} \quad j = 1, 2, \ldots, b-1$$
$$-\sum_{i=1}^{k-1} \hat{\gamma}_i = \overline{y}_{k.} - \overline{y}_{..}$$
$$-\sum_{j=1}^{b-1} \hat{\beta}_j = \overline{y}_{.b} - \overline{y}_{..}$$

In practice, when we encounter similar means in an equation for any design, we replace them with sample equivalents to carry out the estimation of the parameters of the regression model.

6.1.3 Application of the RCBD Regression Model 6.1 to Example 5.1 (Effects Coding Method)

Since the RCBD of Example 5.1 has three levels of block, and five levels of treatment, the specific form of the regression model that applies is

$$y_{ij} = \mu + \sum_{i=1}^{4} \gamma_i X_i + \sum_{j=1}^{2} \beta_j Z_j + \varepsilon \tag{6.2}$$

$$X_i = \begin{cases} -1 & \text{if } y_{ij} \text{ is under the 5th treatment} \\ 1 & \text{if } y_{ij} \text{ is under the } i\text{th treatment} \\ 0 & \text{otherwise} \end{cases}$$

for $i = 1, 2, 3, 4$

$$Z_j = \begin{cases} -1 & \text{if } y_{ij} \text{ is under the 3rd block} \\ 1 & \text{if } y_{ij} \text{ is under the } j\text{th block} \\ 0 & \text{otherwise} \end{cases}$$

for $j = 1, 2$

6.1.4 Estimation of Model Parameters for Example 5.1

Based on the foregoing discussion, and using the response data for this experiment, we manually estimate the parameters of the regression model. First, we calculate the sample equivalents of the means mentioned in the prior formula. Thereafter, we use them to replace the means in the formula to obtain the estimates of the parameters as follows:

$$\overline{y}_{..} = \frac{522}{15} = 34.8 \quad \overline{y}_{1.} = \frac{76}{3} = 25.33$$

$$\overline{y}_{2.} = \frac{90}{3} = 30 \qquad \overline{y}_{3.} = \frac{137}{3} = 45.67$$

$$\overline{y}_{4.} = \frac{102}{3} = 34 \qquad \overline{y}_{5.} = \frac{117}{3} = 39$$

$$\overline{y}_{.1} = \frac{180}{5} = 36 \quad \overline{y}_{.2} = \frac{224}{5} = 44.8$$

$$\overline{y}_{.3} = \frac{118}{5} = 23.6 \Rightarrow \hat{\mu} = 34.80; \;\; \hat{\gamma}_1 = 25.33 - 34.80 = -9.47$$

$$\hat{\gamma}_2 = 30 - 34.80 = -4.8$$

$$\hat{\gamma}_3 = 45.67 - 34.80 = 10.87 \quad \hat{\gamma}_4 = 34 - 34.80 = -0.8$$

$$\overline{y}_{5.} = \frac{117}{3} = 39 \quad \hat{\beta}_1 = 36 - 34.8 = 1.20 \quad \hat{\beta}_2 = 44.8 - 34.8 = 10.00$$

$$\overline{y}_{.3} - \overline{y}_{..} = 23.6 - 34.8 = -11.2 = -(\beta_1 + \beta_2)$$

$$\overline{y}_{5.} - \overline{y} = 39 - 34.8 = 4.2 = -(\hat{\gamma}_1 + \hat{\gamma}_2 + \hat{\gamma}_3 + \hat{\gamma}_4)$$

The above analysis indicates that a typical response for this experiment can be estimated by using the following equation:

$$\hat{y} = 34.8 - 9.47x_1 - 4.8x_2 + 10.87x_3 - 0.8x_4 + 1.2z_1 + 10z_2 \tag{6.3}$$

With this fitted model, we can obtain fitted values for responses and extract residuals. Thereafter, we can carry out some model diagnostics described in previous chapters. Here, we present a table of the estimates and residuals. We leave model diagnostics that involves some graphical analysis of residuals (demonstrated elsewhere in the book) as an exercise for the reader.

TABLE 6.2: Variables, Estimated Responses, and Residuals for Regression Model for Example 5.1 (Effects Coding Method)

x_1	x_2	x_3	x_4	z_1	z_2	y	$\hat{y}$	$(y-\hat{y})$	$(y-\hat{y})^2$
1	0	0	0	1	0	23	26.5333	−3.5333	12.4842
1	0	0	0	0	1	34	35.3333	−1.3333	1.77769
1	0	0	0	−1	−1	19	14.1333	4.8667	23.6848
0	1	0	0	1	0	31	31.2	−0.2	0.04
0	1	0	0	0	1	37	40	−3	9
0	1	0	0	−1	−1	22	18.8	3.2	10.24
0	0	1	0	1	0	52	46.8667	5.1333	26.3508
0	0	1	0	0	1	69	55.6667	13.3333	177.777
0	0	1	0	−1	−1	16	34.4667	−18.4667	341.019
0	0	0	1	1	0	35	35.2	−0.2	0.04
0	0	0	1	0	1	37	44	−7	49
0	0	0	1	−1	−1	30	22.8	7.2	51.84
−1	−1	−1	−1	1	0	39	40.2	−1.2	1.44
−1	−1	−1	−1	0	1	47	49	−2	4
−1	−1	−1	−1	−1	−1	31	27.8	3.2	10.24
Total						522	522	0	718.933

As we can see from Table 6.2, the sum of the estimated responses $\hat{y}$ is the same as the sum of original responses y. The sum of the extracted residuals $\sum_{i=1}^{N}(y_i - \hat{y}_i) = 0$ is as expected.

Using regression theory, we can carry out ANOVA to test for the significance of the fitted model. We know that if a regression analysis uses N responses arising from corresponding N sets of values of the explanatory variables, then the total variation is

$$S_{yy} = \sum_{i=1}^{N} y_i^2 - \frac{\left(\sum_{i=1}^{N} y_i\right)^2}{N} \tag{6.4}$$

The component of the total variation due to regression is

$$SS_{\mathrm{R}} = \sum_{i=1}^{N} (\hat{y}_i - \overline{y})^2 \tag{6.5}$$

while the component due to residual variation is

$$SS_{\mathrm{E}} = \sum_{i=1}^{N} (y_i - \hat{y}_i)^2 \tag{6.6}$$

Thus,

$$S_{yy} = SS_{\mathrm{R}} + SS_{\mathrm{E}} \tag{6.7}$$

Once we know these quantities, we can perform ANOVA to test for the significance of the fitted model. Since SS_{R} is the sum of squares due to the fitted regression model, while SS_{E} is the error sum of squares and $N = bk = 5(3) = 15$, then from the data available in Table 6.2 we can see that

$$\begin{aligned} SS_{\mathrm{E}} &= 718.9333 \\ S_{yy} &= 20766 - \frac{522^2}{15} = 2600.40 \Rightarrow SS_{\mathrm{R}} = 2600.40 - 718.9333 = 1881.4667 \end{aligned}$$

TABLE 6.3: Testing for Significance of the Fitted Regression Model for Example 5.1

Source	Sum of squares	Degrees of freedom	Mean square	*F*-ratio
Regression	1881.4667	6	313.5778	3.489
Error	718.9333	8	89.8666	
Total	2600.40	14		

From Tables 6.2 and 6.3, we see that $F(0.05, 6, 8) = 3.58$ and $F(0.06, 6, 8) = 3.33$. The model is significant at 6% level of significance.

6.2 SAS Analysis of Responses of Example 5.1 Using Dummy Regression (Effects Coding Method)

Next, we use statistical software to analyze the data for the experiment of Example 5.1 and estimate the parameters of the model 6.2 that was fitted to them. We write a SAS program to carry out the analysis. The REG and GLM are the major procedures that we can invoke within SAS for the analysis of the responses of the experiment. Recently, the MIXED procedure has been introduced by SAS to cater for analyses that are not adequately handled by the GLM procedure. We shall use any of them that we deem appropriate for the analysis of the responses of each design we treat in this book. The REG and GLM procedures will suffice for some analyses. However, there are many differences between the two procedures and what they can do. A good primer on both procedures can be found in Onyiah (2005). A more elaborate discussion of the GLM procedure can be found in SAS/STAT 8 Users Guide (1999, Volume 2, pp. 1465–1636).

Suffice to say that if we replace REG in the program below with GLM, our output in SAS will include extra information related to type I and type III sums of squares, associated with parameters in the model, which test for their significance. If the program is left as it is, then SAS will output the estimates of parameters of the model and perform ANOVA (a global test) for the entire model without separating and carrying out analysis for the component parts that make up the total sum of squares for regression.

6.2.1 SAS Program for Regression Analysis of Data of Example 5.1 (Effects Coding Method for Dummy Variables)

We entered the data according to the definition of the dummy variables in the model, and called the REG procedure to carry out the analysis. The fitted model 6.2 is defined in the model statement under this procedure. We should note that the REG procedure allows for many other options. In our analysis, we do not explore many of the options to keep the output as simple as possible.

SAS PROGRAM

```
data rcbdcode1;
input X1-X4 Z1-Z2 Y@@;
datalines;
1 0 0 0 1 0 23 1 0 0 0 0 1 34 1 0 0 0 -1 -1 19
0 1 0 0 1 0 31 0 1 0 0 0 1 37 0 1 0 0 -1 -1 22
0 0 1 0 1 0 52 0 0 1 0 0 1 69 0 0 1 0 -1 -1 16
0 0 0 1 1 0 35 0 0 0 1 0 1 37 0 0 0 1 -1 -1 30
-1 -1 -1 -1 1 0 39 -1 -1 -1 -1 0 1 47 -1 -1 -1 -1 -1 -1 31
;
proc print data=rcbdcode1;
proc reg data=rcbdcode1;
model y=x1-x4 z1-z2; run;
```

SAS OUTPUT

Obs	X1	X2	X3	X4	Z1	Z2	Y
1	1	0	0	0	1	0	23
2	1	0	0	0	0	1	34
3	1	0	0	0	-1	-1	19
4	0	1	0	0	1	0	31
5	0	1	0	0	0	1	37
6	0	1	0	0	-1	-1	22
7	0	0	1	0	1	0	52

Obs	X1	X2	X3	X4	Z1	Z2	Y
8	0	0	1	0	0	1	69
9	0	0	1	0	-1	-1	16
10	0	0	0	1	1	0	35
11	0	0	0	1	0	1	37
12	0	0	0	1	-1	-1	30
13	-1	-1	-1	-1	1	0	39
14	-1	-1	-1	-1	0	1	47
15	-1	-1	-1	-1	-1	-1	31

Analysis of Variance

Source	DF	Sum of Squares	Mean Square	F Value	Pr > F
Model	6	1881.46667	313.57778	3.49	0.0533
Error	8	718.93333	89.86667		
Corrected Total	14	2600.40000			

Root MSE	9.47980	R-Square	0.7235
Dependent Mean	34.80000	Adj R-Sq	0.5162
Coeff Var	27.24081		

Parameter Estimates

Variable	DF	Parameter Estimate	Standard Error	t-Value	Pr > |t|
Intercept	1	34.80000	2.44767	14.22	<.0001
X1	1	-9.46667	4.89535	-1.93	0.0892
X2	1	-4.80000	4.89535	-0.98	0.3555
X3	1	10.86667	4.89535	2.22	0.0572
X4	1	-0.80000	4.89535	-0.16	0.8742
Z1	1	1.20000	3.46153	0.35	0.7378
Z2	1	10.00000	3.46153	2.89	0.0202

It can be seen that SAS estimates for the parameters agree with the manual estimates obtained earlier. The fitted model under SAS is same as Equation 6.3. The SAS findings for the regression model also agrees with the findings that we obtained earlier in the book, by using classical approaches for the analysis of the responses of Example 5.1.

6.3 Dummy Variables Regression Model for the RCBD (Reference Cell Method)

The regression model for RCBD in Equation 6.2 comes out as the same when either the effects coding method or the reference cell method is used for defining the dummy variables. In Section 6.2, we used the effects coding method to define the dummy variables and to analyze the responses of the RCBD experiments presented in Example 5.1. Here, we use the reference cell method to define the dummy variables to be used in a typical RCBD model. As stated under the effects coding method, there are $k-1$ dummy variables defined for treatment that has k levels, while $b-1$ dummy variables are defined for block that has b levels. Thus, the model for the reference cell method, is

$$y_{ij} = \mu + \sum_{i=1}^{k-1} \gamma_i X_i + \sum_{j=1}^{b-1} \beta_j Z_j + \varepsilon \tag{6.8}$$

where

$$X_i = \begin{cases} 1 & \text{if } y_{ij} \text{ is under the } i\text{th treatment} \\ 0 & \text{otherwise} \end{cases}$$

for $i = 1, 2, \ldots, k-1$

$$Z_j = \begin{cases} 1 & \text{if } y_{ij} \text{ is under the } j\text{th block} \\ 0 & \text{otherwise} \end{cases}$$

for $j = 1, 2, \ldots, b-1$

Under this model, a particular cell is chosen as the reference cell for the purpose of defining the dummy variables to be used. Under the unreplicated RCBD, the value of the chosen cell is the value of μ in the regression model 6.8. Suppose we use cell (k, b) as the reference cell, then under the reference cell method, the following relationships exist among cell means, and related quantities (in Table 6.1), and parameters in the model can be identified as follows:

$$\mu_{i.} = \mu + \gamma_i + \frac{1}{b}\sum_{j=1}^{b-1} \beta_j \quad i = 1, 2, \ldots, k-1$$

$$\mu_{k.} = \mu + \frac{1}{b}\sum_{j=1}^{b-1} \beta_j$$

$$\mu_{.j} = \mu + \beta_j + \frac{1}{k}\sum_{i=1}^{k-1} \gamma_i \quad j = 1, 2, \ldots, b-1$$

$$\mu_{.b} = \mu + \frac{1}{k}\sum_{i=1}^{k-1} \gamma_i$$

$$\mu_{kj} = \mu + \beta_j \quad j = 1, 2, \ldots, b-1$$

$$\mu_{ib} = \mu + \gamma_i \quad i = 1, 2, \ldots, k-1$$

$$\mu_{ij} = \mu + \gamma_i + \beta_j \quad i = 1, 2, \ldots, k-1 \quad j = 1, 2, \ldots, b-1$$

6.3.1 Regression Model for RCBD of Example 5.1 (Reference Cell Coding)

With the above relationships, we can state the specific forms of the dummy variables required to obtain the reference cell model for the responses of the RCBD of Example 5.1. If we use cell (5, 3) as the reference cell for the analysis of the data of Example 5.1 for which there are three blocks and five treatments, then the specific model that applies is

$$y_{ij} = \mu + \sum_{i=1}^{4} \gamma_i X_i + \sum_{j=1}^{2} \beta_j Z_j + \varepsilon \tag{6.9}$$

$$X_i = \begin{cases} 1 & \text{if } y_{ij} \text{ is under the } i\text{th treatment} \\ 0 & \text{otherwise} \end{cases}$$

for $i = 1, 2, 3, 4$

and

$$Z_j = \begin{cases} 1 & \text{if } y_{ij} \text{ is under the } j\text{th block} \\ 0 & \text{otherwise} \end{cases}$$

for $j = 1, 2$

$$\mu_{i.} = \mu + \gamma_i + \frac{1}{3}\sum_{j=1}^{2} \beta_j \quad i = 1, 2, 3, 4$$

$$\mu_{k.} = \mu + \frac{1}{3}\sum_{j=1}^{2} \beta_j$$

$$\mu_{.j} = \mu + \beta_j + \frac{1}{5}\sum_{i=1}^{4} \gamma_i \quad j = 1, 2$$

$$\mu_{.b} = \mu + \frac{1}{5}\sum_{i=1}^{4} \gamma_i$$

$$\mu_{kj} = \mu + \beta_j \quad j = 1, 2$$

$$\mu_{ib} = \mu + \gamma_i \quad i = 1, 2, 3, 4$$

$$\mu_{ij} = \mu + \gamma_i + \beta_j \quad i = 1, 2, 3, 4 \quad j = 1, 2$$

6.3.2 Estimation of Parameters of the Model Fitted to Responses of Example 5.1 (Reference Cell Method)

Here, we show how we can manually obtain estimates for the parameters of the model stated in Equation 6.9 and ancillary equations. Using the data for

Example 5.1, the following are the estimates of the parameters of the regression model:

$$\mu_{5.} = \mu + \frac{1}{3}\sum_{j=1}^{2} \beta_j \Rightarrow \mu_{i.} = \mu_{5.} + \gamma_i \quad i = 1, 2, 3, 4$$

$$\Rightarrow \hat{\mu}_{1.} = \frac{76}{3} = \frac{117}{3} + \gamma_1 \Rightarrow \hat{\gamma}_1 = \frac{76}{3} - \frac{117}{3} = -13.6667$$

$$\hat{\mu}_{2.} = \frac{90}{3} = \frac{117}{3} + \gamma_2 \Rightarrow \hat{\gamma}_2 = \frac{90}{3} - \frac{117}{3} = -9.0$$

$$\hat{\mu}_{3} = \frac{137}{3} = \frac{117}{3} + \gamma_3 \Rightarrow \hat{\gamma}_3 = \frac{137}{3} - \frac{117}{3} = 6.6667$$

$$\hat{\mu}_{4.} = \frac{102}{3} = \frac{117}{3} + \gamma_4 \Rightarrow \hat{\gamma}_4 = \frac{102}{3} - \frac{117}{3} = -5.0$$

The above estimates of parameters are related to the treatments in the design. In a similar way, we obtain estimates for block-related parameters as

$$\mu_{.j} = \mu_{.3} + \beta_j \quad j = 1, 2$$

$$\Rightarrow \hat{\mu}_{.1} = \frac{180}{5} = \frac{118}{5} + \beta_1 \Rightarrow \hat{\beta}_1 = \frac{180}{5} - \frac{118}{5} = 12.4$$

$$\hat{\mu}_{.2} - \frac{224}{5} = \frac{118}{5} + \beta_2 \Rightarrow \hat{\beta}_2 = \frac{224}{5} - \frac{118}{5} = 21.2$$

Solving for the intercept of the model, we see that

$$\frac{1}{5}\sum_{i=1}^{4} \hat{\gamma}_i = \left(\frac{-13.667 - 9 + 6.6667 - 5}{5}\right) = -4.2$$

and

$$\frac{1}{3}\sum_{j=1}^{2} \hat{\beta}_j = \left(\frac{12.4 + 21.2}{3}\right) = 11.2$$

so that

$$\frac{118}{5} = \mu + \frac{1}{5}\sum_{i=1}^{4} \hat{\gamma}_i \Rightarrow \mu = \frac{118}{5} + 4.2 = 27.8$$

or

$$\frac{117}{3} = \mu + \frac{1}{3}\sum_{j=1}^{2} \hat{\beta}_j \Rightarrow \mu = \frac{117}{3} - 11.2 = 27.8$$

The above estimations of the parameters of the model indicate that any response of the experiment can be estimated by using the following equation:

$$\hat{y} = 27.8 - 13.6667x_1 - 9.0x_2 + 6.6667x_3 - 5x_4 + 12.4z_1 + 21.2z_2 \quad (6.10)$$

Again, with the estimated model of Equation 6.10, we obtain response estimates and attendant residuals, carry out ANOVA for the fitted model, and leave model diagnostics as an exercise for the reader (Table 6.4).

We can test for significance of fitted model. By the theory of regression, we know that if a regression has N responses arising from n sets of explanatory variables,

$$S_{yy} = \sum_{i=1}^{N} y_i^2 - \frac{\left(\sum_{i=1}^{N} y_i\right)^2}{N}$$

$$SS_{\mathrm{R}} = \sum_{i=1}^{N} (\hat{y}_i - \overline{y})^2$$

$$SS_{\mathrm{E}} = \sum_{i=1}^{N} (y_i - \hat{y}_i)^2$$

then $S_{yy} = SS_{\mathrm{R}} + SS_{\mathrm{E}}$. In this case $N = bk = 5(3) = 15$. Thus, as explained earlier, once we know these quantities, we can perform ANOVA to test for the

TABLE 6.4: Variables, Estimated Responses, and Residuals for Regression Model for Example 5.1 (Reference Cell Coding Method)

x_1	x_2	x_3	x_4	z_1	z_2	y_i	$\hat{y}_i$	$y_i - \hat{y}_i$	$(y_i - \hat{y}_i)^2$
1	0	0	0	1	0	23	26.5333	−3.5333	12.48421
1	0	0	0	0	1	34	35.3333	−1.3333	1.777689
1	0	0	0	0	0	19	14.1333	4.8667	23.68477
0	1	0	0	1	0	31	31.2	−0.2	0.04
0	1	0	0	0	1	37	40	−3	9
0	1	0	0	0	0	22	18.8	3.2	10.24
0	0	1	0	1	0	52	46.8667	5.1333	26.35077
0	0	1	0	0	1	69	55.6667	13.3333	177.7769
0	0	1	0	0	0	16	34.4667	−18.4667	341.019
0	0	0	1	1	0	35	35.2	−0.2	0.04
0	0	0	1	0	1	37	44	−7	49
0	0	0	1	0	0	30	22.8	7.2	51.84
0	0	0	0	1	0	39	40.2	−1.2	1.44
0	0	0	0	0	1	47	49	−2	4
0	0	0	0	0	0	31	27.8	3.2	10.24
Total							522	0	718.9333

significance of the fitted model. SS_{R} is the sum of squares due to the fitted regression model, while SS_{E} is the error sum of squares. From the above table, we can see that

$$SS_{\mathrm{E}} = 718.9333$$

$$S_{yy} = 20766 - \frac{522^2}{15} = 2600.40$$

$$\Rightarrow SS_{\mathrm{R}} = 2600.40 - 718.9333 = 1881.4667$$

TABLE 6.5: Testing for Significance of the Fitted Regression Model for Example 5.1

Source	Sum of squares	Degrees of freedom	Mean square	*F*-ratio
Regression	1881.4667	6	313.5778	3.489
Error	718.9333	8	89.8666	
Total	2600.40	14		

From Tables A5–A7, we see that $F(0.05, 6, 8) = 3.58$ and $F(0.06, 6, 8) = 3.33$. The model is significant at 6% level of significance. Similar result obtained under the effect coding method.

6.3.3 SAS Program for Regression Analysis of Responses of Example 5.1 (Reference Cell Coding for Dummy Variables)

Using the reference cell model, and the REG procedure, we write a SAS program that would analyze the responses of the RCBD of Example 5.1.

SAS PROGRAM

```
data rcbdref1;
input X1-X4 Z1-Z2 Y@@;
datalines;
1 0 0 0 1 0 23 1 0 0 0 0 1 34 1 0 0 0 0 0 19
0 1 0 0 1 0 31 0 1 0 0 0 1 37 0 1 0 0 0 0 22
0 0 1 0 1 0 52 0 0 1 0 0 1 69 0 0 1 0 0 0 16
0 0 0 1 1 0 35 0 0 0 1 0 1 37 0 0 0 1 0 0 30
0 0 0 0 1 0 39 0 0 0 0 0 1 47 0 0 0 0 0 0 31
;
```

```
proc print data=rcbdref1;
proc reg data=rcbdref1;model y=x1-x4 z1-z2; run;
```

SAS OUTPUT

Obs	X1	X2	X3	X4	Z1	Z2	Y
1	1	0	0	0	1	0	23
2	1	0	0	0	0	1	34
3	1	0	0	0	0	0	19
4	0	1	0	0	1	0	31
5	0	1	0	0	0	1	37
6	0	1	0	0	0	0	22
7	0	0	1	0	1	0	52
8	0	0	1	0	0	1	69
9	0	0	1	0	0	0	16
10	0	0	0	1	1	0	35
11	0	0	0	1	0	1	37
12	0	0	0	1	0	0	30
13	0	0	0	0	1	0	39
14	0	0	0	0	0	1	47
15	0	0	0	0	0	0	31

Analysis of Variance

Source	DF	Sum of Squares	Mean Square	F Value	Pr > F
Model	6	1881.46667	313.57778	3.49	0.0533
Error	8	718.93333	89.86667		
Corrected Total	14	2600.40000			

Root MSE	9.47980	R-Square	0.7235
Dependent Mean	34.80000	Adj R-Sq	0.5162
Coeff Var	27.24081		

Parameter Estimates

Variable	DF	Parameter Estimate	Standard Error	t-Value	Pr > |t|
Intercept	1	27.80000	6.47594	4.29	0.0026
X1	1	-13.66667	7.74023	-1.77	0.1154
X2	1	-9.00000	7.74023	-1.16	0.2784
X3	1	6.66667	7.74023	0.86	0.4141
X4	1	-5.00000	7.74023	-0.65	0.5364
Z1	1	12.40000	5.99555	2.07	0.0724
Z2	1	21.20000	5.99555	3.54	0.0077

Again, all the SAS estimates of model parameters using the data agree with estimates obtained by manual calculations. The ANOVA carried out agrees with the results from classical analysis used in Chapter 5 for the responses of Example 5.1.

6.4 Application of Dummy Variables Regression Model to Example 5.2 (Effects Coding Method)

Next, we reanalyze the Uzo-Uwani Cassava data (Example 5.2) using a regression approach. This is another example of analysis of RCBD by using a regression model. First, we start by using the effects coding method to define the dummy variables in the model. There are eight blocks and four treatments, so that the regression model that applies is

$$y_{ij} = \mu + \sum_{i=1}^{3} \gamma_i X_i + \sum_{j=1}^{7} \beta_j Z_j + \varepsilon \tag{6.11}$$

$$X_i = \begin{cases} -1 & \text{if } y_{ij} \text{ is under the 4th treatment} \\ 1 & \text{if } y_{ij} \text{ is under the } i\text{th treatment} \\ 0 & \text{otherwise} \end{cases}$$

for $i = 1, 2, 3$

$$Z_j = \begin{cases} -1 & \text{if } y_{ij} \text{ is under the 8th block} \\ 1 & \text{if } y_{ij} \text{ is under the } j\text{th block} \\ 0 & \text{otherwise} \end{cases}$$

for $j = 1, 2, \ldots, 7$

6.4.1 Estimation of Parameters for the Regression Model for Responses of Example 5.2 (Effects Coding Method)

Again, we use the theory we discussed earlier to manually estimate the parameters of the regression model for the experiment on cassava as follows:

$$\hat{\mu}_{..} = \overline{y}_{..} = 6.78125 \quad \overline{y}_{1.} = 7.75 \quad \overline{y}_{2.} = 6.75 \quad \overline{y}_{3.} = 6.75 \quad \overline{y}_{4.} = 5.875$$

$$\hat{\gamma}_1 = 7.75 - 6.78125 = 0.9688 \quad \hat{\gamma}_2 = 6.75 - 6.78125 = -0.03125$$

$$\hat{\gamma}_3 = 6.75 - 6.78125 = -0.03125 \qquad -\sum_{i=1}^{3}\hat{\gamma}_i = 5.875 - 6.78125 = 0.9064$$

$$\overline{y}_{.1} = 8.5 \quad \overline{y}_{.2} = 3.5 \quad \overline{y}_{.3} = 10.5 \quad \overline{y}_{.4} = 7.75 \quad \overline{y}_{.5} = 3.5 \quad \overline{y}_{.6} = 7.25$$

$$\overline{y}_{.7} = 6.75 \quad \overline{y}_{.8} = 6.5$$

$$\hat{\beta}_1 = 8.5 - 6.78125 = 1.71875 \quad \hat{\beta}_2 = 3.5 - 6.78125 = -3.28125$$

$$\hat{\beta}_3 = 10.5 - 6.78125 = 3.71825 \quad \hat{\beta}_4 = 7.75 - 6.78125 = 0.96875$$

$$\hat{\beta}_5 = 3.5 - 6.78125 = -3.28125 \quad \hat{\beta}_6 = 7.25 - 6.78125 = 0.46875$$

$$\hat{\beta}_7 = 6.75 - 6.78125 = -0.03125 \qquad -\sum_{j=1}^{7}\beta_j = 6.5 - 6.78125 = 0.28125$$

The estimated response for this model is

$$\hat{y} = 6.78125 + 0.9688x_1 - 0.03125x_2 - 0.03125x_3 + 1.71875z_1 - 3.28125z_2 + 3.71825z_3 + 0.96875z_4 - 3.28125z_5 + 0.46875z_6 - 0.03125z_7 \tag{6.12}$$

With the above estimated response (Equation 6.12), we obtain both the estimated reponses and residuals and leave the analysis for checking model adequacy as an exercise for the reader.

As stated earlier, if $S_{yy} = \sum_{i=1}^{N} y_i^2 - \left(\sum_{i=1}^{N} y_i\right)^2/N$, $SS_{\text{R}} = \sum_{i=1}^{N}(\hat{y}_i - \overline{y})^2$, and $SS_{\text{E}} = \sum_{i=1}^{N}(y_i - \hat{y}_i)^2$ then $S_{yy} = SS_{\text{R}} + SS_{\text{E}}$. For this RCBD, $N = bk = 4(8) = 32$.

$$SS_{\text{E}} = 153.1563$$

$$S_{yy} = 1797 - \frac{217^2}{32} = 325.4688$$

$$\Rightarrow SS_{\text{R}} = 325.4688 - 153.1563 = 172.3125$$

TABLE 6.6: Variables, Estimated Responses, and Residuals for Regression Model for Example 5.2 (Effects Coding Method)

x_1	x_2	x_3	z_1	z_2	z_3	z_4	z_5	z_6	z_7	y	$\hat{y}_i$	$y-\hat{y}_i$	$(y-\hat{y_i})^2$
1	0	0	1	0	0	0	0	0	0	8	9.4688	−1.4688	2.157373
1	0	0	0	1	0	0	0	0	0	6	4.4688	1.5312	2.344573
1	0	0	0	0	1	0	0	0	0	13	11.4683	1.5317	2.346105
1	0	0	0	0	0	1	0	0	0	7	8.7188	−1.7188	2.954273
1	0	0	0	0	0	0	1	0	0	4	4.4688	−0.4688	0.219773
1	0	0	0	0	0	0	0	1	0	13	8.2188	4.7812	22.85987
1	0	0	0	0	0	0	0	0	1	2	7.7188	−5.7188	32.70467
1	0	0	−1	−1	−1	−1	−1	−1	−1	9	7.4693	1.5307	2.343042
0	1	0	1	0	0	0	0	0	0	8	8.46875	−0.46875	0.219727
0	1	0	0	1	0	0	0	0	0	3	3.46875	−0.46875	0.219727
0	1	0	0	0	1	0	0	0	0	10	10.46825	−0.46825	0.219258
0	1	0	0	0	0	1	0	0	0	8	7.71875	0.28125	0.079102
0	1	0	0	0	0	0	1	0	0	5	3.46875	1.53125	2.344727
0	1	0	0	0	0	0	0	1	0	10	7.21875	2.78125	7.735352
0	1	0	0	0	0	0	0	0	1	7	6.71875	0.28125	0.079102
0	1	0	−1	−1	−1	−1	−1	−1	−1	3	6.46925	−3.46925	12.0357
0	0	1	1	0	0	0	0	0	0	10	8.46875	1.53125	2.344727
0	0	1	0	1	0	0	0	0	0	2	3.46875	−1.46875	2.157227
0	0	1	0	0	1	0	0	0	0	9	10.46825	−1.46825	2.155758
0	0	1	0	0	0	1	0	0	0	10	7.71875	2.28125	5.204102
0	0	1	0	0	0	0	1	0	0	2	3.46875	−1.46875	2.157227
0	0	1	0	0	0	0	0	1	0	3	7.21875	−4.21875	17.79785
0	0	1	0	0	0	0	0	0	1	10	6.71875	3.28125	10.7666
0	0	1	−1	−1	−1	−1	−1	−1	−1	8	6.46925	1.53075	2.343196
−1	−1	−1	1	0	0	0	0	0	0	8	7.5937	0.4063	0.16508
−1	−1	−1	0	1	0	0	0	0	0	3	2.5937	0.4063	0.16508
−1	−1	−1	0	0	1	0	0	0	0	10	9.5932	0.4068	0.165486
−1	−1	−1	0	0	0	1	0	0	0	6	6.8437	−0.8437	0.71183
−1	−1	−1	0	0	0	0	1	0	0	3	2.5937	0.4063	0.16508
−1	−1	−1	0	0	0	0	0	1	0	3	6.3437	−3.3437	11.18033
−1	−1	−1	0	0	0	0	0	0	1	8	5.8437	2.1563	4.64963
−1	−1	−1	−1	−1	−1	−1	−1	−1	−1	6	5.5942	0.4058	0.164674
Total											217	0	153.1563

TABLE 6.7: ANOVA to Test for Significance of the Fitted Regression Model for Example 5.2

Source	Sum of squares	Degrees of freedom	Mean square	F-ratio
Regression	172.3125	10	17.2312	2.3627
Error	153.1563	21	7.2932	
Total	325.4688	31		

From Tables A5–A7, we see that $F(0.05,\ 10,\ 21) = 2.32$. The model is significant at 5% level of significance, echoing the findings of previous analysis in Chapter 5 using classical methods.

6.4.2 SAS Program for Example 5.2 for the Regression Model with Effects Coding for Dummy Variables

We write a SAS program to analyze the data for Example 5.2 when the effects coding method is used to define the dummy variables in the model. As usual, we call the REG or the GLM procedure for the analysis.

SAS PROGRAM

```
data rcbdcode2;input X1-X3 Z1-Z7 Y@@;
datalines;
1 0 0 1 0 0 0 0 0 0 8 1 0 0 0 1 0 0 0 0 0 6
1 0 0 0 0 1 0 0 0 0 13 1 0 0 0 0 0 1 0 0 0 7
1 0 0 0 0 0 0 1 0 0 4 1 0 0 0 0 0 0 0 1 0 13
1 0 0 0 0 0 0 0 0 1 2 1 0 0 -1 -1 -1 -1 -1 -1 -1 9
0 1 0 1 0 0 0 0 0 0 8 0 1 0 0 1 0 0 0 0 0 3
0 1 0 0 0 1 0 0 0 0 10 0 1 0 0 0 0 1 0 0 0 8
0 1 0 0 0 0 0 1 0 0 5 0 1 0 0 0 0 0 0 1 0 10
0 1 0 0 0 0 0 0 0 1 7 0 1 0 -1 -1 -1 -1 -1 -1 -1 3
0 0 1 1 0 0 0 0 0 0 10 0 0 1 0 1 0 0 0 0 0 2
0 0 1 0 0 1 0 0 0 0 9 0 0 1 0 0 0 1 0 0 0 10
0 0 1 0 0 0 0 1 0 0 2 0 0 1 0 0 0 0 0 1 0 3
0 0 1 0 0 0 0 0 0 1 10 0 0 1 -1 -1 -1 -1 -1 -1 -1 8
-1 -1 -1 1 0 0 0 0 0 0 8 -1 -1 -1 0 1 0 0 0 0 0 3
-1 -1 -1 0 0 1 0 0 0 0 10 -1 -1 -1 0 0 0 1 0 0 0 6
-1 -1 -1 0 0 0 0 1 0 0 3 -1 -1 -1 0 0 0 0 0 1 0 3
-1 -1 -1 0 0 0 0 0 0 1 8 -1 -1 -1 -1 -1 -1 -1 -1 -1 -1 6
;

proc print data=rcbdcode2;
proc reg data=rcbdcode2; model y=x1-x3 z1-z7;
run;
```

SAS OUTPUT

Obs	X1	X2	X3	Z1	Z2	Z3	Z4	Z5	Z6	Z7	Y
1	1	0	0	1	0	0	0	0	0	0	8
2	1	0	0	0	1	0	0	0	0	0	6
3	1	0	0	0	0	1	0	0	0	0	13
4	1	0	0	0	0	0	1	0	0	0	7
5	1	0	0	0	0	0	0	1	0	0	4
6	1	0	0	0	0	0	0	0	1	0	13
7	1	0	0	0	0	0	0	0	0	1	2
8	1	0	0	-1	-1	-1	-1	-1	-1	-1	9
9	0	1	0	1	0	0	0	0	0	0	8
10	0	1	0	0	1	0	0	0	0	0	3
11	0	1	0	0	0	1	0	0	0	0	10
12	0	1	0	0	0	0	1	0	0	0	8
13	0	1	0	0	0	0	0	1	0	0	5
14	0	1	0	0	0	0	0	0	1	0	10
15	0	1	0	0	0	0	0	0	0	1	7
16	0	1	0	-1	-1	-1	-1	-1	-1	-1	3
17	0	0	1	1	0	0	0	0	0	0	10
18	0	0	1	0	1	0	0	0	0	0	2
19	0	0	1	0	0	1	0	0	0	0	9
20	0	0	1	0	0	0	1	0	0	0	10
21	0	0	1	0	0	0	0	1	0	0	2
22	0	0	1	0	0	0	0	0	1	0	3
23	0	0	1	0	0	0	0	0	0	1	10
24	0	0	1	-1	-1	-1	-1	-1	-1	-1	8
25	-1	-1	-1	1	0	0	0	0	0	0	8
26	-1	-1	-1	0	1	0	0	0	0	0	3
27	-1	-1	-1	0	0	1	0	0	0	0	10
28	-1	-1	-1	0	0	0	1	0	0	0	6
29	-1	-1	-1	0	0	0	0	1	0	0	3
30	-1	-1	-1	0	0	0	0	0	1	0	3
31	-1	-1	-1	0	0	0	0	0	0	1	8
32	-1	-1	-1	-1	-1	-1	-1	-1	-1	-1	6

Analysis of Variance

Source	DF	Sum of Squares	Mean Square	F Value	Pr > F
Model	10	172.31250	17.23125	2.36	0.0466
Error	21	153.15625	7.29315		
Corrected Total	31	325.46875			

Root MSE	2.70058	R-Square	0.5294
Dependent Mean	6.78125	Adj R-Sq	0.3053
Coeff Var	39.82428		

Parameter Estimates

Variable	DF	Parameter Estimate	Standard Error	t-Value	Pr > |t|
Intercept	1	6.78125	0.47740	14.20	<.0001
X1	1	0.96875	0.82688	1.17	0.2545
X2	1	-0.03125	0.82688	-0.04	0.9702
X3	1	-0.03125	0.82688	-0.04	0.9702
Z1	1	1.71875	1.26308	1.36	0.1880
Z2	1	-3.28125	1.26308	-2.60	0.0168
Z3	1	3.71875	1.26308	2.94	0.0077
Z4	1	0.96875	1.26308	0.77	0.4516
Z5	1	-3.28125	1.26308	-2.60	0.0168
Z6	1	0.46875	1.26308	0.37	0.7143
Z7	1	-0.03125	1.26308	-0.02	0.9805

It is easy to see that all the SAS estimates agree with the previously obtained manual estimates. Also, if we compare the model ANOVA with previous analyses carried out both by SAS and manual methods for the classical approach, we see that there is an agreement. Thus, this regression model represents an equivalent approach to the analysis of responses of the designed experiment by classical methods.

6.5 Regression Model for RCBD of Example 5.2 (Reference Cell Coding)

Next, we use the reference cell method to obtain the model for analysis of data of Example 5.2. There are eight blocks and four treatments, so we define two sets of dummy variables, one with seven variables for block, and the other with three dummy variables for treatment. We use cell (4, 8) as the reference for this model. The specific model that applies is the same as that mentioned in Equation 6.11, except that the definitions of the dummy variables are different. Here,we restate the model along with the definitions of the dummy variables as

$$y_{ij} = \mu + \sum_{i=1}^{3} \gamma_i X_i + \sum_{j=1}^{7} \beta_j Z_j + \varepsilon$$

$$X_i = \begin{cases} 1 & \text{if } y_{ij} \text{ is under the 4th treatment} \\ 0 & \text{otherwise} \end{cases}$$

for $i = 1, 2, 3$

$$Z_i = \begin{cases} 1 & \text{if } y_{ij} \text{ is under the 8th block} \\ 0 & \text{otherwise} \end{cases}$$

for $j = 1, 2, \ldots, 7$

Again, we use the data to illustrate how manual estimation of the parameters for this model is effected.

$$\overline{y}_{1.} = 7.75 \quad \overline{y}_{2.} = 6.75 \quad \overline{y}_{3.} = 6.75 \quad \overline{y}_{4.} = 5.875 \quad \overline{y}_{.1} = 8.5 \quad \overline{y}_{.2} = 3.5$$

$$\overline{y}_{.3} = 10.5 \quad \overline{y}_{.4} = 7.75 \quad \overline{y}_{.5} = 3.5 \quad \overline{y}_{.6} = 7.25 \quad \overline{y}_{.7} = 6.75 \quad \overline{y}_{.8} = 6.5$$

$$\mu_{i.} = \mu_{4.} + \gamma_i \quad i = 1, 2, 3$$

$$\Rightarrow \mu_{1.} = 7.75 = 5.875 + \gamma_1 \Rightarrow \hat{\gamma}_1 = 7.75 - 5.875 = 1.875$$

$$\mu_{2.} = 6.75 = 5.875 + \gamma_2 \Rightarrow \hat{\gamma}_2 = 6.75 - 5.875 = 0.875$$

$$\mu_{3.} = 6.75 = 5.875 + \gamma_3 \Rightarrow \hat{\gamma}_3 = 6.75 - 5.875 = 0.875$$

$$\mu_{.j} = \mu_{.8} + \beta_j \quad j = 1, 2, \ldots, 7$$

$$\mu_{.1} = 8.5 = 6.5 + \beta_1 \Rightarrow \hat{\beta}_1 = 8.5 - 6.5 = 2.0$$

$$\mu_{.2} = 3.5 = 6.5 + \beta_2 \Rightarrow \hat{\beta}_2 = 3.5 - 6.5 = -3.0$$

$$\mu_{.3} = 10.5 = 6.5 + \beta_3 \Rightarrow \hat{\beta}_3 = 10.5 - 6.5 = 4.0$$

$$\mu_{.4} = 7.75 = 6.5 + \beta_4 \Rightarrow \hat{\beta}_4 = 7.75 - 6.5 = 1.25$$

$$\mu_{.5} = 3.5 = 6.5 + \beta_5 \Rightarrow \hat{\beta}_5 = 3.5 - 6.5 = -3.0$$

$$\mu_{.6} = 7.25 = 6.5 + \beta_6 \Rightarrow \hat{\beta}_6 = 7.25 - 6.5 = 0.75$$

$$\mu_{.7} = 6.75 = 6.5 + \beta_7 \Rightarrow \hat{\beta}_7 = 6.75 - 6.5 = 0.25$$

$$\frac{1}{4}\sum_{i=1}^{3}\hat{\gamma}_i = \frac{1.875+0.875+0.875}{4} = 0.90625$$

$$\frac{1}{8}\sum_{j=1}^{7}\hat{\beta}_j = \frac{2.0-3.0+4.0+1.25-3.0+0.75+0.25}{8} = 0.28125$$

$$6.5 = \mu + \frac{1}{4}\sum_{i=1}^{3}\hat{\gamma}_i \Rightarrow \mu = 6.5 - 0.90625 = 5.59375$$

or

$$5.875 = \mu + \frac{1}{8}\sum_{j=1}^{7}\hat{\beta}_j \Rightarrow \mu = 5.875 - 0.28125 = 5.59375$$

The above estimation leads to the following fitted model:

$$\begin{aligned}\hat{y} &= 5.59375 + 1.875x_1 + 0.875x_2 + 0.875x_3 + 2z_1 - 3z_2 + 4z_3 \\ &\quad + 1.25z_4 - 3z_5 + 0.75z_6 + 0.25z_7 \end{aligned} \tag{6.13}$$

The estimates and residuals for this model are given in Tables 6.8 and 6.9.

Again, it is easy to see that if $S_{yy} = \sum_{i=1}^{N} y_i^2 - \left(\sum_{i=1}^{N} y_i\right)^2/N$, $SS_{\mathrm{R}} = \sum_{i=1}^{N}(\hat{y}_i - \overline{y})^2$, and $SS_{\mathrm{E}} = \sum_{i=1}^{N}(y_i - \hat{y}_i)^2$ then $S_{yy} = SS_{\mathrm{R}} + SS_{\mathrm{E}}$. $N = bk = 4(8) = 32$.

$$SS_{\mathrm{E}} = 153.1563$$

$$S_{yy} = 1797 - \frac{217^2}{32} = 325.4688 \Rightarrow SS_{\mathrm{R}} = 325.4688 - 153.1563 = 172.3125$$

This is the same as the result obtained for the ANOVA (Table 6.9, page 364) for the classical model both by manual and SAS analyses. It is also the same ANOVA we obtained by using effects coding method for the responses of the experiment. This indicates that although the effects coding and reference cell coding methods lead to different model equations and parameter estimates, the ANOVA for both models shows that the same amount of variation is explained by the regression.

6.5.1 SAS Analysis of Data of Example 5.2 Using Regression Model with Reference Cell Coding for Dummy Variables

Further, we analyze the data by writing a SAS program. Here, again, we use the REG procedure as we prefer the briefer output that it produces. To obtain the more elaborate output, the reader is free to replace PROC REG with PROC GLM in the program and run it.

TABLE 6.8: Variables, Estimated Responses, and Residuals for Regression Model for Example 5.2 (Reference Cell Coding Method)

x_1	x_2	x_3	z_1	z_2	z_3	z_4	z_5	z_6	z_7	y	$\hat{y}_i$	$y-\hat{y}_i$	$(y-\hat{y}_i)^2$
1	0	0	1	0	0	0	0	0	0	8	9.46875	−1.46875	2.157227
1	0	0	0	1	0	0	0	0	0	6	4.46875	1.53125	2.344727
1	0	0	0	0	1	0	0	0	0	13	11.46875	1.53125	2.344727
1	0	0	0	0	0	1	0	0	0	7	8.71875	−1.71875	2.954102
1	0	0	0	0	0	0	1	0	0	4	4.46875	−0.46875	0.219727
1	0	0	0	0	0	0	0	1	0	13	8.21875	4.78125	22.86035
1	0	0	0	0	0	0	0	0	1	2	7.71875	−5.71875	32.7041
1	0	0	0	0	0	0	0	0	0	9	7.46875	1.53125	2.344727
0	1	0	1	0	0	0	0	0	0	8	8.46875	−0.46875	0.219727
0	1	0	0	1	0	0	0	0	0	3	3.46875	−0.46875	0.219727
0	1	0	0	0	1	0	0	0	0	10	10.46875	−0.46875	0.219727
0	1	0	0	0	0	1	0	0	0	8	7.71875	0.28125	0.079102
0	1	0	0	0	0	0	1	0	0	5	3.46875	1.53125	2.344727
0	1	0	0	0	0	0	0	1	0	10	7.21875	2.78125	7.735352
0	1	0	0	0	0	0	0	0	1	7	6.71875	0.28125	0.079102
0	1	0	0	0	0	0	0	0	0	3	6.46875	−3.46875	12.03223
0	0	1	1	0	0	0	0	0	0	10	8.46875	1.53125	2.344727
0	0	1	0	1	0	0	0	0	0	2	3.46875	−1.46875	2.157227
0	0	1	0	0	1	0	0	0	0	9	10.46875	−1.46875	2.157227
0	0	1	0	0	0	1	0	0	0	10	7.71875	2.28125	5.204102
0	0	1	0	0	0	0	1	0	0	2	3.46875	−1.46875	2.157227
0	0	1	0	0	0	0	0	1	0	3	7.21875	−4.21875	17.79785
0	0	1	0	0	0	0	0	0	1	10	6.71875	3.28125	10.7666
0	0	1	0	0	0	0	0	0	0	8	6.46875	1.53125	2.344727
0	0	0	1	0	0	0	0	0	0	8	7.59375	0.40625	0.165039
0	0	0	0	1	0	0	0	0	0	3	2.59375	0.40625	0.165039
0	0	0	0	0	1	0	0	0	0	10	9.59375	0.40625	0.165039
0	0	0	0	0	0	1	0	0	0	6	6.84375	−0.84375	0.711914
0	0	0	0	0	0	0	1	0	0	3	2.59375	0.40625	0.165039
0	0	0	0	0	0	0	0	1	0	3	6.34375	−3.34375	11.18066
0	0	0	0	0	0	0	0	0	1	8	5.84375	2.15625	4.649414
0	0	0	0	0	0	0	0	0	0	6	5.59375	0.40625	0.165039
Total											217	0	153.1563

TABLE 6.9: ANOVA to Test for Significance of the Fitted Regression Model for Example 5.2

Source	Sum of squares	Degrees of freedom	Mean square	F-ratio
Regression	172.3125	10	17.2312	2.3627
Error	153.1563	21	7.2932	
Total	325.4688	31		

SAS PROGRAM

```
data rcbdref2;
input X1-X3 Z1-Z7 Y@@;
datalines;
1 0 0 1 0 0 0 0 0 0 8 1 0 0 0 1 0 0 0 0 0 6
1 0 0 0 0 1 0 0 0 0 13 1 0 0 0 0 0 1 0 0 0 7
1 0 0 0 0 0 0 1 0 0 4 1 0 0 0 0 0 0 0 1 0 13
1 0 0 0 0 0 0 0 0 1 2 1 0 0 0 0 0 0 0 0 0 9
0 1 0 1 0 0 0 0 0 0 8 0 1 0 0 1 0 0 0 0 0 3
0 1 0 0 0 1 0 0 0 0 10 0 1 0 0 0 0 1 0 0 0 8
0 1 0 0 0 0 0 1 0 0 5 0 1 0 0 0 0 0 0 1 0 10
0 1 0 0 0 0 0 0 0 1 7 0 1 0 0 0 0 0 0 0 0 3
0 0 1 1 0 0 0 0 0 0 10 0 0 1 0 1 0 0 0 0 0 2
0 0 1 0 0 1 0 0 0 0 9 0 0 1 0 0 0 1 0 0 0 10
0 0 1 0 0 0 0 1 0 0 2 0 0 1 0 0 0 0 0 1 0 3
0 0 1 0 0 0 0 0 0 1 10 0 0 1 0 0 0 0 0 0 0 8
0 0 0 1 0 0 0 0 0 0 8 0 0 0 0 1 0 0 0 0 0 3
0 0 0 0 0 1 0 0 0 0 10 0 0 0 0 0 0 1 0 0 0 6
0 0 0 0 0 0 0 1 0 0 3 0 0 0 0 0 0 0 0 1 0 3
0 0 0 0 0 0 0 0 0 1 8 0 0 0 0 0 0 0 0 0 0 6
;
proc print data=rcbdref2;
proc reg data=rcbdref2;
model y=x1-x3 z1-z7; run;
```

SAS OUTPUT

Obs	X1	X2	X3	Z1	Z2	Z3	Z4	Z5	Z6	Z7	Y
1	1	0	0	1	0	0	0	0	0	0	8
2	1	0	0	0	1	0	0	0	0	0	6
3	1	0	0	0	0	1	0	0	0	0	13
4	1	0	0	0	0	0	1	0	0	0	7
5	1	0	0	0	0	0	0	1	0	0	4
6	1	0	0	0	0	0	0	0	1	0	13
7	1	0	0	0	0	0	0	0	0	1	2
8	1	0	0	0	0	0	0	0	0	0	9
9	0	1	0	1	0	0	0	0	0	0	8
10	0	1	0	0	1	0	0	0	0	0	3
11	0	1	0	0	0	1	0	0	0	0	10

Obs	X1	X2	X3	Z1	Z2	Z3	Z4	Z5	Z6	Z7	Y
12	0	1	0	0	0	0	1	0	0	0	8
13	0	1	0	0	0	0	0	1	0	0	5
14	0	1	0	0	0	0	0	0	1	0	10
15	0	1	0	0	0	0	0	0	0	1	7
16	0	1	0	0	0	0	0	0	0	0	3
17	0	0	1	1	0	0	0	0	0	0	10
18	0	0	1	0	1	0	0	0	0	0	2
19	0	0	1	0	0	1	0	0	0	0	9
20	0	0	1	0	0	0	1	0	0	0	10
21	0	0	1	0	0	0	0	1	0	0	2
22	0	0	1	0	0	0	0	0	1	0	3
23	0	0	1	0	0	0	0	0	0	1	10
24	0	0	1	0	0	0	0	0	0	0	8
25	0	0	0	1	0	0	0	0	0	0	8
26	0	0	0	0	1	0	0	0	0	0	3
27	0	0	0	0	0	1	0	0	0	0	10
28	0	0	0	0	0	0	1	0	0	0	6
29	0	0	0	0	0	0	0	1	0	0	3
30	0	0	0	0	0	0	0	0	1	0	3
31	0	0	0	0	0	0	0	0	0	1	8
32	0	0	0	0	0	0	0	0	0	0	6

Analysis of Variance

Source	DF	Sum of Squares	Mean Square	F Value	Pr > F
Model	10	172.31250	17.23125	2.36	0.0466
Error	21	153.15625	7.29315		
Corrected Total	31	325.46875			

Root MSE	2.70058	R-Square	0.5294
Dependent Mean	6.78125	Adj R-Sq	0.3053
Coeff Var	39.82428		

Parameter Estimates

Variable	DF	Parameter Estimate	Standard Error	t-Value	Pr > |t|
Intercept	1	5.59375	1.58336	3.53	0.0020
X1	1	1.87500	1.35029	1.39	0.1795
X2	1	0.87500	1.35029	0.65	0.5240
X3	1	0.87500	1.35029	0.65	0.5240
Z1	1	2.00000	1.90960	1.05	0.3068
Z2	1	-3.00000	1.90960	-1.57	0.1311
Z3	1	4.00000	1.90960	2.09	0.0485
Z4	1	1.25000	1.90960	0.65	0.5198
Z5	1	-3.00000	1.90960	-1.57	0.1311
Z6	1	0.75000	1.90960	0.39	0.6985
Z7	1	0.25000	1.90960	0.13	0.8971

The estimates for the parameters of the model obtained as part of SAS output agree with the values obtained for the same estimates by manual calculation. The regression ANOVA carried out by SAS corroborates results from previous analysis of the same data that we have presented elsewhere in the book.

6.6 Regression Models for the Latin Square Design

Whether replication occurs or not in a Latin square design, we should be able to obtain cell, row, column, and treatment means for the responses of a Latin squares experiment, which could be presented as in Table 6.10. When there is no replication, the single observation in the cell becomes the cell mean.

In Table 6.10, the subscript k is used in every cell to indicate that any of the r treatments (Latin letters) can appear in any cell following the randomization restrictions, which characterize the Latin square design. The method of randomization for this design had been previously described. The table below contains the means of the r treatments in the design.

Treatment	1	2	...	r	
Means	$\mu_{..1}$	$\mu_{..2}$	...	$\mu_{..r}$	$\mu_{...}$

TABLE 6.10: Typical Cell Means in the Latin Square Design

	Columns				
Rows	**1**	**2**	...	r	**Row means**
1	μ_{11k}	μ_{11k}		μ_{11k}	$\mu_{1..}$
2	μ_{21k}	μ_{22k}		μ_{2rk}	$\mu_{2..}$
$\vdots$	$\vdots$	$\vdots$	...	$\vdots$	$\vdots$
r	μ_{r1k}	μ_{r2k}		μ_{rrk}	$\mu_{r..}$
Column means	$\mu_{.1.}$	$\mu_{.2.}$		$\mu_{.5.}$	$\mu_{...}$

6.6.1 Regression Model for the Latin Square Design Using Effects Coding Method to Define Dummy Variables

Three sets of dummy variables need to be defined to fit a regression model to the typical Latin square design. In the next section, we will use the effects coding method to define the required dummy variables. Each set of dummy variables contains $r-1$ units corresponding to r rows, r columns and r Latin letters. Thus, the model to be fitted to an $r \times r$ Latin square design is of the form

$$y_{ijk} = \mu + \sum_{i=1}^{r-1} \alpha_i X_i + \sum_{j=1}^{r-1} \beta_j Z_j + \sum_{k=1}^{r-1} \gamma_k C_k + \varepsilon \tag{6.14}$$

where

$$X_i = \begin{cases} -1 & \text{if } y_{ijk} \text{ is under the } r\text{th row} \\ 1 & \text{if } y_{ijk} \text{ is under the } i\text{th row} \\ 0 & \text{otherwise} \end{cases}$$

for $i = 1, 2, \ldots, r-1$

$$Z_j = \begin{cases} -1 & \text{if } y_{ijk} \text{ is under the } r\text{th column} \\ 1 & \text{if } y_{ijk} \text{ is under the } j\text{th column} \\ 0 & \text{otherwise} \end{cases}$$

for $j = 1, 2, \ldots, r-1$

$$C_k = \begin{cases} -1 & \text{if } y_{ijk} \text{ is under the } r\text{th treatment or the } r\text{th Latin letter} \\ 1 & \text{if } y_{ijk} \text{ is under the } k\text{th treatment} \\ 0 & \text{otherwise} \end{cases}$$

for $k = 1, 2, \ldots, r-1$

6.6.1.1 Relations for Estimating Model Parameters (Effects Coding Method)

To estimate parameters, we use the following relationships among the cell, row, column, and treatment means with parameters of the regression model 6.14 that help us to estimate the parameters as

$$\mu_{...} = \mu$$

$$-\sum_{i=1}^{r-1} \alpha_i = \mu_{r..} - \mu_{...}$$

$$-\sum_{j=1}^{r-1} \beta_j = \mu_{.r.} - \mu_{...}$$

$$-\sum_{k=1}^{r-1} \gamma_k = \mu_{..r} - \mu_{...}$$

$$\begin{aligned} \mu_{ijk} &= \mu + \alpha_i + \beta_j + \gamma_k \quad i, j, k = 1, 2, \ldots, r-1 \\ \alpha_i &= \mu_{i..} - \mu_{...} \quad i = 1, 2, \ldots, r-1 \\ \beta_j &= \mu_{.j.} - \mu_{...} \quad j = 1, 2, \ldots, r-1 \\ \gamma_k &= \mu_{..k} - \mu_{...} \quad k = 1, 2, \ldots, r-1 \end{aligned}$$

6.6.1.2 Estimation of Parameters of Regression Model for Example 5.5 (Effects Coding Method)

To estimate the parameters for the regression model 6.14 for Example 5.5, we note that $r = 5$ for the design.

We use the following sample values obtained from data to estimate the parameters of the model as follows:

$$\overline{y}_{...} = \frac{1519}{25} = 60.76 \quad \overline{y}_{1..} = \frac{309}{5} = 61.8 \quad \overline{y}_{2..} = \frac{300}{5} = 60$$

$$\overline{y}_{3..} = \frac{303}{5} = 60.6 \quad \overline{y}_{4..} = \frac{297}{5} = 59.4 \quad \overline{y}_{5..} = \frac{310}{5} = 62$$

$$\overline{y}_{.1.} = \frac{288}{5} = 57.6 \quad \overline{y}_{.2.} = \frac{322}{5} = 64.4 \quad \overline{y}_{.3.} = \frac{326}{5} = 65.2$$

$$\overline{y}_{.4.} = \frac{287}{5} = 57.4 \quad \overline{y}_{.1.} = \frac{296}{5} = 59.2 \quad \overline{y}_{..1} = \frac{349}{5} = 69.8$$

$$\overline{y}_{..2} = \frac{318}{5} = 63.6 \quad \overline{y}_{..3} = \frac{283}{5} = 56.6 \quad \overline{y}_{..4} = \frac{311}{5} = 62.2$$

$$\overline{y}_{..5} = \frac{258}{5} = 51.6$$

$$\hat{\alpha}_1 = \overline{y}_{1..} - \overline{y}_{...} = 61.8 - 60.76 = 1.04 \quad \hat{\alpha}_2 = \overline{y}_{2..} - \overline{y}_{...} = 60 - 60.76 = -0.76$$

$$\hat{\alpha}_3 = \overline{y}_{3..} - \overline{y}_{...} = 60.6 - 60.76 = -0.16 \quad \hat{\alpha}_4 = \overline{y}_{4..} - \overline{y}_{...} = 59.4 - 60.76 = -1.36$$

$$\hat{\beta}_1 = \overline{y}_{.1.} - \overline{y}_{...} = 57.6 - 60.76 = -3.16 \quad \hat{\beta}_2 = \overline{y}_{.2.} - \overline{y}_{...} = 64.4 - 60.76 = 3.64$$

$$\hat{\beta}_3 = \overline{y}_{.3.} - \overline{y}_{...} = 65.2 - 60.76 = 4.44 \quad \hat{\beta}_4 = \overline{y}_{.4.} - \overline{y}_{...} = 57.4 - 60.76 = -3.36$$

$$\hat{\gamma}_1 = \overline{y}_{..1} - \overline{y}_{...} = 69.8 - 60.76 = 9.06 \quad \hat{\gamma}_2 = \overline{y}_{..2} - \overline{y}_{...} = 63.6 - 60.76 = 2.84$$

$$\hat{\gamma}_3 = \overline{y}_{..3} - \overline{y}_{...} = 56.6 - 60.76 = -4.16 \quad \hat{\gamma}_4 = \overline{y}_{..4} - \overline{y}_{...} = 62.2 - 60.76 = 1.44$$

The fitted model for the Latin square design data in Example 5.5 is

$$\begin{aligned}\hat{y} = 60.76 &+ 1.04x_1 - 0.76x_2 - 0.16x_3 - 1.36x_4 - 3.16z_1 + 3.64z_2 \\ &+ 4.44z_3 - 3.36z_4 + 9.06c_1 + 2.84c_2 - 4.16c_3 + 1.44c_4 \end{aligned} \quad (6.15)$$

6.6.2 Extracting Residuals and Testing for Significance of the Model

Using this fitted model, we obtain the estimates for the responses, and calculate the residuals for each fitted value (Table 6.11). Thereafter, we test hypothesis that the fitted model is significant.

TABLE 6.11: Fitted Values and Residuals for Example 5.5 Using Effects Coding Method

x_1	x_2	x_3	x_4	z_1	z_2	z_3	z_4	c_1	c_2	c_3	c_4	y	$\hat{y}$	$y-\hat{y}$	$(y-\hat{y})^2$
1	0	0	0	1	0	0	0	−1	−1	−1	−1	54	49.48	4.52	20.4304
1	0	0	0	0	1	0	0	1	0	0	0	76	74.48	1.52	2.3104
1	0	0	0	0	0	1	0	0	0	1	0	60	62.08	−2.08	4.3264
1	0	0	0	0	0	0	1	0	0	0	1	59	59.88	−0.88	0.7744
1	0	0	0	−1	−1	−1	−1	0	1	0	0	60	63.08	−3.08	9.4864
0	1	0	0	1	0	0	0	0	0	0	1	57	58.28	−1.28	1.6384
0	1	0	0	0	1	0	0	−1	−1	−1	−1	50	54.48	−4.48	20.0704
0	1	0	0	0	0	1	0	0	1	0	0	69	67.28	1.72	2.9584
0	1	0	0	0	0	0	1	0	0	1	0	54	52.48	1.52	2.3104
0	1	0	0	−1	−1	−1	−1	1	0	0	0	70	67.48	2.52	6.3504
0	0	1	0	1	0	0	0	0	0	1	0	53	53.28	−0.28	0.0784
0	0	1	0	0	1	0	0	0	0	0	1	66	65.68	0.32	0.1024
0	0	1	0	0	0	1	0	1	0	0	0	75	74.08	0.92	0.8464
0	0	1	0	0	0	0	1	0	1	0	0	59	60.08	−1.08	1.1664
0	0	1	0	−1	−1	−1	−1	−1	−1	−1	−1	50	49.88	0.12	0.0144
0	0	0	1	1	0	0	0	0	1	0	0	60	59.08	0.92	0.8464
0	0	0	1	0	1	0	0	0	0	1	0	60	58.88	1.12	1.2544
0	0	0	1	0	0	1	0	−1	−1	−1	−1	53	54.68	−1.68	2.8224
0	0	0	1	0	0	0	1	1	0	0	0	64	65.08	−1.08	1.1664
0	0	0	1	−1	−1	−1	−1	0	0	0	1	60	59.28	0.72	0.5184
−1	−1	−1	−1	1	0	0	0	1	0	0	0	64	67.88	−3.88	15.0544
−1	−1	−1	−1	0	1	0	0	0	1	0	0	70	68.48	1.52	2.3104
−1	−1	−1	−1	0	0	1	0	0	0	0	1	69	67.88	1.12	1.2544
−1	−1	−1	−1	0	0	0	1	−1	−1	−1	−1	51	49.48	1.52	2.3104
−1	−1	−1	−1	−1	−1	−1	−1	0	0	1	0	56	56.28	−0.28	0.0784
Total												1519	1519	4.26E − 14	100.48

If $S_{yy} = \sum_{i=1}^{N} y_i^2 - \left(\sum_{i=1}^{N} y_i\right)^2 / N$, $SS_{\mathrm{R}} = \sum_{i=1}^{N} (\hat{y}_i - \overline{y})^2$, and $SS_{\mathrm{E}} = \sum_{i=1}^{N} (y_i - \hat{y}_i)^2$ then $S_{yy} = SS_{\mathrm{R}} + SS_{\mathrm{E}}$. For the Latin square design, $N = r^2$. In particular, $N = 5(5) = 25$ for Example 5.5 (Table 6.12).

$$SS_{\mathrm{E}} = 100.48$$

$$S_{yy} = 93669 - \frac{1519^2}{25} = 1374.56$$

$$\Rightarrow SS_{\mathrm{R}} = 1374.56 - 100.48 = 1274.08$$

TABLE 6.12: ANOVA to Test for Significance of the Fitted Regression Model for Example 5.5

Source	Sum of squares	Degrees of freedom	Mean square	F-ratio
Regression	1274.08	12	106.1733	12.6799
Error	100.48	12	8.3733	
Total	1374.56	24		

We see that $F(12, 12, 0.01) = 4.16$, indicating that the fitted model is highly significant at 1% level of significance.

6.6.3 SAS Program for the Analysis of Responses of Example 5.5 (Effects Coding Method)

We write a SAS program to carry out the analysis of the data of Example 5.5, using effects coding method. The SAS program will use the above relationships to estimate the parameters of the regression model. The program is presented as follows:

SAS PROGRAM

```
data latin1;
input x1-x4 z1-z4 c1-c4 y @@;
datalines;
1 0 0 0 1 0 0 0 -1 -1 -1 -1 54 1 0 0 0 0 1 0 0 1 0 0 0 76
1 0 0 0 0 0 1 0 0 0 1 0 60 1 0 0 0 0 0 0 1 0 0 0 1 59
1 0 0 0 -1 -1 -1 -1 0 1 0 0 60 0 1 0 0 1 0 0 0 0 0 0 1 57
0 1 0 0 0 1 0 0 -1 -1 -1 -1 50 0 1 0 0 0 0 1 0 0 1 0 0 69
0 1 0 0 0 0 0 1 0 0 1 0 54 0 1 0 0 -1 -1 -1 -1 1 0 0 0 70
0 0 1 0 1 0 0 0 0 0 1 0 53 0 0 1 0 0 1 0 0 0 0 0 1 66
0 0 1 0 0 0 1 0 1 0 0 0 75 0 0 1 0 0 0 0 1 0 1 0 0 59
0 0 1 0 -1 -1 -1 -1 -1 -1 -1 -1 50 0 0 0 1 1 0 0 0 0 1 0 0 60
0 0 0 1 0 1 0 0 0 0 1 0 60 0 0 0 1 0 0 1 0 -1 -1 -1 -1 53
0 0 0 1 0 0 0 1 1 0 0 0 64 0 0 0 1 -1 -1 -1 -1 0 0 0 1 60
-1 -1 -1 -1 1 0 0 0 1 0 0 0 64 -1 -1 -1 -1 0 1 0 0 0 1 0 0 70
-1 -1 -1 -1 0 0 1 0 0 0 0 1 69 -1 -1 -1 -1 0 0 0 1 -1 -1 -1 -1 51
-1 -1 -1 -1 -1 -1 -1 -1 0 0 1 0 56
;
```

```
proc print data=latin1; sum x1-x4 z1-z4 c1-c4;
proc reg data=latin1;
model y=x1-x4 z1-z4 c1-c4; run;
```

SAS OUTPUT

Obs	x1	x2	x3	x4	z1	z2	z3	z4	c1	c2	c3	c4	y
1	1	0	0	0	1	0	0	0	-1	-1	-1	-1	54
2	1	0	0	0	0	1	0	0	1	0	0	0	76
3	1	0	0	0	0	0	1	0	0	0	1	0	60
4	1	0	0	0	0	0	0	1	0	0	0	1	59
5	1	0	0	0	-1	-1	-1	-1	0	1	0	0	60
6	0	1	0	0	1	0	0	0	0	0	0	1	57
7	0	1	0	0	0	1	0	0	-1	-1	-1	-1	50
8	0	1	0	0	0	0	1	0	0	1	0	0	69
9	0	1	0	0	0	0	0	1	0	0	1	0	54
10	0	1	0	0	-1	-1	-1	-1	1	0	0	0	70
11	0	0	1	0	1	0	0	0	0	0	1	0	53
12	0	0	1	0	0	1	0	0	0	0	0	1	66
13	0	0	1	0	0	0	1	0	1	0	0	0	75
14	0	0	1	0	0	0	0	1	0	1	0	0	59
15	0	0	1	0	-1	-1	-1	-1	-1	-1	-1	-1	50
16	0	0	0	1	1	0	0	0	0	1	0	0	60
17	0	0	0	1	0	1	0	0	0	0	1	0	60
18	0	0	0	1	0	0	1	0	-1	-1	-1	-1	53
19	0	0	0	1	0	0	0	1	1	0	0	0	64
20	0	0	0	1	-1	-1	-1	-1	0	0	0	1	60
21	-1	-1	-1	-1	1	0	0	0	1	0	0	0	64
22	-1	-1	-1	-1	0	1	0	0	0	1	0	0	70
23	-1	-1	-1	-1	0	0	1	0	0	0	0	1	69
24	-1	-1	-1	-1	0	0	0	1	-1	-1	-1	-1	51
25	-1	-1	-1	-1	-1	-1	-1	-1	0	0	1	0	56
	==	==	==	==	==	==	==	==	==	==	==	==	
	0	0	0	0	0	0	0	0	0	0	0	0	

Analysis of Variance

Source	DF	Sum of Squares	Mean Square	F Value	Pr > F
Model	12	1274.08000	106.17333	12.68	<.0001
Error	12	100.48000	8.37333		
Corrected Total	24	1374.56000			

Root MSE	2.89367	R-Square	0.9269
Dependent Mean	60.76000	Adj R-Sq	0.8538
Coeff Var	4.76246		

Parameter Estimates

Variable	DF	Parameter Estimate	Standard Error	t-Value	Pr > |t|
Intercept	1	60.76000	0.57873	104.99	<.0001
x1	1	1.04000	1.15747	0.90	0.3866
x2	1	-0.76000	1.15747	-0.66	0.5238
x3	1	-0.16000	1.15747	-0.14	0.8923
x4	1	-1.36000	1.15747	-1.17	0.2628
z1	1	-3.16000	1.15747	-2.73	0.0183
z2	1	3.64000	1.15747	3.14	0.0085
z3	1	4.44000	1.15747	3.84	0.0024
z4	1	-3.36000	1.15747	-2.90	0.0133
c1	1	9.04000	1.15747	7.81	<.0001
c2	1	2.84000	1.15747	2.45	0.0304
c3	1	-4.16000	1.15747	-3.59	0.0037
c4	1	1.44000	1.15747	1.24	0.2372

The results of SAS analysis agree with our manual analysis carried out elsewhere in the book.

6.7 Dummy Variables Regression Analysis for Example 5.5 (Reference Cell Method)

When we use the reference cell method for defining the dummy variables in the regression analysis, the model that serves as an analog to the classical linear model used in the ANOVA for Latin Square designs (Equation 5.15) is

$$y_{ijk} = \mu + \sum_{i=1}^{r-1} \alpha_i X_i + \sum_{j=1}^{r-1} \beta_j Z_j + \sum_{k=1}^{r-1} \gamma_k C_k + \varepsilon \tag{6.16}$$

where

$$X_i = \begin{cases} 1 & \text{if } y_{ijk} \text{ is under } i\text{th row} \quad i = 1, 2, \ldots, r-1 \\ 0 & \text{otherwise} \end{cases}$$

$$Z_j = \begin{cases} 1 & \text{if } y_{ijk} \text{ is under } j\text{th column} \quad j = 1, 2, \ldots, r-1 \\ 0 & \text{otherwise} \end{cases}$$

$$C_k = \begin{cases} 1 & \text{if } y_{ijk} \text{ is under } k\text{th Latin letter} \quad k = 1, 2, \ldots, r-1 \\ 0 & \text{otherwise} \end{cases}$$

Under this model, we can reanalyze the data for Example 5.5.

For estimation purposes, we note the following relationships among the cell, row, column, and treatment means for a typical $r \times r$ Latin square design.

$$\mu_{ijk} = \mu + \alpha_i + \beta_j + \gamma_j \quad i, j, k = 1, 2, \ldots, r-1$$

$$\mu_{r..} = \mu + \frac{1}{r}\sum_{j=1}^{r-1} \beta_j + \frac{1}{r}\sum_{k=1}^{r-1} \gamma_k$$

$$\mu_{.r.} = \mu + \frac{1}{r}\sum_{i=1}^{r-1}\alpha_i + \frac{1}{r}\sum_{k=1}^{r-1}\gamma_k$$

$$\mu_{..r} = \mu + \frac{1}{r}\sum_{i=1}^{r-1}\alpha_i + \frac{1}{r}\sum_{j=1}^{r-1}\beta_j \tag{6.17}$$

$$\mu_{i..} = \mu + \alpha_i + \frac{1}{r}\sum_{j=1}^{r-1}\beta_j + \frac{1}{r}\sum_{k=1}^{r-1}\gamma_k \quad i = 1, 2, \ldots, r-1$$

$$\mu_{.j.} = \mu + \beta_j + \frac{1}{r}\sum_{i=1}^{r-1}\alpha_i + \frac{1}{r}\sum_{k=1}^{r-1}\gamma_k \quad j = 1, 2, \ldots, r-1$$

$$\mu_{..k} = \mu + \gamma_k + \frac{1}{r}\sum_{i=1}^{r-1}\alpha_i + \frac{1}{r}\sum_{j=1}^{r-1}\beta_j \quad k = 1, 2, \ldots, r-1$$

$$\mu = \mu_{...} - \frac{1}{r}\sum_{i=1}^{r-1}\alpha_i - \frac{1}{r}\sum_{j=1}^{r-1}\beta_j - \frac{1}{r}\sum_{k=1}^{r-1}\gamma_k \tag{6.18}$$

Using the data and the randomization restriction, that occurred in Example 5.5, we can rewrite the table of means to make them specifically apply to the design as in Table 6.13.

Treatment	A	B	C	D	E	Overall
Means	$\mu_{..1}$	$\mu_{..2}$	$\mu_{..3}$	$\mu_{..4}$	$\mu_{..5}$	$\mu_{...}$

TABLE 6.13: Table of Means for Latin Square Design in Example 5.5

	Columns					
Rows	**1**	**2**	**3**	**4**	**5**	**Means**
1	μ_{115}	μ_{121}	μ_{133}	μ_{144}	μ_{112}	$\mu_{1..}$
2	μ_{214}	μ_{225}	μ_{232}	μ_{243}	μ_{251}	$\mu_{2..}$
3	μ_{313}	μ_{324}	μ_{331}	μ_{342}	μ_{355}	$\mu_{3..}$
4	μ_{412}	μ_{423}	μ_{435}	μ_{441}	μ_{454}	$\mu_{4..}$
5	μ_{511}	μ_{522}	μ_{522}	μ_{545}	μ_{553}	$\mu_{5..}$
Means	$\mu_{.1.}$	$\mu_{.2.}$	$\mu_{.3.}$	$\mu_{.4.}$	$\mu_{.5.}$	$\mu_{...}$

6.7.1 Estimation of Model Parameters (Reference Cell Method)

In the above analysis, cell (5, 5) containing treatment 3, that is, the cell with observation y_{553}, was used as the reference cell for the regression model. This is equivalent to using observation in row 5, column 5, and treatment denoted as Latin letter C. By replacing the means in Table 6.13 with their sample equivalents, we can estimate the parameters of the regression model 6.16. The parameters to be estimated are $\alpha_1, \alpha_2, \alpha_3, \alpha_4; \beta_1, \beta_2, \beta_3, \beta_4; \gamma_1, \gamma_2, \gamma_3, \gamma_4$.

From Equations 6.17 and 6.18, using the reference cell above, sample means previously obtained under effects coding method, and making appropriate substitution in the equations, we see that

$$\mu_{i..} = \mu_{5..} + \alpha_i \quad i = 1, 2, 3, 4$$
$$\mu_{.j.} = \mu_{.5.} + \beta_j \quad j = 1, 2, 3, 4$$
$$\mu_{..k} = \mu_{..3} + \gamma_k \quad k = 1, 2, 3, 4$$

Thus,

$$\bar{y}_{5..} = 62$$
$$\bar{y}_{1..} = \bar{y}_{5..} + \alpha_1 \Rightarrow \hat{\alpha}_1 = \bar{y}_{1..} - \bar{y}_{5..} = 61.8 - 62 = -0.2$$
$$\bar{y}_{2..} = \bar{y}_{5..} + \alpha_2 \Rightarrow \hat{\alpha}_2 = \bar{y}_{2..} - \bar{y}_{5..} = 60 - 62 = -2.0$$
$$\bar{y}_{3..} = \bar{y}_{5..} + \alpha_3 \Rightarrow \hat{\alpha}_3 = \bar{y}_{3..} - \bar{y}_{5..} = 60.6 - 62 = -1.4$$
$$\bar{y}_{4..} = \bar{y}_{5..} + \alpha_4 \Rightarrow \hat{\alpha}_4 = \bar{y}_{4..} - \bar{y}_{5..} = 59.4 - 62 = -2.6$$

$$\bar{y}_{.5.} = 59.2$$
$$\bar{y}_{1..} = \bar{y}_{.5.} + \beta_1 \Rightarrow \hat{\beta}_1 = \bar{y}_{.1.} - \bar{y}_{.5.} = 57.6 - 59.2 = -1.6$$
$$\bar{y}_{.2.} = \bar{y}_{.5.} + \beta_2 \Rightarrow \hat{\beta}_2 = \bar{y}_{.2.} - \bar{y}_{.5.} = 64.4 - 59.2 = 5.2$$
$$\bar{y}_{.3.} = \bar{y}_{.5.} + \beta_3 \Rightarrow \hat{\beta}_3 = \bar{y}_{.3.} - \bar{y}_{.5.} = 65.2 - 59.2 = 6.0$$
$$\bar{y}_{.4.} = \bar{y}_{.5.} + \beta_4 \Rightarrow \hat{\beta}_4 = \bar{y}_{.4.} - \bar{y}_{5..} = 57.4 - 59.2 = -1.8$$

$$\bar{y}_{..3} = 56.6$$
$$\bar{y}_{..1} = \bar{y}_{..3} + \gamma_1 \Rightarrow \hat{\gamma}_1 = \bar{y}_{..1} - \bar{y}_{..3} = 69.8 - 56.6 = 13.2$$
$$\bar{y}_{..2} = \bar{y}_{..3} + \gamma_2 \Rightarrow \hat{\gamma}_2 = \bar{y}_{..2} - \bar{y}_{..3} = 63.6 - 56.6 = 7.0$$
$$\bar{y}_{..3} = \bar{y}_{..3} + \gamma_3 \Rightarrow \hat{\gamma}_3 = \bar{y}_{..3} - \bar{y}_{..3} = 62.2 - 56.6 = 5.6$$
$$\bar{y}_{..4} = \bar{y}_{..3} + \gamma_4 \Rightarrow \hat{\gamma}_4 = \bar{y}_{..4} - \bar{y}_{..3} = 51.6 - 56.6 = -5.0$$

Lastly, the intercept is obtained as follows:

$$\mu = \mu_{...} - \frac{1}{r}\sum_{i=1}^{r-1} \alpha_i - \frac{1}{r}\sum_{j=1}^{r-1} \beta_j - \frac{1}{r}\sum_{i=1}^{r-1} \gamma_k$$

$$= 60.76 - \frac{(-0.2 - 2.0 - 1.4 - 2.6)}{5} - \frac{(-1.6 + 5.2 + 6 - 1.8)}{5}$$
$$- \frac{(13.2 + 7 + 5.6 - 5.0)}{5} = 56.28$$

The fitted model is

$$\hat{y} = 56.28 - 0.2x_1 - 2x_2 - 1.4x_3 - 2.6x_4 - 1.6z_1 + 5.2z_2$$
$$+ 6z_3 - 1.8z_4 - 13.2c_1 + 7c_2 + 5.6c_3 - 5c_4 \tag{6.19}$$

Using this fitted model, we obtain the estimated responses and residuals resulting from the estimated responses. We also carry out ANOVA to show that the fitted regression model is significant (Tables 6.14 and 6.15).

TABLE 6.14: Fitted Values, Residuals for Example 5.5 (Reference Cell Method)

x_1	x_2	x_3	x_4	z_1	z_2	z_3	z_4	c_1	c_2	c_3	c_4	y	$\hat{y}$	$y-\hat{y}$	$(y-\hat{y})^2$
1	0	0	0	1	0	0	0	0	0	0	1	54	49.5	4.52	20.4304
1	0	0	0	0	1	0	0	1	0	0	0	76	74.5	1.52	2.3104
1	0	0	0	0	0	1	0	0	0	0	0	60	62.1	−2.1	4.3264
1	0	0	0	0	0	0	1	0	0	1	0	59	59.9	−0.9	0.7744
1	0	0	0	0	0	0	0	0	1	0	0	60	63.1	−3.1	9.4864
0	1	0	0	1	0	0	0	0	0	1	0	57	58.3	−1.3	1.6384
0	1	0	0	0	1	0	0	0	0	0	1	50	54.5	−4.5	20.0704
0	1	0	0	0	0	1	0	0	1	0	0	69	67.3	1.72	2.9584
0	1	0	0	0	0	0	1	0	0	0	0	54	52.5	1.52	2.3104
0	1	0	0	0	0	0	0	1	0	0	0	70	67.5	2.52	6.3504
0	0	1	0	1	0	0	0	0	0	0	0	53	53.3	−0.3	0.0784
0	0	1	0	0	1	0	0	0	0	1	0	66	65.7	0.32	0.1024
0	0	1	0	0	0	1	0	1	0	0	0	75	74.1	0.92	0.8464
0	0	1	0	0	0	0	1	0	1	0	0	59	60.1	−1.1	1.1664
0	0	1	0	0	0	0	0	0	0	0	1	50	49.9	0.12	0.0144
0	0	0	1	1	0	0	0	0	1	0	0	60	59.1	0.92	0.8464
0	0	0	1	0	1	0	0	0	0	0	0	60	58.9	1.12	1.2544
0	0	0	1	0	0	1	0	0	0	0	1	53	54.7	−1.7	2.8224
0	0	0	1	0	0	0	1	1	0	0	0	64	65.1	−1.1	1.1664
0	0	0	1	0	0	0	0	0	0	1	0	60	59.3	0.72	0.5184
0	0	0	0	1	0	0	0	1	0	0	0	64	67.9	−3.9	15.0544
0	0	0	0	0	1	0	0	0	1	0	0	70	68.5	1.52	2.3104
0	0	0	0	0	0	1	0	0	0	1	0	69	67.9	1.12	1.2544
0	0	0	0	0	0	0	1	0	0	0	1	51	49.5	1.52	2.3104
0	0	0	0	0	0	0	0	0	0	0	0	56	56.3	−0.3	0.0784
Total												1519	1519	−0	100.48

If $S_{yy} = \sum_{i=1}^{N} y_i^2 - \left(\sum_{i=1}^{N} y_i\right)^2/N$, $SS_R = \sum_{i=1}^{N} (\hat{y}_i - \overline{y})^2$, and $SS_E = \sum_{i=1}^{N} (y_i - \hat{y}_i)^2$ then $S_{yy} = SS_R + SS_E$.

$$SS_E = 100.48$$

$$S_{yy} = 93669 - \frac{1519^2}{25} = 1374.56 \Rightarrow SS_R = 1374.56 - 100.48 = 1274.08$$

TABLE 6.15: ANOVA to Test for Significance of the Fitted Regression Model for Example 5.5

Source	Sum of squares	Degrees of freedom	Mean square	*F*-ratio
Regression	1274.08	12	106.1733	12.6799
Error	100.48	12	8.3733	
Total	1374.56	24		

From the tables, we see that $F(12, 12, 0.01) = 4.16$, indicating that the fitted model is highly significant.

6.7.2 SAS Program for Fitting Model 6.16 (Reference Cell Method) for Experiment in Example 5.5

We write a SAS program to analyze the data on the basis of the model 6.16 as follows:

SAS PROGRAM

```
data latin2;
input x1-x4 z1-z4 c1-c4 y @@;
datalines;
1 0 0 0 1 0 0 0 0 0 0 1 54 1 0 0 0 0 1 0 0 1 0 0 0 76
1 0 0 0 0 0 1 0 0 0 0 0 60 1 0 0 0 0 0 0 1 0 0 1 0 59
1 0 0 0 0 0 0 0 0 1 0 0 60 0 1 0 0 1 0 0 0 0 0 1 0 57
0 1 0 0 0 1 0 0 0 0 0 1 50 0 1 0 0 0 0 1 0 0 1 0 0 69
0 1 0 0 0 0 0 1 0 0 0 0 54 0 1 0 0 0 0 0 0 1 0 0 0 70
0 0 1 0 1 0 0 0 0 0 0 0 53 0 0 1 0 0 1 0 0 0 0 1 0 66
0 0 1 0 0 0 1 0 1 0 0 0 75 0 0 1 0 0 0 0 1 0 1 0 0 59
0 0 1 0 0 0 0 0 0 0 0 1 50 0 0 0 1 1 0 0 0 0 1 0 0 60
0 0 0 1 0 1 0 0 0 0 0 0 60 0 0 0 1 0 0 1 0 0 0 0 1 53
0 0 0 1 0 0 0 1 1 0 0 0 64 0 0 0 1 0 0 0 0 0 0 1 0 60
0 0 0 0 1 0 0 0 1 0 0 0 64 0 0 0 0 0 1 0 0 0 1 0 0 70
0 0 0 0 0 0 1 0 0 0 1 0 69 0 0 0 0 0 0 0 1 0 0 0 1 51
0 0 0 0 0 0 0 0 0 0 0 0 56
;
proc print data=latin2; sum x1-x4 z1-z4 c1-c4;
proc reg data=latin2; model y=x1-x4 z1-z4 c1-c4;
run;
```

SAS OUTPUT

Obs	x1	x2	x3	x4	z1	z2	z3	z4	c1	c2	c3	c4	y
1	1	0	0	0	1	0	0	0	0	0	0	1	54
2	1	0	0	0	0	1	0	0	1	0	0	0	76
3	1	0	0	0	0	0	1	0	0	0	0	0	60
4	1	0	0	0	0	0	0	1	0	0	1	0	59
5	1	0	0	0	0	0	0	0	0	1	0	0	60
6	0	1	0	0	1	0	0	0	0	0	1	0	57
7	0	1	0	0	0	1	0	0	0	0	0	1	50
8	0	1	0	0	0	0	1	0	0	1	0	0	69
9	0	1	0	0	0	0	0	1	0	0	0	0	54
10	0	1	0	0	0	0	0	0	1	0	0	0	70
11	0	0	1	0	1	0	0	0	0	0	0	0	53
12	0	0	1	0	0	1	0	0	0	0	1	0	66
13	0	0	1	0	0	0	1	0	1	0	0	0	75
14	0	0	1	0	0	0	0	1	0	1	0	0	59
15	0	0	1	0	0	0	0	0	0	0	0	1	50
16	0	0	0	1	1	0	0	0	0	1	0	0	60
17	0	0	0	1	0	1	0	0	0	0	0	0	60
18	0	0	0	1	0	0	1	0	0	0	0	1	53
19	0	0	0	1	0	0	0	1	1	0	0	0	64
20	0	0	0	1	0	0	0	0	0	0	1	0	60
21	0	0	0	0	1	0	0	0	1	0	0	0	64
22	0	0	0	0	0	1	0	0	0	1	0	0	70
23	0	0	0	0	0	0	1	0	0	0	1	0	69
24	0	0	0	0	0	0	0	1	0	0	0	1	51
25	0	0	0	0	0	0	0	0	0	0	0	0	56
	==	==	==	==	==	==	==	==	==	==	==	==	
	5	5	5	5	5	5	5	5	5	5	5	5	

Analysis of Variance

Source	DF	Sum of Squares	Mean Square	F Value	Pr > F
Model	12	1274.08000	106.17333	12.68	<.0001
Error	12	100.48000	8.37333		
Corrected Total	24	1374.56000			

Root MSE	2.89367	R-Square	0.9269
Dependent Mean	60.76000	Adj R-Sq	0.8538
Coeff Var	4.76246		

Parameter Estimates

Variable	DF	Parameter Estimate	Standard Error	t-Value	Pr > |t|
Intercept	1	56.28000	2.08666	26.97	<.0001
x1	1	-0.20000	1.83012	-0.11	0.9148

Variable	DF	Parameter Estimate	Standard Error	t-Value	Pr > |t|
x2	1	-2.00000	1.83012	-1.09	0.2959
x3	1	-1.40000	1.83012	-0.76	0.4591
x4	1	-2.60000	1.83012	-1.42	0.1809
z1	1	-1.60000	1.83012	-0.87	0.3991
z2	1	5.20000	1.83012	2.84	0.0149
z3	1	6.00000	1.83012	3.28	0.0066
z4	1	-1.80000	1.83012	-0.98	0.3448
c1	1	13.20000	1.83012	7.21	<.0001
c2	1	7.00000	1.83012	3.82	0.0024
c3	1	5.60000	1.83012	3.06	0.0099
c4	1	-5.00000	1.83012	-2.73	0.0182

The outcome of SAS analysis is in agreement with the manual analysis carried out before.

6.8 Regression Model for Example 5.7 Using Effects Coding Method to Define Dummy Variables

Next, we define a set of dummy variables to respresent the row, column, Latin factors, and the replicates in the replicated Latin square design in Example 5.7. For the effects coding method, the relevant regression model is

$$y_{ijkl} = \mu + \sum_{i=1}^{r-1} \alpha_i X_i + \sum_{j=1}^{r-1} \beta_j Z_j + \sum_{k=1}^{r-1} \gamma_k C_k + \sum_{l=1}^{n-1} \theta_l H_l + \varepsilon \tag{6.20}$$

where

$$X_i = \begin{cases} -1 & \text{if } y_{ijkl} \text{ is under the } r\text{th row} \\ 1 & \text{if } y_{ijkl} \text{ is under the } i\text{th row} \\ 0 & \text{otherwise} \end{cases}$$

for $i = 1, 2, \ldots, r-1$

$$Z_j = \begin{cases} -1 & \text{if } y_{ijkl} \text{ is under the } r\text{th column} \\ 1 & \text{if } y_{ijkl} \text{ is under the } j\text{th column} \\ 0 & \text{otherwise} \end{cases}$$

for $j = 1, 2, \ldots, r-1$

$$C_k = \begin{cases} -1 & \text{if } y_{ijkl} \text{ is under the } r\text{th treatment (Latin letter)} \\ 1 & \text{if } y_{ijkl} \text{ is under the } k\text{th treatment (Latin letter)} \\ 0 & \text{otherwise} \end{cases}$$

for $k = 1, 2, \ldots, r-1$

and

$$H_l = \begin{cases} -1 & \text{if } y_{ijkl} \text{ is under the } n\text{th replicate} \\ 1 & \text{if } y_{ijkl} \text{ is under the } l\text{th replicate} \\ 0 & \text{otherwise} \end{cases}$$

for $l = 1, 2, \ldots, n-1$

6.8.1 Estimation of Dummy Variables Regression Model Parameters for Example 5.7 (Effects Coding Method)

To manually estimate parameters of the regression model for Example 5.7 when the effects coding method is used to define the dummy variables, we need to extend the estimation theory used for Example 5.5 (Section 6.6.1.2) as follows:

$$\mu_{....} = \mu$$

$$-\sum_{i=1}^{r-1} \alpha_i = \mu_{r...} - \mu_{....}$$

$$-\sum_{j=1}^{r-1} \beta_j = \mu_{.r..} - \mu_{....}$$

$$-\sum_{k=1}^{r-1} \gamma_k = \mu_{..r.} - \mu_{....}$$

$$-\sum_{l=1}^{n-1} \theta_l = \mu_{...r} - \mu_{....}$$

$$\mu_{ijkl} = \mu + \alpha_i + \beta_j + \gamma_k + \theta_l \quad i, j, k = 1, 2, \ldots, r-1 \;\; l = 1, 2, \ldots, n-1$$

$$\begin{aligned} \alpha_i &= \mu_{i...} - \mu_{....} \quad i = 1, 2, \ldots, r-1 \\ \beta_j &= \mu_{.j..} - \mu_{....} \quad j = 1, 2, \ldots, r-1 \\ \gamma_k &= \mu_{..k.} - \mu_{....} \quad k = 1, 2, \ldots, r-1 \\ \theta_l &= \mu_{...l} - \mu_{....} \quad l = 1, 2, \ldots, n-1 \end{aligned} \tag{6.21}$$

To carry out the estimation of parameters of the model, we use the data of Example 5.7 to obtain the sample means as follows:

$$\text{Overall: } \overline{y}_{....} = \frac{357}{75} = 4.76$$

$$\text{Mines: } \overline{y}_{1...} = \frac{49}{15} = 3.2667 \quad \overline{y}_{2...} = \frac{69}{15} = 4.6$$

$$\overline{y}_{3...} = \frac{53}{15} = 3.533 \quad \overline{y}_{4...} = \frac{83}{15} = 5.533$$

$$\overline{y}_{5...} = \frac{103}{15} = 6.8667$$

$$\text{Laboratories: } \overline{y}_{.1..} = \frac{6}{15} = 0.4 \quad \overline{y}_{.2..} = \frac{91}{15} = 6.0667$$

$$\overline{y}_{.3..} = \frac{74}{15} = 4.993 \quad \overline{y}_{.4..} = \frac{105}{15} = 7.0$$

$$\overline{y}_{.5..} = \frac{81}{15} = 5.4$$

$$\text{Catalysts: } \overline{y}_{..1.} = \frac{97}{15} = 6.4667 \quad \overline{y}_{..2.} = \frac{26}{15} = 1.733$$

$$\overline{y}_{..3.} = \frac{44}{15} = 2.933 \quad \overline{y}_{..4.} = \frac{110}{15} = 7.333$$

$$\overline{y}_{..5.} = \frac{80}{15} = 5.333$$

$$\text{Replicates: } \overline{y}_{...1} = \frac{134}{25} = 5.36 \quad \overline{y}_{...2} = \frac{106}{25} = 4.24 \quad \overline{y}_{...3} = \frac{117}{25} = 4.68$$

Using the above relations among the data of Example 5.7, and replacing the row, column, Latin letter, and replicate means with the above sample means, we obtain the following estimates for the parameters of the regression model:

$$\hat{\alpha}_1 = \overline{y}_{1...} - \overline{y}_{....} = 3.2667 - 4.76 = -1.4933$$

$$\hat{\alpha}_2 = \overline{y}_{2...} - \overline{y}_{....} = 4.6 - 4.76 = -0.16$$

$$\hat{\alpha}_3 = \overline{y}_{3...} - \overline{y}_{....} = 3.533 - 4.76 = -1.227$$

$$\hat{\alpha}_4 = \overline{y}_{4...} - \overline{y}_{....} = 5.533 - 4.76 = 0.773$$

$$\hat{\beta}_1 = \overline{y}_{.1..} - \overline{y}_{....} = 0.4 - 4.76 = -4.36$$

$$\hat{\beta}_2 = \overline{y}_{.2..} - \overline{y}_{....} = 6.0667 - 4.76 = 1.3067$$

$$\hat{\beta}_3 = \overline{y}_{.3..} - \overline{y}_{....} = 4.993 - 4.76 = 0.1733$$

$$\hat{\beta}_4 = \overline{y}_{.4..} - \overline{y}_{....} = 7.0 - 4.76 = 2.24$$

$$\hat{\gamma}_1 = \overline{y}_{..1.} - \overline{y}_{....} = 6.4667 - 4.76 = 1.7067$$

$$\hat{\gamma}_2 = \overline{y}_{..2.} - \overline{y}_{....} = 1.733 - 4.76 = -3.027$$

$$\hat{\gamma}_3 = \overline{y}_{..3.} - \overline{y}_{....} = 2.933 - 4.76 = -1.827$$

$$\hat{\gamma}_4 = \overline{y}_{..4.} - \overline{y}_{....} = 7.333 - 4.76 = 2.573$$

$$\hat{\theta}_1 = \overline{y}_{...1} - \overline{y}_{....} = 5.36 - 4.76 = 0.60$$

$$\hat{\theta}_2 = \overline{y}_{...2} - \overline{y}_{....} = 4.24 - 4.76 = -0.52$$

$$\mu = \mu_{....} = 4.76$$

Based on the above values, the estimated model is

$$\begin{aligned}\hat{y} = 4.76 - 1.4933x_1 - 0.16x_2 - 1.227x_3 + 0.773x_4 - 4.36z_1 \\ + 1.3067z_2 + 0.1733z_3 + 2.24z_4 + 1.7067c_1 - 3.027c_2 \\ - 1.827c_3 + 2.573c_4 + 0.6h_1 - 0.52h_2\end{aligned} \tag{6.22}$$

TABLE 6.16: Estimated Responses and Residuals for Example 5.7 (Effects Coding Method)

x_1	x_2	x_3	x_4	z_1	z_2	z_3	z_4	c_1	c_2	c_3	c_4	h_1	h_2	y	$\hat{y}$	$y-\hat{y}$	$(y-\hat{y})^2$
1	0	0	0	1	0	0	0	1	0	0	0	1	0	3	1.21334	1.78666	3.192154
1	0	0	0	0	1	0	0	0	1	0	0	1	0	1	2.14667	−1.14667	1.314852
1	0	0	0	0	0	1	0	0	0	1	0	1	0	2	2.21333	−0.21333	0.04551
1	0	0	0	0	0	0	1	0	0	0	1	1	0	5	8.68	−3.68	13.5424
1	0	0	0	−1	−1	−1	−1	−1	−1	−1	−1	1	0	6	5.08001	0.91999	0.846382
0	1	0	0	1	0	0	0	0	1	0	0	1	0	−4	−2.18667	−1.81333	3.288166
0	1	0	0	0	1	0	0	0	0	1	0	1	0	3	4.68	−1.68	2.8224
0	1	0	0	0	0	1	0	0	0	0	1	1	0	11	7.94666	3.05334	9.322885
0	1	0	0	0	0	0	1	−1	−1	−1	−1	1	0	8	8.01334	−0.01334	0.000178
0	1	0	0	−1	−1	−1	−1	1	0	0	0	1	0	8	7.54667	0.45333	0.205508
0	0	1	0	1	0	0	0	0	0	1	0	1	0	−3	−2.05334	−0.94666	0.896165
0	0	1	0	0	1	0	0	0	0	0	1	1	0	7	8.01333	−1.01333	1.026838
0	0	1	0	0	0	1	0	−1	−1	−1	−1	1	0	5	4.88	0.12	0.0144
0	0	1	0	0	0	0	1	1	0	0	0	1	0	9	8.08	0.92	0.8464
0	0	1	0	−1	−1	−1	−1	0	1	0	0	1	0	3	1.74666	1.25334	1.570861
0	0	0	1	1	0	0	0	0	0	0	1	1	0	5	4.34666	0.65334	0.426853
0	0	0	1	0	1	0	0	−1	−1	−1	−1	1	0	12	8.01334	3.98666	15.89346
0	0	0	1	0	0	1	0	1	0	0	0	1	0	7	8.01333	−1.01333	1.026838
0	0	0	1	0	0	0	1	0	1	0	0	1	0	5	5.34666	−0.34666	0.120173
0	0	0	1	−1	−1	−1	−1	0	0	1	0	1	0	4	4.94666	−0.94666	0.896165
−1	−1	−1	−1	1	0	0	0	−1	−1	−1	−1	1	0	1	3.68001	−2.68001	7.182454
−1	−1	−1	−1	0	1	0	0	1	0	0	0	1	0	11	10.48001	0.51999	0.27039
−1	−1	−1	−1	0	0	1	0	0	1	0	0	1	0	1	4.61333	−3.61333	13.05615
−1	−1	−1	−1	0	0	0	1	0	0	1	0	1	0	11	7.88	3.12	9.7344
−1	−1	−1	−1	−1	−1	−1	−1	0	0	0	1	1	0	13	10.68	2.32	5.3824
1	0	0	0	1	0	0	0	1	0	0	0	0	1	2	0.09334	1.90666	3.635352
1	0	0	0	0	1	0	0	0	1	0	0	0	1	4	1.02667	2.97333	8.840691
1	0	0	0	0	0	1	0	0	0	1	0	0	1	2	1.09333	0.90667	0.82205
1	0	0	0	0	0	0	1	0	0	0	1	0	1	5	7.56	−2.56	6.5536
1	0	0	0	−1	−1	−1	−1	−1	−1	−1	−1	0	1	3	3.96001	−0.96001	0.921619
0	1	0	0	1	0	0	0	0	1	0	0	0	1	−2	−3.30667	1.30667	1.707386
0	1	0	0	0	1	0	0	0	0	1	0	0	1	3	3.56	−0.56	0.3136
0	1	0	0	0	0	1	0	0	0	0	1	0	1	10	6.82666	3.17334	10.07009
0	1	0	0	0	0	0	1	−1	−1	−1	−1	0	1	5	6.89334	−1.89334	3.584736
0	1	0	0	−1	−1	−1	−1	1	0	0	0	0	1	4	6.42667	−2.42667	5.888727
0	0	1	0	1	0	0	0	0	0	1	0	0	1	−5	−3.17334	−1.82666	3.336687
0	0	1	0	0	1	0	0	0	0	0	1	0	1	4	6.89333	−2.89333	8.371358

0	0	1	0	0	0	1	0	−1	−1	−1	−1	0	1	3	3.76	−0.76	0.5776
0	0	1	0	0	0	0	1	1	0	0	0	0	1	8	6.96	1.04	1.0816
0	0	1	0	−1	−1	−1	−1	0	1	0	0	0	1	3	0.62666	2.37334	5.632743
0	0	0	1	1	0	0	0	0	0	0	1	0	1	4	3.22666	0.77334	0.598055
0	0	0	1	0	1	0	0	−1	−1	−1	−1	0	1	9	6.89334	2.10666	4.438016
0	0	0	1	0	0	1	0	1	0	0	0	0	1	6	6.89333	−0.89333	0.798038
0	0	0	1	0	0	0	1	0	1	0	0	0	1	3	4.22666	−1.22666	1.504695
0	0	0	1	−1	−1	−1	−1	0	0	1	0	0	1	4	3.82666	0.17334	0.030047
−1	−1	−1	−1	1	0	0	0	−1	−1	−1	−1	0	1	2	2.56001	−0.56001	0.313611
−1	−1	−1	−1	0	1	0	0	1	0	0	0	0	1	8	9.36001	−1.36001	1.849627
−1	−1	−1	−1	0	0	1	0	0	1	0	0	0	1	3	3.49333	−0.49333	0.243374
−1	−1	−1	−1	0	0	0	1	0	0	1	0	0	1	10	6.76	3.24	10.4976
−1	−1	−1	−1	−1	−1	−1	−1	0	0	0	1	0	1	8	9.56	−1.56	2.4336
1	0	0	0	1	0	0	0	1	0	0	0	−1	−1	3	0.53334	2.46666	6.084412
1	0	0	0	0	1	0	0	0	1	0	0	−1	−1	2	1.46667	0.53333	0.284441
1	0	0	0	0	0	1	0	0	0	1	0	−1	−1	2	1.53333	0.46667	0.217781
1	0	0	0	0	0	0	1	0	0	0	1	−1	−1	5	8	−3	9
1	0	0	0	−1	−1	−1	−1	−1	−1	−1	−1	−1	−1	4	4.40001	−0.40001	0.160008
0	1	0	0	1	0	0	0	0	1	0	0	−1	−1	−4	−2.86667	−1.13333	1.284437
0	1	0	0	0	1	0	0	0	0	1	0	−1	−1	3	4	−1	1
0	1	0	0	0	0	1	0	0	0	0	1	−1	−1	11	7.26666	3.73334	13.93783
0	1	0	0	0	0	0	1	−1	−1	−1	−1	−1	−1	8	7.33334	0.66666	0.444436
0	1	0	0	−1	−1	−1	−1	1	0	0	0	−1	−1	5	6.86667	−1.86667	3.484457
0	0	1	0	1	0	0	0	0	0	1	0	−1	−1	−4	−2.73334	−1.26666	1.604428
0	0	1	0	0	1	0	0	0	0	0	1	−1	−1	7	7.33333	−0.33333	0.111109
0	0	1	0	0	0	1	0	−1	−1	−1	−1	−1	−1	4	4.2	−0.2	0.04
0	0	1	0	0	0	0	1	1	0	0	0	−1	−1	9	7.4	1.6	2.56
0	0	1	0	−1	−1	−1	−1	0	1	0	0	−1	−1	3	1.06666	1.93334	3.737804
0	0	0	1	1	0	0	0	0	0	0	1	−1	−1	5	3.66666	1.33334	1.777796
0	0	0	1	0	1	0	0	−1	−1	−1	−1	−1	−1	7	7.33334	−0.33334	0.111116
0	0	0	1	0	0	1	0	1	0	0	0	−1	−1	4	7.33333	−3.33333	11.11109
0	0	0	1	0	0	0	1	0	1	0	0	−1	−1	5	4.66666	0.33334	0.111116
0	0	0	1	−1	−1	−1	−1	0	0	1	0	−1	−1	3	4.26666	−1.26666	1.604428
−1	−1	−1	−1	1	0	0	0	−1	−1	−1	−1	−1	−1	3	3.00001	−1E−05	1E−10
−1	−1	−1	−1	0	1	0	0	1	0	0	0	−1	−1	10	9.80001	0.19999	0.039996
−1	−1	−1	−1	0	0	1	0	0	1	0	0	−1	−1	3	3.93333	−0.93333	0.871105
−1	−1	−1	−1	0	0	0	1	0	0	1	0	−1	−1	9	7.2	1.8	3.24
−1	−1	−1	−1	−1	−1	−1	−1	0	0	0	1	−1	−1	10	10	0	0
Total														357	357	6.66E − 15	239.7867

If $S_{yy} = \sum_{i=1}^{N} y_i^2 - (\sum_{i=1}^{N} y_i)^2/N$, $SS_{\mathrm{R}} = \sum_{i=1}^{N} (\hat{y}_i - \overline{y})^2$, and $SS_{\mathrm{E}} = \sum_{i=1}^{N} (y_i - \hat{y}_i)^2$ then $S_{yy} = SS_{\mathrm{R}} + SS_{\mathrm{E}}$. In this case, $N = nr^2 = 3(5^2) = 75$.

$$SS_{\mathrm{E}} = 239.7867$$

$$S_{yy} = 2816 - \frac{357^2}{75} = 1115.68 \Rightarrow SS_{\mathrm{R}} = 1115.68 - 239.7867 = 875.8933$$

TABLE 6.17: ANOVA to Test for Significance of the Fitted Regression Model for Example 5.7

Source	Sum of squares	Degrees of freedom	Mean square	*F*-ratio
Regression	875.8933	14	62.56381	15.68
Error	239.7867	60	3.99644	
Total	1115.68	74		

From the Tables (A5–A7) we see that $F(14, 60, 0.01) = 2.394$, indicating that the fitted model is highly significant at 1% level of significance.

6.8.2 SAS Program for Regression Analysis of Data of Example 5.7 (Effects Coding Method)

We write a SAS program to analyze the responses of Example 5.7, using effects coding definition of the dummy variables in the regression model.

SAS PROGRAM

```
data latmines;
input x1-x4 z1-z4 c1-c4 h1-h2 y @@;
datalines;
1 0 0 0 1 0 0 0 1 0 0 0 1 0 3 1 0 0 0 0 1 0 0 0 1 0 0 1 0 1
1 0 0 0 0 0 1 0 0 0 1 0 1 0 2 1 0 0 0 0 0 0 1 0 0 0 1 1 0 5
1 0 0 0 -1 -1 -1 -1 -1 -1 -1 -1 1 0 6 0 1 0 0 1 0 0 0 0 1 0 0 1 0 -4
0 1 0 0 0 1 0 0 0 0 1 0 1 0 3 0 1 0 0 0 0 1 0 0 0 0 1 1 0 11
0 1 0 0 0 0 0 1 -1 -1 -1 -1 1 0 8 0 1 0 0 -1 -1 -1 -1 1 0 0 0 1 0 8
0 0 1 0 1 0 0 0 0 0 1 0 1 0 -3 0 0 1 0 0 1 0 0 0 0 0 1 1 0 7
0 0 1 0 0 0 1 0 -1 -1 -1 -1 1 0 5 0 0 1 0 0 0 0 1 1 0 0 0 1 0 9
0 0 1 0 -1 -1 -1 -1 0 1 0 0 1 0 3 0 0 0 1 1 0 0 0 0 0 0 1 1 0 5
0 0 0 1 0 1 0 0 -1 -1 -1 -1 1 0 12 0 0 0 1 0 0 1 0 1 0 0 0 1 0 7
0 0 0 1 0 0 0 1 0 1 0 0 1 0 5 0 0 0 1 -1 -1 -1 -1 0 0 1 0 1 0 4
-1 -1 -1 -1 1 0 0 0 -1 -1 -1 -1 1 0 1 -1 -1 -1 -1 0 1 0 0 1 0 0 0 1 0 11
-1 -1 -1 -1 0 0 1 0 0 1 0 0 1 0 1 -1 -1 -1 -1 0 0 0 1 0 0 1 0 1 0 11
-1 -1 -1 -1 -1 -1 -1 -1 0 0 0 1 1 0 13 1 0 0 0 1 0 0 0 1 0 0 0 0 1 2
1 0 0 0 0 1 0 0 0 1 0 0 0 1 4 1 0 0 0 0 0 1 0 0 0 1 0 0 1 2
1 0 0 0 0 0 0 1 0 0 0 1 0 1 5 1 0 0 0 -1 -1 -1 -1 -1 -1 -1 -1 0 1 3
0 1 0 0 1 0 0 0 0 1 0 0 0 1 -2 0 1 0 0 0 1 0 0 0 0 1 0 0 1 3
0 1 0 0 0 0 1 0 0 0 0 1 0 1 10 0 1 0 0 0 0 0 1 -1 -1 -1 -1 0 1 5
0 1 0 0 -1 -1 -1 -1 1 0 0 0 0 1 4 0 0 1 0 1 0 0 0 0 0 1 0 0 1 -5
```

```
0 0 1 0 0 1 0 0 0 0 0 1 0 1 4 0 0 1 0 0 0 1 0 -1 -1 -1 -1 0 1 3
0 0 1 0 0 0 0 1 1 0 0 0 0 1 8 0 0 1 0 -1 -1 -1 -1 0 1 0 0 0 1 3
0 0 0 1 1 0 0 0 0 0 0 1 0 1 4 0 0 0 1 0 1 0 0 -1 -1 -1 -1 0 1 9
0 0 0 1 0 0 1 0 1 0 0 0 0 1 6 0 0 0 1 0 0 0 1 0 1 0 0 0 1 3
0 0 0 1 -1 -1 -1 -1 0 0 1 0 0 1 4 -1 -1 -1 -1 1 0 0 0 -1 -1 -1 -1 0 1 2
-1 -1 -1 -1 0 1 0 0 1 0 0 0 0 1 8 -1 -1 -1 -1 0 0 1 0 0 1 0 0 0 1 3
-1 -1 -1 -1 0 0 0 1 0 0 1 0 0 1 10 -1 -1 -1 -1 -1 -1 -1 -1 0 0 0 1 0 1 8
1 0 0 0 1 0 0 0 1 0 0 0 -1 -1 3 1 0 0 0 0 1 0 0 0 1 0 0 -1 -1 2
1 0 0 0 0 0 1 0 0 0 1 0 -1 -1 2 1 0 0 0 0 0 0 1 0 0 0 1 -1 -1 5
1 0 0 0 -1 -1 -1 -1 -1 -1 -1 -1 -1 -1 4 0 1 0 0 1 0 0 0 0 1 0 0 -1 -1 -4
0 1 0 0 0 1 0 0 0 0 1 0 -1 -1 3 0 1 0 0 0 0 1 0 0 0 0 1 -1 -1 11
0 1 0 0 0 0 0 1 -1 -1 -1 -1 -1 -1 8 0 1 0 0 -1 -1 -1 -1 1 0 0 0 -1 -1 5
0 0 1 0 1 0 0 0 0 0 1 0 -1 -1 -4 0 0 1 0 0 1 0 0 0 0 0 1 -1 -1 7
0 0 1 0 0 0 1 0 -1 -1 -1 -1 -1 -1 4 0 0 1 0 0 0 0 1 1 0 0 0 -1 -1 9
0 0 1 0 -1 -1 -1 -1 0 1 0 0 -1 -1 3 0 0 0 1 1 0 0 0 0 0 0 1 -1 -1 5
0 0 0 1 0 1 0 0 -1 -1 -1 -1 -1 -1 7 0 0 0 1 0 0 1 0 1 0 0 0 -1 -1 4
0 0 0 1 0 0 0 1 0 1 0 0 -1 -1 5 0 0 0 1 -1 -1 -1 -1 0 0 1 0 -1 -1 3
-1 -1 -1 -1 1 0 0 0 -1 -1 -1 -1 -1 -1 3 -1 -1 -1 -1 0 1 0 0 1 0 0 0 -1 -1 10
-1 -1 -1 -1 0 0 1 0 0 1 0 0 -1 -1 3 -1 -1 -1 -1 0 0 0 1 0 0 1 0 -1 -1 9
-1 -1 -1 -1 -1 -1 -1 -1 0 0 0 1 -1 -1 10
;
proc print data=latmines; sum x1-x4 z1-z4 c1-c4
h1-h2 y;
proc reg data=latmines;
model y=x1-x4 z1-z4 c1-c4 h1-h2;
run;
```

SAS OUTPUT

Obs	x1	x2	x3	x4	z1	z2	z3	z4	c1	c2	c3	c4	h1	h2	y
1	1	0	0	0	1	0	0	0	1	0	0	0	1	0	3
2	1	0	0	0	0	1	0	0	0	1	0	0	1	0	1
3	1	0	0	0	0	0	1	0	0	0	1	0	1	0	2
4	1	0	0	0	0	0	0	1	0	0	0	1	1	0	5
5	1	0	0	0	-1	-1	-1	-1	-1	-1	-1	-1	1	0	6
6	0	1	0	0	1	0	0	0	0	1	0	0	1	0	-4
7	0	1	0	0	0	1	0	0	0	0	1	0	1	0	3
8	0	1	0	0	0	0	1	0	0	0	0	1	1	0	11
9	0	1	0	0	0	0	0	1	-1	-1	-1	-1	1	0	8
10	0	1	0	0	-1	-1	-1	-1	1	0	0	0	1	0	8
11	0	0	1	0	1	0	0	0	0	0	1	0	1	0	-3
12	0	0	1	0	0	1	0	0	0	0	0	1	1	0	7
13	0	0	1	0	0	0	1	0	-1	-1	-1	-1	1	0	5
14	0	0	1	0	0	0	0	1	1	0	0	0	1	0	9
15	0	0	1	0	-1	-1	-1	-1	0	1	0	0	1	0	3
16	0	0	0	1	1	0	0	0	0	0	0	1	1	0	5
17	0	0	0	1	0	1	0	0	-1	-1	-1	-1	1	0	12
18	0	0	0	1	0	0	1	0	1	0	0	0	1	0	7
19	0	0	0	1	0	0	0	1	0	1	0	0	1	0	5
20	0	0	0	1	-1	-1	-1	-1	0	0	1	0	1	0	4
21	-1	-1	-1	-1	1	0	0	0	-1	-1	-1	-1	1	0	1

Obs	x1	x2	x3	x4	z1	z2	z3	z4	c1	c2	c3	c4	h1	h2	y
22	-1	-1	-1	-1	0	1	0	0	1	0	0	0	1	0	11
23	-1	-1	-1	-1	0	0	1	0	0	1	0	0	1	0	1
24	-1	-1	-1	-1	0	0	0	1	0	0	1	0	1	0	11
25	-1	-1	-1	-1	-1	-1	-1	-1	0	0	0	1	1	0	13
26	1	0	0	0	1	0	0	0	1	0	0	0	0	1	2
27	1	0	0	0	0	1	0	0	0	1	0	0	0	1	4
28	1	0	0	0	0	0	1	0	0	0	1	0	0	1	2
29	1	0	0	0	0	0	0	1	0	0	0	1	0	1	5
30	1	0	0	0	-1	-1	-1	-1	-1	-1	-1	-1	0	1	3
31	0	1	0	0	1	0	0	0	0	1	0	0	0	1	-2
32	0	1	0	0	0	1	0	0	0	0	1	0	0	1	3
33	0	1	0	0	0	0	1	0	0	0	0	1	0	1	10
34	0	1	0	0	0	0	0	1	-1	-1	-1	-1	0	1	5
35	0	1	0	0	-1	-1	-1	-1	1	0	0	0	0	1	4
36	0	0	1	0	1	0	0	0	0	0	1	0	0	1	-5
37	0	0	1	0	0	1	0	0	0	0	0	1	0	1	4
38	0	0	1	0	0	0	1	0	-1	-1	-1	-1	0	1	3
39	0	0	1	0	0	0	0	1	1	0	0	0	0	1	8
40	0	0	1	0	-1	-1	-1	-1	0	1	0	0	0	1	3
41	0	0	0	1	1	0	0	0	0	0	0	1	0	1	4
42	0	0	0	1	0	1	0	0	-1	-1	-1	-1	0	1	9
43	0	0	0	1	0	0	1	0	1	0	0	0	0	1	6
44	0	0	0	1	0	0	0	1	0	1	0	0	0	1	3
45	0	0	0	1	-1	-1	-1	-1	0	0	1	0	0	1	4
46	-1	-1	-1	-1	1	0	0	0	-1	-1	-1	-1	0	1	2
47	-1	-1	-1	-1	0	1	0	0	1	0	0	0	0	1	8
48	-1	-1	-1	-1	0	0	1	0	0	1	0	0	0	1	3
49	-1	-1	-1	-1	0	0	0	1	0	0	1	0	0	1	10
50	-1	-1	-1	-1	-1	-1	-1	-1	0	0	0	1	0	1	8
51	1	0	0	0	1	0	0	0	1	0	0	0	-1	-1	3
52	1	0	0	0	0	1	0	0	0	1	0	0	-1	-1	2
53	1	0	0	0	0	0	1	0	0	0	1	0	-1	-1	2
54	1	0	0	0	0	0	0	1	0	0	0	1	-1	-1	5
55	1	0	0	0	-1	-1	-1	-1	-1	-1	-1	-1	-1	-1	4
56	0	1	0	0	1	0	0	0	0	1	0	0	-1	-1	-4
57	0	1	0	0	0	1	0	0	0	0	1	0	-1	-1	3
58	0	1	0	0	0	0	1	0	0	0	0	1	-1	-1	11
59	0	1	0	0	0	0	0	1	-1	-1	-1	-1	-1	-1	8
60	0	1	0	0	-1	-1	-1	-1	1	0	0	0	-1	-1	5
61	0	0	1	0	1	0	0	0	0	0	1	0	-1	-1	-4
62	0	0	1	0	0	1	0	0	0	0	0	1	-1	-1	7
63	0	0	1	0	0	0	1	0	-1	-1	-1	-1	-1	-1	4
64	0	0	1	0	0	0	0	1	1	0	0	0	-1	-1	9
65	0	0	1	0	-1	-1	-1	-1	0	1	0	0	-1	-1	3
66	0	0	0	1	1	0	0	0	0	0	0	1	-1	-1	5
67	0	0	0	1	0	1	0	0	-1	-1	-1	-1	-1	-1	7
68	0	0	0	1	0	0	1	0	1	0	0	0	-1	-1	4
69	0	0	0	1	0	0	0	1	0	1	0	0	-1	-1	5
70	0	0	0	1	-1	-1	-1	-1	0	0	1	0	-1	-1	3
71	-1	-1	-1	-1	1	0	0	0	-1	-1	-1	-1	-1	-1	3
72	-1	-1	-1	-1	0	1	0	0	1	0	0	0	-1	-1	10
73	-1	-1	-1	-1	0	0	1	0	0	1	0	0	-1	-1	3

```
Obs  x1  x2  x3  x4  z1  z2  z3  z4  c1  c2  c3  c4   h1  h2    y

 74  -1  -1  -1  -1   0   0   0   1   0   0   1   0   -1  -1    9
 75  -1  -1  -1  -1  -1  -1  -1  -1   0   0   0   1   -1  -1   10
     ==  ==  ==  ==   =  ==  ==  ==  ==  ==  ==  ==   ==  == ====
      0   0   0   0   0   0   0   0   0   0   0   0    0   0  357

                          Analysis of Variance

                                Sum of           Mean
Source                  DF     Squares         Square    F Value      Pr > F

Model                   14   875.89333       62.56381      15.65      <.0001
Error                   60   239.78667        3.99644
Corrected Total         74  1115.68000

         Root MSE             1.99911     R-Square      0.7851
         Dependent Mean       4.76000     Adj R-Sq      0.7349
         Coeff Var           41.99813

                     Parameter Estimates
                            Parameter     Standard
Variable                DF   Estimate        Error    t-Value     Pr > |t|

Intercept               1     4.76000      0.23084      20.62       <.0001
x1                      1    -1.49333      0.46167      -3.23       0.0020
x2                      1    -0.16000      0.46167      -0.35       0.7301
x3                      1    -1.22667      0.46167      -2.66       0.0101
x4                      1     0.77333      0.46167       1.68       0.0991
z1                      1    -4.36000      0.46167      -9.44       <.0001
z2                      1     1.30667      0.46167       2.83       0.0063
z3                      1     0.17333      0.46167       0.38       0.7087
z4                      1     2.24000      0.46167       4.85       <.0001
c1                      1     1.70667      0.46167       3.70       0.0005
c2                      1    -3.02667      0.46167      -6.56       <.0001
c3                      1    -1.82667      0.46167      -3.96       0.0002
c4                      1     2.57333      0.46167       5.57       <.0001
h1                      1     0.60000      0.32645       1.84       0.0710
h2                      1    -0.52000      0.32645      -1.59       0.1164
```

The results of this analysis agree with those obtained previously by manual analysis. The ANOVA agrees with that obtained in Chapter 5 using classical methods.

6.9 Dummy Variables Regression Model for Example 5.7 (Reference Cell Coding Method)

The relevant regression model obtained by using reference cell coding method to model the experiment in Example 5.7 is

$$y_{ijkl} = \mu + \sum_{i=1}^{r-1} \alpha_i X_i + \sum_{j=1}^{r-1} \beta_j Z_j + \sum_{k=1}^{r-1} \gamma_k C_k + \sum_{l=1}^{n-1} \theta_l H_l + \varepsilon \tag{6.23}$$

where

$$X_i = \begin{cases} 1 & \text{if } y_{ijkl} \text{ is under the } i\text{th row} \\ 0 & \text{otherwise} \end{cases}$$

for $i = 1, 2, \ldots, r-1$

$$Z_j = \begin{cases} 1 & \text{if } y_{ijkl} \text{ is under the } j\text{th column} \\ 0 & \text{otherwise} \end{cases}$$

for $j = 1, 2, \ldots, r-1$

$$C_k = \begin{cases} 1 & \text{if } y_{ijkl} \text{ is under the } k\text{th Latin letter (treatment)} \\ 0 & \text{otherwise} \end{cases}$$

for $k = 1, 2, \ldots, r-1$

$$H_l = \begin{cases} 1 & \text{if } y_{ijkl} \text{ is under the } l\text{th replicate} \\ 0 & \text{otherwise} \end{cases}$$

for $l = 1, 2, \ldots, n-1$

With this model, we estimate the parameters of the model as stated below, which is an extension of a model with three sets of dummy variables to a model with four sets of dummy variables to account for replication in the designed Latin squares experiment.

$$\mu_{ijkl} = \mu + \alpha_i + \beta_j + \gamma_j + \theta_l \quad i, j, k = 1, 2, \ldots, r-1 \quad l = 1, 2, \ldots, n-1$$

$$\mu_{r...} = \mu + \frac{1}{r}\sum_{j=1}^{r-1}\beta_j + \frac{1}{r}\sum_{k=1}^{r-1}\gamma_k + \frac{1}{n}\sum_{l=1}^{n-1}\theta_l H_l$$

$$\mu_{.r..} = \mu + \frac{1}{r}\sum_{i=1}^{r-1}\alpha_i + \frac{1}{r}\sum_{k=1}^{r-1}\gamma_k + \frac{1}{n}\sum_{l=1}^{n-1}\theta_l H_l$$

$$\mu_{..r.} = \mu + \frac{1}{r}\sum_{i=1}^{r-1}\alpha_i + \frac{1}{r}\sum_{j=1}^{r-1}\beta_j + \frac{1}{n}\sum_{l=1}^{n-1}\theta_l H_l$$

$$\mu = \mu_{....} - \frac{1}{r}\sum_{i=1}^{r-1}\alpha_i - \frac{1}{r}\sum_{j=1}^{r-1}\beta_j - \frac{1}{r}\sum_{k=1}^{r-1}\gamma_k - \frac{1}{n}\sum_{l=1}^{n-1}\theta_l \qquad (6.24)$$

Here is the manual analysis of the data based on a reference cell containing y_{5543}. The reference cell for this model is obtained by considering the observation in replicate 3, belonging to row 5, column 5, with the Latin letter D. In this case, we rewrite the definitions of the set of dummy variables as follows:

$$X_i = \begin{cases} 1 & \text{if } y_{ijkl} \text{ is under the } i\text{th row} \\ 0 & \text{otherwise} \end{cases}$$

for $i = 1, 2, 3, 4$

$$Z_j = \begin{cases} 1 & \text{if } y_{ijkl} \text{ is under the } j\text{th column} \\ 0 & \text{otherwise} \end{cases}$$

for $j = 1, 2, 3, 4$

$$C_k = \begin{cases} 1 & \text{if } y_{ijkl} \text{ is under the } k\text{th Latin letter (treatment)} \\ 0 & \text{otherwise} \end{cases}$$

for $k = 1, 2, 3, 5$

$$H_l = \begin{cases} 1 & \text{if } y_{ijkl} \text{ is under the } l\text{th replicate} \\ 0 & \text{otherwise} \end{cases}$$

for $l = 1, 2$

6.9.1 Estimation of Parameters of Regression Model for Example 5.7 (Reference Cell Coding)

To estimate the parameters for the model, we use sample means obtained in Section 6.8 when we used the effects coding method. The ensuing estimates are

$$\overline{y}_{5...} = 6.8667$$

$$\hat{\alpha}_1 = \overline{y}_{1...} - \overline{y}_{5...} = 3.2667 - 6.8667 = -3.6$$

$$\hat{\alpha}_2 = \overline{y}_{2...} - \overline{y}_{5...} = 4.6 - 6.8667 = -2.2667$$

$$\hat{\alpha}_3 = \overline{y}_{3...} - \overline{y}_{5...} = 3.533 - 6.8667 = -3.3333$$

$$\hat{\alpha}_4 = \overline{y}_{4...} - \overline{y}_{5...} = 5.533 - 6.8667 = -1.3333$$

$$\overline{y}_{.5..} = 5.4$$

$$\hat{\beta}_1 = \overline{y}_{.1..} - \overline{y}_{.5..} = 0.4 - 5.4 = -5.0$$

$$\hat{\beta}_2 = \overline{y}_{.2..} - \overline{y}_{.5..} = 6.0667 - 5.4 = 0.6667$$

$$\hat{\beta}_3 = \overline{y}_{.3..} - \overline{y}_{.5..} = 4.993 - 5.4 = -0.46667$$

$$\hat{\beta}_4 = \overline{y}_{.4..} - \overline{y}_{.5..} = 7.0 - 5.4 = 1.6$$

$$\overline{y}_{..4.} = 7.3333$$

$$\hat{\gamma}_1 = \overline{y}_{..1.} - \overline{y}_{..4.} = 6.4667 - 7.3333 = -0.8667$$

$$\hat{\gamma}_2 = \overline{y}_{..2.} - \overline{y}_{..4.} = 1.7333 - 7.3333 = -5.6$$

$$\hat{\gamma}_3 = \overline{y}_{..3.} - \overline{y}_{..4.} = 2.9333 - 7.3333 = -4.4$$

$$\hat{\gamma}_4 = \overline{y}_{..4.} - \overline{y}_{..4.} = 5.3333 - 7.3333 = -2.0$$

$$\hat{\theta}_1 = \overline{y}_{...1} - \overline{y}_{...3} = 5.36 - 4.68 = 0.68$$

$$\hat{\theta}_2 = \overline{y}_{...2} - \overline{y}_{...3} = 4.24 - 4.68 = -0.44$$

$$\mu = \mu_{....} - \frac{1}{r}\sum_{i=1}^{r-1}\alpha_i - \frac{1}{r}\sum_{j=1}^{r-1}\beta_j - \frac{1}{r}\sum_{k=1}^{r-1}\gamma_k - \frac{1}{n}\sum_{l=1}^{n-1}\theta_l$$

$$\mu = 4.76 - \frac{(-3.6 - 2.2667 - 3.3333 - 1.3333)}{5} - \frac{(-5 + 0.6667 - 0.46667 + 1.6)}{5}$$

$$- \frac{(-0.86667 - 5.6 - 4.4 - 2.0)}{5} - \frac{(0.68 - 0.44)}{3} = 10.00$$

With the above results, the estimated model is

$$\begin{aligned}\hat{y} = 10.00 &- 3.6x_1 - 2.2667x_2 - 3.3333x_3 - 1.3333x_4 - 5.0z_1 + 0.6667z_2\\ &- 0.46667z_3 + 1.6z_4 - 0.8667c_1 - 5.6c_2 - 4.4c_3 - 2.0c_4\\ &+ 0.68h_1 - 0.44h_2 \qquad (6.25)\end{aligned}$$

Using the model, we again obtain the fitted values, extract residuals, and perform ANOVA to test the significance of the fitted model parameters (Tables 6.18 and 6.19).

TABLE 6.18: Estimated Responses and Residuals for Example 5.7 (Reference Cell Coding Method)

x_1	x_2	x_3	x_4	z_1	z_2	z_3	z_4	c_1	c_2	c_3	c_4	h_1	h_2	y	$\hat{y}$	$y-\hat{y}$	$(y-\hat{y})^2$
1	0	0	0	1	0	0	0	1	0	0	0	1	0	3	1.2133	1.7867	3.192297
1	0	0	0	0	1	0	0	0	1	0	0	1	0	1	2.1467	−1.1467	1.314921
1	0	0	0	0	0	1	0	0	0	1	0	1	0	2	2.21333	−0.21333	0.04551
1	0	0	0	0	0	0	1	0	0	0	0	1	0	5	8.68	−3.68	13.5424
1	0	0	0	0	0	0	0	0	0	0	1	1	0	6	5.08	0.92	0.8464
0	1	0	0	1	0	0	0	0	1	0	0	1	0	−4	−2.1867	−1.8133	3.288057
0	1	0	0	0	1	0	0	0	0	1	0	1	0	3	4.68	−1.68	2.8224
0	1	0	0	0	0	1	0	0	0	0	0	1	0	11	7.94663	3.05337	9.323068
0	1	0	0	0	0	0	1	0	0	0	1	1	0	8	8.0133	−0.0133	0.000177
0	1	0	0	0	0	0	0	1	0	0	0	1	0	8	7.5466	0.4534	0.205572
0	0	1	0	1	0	0	0	0	0	1	0	1	0	−3	−2.0533	−0.9467	0.896241
0	0	1	0	0	1	0	0	0	0	0	0	1	0	7	8.0134	−1.0134	1.02698
0	0	1	0	0	0	1	0	0	0	0	1	1	0	5	4.88003	0.11997	0.014393
0	0	1	0	0	0	0	1	1	0	0	0	1	0	9	8.08	0.92	0.8464
0	0	1	0	0	0	0	0	0	1	0	0	1	0	3	1.7467	1.2533	1.570761
0	0	0	1	1	0	0	0	0	0	0	0	1	0	5	4.3467	0.6533	0.426801
0	0	0	1	0	1	0	0	0	0	0	1	1	0	12	8.0134	3.9866	15.89298
0	0	0	1	0	0	1	0	1	0	0	0	1	0	7	8.01333	−1.01333	1.026838
0	0	0	1	0	0	0	1	0	1	0	0	1	0	5	5.3467	−0.3467	0.120201
0	0	0	1	0	0	0	0	0	0	1	0	1	0	4	4.9467	−0.9467	0.896241
0	0	0	0	1	0	0	0	0	0	0	1	1	0	1	3.68	−2.68	7.1824
0	0	0	0	0	1	0	0	1	0	0	0	1	0	11	10.48	0.52	0.2704
0	0	0	0	0	0	1	0	0	1	0	0	1	0	1	4.61333	−3.61333	13.05615
0	0	0	0	0	0	0	1	0	0	1	0	1	0	11	7.88	3.12	9.7344
0	0	0	0	0	0	0	0	0	0	0	0	1	0	13	10.68	2.32	5.3824
1	0	0	0	1	0	0	0	1	0	0	0	0	1	2	0.0933	1.9067	3.635505
1	0	0	0	0	1	0	0	0	1	0	0	0	1	4	1.0267	2.9733	8.840513
1	0	0	0	0	0	1	0	0	0	1	0	0	1	2	1.09333	0.90667	0.82205
1	0	0	0	0	0	0	1	0	0	0	0	0	1	5	7.56	−2.56	6.5536
1	0	0	0	0	0	0	0	0	0	0	1	0	1	3	3.96	−0.96	0.9216
0	1	0	0	1	0	0	0	0	1	0	0	0	1	−2	−3.3067	1.3067	1.707465
0	1	0	0	0	1	0	0	0	0	1	0	0	1	3	3.56	−0.56	0.3136
0	1	0	0	0	0	1	0	0	0	0	0	0	1	10	6.82663	3.17337	10.07028
0	1	0	0	0	0	0	1	0	0	0	1	0	1	5	6.8933	−1.8933	3.584585
0	1	0	0	0	0	0	0	1	0	0	0	0	1	4	6.4266	−2.4266	5.888388
0	0	1	0	1	0	0	0	0	0	1	0	0	1	−5	−3.1733	−1.8267	3.336833

(Continued)

TABLE 6.18 Continued

x_1	x_2	x_3	x_4	z_1	z_2	z_3	z_4	c_1	c_2	c_3	c_4	h_1	h_2	y	$\hat{y}$	$y-\hat{y}$	$(y-\hat{y})^2$
0	0	1	0	0	1	0	0	0	0	0	0	0	1	4	6.8934	−2.8934	8.371764
0	0	1	0	0	0	1	0	0	0	0	1	0	1	3	3.76003	−0.76003	0.577646
0	0	1	0	0	0	0	1	1	0	0	0	0	1	8	6.96	1.04	1.0816
0	0	1	0	0	0	0	0	0	1	0	0	0	1	3	0.6267	2.3733	5.632553
0	0	0	1	1	0	0	0	0	0	0	0	0	1	4	3.2267	0.7733	0.597993
0	0	0	1	0	1	0	0	0	0	0	1	0	1	9	6.8934	2.1066	4.437764
0	0	0	1	0	0	1	0	1	0	0	0	0	1	6	6.89333	−0.89333	0.798038
0	0	0	1	0	0	0	1	0	1	0	0	0	1	3	4.2267	−1.2267	1.504793
0	0	0	1	0	0	0	0	0	0	1	0	0	1	4	3.8267	0.1733	0.030033
0	0	0	0	1	0	0	0	0	0	0	1	0	1	2	2.56	−0.56	0.3136
0	0	0	0	0	1	0	0	1	0	0	0	0	1	8	9.36	−1.36	1.8496
0	0	0	0	0	0	1	0	0	1	0	0	0	1	3	3.49333	−0.49333	0.243374
0	0	0	0	0	0	0	1	0	0	1	0	0	1	10	6.76	3.24	10.4976
0	0	0	0	0	0	0	0	0	0	0	0	0	1	8	9.56	−1.56	2.4336
1	0	0	0	1	0	0	0	1	0	0	0	0	0	3	0.5333	2.4667	6.084609
1	0	0	0	0	1	0	0	0	1	0	0	0	0	2	1.4667	0.5333	0.284409
1	0	0	0	0	0	1	0	0	0	1	0	0	0	2	1.53333	0.46667	0.217781
1	0	0	0	0	0	0	1	0	0	0	0	0	0	5	8	−3	9
1	0	0	0	0	0	0	0	0	0	0	1	0	0	4	4.4	−0.4	0.16
0	1	0	0	1	0	0	0	0	1	0	0	0	0	−4	−2.8667	−1.1333	1.284369
0	1	0	0	0	1	0	0	0	0	1	0	0	0	3	4	−1	1
0	1	0	0	0	0	1	0	0	0	0	0	0	0	11	7.26663	3.73337	13.93805
0	1	0	0	0	0	0	1	0	0	0	1	0	0	8	7.3333	0.6667	0.444489
0	1	0	0	0	0	0	0	1	0	0	0	0	0	5	6.8666	−1.8666	3.484196
0	0	1	0	1	0	0	0	0	0	1	0	0	0	−4	−2.7333	−1.2667	1.604529
0	0	1	0	0	1	0	0	0	0	0	0	0	0	7	7.3334	−0.3334	0.111156
0	0	1	0	0	0	1	0	0	0	0	1	0	0	4	4.20003	−0.20003	0.040012
0	0	1	0	0	0	0	1	1	0	0	0	0	0	9	7.4	1.6	2.56
0	0	1	0	0	0	0	0	0	1	0	0	0	0	3	1.0667	1.9333	3.737649
0	0	0	1	1	0	0	0	0	0	0	0	0	0	5	3.6667	1.3333	1.777689
0	0	0	1	0	1	0	0	0	0	0	1	0	0	7	7.3334	−0.3334	0.111156
0	0	0	1	0	0	1	0	1	0	0	0	0	0	4	7.33333	−3.33333	11.11109
0	0	0	1	0	0	0	1	0	1	0	0	0	0	5	4.6667	0.3333	0.111089
0	0	0	1	0	0	0	0	0	0	1	0	0	0	3	4.2667	−1.2667	1.604529
0	0	0	0	1	0	0	0	0	0	0	1	0	0	3	3	0	0
0	0	0	0	0	1	0	0	1	0	0	0	0	0	10	9.8	0.2	0.04
0	0	0	0	0	0	1	0	0	1	0	0	0	0	3	3.93333	−0.93333	0.871105
0	0	0	0	0	0	0	1	0	0	1	0	0	0	9	7.2	1.8	3.24
0	0	0	0	0	0	0	0	0	0	0	0	0	0	10	10	0	0
Total														357	357.0005	−0.00045	239.7867

If $S_{yy} = \sum_{i=1}^{N} y_i^2 - \left(\sum_{i=1}^{N} y_i\right)^2 / N$, $SS_{\text{R}} = \sum_{i=1}^{N} (\hat{y}_i - \overline{y})^2$, and $SS_{\text{E}} = \sum_{i=1}^{N} (y_i - \hat{y}_i)^2$ then $S_{yy} = SS_{\text{R}} + SS_{\text{E}}$ and $N = 3(5^2) = 75$.

$$SS_{\text{E}} = 239.7867$$

$$S_{yy} = 2816 - \frac{357^2}{75} = 1115.68$$

$$\Rightarrow SS_{\text{R}} = 1115.68 - 239.7867 = 875.8933$$

TABLE 6.19: ANOVA to Test for Significance of the Fitted Regression Model for Example 5.7

Source	Sum of squares	Degrees of freedom	Mean square	F-ratio
Regression	875.8933	14	62.56381	15.68
Error	239.7867	60	3.99644	
Total	1115.68	74		

We find that $F(14, 60, 0.01) = 2.394$, indicating that the fitted model is highly significant at 1% level of significance.

6.9.2 SAS Program for Analysis of Responses of Example 5.7 (Reference Cell Coding Method)

We write a SAS program to analyze the data under the above model. This means that the Latin letters are coded as 1 = A, 2 = B, 3 = C, 4 = E, D = 0. The data were originally coded in Microsoft Excel and imported into SAS before analysis. In the program, we include the SAS code that illustrates the importation of data from another application (in this case Excel) into SAS. Importation of data from an application into SAS for further analysis is treated in detail in Chapter 6 of Onyiah (2005).

SAS PROGRAM

```
PROC IMPORT  OUT= WORK.MINES1
             DATAFILE= "e:  blexamp4_3.xls"
             DBMS=EXCEL REPLACE;
      SHEET="Sheet1$";
      GETNAMES=YES;
      MIXED=NO;
      SCANTEXT=YES;
      USEDATE=YES;
      SCANTIME=YES;
RUN;
data latmines2;
set mines1;
```

```
proc print data=latmines2;
sum x1-x4 z1-z4 c1-c4 h1-h2 y;
proc reg data=latmines2;
model y=x1-x4 z1-z4 c1-c4 h1-h2; run;
```

SAS OUTPUT

Obs	x1	x2	x3	x4	z1	z2	z3	z4	c1	c2	c3	c4	h1	h2	y
1	1	0	0	0	1	0	0	0	1	0	0	0	1	0	3
2	1	0	0	0	0	1	0	0	0	1	0	0	1	0	1
3	1	0	0	0	0	0	1	0	0	0	1	0	1	0	2
4	1	0	0	0	0	0	0	1	0	0	0	0	1	0	5
5	1	0	0	0	0	0	0	0	0	0	0	1	1	0	6
6	0	1	0	0	1	0	0	0	0	1	0	0	1	0	-4
7	0	1	0	0	0	1	0	0	0	0	1	0	1	0	3
8	0	1	0	0	0	0	1	0	0	0	0	0	1	0	11
9	0	1	0	0	0	0	0	1	0	0	0	1	1	0	8
10	0	1	0	0	0	0	0	0	1	0	0	0	1	0	8
11	0	0	1	0	1	0	0	0	0	0	1	0	1	0	-3
12	0	0	1	0	0	1	0	0	0	0	0	0	1	0	7
13	0	0	1	0	0	0	1	0	0	0	0	1	1	0	5
14	0	0	1	0	0	0	0	1	1	0	0	0	1	0	9
15	0	0	1	0	0	0	0	0	0	1	0	0	1	0	3
16	0	0	0	1	1	0	0	0	0	0	0	0	1	0	5
17	0	0	0	1	0	1	0	0	0	0	0	1	1	0	12
18	0	0	0	1	0	0	1	0	1	0	0	0	1	0	7
19	0	0	0	1	0	0	0	1	0	1	0	0	1	0	5
20	0	0	0	1	0	0	0	0	0	0	1	0	1	0	4
21	0	0	0	0	1	0	0	0	0	0	0	1	1	0	1
22	0	0	0	0	0	1	0	0	1	0	0	0	1	0	11
23	0	0	0	0	0	0	1	0	0	1	0	0	1	0	1
24	0	0	0	0	0	0	0	1	0	0	1	0	1	0	11
25	0	0	0	0	0	0	0	0	0	0	0	0	1	0	13
26	1	0	0	0	1	0	0	0	1	0	0	0	0	1	2
27	1	0	0	0	0	1	0	0	0	1	0	0	0	1	4
28	1	0	0	0	0	0	1	0	0	0	1	0	0	1	2
29	1	0	0	0	0	0	0	1	0	0	0	0	0	1	5
30	1	0	0	0	0	0	0	0	0	0	0	1	0	1	3
31	0	1	0	0	1	0	0	0	0	1	0	0	0	1	-2
32	0	1	0	0	0	1	0	0	0	0	1	0	0	1	3
33	0	1	0	0	0	0	1	0	0	0	0	0	0	1	10
34	0	1	0	0	0	0	0	1	0	0	0	1	0	1	5
35	0	1	0	0	0	0	0	0	1	0	0	0	0	1	4
36	0	0	1	0	1	0	0	0	0	0	1	0	0	1	-5
37	0	0	1	0	0	1	0	0	0	0	0	0	0	1	4
38	0	0	1	0	0	0	1	0	0	0	0	1	0	1	3
39	0	0	1	0	0	0	0	1	1	0	0	0	0	1	8
40	0	0	1	0	0	0	0	0	0	1	0	0	0	1	3
41	0	0	0	1	1	0	0	0	0	0	0	0	0	1	4
42	0	0	0	1	0	1	0	0	0	0	0	1	0	1	9
43	0	0	0	1	0	0	1	0	1	0	0	0	0	1	6
44	0	0	0	1	0	0	0	1	0	1	0	0	0	1	3

Obs	x1	x2	x3	x4	z1	z2	z3	z4	c1	c2	c3	c4	h1	h2	y
45	0	0	0	1	0	0	0	0	0	0	1	0	0	1	4
46	0	0	0	0	1	0	0	0	0	0	0	1	0	1	2
47	0	0	0	0	0	1	0	0	1	0	0	0	0	1	8
48	0	0	0	0	0	0	1	0	0	1	0	0	0	1	3
49	0	0	0	0	0	0	0	1	0	0	1	0	0	1	10
50	0	0	0	0	0	0	0	0	0	0	0	0	0	1	8
51	1	0	0	0	1	0	0	0	1	0	0	0	0	0	3
52	1	0	0	0	0	1	0	0	0	1	0	0	0	0	2
53	1	0	0	0	0	0	1	0	0	0	1	0	0	0	2
54	1	0	0	0	0	0	0	1	0	0	0	0	0	0	5
55	1	0	0	0	0	0	0	0	0	0	0	1	0	0	4
56	0	1	0	0	1	0	0	0	0	1	0	0	0	0	-4
57	0	1	0	0	0	1	0	0	0	0	1	0	0	0	3
58	0	1	0	0	0	0	1	0	0	0	0	0	0	0	11
59	0	1	0	0	0	0	0	1	0	0	0	1	0	0	8
60	0	1	0	0	0	0	0	0	1	0	0	0	0	0	5
61	0	0	1	0	1	0	0	0	0	0	1	0	0	0	-4
62	0	0	1	0	0	1	0	0	0	0	0	0	0	0	7
63	0	0	1	0	0	0	1	0	0	0	0	1	0	0	4
64	0	0	1	0	0	0	0	1	1	0	0	0	0	0	9
65	0	0	1	0	0	0	0	0	0	1	0	0	0	0	3
66	0	0	0	1	1	0	0	0	0	0	0	0	0	0	5
67	0	0	0	1	0	1	0	0	0	0	0	1	0	0	7
68	0	0	0	1	0	0	1	0	1	0	0	0	0	0	4
69	0	0	0	1	0	0	0	1	0	1	0	0	0	0	5
70	0	0	0	1	0	0	0	0	0	0	1	0	0	0	3
71	0	0	0	0	1	0	0	0	0	0	0	1	0	0	3
72	0	0	0	0	0	1	0	0	1	0	0	0	0	0	10
73	0	0	0	0	0	0	1	0	0	1	0	0	0	0	3
74	0	0	0	0	0	0	0	1	0	0	1	0	0	0	9
75	0	0	0	0	0	0	0	0	0	0	0	0	0	0	10
	==	==	==	==	==	==	==	==	==	==	==	==	==	==	====
	15	15	15	15	15	15	15	15	15	15	15	15	25	25	357

```
                              The GLM Procedure

                              Number of Observations Read    75
                              Number of Observations Used    75

Dependent Variable: y    y

                                     Sum of
  Source                  DF        Squares    Mean Square    F Value    Pr > F

  Model                   14     875.893333      62.563810      15.65    <.0001

  Error                   60     239.786667       3.996444

  Corrected Total         74    1115.680000


          R-Square     Coeff Var      Root MSE        y Mean

          0.785076      41.99813      1.999111      4.760000
```

Source	DF	Type I SS	Mean Square	F Value	Pr > F
x1	1	41.8133333	41.8133333	10.46	0.0020
x2	1	5.6888889	5.6888889	1.42	0.2375
x3	1	71.1111111	71.1111111	17.79	<.0001
x4	1	13.3333333	13.3333333	3.34	0.0727
z1	1	356.4300000	356.4300000	89.19	<.0001
z2	1	0.9388889	0.9388889	0.23	0.6297
z3	1	16.0444444	16.0444444	4.01	0.0496
z4	1	19.2000000	19.2000000	4.80	0.0323
c1	1	54.6133333	54.6133333	13.67	0.0005
c2	1	135.2000000	135.2000000	33.83	<.0001
c3	1	115.6000000	115.6000000	28.93	<.0001
c4	1	30.0000000	30.0000000	7.51	0.0081
h1	1	13.5000000	13.5000000	3.38	0.0710
h2	1	2.4200000	2.4200000	0.61	0.4395

Source	DF	Type III SS	Mean Square	F Value	Pr > F
x1	1	97.2000000	97.2000000	24.32	<.0001
x2	1	38.5333333	38.5333333	9.64	0.0029
x3	1	83.3333333	83.3333333	20.85	<.0001
x4	1	13.3333333	13.3333333	3.34	0.0727
z1	1	187.5000000	187.5000000	46.92	<.0001
z2	1	3.3333333	3.3333333	0.83	0.3648
z3	1	1.6333333	1.6333333	0.41	0.5251
z4	1	19.2000000	19.2000000	4.80	0.0323
c1	1	5.6333333	5.6333333	1.41	0.2398
c2	1	235.2000000	235.2000000	58.85	<.0001
c3	1	145.2000000	145.2000000	36.33	<.0001
c4	1	30.0000000	30.0000000	7.51	0.0081
h1	1	5.7800000	5.7800000	1.45	0.2338
h2	1	2.4200000	2.4200000	0.61	0.4395

Dependent Variable: y y

Parameter	Estimate	Standard Error	t-Value	Pr > |t|
Intercept	10.00000000	0.89402958	11.19	<.0001
x1	-3.60000000	0.72997209	-4.93	<.0001
x2	-2.26666667	0.72997209	-3.11	0.0029
x3	-3.33333333	0.72997209	-4.57	<.0001
x4	-1.33333333	0.72997209	-1.83	0.0727
z1	-5.00000000	0.72997209	-6.85	<.0001
z2	0.66666667	0.72997209	0.91	0.3648
z3	-0.46666667	0.72997209	-0.64	0.5251
z4	1.60000000	0.72997209	2.19	0.0323
c1	-0.86666667	0.72997209	-1.19	0.2398
c2	-5.60000000	0.72997209	-7.67	<.0001
c3	-4.40000000	0.72997209	-6.03	<.0001
c4	-2.00000000	0.72997209	-2.74	0.0081
h1	0.68000000	0.56543395	1.20	0.2338
h2	-0.44000000	0.56543395	-0.78	0.4395

The regression analysis reproduces the results that have already been obtained by manual analysis. This confirms that the regression approach is as good as the classical linear model approach.

6.10 Regression Model for the Graeco-Latin Square Design

To obtain a regression equivalent of the general linear model ANOVA for the Graeco-Latin squares model, we need to define four sets of dummy variables. We can define the dummy variables by effects coding method or by the reference cell method. For a Graeco-Latin square design containing r^2 units, each of the four sets of dummy variables to be defined must contain $r-1$ variables.

6.10.1 Regression Model for Graeco-Latin Squares Using Effects Coding Method

If we use the effects coding method to define the dummy variables used, then we would typically represent the regression analog of the classical linear model $r \times r$ Graeco-Latin square design ANOVA (Equation 5.20) by

$$y_{ijkl} = \mu + \sum_{i=1}^{r-1} \alpha_i X_i + \sum_{j=1}^{r-1} \beta_j Z_j + \sum_{k=1}^{r-1} \gamma_k C_k + \sum_{l=1}^{r-1} \lambda_l H_l + \varepsilon \qquad (6.26)$$

where

$$X_i = \begin{cases} -1 & \text{if } y_{ijkl} \text{ is under the } r\text{th row} \\ 1 & \text{if } y_{ijkl} \text{ is under the } i\text{th row} \\ 0 & \text{otherwise} \end{cases}$$

for $i = 1, 2, \ldots, r-1$

$$Z_j = \begin{cases} -1 & \text{if } y_{ijkl} \text{ is under the } r\text{th column} \\ 1 & \text{if } y_{ijkl} \text{ is under the } j\text{th column} \\ 0 & \text{otherwise} \end{cases}$$

for $j = 1, 2, \ldots, r-1$

$$C_k = \begin{cases} -1 & \text{if } y_{ijkl} \text{ is under the } r\text{th Latin letter} \\ 1 & \text{if } y_{ijkl} \text{ is under the } k\text{th Latin letter} \\ 0 & \text{otherwise} \end{cases}$$

for $k = 1, 2, \ldots, r-1$

$$H_l = \begin{cases} -1 & \text{if } y_{ijkl} \text{ is under the } r\text{th Greek letter} \\ 1 & \text{if } y_{ijkl} \text{ is under the } l\text{th Greek letter} \\ 0 & \text{otherwise} \end{cases}$$

for $l = 1, 2, \ldots, r-1$

For estimation purposes, consider the following relationships among the row, column, Latin, Greek, and overall means and the parameters of the regression model:

$$\mu_{....} = \mu$$

$$-\sum_{i=1}^{r-1} \alpha_i = \mu_{r...} - \mu_{....}$$

$$-\sum_{j=1}^{r-1} \beta_j = \mu_{.r..} - \mu_{....}$$

$$-\sum_{k=1}^{r-1} \gamma_k = \mu_{..r.} - \mu_{....}$$

$$-\sum_{l=1}^{r-1} \lambda_l = \mu_{...r} - \mu_{....}$$

$$\begin{aligned} \mu_{ijkl} &= \mu + \alpha_i + \beta_j + \gamma_k + \lambda_l \quad i,j,k,l = 1,2,\ldots,r-1 \\ \alpha_i &= \mu_{i...} - \mu_{....} \quad i = 1,2,\ldots,r-1 \\ \beta_j &= \mu_{.j..} - \mu_{....} \quad j = 1,2,\ldots,r-1 \\ \gamma_k &= \mu_{..k.} - \mu_{....} \quad k = 1,2,\ldots,r-1 \\ \lambda_l &= \mu_{...l} - \mu_{....} \quad l = 1,2,\ldots,r-1 \end{aligned} \tag{6.27}$$

This time, we analyze the data for Example 5.9 using the above model. Since the design involves a 5×5 Graeco-Latin square design, then $r = 5$, each of the four sets contain only $r - 1 = 4$ dummy variables.

6.10.2 Estimation of Regression Parameters for Example 5.9 for the Effects Coding Model

$$\overline{y}_{....} = \frac{1087}{25} = 43.48 \quad \overline{y}_{1...} = \frac{223}{5} = 44.6 \quad \overline{y}_{2...} = \frac{230}{5} = 46.0$$

$$\overline{y}_{3...} = \frac{224}{5} = 44.8 \quad \overline{y}_{4...} = \frac{179}{5} = 35.8 \quad \overline{y}_{5...} = \frac{231}{5} = 46.2$$

$$\overline{y}_{.1..} = \frac{233}{5} = 46.6 \quad \overline{y}_{.2..} = \frac{205}{5} = 41 \quad \overline{y}_{.3..} = \frac{218}{5} = 43.6$$

$$\overline{y}_{.4..} = \frac{220}{5} = 44.0 \quad \overline{y}_{.5..} = \frac{211}{5} = 42.2 \quad \overline{y}_{..1.} = \frac{90}{5} = 18.0$$

$$\overline{y}_{..2.} = \frac{226}{5} = 45.2 \quad \overline{y}_{..3.} = \frac{271}{5} = 54.2 \quad \overline{y}_{..4.} = \frac{164}{5} = 32.8$$

$$\overline{y}_{..5.} = \frac{336}{5} = 67.2 \quad \overline{y}_{...1} = \frac{214}{5} = 42.8 \quad \overline{y}_{...2} = \frac{261}{5} = 52.2$$

$$\overline{y}_{...3} = \frac{198}{5} = 39.6 \quad \overline{y}_{...4} = \frac{197}{5} = 39.4 \quad \overline{y}_{...5} = \frac{217}{5} = 43.4$$

$$\hat{\alpha}_1 = \overline{y}_{1...} - \overline{y}_{....} = 44.6 - 43.48 = 1.12$$

$$\hat{\alpha}_2 = \overline{y}_{2...} - \overline{y}_{....} = 46.0 - 43.48 = 2.52$$

$$\hat{\alpha}_3 = \overline{y}_{3...} - \overline{y}_{....} = 44.8 - 43.48 = 1.32$$

$$\hat{\alpha}_4 = \overline{y}_{4...} - \overline{y}_{....} = 35.8 - 43.48 = -7.68$$

$$\hat{\beta}_{.1..} = \overline{y}_{.1..} - \overline{y}_{....} = 46.6 - 43.48 = 3.12$$

$$\hat{\beta}_{.2..} = \overline{y}_{.2..} - \overline{y}_{....} = 41.0 - 43.48 = -2.48$$

$$\hat{\beta}_{.3..} = \overline{y}_{.3..} - \overline{y}_{....} = 43.6 - 43.48 = 0.12$$

$$\hat{\beta}_{.4..} = \overline{y}_{.4..} - \overline{y}_{....} = 44.0 - 43.48 = 0.52$$

$$\hat{\gamma}_{..1.} = \overline{y}_{..1.} - \overline{y}_{....} = 18 - 43.48 = -25.48$$

$$\hat{\gamma}_{..2.} = \overline{y}_{..2.} - \overline{y}_{....} = 45.2 - 43.48 = 1.72$$

$$\gamma_{..3.} = \overline{y}_{..3.} - \overline{y}_{....} = 54.2 - 43.48 = 10.72$$

$$\gamma_{..4.} = \overline{y}_{..4.} - \overline{y}_{....} = 32.8 - 43.48 = -10.68$$

$$\hat{\lambda}_{...1} = \overline{y}_{...1} - \overline{y}_{....} = 42.8 - 43.48 = -0.68$$

$$\hat{\lambda}_{...2} = \overline{y}_{...2} - \overline{y}_{....} = 52.2 - 43.48 = 8.72$$

$$\hat{\lambda}_{...3} = \overline{y}_{...3} - \overline{y}_{....} = 39.6 - 43.48 = -3.88$$

$$\hat{\lambda}_{...4} = \overline{y}_{...4} - \overline{y}_{....} = 39.4 - 43.48 = -4.08$$

The fitted model is

$$\begin{aligned} \hat{y} = {} & 43.48 + 1.12x_1 + 2.52x_2 + 1.32x_3 - 7.68x_4 + 3.12z_1 - 2.48z_2 \\ & + 0.12z_3 + 0.52z_4 - 25.48c_1 + 1.72c_2 + 10.72c_3 - 10.68c_4 \\ & - 0.68h_1 + 8.72h_2 - 3.88h_3 - 4.08h_4 \end{aligned} \tag{6.28}$$

Based on this model, we obtain the fitted values (estimated responses) for the responses of Example 5.9 and the attendant residuals. Thereafter, we carry out ANOVA to check for the significance of the fitted regression model. The results of this analysis are shown in Tables 6.20 and 6.21.

If $S_{yy} = \sum_{i=1}^{N} y_i^2 - (\sum_{i=1}^{N} y_i)^2/N$, $SS_{\text{R}} = \sum_{i=1}^{N} (\hat{y}_i - \overline{y})^2$, and $SS_{\text{E}} = \sum_{i=1}^{N} (y_i - \hat{y}_i)^2$ then $S_{yy} = SS_{\text{R}} + SS_{\text{E}}$. In this design $N = r^2 = 5^2$.

$$SS_{\text{E}} = 450.48$$

$$S_{yy} = 55941 - \frac{1087^2}{25} = 8678.24 \Rightarrow SS_{\text{R}} = 8678.24 - 450.48 = 8227.76$$

TABLE 6.20: Estimated Responses and Residuals for Example 5.9 (Effects Coding Method)

x_1	x_2	x_3	x_4	z_1	z_2	z_3	z_4	c_1	c_2	c_3	c_4	h_1	h_2	h_3	h_4	y	$\hat{y}$	$y-\hat{y}$	$(y-\hat{y})^2$
1	0	0	0	1	0	0	0	0	0	1	0	1	0	0	0	57	57.76	−0.76	0.5776
1	0	0	0	0	1	0	0	0	0	0	1	0	1	0	0	47	40.16	6.84	46.7856
1	0	0	0	0	0	1	0	−1	−1	−1	−1	0	0	1	0	60	64.56	−4.56	20.7936
1	0	0	0	0	0	0	1	0	1	0	0	0	0	0	1	40	42.76	−2.76	7.6176
1	0	0	0	−1	−1	−1	−1	1	0	0	0	−1	−1	−1	−1	19	17.76	1.24	1.5376
0	1	0	0	1	0	0	0	0	1	0	0	−1	−1	−1	−1	55	50.76	4.24	17.9776
0	1	0	0	0	1	0	0	0	0	1	0	0	0	0	1	43	50.16	−7.16	51.2656
0	1	0	0	0	0	1	0	0	0	0	1	1	0	0	0	38	34.76	3.24	10.4976
0	1	0	0	0	0	0	1	1	0	0	0	0	0	1	0	23	17.16	5.84	34.1056
0	1	0	0	−1	−1	−1	−1	−1	−1	−1	−1	0	1	0	0	71	77.16	−6.16	37.9456
0	0	1	0	1	0	0	0	−1	−1	−1	−1	0	0	0	1	74	67.56	6.44	41.4736
0	0	1	0	0	1	0	0	1	0	0	0	1	0	0	0	14	16.16	−2.16	4.6656
0	0	1	0	0	0	1	0	0	1	0	0	0	1	0	0	56	55.36	0.64	0.4096
0	0	1	0	0	0	0	1	0	0	0	1	−1	−1	−1	−1	26	34.56	−8.56	73.2736
0	0	1	0	−1	−1	−1	−1	0	0	1	0	0	0	1	0	54	50.36	3.64	13.2496
0	0	0	1	1	0	0	0	1	0	0	0	0	1	0	0	17	22.16	−5.16	26.6256
0	0	0	1	0	1	0	0	0	1	0	0	0	0	1	0	31	31.16	−0.16	0.0256
0	0	0	1	0	0	1	0	0	0	1	0	−1	−1	−1	−1	47	46.56	0.44	0.1936
0	0	0	1	0	0	0	1	−1	−1	−1	−1	1	0	0	0	61	59.36	1.64	2.6896
0	0	0	1	−1	−1	−1	−1	0	0	0	1	0	0	0	1	23	19.76	3.24	10.4976
−1	−1	−1	−1	1	0	0	0	0	0	0	1	0	0	1	0	30	34.76	−4.76	22.6576
−1	−1	−1	−1	0	1	0	0	−1	−1	−1	−1	−1	−1	−1	−1	70	67.36	2.64	6.9696
−1	−1	−1	−1	0	0	1	0	1	0	0	0	0	0	0	1	17	16.76	0.24	0.0576
−1	−1	−1	−1	0	0	0	1	0	0	1	0	0	1	0	0	70	66.16	3.84	14.7456
−1	−1	−1	−1	−1	−1	−1	−1	0	1	0	0	1	0	0	0	44	45.96	−1.96	3.8416
Total																1087	1087		450.48

It is seen that $F(16, 8, 0.01) = 5.47655$, indicating that the fitted model is highly significant at 1% level of significance.

TABLE 6.21: ANOVA to Test for Significance of the Fitted Regression Model for Example 5.9

Source	**Sum of squares**	**Degrees of freedom**	**Mean square**	***F*-ratio**
Regression	8227.76	16	514.235	9.13
Error	450.48	8	56.31	
Total	8678.24	24		

6.10.3 SAS Program for Responses of Example 5.9 Using Effects Coding Model

As usual, we write a SAS program, carry out the analysis on the basis of the regression model, and present the ensuing output. Additionally, to illustrate the use of the relationships among the row, column, Latin letter, Greek letter, overall mean, and the parameters of the model, we use their sample values to estimate the parameters. We present the key to the coding of Latin and Greek letters in this program in the following table:

Latin letters	A	B	C	D	E
Greek letters	ω	ψ	θ	ρ	ϕ
Position in the design	1	2	3	4	5

SAS PROGRAM

```
data greeklat1;
input x1-x4 z1-z4 c1-c4 h1-h4 y;
datalines;
1 0 0 0 1 0 0 0 0 0 1 0 1 0 0 0 57
1 0 0 0 0 1 0 0 0 0 0 1 0 1 0 0 47
1 0 0 0 0 0 1 0 -1 -1 -1 -1 0 0 1 0 60
1 0 0 0 0 0 0 1 0 1 0 0 0 0 0 1 40
1 0 0 0 -1 -1 -1 -1 1 0 0 0 -1 -1 -1 -1 19
0 1 0 0 1 0 0 0 0 1 0 0 -1 -1 -1 -1 55
0 1 0 0 0 1 0 0 0 0 1 0 0 0 0 1 43
0 1 0 0 0 0 1 0 0 0 0 1 1 0 0 0 38
0 1 0 0 0 0 0 1 1 0 0 0 0 0 1 0 23
0 1 0 0 -1 -1 -1 -1 -1 -1 -1 -1 0 1 0 0 71
0 0 1 0 1 0 0 0 -1 -1 -1 -1 0 0 0 1 74
0 0 1 0 0 1 0 0 1 0 0 0 1 0 0 0 14
0 0 1 0 0 0 1 0 0 1 0 0 0 1 0 0 56
0 0 1 0 0 0 0 1 0 0 0 1 -1 -1 -1 -1 26
0 0 1 0 -1 -1 -1 -1 0 0 1 0 0 0 0 1 0 54
0 0 0 1 1 0 0 0 1 0 0 0 0 1 0 0 17
0 0 0 1 0 1 0 0 0 1 0 0 0 0 1 0 31
0 0 0 1 0 0 1 0 0 0 1 0 -1 -1 -1 -1 47
```

```
0 0 0 1 0 0 0 1 -1 -1 -1 -1 1 0 0 0 61
0 0 0 1 -1 -1 -1 -1 0 0 0 1 0 0 0 1 23
-1 -1 -1 -1 1 0 0 0 0 0 0 1 0 0 1 0 30
-1 -1 -1 -1 0 1 0 0 -1 -1 -1 -1 -1 -1 -1 -1 70
-1 -1 -1 -1 0 0 1 0 1 0 0 0 0 0 0 1 17
-1 -1 -1 -1 0 0 0 1 0 0 1 0 0 1 0 0 70
-1 -1 -1 -1 -1 -1 -1 -1 0 1 0 0 1 0 0 0 44
;
proc print data=greeklat1; sum x1-x4 z1-z4 c1-c4 h1-h4;
proc reg data=greeklat1; model y=x1-x4 z1-z4 c1-c4 h1-h4; run;
```

SAS OUTPUT

Obs	x1	x2	x3	x4	z1	z2	z3	z4	c1	c2	c3	c4	h1	h2	h3	h4	y
1	1	0	0	0	1	0	0	0	0	0	1	0	1	0	0	0	57
2	1	0	0	0	0	1	0	0	0	0	0	1	0	1	0	0	47
3	1	0	0	0	0	0	1	0	-1	-1	-1	-1	0	0	1	0	60
4	1	0	0	0	0	0	0	1	0	1	0	0	0	0	0	1	40
5	1	0	0	0	-1	-1	-1	-1	1	0	0	0	-1	-1	-1	-1	19
6	0	1	0	0	1	0	0	0	0	1	0	0	-1	-1	-1	-1	55
7	0	1	0	0	0	1	0	0	0	0	1	0	0	0	0	1	43
8	0	1	0	0	0	0	1	0	0	0	0	1	1	0	0	0	38
9	0	1	0	0	0	0	0	1	1	0	0	0	0	0	1	0	23
10	0	1	0	0	-1	-1	-1	-1	-1	-1	-1	-1	0	1	0	0	71
11	0	0	1	0	1	0	0	0	-1	-1	-1	-1	0	0	0	1	74
12	0	0	1	0	0	1	0	0	1	0	0	0	1	0	0	0	14
13	0	0	1	0	0	0	1	0	0	1	0	0	0	1	0	0	56
14	0	0	1	0	0	0	0	1	0	0	0	1	-1	-1	-1	-1	26
15	0	0	1	0	-1	-1	-1	-1	0	0	1	0	0	0	1	0	54
16	0	0	0	1	1	0	0	0	1	0	0	0	0	1	0	0	17
17	0	0	0	1	0	1	0	0	0	1	0	0	0	0	1	0	31
18	0	0	0	1	0	0	1	0	0	0	1	0	-1	-1	-1	-1	47
19	0	0	0	1	0	0	0	1	-1	-1	-1	-1	1	0	0	0	61
20	0	0	0	1	-1	-1	-1	-1	0	0	0	1	0	0	0	1	23
21	-1	-1	-1	-1	1	0	0	0	0	0	0	1	0	0	1	0	30
22	-1	-1	-1	-1	0	1	0	0	-1	-1	-1	-1	-1	-1	-1	-1	70
23	-1	-1	-1	-1	0	0	1	0	1	0	0	0	0	0	0	1	17
24	-1	-1	-1	-1	0	0	0	1	0	0	1	0	0	1	0	0	70
25	-1	-1	-1	-1	-1	-1	-1	-1	0	1	0	0	1	0	0	0	44
	==	==	==	==	==	==	==	==	==	==	==	==	==	==	==	==	
	0	0	0	0	0	0	0	0	0	0	0	0	0	0	0	0	

Analysis of Variance

Source	DF	Sum of Squares	Mean Square	F Value	Pr > F
Model	16	8227.76000	514.23500	9.13	0.0018
Error	8	450.48000	56.31000		
Corrected Total	24	8678.24000			

Root MSE	7.50400	R-Square	0.9481
Dependent Mean	43.48000	Adj R-Sq	0.8443
Coeff Var	17.25851		

Parameter Estimates

Variable	DF	Parameter Estimate	Standard Error	t-Value	Pr > |t|
Intercept	1	43.48000	1.50080	28.97	<.0001
x1	1	1.12000	3.00160	0.37	0.7187
x2	1	2.52000	3.00160	0.84	0.4255
x3	1	1.32000	3.00160	0.44	0.6717
x4	1	-7.68000	3.00160	-2.56	0.0337
z1	1	3.12000	3.00160	1.04	0.3290
z2	1	-2.48000	3.00160	-0.83	0.4326
z3	1	0.12000	3.00160	0.04	0.9691
z4	1	0.52000	3.00160	0.17	0.8668
c1	1	-25.48000	3.00160	-8.49	<.0001
c2	1	1.72000	3.00160	0.57	0.5824
c3	1	10.72000	3.00160	3.57	0.0073
c4	1	-10.68000	3.00160	-3.56	0.0074
h1	1	-0.68000	3.00160	-0.23	0.8265
h2	1	8.72000	3.00160	2.91	0.0197
h3	1	-3.88000	3.00160	-1.29	0.2322
h4	1	-4.08000	3.00160	-1.36	0.2111

The result of SAS analysis is the same as that obtained from manual analysis.

6.11 Regression Model for Graeco-Latin Squares (Reference Cell Method)

Next, we define the dummy variables for the regression model equivalent of the classical model for Graeco-Latin square design (Equation 6.20). To do this, we have to choose a particular cell by which we carry out the definition of the dummy variables in the model. The relevant regression model for the design is obtained as follows:

$$y_{ijkl} = \mu + \sum_{i=1}^{r-1} \alpha_i X_i + \sum_{j=1}^{r-1} \beta_j Z_j + \sum_{k=1}^{r-1} \gamma_k C_k + \sum_{l=1}^{r-1} \lambda_l H_l + \varepsilon \tag{6.29}$$

where

$$X_i = \begin{cases} 1 & \text{if } y_{ijkl} \text{ is under the } i\text{th row} \\ 0 & \text{otherwise} \end{cases}$$

for $i = 1, 2, \ldots, r-1$

$$Z_j = \begin{cases} 1 & \text{if } y_{ijkl} \text{ is under the } j\text{th column} \\ 0 & \text{otherwise} \end{cases}$$

for $j = 1, 2, \ldots, r-1$

$$C_k = \begin{cases} 1 & \text{if } y_{ijkl} \text{ is under the } k\text{th Latin letter} \\ 0 & \text{otherwise} \end{cases}$$

for $k = 1, 2, \ldots, r-1$

$$H_l = \begin{cases} 1 & \text{if } y_{ijkl} \text{ is under the } l\text{th Greek letter} \\ 0 & \text{otherwise} \end{cases}$$

for $l = 1, 2, \ldots, r-1$

6.11.1 Estimation of Model Parameters for Reference Cell Method

The parameters of the regression model can be estimated from sample data by using the following relationships:

$$\begin{aligned}
\mu_{ijkl} &= \mu + \alpha_i + \beta_j + \gamma_k + \lambda_l \quad i, j, k, l = 1, 2, \ldots, r-1 \\
\mu_{r...} &= \mu + \frac{1}{r}\sum_{j=1}^{r-1} \beta_j + \frac{1}{r}\sum_{k=1}^{r-1} \gamma_k + \frac{1}{r}\sum_{l=1}^{r-1} \lambda_l \\
\mu_{.r..} &= \mu + \frac{1}{r}\sum_{i=1}^{r-1} \alpha_i + \frac{1}{r}\sum_{k=1}^{r-1} \gamma_k + \frac{1}{r}\sum_{l=1}^{r-1} \lambda_l \\
\mu_{..r.} &= \mu + \frac{1}{r}\sum_{i=1}^{r-1} \alpha_i + \frac{1}{r}\sum_{j=1}^{r-1} \beta_j + \frac{1}{r}\sum_{l=1}^{r-1} \lambda_l \\
\mu_{...r} &= \mu + \frac{1}{r}\sum_{i=1}^{r-1} \alpha_i + \frac{1}{r}\sum_{j=1}^{r-1} \beta_j + \frac{1}{r}\sum_{k=1}^{r-1} \gamma_k \qquad (6.30) \\
\mu_{i...} &= \mu + \alpha_i + \frac{1}{r}\sum_{j=1}^{r-1} \beta_j + \frac{1}{r}\sum_{k=1}^{r-1} \gamma_k + \frac{1}{r}\sum_{l=1}^{r-1} \lambda_l \quad i = 1, 2, \ldots, r-1 \\
\mu_{.j..} &= \mu + \beta_j + \frac{1}{r}\sum_{i=1}^{r-1} \alpha_i + \frac{1}{r}\sum_{k=1}^{r-1} \gamma_k + \frac{1}{r}\sum_{l=1}^{r-1} \lambda_l \quad j = 1, 2, \ldots, r-1 \\
\mu_{..k.} &= \mu + \gamma_k + \frac{1}{r}\sum_{i=1}^{r-1} \alpha_i + \frac{1}{r}\sum_{j=1}^{r-1} \beta_j + \frac{1}{r}\sum_{l=1}^{r-1} \lambda_l \quad k = 1, 2, \ldots, r-1 \\
\mu_{...l} &= \mu + \lambda_l + \frac{1}{r}\sum_{i=1}^{r-1} \alpha_i + \frac{1}{r}\sum_{j=1}^{r-1} \beta_j + \frac{1}{r}\sum_{k=1}^{r-1} \gamma_k \quad l = 1, 2, \ldots, r-1
\end{aligned}$$

$$\mu = \mu_{....} - \frac{1}{r}\sum_{i=1}^{r-1} \alpha_i - \frac{1}{r}\sum_{j=1}^{r-1} \beta_j - \frac{1}{r}\sum_{k=1}^{r-1} \gamma_k - \frac{1}{r}\sum_{l=1}^{r-1} \lambda_l \qquad (6.31)$$

We reanalyze the data for Example 5.9 using the above model. As previously stated, this is a 5×5 Graeco-Latin square design. The model is based on the reference cell with the value y_{5521}, the observation belonged to row 5, column 5, Latin letter A, and Greek letter ψ. Estimating the parameters using the relations stated (Equations 6.30 and 6.31) for the model, we can easily show that

$$\begin{aligned}
\mu_{i...} &= \alpha_i + \mu_{r...} \quad i = 1,2,3,4 \\
\mu_{.j..} &= \beta_j + \mu_{.r..} \quad j = 1,2,3,4 \\
\mu_{..k.} &= \gamma_k + \mu_{..r.} \quad k = 1,3,4,5 \\
\mu_{...l} &= \lambda_l + \mu_{...r} \quad l = 2,3,4,5
\end{aligned}$$

6.11.2 Estimation of the Parameters of the Model Applied to Example 5.9

Thus, using the sample means in place of the cell, row, column, Latin letter, and Greek letter means, we can estimate the parameters as follows:

$$\begin{aligned}
&\bar{y}_{5...} = 46.2 \\
&\bar{y}_{1...} = \alpha_1 + \bar{y}_{5...} \Rightarrow 44.6 = \alpha_1 + 46.2 \Rightarrow \hat{\alpha}_1 = 44.6 - 46.2 = -1.60 \\
&\bar{y}_{2...} = \alpha_2 + \bar{y}_{5...} \Rightarrow 46 = \alpha_2 + 46.2 \Rightarrow \hat{\alpha}_2 = 46 - 46.2 = -0.20 \\
&\bar{y}_{3...} = \alpha_3 + \bar{y}_{5...} \Rightarrow 44.8 = \alpha_3 + 46.2 \Rightarrow \hat{\alpha}_3 = 44.8 - 46.2 = -1.40 \\
&\bar{y}_{4...} = \alpha_4 + \bar{y}_{5...} \Rightarrow 35.8 = \alpha_4 + 46.2 \Rightarrow \hat{\alpha}_4 = 35.8 - 46.2 = -10.40
\end{aligned}$$

$$\begin{aligned}
&\bar{y}_{.5..} = 42.2 \\
&\bar{y}_{.1..} = \beta_1 + \bar{y}_{.5..} \Rightarrow 46.6 = \beta_1 + 42.2 \Rightarrow \hat{\beta}_1 = 46.6 - 42.2 = 4.4 \\
&\bar{y}_{.2..} = \beta_2 + \bar{y}_{.5..} \Rightarrow 41 = \beta_2 + 42.2 \Rightarrow \hat{\beta}_2 = 41 - 42.2 = -1.20 \\
&\bar{y}_{.3..} = \beta_3 + \bar{y}_{.5..} \Rightarrow 43.6 = \beta_3 + 42.2 \Rightarrow \hat{\beta}_3 = 43.6 - 42.2 = 1.40 \\
&\bar{y}_{.4..} = \beta_4 + \bar{y}_{.5..} \Rightarrow 44 = \beta_4 + 42.2 \Rightarrow \hat{\beta}_4 = 44 - 42.2 = 1.80
\end{aligned}$$

$$\begin{aligned}
&\bar{y}_{..2.} = 45.2 \\
&\bar{y}_{..1.} = \gamma_1 + \bar{y}_{..2.} \Rightarrow 18 = \gamma_1 + 45.2 \Rightarrow \hat{\gamma}_1 = 18 - 45.2 = -27.2 \\
&\bar{y}_{..3.} = \gamma_3 + \bar{y}_{..2.} \Rightarrow 54.2 = \gamma_3 + 45.2 \Rightarrow \hat{\gamma}_3 = 54.2 - 45.2 = 9.0 \\
&\bar{y}_{..4.} = \gamma_4 + \bar{y}_{..2.} \Rightarrow 32.8 = \gamma_4 + 45.2 \Rightarrow \hat{\gamma}_4 = 32.8 - 45.2 = -12.40 \\
&\bar{y}_{..5.} = \gamma_5 + \bar{y}_{..2.} \Rightarrow 67.2 = \gamma_5 + 45.2 \Rightarrow \hat{\gamma}_5 = 67.2 - 45.2 = 22.0
\end{aligned}$$

$$\begin{aligned}
&\bar{y}_{...1} = 42.8 \\
&\bar{y}_{...2} = \lambda_2 + \bar{y}_{...1} \Rightarrow 52.2 = \lambda_2 + 42.8 \Rightarrow \hat{\lambda}_2 = 18 - 42.8 = 9.4 \\
&\bar{y}_{...3} = \lambda_3 + \bar{y}_{...1} \Rightarrow 39.6 = \lambda_3 + 42.8 \Rightarrow \hat{\lambda}_3 = 54.2 - 42.8 = -3.2 \\
&\bar{y}_{...4} = \lambda_4 + \bar{y}_{...1} \Rightarrow 39.4 = \lambda_4 + 42.8 \Rightarrow \hat{\lambda}_4 = 32.8 - 42.8 = -3.40 \\
&\bar{y}_{...5} = \lambda_5 + \bar{y}_{...1} \Rightarrow 43.4 = \lambda_5 + 42.8 \Rightarrow \hat{\lambda}_5 = 67.2 - 42.8 = 0.60
\end{aligned}$$

The above estimation was based on the assumption that the reference cell was the one with the observation y_{5521}.

The fitted regression model for the data of Example 5.5 is

$$\begin{aligned}
\hat{y} = 45.96 &- 1.6x_1 - 0.2x_2 - 1.4x_3 - 10.4x_4 + 4.4z_1 - 1.2z_2 + 1.4z_3 + 1.8z_4 \\
&- 27.2c_1 + 9c_2 - 12.4c_3 + 22c_4 + 9.4h_1 - 3.2h_2 - 3.4h_3 + 0.6h_4 \quad (6.32)
\end{aligned}$$

TABLE 6.22: Estimated Responses and Residuals for Example 5.9 (Reference Cell Coding Method)

x_1	x_2	x_3	x_4	z_1	z_2	z_3	z_4	c_1	c_2	c_3	c_4	h_1	h_2	h_3	h_4	y	$\hat{y}$	$y-\hat{y}$	$(y-\hat{y})^2$
1	0	0	0	1	0	0	0	0	1	0	0	0	0	0	0	57	57.76	−0.76	0.5776
1	0	0	0	0	1	0	0	0	0	1	0	1	0	0	0	47	40.16	6.84	46.7856
1	0	0	0	0	0	1	0	0	0	0	1	0	1	0	0	60	64.56	−4.56	20.7936
1	0	0	0	0	0	0	1	0	0	0	0	0	0	1	0	40	42.76	−2.76	7.6176
1	0	0	0	0	0	0	0	1	0	0	0	0	0	0	1	19	17.76	1.24	1.5376
0	1	0	0	1	0	0	0	0	0	0	0	0	0	0	1	55	50.76	4.24	17.9776
0	1	0	0	0	1	0	0	0	1	0	0	0	0	1	0	43	50.16	−7.16	51.2656
0	1	0	0	0	0	1	0	0	0	1	0	0	0	0	0	38	34.76	3.24	10.4976
0	1	0	0	0	0	0	1	1	0	0	0	0	1	0	0	23	17.16	5.84	34.1056
0	1	0	0	0	0	0	0	0	0	0	1	1	0	0	0	71	77.16	−6.16	37.9456
0	0	1	0	1	0	0	0	0	0	0	1	0	0	1	0	74	67.56	6.44	41.4736
0	0	1	0	0	1	0	0	1	0	0	0	0	0	0	0	14	16.16	−2.16	4.6656
0	0	1	0	0	0	1	0	0	0	0	0	1	0	0	0	56	55.36	0.64	0.4096
0	0	1	0	0	0	0	1	0	0	1	0	0	0	0	1	26	34.56	−8.56	73.2736
0	0	1	0	0	0	0	0	0	1	0	0	0	1	0	0	54	50.36	3.64	13.2496
0	0	0	1	1	0	0	0	1	0	0	0	1	0	0	0	17	22.16	−5.16	26.6256
0	0	0	1	0	1	0	0	0	0	0	0	0	1	0	0	31	31.16	−0.16	0.0256
0	0	0	1	0	0	1	0	0	1	0	0	0	0	0	1	47	46.56	0.44	0.1936
0	0	0	1	0	0	0	1	0	0	0	1	0	0	0	0	61	59.36	1.64	2.6896
0	0	0	1	0	0	0	0	0	0	1	0	0	0	1	0	23	19.76	3.24	10.4976
0	0	0	0	1	0	0	0	0	0	1	0	0	1	0	0	30	34.76	−4.76	22.6576
0	0	0	0	0	1	0	0	0	0	0	1	0	0	0	1	70	67.36	2.64	6.9696
0	0	0	0	0	0	1	0	1	0	0	0	0	0	1	0	17	16.76	0.24	0.0576
0	0	0	0	0	0	0	1	0	1	0	0	1	0	0	0	70	66.16	3.84	14.7456
0	0	0	0	0	0	0	0	0	0	0	0	0	0	0	0	44	45.96	−1.96	3.8416
Total																1087	1087	0	450.48

Table 6.22 shows the estimated responses, and the extracted residuals from which we calculate some quantities, which we use later for testing the hypothesis that the fitted regression model is significant (Table 6.23).

If $S_{yy} = \sum_{i=1}^{N} y_i^2 - \left(\sum_{i=1}^{N} y_i\right)^2/N$, $SS_{\mathrm{R}} = \sum_{i=1}^{N} (\hat{y}_i - \overline{y})^2$, and $SS_{\mathrm{E}} = \sum_{i=1}^{N} (y_i - \hat{y}_i)^2$ then $S_{yy} = SS_{\mathrm{R}} + SS_{\mathrm{E}}$.

$$SS_{\mathrm{E}} = 450.48$$
$$S_{yy} = 55941 - \frac{1087^2}{25} = 8678.24 \Rightarrow SS_{\mathrm{R}} = 8678.24 - 450.48 = 8227.76$$

TABLE 6.23: ANOVA to Test for Significance of the Fitted Regression Model for Example 5.9

Source	**Sum of squares**	**Degrees of freedom**	**Mean square**	***F*-ratio**
Regression	8227.76	16	514.235	9.13
Error	450.48	8	56.31	
Total	8678.24	24		

We see that $F(16, 8, 0.01) = 5.47655$, indicating that the fitted model is highly significant at 1% level of significance.

6.11.3 SAS Program for Dummy Regression Analysis of Responses of Example 5.9 Using Reference Cell Method

If we choose the reference cell as the one with the observation y_{5521}, then four sets of dummy variables are defined, each with $r - 1 = 4$ dummy variables. Under this definition we code the variables as follows:

Latin letters	A	B	C	D	E
Code	1 0 0 0	0 0 0 0	0 1 0 0	0 0 1 0	0 0 0 1
Greek letters	ω	ψ	θ	ρ	ϕ
Code	1 0 0 0	0 0 0 0	0 1 0 0	0 0 1 0	0 0 0 1

With the above codes, we write a SAS program to analyze the data. This time, we store the residuals for the fitted responses, and obtain their normal probability plot to test the normality of the errors.

SAS PROGRAM

```
data grklat2;
input x1-x4 z1-z4 c1-c4 h1-h4 y@@;
datalines;
1 0 0 0 1 0 0 0 0 1 0 0 0 0 0 0 57 1 0 0 0 0 1 0 0 0 0 1 0 1 0 0 0 47
1 0 0 0 0 0 1 0 0 0 0 1 0 1 0 0 60 1 0 0 0 0 0 0 1 0 0 0 0 0 0 1 0 40
1 0 0 0 0 0 0 0 1 0 0 0 0 0 0 1 19 0 1 0 0 1 0 0 0 0 0 0 0 0 0 0 1 55
0 1 0 0 0 1 0 0 0 1 0 0 0 0 1 0 43 0 1 0 0 0 0 1 0 0 0 1 0 0 0 0 0 38
0 1 0 0 0 0 0 1 1 0 0 0 0 1 0 0 23 0 1 0 0 0 0 0 0 0 0 0 1 1 0 0 0 71
0 0 1 0 1 0 0 0 0 0 0 1 0 0 1 0 74 0 0 1 0 0 1 0 0 1 0 0 0 0 0 0 0 14
0 0 1 0 0 0 1 0 0 0 0 0 1 0 0 0 56 0 0 1 0 0 0 0 1 0 0 1 0 0 0 0 1 26
0 0 1 0 0 0 0 0 0 1 0 0 0 1 0 0 54 0 0 0 1 1 0 0 0 1 0 0 0 1 0 0 0 17
0 0 0 1 0 1 0 0 0 0 0 0 0 1 0 0 31 0 0 0 1 0 0 1 0 0 1 0 0 0 0 0 1 47
0 0 0 1 0 0 0 1 0 0 0 1 0 0 0 0 61 0 0 0 1 0 0 0 0 0 0 1 0 0 0 1 0 23
0 0 0 0 1 0 0 0 0 0 1 0 0 1 0 0 30 0 0 0 0 0 1 0 0 0 0 0 1 0 0 0 1 70
0 0 0 0 0 0 1 0 1 0 0 0 0 0 1 0 17 0 0 0 0 0 0 0 1 0 1 0 0 1 0 0 0 70
0 0 0 0 0 0 0 0 0 0 0 0 0 0 0 0 44
;
proc print data=grklat2; sum x1-x4 z1-z4 c1-c4 h1-h4;
proc reg data=grklat2;
model y=x1-x4 z1-z4 c1-c4 h1-h4;output out=examp55 r=resid
p=pred;run;
ods;
proc univariate data=examp55 normaltest;var resid;
        probplot resid / normal (mu=est sigma=est)square;
        label resid = 'Residuals';
        inset mean std / format=6.4;
      run;
```

The ensuing output from the SAS program is presented below:

SAS OUTPUT

Obs	x1	x2	x3	x4	z1	z2	z3	z4	c1	c2	c3	c4	h1	h2	h3	h4	y
1	1	0	0	0	1	0	0	0	0	1	0	0	0	0	0	0	57
2	1	0	0	0	0	1	0	0	0	0	1	0	1	0	0	0	47
3	1	0	0	0	0	0	1	0	0	0	0	1	0	1	0	0	60
4	1	0	0	0	0	0	0	1	0	0	0	0	0	0	1	0	40
5	1	0	0	0	0	0	0	0	1	0	0	0	0	0	0	1	19
6	0	1	0	0	1	0	0	0	0	0	0	0	0	0	0	1	55
7	0	1	0	0	0	1	0	0	0	1	0	0	0	0	1	0	43
8	0	1	0	0	0	0	1	0	0	0	1	0	0	0	0	0	38
9	0	1	0	0	0	0	0	1	1	0	0	0	0	1	0	0	23
10	0	1	0	0	0	0	0	0	0	0	0	1	1	0	0	0	71
11	0	0	1	0	1	0	0	0	0	0	0	1	0	0	1	0	74
12	0	0	1	0	0	1	0	0	1	0	0	0	0	0	0	0	14
13	0	0	1	0	0	0	1	0	0	0	0	0	1	0	0	0	56
14	0	0	1	0	0	0	0	1	0	0	1	0	0	0	0	1	26

Obs	x1	x2	x3	x4	z1	z2	z3	z4	c1	c2	c3	c4	h1	h2	h3	h4	y
15	0	0	1	0	0	0	0	0	0	1	0	0	0	1	0	0	54
16	0	0	0	1	1	0	0	0	1	0	0	0	1	0	0	0	17
17	0	0	0	1	0	1	0	0	0	0	0	0	0	1	0	0	31
18	0	0	0	1	0	0	1	0	0	1	0	0	0	0	0	1	47
19	0	0	0	1	0	0	0	1	0	0	0	1	0	0	0	0	61
20	0	0	0	1	0	0	0	0	0	0	1	0	0	0	1	0	23
21	0	0	0	0	1	0	0	0	0	0	1	0	0	1	0	0	30
22	0	0	0	0	0	1	0	0	0	0	0	1	0	0	0	1	70
23	0	0	0	0	0	0	1	0	1	0	0	0	0	0	1	0	17
24	0	0	0	0	0	0	0	1	0	1	0	0	1	0	0	0	70
25	0	0	0	0	0	0	0	0	0	0	0	0	0	0	0	0	44
	==	==	==	==	==	==	==	==	==	==	==	==	==	==	==	==	
	5	5	5	5	5	5	5	5	5	5	5	5	5	5	5	5	

Analysis of Variance

Source	DF	Sum of Squares	Mean Square	F Value	Pr > F
Model	16	8227.76000	514.23500	9.13	0.0018
Error	8	450.48000	56.31000		
Corrected Total	24	8678.24000			

Root MSE	7.50400	R-Square	0.9481
Dependent Mean	43.48000	Adj R-Sq	0.8443
Coeff Var	17.25851		

Parameter Estimates

Variable	DF	Parameter Estimate	Standard Error	t-Value	Pr > |t|
Intercept	1	45.96000	6.18796	7.43	<. 0001
x1	1	-1.60000	4.74595	-0.34	0.7447
x2	1	-0.20000	4.74595	-0.04	0.9674
x3	1	-1.40000	4.74595	-0.29	0.7755
x4	1	-10.40000	4.74595	-2.19	0.0598
z1	1	4.40000	4.74595	0.93	0.3810
z2	1	-1.20000	4.74595	-0.25	0.8068
z3	1	1.40000	4.74595	0.29	0.7755
z4	1	1.80000	4.74595	0.38	0.7144
c1	1	-27.20000	4.74595	-5.73	0.0004
c2	1	9.00000	4.74595	1.90	0.0945
c3	1	-12.40000	4.74595	-2.61	0.0310
c4	1	22.00000	4.74595	4.64	0.0017
h1	1	9.40000	4.74595	1.98	0.0830
h2	1	-3.20000	4.74595	-0.67	0.5192
h3	1	-3.40000	4.74595	-0.72	0.4941
h4	1	0.60000	4.74595	0.13	0.9025

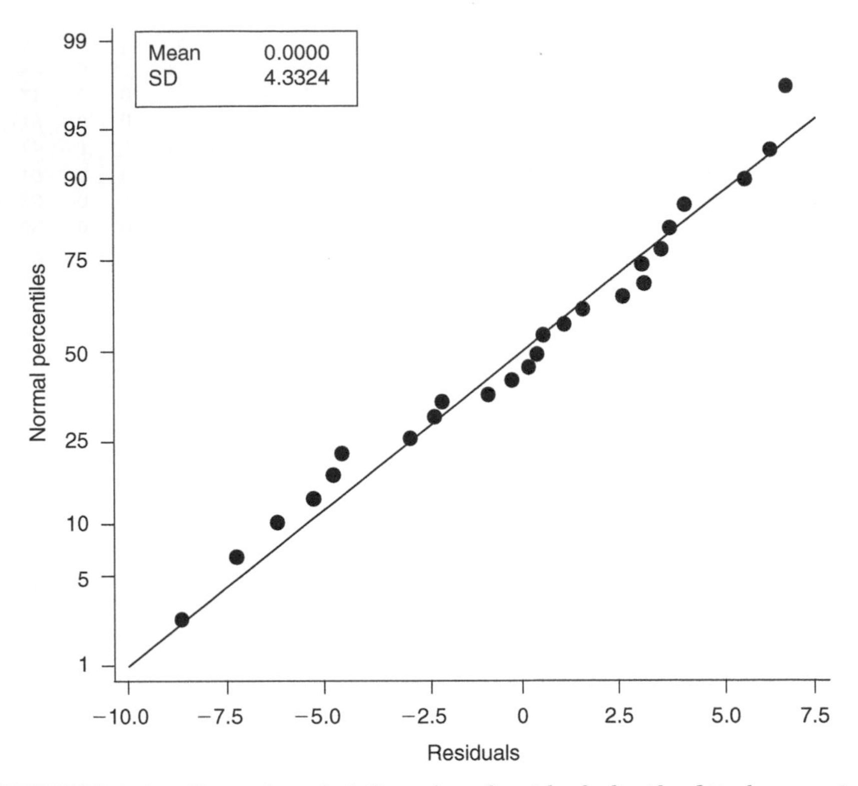

FIGURE 6.1: Normal probability plot of residuals for the fitted regression model for data of Example 5.9 (produced by SAS).

From the normal probability plot for the residuals (Figure 6.1), it appears that there is no violation of the assumption that the residuals are normally independently distributed with zero mean.

6.12 Regression Model for the RCBD Using Orthogonal Contrasts (Example 5.1)

In the regression model for the ANOVA for the RCBD, we can use orthogonal contrasts as weighting coefficients to leave out some levels of a variable, separate some levels of a variable for comparison, or divide the levels of a factor into groups in order to compare their means. The form of the model is

$$Y = \mu + \sum_{i=1}^{k-1} \alpha_i X_i + \sum_{j=1}^{b-1} \beta_j Z_j + \varepsilon \tag{6.33}$$

if there are k levels of treatment and b levels of block in the designed experiment. Thus, for k levels of treatment and b levels of block, $k-1$ and $b-1$ orthogonal contrasts, respectively, need to be defined for a regression model. The number of orthogonal contrasts to be defined for each factor is equal to the number of degrees of freedom for that factor minus one.

We can estimate the parameters of the regression model using the following relations:

$$\hat{\alpha}_i = \frac{\sum X_i Y_i}{\sum X_i^2} \quad \hat{\beta}_j = \frac{\sum Z_j Y_i}{\sum Z_j^2} \tag{6.34}$$

With these, we estimate the parameters for the model 6.33 as follows:

$$\hat{\alpha}_1 = \frac{-142}{60} = -2.3667 \quad \hat{\alpha}_2 = \frac{-86}{36} = -2.38889$$

$$\hat{\alpha}_3 = \frac{55}{18} = 3.05556 \quad \hat{\alpha}_4 = \frac{-15}{6} = 2.5$$

$$\hat{\beta}_1 = \frac{18}{30} = 0.6 \quad \hat{\beta}_2 = \frac{106}{10} = 10.6$$

The estimated regression equation for the responses of this experiment using the orthogonal contrasts (x_1–x_4) and (z_1–z_2) in Table 6.24 is

$$\hat{y} = 34.8 - 2.3667x_1 - 2.38889x_2 + 3.05556x_3 - 2.5x_4 + 0.6z_1 + 10.6z_2 \tag{6.35}$$

With the help of this estimated equation, we obtain the estimated responses, and extract the residuals for the fitted model.

TABLE 6.24: Fitted Values and Residuals for Model 6.33 for Example 5.1

x_1	x_2	x_3	x_4	z_1	z_2	y	yx_1	yx_2	yx_3	yx_4	yz_1	yz_2	d_1	d_2	d_3	d_4	h_1	h_2	$\hat{y}$	$(y-\hat{y})$	$(y-\hat{y})^2$
4	0	0	0	1	0	23	92	0	0	0	23	0	16	0	0	0	1	0	26.5	−3.5333	12.484
4	0	0	0	0	1	34	136	0	0	0	0	34	16	0	0	0	0	1	35.3	−1.3333	1.7777
4	0	0	0	−1	−1	19	76	0	0	0	−19	−19	16	0	0	0	1	1	14.1	4.8667	23.684
−1	3	0	0	1	0	31	−31	93	0	0	31	0	1	9	0	0	1	0	31.2	−0.2	0.04
−1	3	0	0	0	1	37	−37	111	0	0	0	37	1	9	0	0	0	1	40	−3	9
−1	3	0	0	−1	−1	22	−22	66	0	0	−22	−22	1	9	0	0	1	1	18.8	3.2	10.24
−1	−1	2	0	1	0	52	−52	−52	104	0	52	0	1	1	4	0	1	0	46.9	5.1333	26.351
−1	−1	2	0	0	1	69	−69	−69	138	0	0	69	1	1	4	0	0	1	55.7	13.333	177.77
−1	−1	2	0	−1	−1	16	−16	−16	32	0	−16	−16	1	1	4	0	1	1	34.5	−18.467	341.01
−1	−1	−1	1	1	0	35	−35	−35	−35	35	35	0	1	1	1	1	1	0	35.2	−0.2	0.04
−1	−1	−1	1	0	1	37	−37	−37	−37	37	0	37	1	1	1	1	0	1	44	−7	49
−1	−1	−1	1	−1	−1	30	−30	−30	−30	30	−30	−30	1	1	1	1	1	1	22.8	7.2	51.84
−1	−1	−1	−1	1	0	39	−39	−39	−39	−39	39	0	1	1	1	1	1	0	40.2	−1.2	1.44
−1	−1	−1	−1	0	1	47	−47	−47	−47	−47	0	47	1	1	1	1	0	1	49	−2	4
−1	−1	−1	−1	−1	−1	31	−31	−31	−31	−31	−31	−31	1	1	1	1	1	1	27.8	3.2	10.24
0	0	0	0	0	0	522	−142	−86	55	−15	62	106	60	36	18	6	10	10	522	0	718.93

With the residuals extracted, we can use different residual plots which we discussed in Chapter 2 to check whether model assumptions are satisfied. Meanwhile, in order to perform ANOVA for the responses of the above experiment, we need the sums of squares for the contrasts.

$$SS_{X_i} = \frac{(\sum X_i Y_i)^2}{\sum X_i^2} \quad SS_{Z_j} = \frac{(\sum Z_j Y_i)^2}{\sum Z_j^2}$$

Similar results for orthogonal contrasts had been stated in Equations 2.40 and 2.41. Their use in ANOVA had been demonstrated in the analysis of data of Examples 2.2 and 2.3. Thus, using figures from Table 6.24, we obtain the sums of squares and cross products needed for ANOVA as shown in Table 6.25.

TABLE 6.25: Sums of Squares and Cross Products for Example 5.1

Sum of squares and cross products	X_1	X_2	X_3	X_4	Z_1	Z_2
$\sum X_i$ or $\sum Z_j$	0	0	0	0	0	0
$\sum X_i Y_i$	−142	−86	55	−15	62	106
$\sum X_i^2$ or $\sum Z_j^2$	60	36	18	6	30	10

We present the values for the sums of squares and cross products in Table 6.26.

From Table 6.26, we obtain the sums of squares of treatment and block in the experiment as follows (Table 6.27):

TABLE 6.26: Use of Orthogonal Contrasts to Obtain Sums of Squares for ANOVA Example 5.1

Contrast	**Sum of squares**	**Contrast**	**Sum of squares**
X_1	$\frac{(-142)^2}{60} = 336.06667$	X_4	$\frac{(-15)^2}{6} = 37.5$
X_2	$\frac{(-86)^2}{36} = 205.4444$	Z_1	$\frac{18^2}{30} = 10.8$
X_3	$\frac{55^2}{18} = 168.05556$	Z_2	$\frac{23^2}{72} = 1123.6$
$SS_{\text{total}} = \sum Y^2 - \frac{(\sum Y)^2}{N}$	$20766 = \frac{522^2}{15} = 2600.4$		

$$\begin{aligned} SS_{\text{temperature(treatment)}} &= SS_{X_1} + SS_{X_2} + SS_{X_3} + SS_{X_4} \\ &= 336.06667 + 205.4444 + 168.05556 + 37.5 \\ &= 747.06663 \end{aligned}$$

$$\begin{aligned} SS_{\text{algae source(blocks)}} &= SS_{Z_1} + SS_{Z_2} = 10.8 + 1123.6 = 1134.4 \\ SS_{\text{E}} &= SS_{\text{total}} - SS_{\text{algae source}} - SS_{\text{temperature}} \\ &= 2600.4 - 1134.4 - 747.0666 = 718.93337 \end{aligned}$$

TABLE 6.27: ANOVA for Regression Model of the Algae Experiment Based on Orthogonal Contrast

Source	Sum of squares	Degrees of freedom	Mean square	F-ratio
Algae source	1134.4	2	567.2	
Temperature	747.06663	4	186.76666	2.0783
Error	718.93337	8	89.866667	
Total	2600.4	14		

From the tables we see that $F(4, 8, 0.05) = 3.84$, indicating that the temperature is not significant at 5% level of significance. We note that the mean square for algae source is relatively large and conclude that blocking was effective.

6.13 Regression Model for RCBD Using Orthogonal Contrasts (Example 5.2)

Here again, we use orthogonal contrasts to obtain a regression model for the responses of the RCBD described in Example 5.2. The specific model that applies is of the form

$$y_{ij} = \mu + \sum_{i=1}^{3} \alpha_i X_i + \sum_{j=1}^{7} \beta_j Z_j + \varepsilon \tag{6.36}$$

The model takes into account four levels of treatment and eight levels of block in the design.

We proceed to fit the model using methods and formulae described for a similar model for Example 5.1 as shown in Table 6.28. The estimates for the parameters of the model are

$$\hat{\alpha}_1 = \frac{31}{96} = 0.3229 \quad \hat{\alpha}_2 = \frac{7}{48} = 0.1458 \quad \hat{\alpha}_3 = \frac{7}{16} = 0.4375$$

$$\hat{\beta}_1 = \frac{55}{224} = 0.2455 \quad \hat{\beta}_2 = \frac{-85}{168} = -0.506 \quad \hat{\beta}_3 = \frac{83}{120} = 0.6917$$

$$\hat{\beta}_4 = \frac{28}{80} = 0.35 \quad \hat{\beta}_5 = \frac{-40}{48} = -0.8333 \quad \hat{\beta}_6 = \frac{5}{24} = 0.2083$$

$$\hat{\beta}_7 = \frac{1}{8} = 0.125$$

The fitted equation relating responses to the orthogonal contrasts is

$$\begin{aligned}\hat{y} = {} & 0.3229x_1 + 0.1458x_2 + 0.4375x_3 + 0.2455z_1 - 0.506z_2 \\ & + 0.6917z_3 + 0.35z_4 - 0.8333z_5 + 0.2083z_6 + 0.125z_7 \end{aligned} \qquad (6.37)$$

TABLE 6.28: Fitting a Regression Model to Responses of Example 5.2 Using Orthogonal Contrasts

x_1	x_2	x_3	z_1	z_2	z_3	z_4	z_5	z_6	z_7	y	yx_1	yx_2	yx_3	yz_1	yz_2
3	0	0	7	0	0	0	0	0	0	8	24	0	0	56	0
3	0	0	−1	6	0	0	0	0	0	6	18	0	0	−6	36
3	0	0	−1	−1	5	0	0	0	0	13	39	0	0	−13	−13
3	0	0	−1	−1	−1	4	0	0	0	7	21	0	0	−7	−7
3	0	0	−1	−1	−1	−1	3	0	0	4	12	0	0	−4	−4
3	0	0	−1	−1	−1	−1	−1	2	0	13	39	0	0	−13	−13
3	0	0	−1	−1	−1	−1	−1	−1	1	2	6	0	0	−2	−2
3	0	0	−1	−1	−1	−1	−1	−1	−1	9	27	0	0	−9	−9
−1	2	0	7	0	0	0	0	0	0	8	−8	16	0	56	0
−1	2	0	−1	6	0	0	0	0	0	3	−3	6	0	−3	18
−1	2	0	−1	−1	5	0	0	0	0	10	−10	20	0	−10	−10
−1	2	0	−1	−1	−1	4	0	0	0	8	−8	16	0	−8	−8
−1	2	0	−1	−1	−1	−1	3	0	0	5	−5	10	0	−5	−5
−1	2	0	−1	−1	−1	−1	−1	2	0	10	−10	20	0	−10	−10
−1	2	0	−1	−1	−1	−1	−1	−1	1	7	−7	14	0	−7	−7
−1	2	0	−1	−1	−1	−1	−1	−1	−1	3	−3	6	0	−3	−3
−1	−1	1	7	0	0	0	0	0	0	10	−10	−10	10	70	0
−1	−1	1	−1	6	0	0	0	0	0	2	−2	−2	2	−2	12
−1	−1	1	−1	−1	5	0	0	0	0	9	−9	−9	9	−9	−9
−1	−1	1	−1	−1	−1	4	0	0	0	10	−10	−10	10	−10	−10
−1	−1	1	−1	−1	−1	−1	3	0	0	2	−2	−2	2	−2	−2
−1	−1	1	−1	−1	−1	−1	−1	2	0	3	−3	−3	3	−3	−3
−1	−1	1	−1	−1	−1	−1	−1	−1	1	10	−10	−10	10	−10	−10
−1	−1	1	−1	−1	−1	−1	−1	−1	−1	8	−8	−8	8	−8	−8
−1	−1	−1	7	0	0	0	0	0	0	8	−8	−8	−8	56	0
−1	−1	−1	−1	6	0	0	0	0	0	3	−3	−3	−3	−3	18
−1	−1	−1	−1	−1	5	0	0	0	0	10	−10	−10	−10	−10	−10
−1	−1	−1	−1	−1	−1	4	0	0	0	6	−6	−6	−6	−6	−6
−1	−1	−1	−1	−1	−1	−1	3	0	0	3	−3	−3	−3	−3	−3
−1	−1	−1	−1	−1	−1	−1	−1	2	0	3	−3	−3	−3	−3	−3
−1	−1	−1	−1	−1	−1	−1	−1	−1	1	8	−8	−8	−8	−8	−8
−1	−1	−1	−1	−1	−1	−1	−1	−1	−1	6	−6	−6	−6	−6	−6
Total										217	31	7	7	55	−85

Continued

TABLE 6.28: Continued

yz_3	yz_4	yz_5	yz_6	yz_7	x_1^2	x_2^2	x_3^2	z_1^2	z_2^2	z_3^2	z_4^2	z_5^2	z_6^2	z_7^2
0	0	0	0	0	9	0	0	49	0	0	0	0	0	0
0	0	0	0	0	9	0	0	1	36	0	0	0	0	0
65	0	0	0	0	9	0	0	1	1	25	0	0	0	0
−7	28	0	0	0	9	0	0	1	1	1	16	0	0	0
−4	−4	12	0	0	9	0	0	1	1	1	1	9	0	0
−13	−13	−13	26	0	9	0	0	1	1	1	1	1	4	0
−2	−2	−2	−2	2	9	0	0	1	1	1	1	1	1	1
−9	−9	−9	−9	−9	9	0	0	1	1	1	1	1	1	1
0	0	0	0	0	1	4	0	49	0	0	0	0	0	0
0	0	0	0	0	1	4	0	1	36	0	0	0	0	0
50	0	0	0	0	1	4	0	1	1	25	0	0	0	0
−8	32	0	0	0	1	4	0	1	1	1	16	0	0	0
−5	−5	15	0	0	1	4	0	1	1	1	1	9	0	0
−10	−10	−10	20	0	1	4	0	1	1	1	1	1	4	0
−7	−7	−7	−7	7	1	4	0	1	1	1	1	1	1	1
−3	−3	−3	−3	−3	1	4	0	1	1	1	1	1	1	1
0	0	0	0	0	1	1	1	49	0	0	0	0	0	0
0	0	0	0	0	1	1	1	1	36	0	0	0	0	0
45	0	0	0	0	1	1	1	1	1	25	0	0	0	0
−10	40	0	0	0	1	1	1	1	1	1	16	0	0	0
−2	−2	6	0	0	1	1	1	1	1	1	1	9	0	0
−3	−3	−3	6	0	1	1	1	1	1	1	1	1	4	0
−10	−10	−10	−10	10	1	1	1	1	1	1	1	1	1	1
−8	−8	−8	−8	−8	1	1	1	1	1	1	1	1	1	1
0	0	0	0	0	1	1	1	49	0	0	0	0	0	0
0	0	0	0	0	1	1	1	1	36	0	0	0	0	0
50	0	0	0	0	1	1	1	1	1	25	0	0	0	0
−6	24	0	0	0	1	1	1	1	1	1	16	0	0	0
−3	−3	9	0	0	1	1	1	1	1	1	1	9	0	0
−3	−3	−3	6	0	1	1	1	1	1	1	1	1	4	0
−8	−8	−8	−8	8	1	1	1	1	1	1	1	1	1	1
−6	−6	−6	−6	−6	1	1	1	1	1	1	1	1	1	1
83	28	−40	5	1	96	48	16	224	168	120	80	48	24	8

To carry out ANOVA, we calculate the sums of squares using values obtained in Table 6.29:

$$
\begin{aligned}
SS_{\text{fertilizer}} &= SS_{X_1} + SS_{X_2} + SS_{X_3} = 10.01 + 1.02 + 3.06 = 14.09 \\
SS_{\text{block}} &= SS_{Z_1} + SS_{Z_2} + \cdots + SS_{Z_7} \\
&= 13.5045 + 43.0 + 57.408 + 9.8 + 33.33 + 1.0417 + 0.125 \\
&= 158.1845 \\
SS_{\text{E}} &= SS_{\text{total}} - SS_{\text{fertilizer}} - SS_{\text{block}} \\
&= 325.4688 - 14.09 - 158.1845 = 153.1943
\end{aligned}
$$

The results of ANOVA (Table 6.30) agree with that of previous analysis carried out on the same data elsewhere in the chapter.

TABLE 6.29: Use of Orthogonal Contrasts to Obtain Sums of Squares for ANOVA Example 5.2

Contrast	Sum of squares	Contrast	Sum of squares
X_1	$\frac{31^2}{96} = 10.01$	Z_3	$\frac{83^2}{120} = 57.408$
X_2	$\frac{7^2}{48} = 1.02$	Z_4	$\frac{28^2}{80} = 9.8$
X_3	$\frac{7^2}{16} = 3.06$	Z_5	$\frac{(-40)^2}{48} = 33.33$
Z_1	$\frac{55^2}{224} = 13.5045$	Z_6	$\frac{5^2}{24} = 1.0417$
Z_2	$\frac{(-85)^2}{168} = 43.00$	Z_7	$\frac{1^2}{8} = 0.125$
$SS_{\text{total}} = \sum Y^2 - \frac{(\sum Y)^2}{N}$	$1797 - \frac{217^2}{32} = 325.4688$		

TABLE 6.30: ANOVA for Regression Model of the Algae Experiment Based on Orthogonal Contrasts

Source	Sum of squares	Degrees of freedom	Mean square	F-ratio
Fertilizer	14.09	3	4.6967	0.644
Block	158.1845	7	22.5978	
Error	153.1943	21	7.29497	
Total	325.4688	31		

6.14 Exercises

Exercise 6.1

Use effects coding method to define the dummy variables and obtain a regression model for the responses of RCBD in Exercise 5.9. Use manual methods to analyze the data, estimate model parameters, and test for significance of the fitted model. □

Exercise 6.2

Use reference cell coding method using cell (4, 4) for reference to define the dummy variables and obtain a regression model for the responses of the RCBD in Exercise 5.9. Use manual methods to analyze the data, estimate model parameters, and test for significance of the fitted model. □

Exercise 6.3
Use orthogonal contrasts to obtain a regression model for the responses of the RCBD in Exercise 5.9 and test for significance of the fitted model. □

Exercise 6.4
Analyze the data of Exercise 5.7 using effects coding with code of −1 for row 4, column 5, and 4th Latin letter. Use both manual methods to analyze the data, estimate model parameters, and test for significance of the fitted model. □

Exercise 6.5
Analyze the data of Exercise 5.7 using reference cell to define the dummy variables using cell (4, 5, 4), that is, observation row 4, column 5 containing the 4th Latin letter. Use manual methods to analyze the data, estimate model parameters, and test for significance of the fitted model. □

Exercise 6.6
Write a SAS program to analyze the data, estimate model parameters, and test for significance of the fitted model for the experiment treated in Exercise 6.4. □

Exercise 6.7
Write a SAS program to analyze the data, estimate model parameters and, test for significance of the fitted model for the experiment treated in Exercise 6.5. □

Exercise 6.8
Use a SAS program to analyze the data for the experiment treated in Exercise 6.3, estimate model parameters, and test for significance of the fitted model. □

Exercise 6.9
Write a SAS program to carry out the analysis of Exercise 6.2, estimate model parameters, and test for significance of the fitted model. □

Exercise 6.10
Use a SAS program to analyze the data of the experiment treated in Exercise 6.1, estimate model parameters, and test for significance of the fitted model. □

Exercise 6.11
Use effects coding to define the indicator variables and carry out the analysis of the data for the RCBD experiment with a missing value treated in Example 5.4 after the missing value was estimated. Estimate model parameters and test for significance of the fitted model. □

Exercise 6.12
Write a SAS program to carry out the analysis of responses as described in Question 6.11. □

References

Onyiah, L.C. (2005) *SAS Programming and Data Analysis: A Theory and Program-Driven Approach*, University Press of America Inc., Lanham, MD.
SAS/STAT 8 User's Guide (1999) 2, SAS Institute Inc., Cary, NC.

Chapter 7

Factorial Designs—The 2^k and 3^k Factorial Designs

In our studies of different experimental layouts, RCBDs the Latin squares, and the Graeco-Latin square designs, we have already learned that experiments are performed to investigate the effects of a factor (an explanatory variable) or a combination of factors on one or more outcome or response variables. However, the experiments did not investigate the effects of all possible combinations of the studied factors on the responses. For instance, we could only use the Graeco-Latin and the Latin square designs to study the effects of factors on responses if we were certain that there were no interactions between the factors in the design. If each of the factors that is investigated in a given design is set at a number of levels, we may wish to study almost every combination of the levels of the factors. A study that aims to investigate the effects of all possible combinations of the levels of the factors in a full design is called a factorial experiment. Of all possible design configurations that could be devised for such a study, a full factorial experiment provides the most efficient way for carrying out such a study. Other ways of carrying out the same investigation may be time-consuming and expensive. They may also require more units to arrive at the same precision as the typical factorial experiment.

For many investigations, there is no precise knowledge of the factors whose effects are primarily responsible for the observed responses. It is, therefore, usual in such studies to include a large number of explanatory variables (factors) at the exploratory stages of the investigation, to eliminate those factors that do not appear to have significant effect on the observed responses. Factorial experiments could be and are often used at different stages of such an exploratory investigation—at the initial stage to eliminate irrelevant factors and at the intermediate stages (after the investigation has identified the relevant factors) as a tool to explore response surfaces.

In exploratory studies, the result of a full factorial experiment is used to determine the treatment combinations to be studied in further experimentation. A factorial experiment is complete if the investigations are carried out in such a way that every combination of levels of factors receives equal number of tests in an experiment. Suppose there are two factors, A and B with levels k and l, respectively, then the total number of combinations of levels of the factors is kl. A full factorial experiment will investigate all the kl combinations. To estimate error in such an experiment, $n \geq 2$ measurements need to be made for each of these kl combinations. The reader may refer to a special case of a full factorial design involving two factors called the two-way ANOVA (with replication), or the two-factor factorial design, which has been discussed in Chapter 3.

7.1 Advantages of Factorial Designs

Apart from being useful for exploratory purposes, factorial experiments have many other advantages, which we have touched upon in the previous section (Section 7.0). To recapitulate (1) Factorial experiments are more efficient than one-factor-at-a-time experiments, since they reduce the number of observations that need to be made to draw a precise conclusion. Decisions reached under the factorial design may be more accurate than those reached under the one-factor-at-a-time experiments. As an illustration, suppose a chemist is studying the effects of factors, namely, temperature and pressure on the melting time of a substance. We denote the levels of temperature by T_1 and T_2, and the pressure levels by P_1 and P_2. If there is no interaction and the chemist is merely interested in studying the main effects of T and P on the melting time in a one-factor-at-a-time experiment, he would need to make at least three measurements, one at each of the points (T_1, P_1), (T_1, P_2), (T_2, P_1) to obtain the estimates of the main effects of P and T as $(t_1p_2 - t_1p_1)$ and $(t_2p_1 - t_1p_1)$, respectively. Only three observations are required to obtain the estimates, but since there may be experimental error, it would be advisable for the chemist to make at least two observations for each combination of factors. Using this process, he can estimate both the effects and the experimental error. He would need at least six observations to perform this experiment and estimate effects and error. Comparatively, if he used a factorial design, he would have needed to make only one extra observation at (T_1, P_2) in addition to the other three observations at the combinations of levels mentioned above to obtain all the estimates of the main effects of T, P, as well as the error. (2) Factorial designs help detect any interaction between factors that could have been neglected in one-factor-at-a-time experiments. When such interactions are large, and therefore significant, the one-factor-at-a-time experimental strategy is bound to reach conclusions that are misleading. (3) Factorial experiments allow effects of a factor to be estimated at a very large number of levels of other factors. Conclusions of such experiments have wider validity than the one-factor-at-a-time experiments.

7.1.1 The Concept of Interaction of Factors

In a typical two-factor factorial experiment involving factors A (temperature) and B (pressure), there are two possible outcomes. These outcomes are depicted in Figures 7.1a and 7.1b.

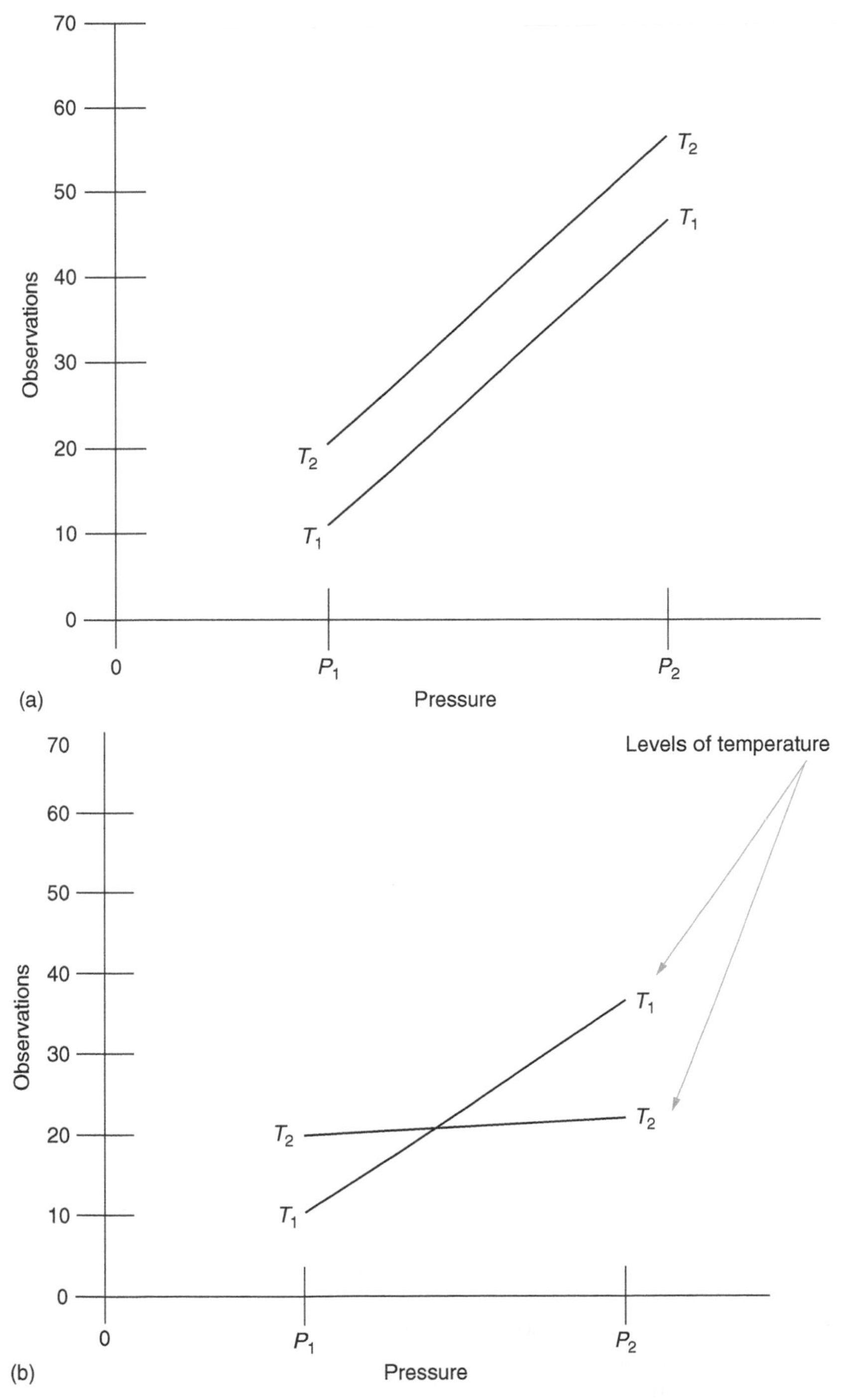

FIGURE 7.1: Two-factor factorial experiment factors P and T (a) without interaction and (b) with interaction.

TABLE 7.1: Responses of a Typical 2^2 Factorial Design

	Factor B	
Factor A	B_1	B_2
A_1	y_{11}	y_{12}
A_2	y_{21}	y_{22}

In the section on two-factor factorial experiments, we dealt with the concept of interaction of factors. Here, we revisit the subject and relate it to factors set at two and three levels only. In the factorial design with factors A and B, each at two levels, the effect of any of these factors is defined as the change in response because of change in the level of that factor. With two factors set at two levels, we have the 2^2 factorial design; the responses of such an experiment are given in Table 7.1.

We could define the main effect of the factor A as the difference between the average response at the first level of A and the average response at the second level of A. This could be shown from our results above as

$$A = \frac{(y_{21} + y_{22})}{2} - \frac{(y_{11} + y_{12})}{2} \tag{7.1}$$

Similarly, the main effect of B is

$$B = \frac{(y_{12} + y_{22})}{2} - \frac{(y_{11} + y_{21})}{2} \tag{7.2}$$

These definitions given for the main effects of A and B would need to be modified if the number of factors becomes larger than two. It has to be borne in mind that if more than one observation at each of these combinations of levels of factors A and B is made, the entries y_{ij} are replaced by the total for each combination of levels of factors i and j. When the response because of a factor is not the same at all levels of another factor, there is an interaction between two factors or more. Figure 7.1a indicates what happens when there is no interaction between the two factors A and B, while Figure 7.1b is a typical case when there is an interaction between the two factors.

Suppose that a factorial experiment is performed to investigate the effect of temperature (T) and pressure (P) on the reaction time of a chemical process, if three levels of these factors were considered. The responses in the experiment could be presented as in Table 7.2.

Three cases representing an overall view of possible outcomes of the experiment are discernible:

TABLE 7.2: Responses of an Experiment with Two Factors at Three Levels

	Temperature levels (T)			
Pressure levels (P)	**1**	**2**	**3**	**Means**
1	y_{11}	y_{12}	y_{13}	$\overline{y}_{1.}$
2	y_{21}	y_{22}	y_{23}	$\overline{y}_{2.}$
3	y_{31}	y_{32}	y_{33}	$\overline{y}_{3.}$
Means	$\overline{y}_{.1}$	$\overline{y}_{.2}$	$\overline{y}_{.3}$	$\overline{y}_{..}$

Case I

If the difference in average response by going from level 1 of P to level 2 of P is $\overline{y}_{2.} - \overline{y}_{1.} = a_1$ and the difference in average response in going from level 2 to level 3 of T is $\overline{y}_{.3} - \overline{y}_{.2} = a_2$ and further, if the difference in responses obtained by going from position determined by level 1 of P and level 2 of T to the position determined by level 2 of P and level 3 of T is $y_{23} - y_{12} = a_1 + a_2$; then there is no interaction between the factors P and T. □

Case II

If $\overline{y}_{2.} - \overline{y}_{1.} = a_1$, $\overline{y}_{.3} - \overline{y}_{.2} = a_2$, but, $y_{23} - y_{12} \neq a_1 + a_2$, then there is an interaction between the factors P and T. □

Case III

If $\overline{y}_{2.} - \overline{y}_{1.} = 0$, $\overline{y}_{.3} - \overline{y}_{.2} = 0$, a special case of zero main effects occurs, and there is no interaction between the factors P and T.

When there is an interaction between factors, we could obtain the effect of interaction. For data in Table 7.3 relating to factors A and B, the interaction would be denoted by AB, and its average effect would be

$$AB = \frac{(y_{22} - y_{21}) - (y_{12} - y_{11})}{2} \tag{7.3}$$

that is, AB is the difference between the effect of A when B is high, and the effect of A when the level of B is low. □

7.2 2^k and 3^k Factorial Designs

In general, factorial experiments can be designed for any number of factors, say b, where factor B_1 has k_1 levels, factor B_2 has k_2 levels, ..., and factor B_b has k_b units for one replicate. For such an experiment, the total number of combinations of levels for all the factors is $L = k_1 \times k_2 \times \cdots \times k_b$. Although it is theoretically possible to construct a factorial design for any finite number of factors, each with any finite number of levels, if interactions between these factors are present, the number of experimental units required to carry out a full experiment quickly becomes large and unwieldy as the number of factors increases. It is precisely for this reason that a class of factorial designs is being very popularly used in many experiments. These designs involve k factors each set at two levels and k factors each set at three levels. These designs are usually referred to as the 2^k and 3^k factorial designs. In the next few sections, we investigate this class of experimental designs.

7.2.1 2^2 Factorial Designs

The simplest form of a factorial design involves two factors, each of whose effects are investigated at two levels. In this design, only four different treatment combinations are investigated to obtain a single replicate of the experiment. This is called the 2^2 factorial design. It is the simplest case of the more general class of experiments referred to as the 2^k factorial designs. A single replicate of this general case, which involves k factors, each set at two levels requires 2^k observations. It is normal when the factors are qualitative (as described in Chapter 2) to describe the levels of the factors in terms of "high" and "low" or "presence" and "absence," whereas if the factors are quantitative, the levels of the factors are easily discernible, and the more apt description would be "high" and "low" (recall examples of quantitative factors described in Chapter 2). The 2^k factorial design is very useful as it could be used in the exploratory stages of an investigation. It becomes even more useful when the random effects model is utilized for the study. When the random effects model is utilized for the study, the 2^k factorial design becomes a way of taking a sample of some factors out of a large population of factors with a view to projecting the results of the experiment to other closely related factors not studied in the experiment. Furthermore, it allows the examination of a fairly large number of combinations of factors whose initial investigation could lead to the elimination of those combinations of factors, which are found not to have significant effects. Sometimes, the purposes of the exploratory analysis could be achieved through fractional replication (particularly when the number of factors to be investigated is large), a technique for dealing with less experimental units than usual for which the 2^k factorial design is well suited. We shall later discuss the procedure for fractional factorial experiments in detail.

7.2.1.1 Standard Order for Treatment Combinations in 2^k Designs

Earlier in this chapter (Section 7.1.1), we used a study of the reaction time of a chemical process at two levels of temperature (T) and pressure (P) to illustrate a typical factorial experiment. This design (the 2^2 factorial design) could be represented geometrically in form of a square (as in Figure 7.2). The vertices of the square represent the treatment combinations in this design. A typical 2^3 factorial experiment can be illustrated by an experiment performed in a wind tunnel to test the stability of a vehicle. In the experiment, three factors are considered to be of importance by the experimenter to his study: wind speed (A), vehicle speed (B), and embankment of the track (C). The factors are all set at two levels. The response is a measure of stability of the vehicle. Figure 7.3 illustrates this design, using the vertices of a cube to represent the treatment combinations. We can also illustrate a 2^4 factorial design as in Figure 7.4.

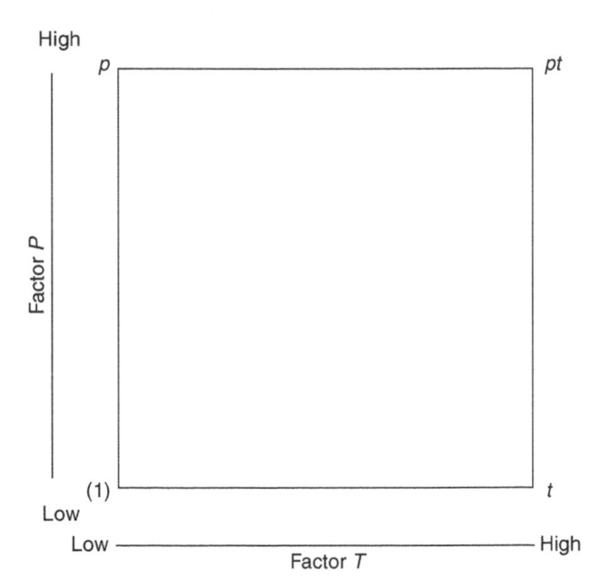

FIGURE 7.2: Typical two-factor factorial experiment with factors P and T at two levels.

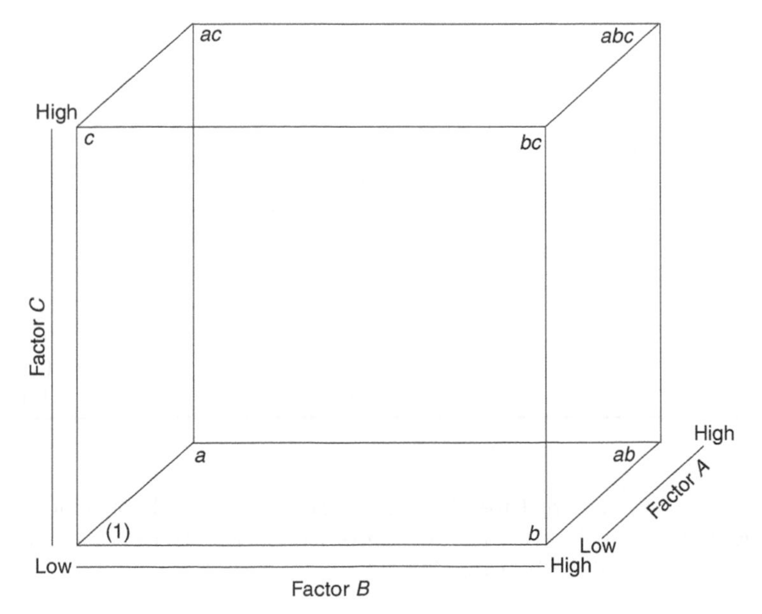

FIGURE 7.3: Typical 2^3 factorial experiment with factors A, B, and C.

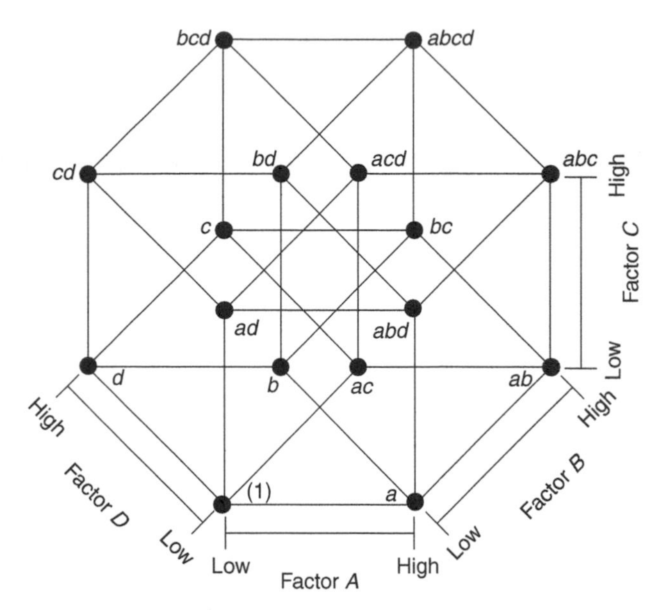

FIGURE 7.4: Typical 2^4 factorial experiment in A, B, C, and D.

In general, if there are two factors A and B each at two levels, the treatment combinations can be written in a standard order or what is sometimes called the "Yates" order, after one of the early investigators of the 2^k factorial design. There are about four conventional methods of listing the treatment combinations. Some of these methods are given in Table 7.3 for the 2^2 factorial design.

TABLE 7.3: Standard Notations for Treatment Combinations in 2^2 Factorial Design

Order of treatment combinations	Standard notations		
	A, B	x_1, x_2	1,2
1	(1)	$--$	0 0
2	a	$+-$	1 0
3	b	$-+$	0 1
4	ab	$++$	1 1

Similarly, if there are three factors A, B, and C, each of which is to be investigated at two levels, then the conventional notations for combinations of treatments are given in Table 7.4.

TABLE 7.4: Standard Notations for Treatment Combinations of 2^3 Factorial Design

Order of treatment combinations	Standard notations		
	A*, *B*, *C	x_1, x_2, x_3	**1, 2, 3**
1	(1)	− − −	0 0 0
2	*a*	+ − −	1 0 0
3	*b*	− + −	0 1 0
4	*ab*	+ + −	1 1 0
5	*c*	− − +	0 0 1
6	*ac*	+ − +	1 0 1
7	*bc*	− + +	0 1 1
8	*abc*	+ + +	1 1 1

When the design contains four factors, *A*, *B*, *C*, and *D*, each set at two levels, then the standard notations are as given in Table 7.5.

TABLE 7.5: Standard Notations for Treatment Combinations of 2^4 Factorial Design

Order of treatment combinations	Standard notations		
	A*, *B*, *C*, *D	x_1, x_2, x_3, x_4	**1, 2, 3, 4**
1	(1)	− − − −	0 0 0 0
2	*a*	+ − − −	1 0 0 0
3	*b*	− + − −	0 1 0 0
4	*ab*	+ + − −	1 1 0 0
5	*c*	− − + −	0 0 1 0
6	*ac*	+ − + −	1 0 1 0
7	*bc*	− + + −	0 1 1 0
8	*abc*	+ + + −	1 1 1 0
9	*d*	− − − +	0 0 0 1
10	*ad*	+ − − +	1 0 0 1
11	*bd*	− + − +	0 1 0 1
12	*abd*	+ + − +	1 1 0 1
13	*cd*	− − + +	0 0 1 1
14	*acd*	+ − + +	1 0 1 1
15	*bcd*	− + + +	0 1 1 1
16	*abcd*	+ + + +	1 1 1 1

It does not matter which of the standard notations is used; however, one may find one notation more convenient in some situations than others. It is precisely for this reason that we shall use the first and third notations in the tables under different circumstances when studying the 2^k and 3^k factorial designs. In Figures 7.2 through 7.4, any of the standard methods for listing treatment combinations could have been used. If we had chosen to use the third standard notation (in order of appearance in the tables), then in Figure 7.2, $(1) = (0\ 0)$; $p = (1\ 0)$, $t = (0\ 1)$, and $pt = (1\ 1)$. Similarly, in Figures 7.3 and

7.4, the notations could be replaced by the corresponding notations shown in the fourth columns of Tables 7.4 and 7.5.

7.3 Contrasts for Factorial Effects in 2^2 and 2^3 Factorial Designs

Consider a 2^2 factorial design, a design in which there are two factors, say A and B, each at two levels. In Table 7.2, we had listed the typical responses from this experiment. We have also shown that the average effect of treatment A could be obtained as

$$A = \frac{(y_{21} + y_{22}) - (y_{11} + y_{12})}{2}$$

If we rewrite the notations in Table 7.2 in terms of the standard notations, $y_{11} = (1)$, $y_{12} = b$, $y_{21} = a$, and $y_{22} = ab$. Equation 7.1 that represents the main effect of A now takes the form

$$A = \frac{[(ab - b) + (a - (1))]}{2} \tag{7.4}$$

If the experiment were replicated n times, then all the replicates will be used to estimate the main effect. This means that each of the treatment combinations (1), a, b, and ab will now be replaced by the total of n responses obtained under it. We would then define the main effect of A as

$$A = \frac{[(ab - b) + (a - (1))]}{2n} \tag{7.5}$$

The terms $(ab - b)$ and $(a - (1))$ are called the simple effects of A. We actually obtained the main effect of A by taking the average of the two simple effects of A; namely, the simple effect of A when B is high and the simple effect of A when B is low. Similarly, by assuming that there are n replicates, we can obtain the main effect of B from Equation 7.2 as

$$B = \frac{[(ab - a) + (b - (1))]}{2n} \tag{7.6}$$

We can obtain the interaction between the factors under the same conditions using Equation 7.3 as

$$AB = \frac{[(ab - b) - (a - (1))]}{2n} \tag{7.7}$$

Another way of deriving these formulae for the factorial effects is to use the geometrical representation of the designs as shown in Figures 7.2 through 7.4. However, one has to be honest and admit that this method becomes less useful as the number of factors becomes large, say beyond 2^4. This is because it is not easy to produce the appropriate figures representing the designs, even when one has an idea of what the figure should be. Use of this method is therefore limited, not because it is not correct, but because it is limited by the logistics. To find the main effects and interactions represented by Figure 7.2, we see that (by analogy with B above) if we take the difference between the treatment combinations on the right hand side of the square and the corresponding treatment combinations on the left hand side of the square, we can obtain the main effect of T as

$$T = \frac{[(pt - p) + (t - (1))]}{2n} \tag{7.8}$$

Similarly, the main effect of P could be obtained as the average of the differences between the treatment combinations on the top of the square and those below as follows

$$P = \frac{[(pt - t) + (p - (1))]}{2n} \tag{7.9}$$

The interaction between P and T could be obtained as the average difference between the treatment combinations that appear on the top vertices of the square and those directly below them

$$PT = \frac{[(pt - t) - (p - (1))]}{2n} \tag{7.10}$$

The main effects and interactions of 2^3 design can be obtained by virtually the same principles which were previously discussed. The main effects for this design are A, B, and C, and the interactions are AB, AC, BC, and ABC. We obtain the main effects of A by calculating the simple effect of A when B and C are high, which is $(abc - bc)$; the simple effect of A when B is high and C is low, which is $(ab - b)$; the simple effect of A when B is low and C is high, which is $(ac - c)$; and the simple effect of A when B and C are low, $(a - (1))$. If there are n replicates (we define the lowercase letters as the sum of all the responses for the treatment combination that it represents), then we take the average of the simple effects for all the n replicates. So now

$$A = \frac{[(abc - bc) + (a - (1)) + (ab - b) + (ac - c)]}{4n}$$

or

$$A = \frac{[(abc + ac + b + (1) - a - ab - c - bc)]}{4n} \tag{7.11}$$

If we consider the interaction AC, we see that the difference between two simple effects makes up its effect when B is high. That is, the difference between the simple effects of A when B and C are high is $(abc - bc)$; and the simple effect of A when B is high and C is low is $(ab - b)$. These give the effect of AC when B is high as $[(abc - bc) - (ab - b)]$. Similarly, the difference between two simple effects makes up AC effect when B is low, which is $[(ac - c) - (a - (1))]$. The full effect of the interaction AC is obtained as the average of the sum of these two quantities (and if there are n replicates), the sum is further divided by n so that

$$AC = \frac{[(abc - bc) - (ab - b) + (ac - c) - (a - (1))]}{4n}$$

or

$$AC = \frac{[abc + ac + b + (1) - a - ab - c - bc]}{4n} \tag{7.12}$$

TABLE 7.6: Factorial Effects and Their Contrasts in a 2^2 Design

Factorial effects	(1)	*a*	*b*	*ab*
I	+	+	+	+
A	−	+	−	+
B	−	−	+	+
AB	+	−	−	+

TABLE 7.7: Factorial Effects and Their Contrasts in a 2^3 Design

Factorial effects	(1)	*a*	*b*	*ab*	*c*	*ac*	*bc*	*abc*
I	+	+	+	+	+	+	+	+
A	−	+	−	+	−	+	−	+
B	−	−	+	+	−	−	+	+
AB	+	−	−	+	+	−	−	+
C	−	−	−	−	+	+	+	+
AC	+	−	+	−	−	+	−	+
BC	+	+	−	−	−	−	+	+
ABC	−	+	+	−	+	−	−	+

Tables 7.6 and 7.7 show all the factorial effects in 2^2 and 2^3 factorial designs as well as their contrasts.

7.3.1 The Models

For the 2^2 factorial design with factors A and B in n replicates, the appropriate model is

$$y_{ijk} = \mu + \alpha_i + \beta_j + (\alpha\beta)_{ij} + \varepsilon_{ijk} \quad i, j = 1, 2 \quad k = 1, \ldots, n \tag{7.13}$$

where μ is the overall mean, α_i is the effect of the ith level of the factor A, β_j is the effect of the jth level of the factor B, $(\alpha\beta)_{ij}$ is the effect of the interaction AB at level i of A and level j of B, and $\varepsilon_{ijk} \sim \text{NID}(0, \sigma^2)$ represents the error. The effects of A, B, and AC could be fixed or random as discussed in previous designs. When they are fixed, then $\alpha_1 + \alpha_2 = 0$; $\beta_1 + \beta_2 = 0$; $\sum_i (\alpha\beta)_{ij} = \sum_j (\alpha\beta)_{ij} = 0$. On the other hand, if they are random, then $\alpha_i \sim \text{NID}(0, \sigma^2_\alpha)$; $\beta_j \sim \text{NID}(0, \sigma^2_\beta)$; and $(\alpha\beta)_{ij} \sim \text{NID}(0, \sigma^2_{\alpha\beta})$. The model 7.13 is a linear model; based on the least square analysis of this model, we calculate the sums of squares to be used in ANOVA as follows:

$$SS_{\text{total}} = \sum_{i=1}^{2}\sum_{j=1}^{2}\sum_{k=1}^{n} y_{ijk}^2 - \frac{y_{\ldots}^2}{4n}$$

$$SS_A = \frac{1}{4n}[(ab - b) + (a - (1))]^2$$

$$SS_B = \frac{1}{4n}[(ab - a) + (b - (1))]^2$$

$$SS_{AB} = \frac{1}{4n}[(ab - b) - (a - (1))]^2$$

$$SS_{\text{E}} = SS_{\text{total}} - SS_A - SS_B - SS_{AB}$$

The ANOVA for this model is shown in Table 7.8.

TABLE 7.8: ANOVA for the 2^2 Factorial Design

Source	Sum of squares	Degrees of freedom	Mean square	F-ratio
A	SS_A	1	$MS_A = SS_A$	MS_A/MS_E
B	SS_B	1	$MS_B = SS_B$	MS_B/MS_E
AB	SS_{AB}	1	$MS_{AB} = SS_{AB}$	MS_{AB}/MS_E
Error	SS_{E}	$2^2(n-1)$	$MS_E = SS_{\text{E}}/2^2(n-1)$	
Total	SS_{total}	$2^2(n-1)$		

For the 2^3 factorial design in A, B, C, with n replicates, the appropriate model is

$$y_{ijkl} = \mu + \alpha_i + \beta_j + (\alpha\beta)_{ij} + \gamma_k + (\alpha\gamma)_{ik} + (\beta\gamma)_{jk} + (\alpha\beta\gamma)_{ijk} + \varepsilon_{ijkl}$$
$$i, j, k = 1, 2; \; l = 1, \ldots, n \qquad (7.14)$$

where μ is the overall mean, α_i is the effect of the ith level of A, β_j is the effect of the jth level of B, γ_k is the effect of the kth level of C, $(\alpha\beta)_{ij}$ is the effect of the interaction of ith level of A with jth level of B, $(\alpha\gamma)_{ik}$ is the effect of the interaction of ith level of A with the kth level of C, and $(\beta\gamma)_{jk}$ is the effect of the interaction of jth level of B with the kth level of C. The interaction of the ith level of A, the jth level of B with the kth level of C is represented by $(\alpha\beta\gamma)_{ijk}$ and $\varepsilon_{ijkl} \sim \text{NID}(0, \sigma^2)$. It is also possible for this design to employ either random or fixed effects. If the random model is chosen, the effects would all be normally and independently distributed random variables with zero mean and appropriate variances as we have illustrated in the 2^2 factorial design (Section 7.3.1), which we discussed in the previous section.

For this design, the sums of squares are

$$SS_{\text{total}} = \sum_{i=1}^{2}\sum_{j=1}^{2}\sum_{k=1}^{2}\sum_{l=1}^{n} y_{ijkl}^2 - \frac{y_{....}^2}{2^3 n}$$

$$SS_A = \frac{1}{8n}[a + ab + ac + abc - (1) - b - c - bc]^2$$

$$SS_B = \frac{1}{8n}[b + ab + bc + abc - (1) - a - c - ac]^2$$

$$SS_{AB} = \frac{1}{8n}[(1) + ab + c + abc - a - b - bc - ac]^2$$

$$SS_C = \frac{1}{8n}[c + ac + bc + abc - (1) - a - b - ab]^2$$

$$SS_{AC} = \frac{1}{8n}[(1) + b + ac + abc - a - ab - c - bc]^2$$

$$SS_{BC} = \frac{1}{8n}[(1) + a + bc + abc - b - ab - c - ac]^2$$

$$SS_{ABC} = \frac{1}{8n}[a + b + bc + abc - (1) - ab - ac - bc]^2$$

$$SS_{\mathrm{E}} = SS_{\mathrm{total}} - SS_A - SS_B - SS_{AB} - SS_C - SS_{AC} - SS_{BC} - SS_{ABC}$$

The ANOVA for a typical 2^3 factorial design is given in Table 7.9.

In addition to the fixed and random models described above, it is possible to have a variety of mixed models in which one or two factors are random and the other(s) is (are) fixed. Tables 7.10 and 7.11 give the expected mean squares for the different configurations of the 2^2 and 2^3 factorial designs.

TABLE 7.9: ANOVA for the 2^3 Factorial Designs

Source	**Sum of squares**	**Degrees of freedom**	**Mean square**	***F*-ratio**
A	SS_A	1	SS_A	MS_A/MS_{E}
B	SS_B	1	SS_B	MS_B/MS_{E}
AB	SS_{AB}	1	SS_{AB}	MS_{AB}/MS_{E}
C	SS_C	1	SS_C	MS_C/MS_{E}
AC	SS_{AC}	1	SS_{AC}	MS_{AC}/MS_{E}
BC	SS_{BC}	1	SS_{BC}	MS_{BC}/MS_{E}
ABC	SS_{ABC}	1	SS_{ABC}	MS_{ABC}/MS_{E}
Residual	SS_{E}	$2^3(n-1)$	$SS_{\mathrm{E}}/2^3(n-1)$	
Total	SS_{total}	$2^3(n-1)$		

TABLE 7.10: Expected Mean Squares in 2^2 Factorial Design

	Expected mean squares		
Factorial effect	**Fixed effects model**	**Random effects model**	**Mixed effects model *B* random**
α_i	$\sigma^2 + 2n\sum_{i=1}^{2}\alpha_i^2$	$\sigma^2 + n\sigma_{\alpha\beta}^2 + 2n\sigma_\alpha^2$	$\sigma^2 + n\sigma_{\alpha\beta}^2 + 2n\sum_{i=1}^{2}\alpha_i^2$
β_j	$\sigma^2 + 2n\sum_{i=1}^{2}\beta_j^2$	$\sigma^2 + n\sigma_{\alpha\beta}^2 + 2n\sigma_\beta^2$	$\sigma^2 + 2n\sigma_\beta^2$
$(\alpha\beta)_{ij}$	$\sigma^2 + n\sum_{i=1}^{2}\sum_{j=1}^{2}(\alpha\beta)_{ij}^2$	$\sigma^2 + n\sigma_{\alpha\beta}^2$	$\sigma^2 + n\sigma_{\alpha\beta}^2$
ε_{ijk}	σ^2	σ^2	σ^2

TABLE 7.11: Expected Mean Squares in 2^3 Factorial Design

	Expected mean squares		
Factorial effect	**Fixed effects model**	**Random effects model**	**Mixed effects (A fixed, B and C random)**
α_i	$\sigma^2 + 4n\sum_{i=1}^{2}\alpha_i^2$	$\sigma^2 + 2n(\sigma^2_{\alpha\beta} + \sigma^2_{\alpha\gamma} + 2\sigma^2_{\alpha}) + n\sigma^2_{\alpha\beta\gamma}$	$\sigma^2 + 4n\sum_{i=1}^{2}\alpha_i^2 + 2n(\sigma^2_{\alpha\beta} + 2\sigma^2_{\alpha\gamma}) + n\sigma^2_{\alpha\beta\gamma}$
β_j	$\sigma^2 + 4n\sum_{i=1}^{2}\beta_j^2$	$\sigma^2 + 2n(\sigma^2_{\alpha\beta} + \sigma^2_{\beta\gamma} + 2\sigma^2_{\beta}) + n\sigma^2_{\alpha\beta\gamma}$	$\sigma^2 + 2n(\sigma^2_{\beta\gamma} + 2\sigma^2_{\beta})$
$(\alpha\beta)_{ij}$	$\sigma^2 + 2n\sum_{i=1}^{2}\sum_{j=1}^{2}(\alpha\beta)^2_{ij}$	$\sigma^2 + 2n(\sigma^2_{\alpha\beta}) + n\sigma^2_{\alpha\beta\gamma}$	$\sigma^2 + 2n\sigma^2_{\alpha\beta} + n\sigma^2_{\alpha\beta\gamma}$
γ_k	$\sigma^2 + 4n\sum_{i=1}^{2}\gamma_k^2$	$\sigma^2 + 2n(\sigma^2_{\beta\gamma} + \sigma^2_{\alpha\gamma} + 2\sigma^2_{\gamma}) + n\sigma^2_{\alpha\beta\gamma}$	$\sigma^2 + 2n(\sigma^2_{\beta\gamma} + 2\sigma^2_{\gamma})$
$(\alpha\gamma)_{ik}$	$\sigma^2 + 2n\sum_{i=1}^{2}\sum_{j=1}^{2}(\alpha\gamma)^2_{ik}$	$\sigma^2 + 2n(\sigma^2_{\alpha\gamma}) + n\sigma^2_{\alpha\beta\gamma}$	$\sigma^2 + 2n\sigma^2_{\alpha\gamma} + n\sigma^2_{\alpha\beta\gamma}$
$(\beta\gamma)_{jk}$	$\sigma^2 + 2n\sum_{i=1}^{2}\sum_{j=1}^{2}(\beta\gamma)^2_{jk}$	$\sigma^2 + 2n(\sigma^2_{\beta\gamma}) + n\sigma^2_{\alpha\beta\gamma}$	$\sigma^2 + 2n\sigma^2_{\beta\gamma}$
$(\alpha\beta\gamma)_{ijk}$	$\sigma^2 + n\sum_{i=1}^{2}\sum_{j=1}^{2}\sum_{k=1}^{2}(\alpha\beta\gamma)^2_{ijk}$	$\sigma^2 + n\sigma^2_{\alpha\beta\gamma}$	$\sigma^2 + n\sigma^2_{\alpha\beta\gamma}$
ε_{ijkl}	σ^2	σ^2	σ^2

Example 7.1

A factorial experiment was designed to study the effects of two factors on the time-to-completion of a chemical reaction. The factors studied were temperature (T) and pressure (P) (each set at two levels). Four replicates of the experiments were performed and the responses obtained were

Temperature	**Low**		**High**	
Pressure	**Low**	**High**	**Low**	**High**
Replicates				
I	78	60	47	40
II	71	69	43	41
III	88	63	53	44
IV	86	64	56	42

We can rewrite the data in terms of the treatment combinations in the design as follows:

Treatment combinations		**Replicates**			**Total**
(1)	78	71	88	86	323
p	60	69	63	64	256
t	47	43	53	56	199
pt	40	41	44	42	167

The sums of squares are obtained as follows:

$$SS_{\text{total}} = 78^2 + 71^2 + 88^2 + \cdots + 44^2 + 42^2 - \frac{945^2}{16} = 3860.94$$

$$SS_P = \frac{[pt + p - (1) - t]^2}{16} = \frac{(167 + 256 - 323 - 256)^2}{16} = 612.56$$

$$SS_T = \frac{[pt + t - (1) - p]^2}{16} = \frac{(167 + 199 - 323 - 256)^2}{16} = 2835.56$$

$$SS_{PT} = \frac{[pt + (1) - p - t]^2}{16} = \frac{(167 + 323 - 256 - 199)^2}{16} = 76.56$$

$$SS_{\text{residual}} = 3860.9 - 612.56 - 2835.56 - 76.56 = 336.22$$

The analysis of variance is presented in Table 7.12.

TABLE 7.12: ANOVA for the Chemical Reaction Experiment.

Source	Sum of squares	Degrees of freedom	Mean square	F-ratio
P	612.56	1	612.56	21.86
T	2835.56	1	2835.56	101.20
PT	76.56	1	76.56	2.73
Residual	336.22	12	336.22	
Total	3860.94	15		

Because $F(1,\, 12,\, 0.01) = 9.33$, we find that the main effects of P and T are significant. Using the methods discussed in the previous chapters, it is possible to carry out the normal probability plot for the residuals of this analysis to check that the normality assumption of the model is satisfied. Other checks for the validity of the model could be carried out.

The next section presents the SAS program for the analysis for the data of Example 7.1 and the subsequent output:

SAS PROGRAM

```
DATA completion;
INPUT p t Y @@;
CARDS;
0 0 78 0 0 71 0 0 88 0 0 86 1 0 60 1 0 69 1 0 63 1 0 64
0 1 47 0 1 43 0 1 53 0 1 56 1 1 40 1 1 41 1 1 44 1 1 42
;
proc print data=completion; sum p t y;
PROC anova DATA=completion;
class p t;
MODEL Y=p|t; RUN;
```

SAS OUTPUT

Obs	p	t	Y
1	0	0	78
2	0	0	71
3	0	0	88
4	0	0	86
5	1	0	60
6	1	0	69
7	1	0	63
8	1	0	64
9	0	1	47
10	0	1	43
11	0	1	53

```
         Obs     p       t       Y

          12     0       1      56
          13     1       1      40
          14     1       1      41
          15     1       1      44
          16     1       1      42
                 =       =     ===
                 8       8     945

              The ANOVA Procedure

            Class Level Information

        Class       Levels        Values
        p                2        0 1

        t                2        0 1

 Number of Observations Read         16
 Number of Observations Used         16

              The ANOVA Procedure
Dependent Variable: Y
                                Sum of
 Source                DF      Squares     Mean Square      F Value      Pr > F

 Model                  3  3524.687500     1174.895833        41.93      <.0001

 Error                 12   336.250000       28.020833

 Corrected Total       15  3860.937500

           R-Square     Coeff Var      Root MSE        Y Mean
           0.912910      8.962490      5.293471      59.06250

Source             DF         Anova SS     Mean Square      F Value      Pr > F

p                   1       612.562500      612.562500        21.86      0.0005
t                   1      2835.562500     2835.562500       101.19      <.0001
p*t                 1        76.562500       76.562500         2.73      0.1242
```

The above SAS program invoked a subroutine within SAS, PROC ANOVA, to analyze the responses of the experiment. The output of the analysis with SAS is in agreement with our manual analysis. In particular, the conclusions we reached in manual ANOVA by using the table of the F-distribution are reached by looking at the column headed Pr > F, in the SAS output, which gives us the p-value associated with the test statistic, the F-value. We usually reject the null hypothesis when the p-value is less than a reasonable level of significance α, or the size of type I error we are willing to allow. ▯

Example 7.2

A 2^3 factorial experiment was performed in three replicates to study the growth of an organism in a medium. The growth of the organism is believed to depend on pH, temperature, and the rate of flow of air into the growth vessel. For convenience, let A = temperature, B = pH, and C = rate of air flow. The response is the number of molds of an organism after 3 days of growth. The observations are as follows:

Temperature	**Low**				**High**			
pH	**Low**		**High**		**Low**		**High**	
Air-flow rate	**Low**	**High**	**Low**	**High**	**Low**	**High**	**Low**	**High**
	42	44	44	46	45	47	47	51
	44	45	45	46	46	46	48	50
	41	43	44	47	47	48	47	52

This can be rewritten for convenience in the standard order for listing treatment combinations as shown below.

Treatment combinations	**Replicates**	**Total**
(1)	42 44 41	127
a	45 46 47	138
b	44 45 44	133
ab	47 48 47	142
c	44 45 43	132
ac	46 46 47	139
bc	47 46 48	141
abc	51 50 52	153

The sums of squares are obtained as

$$SS_{\text{total}} = 42^2 + 44^2 + 41^2 + \cdots + 51^2 + 50^2 + 52^2 - \frac{1105^2}{24} = 158.96$$

$$SS_A = \frac{[153 + 138 + 142 + 141 - 127 - 133 - 132 - 139]^2}{24} = 77.04$$

$$SS_B = \frac{[153 + 142 + 133 + 139 - 127 - 138 - 132 - 141]^2}{24} = 35.04$$

$$SS_{AB} = \frac{[153 + 132 + 142 + 127 - 133 - 138 - 139 - 141]^2}{24} = 0.375$$

$$SS_C = \frac{[153 + 141 + 139 + 132 - 142 - 133 - 127 - 138]^2}{24} = 26.04$$

$$SS_{AC} = \frac{[153 + 133 + 141 + 127 - 142 - 138 - 133 - 132]^2}{24} = 0.375$$

$$SS_{BC} = \frac{[153 + 138 + 139 + 127 - 142 - 141 - 133 - 132]^2}{24} = 3.375$$

$$SS_{ABC} = \frac{[153 + 132 + 138 + 133 - 142 - 141 - 127 - 139]^2}{24} = 2.042$$

$$SS_{\text{residual}} = 158.96 - 77.04 - 35.04 - 0.375 - 26.04 - 0.375 - 3.375 - 2.042$$
$$= 14.673$$

The ANOVA for the experiment is presented in Table 7.13.

TABLE 7.13: ANOVA for the Growth of Organisms

Source	Sum of squares	Degrees of freedom	Mean square	*F*-ratio
A	77.04	1	77.04	84.02
B	35.04	1	35.04	38.21
AB	0.375	1	0.375	0.41
C	26.04	1	26.04	28.40
AC	0.375	1	0.375	0.41
BC	3.375	1	3.375	3.68
ABC	2.042	1	2.04	2.22
Error	14.673	16	0.917	
Total	158.960	23		

From the tables, $F(1, 16, 0.05) = 4.49$ and $F(1, 16, 0.01) = 8.53$; only the main effects are significant. None of the two-factor and three-factor effects is significant either at $\alpha = 0.01$ or at $\alpha = 0.05$. As usual, examination of residuals to check the validity of the model should follow. Analysis of the residuals in this design should reveal that there is no serious violation of basic assumptions underlying this model. Our conclusion is that the growth of this organism is influenced independently by temperature, pH, and rate of flow of air.

Next, we present the SAS program and output for the analysis of data of Example 7.2.

SAS PROGRAM

```
data organism;
INPUT a b c Y @@;
CARDS;
0 0 0 42 0 0 0 44 0 0 0 41 1 0 0 45 1 0 0 46 1 0 0 47
0 1 0 44 0 1 0 45 0 1 0 44 1 1 0 47 1 1 0 48 1 1 0 47
0 0 1 44 0 0 1 45 0 0 1 43 1 0 1 47 1 0 1 46 1 0 1 48
0 1 1 46 0 1 1 46 0 1 1 47 1 1 1 51 1 1 1 50 1 1 1 52
;
proc print data=organism; sum a b c y;
PROC anova DATA=organism;
class a b c;
MODEL Y=a|b|c; RUN;
```

SAS OUTPUT

```
 Obs     a    b    c       Y
   1     0    0    0      42
   2     0    0    0      44
   3     0    0    0      41
   4     1    0    0      45
   5     1    0    0      46
   6     1    0    0      47
   7     0    1    0      44
   8     0    1    0      45
   9     0    1    0      44
  10     1    1    0      47
  11     1    1    0      48
  12     1    1    0      47
  13     0    0    1      44
  14     0    0    1      45
  15     0    0    1      43
  16     1    0    1      47
  17     1    0    1      46
  18     1    0    1      48
  19     0    1    1      46
  20     0    1    1      46
  21     0    1    1      47
  22     1    1    1      51
  23     1    1    1      50
  24     1    1    1      52
        ==   ==   ==    ====
        12   12   12    1105

              The ANOVA Procedure

            Class Level Information

Class          Levels         Values

a                   2         0 1

b                   2         0 1

c                   2         0 1

Number of Observations Read     24
Number of Observations Used     24

              The ANOVA Procedure

Dependent Variable: Y
                                      Sum of
   Source               DF           Squares     Mean Square  F Value  Pr > F

   Model                 7       144.2916667      20.6130952    22.49  <.0001

   Error                16        14.6666667       0.9166667

   Corrected Total      23       158.9583333
```

R-Square	Coeff Var	Root MSE	Y Mean
0.907733	2.079480	0.957427	46.04167

Source	DF	Anova SS	Mean Square	F Value	Pr > F
a	1	77.04166667	77.04166667	84.05	<.0001
b	1	35.04166667	35.04166667	38.23	<.0001
a*b	1	0.37500000	0.37500000	0.41	0.5315
c	1	26.04166667	26.04166667	28.41	<.0001
a*c	1	0.37500000	0.37500000	0.41	0.5315
b*c	1	3.37500000	3.37500000	3.68	0.0730
a*b*c	1	2.04166667	2.04166667	2.23	0.1550

□

7.4 General 2^k Factorial Design

So far, we have discussed the 2^k factorial design for $k = 1, 2, 3$ only. It is possible, however, to carry out the 2^k factorial experiment for any number of factors k, limited only by the capacity of the experimenter to fund, manage, and carry out the investigations of a large number of treatment combinations. It is clear that as the number of factors becomes large, the design becomes difficult to handle since the number of treatment combinations grows out of hand.

7.4.1 Factorial Contrasts in 2^k Factorial Designs

As we have discussed, there are a number of ways of obtaining the factorial effects contrasts in 2^k factorial design. By way of summary, we mention them here in addition to including another method that we have not discussed previously. Once we obtain these contrasts, estimation of the factorial effect or the sum of squares pertaining to it is simplified.

7.4.1.1 Method I

Each main effect is obtained by assigning a + sign to those treatment combinations in which that main effect appears. To all other combinations of treatments, a − sign is assigned. Signs for the interactions could be obtained by the product of the signs of the main effects whose letters appear in the interaction.

7.4.1.2 Method II

It is known that half of all the contrasts of a given main effect or interaction receive + signs while the rest receive − signs. Treatment combinations containing even number of letters that appear in the factorial effect (for this purpose, the value 0 is considered as even) receive one of these signs while the others receive the second (and opposite) sign. Only those treatment combinations that contain all letters in the factorial effect receive the + sign.

7.4.1.3 Method III

In general, the polynomial $(a\pm 1)(b\pm 1)\cdots(p\pm 1)(q\pm 1)$ is evaluated to obtain the contrasts for a factorial effect. For any letter that appears in the factorial effect being considered, the sign in the bracket where it appears is replaced by $-$, while the signs in the brackets containing those letters which do not appear is replaced by $+$. The polynomial is then evaluated, leading to the calculation of the treatment effect by dividing the expanded polynomial equation by $2^k(n)$, while the sums of squares of the effects are obtained by dividing the square of the polynomial by $2^k(n)$, where n is the number of replicates. This means that in a 2^k factorial design, the factorial effect of AB is given by

$AB=$ (contrasts of AB in 2^k factorial design)$/2^{k-1}(n)$; while its sum of squares is

$SS_{AB}=$ (contrasts of AB in 2^k factorial design)$^2/2^k(n)$

For example, in the 2^2 factorial design with n replicates,
Contrast $A=[(a-1)(b+1)]$; the effect of A is obtained as

$$A=[(a-1)(b+1)]/2n=[ab+a-b-(1)]/2n$$
$$SS_A=[ab+a-b-(1)]^2/4n$$

Contrast $AB=[(a-1)(b-1)]/4n$. The effect of AB is obtained as

$$AB=[(a-1)(b-1)]/4n=[ab+(1)-a-b]/2n$$
$$SS_{AB}=[ab+(1)-a-b]^2/4n$$

In the 2^3 factorial design with n replicates, the effect and the sum of squares for the main effect B are obtained using
Contrast $B=[(a+1)(b-1)(c+1)]$ so that the factorial effect of B is obtained as

$$B=[(a+1)(b-1)(c+1)]/4n=[abc+ab+bc+b-ac-c-a-(1)]/4n$$
$$SS_B=[(a+1)(b-1)(c+1)]^2/8n=[abc+ab+bc+b-ac-c-a-(1)]^2/8n$$

In the 2^4 factorial design with n replicates, the main effect C and its sums of squares are obtained using
Contrast $C=[(a+1)(b+1)(c-1)(d+1)]$ as $C=[(a+1)(b+1)(c-1)(d+1)]/8n$; and the effect C is obtained as

$$C=[abcd+abc+acd+ac+bcd+bc+cd+c-abd-ab-ad-a-bd-b-d-(1)]/8n$$
$$SS_C=[abcd+abc+acd+ac+bcd+bc+cd+c-abd-ab-ad-a-bd-b-d-(1)]^2/16n$$

For the 2^5 factorial designs with n replicates, the contrast of the factorial effect CE is

Contrast $CE = (a+1)(b+1)(c-1)(d+1)(e-1)$. Consequently, the effect CE and the corresponding sum of squares are

$$\begin{aligned} CE &= [(a+1)(b+1)(c-1)(d+1)(e-1)]/16n \\ &= [abcde + abce + acde + ace + bcde + bce + cde + ce + abd + ab + ad \\ &\quad + a + bd + b + d + (1) - abcd - abc - acd - ac - bcd - bc \\ &\quad - cd - c - abde - abe - ade - ae - bde - be - de - e]/16n \end{aligned}$$

$$\begin{aligned} SS_{CE} &= [abcde + abce + acde + ace + bcde + bce + cde + ce + abd + ab \\ &\quad + ad + a + bd + b + d + (1) - abcd - abc - acd - ac - bcd \\ &\quad - bc - cd - c - abde - abe - ade - ae - bde - be - de - e]^2/32n \end{aligned}$$

We can obtain the contrasts of any factorial effect in a 2^k factorial design and subsequently its effect and its sum of squares by using the above method.

Another method for calculating contrasts of effects using algorithms due to Yates (1937), which we do not discuss in this book, can be found in the classical books by Kempthorne (1952), Davies (1955, 1958), and Rayner (1967).

7.4.2 Link between Factorial Effects and Group Theory

Underlying all the methods for calculating the contrasts of a factorial effect is *group theory*. The factorial effects group is generated by the factors of the design $(a, b, c, \ldots, p, q, \ldots)$ and their product is abelian; that is, commutative, so that $ca = ac$; $ba = ab$, and so on. Furthermore, $a^2 = b^2 = \cdots = p^2 = q^2 = (1)$. The square of an effect is equivalent to *switching that factor at high level to high level, which in effect, restores it to the low level.* We shall find these concepts useful when we discuss fractional factorial designs and the consequent aliasing of factorial effects.

7.5 Factorial Effects in 2^k Factorial Designs

In a typical 2^k factorial design, there are k main effects, ${}^kC_2$ two-factor interactions, ${}^kC_3$ three-factor interactions, ..., $k(k-1)$-factor interactions and one k-factor interaction in a single replicate. It is obvious that as the number of factors in the design increases, so will the number of experiments and experimental units required, as well as every other resource required for successful completion of the experiments. If resources are available, then it will be possible to replicate the experiments and perform the necessary ANOVA in order to find out which factorial effects significantly contribute to the observations. Just to illustrate what is involved, in Table 7.14 we present the full ANOVA for a typical 2^5 factorial design.

7.5.1 2^k Factorial Designs—A Single Replicate

As we observed in the previous Section 7.4, there are 32 separate experiments in a single replicate of a 2^5 factorial design. This means that to be able

TABLE 7.14: ANOVA of the Typical 2^5 Factorial Design

Source	Sum of squares	Degrees of freedom
A	SS_A	1
B	SS_B	1
AB	SS_{AB}	1
C	SS_C	1
AC	SS_{AC}	1
BC	SS_{BC}	1
ABC	SS_{ABC}	1
D	SS_D	1
AD	SS_{AD}	1
BD	SS_{BD}	1
ABD	SS_{ABD}	1
CD	SS_{AD}	1
ACD	SS_{ACD}	1
BCD	SS_{BCD}	1
$ABCD$	SS_{ABCD}	1
E	SS_E	1
AE	SS_{AE}	1
BE	SS_{BE}	1
ABE	SS_{ABE}	1
CE	SS_{CE}	1
ACE	SS_{ACE}	1
BCE	SS_{BCE}	1
$ABCE$	SS_{ABCE}	1
DE	SS_{DE}	1
ADE	SS_{ADE}	1
BDE	SS_{BDE}	1
$ABDE$	SS_{ABDE}	1
CDE	SS_{CDE}	1
$ACDE$	SS_{ACDE}	1
$BCDE$	SS_{BCDE}	1
$ABCDE$	SS_{ABCDE}	1
Error	SS_{E}	$3^5(n-1)$
Total	SS_{total}	$3^5(n-1)$

to perform a full factorial experiment and estimate all the factorial effects as well as error, at least two replicates of the design are required, that means 64 separate experiments. To obtain similar information from 2^6 and 2^7 factorial designs, at least 128 and 256 separate observations are required, respectively. The task will continue to be even more daunting as the number of factors is increased above seven. Because of this problem, other ways had to be found for making reasonable inferences on factors using the 2^k factorial designs, without having to deal with the increasing workload, costs, and complications, which arise with increasing factors. It is known from experience that the higher order interactions are rarely significant in factorial experiments. Furthermore, since theoretically the expected values of each of the sum of squares for each of these interaction effects is σ^2, the sums of squares of a number of such higher order interactions could be pooled together to estimate the experimental error. It is therefore possible to carry out a 2^k factorial experiment, which involves only a single replicate of the design and use the higher order interactions in the design to estimate the error. We illustrate this with an example.

Example 7.3

A manufacturing plant has designed a vehicle hoping to market it for mass consumption. An experiment was performed to study the effects of four factors on the stability of the vehicle. The four factors were set at two levels and the stability of the vehicle was studied using a 2^4 factorial design. The vehicle is tested in a tunnel where wind speed and embankment of track could be controlled. The speed of the vehicle on the track and the weight of load carried by the vehicle were also studied. The responses are measures of instability. Only one replicate of the experiment could be performed because of resource-related constraints. It was assumed that three-factor and higher order interactions are not significant. They were used to estimate the error and analyze the responses of the study. For convenience, we assume A as vehicle speed, B as air speed, C as level of embankment of track, D as load of vehicle, Lo for low, and Hi for high.

A	Lo								Hi							
B	Lo				Hi				Lo				Hi			
C	Lo		Hi		Lo		Hi		Lo		Hi		Lo		Hi	
D	Lo	Hi	Lo	Hi	Lo	Hi	Lo	Hi	Lo	Hi	Lo	Hi	Lo	Hi	Lo	Hi
Response	38	59	58	79	27	53	30	53	40	62	55	75	30	50	32	54

$$\text{Contrast of } A = (a-1)(b+1)(c+1)(d+1)$$

$$SS_A = \frac{(\text{contrast of } A)^2}{2^4} = \frac{1^2}{16} = 0.06$$

$$\text{Contrast of } B = (a+1)(b-1)(c+1)(d+1)$$

$$SS_B = \frac{(\text{contrast of } B)^2}{2^4} = \frac{-137^2}{16} = 1173.06$$

$$\text{Contrast of } AB = (a-1)(b-1)(c+1)(d+1)$$

$$SS_{AB} = \frac{(\text{contrast of } AB)^2}{2^4} = \frac{5^2}{16} = 1.56$$

$$\text{Contrast of } C = (a+1)(b+1)(c-1)(d+1)$$

$$SS_C = \frac{(\text{contrast of } C)^2}{2^4} = \frac{77^2}{16} = 370.56$$

$$\text{Contrast of } AC = (a-1)(b+1)(c-1)(d+1)$$

$$SS_{AC} = \frac{(\text{contrast of } AC)^2}{2^4} = \frac{-9^2}{16} = 5.06$$

$$\text{Contrast of } BC = (a+1)(b-1)(c-1)(d+1)$$

$$SS_{BC} = \frac{(\text{contrast of } BC)^2}{2^4} = \frac{-59^2}{16} = 217.56$$

$$\text{Contrast of } ABC = (a-1)(b-1)(c-1)(d+1)$$

$$SS_{ABC} = \frac{(\text{contrast of } ABC)^2}{2^4} = \frac{15^2}{16} = 14.06$$

$$\text{Contrast of } D = (a+1)(b+1)(c+1)(d-1)$$

$$SS_{D} = \frac{(\text{contrast of } D)^2}{2^4} = \frac{175^2}{16} = 1914.06$$

$$\text{Contrast of } AD = (a-1)(b+1)(c+1)(d-1)$$

$$SS_{AD} = \frac{(\text{contrast of } AD)^2}{2^4} = \frac{-7^2}{16} = 3.06$$

$$\text{Contrast of } BD = (a+1)(b-1)(c+1)(d-1)$$

$$SS_{BD} = \frac{(\text{contrast of } BD)^2}{2^4} = \frac{7^2}{16} = 3.06$$

$$\text{Contrast of } ABD = (a-1)(b-1)(c+1)(d-1)$$

$$SS_{ABD} = \frac{(\text{contrast of } ABD)^2}{2^4} = \frac{-7^2}{16} = 3.06$$

$$\text{Contrast of } CD = (a+1)(b+1)(c-1)(d-1)$$

$$SS_{CD} = \frac{(\text{contrast of } CD)^2}{2^4} = \frac{-3^2}{16} = 0.56$$

$$\text{Contrast of } ACD = (a-1)(b+1)(c-1)(c-1)$$

$$SS_{ACD} = \frac{(\text{contrast of } ACD)^2}{2^4} = \frac{3^2}{16} = 0.56$$

$$\text{Contrast of } BCD = (a+1)(b-1)(c-1)(d-1)$$

$$SS_{BCD} = \frac{(\text{contrast of } BCD)^2}{2^4} = \frac{1^2}{16} = 0.06$$

$$\text{Contrast of } ABCD = (a-1)(b-1)(c-1)(d-1)$$

$$SS_{ABCD} = \frac{(\text{contrast of } ABCD)^2}{2^4} = \frac{7^2}{16} = 3.06$$

By combining the sums of squares for *ABC*, *ABD*, *ACD*, *BCD*, and *ABCD* interactions, we estimate the error sum of squares as follows:

$$SS_{ABC} + SS_{ABD} + SS_{ACD} + SS_{BCD} + SS_{ABCD}$$
$$= 14.06 + 3.06 + 0.56 + 0.06 + 3.06 = 20.80$$

We perform ANOVA as displayed in Table 7.15.

TABLE 7.15: ANOVA for the Stability of Vehicle Experiment

Source	Sum of squares	Degrees of freedom	Mean square	F-ratio
A	0.06	1	0.06	0.0144
B	1173.06	1	1173.06	281.985
AB	1.56	1	1.56	0.375
C	370.56	1	370.56	89.077
AC	5.06	1	5.06	1.216
BC	217.56	1	217.56	52.298
D	1914.06	1	1914.06	460.110
AD	3.06	1	3.06	0.735
BD	3.06	1	3.06	0.735
CD	0.56	1	0.56	0.135
Error	20.80	5	4.160	
Total	3709.40	15		

Because $F(1, 5, 0.01) = 16.26$, only B, C, BC, and D factorial effects are significant. Since BC is significant, the significance of the main effects of B and C has very little meaning because they were not acting independently of each other. ∎

7.5.2 Obtaining Estimates of Responses

Since only B, C, BC, and D are significant, we can represent the estimates of each response in terms of these significant factorial effects. Using methods discussed previously, we have obtained the estimates as $B = -17.125$, $C = 9.625$, $BC = -7.375$, and $D = 21.875$. The estimates of the measure of instability will usually occur at the vertices of the design, and this can be represented in terms of the equation

$$y = 49.6875 - \left(\frac{17.125}{2}\right) x_1 + \left(\frac{9.625}{2}\right) x_2 - \left(\frac{7.375}{2}\right) x_1 x_2 + \left(\frac{21.875}{2}\right) x_3 \tag{7.15}$$

where 49.6875 is the overall average response, and x_1, x_2, and x_3 are variables associated with the factors, B, C, and D. Each of the x's takes values -1 or $+1$ depending on whether the factor with which it is associated is present or absent for the particular response being estimated using Equation 7.13. This regression equation was supposed to measure the effect on response of unit change in the level of the factor, but since the effects we have calculated are for change from -1 to $+1$, we have actually had two units of change. So, we divide each estimate by 2. As an illustration, the estimate for the treatment combination "a" is

$$y = 49.6875 - \left(\frac{17.125}{2}\right)(-1) + \left(\frac{9.625}{2}\right)(-1) - \left(\frac{7.375}{2}\right)(-1)(-1) + \left(\frac{21.875}{2}\right)(-1) = 38.8125$$

The true response for that treatment combination is 40.00 so that we have the residual as +1.1875; similar substitution for bc that has a response of 30.00 would lead to

$$y = 49.6875 - \left(\frac{17.125}{2}\right)(+1) + \left(\frac{9.625}{2}\right)(+1) - \left(\frac{7.375}{2}\right)(+1)(+1) + \left(\frac{21.875}{2}\right)(-1) = 31.3125$$

The residual for this estimate is −1.3125. In this way, it is possible to estimate all the treatment effects and obtain the appropriate residuals attached to each estimate. Only after these calculations can residual analysis (discussed in detail in Chapter 2) be carried out to check the validity of the model. For the design represented in this problem, we present estimates of factorial effects, residual, and ordered residuals in Tables 7.16 and 7.17. The residual and normal probability plots are shown in Figures 7.5 and 7.6.

TABLE 7.16: Estimates of Factorial Effects/Residuals (Example 7.3)

Treatment combinations	Responses	Estimates or fitted values	Residuals
a	40	38.8125	1.1875
b	27	29.0625	−2.0625
ab	30	29.0625	0.9375
c	58	55.8125	2.1875
ac	55	55.8125	0.1875
bc	30	31.3125	−1.3125
abc	32	31.3125	0.6875
d	59	60.6875	−1.6875
ad	62	60.6875	1.3175
bd	53	50.9375	2.0625
abd	50	50.9375	0.9375
cd	79	77.6875	1.3125
acd	75	77.6875	−2.6875
bcd	53	53.1875	−0.1875
abcd	54	53.1875	0.8125

TABLE 7.17: Ordered Residuals and Their Cumulative Probabilities

Order	Residuals	Cumulative probability
15	2.1875	.9667
14	2.0625	.9
13	1.3175	.8333
12	1.3125	.7667
11	1.1875	.7
10	0.9375	.6333
9	0.9375	.5667
8	0.8125	.5
7	0.6875	.4333
6	0.1875	.3667
5	−0.1875	.3
4	−1.3125	.2333
3	−1.6875	.1667
2	−2.0625	.1
1	−2.6875	.0333

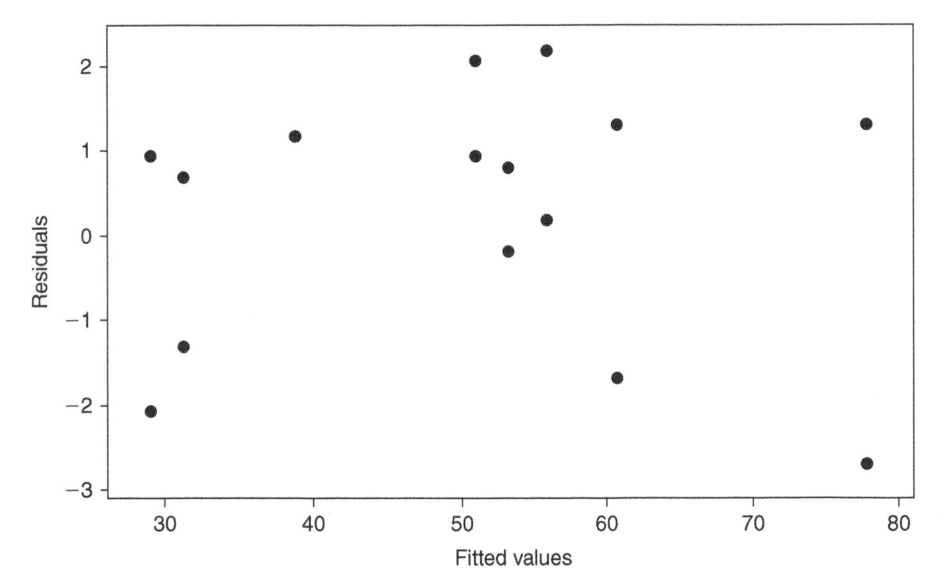

FIGURE 7.5: Plot of residuals against fitted values for Example 7.3.

No gross violation of model assumptions can be inferred from the plot of the residuals against fitted values as indicated in Figure 7.5. There seems to be some concern about normality of the data as evidenced by all the points not being on the straight line in Figure 7.6. The concern is, however, mitigated when we take into account the size of the sample, which is small. The p-value for the Anderson–Darling test is 0.076 which would not be significant for 5% level of significance or less. Although the p-value is relatively low, we

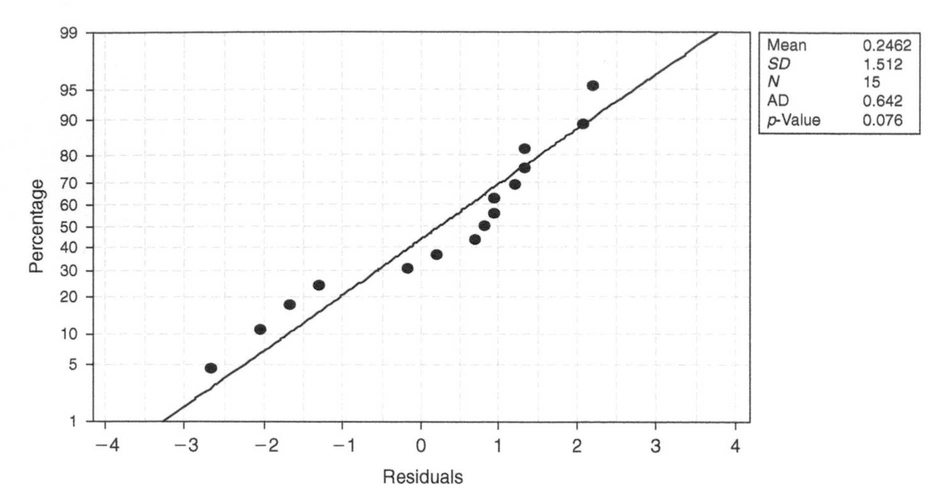

FIGURE 7.6: Normal probability plot for residuals of the vehicle stability experiment (Example 7.3).

note that p-value is based on only 15 residuals. A larger sample could have led to a clearer indication that the data are normal. We would, therefore, not be overly concerned about the normality assumption as it pertains to this single replicate of the design.

Although in Example 7.3 we were running a single replicate of the 2^4 factorial design, using the higher order interactions to estimate error as is usual, we can still write a SAS program for that purpose. This time, we instruct SAS not to fit the higher order interactions used for the estimation of error, thereby ensuring that their sums of squares are considered by SAS as error. Next, we present the SAS program for analyzing the data of Example 7.3 and the output.

SAS PROGRAM

```
DATA single1;
INPUT a b c d Y @@;
CARDS;
0 0 0 0 38 1 0 0 0 40 0 1 0 0 27 1 1 0 0 30 0 0 1 0 58 1 0 1 0 55
0 1 1 0 30 1 1 1 0 32 0 0 0 1 59 1 0 0 1 62 0 1 0 1 53 1 1 0 1 50
0 0 1 1 79 1 0 1 1 75 0 1 1 1 53 1 1 1 1 54
;
proc print data=single1; sum a b c d y;
PROC anova DATA=single1;
class a b c d;
MODEL Y=a b a*b c a*c b*c d a*d b*d c*d;
RUN;
```

SAS OUTPUT

```
Obs   a   b   c   d    Y
  1   0   0   0   0   38
  2   1   0   0   0   40
```

Obs	a	b	c	d	Y
3	0	1	0	0	27
4	1	1	0	0	30
5	0	0	1	0	58
6	1	0	1	0	55
7	0	1	1	0	30
8	1	1	1	0	32
9	0	0	0	1	59
10	1	0	0	1	62
11	0	1	0	1	53
12	1	1	0	1	50
13	0	0	1	1	79
14	1	0	1	1	75
15	0	1	1	1	53
16	1	1	1	1	54
	=	=	=	=	===
	8	8	8	8	795

The ANOVA Procedure

Class Level Information

Class	Levels	Values
a	2	0 1
b	2	0 1
c	2	0 1
d	2	0 1

Number of Observations Read 16
Number of Observations Used 16

Dependent Variable: Y

Source	DF	Sum of Squares	Mean Square	F Value	Pr > F
Model	10	3688.625000	368.862500	88.62	<.0001
Error	5	20.812500	4.162500		
Corrected Tota	15	3709.437500			

R-Square	Coeff Var	Root MSE	Y Mean
0.994389	4.106104	2.040221	49.68750

Source	DF	Anova SS	Mean Square	F Value	Pr > F
a	1	0.062500	0.062500	0.02	0.9072
b	1	1173.062500	1173.062500	281.82	<.0001
a*b	1	1.562500	1.562500	0.38	0.5669
c	1	370.562500	370.562500	89.02	0.0002
a*c	1	5.062500	5.062500	1.22	0.3203
b*c	1	217.562500	217.562500	52.27	0.0008
d	1	1914.062500	1914.062500	459.83	<.0001
a*d	1	3.062500	3.062500	0.74	0.4302
b*d	1	3.062500	3.062500	0.74	0.4302
c*d	1	0.562500	0.562500	0.14	0.7282

7.5.3 Dealing with Significant Higher Order Interactions in the Single and Half Replicates of a 2^k Factorial Design

The practice of using three-factor or higher order interactions for estimating error in the single or half replicates of the 2^k factorial design is based on the assumption that the higher order effects are negligible and that under such circumstances, theory indicates that the expected value of the sums of squares for each of the higher order interaction effect is the same as the variance of error σ^2. These assumptions clearly do not hold when the higher order interaction is significant. Since the rules are violated if they are significant, we need to make sure that when this happens, these interaction effects are not employed in ANOVA. A method for dealing with this involves the screening of all the factorial effects before choosing those whose contributions are negligible; these sums of squares should be pooled together to estimate error for use in ANOVA. The process of screening involves the ordering of estimates of the effects according to their magnitudes and calculating their cumulative probabilities. These effects are then plotted on a normal probability paper. Daniel (1959) suggests that better results could be obtained by plotting the effects on "half normal" paper. It is expected that when the plot is carried out, those estimates of factorial effects that are significant will stand out while the rest that cannot be distinguished from error will all stand in a straight line. In this way, it is possible to identify the higher order interactions that are significant and thus avoid using them as estimates of the common error variance σ^2. This method is illustrated with the results of Example 7.3 and the ordered estimates are in Table 7.18 while the plot is shown in Figure 7.7.

TABLE 7.18: Ordered Effects for Problem 7.3 and Their Cumulative Probabilities

Order	Effects	Cumulative probability
15	21.875	.9667
14	9.625	.9
13	1.375	.8333
12	0.875	.7667
11	0.875	.7
10	0.625	.6333
9	0.375	.5667
8	0.125	.5
7	0.125	.4333
6	−0.375	.3667
5	−0.875	.3
4	−0.875	.2333
3	−1.125	.1667
2	−7.375	.1
1	−17.125	.0333

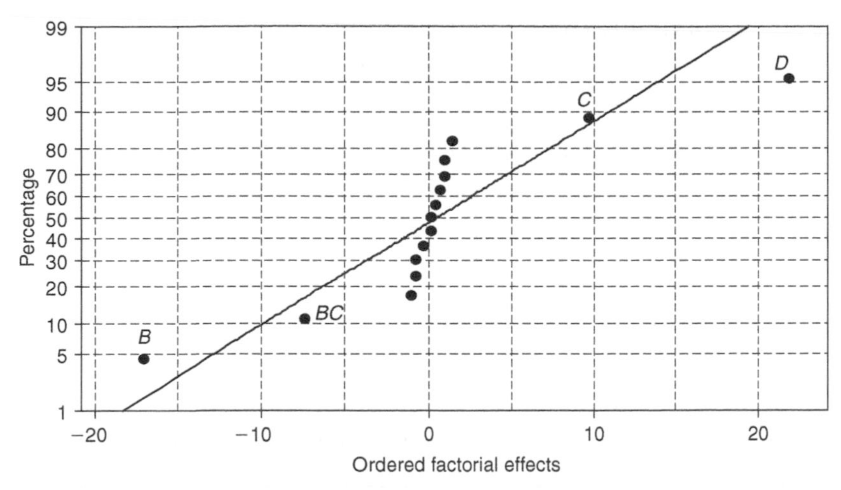

FIGURE 7.7: Cumulative probability plot of ordered effects for the design of Example 7.3.

7.5.4 Collapsing a Single Replicate of 2^k Factorial Design into a Full 2^{k-1} Factorial Design

The result of the analysis of data of Example 7.3 indicates that only B, C, BC, and D are significant. Because neither A nor any other interaction involving A is significant, we notice that we could actually have analyzed the data as if they were coming from a 2^3 (i.e., 2^{4-1}) instead of the single replicate of the 2^3 factorial design used in the previous analysis. If this method of analysis were adopted, A would be disregarded and treatment combinations such as a, ab, ac, etc. would be listed as (1), b, c, respectively. The reader should check that this would result in a 2^3 factorial design in two replicates; this in turn means that we can estimate treatment effects as well as error. This is a method for treating responses of an experiment when the initial analysis indicates that some factors are not significant. This analysis was carried out for the responses of Example 7.3. The ANOVA is presented in Table 7.19.

$$\text{Contrast of } B = (b-1)(c+1)(d+1)$$

$$SS_B = \frac{(\text{contrast of } B)^2}{2^4} = \frac{137^2}{16} = 1173.0625$$

$$\text{Contrast of } C = (b+1)(c-1)(d+1)$$

$$SS_C = \frac{(\text{contrast of } C)^2}{2^4} = \frac{77^2}{16} = 370.5625$$

$$\text{Contrast of } BC = (b-1)(c-1)(d+1)$$

$$SS_{BC} = \frac{(\text{contrast of } BC)^2}{2^4} = \frac{(-59)^2}{16} = 217.5625$$

$$\text{Contrast of } D = (b+1)(c+1)(d-1)$$

$$SS_D = \frac{(\text{contrast of } D)^2}{2^4} = \frac{175^2}{16} = 1914.0625$$

$$\text{Contrast of } BD = (b-1)(c+1)(d-1)$$

$$SS_{BD} = \frac{(\text{contrast of } BD)^2}{2^4} = \frac{7^2}{16} = 3.0625$$

$$\text{Contrast of } CD = (b+1)(c-1)(d-1)$$

$$SS_{CCD} = \frac{(\text{contrast of } CD)^2}{2^4} = \frac{(-3)^2}{16} = 0.5625$$

$$\text{Contrast of } BCD = (b-1)(c-1)(d-1)$$

$$SS_{BCD} = \frac{(\text{contrast of } BCD)^2}{2^4} = \frac{1^2}{16} = 0.0625$$

TABLE 7.19: ANOVA for the 2^3 Factorial Design in B,C, and D

Sources	Sum of squares	Degrees of freedom	Mean square	F-ratio
B	1173.0625	1	1173.0625	308.0673
C	370.5625	1	370.0625	97.3164
BC	217.5625	1	217.5625	57.1358
D	1914.0625	1	1914.0625	502.667
BD	3.0625	1	3.0625	0.804
CD	0.5625	1	0.5625	0.1477
BCD	0.0625	1	0.0625	0.0164
Residual	30.4625	8	3.807813	
Total	3709.400	15		

From the tables, $F(1, 8, 0.01) = 11.26$. We see that the results of this analysis merely confirm what we had discovered by studying the stability of the vehicle using a single replicate of the 2^4 factorial design: that only B, C, BC, and D are significant. This is an obvious example where a single replicate of the 2^4 factorial design has served as an exploratory study and indicated which factors are significant.

7.6 3^k Factorial Designs

So far in this chapter, we have discussed factorial designs with factors set at two levels only. It is clear, however, that factors could assume more than two levels in an experiment. The simplest case when all the levels of the factors are above two is the 3^2 factorial design. This factorial design requires nine units to obtain a single replicate of the experiment. Since each of the two factors involved in this design has three levels, we denote these levels by 0, 1, and 2 (corresponding to the system using the 2^k factorial design). Suppose that the factors involved in our design are A and B. Let the levels of A be represented by x_1 and those of B by x_2. When $x_3 = 0$, the level of A is low; $x_1 = 1$ refers to an intermediate level of A; while $x_1 = 2$ indicates that A is at high level. Similarly, when $x_2 = 0$, 1, and 2, then the levels of B is at low, intermediate, and high levels, respectively. By following the convention of the 2^k factorial design which we discussed earlier, we can list the nine treatment combinations in this design as

$$\begin{aligned}(a_0b_0 \quad a_0b_1 \quad a_0b_2) &\equiv [(0,0),(0,1),(0,2)]\\ (a_1b_0 \quad a_1b_1 \quad a_1b_2) &\equiv [(1,0),(1,1),(1,2)]\\ (a_2b_0 \quad a_2b_1 \quad a_2b_2) &\equiv [(2,0),(2,1),(2,2)]\end{aligned}$$

From Figure 7.8 (below), we can see that it is possible to use modulo arithmetic to obtain the levels for the main effects of this design. The main effect of A represents the comparison of the following groups of points:

(0, 0), (0, 1), (0, 2) for each of which $x_1 = 0$ (modulo 3)
(1, 0), (1, 2), (1, 2) for each of which $x_2 = 1$ (modulo 3)
(2, 0), (2, 1), (2, 2) for each of which $x_3 = 3$ (modulo 3)

Similarly, the main effects of B are obtained as a result of the comparison of points for which $x_2 = 0$, 1, 2 (modulo 3), respectively. Reduction of numbers modulo n means that any number divided by n is replaced by the remainder after the division (e.g., 4 modulo 3 = 1; 3 modulo 3 = 0; 0 modulo 3 = 0, and so on).

Thus, each main effect is the comparison of the effect of the factor at three different levels, $x_i = 0$, 1, 2. Each main effect has 2 degrees of freedom. The interaction AB has 4 degrees of freedom. The degrees of freedom

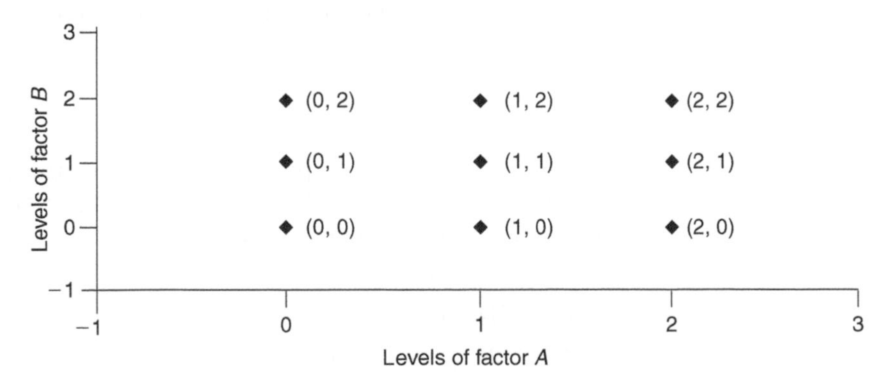

FIGURE 7.8: The nine points of the 3^2 factorial design.

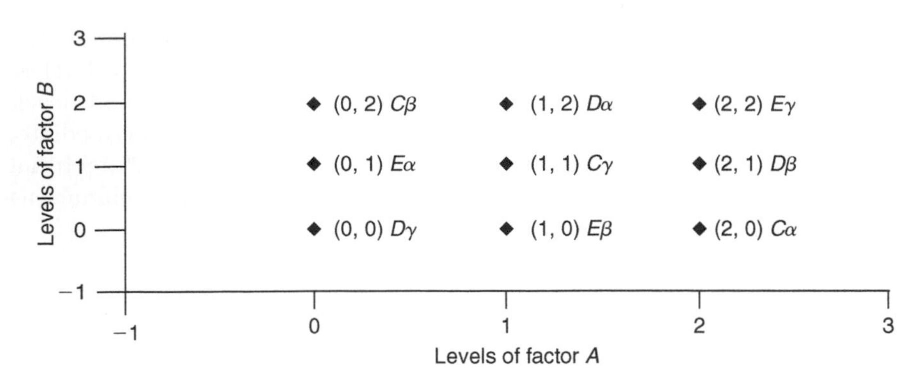

FIGURE 7.9: The decomposition of the AB interaction in a 3^2 factorial experiment into its component parts (Latin letters refer to factor A; Greek letters refer to factor B).

of AB can be decomposed into its component parts to reveal how they arose. This decomposition of the AB into its component parts is best illustrated by considering it as arising from a superposition of a 3×3 Latin square design on another 3×3 Latin square design to produce an orthogonal design or a Graeco-Latin square design. The proposed design is shown in Figure 7.9 where Latin letters represent the factor A while Greek letters represent the factor B.

We consider those treatments represented by the Latin Letters:

$$C\text{: } (0,\ 2),\ (1,\ 1),\ (2,\ 0)$$
$$D\text{: } (0,\ 0),\ (1,\ 1),\ (2,\ 1)$$
$$E\text{: } (0,\ 1),\ (1,\ 0),\ (2,\ 2)$$

For those treatment combinations in C, $x_1 + x_2 = 2,\ 2,\ 2 = 2$ (modulo 3)

For those treatment combinations in D, $x_1 + x_2 = 0,\ 3,\ 3 = 0$ (modulo 3)

For those in E, $x_1 + x_2 = 1,\ 1,\ 4 = 1$ (modulo 3)

We can see that the Latin letters represent comparisons between treatment combinations for which the value of $x_1 + x_2 = 0$, 1, and 2. If we consider the treatment combinations represented by the Greek letters:

$$\alpha\text{: } (0,\ 1),\ (1,\ 2),\ (2,\ 0)$$
$$\beta\text{: } (0,\ 2),\ (1,\ 0),\ (2,\ 1)$$
$$\gamma\text{: } (0,\ 0),\ (1,\ 1),\ (2,\ 2)$$

For those treatment combinations under α: $x_1 + 2x_2 = 2,\ 5,\ 2 = 2$ modulo 3

For those treatments under β: $x_1 + 2x_2 = 4,\ 1,\ 4 = 1$ modulo 3

For those treatments under γ: $x_1 + 2x_2 = 0,\ 3,\ 6 = 0$ modulo 3

We have shown that each of the main effects of A and B is obtained as a comparison with treatments for which $x_i = 0,\ 1,\ 2$ modulo 3; $(i = 1, 2)$. We have further shown that the AB interaction is composed of two parts AB and AB^2 whose treatment combinations have values 0, 1, 2 modulo 3, obtained by solving $x_1 + x_2$ modulo 3 and $x_1 + 2x_2$ modulo 3. Each of these factorial effects has two degrees of freedom associated with it, and since the AB interaction is made up AB and AB^2, there is no special gain in separately testing for the interactions AB and AB^2, in ANOVA. We will notice that the component parts of interactions between factors, which increase in number and complexity as the increase in number of factors in the design, do not appear in the ANOVA. The splitting of the interactions finds use in confounding and fractional replication of the 3^k factorial designs.

Since we have shown that both the main effects and the interactions have levels 0, 1, and 2 modulo 3, we could (as a notation) use these levels as subscripts to denote which level of the factorial effect we are considering. Using this, for the 3^2 factorial design with n replicates, the sums of squares for the main effects and interactions could be calculated as:

$$SS_A = \frac{[A_0^2 + A_1^2 + A_2^2]}{3n} - \frac{G^2}{3^2 n} = \frac{\sum_{i=0}^{2} A_i^2}{3n} - \frac{G^2}{3^2 n}$$

$$SS_B = \frac{[B_0^2 + B_1^2 + B_2^2]}{3n} - \frac{G^2}{3^2 n} = \frac{\sum_{i=0}^{2} B_i^2}{3n} - \frac{G^2}{3^2 n}$$

$$SS_{AB} = \frac{[(AB)_0^2 + (AB)_1^2 + (AB)_2^2]}{3n} - \frac{G^2}{3^2 n} = \frac{\sum_{i=0}^{2} (AB)_i^2}{3n} - \frac{G^2}{3^2 n}$$

$$SS_{AB^2} = \frac{[(AB^2)_0^2 + (AB^2)_1^2 + (AB^2)_2^2]}{3n} - \frac{G^2}{3^2 n} = \frac{\sum_{i=0}^{2} (AB^2)_i^2}{3n} - \frac{G^2}{3^2 n}$$

Here, A_1 is the sum of all observations when the factor A is at low level, A_2 is the sum of all observations when A is at intermediate level, and A_3 is the sum of all observations when A is at high level. Similar interpretation can be applied to other effects that appear in the formulae above. This method identifies the sums of squares of components of each factorial effect. For instance, AB is divided into AB and AB^2. But we note that this design simply involves two factors each set at three levels so that the model that applies is

$$y_{ijk} = \mu + \alpha_i + \beta_j + (\alpha\beta)_{ij} + \varepsilon_{ijk} \quad i, j = 0, 1, 2;\; k = 1, \ldots, n \tag{7.16}$$

This means that we can also obtain the above sums of squares as usual by using the following formulae:

$$SS_A = \sum_{i=0}^{2} \frac{y_{i..}^2}{3n} - \frac{G^2}{3^2 n}$$

$$SS_B = \sum_{i=0}^{2} \frac{y_{.j.}^2}{3n} - \frac{G^2}{3^2 n}$$

$$SS_{AB} = \sum_{i=0}^{2}\sum_{j=0}^{2} \frac{y_{ij.}^2}{3n} - \frac{G^2}{3^2 n} - SS_A - SS_B \tag{7.17a}$$

$$SS_{\text{total}} = \sum_{i=0}^{2}\sum_{j=0}^{2}\sum_{k=1}^{n} y_{ijk}^2 - \frac{G^2}{3^2 n}$$

$$SS_{\text{E}} = SS_{\text{total}} - SS_A - SS_B - SS_{AB}$$

The sum of squares for treatment is easily obtained by using the sum of responses for each treatment combination from all replicates as

$$SS_{\text{treatment}} = \sum_{i=0}^{2}\sum_{j=0}^{2} \frac{y_{ij.}^2}{n} - \frac{G^2}{3^2 n} \tag{7.17b}$$

In this design, the residual sum of squares is associated with $3^2(n-1)$ degrees of freedom. G is used in the above formulae as the grand total. We present an example to illustrate this design.

Example 7.4

A company that was carrying out research to improve the quality of beer produced from guinea corn was studying the alcohol content of fermented guinea corn. Since it seemed that the fermentation period and temperature at which the fermented corn was dried would impact the alcohol content of the beer, these factors were studied. Time of fermentation, A, was set at three levels (2, 3, and 4 days) while drying temperature, B was also set at three levels (20°C, 25°C, and 30°C). Only two replicates of the experiments were performed, and the responses (percentage alcohol) are given as follows:

		Fermentation period (days)			
Drying temperature (°C)		**2**	**3**	**4**	**Total *(B)***
(C)	Levels	0	1	2	
20	0	44, 45.8 (89.8)	44.5, 38.1 (82.6)	32, 35.20 (67.2)	239.6
25	1	39.1, 42.7 (81.8)	34.6, 29.1 (63.7)	31.1, 37.0 (68.1)	213.6
30	2	34, 36.2 (70.2)	26.9, 30.7 (57.6)	30.9, 33.5 (64.4)	192.2
Total *(A)*		241.8	203.9	199.7	645.4

The figures in the parentheses represent the totals for the cells. Is any of the factors significant at $\alpha = 1\%$?

The total sum of squares for this experiment is

$$SS_{\text{total}} = \sum_{i=0}^{2}\sum_{j=0}^{2}\sum_{k=1}^{n} y_{ijk}^2 - \frac{G^2}{3^2 n} = [(44)^2 + (45.84)^2 + \cdots + (33.5)^2]$$

$$- \frac{(645.4)^2}{18} = 529.244$$

With the sums of the replicates in parentheses, we can calculate the treatment sum of squares as

$$SS_{\text{treatment}} = \sum_{i=0}^{2}\sum_{j=0}^{2} \frac{y_{ij.}^2}{n} - \frac{G^2}{3^2 n} = \frac{[(89.8)^2 + (82.56)^2 + \cdots + (62.4)^2]}{2}$$

$$- \frac{(645.4)^2}{18} = 449.994$$

$$SS_{\text{E}} = 529.244 - 449.994 = 79.25$$

We now show that the treatment sum of squares is subdivided according to factorial effects as

$$SS_A = \frac{[(239.6)^2 + (213.6)^2 + (192.2)^2]}{6} - \frac{(645.4)^2}{18} = 187.818$$

$$SS_B = \frac{[(241.8)^2 + (203.9)^2 + (199.7)^2]}{6} - \frac{(645.4)^2}{18} = 179.248$$

As we stated in earlier discussion, the AB interaction has two parts, AB and AB^2. When calculating the sums of squares manually, these effects are obtained separately, and their sum is used as the AB interaction in ANOVA. Most computer packages will simply give the sum of squares for the AB interaction, which is the total sum of squares for all effects under AB. To obtain the AB part of the interaction, we consider the points for which $x_1+x_2=0$, 1, 2 (modulo 3).

Sum of responses for which $x_1+x_2=0$ is $89.8+68.1+57.6=215.5$

Sum of responses for which $x_1+x_2=1$ is $82.6+81.8+64.4=228.80$

Sum of responses for which $x_1+x_2=2$ is $67.2+63.7+70.2=201.10$

This means that according to our notation in Equation 6.14, $(AB)_0=215.5$, $(AB)_1=228.80$, and $(AB)_2=201.1$ and we obtain the sum of squares as

$$SS_{AB} = \frac{[(215.5)^2+(228.8)^2+(201.1)^2]}{6} - \frac{(645.4)^2}{18} = 63.9744$$

Similarly, we can obtain all the sums for which $x_1+2x_2=0$, 1, 2 (modulo 3) to calculate the sum of squares for AB^2 interaction. The sums are

For $x_1+2x_2=0: 89.8+63.7+64.4=217.9$

For $x_1+2x_2=1: 67.2+81.8+57.6=206.6$

For $x_1+2x_2=2: 82.6+68.1+70.2=220.9$

So that

$$SS_{AB^2} = \frac{[(217.9)^2+(206.6)^2+(220.9)^2]}{6} - \frac{(645.4)^2}{18} = 18.9544$$

The AB interaction has the total sum of squares, $SS_{AB(\text{inter})}=82.929$. As a check that our calculations are accurate, we note that

$$SS_A + SS_B + SS_{AB(\text{inter})} = 187.818 + 179.248 + 82.929 = 449.994$$

which is exactly the same value as the treatment sum of squares that we had earlier obtained. We now perform the analysis of variance as usual; the ANOVA for this experiment is presented in Table 7.20. Before the ANOVA is carried out for this design, it is important to recall that the ANOVA being carried out here is essentially the same as that for the two-way classification with replication which we discussed in Chapter 2. Here, each factor was held at only three levels with two replicates. We therefore state the hypotheses being tested as follows:
For factor A,

$$H_0 : \alpha_0 = \alpha_1 = \alpha_2 = 0 \text{ vs } H_1 : \alpha_i \neq 0 \quad \text{for at least one } i$$

$$H_0' : \beta_0 = \beta_1 = \beta_2 = 0 \text{ vs } H_1' : \beta_j \neq 0 \quad \text{for at least one } j$$

$$H_0'' : \alpha_i\beta_0 = \alpha_i\beta_1 = \alpha_i\beta_2 = 0 \text{ vs } H_0'' : (\alpha\beta)_{ij} \neq 0 \quad \text{for a fixed } i \text{ and some } j$$

TABLE 7.20: ANOVA for the Guinea-Corn Fermentation Problem

Sources	Sum of squares	Degrees of freedom	Mean square	*F*-ratio
A	187.818	2	93.909	10.66
B	179.248	2	89.624	10.15
AB (inter)	82.929	4	20.732	2.35
Residual	79.250	9	8.806	
Total	529.244	17		

Because $F(2, 9, 0.01) = 8.02$ and $F(4, 9, 0.01) = 6.42$, we see that only the main effects of A and B are significant at 1% level of significance. As usual, residual analysis, which we discussed in Chapter 2, should follow to ensure that no gross violation of the basic assumptions of the model has taken place. This is left as an exercise for the reader. The analysis should be carried out, and the extent to which the assumptions are satisfied should be discussed. Next, we present a SAS program for analyzing the data and the ensuing output for Example 7.4. □

SAS PROGRAM

```
DATA ferment;
INPUT temp days Y @@;
CARDS;
0 0 44 0 0 45.8 0 1 44.5 0 1 38.1 0 2 32 0 2 35.2
1 0 39.1 1 0 42.7 1 1 34.6 1 1 29.1 1 2 31.1 1 2 37
2 0 34 2 0 36.2 2 1 26.9 2 1 30.7 2 2 30.9 2 2 33.5
;
proc print data=ferment; sum temp days y;
PROC anova DATA=ferment;
class temp days;
MODEL Y=temp|days; RUN;
```

SAS OUTPUT

```
 Obs    temp    days      Y

   1      0       0     44.0
   2      0       0     45.8
   3      0       1     44.5
   4      0       1     38.1
   5      0       2     32.0
   6      0       2     35.2
   7      1       0     39.1
   8      1       0     42.7
   9      1       1     34.6
```

Obs	temp	days	Y
10	1	1	29.1
11	1	2	31.1
12	1	2	37.0
13	2	0	34.0
14	2	0	36.2
15	2	1	26.9
16	2	1	30.7
17	2	2	30.9
18	2	2	33.5
	====	====	=====
	18	18	645.4

The ANOVA Procedure

Class Level Information

Class	Levels	Values
temp	3	0 1 2
days	3	0 1 2

Number of Observations Read 18
Number of Observations Used 18

The ANOVA Procedure

Dependent Variable: Y

Source	DF	Sum of Squares	Mean Square	F Value	Pr > F
Model	8	449.9944444	56.2493056	6.39	0.0059
Error	9	79.2500000	8.8055556		
Corrected Total	17	529.2444444			

R-Square	Coeff Var	Root MSE	Y Mean
0.850258	8.276027	2.967416	35.85556

Source	DF	Anova SS	Mean Square	F Value	Pr > F
temp	2	187.8177778	93.9088889	10.66	0.0042
days	2	179.2477778	89.6238889	10.18	0.0049
temp*days	4	82.9288889	20.7322222	2.35	0.1316

7.6.1 3^3 Factorial Design

When there are three factors, each at three levels, to obtain a single replicate of the experiment, 3^3, that is, 27 observations are required. This is the number of treatment combinations that make up this design. Each of the two main effects has 2 degrees of freedom; each of the 2 two-factor interactions has 4 degrees of freedom; there is only 1 three-factor interaction that has 8 degrees of freedom. There are only 26 degrees of freedom in one replicate of the design that is exhausted by the main and the interaction effects. To perform a full experiment of this design and estimate all effects and error, we need at least two replicates. When n replicates of the design are used, the total number of degrees of freedom is $3^3(n-1)$; in such situations, the degrees of freedom left for estimation of error is $3^3(n-1)$. The model for this design is

$$y_{ijkl} = \mu + \alpha_i + \beta_j + (\alpha\beta)_{ij} + \gamma_k + (\alpha\gamma)_{ik} + (\beta\gamma)_{jk} + (\alpha\beta\gamma)_{ijk} + \varepsilon_{ijkl}$$
$$i, j, k = 0, 1, 2 \quad l = 1, \ldots, n \tag{7.18}$$

so that taking a cue from Equation 7.17a, we obtain the sums of squares for the factorial effects without splitting them into their components as

$$\begin{aligned}
SS_A &= \sum_{i=0}^{2} \frac{y_{i\ldots}^2}{3^2 n} - \frac{G^2}{3^3 n} \\
SS_B &= \sum_{j=0}^{2} \frac{y_{.j..}^2}{3^2 n} - \frac{G^2}{3^3 n} \\
SS_{AB} &= \sum_{i=0}^{2} \sum_{j=0}^{2} \frac{y_{ij.}^2}{3n} - \frac{G^2}{3^3 n} - SS_A - SS_B \\
SS_C &= \sum_{k=0}^{2} \frac{y_{..k.}^2}{3^2 n} - \frac{G^2}{3^3 n} \\
SS_{AC} &= \sum_{i=0}^{2} \sum_{k=0}^{2} \frac{y_{i.k.}^2}{3n} - \frac{G^2}{3^3 n} - SS_A - SS_C \\
SS_{BC} &= \sum_{j=0}^{2} \sum_{k=0}^{2} \frac{y_{.jk.}^2}{3n} - \frac{G^2}{3^3 n} - SS_B - SS_C \\
SS_{ABC} &= \sum_{i=0}^{2} \sum_{j=0}^{2} \sum_{k=0}^{2} \frac{y_{ijk.}^2}{n} - \frac{G^2}{3^3 n} - SS_A - SS_B - SS_{AB} \\
&\quad - SS_C - SS_{AC} - SS_{BC} \\
SS_{\text{total}} &= \sum_{i=0}^{2} \sum_{j=0}^{2} \sum_{k=0}^{2} \sum_{l=1}^{n} y_{ijkl}^2 - \frac{G^2}{3^3 n} \\
SS_{\text{E}} &= SS_{\text{total}} - SS_A - SS_B - SS_{AB} - SS_C \\
&\quad - SS_{AC} - SS_{BC} - SS_{ABC}
\end{aligned} \tag{7.19}$$

Suppose that the three factors involved in this design are A, B, and C, then all treatment combinations in a full replicate of the design could be listed in terms of the levels 0, 1, 2 (where 0, 1, 2 represent low, intermediate, and high), as follows:

0	**1**	**2**
0 0 0	0 0 1	0 0 2
0 1 0	0 1 1	0 1 2
0 2 0	0 2 1	0 2 2
1 0 0	1 0 1	1 0 2
1 1 0	1 1 1	1 1 2
1 2 0	1 2 1	1 2 2
2 0 0	2 0 1	2 0 2
2 1 0	2 1 1	2 1 2
2 2 0	2 2 1	2 2 2

In addition, the two-factor and three-factor interactions in 3^3 factorial design could further be partitioned into their component parts. This splitting of the interactions into their component parts is found useful in confounding and fractional replication. If we represent the levels of A by x_1, the levels of B by x_2, and those of C by x_3, then the observations that make up each factorial effect is obtained by the appropriate sum of levels of each factors whose letters appear in that effect (modulo 3). For each factorial effect, only those observations whose sum of levels (modulo 3) satisfies values 0, 1, and 2 for each component are included under that component. The factorial effects for this design, their degrees of freedom, and how to identify the responses belonging to them are listed in Table 7.21.

We illustrate how to obtain the sums of squares of this design using an example.

TABLE 7.21: Factorial Effects and Their Components in 3^k Factorial Design

Factorial effect in 3^k factorial design	Components of factorial effect	Degrees of freedom	How to identify responses in factorial effect
A		2	$x_1 = 0, 1, 2$ (modulo 3)
B		2	$x_2 = 0, 1, 2$ (modulo 3)
AB	AB	2	$x_1 + x_2 = 0, 1, 2$ (modulo 3)
	AB^2	2	$x_1 + 2x_2 = 0, 1, 2$ (modulo 3)
C		2	$x_3 = 0, 1, 2$ (modulo 3)
AC	AC	2	$x_1 + x_3 = 0, 1, 2$ (modulo 3)
	AC^2	2	$x_1 + 2x_3 = 0, 1, 2$ (modulo 3)
BC	BC	2	$x_2 + x_3 = 0, 1, 2$ (modulo 3)
	BC^2	2	$x_2 + 2x_3 = 0, 1, 2$ (modulo 3)
ABC	ABC	2	$x_1 + x_2 + x_3 = 0, 1, 2$ (modulo 3)
	ABC^2	2	$x_1 + x_2 + 2x_3 = 0, 1, 2$ (modulo 3)
	AB^2C	2	$x_1 + 2x_2 + x_3 = 0, 1, 2$ (modulo 3)
	AB^2C^2	2	$x_1 + 2x_2 + 2x_3 = 0, 1, 2$ (modulo 3)

Example 7.5
The yield of a chemical product from a pilot plant is thought to depend on temperature, catalyst concentration, and time of reaction. Each of these three factors was set at three levels, and the yields of the plant for the combinations of factors were studied. The following responses were obtained for two replicates of the experiment. The temperature levels were 25°C, 27.5°C, and 30°C. Catalyst concentration levels were 10%, 12.5%, and 15%; the yields of the plant for 10, 15, and 20 min were studied. For convenience, let temperature be A, catalyst concentration be B, and time be C. The yields of the experiment in kilograms are presented as follows:

		C					
		Replicate I			Replicate II		
B	***A***	**0**	**1**	**2**	**0**	**1**	**2**
0	0	4.20	9.8	14.26	4.93	10.10	13.79
	1	12.50	15.51	18.70	13.00	15.22	17.98
	2	16.80	19.00	24.20	15.96	18.60	22.58
1	0	5.11	11.90	13.30	5.31	11.34	15.64
	1	14.30	18.80	21.60	15.35	20.96	24.27
	2	19.86	26.37	29.72	20.88	24.38	29.64
2	0	6.70	12.40	16.20	7.00	13.15	17.10
	1	15.38	19.20	22.40	14.90	20.10	23.40
	2	21.30	27.30	31.40	24.00	27.20	29.64

We find the sum of the responses for the two replicates to use them for the calculation of the treatment sums of squares.

		C					
B	***A***	**0**	**1**	**2**	***A* (total)**		***B* (total)**
0	0	9.13	19.90	28.39			
	1	25.50	30.73	36.68	192.57	0	267.47
	2	32.76	37.60	46.78			
1	0	10.42	23.24	28.94			
	1	29.65	39.76	45.87	323.57	1	328.73
	2	40.74	50.75	59.36			
2	0	13.70	25.55	33.30			
	1	30.28	39.30	45.80	428.83	2	348.77
	2	45.30	54.50	61.04			
Totals		237.48	321.32	386.16			

Using the methods shown in Table 7.21, we identify the responses belonging to each of the main effects, we also identify those responses belonging to the interaction by obtaining the responses that belong to their component parts; we then use the sums of these responses to calculate the sums of squares corresponding to each factorial effect. The sums of the responses are shown in the table above. We obtain the sums of the responses for the interactions; first, we describe how they are obtained before presenting the results. For each factorial effect, its effect when its level is either 0, 1, or 2 is obtained by summing the nine responses that fall under each contrast value. For instance, consider the factorial effect AB^2C^2. The complete listing of all the treatment combinations belonging to this factorial effect, which fall under different levels of its contrasts, is given as follows:

AB^2C^2 Contrasts: $x_1 + 2x_2 + 2x_3 = 0\ 1\ 2$

0	**1**	**2**
0 0 0	0 0 2	0 0 1
0 1 2	0 1 1	0 1 0
0 2 1	0 2 0	0 2 2
1 1 0	1 0 0	1 0 2
1 0 1	1 1 2	1 1 1
1 2 2	1 2 1	1 2 0
2 2 0	2 0 1	2 0 0
2 0 2	2 1 0	2 1 2
2 1 1	2 2 2	2 2 1

By summing all the responses under 0, 1, 2, the total responses when this treatment combination (modulo 3) has these levels are obtained and used to calculate the sum of squares. The sums of responses for this and other factorial effects in the ABC interaction effects are presented below. Meanwhile, we have obtained the totals for other treatment combinations using the same method.

AB	$(x_1+x_2=0)$ 323.65	AB^2	$(x_1+2x_2=0)$ 333.54
	$(x_1+x_2=1)$ 316.35	AB^2	$(x_1+2x_2=1)$ 295.12
	$(x_1+x_2=2)$ 304.97	AB^2	$(x_1+2x_2=2)$ 316.31
AC	$(x_1+x_3=0)$ 304.45	AC^2	$(x_1+2x_3=0)$ 310.22
	$(x_1+x_3=1)$ 321.30	AC^2	$(x_1+2x_2=1)$ 318.91
	$(x_1+x_3=2)$ 319.22	AC^2	$(x_1+2x_3=2)$ 315.84
BC	$(x_2+x_3=0)$ 320.91	BC^2	$(x_2+2x_3=0)$ 321.28
	$(x_2+x_3=1)$ 309.18	BC^2	$(x_2+2x_3=1)$ 311.68
	$(x_2+x_3=2)$ 314.88	BC^2	$(x_2+2x_3=2)$ 312.01

By the same token, the total responses for different levels of the components of ABC interaction are

	Levels of ABC		
Components of ABC	**0**	**1**	**2**
ABC	309.72	317.12	318.13
AB^2C	313.56	315.62	315.79
ABC^2	314.57	317.68	312.72
AB^2C^2	312.63	315.38	316.96

The sum of all the responses obtained in the experiment is

$$G = (4.2 + 9.8 + \cdots + 27.2 + 29.64) = 944.97$$

We can obtain the total sum of squares by the usual method as

$$\begin{aligned} SS_{\text{total}} &= \sum_{i=0}^{2}\sum_{j=0}^{2}\sum_{k=0}^{2}\sum_{l=1}^{n} y_{ijkl}^2 - \frac{G^2}{3^3 n} \\ &= [(4.2)^2 + (9.8)^2 + \cdots + (27.2)^2 + (29.64)^2] - \frac{(944.97)^2}{54} \\ &= 2468.46 \end{aligned}$$

We can also obtain the sum of squares for the treatments by using the sum of the responses for the replicates as

$$\begin{aligned} SS_{\text{treatment}} &= \sum_{i=0}^{2}\sum_{j=0}^{2}\sum_{k=0}^{2} \frac{y_{ijk.}^2}{n} - \frac{G^2}{3^3 n} \\ &= \frac{[(9.13)^2 + (19.9)^2 + \cdots + (59.36)^2]}{2} - \frac{(944.97)^2}{54} = 2446.788 \end{aligned}$$

The treatment sum of squares is made up of the sums of squares of all the factorial effects in the design. Splitting these sums of squares into their components, we obtain the following (Table 7.22):

$$SS_A = \frac{1}{18}[(192.57)^2 + (323.57)^2 + (428.83)^2] - \frac{(944.97)^2}{54} = 1556.66$$

$$SS_B = \frac{1}{18}[(267.47)^2 + (328.73)^2 + (348.7)^2] - \frac{(944.97)^2}{54} = 199.33$$

$$SS_{AB} = \frac{1}{18}[(323.65)^2 + (316.35)^2 + (304.97)^2] - \frac{(944.97)^2}{54} = 9.85$$

$$SS_{AB^2} = \frac{1}{18}[(333.54)^2 + (316.31)^2 + (295.12)^2] - \frac{(944.97)^2}{54} = 41.148$$

$$SS_C = \frac{1}{18}[(237.48)^2 + (321.33)^2 + (386.16)^2] - \frac{(944.97)^2}{54} = 617.40$$

$$SS_{AC} = \frac{1}{18}[(304.45)^2 + (321.30)^2 + (319.22)^2] - \frac{(944.97)^2}{54} = 9.38$$

$$SS_{AC^2} = \frac{1}{18}[(310.22)^2 + (318.91)^2 + (315.84)^2] - \frac{(944.97)^2}{54} = 2.16$$

$$SS_{BC} = \frac{1}{18}[(320.91)^2 + (309.18)^2 + (314.88)^2] - \frac{(944.97)^2}{54} = 3.82$$

$$SS_{BC^2} = \frac{1}{18}[(321.28)^2 + (312.01)^2 + (311.68)^2] - \frac{(944.97)^2}{54} = 3.30$$

$$SS_{ABC} = \frac{1}{18}[(309.72)^2 + (317.12)^2 + (318.13)^2] - \frac{(944.97)^2}{54} = 2.34$$

$$SS_{ABC^2} = \frac{1}{18}[(314.57)^2 + (317.68)^2 + (312.72)^2] - \frac{(944.97)^2}{54} = 0.698$$

$$SS_{AB^2C} = \frac{1}{18}[(313.56)^2 + (315.62)^2 + (315.79)^2] - \frac{(944.97)^2}{54} = 0.171$$

$$SS_{AB^2C^2} = \frac{1}{18}[(312.63)^2 + (315.38)^2 + (316.96)^2] - \frac{(944.97)^2}{54} = 0.533$$

$$SS_{\text{residual}} = 2468.46 - 2446.788 = 21.672$$

TABLE 7.22: ANOVA for the Pilot Plant Problem

Source	Sum of squares	Degrees of freedom	Mean square	F-ratio
A	1556.66	2	778.33	969.6802
B	199.33	2	99.665	124.1674
AB interaction	50.998	4	12.7495	15.8839
C	617.400	2	308.7000	384.5930
AC interaction	11.540	4	2.8850	3.5943
BC interaction	7.120	4	1.7800	2.2176
ABC interaction	3.742	8	0.4677	0.5827
Residual	21.672	27	0.8027	
Total	2468.4160	53		

Because $F(2, 27, 0.01) = 5.49$, $F(4, 27, 0.01) = 4.11$, and $F(8, 27, 0.01) = 3.56$ only A, B, AB, and C are significant. Other analyses that are performed for the validation of model assumptions should follow. Next, we present a SAS program for analyzing this data, and the output for the analysis.

SAS PROGRAM

```
DATA EXAMP7_5regular;
INPUT a b c Y @@;
CARDS;
0 0 0 4.2 0 0 0 4.93 0 0 1 9.8 0 0 1 10.10
1 0 0 12.5 1 0 0 13.0 1 0 1 15.51 1 0 1 15.22
0 0 2 14.6 0 0 2 13.79 1 0 2 18.7 1 0 2 17.98
2 0 0 16.8 2 0 0 15.96 2 0 1 19 2 0 1 18.6
2 0 2 24.2 2 0 2 22.58 0 1 0 5.11 0 1 0
5.31 0 1 1 11.9 0 1 1 11.34 1 1 0 14.3 1 1 0 15.35 1 1 1 18.8 1 1 1 20.96 0 1
2 13.3 0 1 2 15.64 1 1 2 21.6 1 1 2 24.27
2 1 0 19.86 2 1 0 20.88 2 1 1 26.37 2 1 1 24.38
2 1 2 29.72 2 1 2 29.64 0 2 0 6.7 0 2 0 7 0 2 1 12.4 0 2 1 13.15
1 2 0 15.38 1 2 0 14.9 1 2 1 19.2 1 2 1 20.1 0 2 2 16.2 0 2 2 17.1 1 2 2 22.4
1 2 2 23.4 2 2 0 21.3 2 2 0 24 2 2 1 27.3 2 2 1 27.2
2 2 2 31.4 2 2 2 29.64
;
```

```
proc print data=examp7_5regular; sum a b c y;
PROC anova DATA=EXAMP7_5regular;
class a b c;
MODEL Y=a|b|c; RUN;
```

SAS OUTPUT

Obs	a	b	c	Y
1	0	0	0	4.20
2	0	0	0	4.93
3	0	0	1	9.80
4	0	0	1	10.10
5	1	0	0	12.50
6	1	0	0	13.00
7	1	0	1	15.51
8	1	0	1	15.22
9	0	0	2	14.60
10	0	0	2	13.79
11	1	0	2	18.70
12	1	0	2	17.98
13	2	0	0	16.80
14	2	0	0	15.96
15	2	0	1	19.00
16	2	0	1	18.60
17	2	0	2	24.20
18	2	0	2	22.58
19	0	1	0	5.11
20	0	1	0	5.31
21	0	1	1	11.90
22	0	1	1	11.34
23	1	1	0	14.30
24	1	1	0	15.35
25	1	1	1	18.80
26	1	1	1	20.96
27	0	1	2	13.30
28	0	1	2	15.64
29	1	1	2	21.60
30	1	1	2	24.27
31	2	1	0	19.86
32	2	1	0	20.88
33	2	1	1	26.37
34	2	1	1	24.38
35	2	1	2	29.72
36	2	1	2	29.64
37	0	2	0	6.70
38	0	2	0	7.00
39	0	2	1	12.40
40	0	2	1	13.15
41	1	2	0	15.38
42	1	2	0	14.90
43	1	2	1	19.20
44	1	2	1	20.10
45	0	2	2	16.20
46	0	2	2	17.10

Obs	a	b	c	Y
47	1	2	2	22.40
48	1	2	2	23.40
49	2	2	0	21.30
50	2	2	0	24.00
51	2	2	1	27.30
52	2	2	1	27.20
53	2	2	2	31.40
54	2	2	2	29.64
	==	==	==	======
	54	54	54	944.97

Class Level Information

Class	Levels	Values
a	3	0 1 2
b	3	0 1 2
c	3	0 1 2

Number of Observations Read 54

The ANOVA Procedure

Dependent Variable: Y

Source	DF	Sum of Squares	Mean Square	F Value	Pr > F
Model	26	2446.788533	94.107251	117.48	<.0001
Error	27	21.627550	0.801020		
Corrected Total	53	2468.416083			

R-Square	Coeff Var	Root MSE	Y Mean
0.991238	5.114433	0.894997	17.49944

Source	DF	Anova SS	Mean Square	F Value	Pr > F
a	2	1556.656578	778.328289	971.67	<.0001
b	2	199.334800	99.667400	124.43	<.0001
a*b	4	50.994856	12.748714	15.92	<.0001
c	2	617.398033	308.699017	385.38	<.0001
a*c	4	11.535689	2.883922	3.60	0.0178
b*c	4	7.123067	1.780767	2.22	0.0931
a*b*c	8	3.745511	0.468189	0.58	0.7817

□

7.7 Extension to k Factors at Three Levels

So far we have dealt with 3^k factorial designs with $k = 2, 3$ only. The principles and methods we have employed in dealing with these designs can easily be extended and applied to the 3^k factorial designs (for $k>3$). We have also shown that for 3^2 factorial design, there are two main effects, each with 2 degrees of freedom, and 1 two-factor interaction with 4 degrees of freedom. For the 3^3 factorial design, there are three main effects, each with 2 degrees of freedom, 3 two-factor interactions, each having 4 degrees of freedom and 1 three-factor interaction with 8 degrees of freedom. For 3^4 factorial design in A, B, C, and D, the factorial effects are A, B, AB, C, AC, BC, ABC, D, AD, BD, ABD, CD, ACD, BCD, and $ABCD$ (4 main effects, 6 two-factor interactions, 4 three-factor interactions, and 1 four-factor interaction). Each of the main effects has 2 degrees of freedom, each two-factor interaction has 4 degrees of freedom, while each of the three-factor interactions has 8 degrees of freedom. The four-factor interaction has 16 degrees of freedom. In general, if there are k factors at three levels, there will be k main effects, ${}^kC_2$ two-factor interactions, ${}^kC_3$ three-factor interactions, ... and one k-factor interaction. For $p \geq 2$, each p-factor interaction has 2^p degrees of freedom.

Furthermore, we recall that each interaction of factors in the 3^2 or 3^3 factorial designs could be broken down into components, each of which had 2 degrees of freedom. In these designs, all the two-factor interactions had two components and the three-factor interactions had four components, each of which had 2 degrees of freedom. In general, in 3^k factorial designs, any interaction of p factors could be broken down into 2^{p-1} factorial effects, each with 2 degrees of freedom. To write down these factorial effects, it is conventional not to allow the first letter in the factorial effect to have an exponent that is greater than one. If this is not the case (which sometimes could arise in fractional replication when the generalized interaction of a factorial effect is obtained), the factorial effect should be squared and each letter reduced (modulo 3) to obtain an effect without an exponent that is greater than one in the first letter (see Chapter 9 for fractional replication in 3^k factorial designs). For instance, suppose we have two factorial effects, $X=AB^2$ and $Y=AC^2$, then $XY=A^2B^2C^2=A^4B^4C^4=ABC$. We had to square $A^2B^2C^2$ to obtain an effect $A^4B^4C^4$, which when reduced (modulo 3) produces ABC that has A with exponent one and is an effect in the design.

In a typical 3^k factorial design with n replicates, there are $3^k(n)-1$ degrees of freedom. As we have demonstrated for the 3^2 and 3^3 factorial designs, the listing of treatment effects is better carried out according to digital notation. For instance, 0 2 0 1 2 is a treatment combination in a typical 3^5 factorial design. In the same design, if we represent the contrasts of the letters in the design, namely, A, B, C, D, E, by x_1, x_2, x_3, x_4, and x_5, respectively, then we can represent the levels of a typical five-factor factorial effect, $ABCDE$, by $x_1+x_2+x_3+x_4+x_5$ (modulo 3). This equation will be used (as we did in the case of the 3^3 design to obtain the responses that fall under different levels of this factor (levels 0, 1, 2)). These responses will in turn be used to estimate the factor and more importantly to calculate the sums of squares that will be used in ANOVA. For the 3^5 factorial design, the treatment combination 0 2 0 1 2 would fall under level 1 of the factorial effect, $ABCDE$.

The same difficulties and complications that arise in 2^k factorial designs also arise in the 3^k factorial designs; indeed, they become more prominent. As k, the number of factors in this design increases, the design and the analysis of the experiment becomes more complicated. Frequently, only one replicate of the 3^k factorial design is used because of these difficulties. In such designs as was the case in the 2^k factorial designs, higher order interactions whose effects are considered insignificant are used to estimate the error. When one replicate becomes more difficult to manage, fractional replication is used. We discuss fractional replication in this design in Chapter 9.

As usual, when calculating the sums of squares, it is advisable (especially when this is carried out manually) to use all the observations to obtain the total sum of squares. Next, the replicates for each treatment combination should be added together, and their sums should be used to estimate treatment sum of squares as we demonstrated in Example 7.5 on 3^3 factorial design. Once we have obtained the total and treatment sums of squares, we can determine the residual sum of squares as the difference between the two. The treatment sums of squares can be broken down to the different sums of squares for the factorial effects; it also helps us check that our calculations are accurate and free of errors, since the sum of all the sums of squares for the factorial effects should be the same as the treatment sum of squares. This check is necessary because of the complicated calculations encountered in this design. Computer packages considerably reduce the amount of computations needed.

7.8 Exercises

Exercise 7.1

A 2^3 factorial experiment was carried out on methods of cultivating cauliflower. The factors were

1. Fertilizers applied in spring or summer
2. Spraying or nonspraying
3. Irrigation or lack of irrigation

The experiment was performed twice, and the percentage of poor quality cauliflower was observed in each case. Analyze the data fully, stating your conclusions clearly.

	Replicate	Spring		Summer	
		Spray	Nonspray	Spray	Nonspray
Irrigated	1	20.5	28.7	26.2	37.5
	2	19.7	31.3	29.9	35
Nonirrigated	1	24.8	21.8	19.7	29.4
	2	26.5	26	27.0	26.6

Suggest the treatment combinations that are significant. □

Exercise 7.2

In a full 2^k factorial experiment for the study of the effects of combinations of three fertilizer types, N, K, and P each at two levels, on the growth of sweet potatoes; the following data show the weights of good sweet potatoes picked for plots of the same size following applications of fertilizer. The experiment was replicated three times:

(1)	39.63	41.27	40.22
k	56.08	54.86	45.37
n	33.75	33.26	38.25
nk	50.83	50.63	51.93
p	42.47	53.94	47.94
kp	59.84	51.43	50.96
np	54.49	53.32	40.31
npk	63.1	49.49	48.97

Obtain the factorial effects and hence analyze the data fully. Check for violations of assumptions. □

Exercise 7.3

State the important features of a full 2^k factorial experiment and write brief notes on each of them. What is meant by main effects and interaction effects? Indicate why factorial experiments are advantageous in this respect. What are the disadvantages of factorial experiments when resources are limited? Describe some of the methods used to overcome each disadvantage you identify. □

Exercise 7.4

The wear index of a metal that is used to make engine blocks is measured under experimental conditions. The wear is affected by the length of time the metal is under stress, the temperature of the metal, and the percentage of the alloy in the metal. Low values of the index indicate little wear. A full 3^3 factorial experiment was performed in three replicates to investigate the effect of the three factors on wear. The factors and their levels are

Temperature (°C)	145	160	175
Time (h)	2	4	6
Alloy (%)	4	7	10

The data below show the responses over the three replicates:

		Temperature (°C)		
	Time (h)	**145**	**160**	**175**
Alloy 4%	2	14, 16.4, 16.5	15.4, 15.5, 16.3	13.1, 12.1, 11.0
	4	16.1, 18.2, 19.5	17, 21.2, 19.4	16.5, 14.8, 15.2
	6	19.2, 19.6, 20.0	21.4, 18.1, 22.3	18.6, 17.1, 17.0
Alloy 7%	2	15.2, 18.3, 21.9	16.8, 18.2, 21.2	14.2, 16.8, 17.2
	4	17.3, 18.4, 22.1	16.4, 24.2, 22.0	17.3, 18.8, 20.5
	6	22.0, 18.6, 19.2	20.4, 18.5, 23.9	18.2, 20.3, 21.8
Alloy 10%	2	18.4, 22.3, 18.2	19.4, 23.1, 17.9	14.9, 20.1, 18.1
	4	21.0, 21.6, 19.7	20.1, 22.0, 23.5	15.5, 16.4, 15.6
	6	22.0, 19.7, 23.4	23.0, 18.7, 26.2	23.2, 20.1, 21.7

Write a report of your analysis of the responses and the conclusions of this experiment. □

Exercise 7.5

It is suspected that the yield of a chemical plant is influenced by five factors: A, temperature; B, pressure; C, operating mode (manual or automatic); D, catalyst concentration; and E, speed. These factors were all set at two levels, and the yields of the plants were studied as 2^5 factorial experiment. However, because of many constraints, only one full replicate of the design could be obtained.

(1) 13	*d* 14	*e* 22	*de* 17
a 17	*ad* 20	*ae* 23	*ade* 24
b 15	*bd* 14	*be* 18	*bde* 19
ab 17	*abd* 22	*abe* 21	*abde* 24
c 16	*cd* 18	*ce* 25	*cde* 16
ac 18	*acd* 19	*ace* 17	*acde* 18
bc 11	*bcd* 17	*bce* 14	*bcde* 22
abc 23	*abcd* 19	*abce* 25	*abcde* 19

Using factorial effects contrasts, obtain the estimates of effects in this design. Obtain the sums of squares for the effects. Assuming that all second- and higher order interactions are negligible, combine them to estimate error and hence draw conclusions on the plant. Plot the estimated effects on a normal probability paper. Can you justify the assumption that all second- and higher order interactions are negligible? □

Exercise 7.6

Collapse the design in Exercise 7.5 into eight replicates of a 2^2 factorial design in the factors that have been found significant following the analysis. Analyze the data fully. Are your conclusions different from those reached in Exercise 7.6 above? □

Exercise 7.7

A 2^4 experiment was performed in three replicates to study a system. The responses of the design are presented below. Analyze the data and draw conclusions regarding the factors A, B, C, and D.

Treatment	**Replicates**			**Treatment**	**Replicates**		
combinations	**I**	**II**	**III**	**combinations**	**I**	**II**	**III**
(1)	7	6	5	d	4	4	6
a	3	4	5	ad	2	2	4
b	6	7	4	bd	5	6	5
ab	4	4	3	abd	3	4	2
c	5	6	4	cd	4	6	4
ac	4	4	4	acd	2	2	4
bc	7	5	7	bcd	4	4	3
abc	4	5	5	$abcd$	2	2	4

□

Exercise 7.8

The distance travelled (in miles) between recharges of JCY batteries used for electric cars, appears to depend on charging time and the speed at which the car is operated. A pilot study of the prototype JCY battery was carried out at three preset charging periods as well as three operating speeds. The batteries were tested in a car running on rails within the factory and robots were used as drivers. The experiment was replicated twice and the responses in miles achieved between charges are presented below:

	Charging period (h)		
Operating speed (km/h)	**18**	**24**	**30**
56	330, 309	355, 370	380, 360
72	335, 327	377, 381	390, 385
88	326, 356	390, 386	388, 396

Analyze the data fully comparing the means as appropriate and write a report to the manager. □

Exercise 7.9
In an experiment to study the yields of a chemical process under four factors A, temperature; B, pressure; C, catalyst concentration; and D, stirring rate, each factor was set at two levels only. Only one replicate of the experiment could be run. The yields are

(1) 20 a 14 b 18 ab 12 c 18 ac 14 bc 22 abc 14 d 16 ad 10 bd 16 abd 10 cd 8 acd 9 bcd 13 $abcd$ 10

Obtain the effects of the treatment combinations. Assuming that all the second- and third-order interactions are not significant, use their combined sums of squares as an estimate of the residual sum of squares and analyze the data. Obtain a normal probability plot of the estimates. Can you confirm that the second- and third-order effects are not significant? □

Exercise 7.10
Collapse the responses in Exercise 7.9 into four replicates of a 2^2 factorial design in two factors: temperature and pressure. Analyze the data fully. Are your conclusions different from those reached in Exercise 7.9? □

Exercise 7.11
Write a SAS program to analyze the data of Exercise 7.4. □

Exercise 7.12
Write a SAS program to analyze the data of Exercise 7.5. □

Exercise 7.13
Write a SAS program to analyze the data of Exercise 7.8. □

References

Daniel, C. (1959) Use of half-normal plots in interpreting two level experiments, *Technometrics*, 1, 311–342.

Davies, O.L. (1955) *Design and Analysis of Industrial Experiments*, Oliver and Boyd, London, England.

Davies, O.L. (1958) *Design and Analysis of Industrial Experiments*, 2nd Ed., Oliver and Boyd, London, England.

Kempthorne, O. (1952) *The Design and Analysis of Experiments*, Wiley, New York.

Rayner, A.A. (1967) The square summing check on the main effects and interactions in a 2^k experiment as calculated by Yates' algorithm. *Biometrics* 23, 571–573.

Yates, F. (1937) *Design and Analysis of Factorial Experiments*, Tech. Comm. No. 35, Imperial Bureau of Soil Sciences, London.

Chapter 8

Regression Models for 2^k and 3^k Factorial Designs

In Chapter 7, we used classical analysis of variance (ANOVA) approaches to analyze data dealing with responses of the typical 2^k and 3^k factorial experiments. For all those designs, we assumed that the effects were fixed for the models and all our analyses; we could also have easily used random effects as described in the earlier chapters. As we have demonstrated in Chapters 4 and 6, we can fit typical regression models to data generated from experimental designs and analyze them from the regression theoretical standpoint. Here, we have developed similar regression models for the 2^k and 3^k factorial designs which were discussed in Chapter 7. We concentrate again on developing regression models for the examples in Chapter 7, which had been previously analyzed by other approaches. We will estimate parameters for the models. In each case, we will endeavor to carry out ANOVA to test for significance of the fitted models. We do, however, recognize that for the single replicate of a design, we will not be in a position to perform ANOVA.

8.1 Regression Models for the 2^2 Factorial Design Using Effects Coding Method (Example 7.1)

In this section, we consider the type of regression model that could be fitted to responses of a typical 2^k factorial experiment. To do this, we consider the simplest form of the 2^k factorial experiment, the 2^2 factorial, which has two factors, each set at two levels. Under the effects coding method, this design can be analyzed by using the regression model of the type

$$Y = \mu + \alpha X_i + \beta Z_j + \gamma_{ij} X_i Z_j + \varepsilon \tag{8.1}$$

where

$$X_i = \begin{cases} -1 & \text{if } i = 2 \\ 1 & \text{if } i = 1 \\ 0 & \text{otherwise} \end{cases}$$

$$Z_j = \begin{cases} -1 & \text{if } j = 2 \\ 1 & \text{if } j = 1 \\ 0 & \text{otherwise} \end{cases}$$

This is because each factor is set at two levels, and we define two sets of dummy variables, each containing only one dummy variable, that is $2-1=1$, corresponding to $k-1$ dummy variables for a factor with k levels, where $k=2$.

8.1.1 Estimates of the Model Parameters

With this model, we estimate the parameters as follows:

$$\begin{aligned}
\hat{\mu} &= \hat{\mu}_{..} \\
-\hat{\alpha}_1 &= \hat{\mu}_{2.} - \hat{\mu}_{..} \\
-\hat{\beta}_1 &= \hat{\mu}_{.2} - \hat{\mu}_{..} \\
\hat{\mu}_{11} &= \hat{\mu} + \hat{\alpha}_1 + \hat{\beta}_1 + \hat{\gamma}_{11} \\
\hat{\alpha}_1 &= \hat{\mu}_{1.} - \hat{\mu}_{..} \\
\hat{\beta}_1 &= \hat{\mu}_{.1} - \hat{\mu}_{..} \\
\hat{\gamma}_{11} &= \hat{\mu}_{11} - \hat{\mu}_{1.} - \hat{\mu}_{.1} + \hat{\mu}_{..}
\end{aligned}$$

From the experiment of Example 7.1, we obtained the data and the means as shown in Table 8.1.

TABLE 8.1: Marginal Means for Data of Example 7.1

	Pressure (P)			
Temperature (T)	**I**	**II**	**Total**	**Mean**
I	78,71,88,86	60,69,63,64	579	$72.375 = \hat{\mu}_{1.}$
II	47,43,53,56	0,41,44,42	366	$45.75 = \hat{\mu}_{2.}$
Total	522	423	945	
Mean	$65.25 = \hat{\mu}_{.1}$	$52.875 = \hat{\mu}_{.2}$	$59.0625 = \hat{\mu}_{..}$	

With the data, we obtain the following statistics, which will be used in model parameter estimation:

$$\begin{aligned}
\overline{y}_{..} &= \frac{945}{16} = 59.0625 = \hat{\mu}_{..} & \overline{y}_{1.} &= \frac{579}{8} = 72.375 = \hat{\mu}_{1.} \\
\overline{y}_{.1} &= \frac{522}{8} = 65.25 = \hat{\mu}_{1.} & \overline{y}_{2.} &= \frac{366}{8} = 45.75 = \hat{\mu}_{2.} \\
\overline{y}_{.2} &= \frac{423}{8} = 52.875 = \hat{\mu}_{.2} & \overline{y}_{11} &= \frac{323}{4} = 80.75 = \hat{\mu}_{11} \\
\overline{y}_{12} &= \frac{256}{4} = 64 = \hat{\mu}_{12} & \overline{y}_{21} &= \frac{199}{4} = 49.75 = \hat{\mu}_{21} \\
\overline{y}_{22} &= \frac{167}{4} = 41.75 = \hat{\mu}_{22}.
\end{aligned}$$

We use the above statistics to estimate the regression model parameters for the data of Example 7.1, which was obtained by using effects coding to define the dummy variables as follows:

$$\hat{\alpha}_1 = \hat{\mu}_{1.} - \hat{\mu}_{..} \Rightarrow \hat{\alpha}_1 = \overline{y}_{1.} - \overline{y}_{..} = 72.375 - 59.0625 = 13.3125$$

$$\hat{\beta}_1 = \hat{\mu}_{.1} - \hat{\mu}_{..} \Rightarrow \hat{\beta}_1 = \overline{y}_{.1} - \overline{y}_{..} = 65.25 - 59.0625 = 6.1825$$

$$\hat{\gamma}_{11} = \hat{\mu}_{11} - \hat{\mu}_{.1} - \hat{\mu}_{1.} + \hat{\mu}_{..} \Rightarrow \hat{\gamma}_{11} = 80.75 - 65.25 - 72.375 + 59.0625 = 2.1875$$

The resulting estimated model for the design of Example 7.1 is

$$\hat{y} = 59.0625 + 13.3125x + 6.1825z + 2.1875xz \tag{8.2}$$

With Equation 8.2, we can obtain the estimated values of the responses and thus find the attendant residuals.

8.1.2 Testing for Significance of the Fitted Model 8.2

Using the fitted model 8.2, we obtain the estimates for each response of the experiment; we extract the residuals as the difference between responses and fitted values as shown in Table 8.2.

As usual, the sum of the squared residuals gives us the sum of squares due to error, which we could utilize in ANOVA to test for significance of the fitted

TABLE 8.2: Fitted Values and Residuals for the Fitted Model 8.2 for Example 7.1 (Effects Coding Method)

	z	x	xz	y	$\hat{y}$	**Residuals** $= y - \hat{y}$	$(y - \hat{y})^2$
	1	1	1	78	80.745	−2.745	7.535025
	1	1	1	71	80.745	−9.745	94.96503
	1	1	1	88	80.745	7.255	52.63502
	1	1	1	86	80.745	5.255	27.61503
	−1	1	−1	60	64.005	−4.005	16.04003
	−1	1	−1	69	64.005	4.995	24.95003
	−1	1	−1	63	64.005	−1.005	1.010025
	−1	1	−1	64	64.005	−0.005	2.5E−05
	1	−1	−1	47	49.745	−2.745	7.535025
	1	−1	−1	43	49.745	−6.745	45.49503
	1	−1	−1	53	49.745	3.255	10.59503
	1	−1	−1	56	49.745	6.255	39.12503
	−1	−1	1	40	41.755	−1.755	3.080025
	−1	−1	1	41	41.755	−0.755	0.570025
	−1	−1	1	44	41.755	2.245	5.040025
	−1	−1	1	42	41.755	0.245	0.060025
Total	0	0	0	945	945	0	336.2504

model. From regression theory, we see that since the total number of units in the designed experiment is $N = 4(2^2) = 16$,

$$SS_{\text{total}} = \sum_{i=1}^{16} y_i^2 - \frac{\left(\sum_{i=1}^{16} y_i\right)^2}{16} = 59675 - \frac{945^2}{16} = 3860.938$$

$$SS_{\text{residual}} = \sum_{i=1}^{16} (y_i - \hat{y}_i)^2 = 336.2504$$

$$SS_{\text{regression}} = SS_{\text{total}} - SS_{\text{residual}} = 3860.938 - 336.2504 = 3524.687$$

TABLE 8.3: ANOVA to Test for Significance of Fitted Model for Data of Example 7.1

Source	Sum of squares	Degrees of freedom	Mean square	F-ratio	p-value
Regression	3524.687	3	1174.896	48.91753	1.1476E−07
Error	336.2504	14	24.01789		
Total	3860.938	17			

We perform ANOVA to test for the significance of the fitted model (Table 8.3):

Clearly, the fitted model of Equation 8.2 is highly significant.

8.1.3 SAS Analysis of Data of Example 7.1 under the Model 8.1

We write a SAS program to carry out the analysis of the data of Example 7.1 using effects coding to define the dummy variables in its model.

SAS PROGRAM

```
data _2fact1a;
input z x y@@;
h=x*z;
datalines;
1 1 78 1 1 71 1 1 88 1 1 86 -1 1 60 -1 1 69 -1 1 63 -1 1 64
1 -1 47 1 -1 43 1 -1 53 1 -1 56 -1 -1 40 -1 -1 41  -1 -1 44 -1 -1 42
;
proc print data=_2fact1a; var x z h y; sum x z h y;
proc reg data=_2fact1a; model y=x z h;
run;
```

SAS OUTPUT

```
Obs    x    z    h    y

  1    1    1    1   78
  2    1    1    1   71
```

Obs	x	z	h	y
3	1	1	1	88
4	1	1	1	86
5	1	-1	-1	60
6	1	-1	-1	69
7	1	-1	-1	63
8	1	-1	-1	64
9	-1	1	-1	47
10	-1	1	-1	43
11	-1	1	-1	53
12	-1	1	-1	56
13	-1	-1	1	40
14	-1	-1	1	41
15	-1	-1	1	44
16	-1	-1	1	42
	==	==	==	===
	0	0	0	945

```
                The REG Procedure
                 Model: MODEL1
              Dependent Variable: y

       Number of Observations Read    16
       Number of Observations Used    16


              Analysis of Variance

                          Sum of          Mean
Source            DF     Squares        Square  F Value  Pr > F

Model              3  3524.68750    1174.89583    41.93  <.0001
Error             12   336.25000      28.02083
Corrected Total   15  3860.93750

        Root MSE            5.29347   R-Square    0.9129
        Dependent Mean     59.06250   Adj R-Sq    0.8911
        Coeff Var           8.96249

              Parameter Estimates

                     Parameter    Standard
     Variable   DF    Estimate       Error    t Value    Pr > |t|

     Intercept   1    59.06250     1.32337      44.63      <.0001
     x           1    13.31250     1.32337      10.06      <.0001
     z           1     6.18750     1.32337       4.68      0.0005
     h           1     2.18750     1.32337       1.65      0.1242
```

The outputs of SAS analysis agree entirely with those of the manual analysis carried out previously. The estimates are the same as those obtained previously.

8.2 Regression Model for Example 7.1 Using Reference Cell to Define Dummy Variables

The regression model which is appropriate for this design is

$$Y = \mu + \alpha_1 X_i + \beta_1 Z_j + \gamma_{11} X_i Z_j + \varepsilon \tag{8.3}$$

where

$$X_i = \begin{cases} 1 & \text{if } i = 1 \\ 0 & \text{otherwise} \end{cases}$$

$$Z_j = \begin{cases} 1 & \text{if } j = 1 \\ 0 & \text{otherwise} \end{cases}$$

provided that we make cell (2, 2) the reference cell, in the design. Cell (2, 2) refers to the position for which both temperature and pressure are set high.

8.2.1 Estimates of the Model Parameters

We can estimate the model parameters for the regression model. Other relevant formulae, which are specific to 2^2 factorial design when cell (2, 2) is chosen as the reference cell are as follows:

$$\begin{aligned}
\mu &= \mu_{22} \\
\mu_{11} &= \mu + \alpha_1 + \beta_1 + \gamma_{11} \\
\mu_{21} &= \mu + \beta_1 \\
\mu_{12} &= \mu + \alpha_1 \\
\mu_{1.} &= \mu + \alpha_1 + \frac{\beta_1 + \gamma_{11}}{2} \\
\mu_{.1} &= \mu + \beta_1 + \frac{\alpha_1 + \gamma_{11}}{2} \\
\Rightarrow \mu_{1.} - \mu_{12} &= \frac{\beta_1 + \gamma_{11}}{2} \Rightarrow \beta_1 = 2(\mu_{1.} - \mu_{12}) - \gamma_{11} \\
\Rightarrow \mu_{.1} - \mu_{21} &= \frac{\alpha_1 + \gamma_{11}}{2} \Rightarrow \alpha_1 = 2(\mu_{.1} - \mu_{21}) - \gamma_{11} \\
\mu_{2.} &= \mu + \frac{\beta_1}{2} \\
\mu_{.2} &= \mu + \frac{\alpha_1}{2}
\end{aligned}$$

From the table of means for Example 7.1 used in Section 8.1 (Table 8.1), we see that because cell (2, 2) is the reference cell in this case, the estimates are as follows:

$$\begin{aligned}
\hat{\mu}_{22} &= \hat{\mu} = 41.75 \\
\hat{\mu}_{21} &= \hat{\mu} + \hat{\beta}_1 \Rightarrow 49.75 = 41.75 + \hat{\beta}_1 \Rightarrow \hat{\beta}_1 = 49.75 - 41.75 = 8 \\
\hat{\mu}_{12} &= \hat{\mu} + \hat{\alpha}_1 \Rightarrow 64 = 41.75 + \hat{\alpha}_1 \Rightarrow \hat{\alpha}_1 = 64.00 - 41.75 = 22.25 \\
\hat{\beta}_1 &= 2(\hat{\mu}_{.1} - \hat{\mu}_{21}) - \hat{\gamma}_{11} = 22.25 = 2(65.25 - 49.75) - \hat{\gamma}_{11} \\
\Rightarrow \hat{\gamma}_{11} &= 2(65.25 - 49.75) - 22.25 = 8.75
\end{aligned}$$

or

$$\begin{aligned}
\hat{\beta}_1 &= 2(\hat{\mu}_{1.} - \hat{\mu}_{12}) - \hat{\gamma}_{11} = 8 = 2(72.375 - 64) - \hat{\gamma}_{11} \\
\Rightarrow \hat{\gamma}_{11} &= 2(72.375 - 64) - 8 = 8.75
\end{aligned}$$

The fitted model for this design under the model proposed in Equation 8.3 is

$$\hat{y} = 41.75 + 22.25x + 8z + 8.75xz \tag{8.4}$$

8.2.2 Testing for Significance of the Fitted Model

We obtained the fitted values for the responses of the 2^2 factorial design in Example 7.1; thereafter we extracted the attendant residuals using the fitted values and the actual responses observed for the experiment. They are shown in Table 8.4.

TABLE 8.4: Fitted Values and Residuals for the Fitted Model 8.4 for Example 7.1 (Reference Cell Coding)

	x	z	xz	y	$\hat{y}$	**Residuals** = $y - \hat{y}$	$(y - \hat{y})^2$
	1	1	1	78	80.75	−2.75	7.5625
	1	1	1	71	80.75	−9.75	95.0625
	1	1	1	88	80.75	7.25	52.5625
	1	1	1	86	80.75	5.25	27.5625
	1	0	0	60	64	−4	16
	1	0	0	69	64	5	25
	1	0	0	63	64	−1	1
	1	0	0	64	64	0	0
	0	1	0	47	49.75	−2.75	7.5625
	0	1	0	43	49.75	−6.75	45.5625
	0	1	0	53	49.75	3.25	10.5625
	0	1	0	56	49.75	6.25	39.0625
	0	0	0	40	41.75	−1.75	3.0625
	0	0	0	41	41.75	−0.75	0.5625
	0	0	0	44	41.75	2.25	5.0625
	0	0	0	42	41.75	0.25	0.0625
Total	8	8	4	945	945	0	336.25

Again, from regression theory, we see that

$$SS_{\text{total}} = \sum_{i=1}^{16} y_i^2 - \frac{\left(\sum_{i=1}^{16} y_i\right)^2}{16} = 59675 - \frac{945^2}{16} = 3860.938$$

$$SS_{\text{residual}} = \sum_{i=1}^{16} (y_i - \hat{y}_i)^2 = 336.2504$$

$$SS_{\text{regression}} = SS_{\text{total}} - SS_{\text{residual}} = 3860.938 - 336.2504 = 3524.687$$

We perform ANOVA to test for significance of the fitted model (Table 8.5).

TABLE 8.5: ANOVA to Test for Significance of Fitted Model 8.4 for Data of Example 7.1

Source	**Sum of squares**	**Degrees of freedom**	**Mean square**	***F*-ratio**	***p*-Value**
Regression	3524.687	3	1174.896	48.91753	1.1476E−07
Error	336.2504	14	24.01789		
Total	3860.938	17			

The result of ANOVA indicates that the fitted model is highly significant. Although we reached this conclusion by using cell (2, 2) as the reference cell for the model 8.3 above, we could have used any cell within the design to carry out the analysis. The estimates for the parameters would be different but conclusion reached in ANOVA would be the same.

8.2.3 SAS Analysis of Data of Example 7.1 under the Model 8.1

Under the model 8.3 with reference cell (2, 2), we present the SAS program that would analyze the data of Example 7.1 and the attendant output.

SAS PROGRAM

```
data _2kfact1;
input x z y@@;
h=x*z;
datalines;
1 1 78 1 1 71 1 1 88 1 1 86 1 0 60 1 0 69 1 0 63 1 0 64
0 1 47 0 1 43 0 1 53 0 1 56 0 0 40 0 0 41 0 0 44 0 0 42
;
proc print data=_2kfact1; var x z h y; sum x z h y;
proc reg data=_2kfact1;
model y=x z h; run;
```

SAS OUTPUT

Obs	x	z	h	y
1	1	1	1	78
2	1	1	1	71
3	1	1	1	88
4	1	1	1	86
5	1	0	0	60
6	1	0	0	69
7	1	0	0	63
8	1	0	0	64
9	0	1	0	47
10	0	1	0	43
11	0	1	0	53
12	0	1	0	56
13	0	0	0	40
14	0	0	0	41
15	0	0	0	44
16	0	0	0	42

The REG Procedure
Model: MODEL1
Dependent Variable: y

Number of Observations Read	16
Number of Observations Used	16

Analysis of Variance

Source	DF	Sum of Squares	Mean Square	F Value	Pr > F
Model	3	3524.68750	1174.89583	41.93	<.0001
Error	12	336.25000	28.02083		
Corrected Total	15	3860.93750			

Root MSE	5.29347	R-Square	0.9129
Dependent Mean	59.06250	Adj R-Sq	0.8911
Coeff Var	8.96249		

Parameter Estimates

Variable	DF	Parameter Estimate	Standard Error	t Value	Pr > |t|
Intercept	1	41.75000	2.64674	15.77	<.0001
x	1	22.25000	3.74305	5.94	<.0001
z	1	8.00000	3.74305	2.14	0.0539
h	1	8.75000	5.29347	1.65	0.1242

Again, we can see that the estimates obtained from SAS analysis agree with the previous estimates obtained by manual calculations. All the conclusions from SAS analysis confirm the method adopted for manual analysis.

8.3 General Regression Models for the Three-Way Factorial Design

Before we consider using dummy variable regression to model and analyze the responses of a three-factor factorial experiment, let us consider the factorial experiment in three factors A, B, and C set at a, b, and c levels, respectively. A full factorial design would study all the factors and the factorial effects resulting from all possible combinations of different levels of three factors. Such a designed experiment is popularly known as the three-way ANOVA. A fixed effects model can easily be written for this design, and typically the model and the attendant assumptions would look somewhat like the following (if the designed experiment is replicated n times):

$$
\begin{aligned}
y_{ijkl} &= \mu + \alpha_i + \beta_j + (\alpha\beta)_{ij} + \gamma_k + (\alpha\gamma)_{ik} + (\beta\gamma)_{jk} \\
&\quad + (\alpha\beta\gamma)_{ijk} + \varepsilon_{ijkl} \\
&\quad i = 1,2,\ldots,a \;\; j = 1,2,\ldots,b \;\; k = 1,2,\ldots,c \;\; l = 1,2,\ldots,n \\
&\sum_{i=1}^{a} \alpha_i = 0 \; \sum_{j=1}^{b} \beta_j = 0 \; \sum_{k=1}^{c} \gamma_k = 0 \; \sum_i (\alpha\beta)_{ij} = \sum_j (\alpha\beta)_{ij} = 0 \\
&\sum_i (\alpha\gamma)_{ik} = \sum_k (\alpha\gamma)_{ik} = 0 \quad \sum_j (\beta\gamma)_{jk} = \sum_k (\beta\gamma)_{jk} = 0 \\
&\sum_{i,j} (\alpha\beta\gamma)_{ijk} = \sum_{i,k} (\alpha\beta\gamma)_{ijk} = \sum_{j,k} (\alpha\beta\gamma)_{ijk} = 0 \quad \varepsilon_{ijkl} \sim \text{NID}(0, \sigma^2)
\end{aligned}
\tag{8.5}
$$

An illustration of the analysis of such a design is given in Example 7.2 where the factors A, B, and C were all set at two levels, leading to a 2^3 factorial experiment. In the next section, we see how this model can be reset to study the three-way ANOVA from the dummy regression perspective. We will also see how the responses are analyzed using regression models.

8.3.1 Modeling the General Three-Way Factorial Design Using Effects Coding Method to Define Dummy Variables

We already know that there are two ways of representing a dummy variable in the regression approach to modeling the responses of a typical experimental design. We use either the effects coding method or the reference cell coding method. When we use the effects coding method for defining dummy variables, we can show that the regression model for a typical three-way ANOVA or three-factor factorial design would be of the form

$$Y = \mu + \sum_{i=1}^{a-1} \alpha_i X_i + \sum_{j=1}^{b-1} \beta_j Z_j + \sum_{i=1}^{a-1}\sum_{j=1}^{b-1} \gamma_{ij} X_i Z_j + \sum_{k=1}^{c-1} \omega_k H_k$$

$$+ \sum_{i=1}^{a-1}\sum_{k=1}^{c-1} \theta_{ik} X_i H_k + \sum_{j=1}^{b-1}\sum_{k=1}^{c-1} \lambda_{jk} Z_j H_k$$

$$+ \sum_{i=1}^{a-1}\sum_{j=1}^{b-1}\sum_{k=1}^{c-1} \delta_{ijk} X_i Z_j H_k + \varepsilon \tag{8.6}$$

where

$$X_i = \begin{cases} -1 & \text{if } i = a \\ 1 & \text{if } i = 1, 2, \ldots, a-1 \\ 0 & \text{otherwise} \end{cases}$$

$$Z_j = \begin{cases} -1 & \text{if } j = b \\ 1 & \text{if } j = 1, 2, \ldots, b-1 \\ 0 & \text{otherwise} \end{cases}$$

$$H_k = \begin{cases} -1 & \text{if } k = c \\ 1 & \text{if } k = 1, 2, \ldots, c-1 \\ 0 & \text{otherwise} \end{cases}$$

and the three factors A, B, and C have levels a, b, and c, respectively. We define three sets of dummy variables containing $a-1$, $b-1$, and $c-1$ dummy variables, respectively. The above is the dummy regression model for a typical three-way ANOVA.

8.3.2 Estimation of Parameters for the General Three-Way Factorial Experiment with Effects Code Definition for Dummy Variables

Under the above model 8.6, which is based on effects coding method, we state the following relations that can be used to estimate the parameters of any regression model in which there are three factors:

$$
\begin{aligned}
\mu_{...} &= \mu \\
-\sum_{i=1}^{a-1} \alpha_i &= \mu_{a..} - \mu_{...} \\
-\sum_{j=1}^{b-1} \beta_j &= \mu_{.b.} - \mu_{...} \\
-\sum_{k=1}^{c-1} \gamma_k &= \mu_{..c} - \mu_{...} \\
\mu_{ijk} &= \mu + \alpha_i + \beta_i + \gamma_{ij} + \omega_k + \theta_{ik} + \lambda_{jk} + \delta_{ijk} \\
\alpha_i &= \mu_{i..} - \mu_{...} \\
\beta_j &= \mu_{.j.} - \mu_{...} \\
\omega_k &= \mu_{..k} - \mu_{...} \\
\gamma_{ij} &= \mu_{ij.} - \mu_{i..} - \mu_{.j.} + \mu_{...} \quad i = 1, 2, \ldots, a-1 \quad j = 1, 2, \ldots, b-1 \\
\theta_{ik} &= \mu_{i.k} - \mu_{i..} - \mu_{..k} + \mu_{...} \quad i = 1, 2, \ldots, a-1 \quad k = 1, 2, \ldots, c-1 \\
\lambda_{jk} &= \mu_{.jk} - \mu_{.j.} - \mu_{..k} + \mu_{...} \quad j = 1, 2, \ldots, b-1 \quad k = 1, 2, \ldots, c-1 \\
\delta_{ijk} &= \mu_{ijk} + \mu_{i..} + \mu_{.j.} + \mu_{..k} - \mu_{ij.} - \mu_{i.k} - \mu_{.jk} - \mu_{...} \\
&\quad i = 1, 2, \ldots, a-1 \;\; j = 1, 2, \ldots, b-1 \;\; k = 1, 2, \ldots, c-1
\end{aligned}
$$

These relations above apply in general to any three-way ANOVA and can be used to estimate parameters of any three-way factorial design. This is achieved by using the sample estimates of the means in the relations to replace the means and solve for estimates of the parameters of the model.

8.3.3 Fitting the Regression Model 8.6 (Effects Coding Method) for the Experimental Design of Example 7.2

We modify the above model and ancillary relations to obtain their equivalents for a typical 2^3 factorial experiment and obtain the estimates of the parameters. Thereafter, we write a SAS program to carry out the same analysis and obtain outputs that should include estimates of the parameters. In Example 7.2, each of the factors, A, B, and C was set at two levels. By modifying the model equations to produce those that apply to the 2^3 factorial design discussed in Example 7.2, we obtain the following results:

$$Y = \mu + \alpha X + \beta Z + \gamma XZ + \omega H + \theta XH + \lambda ZH + \delta XZH + \varepsilon \tag{8.7}$$

$$
\begin{aligned}
\mu_{...} &= \mu \\
\mu_{ijk} &= \mu + \alpha + \beta + \gamma + \omega + \theta + \lambda + \delta
\end{aligned}
$$

$$\alpha = \mu_{1..} - \mu_{...}$$

$$\beta = \mu_{.1.} - \mu_{...}$$

$$\gamma = \mu_{..1} - \mu_{...}$$

$$\omega = \mu_{11.} - \mu_{1..} - \mu_{.1.} + \mu_{...}$$

$$\theta = \mu_{1.1} - \mu_{1..} - \mu_{..1} + \mu_{...}$$

$$\lambda = \mu_{.11} - \mu_{.1.} - \mu_{..1} + \mu_{...}$$

$$\delta = \mu_{111} + \mu_{1..} + \mu_{.1.} + \mu_{..1} - \mu_{11.} - \mu_{1.1} - \mu_{.11} - \mu_{...}$$

We use the statistics (means) obtained for the data of Example 7.2 to show how estimation of the parameters can be carried out manually. The means from the data are

$$\overline{y}_{...} = \frac{1105}{24} = 46.0416 \quad \overline{y}_{1..} = \frac{531}{12} = 44.25 \quad \overline{y}_{.1.} = \frac{538}{12} = 44.833$$

$$\overline{y}_{..1} = \frac{540}{12} = 45 \quad \overline{y}_{11.} = \frac{259}{6} = 43.1667 \quad \overline{y}_{1.1} = \frac{260}{6} = 43.3333$$

$$\overline{y}_{.11} = \frac{265}{6} = 44.1667 \quad \overline{y}_{111} = \frac{127}{3} = 42.3333$$

$$\hat{\mu} = \overline{y}_{...} = 46.04167$$

$$\hat{\alpha} = \overline{y}_{1..} - \overline{y}_{...} = 44.25 - 46.04167 = -1.79167$$

$$\hat{\beta} = \overline{y}_{.1.} - \overline{y}_{...} = 44.8333 - 46.04167 = -1.20867$$

$$\hat{\gamma} = \overline{y}_{..1} - \overline{y}_{...} = 45 - 46.04167 = -1.04167$$

$$\hat{\omega} = \overline{y}_{11.} - \overline{y}_{1..} - \overline{y}_{.1.} + \overline{y}_{...} \Rightarrow \hat{\omega} = 43.1667 - 44.25 - 44.8333 + 46.04167 = 0.125$$

$$\hat{\theta} = \overline{y}_{1.1} - \overline{y}_{1..} - \overline{y}_{..1} + \overline{y}_{...} \Rightarrow \hat{\theta} = 43.3333 - 45 - 44.25 + 46.04167 = 0.125$$

$$\hat{\lambda} = \overline{y}_{11} - \overline{y}_{.1.} - \overline{y}_{..1} + \overline{y}_{...} \Rightarrow \hat{\lambda} = 44.1667 - 44.83330 - 45 + 46.04167 = 0.375$$

$$\hat{\delta} = \overline{y}_{111} + \overline{y}_{1..} + \overline{y}_{.1.} + \overline{y}_{..1} - \overline{y}_{11.} - \overline{y}_{1.1} - \overline{y}_{11.} - \overline{y}_{...}$$

$$\Rightarrow \hat{\delta} = 42.3333 + 44.25 + 44.8333 + 45 - 43.1667 - 43.3333 - 44.1667 - 46.04167 = -0.29167$$

The fitted model is

$$\hat{y} = 46.04167 - 1.79167x - 1.20867z + 0.125xz - 1.04167h + 0.125xh + 0.375zh - 0.29167xzh \quad (8.8)$$

8.3.4 Testing for Significance of the Fitted Model

We obtained the fitted values for the responses of the 2^3 factorial design in Example 7.2. Thereafter we extracted the attendant residuals using the fitted values and the responses of the experiment. They are shown in Table 8.6.

TABLE 8.6: Fitted Values and Residuals for the Fitted Model 8.5 for Example 7.2 (Effects Coding Method)

	x	z	h	xz	xh	zh	xzh	y	$\hat{y}$	Residuals = $y-\hat{y}$	$(y-\hat{y})^2$
	1	1	1	1	1	1	1	42	42.33296	−0.33296	0.110862
	1	1	1	1	1	1	1	44	42.33296	1.66704	2.779022
	1	1	1	1	1	1	1	41	42.33296	−1.33296	1.776782
	−1	1	1	−1	1	−1	−1	45	46.00044	−1.00044	1.00088
	−1	1	1	−1	1	−1	−1	46	46.00044	−0.00044	1.94E−07
	−1	1	1	−1	1	−1	−1	47	46.00044	0.99956	0.99912
	1	−1	1	−1	−1	1	−1	44	44.3337	−0.3337	0.111356
	1	−1	1	−1	−1	1	−1	45	44.3337	0.6663	0.443956
	1	−1	1	−1	−1	1	−1	44	44.3337	−0.3337	0.111356
	−1	−1	1	1	−1	−1	1	47	47.3329	−0.3329	0.110822
	−1	−1	1	1	−1	−1	1	48	47.3329	0.6671	0.445022
	−1	−1	1	1	−1	−1	1	47	47.3329	−0.3329	0.110822
	1	1	−1	1	−1	−1	−1	44	43.9997	0.0003	9E−08
	1	1	−1	1	−1	−1	−1	45	43.9997	1.0003	1.0006
	1	1	−1	1	−1	−1	−1	43	43.9997	−0.9997	0.9994
	−1	1	−1	−1	−1	1	1	47	46.9989	0.0011	1.21E−06
	−1	1	−1	−1	−1	1	1	46	46.9989	−0.9989	0.997801
	−1	1	−1	−1	−1	1	1	48	46.9989	1.0011	1.002201
	1	−1	−1	−1	1	−1	1	46	46.33364	−0.33364	0.111316
	1	−1	−1	−1	1	−1	1	46	46.33364	−0.33364	0.111316
	1	−1	−1	−1	1	−1	1	47	46.33364	0.66636	0.444036
	−1	−1	−1	1	1	1	−1	51	51.00112	−0.00112	1.25E−06
	−1	−1	−1	1	1	1	−1	50	51.00112	−1.00112	1.002241
	−1	−1	−1	1	1	1	−1	52	51.00112	0.99888	0.997761
Total	0	0	0	0	0	0	0	1105	1105	0	14.66668

To test for the significance of the fitted model, we calculate the sums of squares as follows (Table 8.7):

$$SS_{\text{total}} = \sum_{i=1}^{24} y_i^2 - \frac{\left(\sum_{i=1}^{24} y_i\right)^2}{24} = 51035 - \frac{1105^2}{24} = 158.9583$$

$$SS_{\text{residual}} = \sum_{i=1}^{24} (y_i - \hat{y}_i)^2 = 14.66668$$

$$SS_{\text{regression}} = SS_{\text{total}} - SS_{\text{residual}} = 158.9583 - 14.6668 = 144.29$$

TABLE 8.7: ANOVA to Test for Significance of Fitted Model 8.8 for Data of Example 7.2

Source	Sum of squares	Degrees of freedom	Mean square	F-ratio	p-Value
Regression	144.29	7	20.6131	22.487	3.76692E−07
Error	14.667	16	0.916668		
Total	158.96	23			

ANOVA shows that the fitted model is significant.

8.3.5 SAS Analysis of Responses of a 2^3 Factorial Design (Example 7.2)

The SAS program which carries out the analysis of data and obtains estimates for the model parameters displayed in its output after execution, is as follows:

SAS PROGRAM

```
data _2fact2A;
input x z h y@@;
t1=x*z; t2=z*h; t3=x*h; T4=X*Z*H;
cards;
1 1 1  42 1 1 1 44 1 1 1 41 -1 1 1 45 -1 1 1 46 -1 1 1 47
1 -1 1 44 1 -1 1  45 1 -1 1 44 -1 -1 1  47 -1 -1 1 48 -1 -1 1 47
1 1 -1 44 1 1 -1  45 1 1 -1 43 -1 1 -1  47 -1 1 -1 46 -1 1 -1 48
1 -1 -1 46 1 -1  -1 46 1 -1 -1 47 -1 -1  -1 51 -1 -1 -1 50 -1 -1 -1 52
;
proc print data=_2fact2A; sum x z h t1-t4 y;
proc REG data=_2fact2A; model y=x z h t1-t4; run;
```

SAS OUTPUT

Obs	x	z	h	y	t1	t2	t3	T4
1	1	1	1	42	1	1	1	1
2	1	1	1	44	1	1	1	1
3	1	1	1	41	1	1	1	1
4	-1	1	1	45	-1	1	-1	-1
5	-1	1	1	46	-1	1	-1	-1
6	-1	1	1	47	-1	1	-1	-1
7	1	-1	1	44	-1	-1	1	-1
8	1	-1	1	45	-1	-1	1	-1
9	1	-1	1	44	-1	-1	1	-1
10	-1	-1	1	47	1	-1	-1	1
11	-1	-1	1	48	1	-1	-1	1
12	-1	-1	1	47	1	-1	-1	1
13	1	1	-1	44	1	-1	-1	-1
14	1	1	-1	45	1	-1	-1	-1
15	1	1	-1	43	1	-1	-1	-1
16	-1	1	-1	47	-1	-1	1	1

```
Obs   x   z   h    y   t1   t2   t3   T4

 17  -1   1  -1   46   -1   -1    1    1
 18  -1   1  -1   48   -1   -1    1    1
 19   1  -1  -1   46   -1    1   -1    1
 20   1  -1  -1   46   -1    1   -1    1
 21   1  -1  -1   47   -1    1   -1    1
 22  -1  -1  -1   51    1    1    1   -1
 23  -1  -1  -1   50    1    1    1   -1
 24  -1  -1  -1   52    1    1    1   -1
     ==  ==  == ====   ==   ==   ==   ==
      0   0   0 1105    0    0    0    0

                The REG Procedure
                 Model: MODEL1
             Dependent Variable: y

         Number of Observations Read    24
         Number of Observations Used    24

              Analysis of Variance

                          Sum of         Mean
Source            DF     Squares       Square   F Value  Pr > F

Model              7   144.29167     20.61310     22.49  <.0001
Error             16    14.66667      0.91667
Corrected Total   23   158.95833

       Root MSE            0.95743    R-Square    0.9077
       Dependent Mean     46.04167    Adj R-Sq    0.8674
       Coeff Var           2.07948

              Parameter Estimates

                   Parameter     Standard
Variable    DF      Estimate        Error   t Value     Pr > |t|

Intercept    1      46.04167      0.19543    235.59       <.0001
x            1      -1.79167      0.19543     -9.17       <.0001
z            1      -1.20833      0.19543     -6.18       <.0001
h            1      -1.04167      0.19543     -5.33       <.0001
t1           1       0.12500      0.19543      0.64       0.5315
t2           1       0.37500      0.19543      1.92       0.0730
t3           1       0.12500      0.19543      0.64       0.5315
T4           1      -0.29167      0.19543     -1.49       0.1550
```

The estimates of the parameters for the model obtained by SAS are in agreement with the theory and estimates obtained by manual calculations; so also the ANOVA.

8.4 The General Regression Model for a Three-Way ANOVA (Reference Cell Coding Method)

If we intend to use reference cell coding method to model a three-factor factorial experiment, the model that is applicable is the same as the one used in Section 8.3 for effects coding method, save the definitions of the indicator variables in the model.

$$
\begin{aligned}
Y = \mu &+ \sum_{i=1}^{a-1} \alpha_i X_i + \sum_{j=1}^{b-1} \beta_j Z_j + \sum_{i=1}^{a-1}\sum_{j=1}^{b-1} \gamma_{ij} X_i Z_j + \sum_{k=1}^{c-1} \omega_k H_k \\
&+ \sum_{i=1}^{a-1}\sum_{k=1}^{c-1} \theta_{ik} X_i H_k + \sum_{j=1}^{b-1}\sum_{k=1}^{c-1} \lambda_{jk} Z_j H_k \\
&+ \sum_{i=1}^{a-1}\sum_{j=1}^{b-1}\sum_{k=1}^{c-1} \delta_{ijk} X_i Z_j H_k + \varepsilon
\end{aligned} \tag{8.9a}
$$

where

$$X_i = \begin{cases} 1 & \text{if } i = 1, 2, \ldots, a-1 \\ 0 & \text{otherwise} \end{cases}$$

$$Z_j = \begin{cases} 1 & \text{if } j = 1, 2, \ldots, b-1 \\ 0 & \text{otherwise} \end{cases}$$

$$H_k = \begin{cases} 1 & \text{if } k = 1, 2, \ldots, c-1 \\ 0 & \text{otherwise} \end{cases}$$

To estimate the parameters of the above model, we first obtain the usual row, column, and cell means, and use them in the following set of equations derived from the model:

$$
\begin{aligned}
\mu_{ijk} &= \mu + \alpha_i + \beta_j + \omega_k + \gamma_{ij} + \theta_{ik} + \lambda_{jk} + \delta_{ijk} \\
&\quad i = 1, 2, \ldots, a-1 \quad j = 1, 2, \ldots, b-1 \quad k = 1, 2, \ldots, c-1 \\
\mu_{ajk} &= \mu + \beta_j + \delta_k + \lambda_{jk} \quad j = 1, 2, \ldots, b-1 \quad k = 1, 2, \ldots, c-1 \\
\mu_{ibk} &= \mu + \alpha_i + \delta_k + \theta_{ik} \quad i = 1, 2, \ldots, a-1 \quad k = 1, 2, \ldots, c-1 \\
\mu_{ijc} &= \mu + \alpha_i + \beta_j + \gamma_{ij} \quad i = 1, 2, \ldots, a-1 \quad j = 1, 2, \ldots, b-1 \\
\mu_{abk} &= \mu + \omega_k \quad k = 1, 2, \ldots, c-1 \\
\mu_{ajc} &= \mu + \beta_j \quad j = 1, 2, \ldots, b-1 \\
\mu_{ibc} &= \mu + \alpha_i \quad i = 1, 2, \ldots, a-1
\end{aligned}
$$

$$\mu_{ij\cdot} = \mu + \alpha_i + \beta_j + \gamma_{ij} + \sum_{k=1}^{c-1}(\omega_k + \theta_{ik} + \lambda_{jk} + \delta_{ijk})/c$$

$$\mu_{i\cdot k} = \mu + \alpha_i + \omega_k + \theta_{ik} + \sum_{j=1}^{b-1}(\beta_j + \gamma_{ij} + \lambda_{jk} + \delta_{ijk})/b$$

$$\mu_{\cdot jk} = \mu + \beta_j + \omega_k + \lambda_{jk} + \sum_{i=1}^{a-1}(\alpha_i + \gamma_{ij} + \theta_{ik} + \delta_{ijk})/a$$

$$\mu_{ab\cdot} = \mu + \sum_{k=1}^{c-1}(\delta_k)/c$$

$$\mu_{a\cdot c} = \mu + \sum_{j=1}^{b-1}(\beta_j)/b$$

$$\mu_{\cdot bc} = \mu + \sum_{i=1}^{a-1}(\alpha_i)/a$$

$$\mu_{abc} = \mu$$

Based on the above relations between the parameters of the model, we simply use sample values to solve for their estimates and to determine the estimated relationship between the variables in the model.

8.4.1 Regression Model for 2^3 Factorial Design and Relations for Estimating Parameters (Reference Cell Method)

In Example 7.2, the levels of the factors A, B, and C were set at $a=b=c=2$. Consequently, we can rewrite the model for the general three-way ANOVA and the estimating equations for parameters, which is specifically tailored for this experiment as follows:

$$Y = \mu + \alpha X + \beta Z + \gamma XZ + \omega H + \theta XH + \lambda ZH + \delta XZH + \varepsilon \qquad (8.9b)$$

where

$$X = \begin{cases} 1 & \text{if level of } A \text{ is } 1 \\ 0 & \text{if level of } A \text{ is } 2 \end{cases}$$

$$Z = \begin{cases} 1 & \text{if level of } B \text{ is } 1 \\ 0 & \text{if level of } B \text{ is } 2 \end{cases}$$

$$H = \begin{cases} 1 & \text{if level of } C \text{ is } 1 \\ 0 & \text{if level of } C \text{ is } 2 \end{cases}$$

$$\begin{aligned}
\mu &= \mu_{222}\\
\mu_{111} &= \mu + \alpha + \beta + \gamma + \omega + \theta + \lambda + \delta\\
\mu_{122} &= \mu + \alpha ...\\
\mu_{212} &= \mu + \beta\\
\mu_{221} &= \mu + \omega\\
\mu_{211} &= \mu + \beta + \omega + \lambda\\
\mu_{121} &= \mu + \alpha + \omega + \theta\\
\mu_{112} &= \mu + \alpha + \beta + \gamma
\end{aligned}$$

8.4.1.1 Application to Example 7.2

From data for Example 7.2, we see that

(1)	42,44,41	$\hat{\mu}_{111} = \frac{42+44+41}{3} = 42.33$
a	45,46,47	$\hat{\mu}_{211} = \frac{45+46+47}{3} = 46$
b	44,45,44	$\hat{\mu}_{121} = \frac{44+45+44}{3} = 44.33$
ab	47,48,47	$\hat{\mu}_{221} = \frac{47+48+47}{3} = 47.33$
c	44,45,43	$\hat{\mu}_{112} = \frac{44+45+43}{3} = 44$
ac	47,46,48	$\hat{\mu}_{212} = \frac{47+46+48}{3} = 47$
bc	46,46,47	$\hat{\mu}_{122} = \frac{46+46+47}{3} = 46.33$
abc	51,50,52	$\hat{\mu}_{222} = \frac{51+50+52}{3} = 51$

The manual estimates for the parameters of the regression model for Example 7.2 based on reference cell coding method are as follows:

$$\begin{aligned}
\hat{\mu}_{222} &= 51 = \hat{\mu}\\
\hat{\mu}_{122} &= \hat{\mu} + \hat{\alpha} \Rightarrow 46.33 = 51 + \hat{\alpha} \Rightarrow \hat{\alpha} = -4.67\\
\hat{\mu}_{212} &= \hat{\mu} + \hat{\beta} \Rightarrow 47 = 51 + \hat{\beta} \Rightarrow \hat{\beta} = -4\\
\hat{\mu}_{221} &= \hat{\mu} + \hat{\omega} \Rightarrow 47.33 = 51 + \hat{\omega} \Rightarrow \hat{\omega} = -3.67\\
\hat{\mu}_{211} &= \hat{\mu} + \hat{\beta} + \hat{\omega} + \hat{\lambda} \Rightarrow 46 = 51 - 4 - 3.67 + \hat{\lambda} \Rightarrow \hat{\lambda} = 2.67\\
\hat{\mu}_{121} &= \hat{\mu} + \hat{\alpha} + \hat{\omega} + \hat{\theta} \Rightarrow 44.33 = 51 - 4.67 - 3.67 + \hat{\theta} \Rightarrow \hat{\theta} = 1.67\\
\hat{\mu}_{112} &= \hat{\mu} + \hat{\alpha} + \hat{\beta} + \hat{\gamma} \Rightarrow 44 = 51 - 4.67 - 4 + \hat{\gamma} \Rightarrow \hat{\gamma} = 1.67\\
\hat{\mu}_{111} &= \hat{\mu} + \hat{\alpha} + \hat{\gamma} + \hat{\omega} + \hat{\theta} + \lambda + \hat{\delta}\\
&\Rightarrow 42.33 = 51 - 4.67 - 4 + 1.67 - 3.67 + 1.67 + 2.67 + \hat{\delta} \Rightarrow \hat{\delta} = -2.34
\end{aligned}$$

The fitted model is

$$\hat{y} = 51 - 4.67x - 4z + 1.67xz - 3.67h + 2.67xh + 1.67zh - 2.34xzh \quad (8.10)$$

8.4.2 Testing for Significance of the Fitted Model 8.6 for Example 7.2

We use the estimated model 8.10 and responses of the experiment to obtain the fitted values and the attendant residual for each response of the experiment as displayed in Table 8.8.

TABLE 8.8: Estimates and Residuals for the Fitted Model 8.6 for Example 7.2 (Reference Cell Coding)

	x	z	h	y	xz	xh	zh	xzh	$\hat{y}$	Residuals = $y - \hat{y}$	$(y - \hat{y})^2$
	1	1	1	42	1	1	1	1	42.3	−0.33	0.1089
	1	1	1	44	1	1	1	1	42.3	1.67	2.7889
	1	1	1	41	1	1	1	1	42.3	−1.33	1.7689
	0	1	1	45	0	1	0	0	46	−1	1
	0	1	1	46	0	1	0	0	46	0	0
	0	1	1	47	0	1	0	0	46	1	1
	1	0	1	44	0	0	1	0	44.3	−0.33	0.1089
	1	0	1	45	0	0	1	0	44.3	0.67	0.4489
	1	0	1	44	0	0	1	0	44.3	−0.33	0.1089
	0	0	1	47	0	0	0	0	47.3	−0.33	0.1089
	0	0	1	48	0	0	0	0	47.3	0.67	0.4489
	0	0	1	47	0	0	0	0	47.3	−0.33	0.1089
	1	1	0	44	1	0	0	0	44	0	0
	1	1	0	45	1	0	0	0	44	1	1
	1	1	0	43	1	0	0	0	44	−1	1
	0	1	0	47	0	0	0	0	47	0	0
	0	1	0	46	0	0	0	0	47	−1	1
	0	1	0	48	0	0	0	0	47	1	1
	1	0	0	46	0	0	0	0	46.3	−0.33	0.1089
	1	0	0	46	0	0	0	0	46.3	−0.33	0.1089
	1	0	0	47	0	0	0	0	46.3	0.67	0.4489
	0	0	0	51	0	0	0	0	51	0	0
	0	0	0	50	0	0	0	0	51	−1	1
	0	0	0	52	0	0	0	0	51	1	1
Total	12	12	12	1105	6	6	6	3	1105	0	14.6668

To test for significance of the fitted model, we calculate the sums of squares as follows:

$$SS_{\text{total}} = \sum_{i=1}^{24} y_i^2 - \frac{\left(\sum_{i=1}^{24} y_i\right)^2}{24} = 51035 - \frac{1105^2}{24} = 158.9583$$

$$SS_{\text{residual}} = \sum_{i=1}^{24} (y_i - \hat{y}_i)^2 = 14.66668$$

$$SS_{\text{regression}} = SS_{\text{total}} - SS_{\text{residual}} = 158.9583 - 14.6668 = 144.29$$

TABLE 8.9: ANOVA to Test for Significance of Fitted Model 8.4 for Data of Example 7.2

Source	Sum of squares	Degrees of freedom	Mean square	F-ratio	p-Value
Regression	144.29	7	20.6131	22.487	3.76692E−07
Error	14.667	16	0.916668		
Total	158.96	23			

ANOVA shows that the fitted model is significant.

8.4.3 SAS Program for Analysis of Data of Example 7.2 (Reference Cell Method)

Next, we carry out the analysis of the data of Example 7.2 using SAS. The SAS program which does the analysis and its output are as follows:

SAS PROGRAM

```
data _2fact2;
input x z h y@@;
t1=x*z; t2=z*h; t3=x*h; T4=X*Z*H; f1=x*y;f2=z*y; f3=h*y;
f4=t1*y;  f5=t2*y;  f6=t3*y;  f7=t4*y;
cards;
1 1 1  42 1 1 1 44 1 1 1 41 0 1 1 45 0 1 1 46 0 1 1 47
1 0 1 44 1 0 1  45 1 0 1 44 0 0 1  47 0 0 1 48 0 0 1 47
1 1 0 44 1 1 0 45 1 1 0 43 0 1 0 47 0 1 0 46 0 1 0 48
1 0 0 46 1 0 0 46 1 0 0 47 0 0 0 51 0 0 0 50 0 0 0 52
;
proc print data=_2fact2; sum x z h t1-t4 y f1-f7;
proc REG data=_2fact2;
model y=x z h t1-t4; run;
```

SAS OUTPUT

Obs	x	z	h	y	t1	t2	t3	T4	f1	f2	f3	f4	f5	f6	f7
1	1	1	1	42	1	1	1	1	42	42	42	42	42	42	42
2	1	1	1	44	1	1	1	1	44	44	44	44	44	44	44
3	1	1	1	41	1	1	1	1	41	41	41	41	41	41	41
4	0	1	1	45	0	1	0	0	0	45	45	0	45	0	0
5	0	1	1	46	0	1	0	0	0	46	46	0	46	0	0
6	0	1	1	47	0	1	0	0	0	47	47	0	47	0	0

Obs	x	z	h	y	t1	t2	t3	T4	f1	f2	f3	f4	f5	f6	f7
7	1	0	1	44	0	0	1	0	44	0	44	0	0	44	0
8	1	0	1	45	0	0	1	0	45	0	45	0	0	45	0
9	1	0	1	44	0	0	1	0	44	0	44	0	0	44	0
10	0	0	1	47	0	0	0	0	0	0	47	0	0	0	0
11	0	0	1	48	0	0	0	0	0	0	48	0	0	0	0
12	0	0	1	47	0	0	0	0	0	0	47	0	0	0	0
13	1	1	0	44	1	0	0	0	44	44	0	44	0	0	0
14	1	1	0	45	1	0	0	0	45	45	0	45	0	0	0
15	1	1	0	43	1	0	0	0	43	43	0	43	0	0	0
16	0	1	0	47	0	0	0	0	0	47	0	0	0	0	0
17	0	1	0	46	0	0	0	0	0	46	0	0	0	0	0
18	0	1	0	48	0	0	0	0	0	48	0	0	0	0	0
19	1	0	0	46	0	0	0	0	46	0	0	0	0	0	0
20	1	0	0	46	0	0	0	0	46	0	0	0	0	0	0
21	1	0	0	47	0	0	0	0	47	0	0	0	0	0	0
22	0	0	0	51	0	0	0	0	0	0	0	0	0	0	0
23	0	0	0	50	0	0	0	0	0	0	0	0	0	0	0
24	0	0	0	52	0	0	0	0	0	0	0	0	0	0	0
	==	==	==	====	==	==	==	==	===	===	===	===	===	===	===
	12	12	12	1105	6	6	6	3	531	538	540	259	265	260	127

Analysis of Variance

Source	DF	Sum of Squares	Mean Square	F Value	Pr > F
Model	7	144.29167	20.61310	22.49	<.0001
Error	16	14.66667	0.91667		
Corrected Total	23	158.95833			

Root MSE	0.95743	R-Square	0.9077
Dependent Mean	46.04167	Adj R-Sq	0.8674
Coeff Var	2.07948		

Parameter Estimates

Variable	DF	Parameter Estimate	Standard Error	t Value	Pr > |t|
Intercept	1	51.00000	0.55277	92.26	<.0001
x	1	-4.66667	0.78174	-5.97	<.0001
z	1	-4.00000	0.78174	-5.12	0.0001
h	1	-3.66667	0.78174	-4.69	0.0002
t1	1	1.66667	1.10554	1.51	0.1512
t2	1	2.66667	1.10554	2.41	0.0282
t3	1	1.66667	1.10554	1.51	0.1512
T4	1	-2.33333	1.56347	-1.49	0.1550

As we can see, SAS estimates of the parameters of the model are the same as those obtained earlier using manual calculations, agreeing with the fitted model 8.10.

8.5 Regression Models for the Four-Factor Factorial Design Using Effects Coding Method

Under the classical analysis of the four-factor factorial experiment, the general model can be written as follows:

$$\begin{aligned} y_{ijklm} &= \mu + \alpha_i + \beta_j + (\alpha\beta)_{ij} + \gamma_k + (\alpha\gamma)_{ik} + (\beta\gamma)_{jk} + (\alpha\beta\gamma)_{ijk} \\ &\quad + \delta_l + (\alpha\delta)_{il} + (\beta\delta)_{jl} + (\alpha\beta\delta)_{ijl} + (\gamma\delta)_{kl} + (\alpha\gamma\delta)_{ikl} \\ &\quad + (\beta\gamma\delta)_{jkl} + (\alpha\beta\gamma\delta)_{ijkl} + \varepsilon_{ijklm} \end{aligned} \tag{8.11}$$

where $i = 1, 2, \ldots, a$ $j = 1, 2, \ldots, b$ $k = 1, 2, \ldots, c$ $l = 1, 2, \ldots, d$
$m = 1, 2, \ldots, n$ $\varepsilon_{ijklm} \sim \text{NID}(0, \sigma^2)$

The responses of the four-factor factorial experiment can be analyzed on the basis of this model, whose assumptions can be augmented by using either fixed, random, or mixed effects.

We can use effects coding method for defining the dummy variables in a typical four-factor design. Under this regression method, the underlying model for the full four-factor factorial experiment is obtained as

$$\begin{aligned} Y &= \mu + \sum_{i=1}^{a-1} \alpha_i X_i + \sum_{j=1}^{b-1} \beta_j Z_j + \sum_{i=1}^{a-1}\sum_{j=1}^{b-1} \gamma_{ij} X_i Z_j + \sum_{k=1}^{c-1} \omega_k H_k \\ &\quad + \sum_{i=1}^{a-1}\sum_{k=1}^{c-1} \theta_{ik} X_i H_k + \sum_{j=1}^{b-1}\sum_{k=1}^{c-1} \lambda_{jk} Z_j H_k + \sum_{i=1}^{a-1}\sum_{j=1}^{b-1}\sum_{k=1}^{c-1} \delta_{ijk} X_i Z_j H_k \\ &\quad + \sum_{l=1}^{d-1} \tau_l M_l + \sum_{i=1}^{a-1}\sum_{l=1}^{d-1} \psi_{il} X_i M_l + \sum_{j=1}^{b-1}\sum_{l=1}^{d-1} \rho_{jl} Z_j M_l \\ &\quad + \sum_{i=1}^{a-1}\sum_{j=1}^{b-1}\sum_{l=1}^{d-1} \phi_{ijl} X_i Z_j M_l + \sum_{k=1}^{c-1}\sum_{l=1}^{d-1} \upsilon_{kl} H_k M_l + \sum_{i=1}^{a-1}\sum_{k=1}^{c-1}\sum_{l=1}^{d-1} \sigma_{ikl} X_i H_k M_l \\ &\quad + \sum_{j=1}^{b-1}\sum_{k=1}^{c-1}\sum_{l=1}^{d-1} \pi_{jkl} Z_j H_k M_l + \sum_{i=1}^{a-1}\sum_{j=1}^{b-1}\sum_{k=1}^{c-1}\sum_{l=1}^{d-1} \eta_{ijkl} X_i Z_j H_k M_l + \varepsilon \end{aligned} \tag{8.12}$$

where

$$X_i = \begin{cases} -1 & \text{if } i = a \\ 1 & \text{if } i = 1, 2, \ldots, a-1 \\ 0 & \text{otherwise} \end{cases}$$

$$Z_j = \begin{cases} -1 & \text{if } j = b \\ 1 & \text{if } j = 1, 2, \ldots, b-1 \\ 0 & \text{otherwise} \end{cases}$$

$$H_k = \begin{cases} -1 & \text{if } k = c \\ 1 & \text{if } k = 1, 2, \ldots, c-1 \\ 0 & \text{otherwise} \end{cases}$$

$$M_l = \begin{cases} -1 & \text{if } l = d \\ 1 & \text{if } l = 1, 2, \ldots, d-1 \\ 0 & \text{otherwise} \end{cases}$$

Estimation of all the parameters of the model can be accomplished using means from the sample substituted in the following relations:

$$\begin{aligned}
\mu &= \mu_{....} \\
\alpha_i &= \mu_{i...} - \mu \quad i = 1, 2, \ldots, a-1 \\
\beta_j &= \mu_{.j..} - \mu \quad j = 1, 2, \ldots, b-1 \\
\omega_k &= \mu_{..k.} - \mu \quad k = 1, 2, \ldots, c-1 \\
\tau_l &= \mu_{...l} - \mu \quad l = 1, 2, \ldots, d-1 \\
\gamma_{ij} &= \mu_{ij..} - \mu_{i...} - \mu_{.j..} + \mu \quad i = 1, 2, \ldots, a-1 \;\; j = 1, 2, \ldots, b-1 \\
\theta_{ik} &= \mu_{i.k.} - \mu_{i...} - \mu_{..k.} + \mu \quad i = 1, 2, \ldots, a-1 \;\; k = 1, 2, \ldots, c-1 \\
\psi_{il} &= \mu_{i..l} - \mu_{i...} - \mu_{...l} + \mu \quad i = 1, 2, \ldots, a-1 \;\; l = 1, 2, \ldots, d-1 \\
\upsilon_{kl} &= \mu_{..kl} - \mu_{..k.} - \mu_{...l} + \mu \quad k = 1, 2, \ldots, c-1 \;\; l = 1, 2, \ldots, d-1 \\
\rho_{jl} &= \mu_{.j.l} - \mu_{.j..} - \mu_{...l} + \mu \quad j = 1, 2, \ldots, b-1 \;\; l = 1, 2, \ldots, d-1 \\
\lambda_{jk} &= \mu_{.jk.} - \mu_{.j..} - \mu_{..k.} + \mu \quad i = 1, 2, \ldots, a-1 \;\; j = 1, 2, \ldots, b-1 \\
\delta_{ijk} &= \mu_{ijk.} + \mu_{i...} + \mu_{.j..} + \mu_{..k.} - \mu_{ij..} - \mu_{i.k.} - \mu_{.jk.} - \mu_{....} \\
&\quad i = 1, 2, \ldots, a-1 \;\; j = 1, 2, \ldots, b-1 \;\; k = 1, 2, \ldots, c-1 \\
\phi_{ijl} &= \mu_{ij.l} + \mu_{i...} + \mu_{.j..} + \mu_{...l} - \mu_{ij..} - \mu_{i..l} - \mu_{.j.l} - \mu_{....} \\
i &= 1, 2, \ldots, a-1 \;\; j = 1, 2, \ldots, b-1 \;\; l = 1, 2, \ldots, d-1 \\
\pi_{jkl} &= \mu_{.jkl} + \mu_{.j..} + \mu_{..k.} + \mu_{...l} - \mu_{.jk.} - \mu_{.j.l} - \mu_{..kl} - \mu_{....} \\
j &= 1, 2, \ldots, b-1 \;\; k = 1, 2, \ldots, c-1 \;\; l = 1, 2, \ldots, d-1 \\
\sigma_{ikl} &= \mu_{i.kl} + \mu_{i...} + \mu_{..k.} + \mu_{...l} - \mu_{ik..} - \mu_{i..l} - \mu_{..kl} - \mu_{....} \\
i &= 1, 2, \ldots, a-1 \;\; k = 1, 2, \ldots, c-1 \;\; l = 1, 2, \ldots, d-1 \\
\eta_{ijkl} &= \mu_{ijkl} + \mu_{ij..} + \mu_{i.k.} + \mu_{i..l} + \mu_{.jk.} + \mu_{.j.l} + \mu_{..kl} + \mu_{....} - \mu_{ijk.} \\
&\quad - \mu_{ij.l} - \mu_{i.kl} - \mu_{.jkl} - \mu_{i...} - \mu_{.j..} - \mu_{..k.} - \mu_{...l} \\
i &= 1, 2, \ldots, a-1 \;\; j = 1, 2, \ldots, b-1 \;\; k = 1, 2, \ldots, c-1
\end{aligned}$$

8.5.1 Regression Model for the 2^4 Factorial Design Using Effects Coding Method for Defining Dummy Variables

Next, we modify the model described above so that it can specifically be applied to a 2^4 factorial design. The model and ancillary relations would then take the form

$$\begin{aligned} Y = \mu &+ \alpha_1 X_i + \beta_1 Z_j + \gamma_{11} X_i Z_j + \omega_1 H_k + \theta_{11} X_i H_k + \lambda_{11} Z_j H_k \\ &+ \delta_{111} X_i Z_j H_k + \tau_1 M_l + \psi_{11} X_i M_l + \rho_{11} Z_j M_l + \phi_{111} X_i Z_j M_l \\ &+ \upsilon_{11} H_k M_l + \sigma_{111} X_i H_k M_l + \pi_{111} Z_j H_k M_l \\ &+ \eta_{1111} X_i Z_j H_k M_l + \varepsilon \end{aligned} \tag{8.13}$$

where

$$X_i = \begin{cases} -1 & \text{if } i = 2 \\ 1 & \text{if } i = 1 \\ 0 & \text{otherwise} \end{cases}$$

$$Z_j = \begin{cases} -1 & \text{if } j = 2 \\ 1 & \text{if } j = 1 \\ 0 & \text{otherwise} \end{cases}$$

$$H_k = \begin{cases} -1 & \text{if } k = 2 \\ 1 & \text{if } k = 1 \\ 0 & \text{otherwise} \end{cases}$$

$$M_l = \begin{cases} -1 & \text{if } l = 2 \\ 1 & \text{if } l = 1 \\ 0 & \text{otherwise} \end{cases}$$

8.5.2 Application of the Model to Example 7.3 (Effects Coding Method)

In order to estimate parameters for the 2^4 factorial experiment, we simply set the values of $a=b=c=d=2$ in the relations presented in Section 8.4. Thereafter, using the data of Example 7.3, we obtain values for the cell means needed to estimate the parameters of the model as follows:

$$\begin{aligned} \hat{\mu}_{1\ldots} &= \frac{[(1) + b + c + bc + d + bd + cd + bcd]}{8} \\ &= \frac{(38 + 27 + 58 + 30 + 59 + 53 + 79 + 53)}{8} = 49.625 \end{aligned}$$

$$\hat{\mu}_{.1..} = \frac{[(1) + a + c + ac + d + ad + cd + acd]}{8}$$
$$= \frac{(38 + 40 + 58 + 55 + 59 + 62 + 79 + 75)}{8} = 58.25$$
$$\hat{\mu}_{..1.} = \frac{[(1) + a + b + ab + d + ad + bd + abd]}{8}$$
$$= \frac{(38 + 40 + 27 + 30 + 59 + 62 + 53 + 50)}{8} = 44.875$$
$$\hat{\mu}_{...1} = \frac{[(1) + a + b + ab + c + ac + bc + abc]}{8}$$
$$= \frac{(38 + 40 + 27 + 30 + 58 + 55 + 30 + 32)}{8} = 38.75$$
$$\hat{\mu}_{11..} = \frac{[(1) + c + d + cd]}{4} = \frac{(38 + 58 + 59 + 79)}{4} = 58.50$$
$$\hat{\mu}_{1.1.} = \frac{[(1) + b + d + bd]}{4} = \frac{(38 + 27 + 59 + 53)}{4} = 44.25$$
$$\hat{\mu}_{1..1} = \frac{[(1) + b + c + bc]}{4} = \frac{(38 + 27 + 58 + 30)}{4} = 38.25$$
$$\hat{\mu}_{.11.} = \frac{[(1) + a + d + ad]}{4} = \frac{(38 + 40 + 59 + 62)}{4} = 49.75$$
$$\hat{\mu}_{..11} = \frac{[(1) + a + b + ab]}{4} = \frac{(38 + 40 + 27 + 30)}{4} = 33.75$$
$$\hat{\mu}_{.1.1} = \frac{[(1) + a + c + ac]}{4} = \frac{(38 + 40 + 58 + 55)}{4} = 47.75$$
$$\hat{\mu}_{111.} = \frac{(1) + d}{2} = \frac{59 + 38}{2} = 48.50$$
$$\hat{\mu}_{.111} = \frac{(1) + a}{2} = \frac{38 + 40}{2} = 39$$
$$\hat{\mu}_{11.1} = \frac{(1) + c}{2} = \frac{38 + 58}{2} = 48$$
$$\hat{\mu}_{1.11} = \frac{(1) + b}{2} = \frac{38 + 27}{2} = 32.50$$
$$\hat{\mu}_{1111} = 38$$

With the above values for the different means, we can estimate the parameters of the regression model as follows (Table 8.10):

$$\hat{\mu}_{....} = \frac{795}{16} = 49.6875$$
$$\hat{\alpha}_1 = \hat{\mu}_{1...} - \hat{\mu}_{....} = 49.625 - 49.6875 = -0.0625$$
$$\hat{\beta}_1 = \hat{\mu}_{.1..} - \hat{\mu}_{....} = 58.25 - 49.6875 = 8.5625$$
$$\hat{\omega}_1 = \hat{\mu}_{..1.} - \hat{\mu}_{....} = 44.875 - 49.6875 = -4.8125$$

$$\hat{\tau}_1 = \hat{\mu}_{...1} - \hat{\mu}_{....} = 38.75 - 49.6875 = -10.9375$$

$$\hat{\gamma}_{11} = \hat{\mu}_{11..} - \hat{\mu}_{1...} - \hat{\mu}_{.1..} + \hat{\mu}_{....} = 58.5 - 49.625 - 58.25 + 49.6875 = 0.3125$$

$$\hat{\theta}_{11} = \hat{\mu}_{1.1.} - \hat{\mu}_{1...} - \hat{\mu}_{..1.} + \hat{\mu}_{....} = 44.25 - 49.625 - 44.875 + 49.6875 = -0.5625$$

$$\hat{\psi}_{11} = \hat{\mu}_{1..1} - \hat{\mu}_{1...} - \hat{\mu}_{...1} + \hat{\mu}_{....} = 38.25 - 49.625 - 38.75 + 49.6875 = -0.4375$$

$$\hat{\upsilon}_{11} = \hat{\mu}_{..11} - \hat{\mu}_{..1.} - \hat{\mu}_{...1} + \hat{\mu}_{....} = 33.75 - 44.875 - 38.75 + 49.6875 = -0.1875$$

$$\hat{\rho}_{11} = \hat{\mu}_{.1.1} - \hat{\mu}_{.1..} - \hat{\mu}_{...1} + \hat{\mu}_{....} = 47.75 - 58.25 - 38.75 + 49.6875 = 0.4345$$

$$\hat{\lambda}_{11} = \hat{\mu}_{.11.} - \hat{\mu}_{.1..} - \hat{\mu}_{..1.} + \hat{\mu}_{....} = 49.75 - 58.25 - 44.875 = 49.6875 = -3.6875$$

$$\begin{aligned}\hat{\delta}_{111} &= \hat{\mu}_{111.} + \hat{\mu}_{1...} + \hat{\mu}_{.1..} + \hat{\mu}_{..1.} - \hat{\mu}_{11..} - \hat{\mu}_{1.1.} - \hat{\mu}_{.11.} - \hat{\mu}_{....} \\ &= 48.5 + 49.625 + 58.25 + 44.875 - 58.5 - 44.25 - 49.75 - 49.6875 \\ &= -0.9375\end{aligned}$$

$$\begin{aligned}\hat{\phi}_{111} &= \mu_{11.1} + \hat{\mu}_{1...} + \hat{\mu}_{.1..} + \hat{\mu}_{...1} - \hat{\mu}_{11..} - \hat{\mu}_{.1.1} - \hat{\mu}_{1..1} - \hat{\mu}_{....} \\ &= 48 + 49.625 + 58.25 + 38.75 - 58.5 - 47.75 - 38.25 - 49.6875 \\ &= 0.4375\end{aligned}$$

$$\begin{aligned}\hat{\pi}_{111} &= \hat{\mu}_{.111} + \hat{\mu}_{.1..} + \hat{\mu}_{..1.} + \hat{\mu}_{...1} - \hat{\mu}_{.11.} - \hat{\mu}_{..11} - \hat{\mu}_{.1.1} - \hat{\mu}_{....} \\ &= 39 + 58.25 + 44.875 + 38.75 - 49.75 - 33.75 - 47.75 - 49.6875 \\ &= 0.0625\end{aligned}$$

$$\begin{aligned}\hat{\sigma}_{111} &= \hat{\mu}_{.111} + \hat{\mu}_{.1..} + \hat{\mu}_{..1.} + \hat{\mu}_{...1} - \hat{\mu}_{.11.} - \hat{\mu}_{..11} - \hat{\mu}_{.1.1} - \hat{\mu}_{....} \\ &= 32.5 + 49.625 + 44.875 + 38.75 - 44.25 - 33.75 - 38.25 - 49.6875 \\ &= -0.1875\end{aligned}$$

$$\begin{aligned}\hat{\eta}_{1111} &= \hat{\mu}_{1111} + \hat{\mu}_{11..} + \hat{\mu}_{1.1.} + \hat{\mu}_{1..1} + \hat{\mu}_{.11.} + \hat{\mu}_{.1.1} + \hat{\mu}_{..11} + \hat{\mu}_{....} \\ &\quad - \hat{\mu}_{111.} - \hat{\mu}_{11.1} - \hat{\mu}_{1.11} - \hat{\mu}_{.111} - \hat{\mu}_{1...} - \hat{\mu}_{.1..} - \hat{\mu}_{..1.} - \hat{\mu}_{...1} \\ &= 38 + 58.5 + 44.25 + 38.25 + 49.75 + 47.75 + 33.75 + 49.6875 \\ &\quad - 48.5 - 48 - 32.5 - 39 - 49.625 - 58.25 - 44.875 - 38.75 = 0.4375\end{aligned}$$

$$\begin{aligned}\hat{y} = \ & 49.6875 - 0.0625x + 8.5625z + 0.3125xz - 4.8125h - 0.5625xh \\ & - 3.6875zh - 0.9375xzh - 10.9375m - 0.4375xm + 0.4345zm \\ & + 0.4375xzm - 0.1875hm - 0.1875xhm + 0.0625zhm + 0.4375xzhm\end{aligned} \tag{8.14}$$

There is no further analysis to test for significance of fitted model because we have a single replicate of the experiment, which leaves us with no estimate for error. Recall that this was true in the traditional ANOVA for this example carried out in Chapter 7.

TABLE 8.10: Estimates and Residuals for Fitted Model 8.13 for Example 7.3 (Effects Coding Method)

x	z	h	m	y	xz	xh	zh	xzh	xm	zm	xzm	xhm	zhm	$xzhm$	hm	$\hat{y}$	$y-\hat{y}$
1	1	1	1	38	1	1	1	1	1	1	1	1	1	1	1	38	0
−1	1	1	1	40	−1	−1	1	−1	−1	1	−1	−1	1	−1	1	40	0
1	−1	1	1	27	−1	1	−1	−1	1	−1	−1	1	−1	−1	1	27	0
−1	−1	1	1	30	1	−1	−1	1	−1	−1	1	−1	−1	1	1	30	0
1	1	−1	1	58	1	−1	−1	−1	1	1	1	−1	−1	−1	−1	58	0
−1	1	−1	1	55	−1	1	−1	1	−1	1	−1	1	−1	1	−1	55	0
1	−1	−1	1	30	−1	−1	1	1	1	−1	−1	−1	1	1	−1	3	0
−1	−1	−1	1	32	1	1	1	−1	−1	−1	1	1	1	−1	−1	32	0
1	1	1	−1	59	1	1	1	1	−1	−1	−1	−1	−1	−1	−1	59	0
−1	1	1	−1	62	−1	−1	1	−1	1	−1	1	1	−1	1	−1	62	0
1	−1	1	−1	53	−1	1	−1	−1	−1	1	1	−1	1	1	−1	53	0
−1	−1	1	−1	50	1	−1	−1	1	1	1	−1	1	1	−1	−1	50	0
1	1	−1	−1	79	1	−1	−1	−1	−1	−1	−1	1	1	1	1	79	0
−1	1	−1	−1	75	−1	1	−1	1	1	−1	1	−1	1	−1	1	75	0
1	−1	−1	−1	53	−1	−1	1	1	−1	1	1	1	−1	−1	1	53	0
−1	−1	−1	−1	54	1	1	1	−1	1	1	−1	−1	−1	1	1	54	0

8.5.3 SAS Program for Analysis of Data of Example 7.3

We write a SAS program to analyze the data of the 2^4 factorial design (Example 6.3) and to estimate the parameters of the regression model when the effects coding method is used to define the dummy variables in the model.

SAS PROGRAM

```
data fourfact2;
input x z h m y @@;
t1=x*z; t2=x*h; t3=z*h; t4=x*z*h; t5=x*m; t6=z*m; t7=x*z*m; t8=x*h*m;
t9=z*h*m; t10=x*z*h*m; t11=h*m;
datalines;
1 1 1 1 38 -1 1 1 1 40 1 -1 1 1 27 -1 -1 1 1 30 1 1 -1 1 58
-1 1 -1 1 55 1 -1 -1 1 30 -1 -1 -1 1 32 1 1 1 -1 59 -1 1 1 -1 62
1 -1 1 -1 53 -1 -1 1 -1 50  1 1 -1 -1 79 -1 1 -1 -1 75
1 -1 -1 -1 53 -1 -1 -1 -1 54
;
proc print data=fourfact2; sum x z h m y t1-t11;
proc reg data=fourfact2;
model y=x z h m t1-t11; run;
```

SAS OUTPUT

Obs	x	z	h	m	y	t1	t2	t3	t4	t5	t6	t7	t8	t9	t10	t11
1	1	1	1	1	38	1	1	1	1	1	1	1	1	1	1	1
2	-1	1	1	1	40	-1	-1	1	-1	-1	1	-1	-1	1	-1	1
3	1	-1	1	1	27	-1	1	-1	-1	1	-1	-1	1	-1	-1	1
4	-1	-1	1	1	30	1	-1	-1	1	-1	-1	1	-1	-1	1	1
5	1	1	-1	1	58	1	-1	-1	-1	1	1	1	-1	-1	-1	-1
6	-1	1	-1	1	55	-1	1	-1	1	-1	1	-1	1	-1	1	-1
7	1	-1	-1	1	30	-1	-1	1	1	1	-1	-1	-1	1	1	-1
8	-1	-1	-1	1	32	1	1	1	-1	-1	-1	1	1	1	-1	-1
9	1	1	1	-1	59	1	1	1	1	-1	-1	-1	-1	-1	-1	-1
10	-1	1	1	-1	62	-1	-1	1	-1	1	-1	1	1	-1	1	-1
11	1	-1	1	-1	53	-1	1	-1	-1	-1	1	1	-1	1	1	-1
12	-1	-1	1	-1	50	1	-1	-1	1	1	1	-1	1	1	-1	-1
13	1	1	-1	-1	79	1	-1	-1	-1	-1	-1	-1	1	1	1	1
14	-1	1	-1	-1	75	-1	1	-1	1	1	-1	1	-1	1	-1	1
15	1	-1	-1	-1	53	-1	-1	1	1	-1	1	1	1	-1	-1	1
16	-1	-1	-1	-1	54	1	1	1	-1	1	1	-1	-1	-1	1	1
	==	==	==	==	===	==	==	==	==	==	==	==	==	==	===	===
	0	0	0	0	795	0	0	0	0	0	0	0	0	0	0	0

```
                    The REG Procedure
                      Model: MODEL1
                  Dependent Variable: y

          Number of Observations Read      16
          Number of Observations Used      16
```

Analysis of Variance

Source	DF	Sum of Squares	Mean Square	F Value	Pr > F
Model	15	3709.43750	247.29583	.	.
Error	0	0	.		
Corrected Total	15	3709.43750			

Root MSE	.	R-Square	1.0000
Dependent Mean	49.68750	Adj R-Sq	.
Coeff Var	.		

Parameter Estimates

Variable	DF	Parameter Estimate	Standard Error	t Value	Pr > |t|
Intercept	1	49.68750	.	.	.
x	1	-0.06250	.	.	.
z	1	8.56250	.	.	.
h	1	-4.81250	.	.	.
m	1	-10.93750	.	.	.
t1	1	0.31250	.	.	.
t2	1	-0.56250	.	.	.
t3	1	-3.68750	.	.	.
t4	1	-0.93750	.	.	.
t5	1	-0.43750	.	.	.
t6	1	0.43750	.	.	.
t7	1	0.43750	.	.	.
t8	1	-0.18750	.	.	.
t9	1	-0.06250	.	.	.
t10	1	0.43750	.	.	.
t11	1	-0.18750	.	.	.

As we can see, the results of SAS estimates agree with the previous manual analysis. SAS could not carry out ANOVA as the experiment was not replicated, so it could not estimate error.

8.6 Regression Analysis for a Four-Factor Factorial Experiment Using Reference Cell Coding to Define Dummy Variables

When we use dummy variable regression to investigate the effects of the factors on the response variable, we can use reference cell method to define the dummy variables in the design. Under this method, the underlying regression model for the full four-factor factorial experiment is obtained as

$$Y = \mu + \sum_{i=1}^{a-1} \alpha_i X_i + \sum_{j=1}^{b-1} \beta_j Z_j + \sum_{i=1}^{a-1}\sum_{j=1}^{b-1} \gamma_{ij} X_i Z_j + \sum_{k=1}^{c-1} \omega_k H_k$$
$$+ \sum_{i=1}^{a-1}\sum_{k=1}^{c-1} \theta_{ik} X_i H_k + \sum_{j=1}^{b-1}\sum_{k=1}^{c-1} \lambda_{jk} Z_j H_k + \sum_{i=1}^{a-1}\sum_{j=1}^{b-1}\sum_{k=1}^{c-1} \delta_{ijk} X_i Z_j H_k$$

$$+\sum_{l=1}^{d-1}\tau_l M_l+\sum_{i=1}^{a-1}\sum_{l=1}^{d-1}\psi_{il}X_iM_l+\sum_{j=1}^{b-1}\sum_{l=1}^{d-1}\rho_{jl}Z_jM_l$$
$$+\sum_{i=1}^{a-1}\sum_{j=1}^{b-1}\sum_{l=1}^{d-1}\phi_{ijl}X_iZ_jM_l+\sum_{k=1}^{c-1}\sum_{l=1}^{d-1}\upsilon_{kl}H_kM_l+\sum_{i=1}^{a-1}\sum_{k=1}^{c-1}\sum_{l=1}^{d-1}\sigma_{ikl}X_iH_kM_l$$
$$+\sum_{j=1}^{b-1}\sum_{k=1}^{c-1}\sum_{l=1}^{d-1}\pi_{jkl}Z_jH_kM_l+\sum_{i=1}^{a-1}\sum_{j=1}^{b-1}\sum_{k=1}^{c-1}\sum_{l=1}^{d-1}\eta_{ijkl}X_iZ_jH_kM_l+\varepsilon \qquad (8.15)$$

where

$$X_i=\begin{cases}1 & \text{if } i=1,2,\ldots,a-1\\ 0 & \text{otherwise}\end{cases}$$

$$Z_j=\begin{cases}1 & \text{if } j=1,2,\ldots,b-1\\ 0 & \text{otherwise}\end{cases}$$

$$H_k=\begin{cases}1 & \text{if } k=1,2,\ldots,c-1\\ 0 & \text{otherwise}\end{cases}$$

$$M_l=\begin{cases}1 & \text{if } l=1,2,\ldots,d-1\\ 0 & \text{otherwise}\end{cases}$$

Estimation of parameters of the model can be accomplished by using the following relations, replacing the means with sample values:

$$\mu_{ijkl}=\mu+\alpha_i+\beta_j+\gamma_{ij}+\omega_k+\theta_{ik}+\lambda_{jk}+\delta_{ijk}+\tau_l+\psi_{il}+\rho_{jl}+\phi_{ijl}$$
$$+\upsilon_{kl}+\sigma_{ikl}+\pi_{jkl}+\eta_{ijkl}$$
$$\mu_{ibcd}=\mu+\alpha_i \quad i=1,2,\ldots,a-1$$
$$\mu_{ajcd}=\mu+\beta_j \quad j=1,2,\ldots,b-1$$
$$\mu_{abkd}=\mu+\omega_k \quad k=1,2,\ldots,c-1$$
$$\mu_{abcl}=\mu+\tau_l \quad l=1,2,\ldots,d-1$$
$$\mu_{ijcd}=\mu+\alpha_i+\beta_j+\gamma_{ij} \quad i=1,2,\ldots,a-1 \quad j=1,2,\ldots,b-1$$
$$\mu_{ibkd}=\mu+\alpha_i+\omega_k+\theta_{ik} \quad i=1,2,\ldots,a-1 \quad k=1,2,\ldots,c-1$$
$$\mu_{ibcl}=\mu+\alpha_i+\tau_l+\psi_{il} \quad i=1,2,\ldots,a-1 \quad l=1,2,\ldots,d-1$$
$$\mu_{abkl}=\mu+\omega_k+\tau_l+\upsilon_{kl} \quad k=1,2,\ldots,c-1 \quad l=1,2,\ldots,d-1$$
$$\mu_{ajcl}=\mu+\beta_j+\tau_l+\rho_{jl} \quad j=1,2,\ldots,b-1 \quad l=1,2,\ldots,d-1$$
$$\mu_{ajkd}=\mu+\beta_j+\omega_k+\lambda_{jk} \quad j=1,2,\ldots,b-1 \quad k=1,2,\ldots,c-1$$
$$\mu_{ijkd}=\mu+\alpha_i+\beta_j+\gamma_{ij}+\omega_k+\theta_{ik}+\lambda_{jk}+\delta_{ijk}$$

$$
\begin{aligned}
i &= 1, 2, \ldots, a-1 \;\; j = 1, 2, \ldots, b-1 \;\; k = 1, 2, \ldots, c-1 \\
\mu_{ijcl} &= \mu + \alpha_i + \beta_j + \gamma_{ij} + \tau_l + \psi_{il} + \rho_{jl} + \phi_{ijl} \\
i &= 1, 2, \ldots, a-1 \;\; j = 1, 2, \ldots, b-1 \;\; l = 1, 2, \ldots, d-1 \\
\mu_{ajkl} &= \mu + \beta_j + \omega_k + \lambda_{jk} + \tau_l + \rho_{jl} + \upsilon_{kl} + \pi_{jkl} \\
j &= 1, 2, \ldots, b-1 \;\; k = 1, 2, \ldots, c-1 \;\; l = 1, 2, \ldots, d-1 \\
\mu_{ibkl} &= \mu + \alpha_i + \omega_k + \theta_{ik} + \tau_l + \psi_{il} + \upsilon_{kl} + \sigma_{ikl} \\
i &= 1, 2, \ldots, a-1 \;\; k = 1, 2, \ldots, c-1 \;\; l = 1, 2, \ldots, d-1 \\
\mu_{ijkl} &= \mu + \alpha_i + \beta_j + \gamma_{ij} + \omega_k + \theta_{ik} + \lambda_{jk} + \delta_{ijk} + \tau_l + \psi_{il} \\
&\quad + \rho_{jl} + \phi_{ijl} + \upsilon_{kl} + \sigma_{ikl} + \pi_{jkl} + \eta_{ijkl} \\
i &= 1, 2, \ldots, a-1 \;\; j = 1, 2, \ldots c-1 \;\; k = 1, 2, \ldots, c-1 \\
l &= 1, 2, \ldots, d-1 \\
\mu_{abcd} &= \mu
\end{aligned}
$$

8.6.1 Application of Regression Model (Using Reference Cell Coding for Dummy Variables) to Example 7.3

For Example 7.3, there was a single replicate of the 2^4 factorial design. Manual analysis depended on the use of higher order interactions to estimate error, and error sum of squares and carry out ANOVA. Here, we simply model the data using a regression approach based on the definition of the dummy variables by making reference to a single cell in the dataset. We will not be able to perform ANOVA precisely because the design was not replicated. Nevertheless, we will be able to estimate the parameters of the model based on the particular cell that was referenced. In this particular analysis, we assume that the referenced cell is the one in which the level for each of A, B, C, and D is 2. Also, because this is a 2^4 factorial design, the number of levels for each of them is 2. Thus, $a = b = c = d = 2$ in the above relations for estimating the parameters of the four-factor factorial design. The model which applies to the design is

$$
\begin{aligned}
Y &= \mu + \alpha_1 X + \beta_1 Z + \gamma_{11} XZ + \omega_1 H + \theta_{11} XH + \lambda_{11} ZH \\
&\quad + \tau_1 M + \psi_{11} XM + \rho_{11} ZM + \nu_{11} HM + \varepsilon
\end{aligned} \tag{8.16}
$$

8.6.1.1 Estimation of Parameters of the Model

Using the data for Example 7.3, we see that the means of the cells have the following values (Table 8.11).

Using the data in Table 8.11, we estimate the parameters of the regression model with the reference cell, as cell $(A, B, C, D) = (2, 2, 2, 2)$. The cell corresponds to treatment combination *abcd* in the normal 2^4 factorial design. The resulting estimates are

$$
\begin{aligned}
\hat{\mu}_{1222} &= 53 = 54 + \hat{\alpha}_1 \Rightarrow \hat{\alpha}_1 = 53 - 54 = -1 \\
\hat{\mu}_{2122} &= 75 = 54 + \hat{\beta}_1 \Rightarrow \hat{\beta}_1 = 75 - 54 = 21 \\
\hat{\mu}_{2212} &= 50 = 54 + \hat{\omega}_1 \Rightarrow \hat{\omega}_1 = 50 - 54 = -4 \\
\hat{\mu}_{2221} &= 32 = 54 + \hat{\tau}_1 \Rightarrow \hat{\tau}_1 = 32 - 54 = -22 \\
\hat{\mu}_{1122} &= 79 = 54 - 1 + 21 + \hat{\gamma}_{11} \Rightarrow \hat{\gamma}_{11} = 5
\end{aligned}
$$

TABLE 8.11: Means for Data of Example 7.3

Treatment combinations	Mean under reference cell regression model	Observed values
(1)	$\hat{\mu}_{1111}$	38
a	$\hat{\mu}_{2111}$	40
b	$\hat{\mu}_{1211}$	27
ab	$\hat{\mu}_{2211}$	30
c	$\hat{\mu}_{1121}$	58
ac	$\hat{\mu}_{2121}$	55
bc	$\hat{\mu}_{1221}$	30
abc	$\hat{\mu}_{2221}$	32
d	$\hat{\mu}_{1112}$	59
ad	$\hat{\mu}_{2112}$	62
bd	$\hat{\mu}_{1212}$	53
abd	$\hat{\mu}_{2212}$	50
cd	$\hat{\mu}_{1122}$	79
acd	$\hat{\mu}_{2122}$	75
bcd	$\hat{\mu}_{1222}$	53
$abcd$	$\hat{\mu}_{2222}$	54

$$
\begin{aligned}
\hat{\mu}_{1212} &= 53 = 54 - 1 - 4 + \hat{\theta}_{11} \Rightarrow \hat{\theta}_{11} = 4 \\
\hat{\mu}_{1221} &= \hat{\mu} + \hat{\alpha}_1 + \hat{\tau}_1 + \hat{\psi}_{11} \Rightarrow 30 = 54 - 22 - 1 + \hat{\psi}_{11} \Rightarrow \hat{\psi}_{11} = -1 \\
\hat{\mu}_{2211} &= \hat{\mu} + \omega_1 + \hat{\tau}_1 + \hat{\upsilon}_{11} \Rightarrow 30 = 54 - 4 - 22 + \hat{\upsilon}_{11} \Rightarrow \hat{\upsilon}_{11} = 2 \\
\hat{\mu}_{2121} &= \hat{\mu} + \hat{\beta}_1 + \hat{\tau}_1 + \hat{\rho}_{11} \Rightarrow 55 = 54 + 21 - 22 + \hat{\rho}_{11} \Rightarrow \hat{\rho}_{11} = 2 \\
\hat{\mu}_{2112} &= \hat{\mu} + \hat{\beta}_1 + \hat{\omega}_1 + \hat{\lambda}_{11} \Rightarrow 62 = 54 + 21 - 4 + \hat{\lambda}_{11} \Rightarrow \hat{\lambda}_{11} = -9 \\
\hat{\mu}_{1112} &= \hat{\mu} + \hat{\alpha}_1 + \hat{\beta}_1 + \hat{\gamma}_{11} + \hat{\omega}_1 + \hat{\theta}_{11} + \hat{\lambda}_{11} + \hat{\delta}_{111} \\
&\Rightarrow 59 = 54 - 1 + 21 + 5 - 4 + 4 - 9 + \hat{\delta}_{111} \Rightarrow \hat{\delta}_{111} = -11 \\
\hat{\mu}_{1121} &= \hat{\mu} + \hat{\alpha}_1 + \hat{\beta}_1 + \hat{\gamma}_{11} + \hat{\tau}_1 + \hat{\psi}_{11} + \hat{\rho}_{11} + \hat{\phi}_{111} \\
&\Rightarrow 58 = 54 - 1 + 21 + 5 - 22 - 1 + 2 + \hat{\phi}_{111} \Rightarrow \hat{\phi}_{111} = 0 \\
\hat{\mu}_{1211} &= \hat{\mu} + \hat{\alpha}_1 + \hat{\omega}_1 + \hat{\theta}_{11} + \hat{\tau}_1 + \hat{\psi}_{11} + \hat{\upsilon}_{11} + \hat{\sigma}_{111} \\
&\Rightarrow 27 = 54 - 1 - 4 + 4 - 22 - 1 + 2 + \hat{\sigma}_{111} \Rightarrow \hat{\sigma}_{111} = -5 \\
\hat{\mu}_{2111} &= \hat{\mu} + \hat{\beta}_1 + \hat{\omega}_1 + \hat{\lambda}_{11} + \hat{\tau}_1 + \hat{\rho}_{11} + \hat{\upsilon}_{11} + \hat{\pi}_{111} \\
&\Rightarrow 40 = 54 + 21 - 4 - 9 - 22 + 2 + 2 + \hat{\pi}_{111} \Rightarrow \hat{\pi}_{111} = -4 \\
\hat{\mu}_{1111} &= \hat{\mu} + \hat{\alpha}_1 + \hat{\beta}_1 + \hat{\gamma}_{11} + \hat{\omega}_1 + \hat{\theta}_{11} + \hat{\lambda}_{11} + \hat{\delta}_{111} + \hat{\tau}_1 \\
&\quad + \hat{\psi}_{11} + \hat{\rho}_{11} + \hat{\phi}_{111} + \hat{\upsilon}_{11} + \hat{\sigma}_{111} + \hat{\pi}_{111} + \hat{\eta}_{1111} \\
\Rightarrow 38 &= 54 - 1 + 21 + 5 - 4 + 4 - 9 - 11 - 22 - 1 + 2 \\
&\quad + 0 + 2 - 5 - 4 + \hat{\eta}_{1111} \Rightarrow \hat{\eta}_{1111} = 7.
\end{aligned}
$$

The estimate for the intercept of this model is the value of the reference cell, $\hat{\mu}_{2222} = 54$. The above estimations of the parameters of the model indicate that the fitted model is

$$
\begin{aligned}
\hat{y} &= 54 - x + 21z + 5xz - 4h + 4xh - 9zh - 11xzh - 22m - xm \\
&\quad + 2zm + 0xzm + 2hm - 5xhm - 4zhm + 7xzhm
\end{aligned} \tag{8.17}
$$

With this result, we can obtain the estimated responses for this design and the accompanying residuals as shown in Table 8.12.

TABLE 8.12: Estimates and Residuals for Fitted Model 8.17 for Example 7.3 (Reference Cell Coding)

x	z	h	m	y	xz	xh	zh	xzh	xm	zm	xzm	xhm	zhm	$xzhm$	hm	$\hat{y}$	$y-\hat{y}$
1	1	1	1	38	1	1	1	1	1	1	1	1	1	1	1	38	0
0	1	1	1	40	0	0	1	0	0	1	0	0	1	0	1	40	0
1	0	1	1	27	0	1	0	0	1	0	0	1	0	0	1	27	0
0	0	1	1	30	0	0	0	0	0	0	0	0	0	0	1	30	0
1	1	0	1	58	1	0	0	0	1	1	1	0	0	0	0	58	0
0	1	0	1	55	0	0	0	0	0	1	0	0	0	0	0	55	0
1	0	0	1	30	0	0	0	0	1	0	0	0	0	0	0	30	0
0	0	0	1	32	0	0	0	0	0	0	0	0	0	0	0	32	0
1	1	1	0	59	1	1	1	1	0	0	0	0	0	0	0	59	0
0	1	1	0	62	0	0	1	0	0	0	0	0	0	0	0	62	0
1	0	1	0	53	0	1	0	0	0	0	0	0	0	0	0	53	0
0	0	1	0	50	0	0	0	0	0	0	0	0	0	0	0	50	0
1	1	0	0	79	1	0	0	0	0	0	0	0	0	0	0	79	0
0	1	0	0	75	0	0	0	0	0	0	0	0	0	0	0	75	0
1	0	0	0	53	0	0	0	0	0	0	0	0	0	0	0	53	0
0	0	0	0	54	0	0	0	0	0	0	0	0	0	0	0	54	0

As we can see the estimates $\hat{y}$ are the same as the responses y, so that the resulting residuals $(y - \hat{y})$ are all zeros. Recall that the ANOVA for this design was only achieved by assuming that the second- and third-order interactions were not significant, and their sums of squares were combined to estimate the error variance mean square.

8.6.2 SAS Program for Analysis of Data of Example 7.3

We write a SAS program to analyze the same data under the reference cell model stated above as follows:

SAS PROGRAM

```
data fourfact2;
input x z h m y @@;
t1=x*z; t2=x*h; t3=z*h; t4=x*z*h; t5=x*m; t6=z*m; t7=x*z*m; t8=x*h*m;
t9=z*h*m; t10=x*z*h*m; t11=h*m;
datalines;
1 1 1 1 38 0 1 1 1 40 1 0 1 1 27 0 0 1 1 30 1 1 0 1 58  0 1 0 1 55 1 0 0 1 30
0 0 0 1 32 1 1 1 0 59 0 1 1 0 62 1 0 1 0 53 0 0 1 0 50  1 1 0 0 79 0 1 0 0 75
1 0 0 0 53 0 0 0 0 54
;
proc print data=fourfact2;
proc reg data=fourfact2;
model y=x z h m t1-t11; run;
```

SAS OUTPUT

Obs	x	z	h	m	y	t1	t2	t3	t4	t5	t6	t7	t8	t9	t10	t11
1	1	1	1	1	38	1	1	1	1	1	1	1	1	1	1	1
2	0	1	1	1	40	0	0	1	0	0	1	0	0	1	0	1
3	1	0	1	1	27	0	1	0	0	1	0	0	1	0	0	1
4	0	0	1	1	30	0	0	0	0	0	0	0	0	0	0	1
5	1	1	0	1	58	1	0	0	0	1	1	1	0	0	0	0
6	0	1	0	1	55	0	0	0	0	0	1	0	0	0	0	0
7	1	0	0	1	30	0	0	0	0	1	0	0	0	0	0	0
8	0	0	0	1	32	0	0	0	0	0	0	0	0	0	0	0
9	1	1	1	0	59	1	1	1	1	0	0	0	0	0	0	0
10	0	1	1	0	62	0	0	1	0	0	0	0	0	0	0	0
11	1	0	1	0	53	0	1	0	0	0	0	0	0	0	0	0
12	0	0	1	0	50	0	0	0	0	0	0	0	0	0	0	0
13	1	1	0	0	79	1	0	0	0	0	0	0	0	0	0	0
14	0	1	0	0	75	0	0	0	0	0	0	0	0	0	0	0
15	1	0	0	0	53	0	0	0	0	0	0	0	0	0	0	0
16	0	0	0	0	54	0	0	0	0	0	0	0	0	0	0	0

```
                    The REG Procedure
                      Model: MODEL1
                  Dependent Variable: y

        Number of Observations Read          16
        Number of Observations Used          16
```

Analysis of Variance

Source	DF	Sum of Squares	Mean Square	F Value	Pr > F
Model	15	3709.43750	247.29583	.	.
Error	0	0	.		
Corrected Total	15	3709.43750			

Root MSE	.	R-Square	1.0000
Dependent Mean	49.68750	Adj R-Sq	.
Coeff Var	.		

Parameter Estimates

Variable	DF	Parameter Estimate	Standard Error	t Value	Pr > |t|
Intercept	1	54.00000	.	.	.
x	1	-1.00000	.	.	.
z	1	21.00000	.	.	.
h	1	-4.00000	.	.	.
m	1	-22.00000	.	.	.
t1	1	5.00000	.	.	.
t2	1	4.00000	.	.	.
t3	1	-9.00000	.	.	.
t4	1	-11.00000	.	.	.
t5	1	-1.00000	.	.	.
t6	1	2.00000	.	.	.
t7	1	4.44089E-16	.	.	.
t8	1	-5.00000	.	.	.
t9	1	-4.00000	.	.	.
t10	1	7.00000	.	.	.
t11	1	2.00000	.	.	.

Clearly, the estimates obtained from SAS agree with the manual estimates obtained previously. However, notice that SAS could not supply any information about standard error of estimates because only one replicate of the experiment was studied. This agrees with the manual analysis of data of Example 7.3 carried out elsewhere in the book. In that case, the second- and third-order interactions were deliberately not considered to be significant in the design and were used to estimate error variance to perform ANOVA.

8.7 Dummy Variables Regression Models for Experiment in 3^k Factorial Designs

As usual, we use indicator variables regression to model the relationship between response variables and different factors in the design of Example 7.4.

The experiment was carried out as a typical two-factor factorial design involving factors A and B, each of which had three levels. We can use either the effects coding method or the reference cell method to define the dummy variables to be included in the model for this design. We discuss both methods in the next few sections.

8.7.1 Dummy Variables Regression Model (Effects Coding Method) for a 3^2 Factorial Design

By modifying the results for the general two-factor factorial experiment, we obtain the regression model for the experimental data of Example 7.4, when effects coding method is used to define the dummy variables utilized. The modified model and ancillary relations are given as follows:

$$Y = \mu + \sum_{i=1}^{2} \alpha_i X_i + \sum_{j=1}^{2} \beta_j Z_j + \sum_{i=1}^{2}\sum_{j=1}^{2} \gamma_{ij} X_i Z_j + \varepsilon \qquad (8.18)$$

where

$$X_i = \begin{cases} 1 & \text{if } i = 1,2 \\ -1 & \text{if } i = 3 \\ 0 & \text{otherwise} \end{cases}$$

$$Z_j = \begin{cases} 1 & \text{if } j = 1,2 \\ -1 & \text{if } j = 3 \\ 0 & \text{otherwise} \end{cases}$$

8.7.2 Relations for Estimating Parameters of the Model

To estimate the parameters for the model, we state the following relations based on the row-factor, column-factor, and cell means for the model, which enable us to carry out the estimation:

$$\begin{aligned}
\mu &= \mu_{..} \\
\alpha_i &= \mu_{i.} - \mu_{..} \\
\beta_j &= \mu_{.j} - \mu_{..} \\
\gamma_{ij} &= \mu_{ij} - \mu_{.j} - \mu_{i.} + \mu_{..} \\
-(\alpha_1 + \alpha_2) &= \mu_{3.} - \mu_{..} \\
-(\beta_1 + \beta_2) &= \mu_{.3} - \mu_{..} \\
-\sum_{j=1}^{2} \gamma_{ij} &= \mu_{3j} - \mu_{3.} - \mu_{.j} + \mu_{..} \quad \text{for } j = 1,2 \\
-\sum_{i=1}^{2} \gamma_{ij} &= \mu_{i3} - \mu_{.3} - \mu_{i.} + \mu_{..} \quad \text{for } i = 1,2
\end{aligned}$$

8.7.3 Fitting the Model to Example 7.4

We use the responses of Example 7.4 to obtain the row-factor, column-factor and cell means and present them in Table 8.13.

TABLE 8.13: Cell, Row, and Column Means for Example 7.4

	Factor B			
Factor A	**I**	**II**	**III**	**Row means**
I	44.9	41.3	33.6	39.933
II	40.9	31.85	34.05	35.6
III	35.1	28.8	32.2	32.033
Column means	40.3	33.983	33.383	35.8555

These cell, row, and column means will be regarded as estimates of the means that appear in the above relations to obtain estimates of the parameters of the model. Thus, using the relations and data from Table 8.13, we obtain the estimates for the model parameters as follows:

$$\begin{aligned}
\hat{\mu} &= \bar{y}_{..} = 35.8555 \\
\hat{\alpha}_1 &= \bar{y}_{1.} - \bar{y}_{..} = 39.93 - 35.8555 = 4.08 \\
\hat{\alpha}_2 &= \bar{y}_{2.} - \bar{y}_{..} = 35.6 - 35.8555 = -0.2555 \\
\hat{\beta}_1 &= \bar{y}_{.1} - \bar{y}_{..} = 40.3 - 35.8555 = 4.444 \\
\hat{\beta}_2 &= \bar{y}_{.2} - \bar{y}_{..} = 33.983 - 35.8555 = -1.8725 \\
\hat{\gamma}_{11} &= \bar{y}_{11} - \bar{y}_{.1} - \bar{y}_{1.} + \bar{y}_{..} = 44.9 - 39.93 - 40.3 + 35.8555 = 0.5221 \\
\hat{\gamma}_{12} &= \bar{y}_{12} - \bar{y}_{.2} - \bar{y}_{1.} + \bar{y}_{..} = 41.3 - 39.93 - 33.983 + 35.8555 = 3.23 \\
\hat{\gamma}_{21} &= \bar{y}_{21} - \bar{y}_{.1} - \bar{y}_{2.} + \bar{y}_{..} = 40.9 - 35.6 - 40.3 + 35.8555 = 0.8556 \\
\hat{\gamma}_{22} &= \bar{y}_{22} - \bar{y}_{.2} - \bar{y}_{2.} + \bar{y}_{..} = 31.85 - 35.6 - 33.983 + 35.8555 \\
&= -1.8777
\end{aligned}$$

Based on the above analysis, we see that the model estimate for a typical response is

$$\begin{aligned}
\hat{y} = {} & 35.8555 + 4.08x_1 - 0.2555x_2 + 4.444z_1 - 1.8722z_2 \\
& + 0.5221x_1z_1 + 3.23x_1z_2 + 0.8556x_2z_1 - 1.8777x_2z_2 \qquad (8.19)
\end{aligned}$$

With this model, we can obtain the fitted values for the responses and extract the residuals as in Table 8.14.

TABLE 8.14: Estimates and Residuals for Fitted Model 8.18 for Example 7.4 (Effects Coding Method)

	x_1	x_2	z_1	z_2	y	x_1z_1	x_1z_2	x_2z_1	x_2z_2	$\hat{y}$	$y-\hat{y}$	$(y-\hat{y})^2$
	1	0	1	0	44	1	0	0	0	44.90166	−0.90166	0.812990756
	1	0	1	0	45.8	1	0	0	0	44.90166	0.89834	0.807014756
	1	0	0	1	44.5	0	1	0	0	41.29306	3.20694	10.28446416
	1	0	0	1	38.1	0	1	0	0	41.29306	−3.19306	10.19563216
	1	0	−1	−1	32	−1	−1	0	0	33.61196	−1.61196	2.598415042
	1	0	−1	−1	35.2	−1	−1	0	0	33.61196	1.58804	2.521871042
	0	1	1	0	39.1	0	0	1	0	40.89961	−1.79961	3.238596152
	0	1	1	0	42.7	0	0	1	0	40.89961	1.80039	3.241404152
	0	1	0	1	34.6	0	0	0	1	31.84981	2.75019	7.563545036
	0	1	0	1	29.1	0	0	0	1	31.84981	−2.74981	7.561455036
	0	1	−1	−1	31.1	0	0	−1	−1	34.05061	−2.95061	8.706099372
	0	1	−1	−1	37	0	0	−1	−1	34.05061	2.94939	8.698901372
	−1	−1	1	0	34	−1	0	−1	0	35.09741	−1.09741	1.204308708
	−1	−1	1	0	36.2	−1	0	−1	0	35.09741	1.10259	1.215704708
	−1	−1	0	1	26.9	0	−1	0	−1	28.80631	−1.90631	3.634017816
	−1	−1	0	1	30.7	0	−1	0	−1	28.80631	1.89369	3.586061816
	−1	−1	−1	−1	30.9	1	1	1	1	32.18961	−1.28961	1.663093952
	−1	−1	−1	−1	33.5	1	1	1	1	32.18961	1.31039	1.717121952
Total					645.4					645.4	0	79.25

As we discussed in Chapter 3, the extracted residuals can be used in a variety of ways to evaluate the validity of various model assumptions, such as normality, equality of variances, etc. This is left as an exercise for the reader.

We test for significance of the fitted model (Table 8.15).

$$SS_{\text{total}} = \sum_{i=1}^{18} y_i^2 - \frac{\left(\sum_{i=1}^{18} y_i\right)^2}{18} = 23670.42 - \frac{645.4^2}{18} = 529.24$$

$$SS_{\text{residual}} = \sum_{i=1}^{18} (y_i - \hat{y}_i)^2 = 79.25$$

$$SS_{\text{regression}} = SS_{\text{total}} - SS_{\text{residual}} = 529.24 - 79.25 = 449.99$$

$$r^2 = \frac{449.99}{529.24} = 0.8503$$

TABLE 8.15: ANOVA to Test for Significance of Fitted Model 8.19

Source	**Sum of squares**	**Degrees of freedom**	**Mean square**	***F*-ratio**	***p*-Value**
Regression	449.99	8	56.2488	6.3879	0.005886
Error	79.25	9	8.8056		
Total	529.24	17			

ANOVA shows that the fitted model is significant.

8.7.4 SAS Program and Analysis of Data for Example 7.4 (Effects Coding Model)

We now analyze the same data fully by writing a SAS program, which should obtain estimates for model parameters, and test the significance of the model.

SAS PROGRAM

```
DATA EXAMP7_4;
INPUT X1-X2 Z1-Z2 y@@;
H1=X1*Z1; H2=X1*Z2; H3=X2*Z1; H4=X2*Z2;
```

```
DATALINES;
1 0 1 0 44 1 0 1 0 45.8 1 0 0 1 44.5 1 0 0 1 38.1 1 0 -1 -1 32 1 0 -1 -1 35.2
0 1 1 0 39.1 0 1 1 0 42.7 0 1 0 1 34.6 0 1 0 1 29.1 0 1 -1 -1 31.1 0 1 -1 -1
37.0 -1 -1 1 0 34 -1 -1 1 0 36.2 -1 -1 0 1 26.9 -1 -1 0 1 30.7 -1 -1 -1 -1
30.9 -1 -1 -1 -1 33.5
;
PROC PRINT DATA=EXAMP7_4;  SUM X1-X2 Z1-Z2 H1-H4;
PROC REG DATA=EXAMP7_4;
MODEL Y=X1 X2 Z1 Z2 H1-H4; RUN;
```

SAS OUTPUT

```
 Obs   X1    X2    Z1    Z2       y    H1    H2    H3    H4
   1    1     0     1     0    44.0    1     0     0     0
   2    1     0     1     0    45.8    1     0     0     0
   3    1     0     0     1    44.5    0     1     0     0
   4    1     0     0     1    38.1    0     1     0     0
   5    1     0    -1    -1    32.0   -1    -1     0     0
   6    1     0    -1    -1    35.2   -1    -1     0     0
   7    0     1     1     0    39.1    0     0     1     0
   8    0     1     1     0    42.7    0     0     1     0
   9    0     1     0     1    34.6    0     0     0     1
  10    0     1     0     1    29.1    0     0     0     1
  11    0     1    -1    -1    31.1    0     0    -1    -1
  12    0     1    -1    -1    37.0    0     0    -1    -1
  13   -1    -1     1     0    34.0   -1     0    -1     0
  14   -1    -1     1     0    36.2   -1     0    -1     0
  15   -1    -1     0     1    26.9    0    -1     0    -1
  16   -1    -1     0     1    30.7    0    -1     0    -1
  17   -1    -1    -1    -1    30.9    1     1     1     1
  18   -1    -1    -1    -1    33.5    1     1     1     1
       ==    ==    ==    ==           ==    ==    ==    ==
        0     0     0     0            0     0     0     0

                      The REG Procedure
                        Model: MODEL1
                   Dependent Variable: y

          Number of Observations Read        18
          Number of Observations Used        18

                    Analysis of Variance

                            Sum of        Mean
Source             DF      Squares      Square    F Value    Pr > F

Model               8    449.99444    56.24931       6.39    0.0059
Error               9     79.25000     8.80556
Corrected Total    17    529.24444

      Root MSE             2.96742  R-Square    0.8503
      Dependent Mean      35.85556  Adj R-Sq    0.7172
      Coeff Var            8.27603
```

Parameter Estimates

Variable	DF	Parameter Estimate	Standard Error	t Value	Pr > |t|
Intercept	1	35.85556	0.69943	51.26	<.0001
X1	1	4.07778	0.98914	4.12	0.0026
X2	1	-0.25556	0.98914	-0.26	0.8019
Z1	1	4.44444	0.98914	4.49	0.0015
Z2	1	-1.87222	0.98914	-1.89	0.0909
H1	1	0.52222	1.39885	0.37	0.7176
H2	1	3.23889	1.39885	2.32	0.0458
H3	1	0.85556	1.39885	0.61	0.5559
H4	1	-1.87778	1.39885	-1.34	0.2124

The output of the SAS program gives the same values for estimates of the model parameters as those previously obtained using manual calculations. SAS test for significance of the model yields a p-value of 0.0059, indicating a highly significant fitted model.

8.7.5 Fitting the Model to Example 7.4 by Reference Cell Method

We can also represent the responses of the experiment in Example 7.4 with a regression model in which reference is made to a cell to define the dummy variables. The model to be used is of the form

$$Y = \mu + \sum_{i=1}^{2} \alpha_i X_i + \sum_{j=1}^{2} \beta_j Z_j + \sum_{i=1}^{2} \sum_{j=1}^{2} \gamma_{ij} X_i Z_j + \varepsilon \tag{8.20}$$

where

$$X_i = \begin{cases} 1 & \text{if } i = 1, 2 \\ 0 & \text{otherwise} \end{cases}$$

$$Z_j = \begin{cases} 1 & \text{if } j = 1, 2 \\ 0 & \text{otherwise} \end{cases}$$

Here, we used cell $(A,\ B) = (3,\ 3)$ as the reference cell, since each of the factors A and B in the design has three levels. Any other cell could have been used, but the parameters would be different for each reference cell chosen. The relevant relations that will help us to estimate the parameters of the model are as follows:

$$
\begin{aligned}
&\mu_{33} = \mu \\
&\mu_{3j} = \mu + \alpha_i \quad i = 1, 2 \\
&\mu_{i3} = \mu + \beta_j \quad j = 1, 2 \\
&\mu_{ij} = \mu + \alpha_i + \beta_j + \gamma_{ij} \quad i = 1, 2 \quad j = 1, 2
\end{aligned}
$$

Thus using the means of Table 8.13, $\hat{\mu} = \bar{y}_{33} = 32.2$

$$
\begin{aligned}
&\bar{y}_{13} = 33.6 = \bar{y}_{33} + \hat{\alpha}_1 \Rightarrow 33.6 = 32.2 + \hat{\beta}_1 \Rightarrow \hat{\alpha}_1 = 33.6 - 32.2 = 1.4 \\
&\bar{y}_{23} = 34.05 = \bar{y}_{33} + \hat{\alpha}_2 \Rightarrow 34.05 = 32.2 + \hat{\alpha}_1 \Rightarrow \hat{\alpha}_1 = 34.05 - 32.2 = 1.85 \\
&\bar{y}_{31} = 35.1 = \bar{y}_{33} + \hat{\beta}_1 \Rightarrow 35.1 = 32.2 + \hat{\beta}_1 \Rightarrow \hat{\beta}_1 = 35.1 - 32.2 = 2.9 \\
&\bar{y}_{32} = 28.8 = \bar{y}_{33} + \hat{\beta}_2 \Rightarrow 28.8 = 32.2 + \hat{\beta}_2 \Rightarrow \hat{\beta}_2 = 28.8 - 32.2 = -3.4 \\
&\bar{y}_{11} = \bar{y}_{33} + \hat{\alpha}_1 + \hat{\beta}_1 + \hat{\gamma}_{11} \Rightarrow 44.9 = 32.2 + 1.3 + 2.9 + \hat{\gamma}_{11} \Rightarrow \hat{\gamma}_{11} = 8.4 \\
&\bar{y}_{12} = \bar{y}_{33} + \hat{\alpha}_1 + \hat{\beta}_2 + \hat{\gamma}_{12} \Rightarrow 41.3 = 32.2 - 3.5 + 1.3 + \hat{\gamma}_{12} \Rightarrow \hat{\gamma}_{12} = 11.1 \\
&\bar{y}_{21} = 40.9 = \bar{y}_{33} + \hat{\alpha}_2 + \hat{\beta}_1 + \hat{\gamma}_{21} \Rightarrow 40.9 = 32.2 + 1.85 + 2.9 + \hat{\gamma}_{21} \Rightarrow \hat{\gamma}_{21} = 3.95 \\
&\bar{y}_{22} = 31.85 = \bar{y}_{33} + \hat{\alpha}_2 + \hat{\beta}_2 + \hat{\gamma}_{22} = 31.85 - 32.2 + 2.2 + 3.4 - 1.85 = 1.2
\end{aligned}
$$

Thus, the estimated equation relating the responses to the dummy variables in the model is

$$
\begin{aligned}
\hat{y} = {} & 32.2 + 1.4x_1 + 1.85x_2 + 2.9z_1 - 3.4z_2 \\
& + 8.4x_1z_1 + 11.1x_1z_2 + 3.95x_2z_1 + 1.2x_2z_2 \qquad (8.21)
\end{aligned}
$$

With this equation we again extract the fitted values and the residuals for this design (Table 8.16).

TABLE 8.16: Estimates and Residuals for Fitted Model 8.21 for Example 7.4 (Reference Cell Coding)

	x_1	x_2	z_1	z_2	y	x_1z_1	x_1z_2	x_2z_1	x_2z_2	$\hat{y}$	$(y^- - \hat{y})$	$(y^- - \hat{y})^2$
	1	0	1	0	44	1	0	0	0	44.9	−0.9	0.81
	1	0	1	0	45.8	1	0	0	0	44.9	0.9	0.81
	1	0	0	1	44.5	0	1	0	0	41.3	3.2	10.24
	1	0	0	1	38.1	0	1	0	0	41.3	−3.2	10.24
	1	0	0	0	32	0	0	0	0	33.6	−1.6	2.56
	1	0	0	0	35.2	0	0	0	0	33.6	1.6	2.56
	0	1	1	0	39.1	0	0	1	0	40.9	−1.8	3.24
	0	1	1	0	42.7	0	0	1	0	40.9	1.8	3.24
	0	1	0	1	34.6	0	0	0	1	31.85	2.75	7.5625
	0	1	0	1	29.1	0	0	0	1	31.85	−2.75	7.5625
	0	1	0	0	31.1	0	0	0	0	34.05	−2.95	8.7025
	0	1	0	0	37	0	0	0	0	34.05	2.95	8.7025
	0	0	1	0	34	0	0	0	0	35.1	−1.1	1.21
	0	0	1	0	36.2	0	0	0	0	35.1	1.1	1.21
	0	0	0	1	26.9	0	0	0	0	28.8	−1.9	3.61
	0	0	0	1	30.7	0	0	0	0	28.8	1.9	3.61
	0	0	0	0	30.9	0	0	0	0	32.2	−1.3	1.69
	0	0	0	0	33.5	0	0	0	0	32.2	1.3	1.69
Total					645.4					645.4	0	79.25

We test for significance of the fitted model (Table 8.17).

$$SS_{\text{total}} = \sum_{i=1}^{18} y_i^2 - \frac{\left(\sum_{i=1}^{18} y_i\right)^2}{18} = 23670.42 - \frac{645.4^2}{18} = 529.24$$

$$SS_{\text{residual}} = \sum_{i=1}^{18} (y_i - \hat{y}_i)^2 = 79.25$$

$$SS_{\text{regression}} = SS_{\text{total}} - SS_{\text{residual}} = 529.24 - 79.25 = 449.99$$

TABLE 8.17: ANOVA to Test for Significance of Fitted Model 8.21 for Data of Example 7.4

Source	Sum of squares	Degrees of freedom	Mean square	F-ratio	p-Value
Regression	449.99	8	56.2488	6.3879	0.005886
Error	79.25	9	8.8056		
Total	529.24	17			

ANOVA shows that the fitted model is significant.

8.7.6 SAS Program and Analysis of Data for Example 7.4 (Reference Cell Model)

Next, we write a SAS program to analyze the data and to estimate parameters for the model.

SAS PROGRAM

```
DATA EXAMP7_4a;
INPUT X1-X2 Z1-Z2 y@@;
H1=X1*Z1; H2=X1*Z2; H3=X2*Z1; H4=X2*Z2;
DATALINES;
1 0 1 0 44 1 0 1 0 45.8 1 0 0 1 44.5 1 0 0 1 38.1 1 0 0 0 32 1 0 0 0 35.2 0 1
1 0 39.1 0 1 1 0 42.7 0 1 0 1 34.6 0 1 0 1 29.1 0 1 0 0 31.1 0 1 0 0 37.0 0 0
1 0 34 0 0 1 0 36.2 0 0 0 1 26.9 0 0 0 1 30.7 0 0 0 0 30.9 0 0 0 0 33.5
;
PROC PRINT DATA=EXAMP7_4A; SUM X1-X2 Z1-Z2 H1-H4;
PROC REG DATA=EXAMP7_4A;
MODEL Y=X1 X2 Z1 Z2 H1-H4; RUN;
```

SAS OUTPUT

Obs	X1	X2	Z1	Z2	y	H1	H2	H3	H4
1	1	0	1	0	44.0	1	0	0	0
2	1	0	1	0	45.8	1	0	0	0
3	1	0	0	1	44.5	0	1	0	0
4	1	0	0	1	38.1	0	1	0	0
5	1	0	0	0	32.0	0	0	0	0
6	1	0	0	0	35.2	0	0	0	0
7	0	1	1	0	39.1	0	0	1	0
8	0	1	1	0	42.7	0	0	1	0
9	0	1	0	1	34.6	0	0	0	1
10	0	1	0	1	29.1	0	0	0	1
11	0	1	0	0	31.1	0	0	0	0
12	0	1	0	0	37.0	0	0	0	0
13	0	0	1	0	34.0	0	0	0	0
14	0	0	1	0	36.2	0	0	0	0
15	0	0	0	1	26.9	0	0	0	0
16	0	0	0	1	30.7	0	0	0	0
17	0	0	0	0	30.9	0	0	0	0
18	0	0	0	0	33.5	0	0	0	0
	==	==	==	==		==	==	==	==
	6	6	6	6		2	2	2	2

The REG Procedure
Model: MODEL1
Dependent Variable: y

Number of Observations Read	18
Number of Observations Used	18

Analysis of Variance

Source	DF	Sum of Squares	Mean Square	F Value	Pr > F
Model	8	449.99444	56.24931	6.39	0.0059
Error	9	79.25000	8.80556		
Corrected Total	17	529.24444			

Root MSE	2.96742	R-Square	0.8503
Dependent Mean	35.85556	Adj R-Sq	0.7172
Coeff Var	8.27603		

Parameter Estimates

Variable	DF	Parameter Estimate	Standard Error	t Value	Pr > |t|
Intercept	1	32.20000	2.09828	15.35	<.0001
X1	1	1.40000	2.96742	0.47	0.6483
X2	1	1.85000	2.96742	0.62	0.5485
Z1	1	2.90000	2.96742	0.98	0.3540
Z2	1	-3.40000	2.96742	-1.15	0.2814

```
                    Parameter     Standard
Variable  DF         Estimate        Error    t Value    Pr > |t|

H1         1          8.40000      4.19656       2.00      0.0764
H2         1         11.10000      4.19656       2.65      0.0267
H3         1          3.95000      4.19656       0.94      0.3712
H4         1          1.20000      4.19656       0.29      0.7814
```

The output of SAS agrees with the ANOVA and the estimates of model parameters previously obtained by manual calculations.

8.8 Fitting Regression Model for Example 7.5 (Effects Coding Method)

We fit a regression model on the data of Example 7.5 based on the theory of effects coding which we discussed earlier. To obtain the estimates of the regression model for Example 7.5, we list the cell means for the data for the experiment in Example 7.5, which is denoted in Table 8.18.

TABLE 8.18: The Cell Means of 3^3 Factorial Designs

		C-levels			
***B*-levels**	***A*-levels**	**1**	**2**	**3**	
	1	$\bar{y}_{111}$	$\bar{y}_{112}$	$\bar{y}_{113}$	
1	2	$\bar{y}_{211}$	$\bar{y}_{212}$	$\bar{y}_{213}$	
	3	$\bar{y}_{311}$	$\bar{y}_{312}$	$\bar{y}_{313}$	
	1	$\bar{y}_{121}$	$\bar{y}_{122}$	$\bar{y}_{123}$	
2	2	$\bar{y}_{221}$	$\bar{y}_{222}$	$\bar{y}_{223}$	
	3	$\bar{y}_{321}$	$\bar{y}_{322}$	$\bar{y}_{323}$	
	1	$\bar{y}_{131}$	$\bar{y}_{132}$	$\bar{y}_{133}$	
3	2	$\bar{y}_{231}$	$\bar{y}_{232}$	$\bar{y}_{233}$	
	3	$\bar{y}_{331}$	$\bar{y}_{332}$	$\bar{y}_{333}$	$\bar{y}_{...}$

We then obtain the corresponding estimates for the cell means (Tables 8.19 through 8.22).

Under the effects coding method for defining dummy variables, the relevant regression model is

$$Y = \mu + \sum_{i=1}^{2}\alpha_i X_i + \sum_{j=1}^{2}\beta_j Z_j + \sum_{i=1}^{2}\sum_{j=1}^{2}\gamma_{ij}X_i Z_j + \sum_{k=1}^{2}\omega_k H_k + \sum_{i=1}^{2}\sum_{k=1}^{2}\theta_{ik}X_i H_k$$
$$+\sum_{j=1}^{2}\sum_{k=1}^{2}\psi_{jk}Z_j H_k + \sum_{i=1}^{2}\sum_{j=1}^{2}\sum_{k=1}^{2}\delta_{ijk}X_i Z_j H_k + \varepsilon \quad (8.22)$$

TABLE 8.19: Cell Means of a 3^3 Factorial Design in Example 7.5

		C-levels			
***B*-levels**	***A*-levels**	**1**	**2**	**3**	
	1	4.565	9.95	14.195	
1	2	12.75	15.365	18.34	
	3	16.38	18.8	23.39	
	1	5.21	11.62	14.47	
2	2	14.825	19.88	22.935	
	3	20.37	25.375	29.68	
	1	6.85	12.775	16.65	
3	2	15.14	19.65	22.9	
	3	22.65	27.25	30.52	17.49944

TABLE 8.20: Means for Levels of Main Effects of the 3^3 Factorial Design in Example 7.5

Combination of levels	**1**	**2**	**3**
A	10.698333	17.976111	23.823889
B	14.8595	18.2628	19.376111
C	13.19333	17.8511	21.453333

TABLE 8.21: Means for Combinations of Levels of 3^3 Factorial Design in Example 7.5

Combination of levels	**(1, 1)**	**(1, 2)**	**(2, 1)**	**(2, 2)**
AB	9.57	10.4333	15.492	19.2133
AC	5.54166	11.4483	14.23833	18.2933
BC	11.231667	14.705	13.46833	18.95833

where

$$X_i = \begin{cases} 1 & \text{if } i = 1, 2 \\ -1 & \text{if } i = 3 \\ 0 & \text{otherwise} \end{cases}$$

$$Z_j = \begin{cases} 1 & \text{if } j = 1, 2 \\ -1 & \text{if } j = 3 \\ 0 & \text{otherwise} \end{cases}$$

TABLE 8.22: Means for Combinations of Levels of 3^3 Factorial Design in Example 7.5

Combination of levels	(1, 1, 1)	(1, 1, 2)	(1, 2, 1)	(1, 2, 2)	(2, 1, 1)	(2, 2, 1)	(2, 1, 2)	(2, 2, 2)
ABC	4.565	9.95	5.21	11.62	12.75	14.825	15.365	19.88

$$H_k = \begin{cases} 1 & \text{if } k = 1,2 \\ -1 & \text{if } k = 3 \\ 0 & \text{otherwise} \end{cases}$$

Based on the above model and Tables 8.19–8.22, we can estimate the parameters of the model by using the following relations:

$$\begin{aligned}
\mu &= \mu_{...} \\
\mu_{ijk} &= \mu + \alpha_i + \beta_j + \gamma_{ij} + \omega_k + \theta_{ik} + \psi_{jk} + \delta_{ijk} \quad i,j,k = 1,2 \\
\mu_{ij.} &= \mu + \alpha_i + \beta_j + \gamma_{ij} \quad j = 1,2 \\
\mu_{.jk} &= \mu + \beta_j + \omega_k + \psi_{jk} \quad j,k = 1,2 \\
\mu_{i.k} &= \mu + \alpha_i + \omega_k + \theta_{ik} \quad i,k = 1,2 \\
\mu_{i..} &= \mu + \alpha_i \quad i = 1,2 \\
\mu_{.j.} &= \mu + \beta_j \quad j = 1,2 \\
\mu_{..k} &= \mu + \omega_k \quad k = 1,2
\end{aligned}$$

Using the data of Example 7.5, we obtain the estimates of the parameters of the regression model by manual calculations as follows:

$$\begin{aligned}
\mu &= \bar{y}_{...} = 17.49944 \\
\hat{\alpha}_1 &= \bar{y}_{1..} - \bar{y}_{...} = 10.698333 - 17.49944 = -6.80111 \\
\hat{\alpha}_2 &= \bar{y}_{2..} - \bar{y}_{...} = 17.976111 - 17.49944 = 0.47667 \\
\hat{\beta}_1 &= \bar{y}_{.1.} - \bar{y}_{...} = 14.859444 - 17.49944 = -2.63999 \\
\hat{\beta}_2 &= \bar{y}_{.2.} - \bar{y}_{...} = 18.262778 - 17.49944 = 0.76333 \\
\hat{\omega}_1 &= \bar{y}_{..1} - \bar{y}_{...} = 13.193333 - 17.49944 = -4.30611 \\
\hat{\omega}_2 &= \bar{y}_{..2} - \bar{y}_{...} = 17.851666 - 17.49944 = 0.35222 \\
\gamma_{ij} &= \bar{y}_{ij.} - \bar{y}_{...} - \hat{\alpha}_i - \hat{\beta}_j \quad i,j = 1,2 \\
\gamma_{11} &= \bar{y}_{11.} - \bar{y}_{...} - \hat{\alpha}_1 - \hat{\beta}_1 \\
&\Rightarrow \hat{\gamma}_{11} = 9.57 - 17.49944 + 6.80111 + 2.63999 = 1.51166 \\
\gamma_{12} &= \bar{y}_{12.} - \bar{y}_{...} - \hat{\alpha}_1 - \hat{\beta}_2 \\
&\Rightarrow \hat{\gamma}_{12} = 10.43333 - 17.49944 + 6.80111 - 0.76333 = -1.02833 \\
\gamma_{21} &= \bar{y}_{21.} - \bar{y}_{...} - \hat{\alpha}_2 - \hat{\beta}_1 \\
&\Rightarrow \hat{\gamma}_{21} = 15.485 - 17.49944 - 0.46667 + 2.63999 = 0.14883
\end{aligned}$$

$$\gamma_{22} = \bar{y}_{22.} - \bar{y}_{...} - \hat{\alpha}_2 - \hat{\beta}_2$$
$$\Rightarrow \hat{\gamma}_{22} = 19.21333 - 17.49944 - 0.47667 - 0.76333 = 0.47389$$
$$\hat{\theta}_{ik} = \bar{y}_{i.k} - \bar{y}_{...} - \hat{\alpha}_i - \hat{\omega}_k \quad i, k = 1, 2$$
$$\hat{\theta}_{11} = \bar{y}_{1.1} - \bar{y}_{...} - \hat{\alpha}_1 - \hat{\omega}_1$$
$$\Rightarrow \hat{\theta}_{11} = 5.541667 - 17.49944 + 6.80111 + 4.30611 = -0.850552$$
$$\hat{\theta}_{12} = \bar{y}_{1.2} - \bar{y}_{...} - \hat{\alpha}_1 - \hat{\omega}_2$$
$$\Rightarrow \hat{\theta}_{12} = 11.448333 - 17.49944 + 6.80111 - 0.352222 = 0.39778$$
$$\hat{\theta}_{21} = \bar{y}_{2.1} - \bar{y}_{...} - \hat{\alpha}_2 - \hat{\omega}_1$$
$$\Rightarrow \hat{\theta}_{21} = 14.238333 - 17.49944 - 0.47667 + 4.306111 = 0.56833$$
$$\hat{\theta}_{22} = \bar{y}_{2.2} - \bar{y}_{...} - \hat{\alpha}_2 - \hat{\omega}_2 \Rightarrow \hat{\theta}_{12} = 18.29833 - 17.49944 - 0.47667 - 0.352222 = -0.029996$$
$$\hat{\psi}_{12} = \bar{y}_{.jk} - \bar{y}_{...} - \hat{\beta}_j - \hat{\omega}_k \quad j, k = 1, 2$$
$$\hat{\psi}_{11} = \bar{y}_{.11} - \bar{y}_{...} - \hat{\beta}_1 - \hat{\omega}_1$$
$$\Rightarrow \hat{\psi}_{11} = 11.231667 - 17.49944 + 2.63999 + 4.306111 = 0.67833$$
$$\hat{\psi}_{12} = \bar{y}_{.12} - \bar{y}_{...} - \hat{\beta}_1 - \hat{\omega}_2$$
$$\Rightarrow \hat{\psi}_{12} = 14.705 - 17.49944 + 2.63999 - 0.35222 = -0.5067$$
$$\hat{\psi}_{21} = \bar{y}_{.21} - \bar{y}_{...} - \hat{\beta}_2 - \hat{\omega}_1$$
$$\Rightarrow \hat{\psi}_{21} = 13.46833 - 17.49944 - 0.76333 + 4.306111 = -0.48833$$
$$\hat{\psi}_{22} = \bar{y}_{.22} - \bar{y}_{...} - \hat{\beta}_2 - \hat{\omega}_2$$
$$\Rightarrow \hat{\psi}_{22} = 18.95833 - 17.49944 - 0.76333 - 0.35222 = 0.3433$$
$$\hat{\delta}_{ijk} = \bar{y}_{ijk} - \bar{y}_{...} - \hat{\alpha}_i - \hat{\beta}_j - \hat{\gamma}_{ij} - \hat{\omega}_k - \hat{\theta}_{ik} - \hat{\psi}_{jk} \quad i, j, k = 1, 2$$
$$\hat{\delta}_{111} = \bar{y}_{111} - \bar{y}_{...} - \hat{\alpha}_1 - \hat{\beta}_1 - \hat{\gamma}_{11} - \hat{\omega}_1 - \hat{\theta}_{11} - \hat{\psi}_{11}$$
$$\Rightarrow \hat{\delta}_{111} = 4.565 - 17.49944 + 6.80111 + 2.639944 - 1.51166 + 4.306111 - 0.850552 - 0.67833 = -0.5266$$
$$\hat{\delta}_{112} = \bar{y}_{112} - \bar{y}_{...} - \hat{\alpha}_1 - \hat{\beta}_1 - \hat{\gamma}_{11} - \hat{\omega}_2 - \hat{\theta}_{12} - \hat{\psi}_{12}$$
$$\Rightarrow \hat{\delta}_{112} = 9.95 - 17.49944 + 6.80111 + 2.639944 - 1.51166 - 0.35222 - 0.39778 + 0.5067 = 0.13667$$
$$\hat{\delta}_{121} = \bar{y}_{121} - \bar{y}_{...} - \hat{\alpha}_1 - \hat{\beta}_2 - \hat{\gamma}_{12} - \hat{\omega}_1 - \hat{\theta}_{11} - \hat{\psi}_{21}$$
$$\Rightarrow \hat{\delta}_{121} = 5.21 - 17.49944 + 6.80111 - 0.76333 + 1.02833 + 4.30611 + 0.850552 + 0.48833 = 0.42166$$
$$\hat{\delta}_{122} = \bar{y}_{122} - \bar{y}_{...} - \hat{\alpha}_1 - \hat{\beta}_2 - \hat{\gamma}_{12} - \hat{\omega}_2 - \hat{\theta}_{12} - \hat{\psi}_{22}$$
$$\Rightarrow \hat{\delta}_{121} = 11.62 - 17.49944 + 6.80111 - 0.76333 + 1.02833 - 0.35222 - 0.39778 - 0.3433 = 0.09333$$

$$\delta_{211} = \bar{y}_{211} - \bar{y}_{...} - \hat{\alpha}_2 - \hat{\beta}_1 - \hat{\gamma}_{21} - \hat{\omega}_1 - \hat{\theta}_{21} - \hat{\psi}_{11}$$
$$\Rightarrow \hat{\delta}_{211} = 12.75 - 17.49944 - 0.47667 + 2.63994 - 0.14883 + 4.30611 - 0.56833 - 0.67833 = 0.32445$$
$$\delta_{212} = \bar{y}_{212} - \bar{y}_{...} - \hat{\alpha}_2 - \hat{\beta}_1 - \hat{\gamma}_{21} - \hat{\omega}_2 - \hat{\theta}_{21} - \hat{\psi}_{12}$$
$$\Rightarrow \hat{\delta}_{212} = 15.365 - 17.49944 - 0.47667 + 2.63994 - 0.14883 - 0.35222 + 0.03 - 0.50667 = 0.06445$$
$$\hat{\delta}_{221} = \bar{y}_{221} - \bar{y}_{...} - \hat{\alpha}_2 - \hat{\beta}_2 - \hat{\gamma}_{22} - \hat{\omega}_1 - \hat{\theta}_{21} - \hat{\psi}_{21}$$
$$\Rightarrow \hat{\delta}_{221} = 14.825 - 17.49944 - 0.47667 - 0.76333 - 0.47389 - 4.30611 - 0.56833 - 0.48833 = -0.16222$$
$$\hat{\delta}_{222} = \bar{y}_{222} - \bar{y}_{...} - \hat{\alpha}_2 - \hat{\beta}_2 - \hat{\gamma}_{22} - \hat{\omega}_2 - \hat{\theta}_{22} - \hat{\psi}_{22}$$
$$\Rightarrow \hat{\delta}_{222} = 19.88 - 17.49944 - 0.47667 - 0.76333 - 0.47389 - 0.35222 + 0.03 - 0.3433 = 0.001146$$

The fitted model that can help us estimate the responses for different combinations of levels of the factors in the design is

$$\begin{aligned}\hat{y} = {} & 17.4994 - 6.80111x_1 + 0.47667x_2 - 2.63999z_1 + 0.766333z_2 \\ & - 4.30611h_1 + 0.35222h_2 + 1.51166c_1 - 1.02833c_2 \\ & - 0.85055c_3 + 0.39778c_4 + 0.14883c_5 + 0.47389c_6 + 0.56833c_7 - 0.02999c_8 \\ & + 0.67833c_9 - 0.5067c_{10} - 0.48833c_{11} + 0.3433c_{12} - 0.5266c_{13} \\ & + 0.13667c_{14} + 0.42166c_{15} + 0.09333c_{16} + 0.32445c_{17} + 0.06445c_{18} \\ & - 0.16222c_{19} + 0.001146c_{20} \end{aligned} \tag{8.23}$$

We would usually test for significance of the fitted model. This part is left as an exercise for the reader.

8.8.1 SAS Analysis of Data of Example 7.5 Using Effects Coding Method

Next, we write a SAS program to analyze the data of Example 7.5 based on the use of effects coding method for defining the dummy variables in the regression model.

SAS PROGRAM

```
DATA EXAMP7_5;
INPUT X1-X2 Z1-Z2 h1-h2 Y @@;
c1=X1*Z1; c2=X1*Z2; c3=X1*h1; c4=X1*h2; c5=X2*Z1; c6=X2*Z2; c7=X2*h1;
c8=X2*h2; c9=Z1*h1;  c10=Z1*h2; c11=z2*h1; c12=z2*h2;  c13=X1*Z1*h1;
c14=X1*Z1*h2; c15=X1*Z2*h1;  c16=X1*Z2*h2; c17=X2*Z1*h1; c18=X2*Z1*h2;
c19=X2*Z2*h1; c20=X2*Z2*h2;
```

```
CARDS;
1 0 1 0 1 0 4.2 1 0 1 0 1 0 4.93 1 0 1 0 0 1 9.8 1 0 1 0 0 1 10.10
0 1 1 0 1 0 12.5 0 1 1 0 1 0 13.0 0 1 1 0 0 1 15.51 0 1 1 0 0 1 15.22
1 0 1 0 -1 -1 14.6 1 0 1 0 -1 -1 13.79 0 1 1 0 -1 -1 18.7 0 1 1 0 -1 -1 17.98
-1 -1 1 0 1 0 16.8 -1 -1 1 0 1 0 15.96 -1 -1 1 0 0 1 19 -1 -1 1 0 0 1 18.6 -1
-1 1 0 -1 -1 24.2 -1 -1 1 0 -1 -1 22.58 1 0 0 1 1 0 5.11 1 0 0 1 1 0 5.31 1 0
0 1 0 1 11.9 1 0 0 1 0 1 11.34 0 1 0 1 1 0 14.3 0 1 0 1 1 0 15.35 0 1 0 1 0 1
18.8 0 1 0 1 0 1 20.96 1 0 0 1 -1 -1 13.3 1 0 0 1 -1 -1 15.64 0 1 0 1 -1 -1
21.6 0 1 0 1 -1 -1 24.27 -1 -1 0 1 1 0 19.86 -1 -1 0 1 1 0 20.88 -1 -1 0 1 0
1 26.37 -1 -1 0 1 0 1 24.38 -1 -1 0 1 -1 -1 29.72 -1 -1 0 1 -1 -1 29.64 1 0
-1 -1 1 0 6.7 1 0 -1 -1 1 0 7 1 0 -1 -1 0 1 12.4 1 0 -1 -1 0 1 13.15  0 1 -1
-1 1 0 15.38 0 1 -1 -1 1 0 14.9 0 1 -1 -1 0 1 19.2 0 1 -1 -1 0 1 20.1 1 0 -1
-1 -1 -1 16.2 1 0 -1 -1 -1 -1 17.1 0 1 -1 -1 -1 -1 22.4 0 1 -1 -1 -1 -1 23.4
-1 -1 -1 -1 1 0 21.3 -1 -1 -1 -1 1 0 24 -1 -1 -1 -1 0 1 27.3 -1 -1 -1 -1 0 1
27.2 -1 -1 -1 -1 -1 -1 31.4 -1 -1 -1 -1 -1 -1 29.64
;
proc print data=examp7_5; sum x1-x2 z1-z2 h1-h2 c1-c20 y;
/* SUMMATION: DATA ENTRY CHECK */
PROC REG DATA=EXAMP7_5;
MODEL Y=X1-X2 Z1-Z2 h1-h2 c1-c20;
RUN;
```

SAS OUTPUT

Obs	X1	X2	Z1	Z2	h1	h2	Y	c1	c2	c3	c4	c5	c6	c7	c8	c9	c10	c11	c12	c13	c14	c15	c16	c17	c18	c19	c20
1	1	0	1	0	1	0	4.20	1	0	1	0	0	0	0	0	1	0	0	0	1	0	0	0	0	0	0	0
2	1	0	1	0	1	0	4.93	1	0	1	0	0	0	0	0	1	0	0	0	1	0	0	0	0	0	0	0
3	1	0	1	0	0	1	9.80	1	0	0	1	0	0	0	0	0	1	0	0	0	1	0	0	0	0	0	0
4	1	0	1	0	0	1	10.10	1	0	0	1	0	0	0	0	0	1	0	0	0	1	0	0	0	0	0	0
5	0	1	1	0	1	0	12.50	0	0	0	0	1	0	1	0	1	0	0	0	0	0	0	0	1	0	0	0
6	0	1	1	0	1	0	13.00	0	0	0	0	1	0	1	0	1	0	0	0	0	0	0	0	1	0	0	0
7	0	1	1	0	0	1	15.51	0	0	0	0	1	0	0	1	0	1	0	0	0	0	0	0	0	1	0	0
8	0	1	1	0	0	1	15.22	0	0	0	0	1	0	0	1	0	1	0	0	0	0	0	0	0	1	0	0
9	1	0	1	0	-1	-1	14.60	1	0	-1	-1	0	0	0	0	-1	-1	0	0	-1	-1	0	0	0	0	0	0
10	1	0	1	0	-1	-1	13.79	1	0	-1	-1	0	0	0	0	-1	-1	0	0	-1	-1	0	0	0	0	0	0
11	0	1	1	0	-1	-1	18.70	0	0	0	0	1	0	-1	-1	-1	-1	0	0	0	0	0	0	-1	-1	0	0
12	0	1	1	0	-1	-1	17.98	0	0	0	0	1	0	-1	-1	-1	-1	0	0	0	0	0	0	-1	-1	0	0
13	-1	-1	1	0	1	0	16.80	-1	0	-1	0	-1	0	-1	0	1	0	0	0	-1	0	0	0	-1	0	0	0
14	-1	-1	1	0	1	0	15.96	-1	0	-1	0	-1	0	-1	0	1	0	0	0	-1	0	0	0	-1	0	0	0
15	-1	-1	1	0	0	1	19.00	-1	0	0	-1	-1	0	0	-1	0	1	0	0	0	-1	0	0	0	-1	0	0
16	-1	-1	1	0	0	1	18.60	-1	0	0	-1	-1	0	0	-1	0	1	0	0	0	-1	0	0	0	-1	0	0
17	-1	-1	1	0	-1	-1	24.20	-1	0	1	1	-1	0	1	1	-1	-1	0	0	1	1	0	0	1	1	0	0
18	-1	-1	1	0	-1	-1	22.58	-1	0	1	1	-1	0	1	1	-1	-1	0	0	1	1	0	0	1	1	0	0
19	1	0	0	1	1	0	5.11	0	1	1	0	0	0	0	0	0	0	1	0	0	0	1	0	0	0	0	0
20	1	0	0	1	1	0	5.31	0	1	1	0	0	0	0	0	0	0	1	0	0	0	1	0	0	0	0	0
21	1	0	0	1	0	1	11.90	0	1	0	1	0	0	0	0	0	0	0	1	0	0	0	1	0	0	0	0
22	1	0	0	1	0	1	11.34	0	1	0	1	0	0	0	0	0	0	0	1	0	0	0	1	0	0	0	0
23	0	1	0	1	1	0	14.30	0	0	0	0	0	1	1	0	0	0	1	0	0	0	0	0	0	0	1	0
24	0	1	0	1	1	0	15.35	0	0	0	0	0	1	1	0	0	0	1	0	0	0	0	0	0	0	1	0
25	0	1	0	1	0	1	18.80	0	0	0	0	0	1	0	1	0	0	0	1	0	0	0	0	0	0	0	1
26	0	1	0	1	0	1	20.96	0	0	0	0	0	1	0	1	0	0	0	1	0	0	0	0	0	0	0	1
27	1	0	0	1	-1	-1	13.30	0	1	-1	-1	0	0	0	0	0	0	-1	-1	0	0	-1	-1	0	0	0	0
28	1	0	0	1	-1	-1	15.64	0	1	-1	-1	0	0	0	0	0	0	-1	-1	0	0	-1	-1	0	0	0	0
29	0	1	0	1	-1	-1	21.60	0	0	0	0	0	1	-1	-1	0	0	-1	-1	0	0	0	0	0	0	-1	-1
30	0	1	0	1	-1	-1	24.27	0	0	0	0	0	1	-1	-1	0	0	-1	-1	0	0	0	0	0	0	-1	-1
31	-1	-1	0	1	1	0	19.86	0	-1	-1	0	0	-1	-1	0	0	0	1	0	0	0	-1	0	0	0	-1	0
32	-1	-1	0	1	1	0	20.88	0	-1	-1	0	0	-1	-1	0	0	0	1	0	0	0	-1	0	0	0	-1	0
33	-1	-1	0	1	0	1	26.37	0	-1	0	-1	0	-1	0	-1	0	0	0	1	0	0	0	-1	0	0	0	-1
34	-1	-1	0	1	0	1	24.38	0	-1	0	-1	0	-1	0	-1	0	0	0	1	0	0	0	-1	0	0	0	-1
35	-1	-1	0	1	-1	-1	29.72	0	-1	1	1	0	-1	1	1	0	0	-1	-1	0	0	1	1	0	0	1	1

Obs	X1	X2	Z1	Z2	h1	h2	Y	c1	c2	c3	c4	c5	c6	c7	c8	c9	c10	c11	c12	c13	c14	c15	c16	c17	c18	c19	c20
36	-1	-1	0	1	-1	-1	29.64	0	-1	1	1	0	-1	1	1	0	0	-1	-1	0	0	1	1	0	0	1	1
37	1	0	-1	-1	1	0	6.70	-1	-1	1	0	0	0	0	0	-1	0	-1	0	-1	0	-1	0	0	0	0	0
38	1	0	-1	-1	1	0	7.00	-1	-1	1	0	0	0	0	0	-1	0	-1	0	-1	0	-1	0	0	0	0	0
39	1	0	-1	-1	0	1	12.40	-1	-1	0	1	0	0	0	0	0	-1	0	-1	0	-1	0	-1	0	0	0	0
40	1	0	-1	-1	0	1	13.15	-1	-1	0	1	0	0	0	0	0	-1	0	-1	0	-1	0	-1	0	0	0	0
41	0	1	-1	-1	1	0	15.38	0	0	0	0	-1	-1	1	0	-1	0	-1	0	0	0	0	0	-1	0	-1	0
42	0	1	-1	-1	1	0	14.90	0	0	0	0	-1	-1	1	0	-1	0	-1	0	0	0	0	0	-1	0	-1	0
43	0	1	-1	-1	0	1	19.20	0	0	0	0	-1	-1	0	1	0	-1	0	-1	0	0	0	0	0	-1	0	-1
44	0	1	-1	-1	0	1	20.10	0	0	0	0	-1	-1	0	1	0	-1	0	-1	0	0	0	0	0	-1	0	-1
45	1	0	-1	-1	-1	-1	16.20	-1	-1	-1	-1	0	0	0	0	1	1	1	1	1	1	1	1	0	0	0	0
46	1	0	-1	-1	-1	-1	17.10	-1	-1	-1	-1	0	0	0	0	1	1	1	1	1	1	1	1	0	0	0	0
47	0	1	-1	-1	-1	-1	22.40	0	0	0	0	-1	-1	-1	-1	1	1	1	1	0	0	0	0	1	1	1	1
48	0	1	-1	-1	-1	-1	23.40	0	0	0	0	-1	-1	-1	-1	1	1	1	1	0	0	0	0	1	1	1	1
49	-1	-1	-1	-1	1	0	21.30	1	1	-1	0	1	1	-1	0	-1	0	-1	0	1	0	1	0	1	0	1	0
50	-1	-1	-1	-1	1	0	24.00	1	1	-1	0	1	1	-1	0	-1	0	-1	0	1	0	1	0	1	0	1	0
51	-1	-1	-1	-1	0	1	27.30	1	1	0	-1	1	1	0	-1	0	-1	0	-1	0	1	0	1	0	1	0	1
52	-1	-1	-1	-1	0	1	27.20	1	1	0	-1	1	1	0	-1	0	-1	0	-1	0	1	0	1	0	1	0	1
53	-1	-1	-1	-1	-1	-1	31.40	1	1	1	1	1	1	1	1	1	1	1	1	-1	-1	-1	-1	-1	-1	-1	1
54	-1	-1	-1	-1	-1	-1	29.64	1	1	1	1	1	1	1	1	1	1	1	1	-1	-1	-1	-1	-1	-1	-1	1
	==	==	==	==	==	==	======																				
	0	0	0	0	0	0	944.97																				

The REG Procedure
Model: MODEL1
Dependent Variable: Y

Number of Observations Read	54
Number of Observations Used	54

Analysis of Variance

Source	DF	Sum of Squares	Mean Square	F Value	Pr > F
Model	26	2446.78853	94.10725	117.48	<.0001
Error	27	21.62755	0.80102		
Corrected Total	53	2468.41608			

Root MSE	0.89500	R-Square	0.9912
Dependent Mean	17.49944	Adj R-Sq	0.9828
Coeff Var	5.11443		

Parameter Estimates

Variable	DF	Parameter Estimate	Standard Error	t Value	Pr > |t|
Intercept	1	17.49944	0.12179	143.68	<.0001
X1	1	-6.80111	0.17224	-39.49	<.0001
X2	1	0.47667	0.17224	2.77	0.0101
Z1	1	-2.64000	0.17224	-15.33	<.0001
Z2	1	0.76333	0.17224	4.43	0.0001
h1	1	-4.30611	0.17224	-25.00	<.0001
h2	1	0.35222	0.17224	2.04	0.0507
c1	1	1.51167	0.24359	6.21	<.0001
c2	1	-1.02833	0.24359	-4.22	0.0002
c3	1	-0.85056	0.24359	-3.49	0.0017

Variable	DF	Parameter Estimate	Standard Error	t Value	Pr > |t|
c4	1	0.39778	0.24359	1.63	0.1141
c5	1	0.14889	0.24359	0.61	0.5462
c6	1	0.47389	0.24359	1.95	0.0622
c7	1	0.56833	0.24359	2.33	0.0273
c8	1	-0.03000	0.24359	-0.12	0.9029
c9	1	0.67833	0.24359	2.78	0.0097
c10	1	-0.50667	0.24359	-2.08	0.0471
c11	1	-0.48833	0.24359	-2.00	0.0551
c12	1	0.34333	0.24359	1.41	0.1701
c13	1	-0.52667	0.34448	-1.53	0.1379
c14	1	0.13667	0.34448	0.40	0.6947
c15	1	0.42167	0.34448	1.22	0.2315
c16	1	0.09333	0.34448	0.27	0.7885
c17	1	0.32444	0.34448	0.94	0.3546
c18	1	0.06444	0.34448	0.19	0.8530
c19	1	-0.16222	0.34448	-0.47	0.6415
c20	1	0.00111	0.34448	0.00	0.9975

The estimates of the model parameters obtained by SAS agree with those previously obtained by manual calculations. With these estimates of parameters, we can write the estimated model for the regression analysis.

8.9 Fitting Regression Model 8.22 (Reference Cell Coding Method) to Responses of Experiment of Example 7.5

To obtain the regression model for Example 7.5, we find the cell means for the experiment responses for Example 7.5, which is presented in Table 8.23.

TABLE 8.23: Means of 3^3 Factorial Design (Example 7.5)

		C-levels		
***B*-levels**	***A*-levels**	**1**	**2**	**3**
	1	μ_{111}	μ_{112}	μ_{113}
1	2	μ_{211}	μ_{212}	μ_{213}
	3	μ_{311}	μ_{312}	μ_{313}
	1	μ_{121}	μ_{122}	μ_{123}
2	2	μ_{221}	μ_{222}	μ_{223}
	3	μ_{321}	μ_{322}	μ_{323}
	1	μ_{131}	μ_{132}	μ_{133}
3	2	μ_{231}	μ_{232}	μ_{233}
	3	μ_{331}	μ_{332}	μ_{333}

We then obtain the corresponding estimates for the above cell means as follows:

		C-levels		
***B*-levels**	***A*-levels**	**1**	**2**	**3**
	1	4.565	9.95	14.195
1	2	12.75	15.365	18.34
	3	16.38	18.8	23.39
	1	5.21	11.62	14.47
2	2	14.825	19.88	22.935
	3	20.37	25.375	29.68
	1	6.85	12.775	16.65
3	2	15.14	19.65	22.9
	3	22.65	27.25	30.52

Under the reference cell method for defining dummy variables, the relevant regression model is:

$$Y = \mu + \sum_{i=1}^{2} \alpha_i X_i + \sum_{j=1}^{2} \beta_j Z_j + \sum_{i=1}^{2}\sum_{j=1}^{2} \gamma_{ij} X_i Z_j + \sum_{k=1}^{2} \omega_k H_k + \sum_{i=1}^{2}\sum_{k=1}^{2} \theta_{ik} X_i H_k$$
$$+ \sum_{j=1}^{2}\sum_{k=1}^{2} \psi_{jk} Z_j H_k + \sum_{i=1}^{2}\sum_{j=1}^{2}\sum_{k=1}^{2} \delta_{ijk} X_i Z_j H_k + \varepsilon \tag{8.24}$$

where

$$X_i = \begin{cases} 1 & \text{if } i = 1, 2 \\ 0 & \text{otherwise} \end{cases}$$

$$Z_j = \begin{cases} 1 & \text{if } j = 1, 2 \\ 0 & \text{otherwise} \end{cases}$$

$$H_k = \begin{cases} 1 & \text{if } k = 1, 2 \\ 0 & \text{otherwise} \end{cases}$$

We use cell (3, 3, 3) as the reference cell, the cell for which the levels of the factors A, B, and C are three. Any other cell in the design could have been used. The estimated parameters will be different for different cells. Based on the use of cell (3, 3, 3) as the reference cell, to estimate the parameters of the model, we use the following relations:

$$\begin{aligned}
\mu_{ijk} &= \mu + \alpha_i + \beta_j + \gamma_{ij} + \omega_k + \theta_{ik} + \psi_{jk} + \delta_{ijk} \quad j, k = 1, 2 \\
\mu_{3jk} &= \mu + \beta_j + \omega_k + \psi_{jk} \quad j, k = 1, 2 \\
\mu_{i3k} &= \mu + \alpha_i + \omega_k + \theta_{ik} \quad i, k = 1, 2 \\
\mu_{ij3} &= \mu + \alpha_i + \beta_j + \gamma_{ij} \quad i, j = 1, 2
\end{aligned}$$

$$\mu_{33k} = \mu + \omega_k \quad k = 1, 2$$
$$\mu_{3j3} = \mu + \beta_j \quad j = 1, 2$$
$$\mu_{i33} = \mu + \alpha_i \quad i = 1, 2$$
$$\mu_{333} = \mu$$

Using the estimated means in the table above, we estimate the parameters of the model as follows:

$$\hat{\mu} = \hat{\mu}_{333} = 30.52$$
$$\hat{\mu}_{133} = \hat{\mu}_{333} + \hat{\alpha}_1 \Rightarrow 16.65 = 30.52 + \hat{\alpha}_1 \Rightarrow \hat{\alpha}_1 = 16.65 - 30.52 = -13.87$$
$$\hat{\mu}_{233} = \hat{\mu}_{333} + \hat{\alpha}_2 \Rightarrow 22.9 = 30.52 + \hat{\alpha}_2 \Rightarrow \hat{\alpha}_2 = 22.9 - 30.52 = -7.62$$
$$\hat{\mu}_{313} = \hat{\mu}_{333} + \hat{\beta}_1 \Rightarrow 23.39 = 30.52 + \hat{\beta}_1 \Rightarrow \hat{\beta}_1 = 23.39 - 30.52 = -7.13$$
$$\hat{\mu}_{323} = \hat{\mu}_{333} + \hat{\beta}_2 \Rightarrow 29.68 = 30.52 + \hat{\beta}_2 \Rightarrow \hat{\beta}_2 = 29.68 - 30.52 = -0.84$$
$$\hat{\mu}_{331} = \hat{\mu}_{333} + \hat{\omega}_1 \Rightarrow 22.65 = 30.52 + \hat{\omega}_1 \Rightarrow \hat{\omega}_1 = 22.65 - 30.52 = -7.87$$
$$\hat{\mu}_{332} = \hat{\mu}_{333} + \hat{\alpha}_1 \Rightarrow 27.25 = 30.52 + \hat{\omega}_2 \Rightarrow \hat{\omega}_2 = 27.25 - 30.52 = -3.27$$
$$\hat{\mu}_{113} = \hat{\mu}_{333} + \hat{\alpha}_1 + \hat{\beta}_1 + \hat{\gamma}_{11} \Rightarrow 14.195 = 30.52 - 13.87 - 7.13 + \hat{\gamma}_{11} \Rightarrow \hat{\gamma}_{11} = 4.675$$
$$\hat{\mu}_{123} = \hat{\mu}_{333} + \hat{\alpha}_1 + \hat{\beta}_2 + \hat{\gamma}_{12} \Rightarrow 14.47 = 30.52 - 13.87 - 0.84 + \hat{\gamma}_{12} \Rightarrow \hat{\gamma}_{12} = -1.34$$
$$\hat{\mu}_{131} = \hat{\mu}_{333} + \hat{\alpha}_1 + \hat{\omega}_1 + \hat{\theta}_{11} \Rightarrow 6.85 = 30.52 - 13.87 - 7.87 + \hat{\theta}_{11} \Rightarrow \hat{\theta}_{11} = -1.93$$
$$\hat{\mu}_{132} = \hat{\mu}_{333} + \hat{\alpha}_1 + \hat{\omega}_2 + \hat{\theta}_{12} \Rightarrow 12.775 = 30.52 - 13.87 - 3.27 + \hat{\theta}_{12} \Rightarrow \hat{\theta}_{12} = -0.605$$
$$\hat{\mu}_{213} = \hat{\mu}_{333} + \hat{\alpha}_2 + \hat{\beta}_1 + \hat{\gamma}_{21} \Rightarrow 18.34 = 30.52 - 7.62 - 7.13 + \hat{\gamma}_{21} \Rightarrow \hat{\gamma}_{21} = 2.57$$
$$\hat{\mu}_{223} = \hat{\mu}_{333} + \hat{\alpha}_2 + \hat{\beta}_2 + \hat{\gamma}_{22} \Rightarrow 22.935 = 30.52 - 7.62 - 0.84 + \hat{\gamma}_{22} \Rightarrow \hat{\gamma}_{22} = 0.875$$
$$\hat{\mu}_{231} = \hat{\mu}_{333} + \hat{\alpha}_2 + \hat{\omega}_1 + \hat{\theta}_{21} \Rightarrow 15.14 = 30.52 - 7.62 - 7.87 + \hat{\theta}_{21} \Rightarrow \hat{\theta}_{21} = -0.11$$
$$\hat{\mu}_{232} = \hat{\mu}_{333} + \hat{\alpha}_2 + \hat{\omega}_2 + \hat{\theta}_{22} \Rightarrow 19.65 = 30.52 - 7.62 - 3.27 + \hat{\theta}_{22} \Rightarrow \hat{\theta}_{22} = 0.02$$
$$\hat{\mu}_{311} = \hat{\mu}_{333} + \hat{\beta}_1 + \hat{\omega}_1 + \hat{\psi}_{11} \Rightarrow 16.38 = 30.52 - 7.13 - 7.87 + \hat{\psi}_{11} \Rightarrow \hat{\psi}_{11} = 0.86$$
$$\hat{\mu}_{312} = \hat{\mu}_{333} + \hat{\beta}_1 + \hat{\omega}_2 + \hat{\psi}_{12} \Rightarrow 18.8 = 30.52 - 0.84 - 3.27 + \hat{\psi}_{12} \Rightarrow \hat{\psi}_{12} = -1.32$$

$$\hat{\mu}_{321} = \hat{\mu}_{333} + \hat{\beta}_2 + \hat{\omega}_1 + \hat{\psi}_{21} \Rightarrow 20.37 = 30.52 - 0.84 - 7.87 + \hat{\psi}_{21} \Rightarrow \hat{\psi}_{21} = -1.44$$

$$\hat{\mu}_{322} = \hat{\mu}_{333} + \hat{\beta}_2 + \hat{\omega}_2 + \hat{\psi}_{22} \Rightarrow 25.375 = 30.52 - 0.84 - 3.27 + \hat{\psi}_{22} \Rightarrow \hat{\psi}_{22} = -1.035$$

$$\hat{\mu}_{111} = \hat{\mu}_{333} + \hat{\alpha}_1 + \hat{\beta}_1 + \hat{\gamma}_{11} + \hat{\omega}_1 + \hat{\theta}_{11} + \hat{\psi}_{11} + \hat{\delta}_{111}$$
$$\Rightarrow 4.565 = 30.52 - 13.87 - 7.13 + 4.675 - 7.87 - 1.93 + 0.86 + \hat{\delta}_{111}$$
$$\Rightarrow \hat{\delta}_{111} = -0.69$$

$$\hat{\mu}_{112} = \hat{\mu}_{333} + \hat{\alpha}_1 + \hat{\beta}_1 + \hat{\gamma}_{11} + \hat{\omega}_2 + \hat{\theta}_{12} + \hat{\psi}_{12} + \hat{\delta}_{112}$$
$$\Rightarrow 9.95 = 30.52 - 13.87 - 7.13 + 4.675 - 3.27 - 0.605 - 1.32 + \hat{\delta}_{112} \Rightarrow \hat{\delta}_{112} = 0.95$$

$$\hat{\mu}_{121} = \hat{\mu}_{333} + \hat{\alpha}_1 + \hat{\beta}_2 + \hat{\gamma}_{12} + \hat{\omega}_1 + \hat{\theta}_{11} + \hat{\psi}_{21} + \hat{\delta}_{121}$$
$$\Rightarrow 5.21 = 30.52 - 13.87 - 0.84 - 1.34 - 7.87 - 1.93 - 1.44 + \hat{\delta}_{121} \Rightarrow \hat{\delta}_{121} = 1.98$$

$$\hat{\mu}_{122} = \hat{\mu}_{333} + \hat{\alpha}_1 + \hat{\beta}_2 + \hat{\gamma}_{12} + \hat{\omega}_2 + \hat{\theta}_{12} + \hat{\psi}_{22} + \hat{\delta}_{122}$$
$$\Rightarrow 11.62 = 30.52 - 13.87 - 0.84 - 1.34 - 3.27 - 0.605 - 1.035 + \hat{\delta}_{122}$$
$$\Rightarrow \hat{\delta}_{122} = 2.06$$

$$\hat{\mu}_{211} = \hat{\mu} + \hat{\alpha}_2 + \hat{\beta}_1 + \hat{\gamma}_{21} + \hat{\omega}_1 + \hat{\theta}_{21} + \hat{\psi}_{11} + \hat{\delta}_{211}$$
$$\Rightarrow 12.75 = 30.52 - 7.62 - 7.13 + 2.57 - 7.87 + 0.11 + 0.86 + \hat{\delta}_{211}$$
$$\Rightarrow \hat{\delta}_{211} = 1.31$$

$$\hat{\mu}_{221} = \hat{\mu}_{333} + \hat{\alpha}_2 + \hat{\beta}_2 + \hat{\gamma}_{22} + \hat{\omega}_1 + \hat{\theta}_{21} + \hat{\psi}_{21} + \hat{\delta}_{221}$$
$$\Rightarrow 14.825 = 30.52 - 7.62 - 0.84 + 0.875 - 7.87 + 0.11 - 1.44 + \hat{\delta}_{221}$$
$$\Rightarrow \hat{\delta}_{221} = 1.09$$

$$\hat{\mu}_{222} = \hat{\mu}_{333} + \hat{\alpha}_2 + \hat{\beta}_2 + \hat{\gamma}_{22} + \hat{\omega}_2 + \hat{\theta}_{22} + \hat{\psi}_{22} + \hat{\delta}_{222}$$
$$\Rightarrow 19.88 = 30.52 - 7.62 - 0.84 + 0.875 - 3.27 + 0.02 - 1.035 + \hat{\delta}_{222}$$
$$\Rightarrow \hat{\delta}_{222} = 1.23$$

$$\hat{\mu}_{212} = \hat{\mu}_{333} + \hat{\alpha}_2 + \hat{\beta}_1 + \hat{\gamma}_{21} + \hat{\omega}_2 + \hat{\theta}_{22} + \hat{\psi}_{12} + \hat{\delta}_{212}$$
$$\Rightarrow 15.365 = 30.52 - 7.62 - 7.13 + 2.57 - 3.27 + 0.02 - 1.32 + \hat{\delta}_{212}$$
$$\Rightarrow \hat{\delta}_{212} = 1.595$$

The values obtained above are the estimates of the parameters of the model when all data are referenced to cell (3, 3, 3). The resulting estimated response under this model is

$$\begin{aligned}\hat{y} = 30.52 - 13.87x_1 - 7.62x_2 - 7.13z_1 - 0.84z_2 - 7.87h_1 - 3.27h_2 \\ + 4.675c_1 - 1.34c_2 - 1.93c_3 - 0.605c_4 + 2.57c_5 + 0.875c_6 + 0.11c_7 \\ + 0.02c_8 + 0.86c_9 - 1.32c_{10} - 1.44c_{11} - 1.035c_{12} - 0.69c_{13} + 0.95c_{14} \\ + 1.98c_{15} + 2.06c_{16} + 1.31c_{17} + 1.595c_{18} + 1.09c_{19} + 1.23c_{20} \qquad (8.25)\end{aligned}$$

With this, we can obtain the estimated responses at each combinations of levels of the factors in the design using the model definitions of x_i, z_j, and c_k. We can also obtain the residual for each fitted value and carry out analysis as described in Chapter 2. Just like in the effects coding model, the reader is welcome to carry out the analysis to test for significance of the fitted model.

8.9.1 SAS Analysis of Data of Example 7.5

We can write a SAS program to analyze the responses of the experiment in Example 7.5 based on the use of reference cell coding method, and the use of cell (3, 3, 3). When we use the reference cell method to define the dummy variables in a three-factor factorial design described above, the relevant SAS program and its output are presented as follows:

SAS PROGRAM

```
DATA EXAMP7_5cell;
INPUT X1-X2 Z1-Z2 H1-H2 Y @@;
c1=X1*Z1; c2=X1*Z2; c3=X1*H1; c4=X1*H2; c5=X2*Z1; c6=X2*Z2; c7=X2*h1;
c8=X2*h2; c9=Z1*H1; C10=Z1*h2; C11=z2*H1; C12=z2*H2;  c13=X1*Z1*H1;
C14=X1*Z1*H2; C15=X1*Z2*H1;  C16=X1*Z2*H2; C17=X2*Z1*H1; C18=X2*Z1*H2;
C19=X2*Z2*H1; C20=X2*Z2*H2;
CARDS;
1 0 1 0 1 0 4.2 1 0 1 0 1 0 4.93 1 0 1 0 0 1 9.8 1 0 1 0 0 1 10.10 0 1 1 0 1
0 12.5 0 1 1 0 1 0 13.0 0 1 1 0 0 1 15.51 0 1 1 0 0 1 15.22 1 0 1 0 0 0 14.6
1 0 1 0 0 0 13.79 0 1 1 0 0 0 18.7 0 1 1 0 0 0 17.98 0 0 1 0 1 0 16.8 0 0 1 0
1 0 15.96 0 0 1 0 0 1 19 0 0 1 0 0 1 18.6 0 0 1 0 0 0 24.2 0 0 1 0 0 0 22.58
1 0 0 1 1 0 5.11 1 0 0 1 1 0 5.31 1 0 0 1 0 1 11.9 1 0 0 1 0 1 11.34 0 1 0 1
1 0 14.3 0 1 0 1 1 0 15.35 0 1 0 1 0 1 18.8 0 1 0 1 0 1 20.96 1 0 0 1 0 0
13.3 1 0 0 1 0 0 15.64 0 1 0 1 0 0 21.6 0 1 0 1 0 0 24.27 0 0 0 1 1 0 19.86 0
0 0 1 1 0 20.88 0 0 0 1 0 1 26.37 0 0 0 1 0 1 24.38 0 0 0 1 0 0 29.72 0 0 0 1
0 0 29.64 1 0 0 0 1 0 6.7 1 0 0 0 1 0 7 1 0 0 0 0 1 12.4 1 0 0 0 0 1 13.15 0
1 0 0 1 0 15.38 0 1 0 0 1 0 14.9 0 1 0 0 0 1 19.2 0 1 0 0 0 1 20.1 1 0 0 0 0
0 16.2 1 0 0 0 0 0 17.1 0 1 0 0 0 0 22.4 0 1 0 0 0 0 23.4 0 0 0 0 1 0 21.3 0
0 0 0 1 0 24 0 0 0 0 0 1 27.3 0 0 0 0 0 1 27.2 0 0 0 0 0 0 31.4 0 0 0 0 0 0
29.64
;
```

```
proc print data=examp6_5cell; sum x1-x2 z1-z2 H1-H2 C1-C20 y;
/* SUMMATION ALLOWS ME TO CHECK THAT DATA WERE ENTERED CORRECTLY*/
PROC REG DATA=EXAMP7_5cell;
MODEL Y=X1-X2 Z1-Z2 H1-H2 C1-C20;
RUN;
```

SAS OUTPUT

Obs	X1	X2	Z1	Z2	C1	C2	Y	C1	C2	C3	C4	C5	C6	C7	C8	C9	C10	C11	C12	C13	C14	C15	C16	C17	C18	C19	C20
1	1	0	1	0	1	0	4.20	1	0	1	0	0	0	0	0	1	0	0	0	1	0	0	0	0	0	0	0
2	1	0	1	0	1	0	4.93	1	0	1	0	0	0	0	0	1	0	0	0	1	0	0	0	0	0	0	0
3	1	0	1	0	0	1	9.80	1	0	0	1	0	0	0	0	0	1	0	0	0	1	0	0	0	0	0	0
4	1	0	1	0	0	1	10.10	1	0	0	1	0	0	0	0	0	1	0	0	0	1	0	0	0	0	0	0
5	0	1	1	0	1	0	12.50	0	0	0	0	1	0	1	0	1	0	0	0	0	0	0	0	1	0	0	0
6	0	1	1	0	1	0	13.00	0	0	0	0	1	0	1	0	1	0	0	0	0	0	0	0	1	0	0	0
7	0	1	1	0	0	1	15.51	0	0	0	0	1	0	0	1	0	1	0	0	0	0	0	0	0	1	0	0
8	0	1	1	0	0	1	15.22	0	0	0	0	1	0	0	1	0	1	0	0	0	0	0	0	0	1	0	0
9	1	0	1	0	0	0	14.60	1	0	0	0	0	0	0	0	0	0	0	0	0	0	0	0	0	0	0	0
10	1	0	1	0	0	0	13.79	1	0	0	0	0	0	0	0	0	0	0	0	0	0	0	0	0	0	0	0
11	0	1	1	0	0	0	18.70	0	0	0	0	1	0	0	0	0	0	0	0	0	0	0	0	0	0	0	0
12	0	1	1	0	0	0	17.98	0	0	0	0	1	0	0	0	0	0	0	0	0	0	0	0	0	0	0	0
13	0	0	1	0	1	0	16.80	0	0	0	0	0	0	0	0	1	0	0	0	0	0	0	0	0	0	0	0
14	0	0	1	0	1	0	15.96	0	0	0	0	0	0	0	0	1	0	0	0	0	0	0	0	0	0	0	0
15	0	0	1	0	0	1	19.00	0	0	0	0	0	0	0	0	0	1	0	0	0	0	0	0	0	0	0	0
16	0	0	1	0	0	1	18.60	0	0	0	0	0	0	0	0	0	1	0	0	0	0	0	0	0	0	0	0
17	0	0	1	0	0	0	24.20	0	0	0	0	0	0	0	0	0	0	0	0	0	0	0	0	0	0	0	0
18	0	0	1	0	0	0	22.58	0	0	0	0	0	0	0	0	0	0	0	0	0	0	0	0	0	0	0	0
19	1	0	0	1	1	0	5.11	0	1	1	0	0	0	0	0	0	0	1	0	0	0	1	0	0	0	0	0
20	1	0	0	1	1	0	5.31	0	1	1	0	0	0	0	0	0	0	1	0	0	0	1	0	0	0	0	0
21	1	0	0	1	0	1	11.90	0	1	0	1	0	0	0	0	0	0	0	1	0	0	0	1	0	0	0	0
22	1	0	0	1	0	1	11.34	0	1	0	1	0	0	0	0	0	0	0	1	0	0	0	1	0	0	0	0
23	0	1	0	1	1	0	14.30	0	0	0	0	0	1	1	0	0	0	1	0	0	0	0	0	0	0	1	0
24	0	1	0	1	1	0	15.35	0	0	0	0	0	1	1	0	0	0	1	0	0	0	0	0	0	0	1	0
25	0	1	0	1	0	1	18.80	0	0	0	0	0	1	0	1	0	0	0	1	0	0	0	0	0	0	0	1
26	0	1	0	1	0	1	20.96	0	0	0	0	0	1	0	1	0	0	0	1	0	0	0	0	0	0	0	1
27	1	0	0	1	0	0	13.30	0	1	0	0	0	0	0	0	0	0	0	0	0	0	0	0	0	0	0	0
28	1	0	0	1	0	0	15.64	0	1	0	0	0	0	0	0	0	0	0	0	0	0	0	0	0	0	0	0
29	0	1	0	1	0	0	21.60	0	0	0	0	0	1	0	0	0	0	0	0	0	0	0	0	0	0	0	0
30	0	1	0	1	0	0	24.27	0	0	0	0	0	1	0	0	0	0	0	0	0	0	0	0	0	0	0	0
31	0	0	0	1	1	0	19.86	0	0	0	0	0	0	0	0	0	0	1	0	0	0	0	0	0	0	0	0
32	0	0	0	1	1	0	20.88	0	0	0	0	0	0	0	0	0	0	1	0	0	0	0	0	0	0	0	0
33	0	0	0	1	0	1	26.37	0	0	0	0	0	0	0	0	0	0	0	1	0	0	0	0	0	0	0	0
34	0	0	0	1	0	1	24.38	0	0	0	0	0	0	0	0	0	0	0	1	0	0	0	0	0	0	0	0
35	0	0	0	1	0	0	29.72	0	0	0	0	0	0	0	0	0	0	0	0	0	0	0	0	0	0	0	0
36	0	0	0	1	0	0	29.64	0	0	0	0	0	0	0	0	0	0	0	0	0	0	0	0	0	0	0	0
37	1	0	0	0	1	0	6.70	0	0	1	0	0	0	0	0	0	0	0	0	0	0	0	0	0	0	0	0
38	1	0	0	0	1	0	7.00	0	0	1	0	0	0	0	0	0	0	0	0	0	0	0	0	0	0	0	0
39	1	0	0	0	0	1	12.40	0	0	0	1	0	0	0	0	0	0	0	0	0	0	0	0	0	0	0	0
40	1	0	0	0	0	1	13.15	0	0	0	1	0	0	0	0	0	0	0	0	0	0	0	0	0	0	0	0
41	0	1	0	0	1	0	15.38	0	0	0	0	0	0	1	0	0	0	0	0	0	0	0	0	0	0	0	0
42	0	1	0	0	1	0	14.90	0	0	0	0	0	0	1	0	0	0	0	0	0	0	0	0	0	0	0	0
43	0	1	0	0	0	1	19.20	0	0	0	0	0	0	0	1	0	0	0	0	0	0	0	0	0	0	0	0
44	0	1	0	0	0	1	20.10	0	0	0	0	0	0	0	1	0	0	0	0	0	0	0	0	0	0	0	0
45	1	0	0	0	0	0	16.20	0	0	0	0	0	0	0	0	0	0	0	0	0	0	0	0	0	0	0	0
46	1	0	0	0	0	0	17.10	0	0	0	0	0	0	0	0	0	0	0	0	0	0	0	0	0	0	0	0
47	0	1	0	0	0	0	22.40	0	0	0	0	0	0	0	0	0	0	0	0	0	0	0	0	0	0	0	0
48	0	1	0	0	0	0	23.40	0	0	0	0	0	0	0	0	0	0	0	0	0	0	0	0	0	0	0	0
49	0	0	0	0	1	0	21.30	0	0	0	0	0	0	0	0	0	0	0	0	0	0	0	0	0	0	0	0
50	0	0	0	0	1	0	24.00	0	0	0	0	0	0	0	0	0	0	0	0	0	0	0	0	0	0	0	0
51	0	0	0	0	0	1	27.30	0	0	0	0	0	0	0	0	0	0	0	0	0	0	0	0	0	0	0	0
52	0	0	0	0	0	1	27.20	0	0	0	0	0	0	0	0	0	0	0	0	0	0	0	0	0	0	0	0
53	0	0	0	0	0	0	31.40	0	0	0	0	0	0	0	0	0	0	0	0	0	0	0	0	0	0	0	0
54	0	0	0	0	0	0	29.64	0	0	0	0	0	0	0	0	0	0	0	0	0	0	0	0	0	0	0	0
	==	==	==	==	==	==	======	==	==	==	==	==	==	==	==	==	===	===	===	===	===	===	===	===	===	===	===
	18	18	18	18	18	18	994.97	6	6	6	6	6	6	6	6	6	6	6	6	2	2	2	2	2	2	2	2

The REG Procedure

Model: MODEL1

Dependent Variable: Y

Number of Observations Read	54
Number of Observations Used	54

Analysis of Variance

Source	DF	Sum of Squares	Mean Square	F Value	Pr > F
Model	26	2446.78853	94.10725	117.48	<.0001
Error	27	21.62755	0.80102		
Corrected Total	53	2468.41608			

Root MSE	0.89500	R-Square	0.9912
Dependent Mean	17.49944	Adj R-Sq	0.9828
Coeff Var	5.11443		

Parameter Estimates

Variable	DF	Parameter Estimate	Standard Error	t Value	Pr > |t|
Intercept	1	30.52000	0.63286	48.23	<.0001
X1	1	-13.87000	0.89500	-15.50	<.0001
X2	1	-7.62000	0.89500	-8.51	<.0001
Z1	1	-7.13000	0.89500	-7.97	<.0001
Z2	1	-0.84000	0.89500	-0.94	0.3563
H1	1	-7.87000	0.89500	-8.79	<.0001
H2	1	-3.27000	0.89500	-3.65	0.0011
C1	1	4.67500	1.26572	3.69	0.0010
C2	1	-1.34000	1.26572	-1.06	0.2991
C3	1	-1.93000	1.26572	-1.52	0.1389
C4	1	-0.60500	1.26572	-0.48	0.6365
C5	1	2.57000	1.26572	2.03	0.0523
C6	1	0.87500	1.26572	0.69	0.4953
C7	1	0.11000	1.26572	0.09	0.9314
C8	1	0.02000	1.26572	0.02	0.9875
C9	1	0.86000	1.26572	0.68	0.5026
C10	1	-1.32000	1.26572	-1.04	0.3062
C11	1	-1.44000	1.26572	-1.14	0.2652
C12	1	-1.03500	1.26572	-0.82	0.4207
C13	1	-0.69000	1.78999	-0.39	0.7029
C14	1	0.95000	1.78999	0.53	0.5999
C15	1	1.98000	1.78999	1.11	0.2784
C16	1	2.06000	1.78999	1.15	0.2599
C17	1	1.31000	1.78999	0.73	0.4706
C18	1	1.59500	1.78999	0.89	0.3808
C19	1	1.09000	1.78999	0.61	0.5477
C20	1	1.23000	1.78999	0.69	0.4978

The estimates of the model parameters obtained by SAS analysis are in agreement with those previously obtained by manual calculations.

8.10 Exercises

Exercise 8.1

For the experiment of Exercise 7.1, obtain a regression model for the responses using reference cell coding to define the indicator variables in the model, with cell (2, 2, 2) as the reference cell. Fit the model and extract the fitted values and residuals. Perform ANOVA and test for significance of the fitted model. □

Exercise 8.2

For the experiment of Exercise 7.1, obtain a regression model for the responses using effects coding method to define the indicator variables in the model. Fit the model, extract the fitted values and residuals. Perform ANOVA and test for significance of the fitted model. □

Exercise 8.3

Suppose that in an experiment involving three factors, A, B, and C, set at two levels, the following data were obtained:

(1)	29.63	21.27	20.22
a	56.08	54.86	45.37
b	33.75	33.26	38.25
ab	30.83	30.63	31.93
c	42.47	53.94	47.94
ac	59.84	51.43	50.96
bc	54.49	53.32	40.31
abc	23.1	29.49	28.97

Obtain a regression model for the responses using reference cell coding to define the indicator variables in the model. Fit the model and extract the fitted values and residuals. Perform ANOVA and test for significance of the fitted model; use cell (2, 2, 2). □

Exercise 8.4

For the experiment of Exercise 8.3, obtain a regression model for the responses using effects coding method to define the indicator variables in the model. Fit the model and extract the fitted values and residuals. Perform ANOVA and test for significance of the fitted model. □

Exercise 8.5

To study the effects of four factors A, B, C, and D on a process, a 2^4 factorial experiment was run in two replicates and the responses are as follows:

(1)	6	7
a	10	12
b	30	33
ab	55	60
c	6	4
ac	15	18
bc	26	23
abc	60	62
d	8	10
ad	12	15
bd	34	32
abd	60	61
cd	16	15
acd	5	6
bcd	35	37
$abcd$	52	53

Write a SAS program to analyze the responses of the experiment by using cell (2, 2, 2, 2) as the reference cell for the regression model. Identify which factorial effects are significant. □

Exercise 8.6

Write a SAS program to analyze the responses for Exercise 8.5, defining the dummy variables by effects coding. □

Exercise 8.7

For data of Example 7.1, obtain the reference cell regression model for the analysis of the experimental responses by assuming that the reference cell is (2, 1). Obtain the estimates for the model and the equation for the fitted model. Obtain the estimated responses, extract the residuals, and perform ANOVA to test whether the fitted model is significant. □

Exercise 8.8

Write SAS program to carry out the analysis required for Exercise 8.1. □

Exercise 8.9

Write a SAS program to carry out the analysis required for Exercise 8.2. □

Exercise 8.10

Write a SAS program to analyze the responses of Example 7.1, according to the regression model suggested in Exercise 8.7. □

Chapter 9

Fractional Replication and Confounding in 2^k and 3^k Factorial Designs

As we pointed out in Chapter 7, when the number of factors k in a 2^k or 3^k factorial design becomes large, the number of treatment combinations becomes large and unwieldy. Even when 2^k factorial experiments involving such number of factors are performed, there is the cumbersome task of identifying and estimating the k main effects, the ${}^kC_2$ two-factor interactions, the ${}^kC_3$ three-factor interactions, ..., and the k one-factor interaction. There is also the task of calculating their sums of squares for use in ANOVA. The same is true of the 3^k factorial design. Resources and convenience may dictate that even a single replicate of the 2^k factorial design, involving a large number of factors, is not a feasible proposition. In such situations, fractional replication is used for the study of the factors. An attempt is then made to design an experiment to use a fraction of a single replicate of the 2^k or 3^k factorial design to elicit information on the effects of the factors on the response of the system. Such studies of a fraction of a 2^k or 3^k design would eventually lead to the whittling down of the number of factors that should be thoroughly investigated. The theory of fractional replication in these designs is based on the fact that sometimes it is possible to identify a high-order interaction of factors, whose effect is known prior to experimentation to be negligible. *This high-order interaction is then used as the identity, I, in the proposed design.* Consider the half-fraction, if the *principal fraction* is chosen, all of its contrasts that appear with the + sign are used to define the fractional replicate of the design to be carried out in actual experimentation. This high-order interaction (I) is used to define the fractional replicate of the design to be carried out in actual experimentation. It is also possible to choose the *complementary fraction* as the defining relation. In this case, all the members of the relation will have contrasts with a negative sign.

9.1 Construction of the 2^{k-1} Fractional Factorial Design

In the 2^k factorial designs, the simplest fractional factorial that could be used to study the effects of factors, on the response of a process, is the half replicate. In such designs, only 2^{k-1} units are required to perform the experiment and elicit information on the k factors. Consequently, only 2^{k-1} treatment combinations are studied in the actual experiment. These designs are very useful for screening and exploratory studies. The 2^{k-1} factorial designs can be constructed by choosing the highest order interaction in the 2^k factorial design, and using its contrasts to split the treatment combinations into two sets. It is usual to use the treatment combinations in the contrasts of the high-order interaction, which have either − sign or + sign to define the half replicate of the design. The half replicate defined by the + sign is called the principal fraction of the 2^k factorial design. The other half fraction that appears with a − sign is called the complementary fraction of the design. To illustrate, we construct a 2^{3-1} factorial design. The highest order interaction for the 2^3 factorial design in A, B, and C is ABC. In the contrast of ABC presented in Chapter 7 (Table 7.7), the treatment combinations with + signs are as follows:

$a \quad b \quad c \quad abc$ (principal fraction)

Treatment combinations with − signs are as follows:

$ab \quad ac \quad bc$ (1) (complementary fraction)

We can use either of these sets of treatment combinations to construct the factorial design. If we choose the treatment combinations with the + sign, we can list all the factorial effects in this design and their contrasts. In this case, $I = +ABC$ is the defining relation for the design. Similarly, $I = -ABC$ is the defining relation, if we choose only those treatment combinations with − signs. The interactions between $+ABC$ and $-ABC$ are called the design generators. We present the 2^3 design with $-ABC$ as the generator (Table 9.1).

From the list of factorial effects and contrasts under the 2^3 factorial design in A, B, and C, the contrasts of the main effects are as follows:

$$A = -BC$$
$$B = -AC$$
$$C = -AB$$

Further, it can be seen that the contrast that estimates A is the same as the contrast that estimates BC, with the signs changed. Similarly, the contrast

TABLE 9.1: Effects and Contrasts of 2^{3-1} Factorial Design with $I = -ABC$

	Contrasts			
Factorial effects	**(1)**	***ab***	***ac***	***bc***
A	−	+	+	−
B	−	+	−	+
AB	+	+	−	−
C	−	−	+	+
AC	+	−	+	−
BC	+	−	−	+
ABC	−	−	−	−

that estimates B is the same as the contrast that estimates AC, but with the signs changed. We say that A is an alias of $-BC$ while B is an alias of $-AC$ in the design. We can obtain the aliases for the factorial effects in this design by using the defining relation, $I = -ABC$ (for which all the contrasts of ABC in 2^3 design are negative); we multiply this identity by the factorial effect of interest. For instance, to obtain the alias of B in the design, the multiplication leads to

$$B = B \cdot I = B(-ABC) = -AB^2C = -AC$$

Note that $B = -AC$, because any effect multiplied by the identity is the same as itself. In this design, $-AC$ is the generalized interaction of B. If we had chosen the other set of treatment combinations whose contrasts had the + sign, the defining relation would be $I = +ABC$. The contrasts of the factorial effects under this design would be as shown in Table 9.2.

Under this design which is based on principal fraction of ABC, we find that the contrasts for the main effects A, B, and C are as follows:

$$A \cdot I = A^2BC = BC$$
$$B \cdot I = AB^2C = AC$$
$$C \cdot I = ABC^2 = AB$$

Interaction AC is called the generalized interaction of B in the design. Any of the two designs, which we have discussed (represented by $I = -ABC$ and $I = +ABC$), could be chosen for the definition of the 2^{3-1} factorial design. The main effects A, B, and C are said to be aliased by BC, AC, and AB, respectively, in the design represented by the principal fraction $+ABC$. The main effects are said to be aliased by the first-order interactions because when we estimate A, it is difficult to say whether we are actually estimating A or BC. In other words, we cannot differentiate between A and BC. The same applies to B and AC, and C and AB. Further, we can say the same for A and $-BC$, B and $-AC$, and C and $-AB$ in the complementary fraction, represented by

TABLE 9.2: Effects and Contrasts of 2^{3-1} Factorial Design with $I = +ABC$

	Contrasts			
Factorial effects	***a***	***b***	***c***	***abc***
A	+	−	−	+
B	−	+	−	+
AB	+	+	−	−
C	−	−	+	+
AC	+	−	+	−
BC	−	+	+	−
ABC	+	+	+	+

$I = -ABC$. Another way of looking at these results is to say that when we estimate A in $I = +ABC$, we are estimating $A + BC$. This way of looking at the alias structure has the benefit of showing that if after we have run one of the fractions, we then run the other (its compliment) we can isolate and estimate the main effects by simply adding the estimates under the two fractions. We usually estimate the main effect of A in $I = -ABC$ as $(A - BC)/2$ and estimate the same effect in $I = +ABC$ as $(A + BC)/2$; it can be seen that once we have these two halves of the fractions, we can estimate A and BC as follows:

$$A = \frac{[(A + BC) + (A - BC)]}{2}$$

$$BC = \frac{[(A + BC) - (A - BC)]}{2}$$

Other effects in the design can be obtained by addition and subtraction of estimates of the effects from the two half fractions of the 2^3 factorial design as follows:

$$B = \frac{[(B + AC) + (B - AC)]}{2}$$

$$AC = \frac{[(B + AC) - (B - AC)]}{2}$$

9.1.1 Some Requirements of Good 2^{k-1} Fractional Factorial Designs

1. In the planning of fractionals of the 2^k factorial designs, when k is fairly large (for instance, $k > 4$), it is desirable that no main effect or low-order interaction should be aliased by any other main effect or low-order interaction. Instead, main effects and low-order interactions should be aliased with higher order interactions.
2. It is necessary (indeed desirable) to have some contrasts whose effects and those of their aliases are known *a priori* to be nonsignificant (or for which it could be assumed, with relative certainty, that they and their aliases are not significant), so that they can be used to obtain an estimate of error sum of squares in ANOVA.

9.2 Contrasts of the 2^{k-1} Fractional Factorial Design

In Section 9.1, we discussed the half fraction of the 2^3 factorial design. In this section, we discuss the general 2^{k-1} fractional factorial design. This is the half fraction of any factorial design with k factors set at two levels. In general, there are two ways of obtaining the estimates of factorial effects in 2^{k-1} fractional factorial design.

9.2.1 Estimation of the Effects of a 2^{k-1} Fractional Design Using a Full 2^{k-1} Factorial Design

We can use the normal treatment combinations in a full 2^{k-1} factorial design (i.e., a factorial design with $k-1$ factors at two levels) to obtain the contrasts for the effects in the 2^{k-1} fractional factorial design. In such situations, the treatments are written down as would be the case in a full 2^{k-1} factorial design, and the kth factor is then associated with the treatment combinations. This kth element is disregarded in the estimation of factorial effects, as the effect is estimated from the full 2^{k-1} factorial experiment. The kth factor is only added (in the treatment combination) to indicate the actual treatment effect in 2^{k-1} fractional factorial, which is being estimated. For instance, in 2^{5-1} fractional factorial design, involving factors A, B, C, D, and E, let $I=-ABCDE$; the treatment effects in this design have the contrasts shown in Table 9.3 (below).

TABLE 9.3: A Typical 2^{5-1} Fractional Factorial with $I=-ABCDE$

Treatment	Factorial effects contrasts				
combinations	A	B	C	D	$E=-ABCD$
(1)	−	−	−	−	−
$a(e)$	+	−	−	−	+
$b(e)$	−	+	−	−	+
ab	+	+	−	−	−
$c(e)$	−	−	+	−	+
ac	+	−	+	−	−
bc	−	+	+	−	−
$abc(e)$	+	+	+	−	+
$d(e)$	−	−	−	+	+
ad	+	−	−	+	−
bd	−	+	−	+	−
$abd(e)$	+	+	−	+	+
cd	−	−	+	+	−
$acd(e)$	+	−	+	+	+
$bcd(e)$	−	+	+	+	+
$abcd$	+	+	+	+	−

For this design, it is possible to estimate the fractional effect AE by estimating the contrast of the main effect A in the 2^4 full factorial design. In the 2^4 design, the main effect A would be estimated as usual, but the letters in the parenthesis are added to indicate which factorial effect in the 2^4 fractional factorial is being estimated. AE can therefore be estimated as follows:

$$A = (1/2^3)[abcd + abd(e) + abc(e) + ab + acd(e) + ad + ac + a(e) - bcd(e) - bd - bc - b(e) - cd - d(e) - c(e) - (1)]$$

Similarly, AB in 2^{5-1} fractional factorial design is obtained through the 2^4 design as follows:

$$AB = (1/2^3)[abcd + ab(e) + abd(e) + ab + d(e) + cd + c(e) + (1) - ac - ad \\ - a(e) - bcd(e) - bc - bd - bcd(e) - b(e)]$$

In a similar manner, we can obtain the contrasts for all the treatments, which appear in the 2^{5-1} fractional factorial design and estimate them. It can be seen that for this design, all the main effects are aliased with third-order interactions while the first-order interactions are all aliased with second-order interactions. The alias structure for the factorial effects is as follows:

$$\begin{aligned} I &= -ABCDE \\ A &= -BCDE \\ B &= -ACDE \\ C &= -ABDE \\ D &= -ABCE \\ E &= -ABCD \\ CD &= -ABE \\ AE &= -BCD \\ BE &= -ACD \\ DE &= -ABC \end{aligned}$$

Here, we further illustrate the methods we described for the estimation of the factorial effects by an example.

Example 9.1

The yield of a chemical reaction is thought to depend on temperature, pressure, reaction time, reactant concentration, and addition of a catalyst. Each of these factors is set at two levels, and a half replicate of this design is constructed to explore which of them is significant. The factors are set as follows:

		Levels	
Temperature	A	30°C	40°C
Pressure	B	1 atm	2 atm
Reaction time	C	2 h	4 h
Reactant concentration	D	30%	40%
Catalyst	E	Absent	Present

The highest order interaction $ABCDE$ is used to define the half replicate of the design. The treatment combinations and the resulting responses are shown below:

(1)	ae	be	ab	ce	ac	bc	$abce$	d	ad	bd	$abde$	cd	$acde$	$bcde$	$abcd$
30.3	27.6	25.8	31.3	24.9	36.3	24.9	28.3	37.9	34.9	35.9	28.3	33.6	37.0	32.2	37.4

If we use the 2^4 effects, as described earlier, we shall obtain the factorial effects of A (which is AE in this design) and AB as follows:

$$AE = \left(\frac{1}{8}\right)(37.4 + 28.3 + 28.3 + 31.3 + 34.9 + 36.3 + 27.6 - 32.2 - 35.9 - 24.9 \\ - 25.8 - 33.6 - 37.9 - 24.9 - 30.3) = 1.95$$

$$AB = \left(\frac{1}{8}\right)(37.4 + 28.3 + 31.3 + 33.6 + 24.9 + 37.9 + 30.3 - 37 - 36.3 - 34.9 \\ - 27.6 - 32.2 - 24.9 - 35.9 - 25.8) = -0.325$$

The values of other estimates, obtained in a similar manner, are presented below without showing all the calculations:

$$BE = -2.30 \quad CE = 0.325 \quad AC = 0.05 \quad ABC = -1.225 \quad DE = 5.975$$

$$AD = -2.45 \quad BD = 0.04 \quad ABD = -0.375 \quad CD = 0.475$$

$$ACDE = 0.90 \quad BCDE = 1.85 \quad ABCD = 2.825$$

We calculate the contrasts for all the effects using the methods described in Chapter 7 and estimate both the factorial effects and their sums of squares in fractional factorial designs to show that both methods yield the same estimates (Table 9.4):

$$\text{Contrast of } A(E) = (a - 1)(b + 1)(c + 1)(d + 1)$$

$$SS_{A(E)} = [\text{contrast of } A(E)]^2/2^4 = \frac{(15.6)^2}{16} = 15.21$$

$$\text{Contrast of } B(E) = (a + 1)(b - 1)(c + 1)(d + 1)$$

$$SS_{B(E)} = [\text{contrast of } B(E)]^2/2^4 = \frac{(-18.6)^2}{16} = 21.60$$

$$\text{Contrast of } AB = (a - 1)(b - 1)(c + 1)(d + 1)$$

$$SS_{AB} = (\text{contrast of } AB)^2/2^4 = \frac{(-2.6^2)}{16} = 0.423$$

$$\text{Contrast of } C(E) = (a + 1)(b + 1)(c - 1)(d + 1)$$

$$SS_{C(E)} = [\text{contrast of } C(E)]^2/2^4 = \frac{(2.6)^2}{16} = 0.423$$

$$\text{Contrast of } AC = (a - 1)(b + 1)(c - 1)(d + 1)$$

$$SS_{AC} = (\text{contrast of } AC)^2/2^4 = \frac{(31.2)^2}{16} = 60.84$$

$$\text{Contrast of } BC = (a + 1)(b - 1)(c - 1)(d + 1)$$

$$SS_{BC} = (\text{contrast of } BC)^2/2^4 = \frac{(0.4)^2}{16} = 0.01$$

$$\text{Contrast of } ABC(E) = (a - 1)(b - 1)(c - 1)(d + 1)$$

$$SS_{ABC(E)} = [\text{contrast of } ABC(E)]^2/2^4 = \frac{(-9.8)^2}{16} = 6.0$$

$$\text{Contrast of } D = (a+1)(b+1)(c+1)(d-1)$$

$$SS_{D(E)} = [\text{contrast of } D(E)]^2/2^4 = \frac{(47.8)^2}{16} = 142.8$$

$$\text{Contrast of } AD = (a-1)(b+1)(c+1)(d-1)$$

$$SS_{AD} = (\text{contrast of } AD)^2/2^4 = \frac{(-19.6)^2}{16} = 24.01$$

$$\text{Contrast of } BD = (a+1)(b-1)(c+1)(d-1)$$

$$SS_{BD} = (\text{contrast of } BD)^2/2^4 = \frac{(-0.8)^2}{16} = 0.04$$

$$\text{Contrast of } ABD(E) = (a-1)(b-1)(c+1)(d-1)$$

$$SS_{ABD(E)} = [\text{contrast of } ABD(E)]^2/2^4 = \frac{(-3)^2}{16} = 0.423$$

$$\text{Contrast of } CD = (a+1)(b+1)(c-1)(d-1)$$

$$SS_{ACD} = (\text{contrast of } CD)^2/2^4 = \frac{(-3)^2}{16} = 0.903$$

$$\text{Contrast of } ACD(E) = (a-1)(b+1)(c-1)(d-1)$$

$$SS_{ACD(E)} = [\text{contrast of } ACD(E)]^2/2^4 = \frac{(7.2)^2}{16} = 3.24$$

$$\text{Contrast of } BCD(E) = (a+1)(b-1)(c-1)(d-1)$$

$$SS_{BCD(E)} = [\text{contrast of } BCD(E)]^2/2^4 = \frac{(14.8)^2}{16} = 13.69$$

$$\text{Contrast of } ABCD = (a-1)(b-1)(c-1)(d-1)$$

$$SS_{ABCD} = (\text{contrast of } ABCD)^2/2^4 = \frac{(22.6)^2}{16} = 31.123$$

TABLE 9.4: Estimation of Factorial Effects in Fractional Factorial Designs Illustrated with the Responses of Example 9.1

Treatment combinations	Responses	Estimates	Sum of squares
(1)	30.3		
a(e)	27.6	1.95	15.21
b(e)	25.8	−2.30	21.60
ab	31.3	−0.325	0.423
c(e)	24.9	0.325	0.423
ac	36.3	3.90	60.84*
bc	24.9	0.05	0.01
abc(e)	28.3	−1.225	6.0
d(e)	37.9	5.975	142.8*
ad	34.9	−2.45	24.01
bd	35.9	−0.10	0.04
abd(e)	28.3	−0.375	0.423
cd	33.6	0.475	0.903
acd(e)	37.0	0.90	3.24
bcd(e)	32.2	1.85	13.69
abcd	37.4	2.825	31.123

We use *ABCD*, *ACDE*, *BCDE*, and *ABCE* interaction effects to estimate the error sum of squares in this design to be used in ANOVA. Only those factorial effects with * are significant at 1% level of significance (i.e., *AC* and *DE*) since $F(1, 4, 0.01) = 7.71$. Only *AC* and *DE* are significant, suggesting that *B* is not an important factor. We could well have run this experiment as a 2^4 factorial design in one replicate. This is left as an exercise for the reader. ▯

9.3 General 2^{k-p} Fractional Factorial Design

Apart from one half replicate of the 2^k factorial design, the next most popular fractional replication is the 1/4 fraction of the 2^k factorial design. This is usually denoted as the 2^{k-2} factorial design and requires 1/4 of the 2^k units needed to achieve a single replicate of a full factorial design with k factors, when each factor is set at two levels. The quarter fractional design is used when the half fraction becomes difficult to manage, due to increase in number of factors that need to be investigated in a design. It is used for preliminary screening for large k, when it is hoped that 1/4 fraction of the designed experiment would be sufficient to help in identifying factors that contribute significantly to the response of the process. The overall aim is to keep the cost and complications of the design to the barest minimum possible, while still achieving the objective of the investigation. The quarter fractional replication requires two factorial effects for the definition of the design. In this design, the defining relation, I, would contain the two contrasts, and the general interaction can be obtained by multiplying the two contrasts with the exponents reduced (modulo 2). It is useful in planning this experiment to ensure that all the requirements of a good fractional design, which were discussed earlier in this chapter, are met. Choice of the defining contrasts should be guided by the principle of ensuring that no main effect is aliased with another; low-order interactions (particularly, the first-order interactions) should as much as possible be aliased with higher order interactions. Of course, this will always depend on how large the number of factors k is. To illustrate the construction of the design, consider a 2^{6-2} factorial design in factors A, B, C, D, E, and F. We could choose the defining contrasts to be *ABCD* and *CDEF*; however, by the choice of these interactions, we have the defining relation as follows:

$$I = ABCD = CDEF = ABEF$$

because $$ABCD \cdot CDEF = ABC^2D^2EF = ABEF$$

The implication of the above defining relation is that we automatically choose a third factorial effect, which is *the generalized interaction of the two original defining contrasts as part of the defining relation.* This is not difficult to understand since we recall that in the 2^{3-1} factorial design, with $I = +ABC$, A is aliased with its generalized interaction in I. The generalized interaction of *ABCD* and *CDEF* is *ABEF*. With *ABCD*, *CDEF*, and *ABEF* in the defining relation, any factorial effect in this design should be aliased with three other factorial effects. It is easy to work out the alias structure of

TABLE 9.5: Alias Structure of 2^{k-2} Factorial Design

Factorial effects	Aliased effects (three in each case)		
I	$ABCD$	$CDEF$	$ABEF$
A	BCD	BEF	$ACDEF$
B	ACD	AEF	$BCDEF$
AB	CD	EF	$ABCDEF$
C	ABD	DEF	$ABCEF$
AC	BD	$ADEF$	$BCEF$
BC	AD	$BDEF$	$ACEF$
D	ABC	CEF	$ABDEF$
E	CDF	ABF	$ABCDE$
F	CDE	ABE	$ABCDEF$
AE	$BCDE$	$ACDF$	BF
AF	$BCDF$	$ACDE$	BE
CF	$ABDF$	$ABCE$	DE
DF	CE	$ABCF$	$ABDE$
ADE	BCE	ACF	BDF
ADF	BCF	ACE	BDE

the design. The structure is such that each main effect is aliased with second-order interactions and higher. On the other hand, the first-order interactions are aliased with other first-order interactions and higher. The alias structure is presented in Table 9.5.

From the above discussion, it should be clear that there are many other ways of constructing the 2^{6-2} design from the six factors, A, B, C, D, E, and F. For instance, $I = ABCD = ADEF = ABCEF$. For this design, main effects are aliased by second- and third-order interactions, whereas the first-order interactions are aliased by first- and second-order interactions. Another design is $I = ABC = DEF = ABCDEF$; the main effects here are aliased by first- and fourth-order interactions. Although a variety of designs could be obtained, they are not always all good enough. For example, consider the last design mentioned above, although there are six factors in the design (large enough by the standards of many experiments), it is found that main effects are aliased by first-order interactions, which should themselves be important in this experiment. On that basis, it is not a good design. Compare the designs represented by $I = ABC = DEF = ABCDEF$ and the other, $I = ABCD = CDEF = ABEF$; obviously, for the reason we have just mentioned, the former is not as good as the latter. In the latter, the number of factors is six, and the first-order interactions are aliased with first-order interactions, which are less important than the main effects. Further, this is also more acceptable because it is the highest level of design efficiency that could be achieved, given all the constraints imposed on the design. In general, if there are k factors in a design, each to be investigated at two levels, and we wish to construct a $1/2^p$ fractional factorial of this design, then we need to choose p factorial effects, which make up the defining contrasts for the design. Since there are p factorial effects, we shall have in the defining relation, p factorial effects called *generators*. Apart from those chosen p factorial effects, we shall have other effects ($2p - p - 1$ of them), which are the generalized interactions of these p factorial effects. They are obtained by multiplying factorial effects, two at a time, three at a time, and so on (without repetition). Suppose we are interested in constructing a 2^{6-3} fractional factorial design, if we choose *ABD*, *ACE*, and *BCF* as our defining contrasts, then we have the defining relation as follows:

$$I = ABD = ACE = BCF = ABD \cdot ACE\; ABD \cdot BCF = ACE \cdot BCF$$
$$= ABD \cdot ACE \cdot BCF$$

i.e.,

$$I = ABD = ACE = BCF = BCDE = ACDF = ABEF = DEF$$

Once the defining contrast of any fractional factorial experiment has been obtained, the alias structure can be found. The structure is simply obtained by multiplying through the defining relation with the factorial effect of interest, bearing in mind that A^2, for instance, is the same as I (recall the relationship between factorial effects and group theory which we discussed in Chapter 7—The Reduction of the Exponent of a Letter by Modulo Arithmetic). In 2^{6-3} design that we have just constructed, the aliases of A are

$$A \cdot I = A \cdot ABD = A \cdot ACE = A \cdot BCF = A \cdot BCDE$$
$$= A \cdot ACDF = A \cdot ABEF = A \cdot DEF$$

which on reduction according to this rule leads to

$$A = BD = CE = ABCF = ABCDE = CDF = BEF = ADEF$$

Clearly, for this design, which is strictly constrained, main effects are aliased with first-order interactions and higher. The aim of every factorial design, within the limits of the constraints, should always be to ensure that no factorial effect of special interest is aliased with another effect, which is of special interest in a study. It is important to recognize that there are other ways of defining the fractional factorial designs we have just considered. If we consider the 2^{6-2}, we should note that all our design contrasts are positive. What we have stated in each case regarding its design, refer only to the *principal fractions* of the designs. Take the 2^{6-2} design, we could have chosen two negative contrasts to make up the defining relation as follows:

$$I = -ABCD = -CDEF = ABEF$$

One of the other possible ways of defining the design, which could have been obtained by using the same set of factorial effects that make up the 2^6 factorial design, is shown as follows:

$$I = -ABCD = CDEF = -ABEF \quad \text{or} \quad I = ABCD = -CDEF = ABEF$$

The 2^{6-2} design has six design configurations that could arise by arrangement of signs of the defining contrasts. In general, p factorial effects are chosen to define a 2^{k-p} fractional factorial design (i.e., those effects which together with their generalized interactions form the defining relation), and the number of possible design configurations is obtained as the sum of ${}^{p}C_1 + {}^{p}C_2 + \cdots + {}^{p}C_p$ for $p > 1$; the number is 2 when $p = 1$.

Once the design is defined and treatment effects listed as stated earlier (i.e., in standard order for the full fractional design with $k - p$ factors, and with missing letters inserted in brackets to indicate the actual factorial effects being estimated in a 2^{k-p} fractional factorial design), the experiment is performed by studying only those treatment combinations that appear in the design. The methods for calculating contrasts (Chapter 7) could then be used to analyze the responses, to estimate the factorial effects, and to calculate their sums of squares. This method had been demonstrated for the 2^{k-1} fractional factorial design. On the other hand, it is possible to estimate the factorial effects, by using formula and by listing the contrasts. Hence, to estimate the factorial effect j, we use the following formula:

$$j = 2(\text{contrast of } j)/N$$

where $N = 2^{k-p}$, so that the main effect A could be estimated as

$$A = 2(\text{contrast of } A)/N$$

Sums of squares are obtained in this case as

$$SS_A = (\text{contrast of } A)^2/N$$

9.4 Resolution of a Fractional Factorial Design

Since exploratory analysis is one of the main uses of fractional factorial designs, it is advisable to run 2^k factorial designs in sequences of half fractional factorials. One of the fractions is then run and its responses are analyzed. Based on the analysis results, it could be decided that some factors should be dropped because they are not significant, or that all the factors be retained because it is believed that they all contribute to the design. The practice is that when such an experiment is carried out, if ambiguities arise in the interpretation of the results, the second half should be run to resolve them. In such situations, the half fractions of the design are run in two blocks, defined by the high-order interaction. Information is lost only for the higher order interaction that was used to define the fractional factorial design. On the other hand, it is possible that enough will be learnt from the one half fraction to enable further experimentation to proceed without running the other half fraction, thus saving resources. Sequential use of fractional factorials is therefore highly recommended.

In the context of sequential use of fractional factorials, to study the effects of factors on a response variable, the *resolution* of a fractional factorial design is important. A fractional factorial design is said to be of *resolution*, r, if no p-factor interaction is aliased with a factorial effect of less than $r-p$ factors. The 2^{6-2} fractional factorial design with $I=ABCD=CDEF=ABEF$ is a design of resolution IV, conventionally written as 2^{6-2}_{IV}, with the subscripts in roman numerals indicating the resolution of the design. The fractional factorial designs in A, B, C with the generators $C=\pm AB$ (which we discussed earlier as part of our study of the 2^{k-1} factorial designs; these are the designs with the defining relations $I=\pm ABC$) are both of resolution III, written as 2^{3-1}_{III}. Because of the properties of designs with resolutions III, IV, and V, they are often used in sequential fractional factorial experiments. Owing to the importance of these designs, we discuss them further in the next few sections.

9.4.1 Resolution III Designs

A fractional factorial design is said to be of resolution III, if the main effects are not aliased with one another but are aliased with first-order interactions, while the first-order interactions are in turn aliased with one another and higher. These designs are constructed for investigating $k=N-1$ factors, while performing only N experiments, where N is a multiple of four. Under this design, it is possible to carry out experiments involving only four combinations of factors to investigate only three factors. It is also possible to run experiments involving eight different combinations of factors to investigate seven factors. Similarly, 16 experiments could be performed to study 15 factors. Such resolution III designs are said to be *saturated* and are highly used in industrial experiments (where the number of factors to be investigated in a designed experiment is often large) as a part of an exploratory analysis. The design 2^{3-1}_{III}, which we discussed earlier, is used for studying three factors while performing only four experiments involving different combinations of factors, A, B, and C, which make up the design. Consider the 2^{6-3}_{III} design; this design can easily be constructed by first writing down the contrasts in the full 2^3 factorial design in A, B, and C, and equating the rest of the main effects D, E, F to first-order interactions of the letters A, B, and C. In this way, we obtain the generators of the design as $D=AB$, $E=AC$, and $F=BC$ (Table 9.6 displays

TABLE 9.6: Resolution III 2^{6-3} Fractional Factorial Design

Factorial effects contrasts						Treatment combinations
A	B	C	$D=AB$	$E=AC$	$F=BC$	
−	−	−	+	+	+	*def*
+	−	−	−	−	+	*af*
−	+	−	−	+	−	*be*
+	+	−	+	−	−	*abd*
−	−	+	+	−	−	*cd*
+	−	+	−	+	−	*ace*
−	+	+	−	−	+	*bcf*
+	+	+	+	+	+	*abcdef*

TABLE 9.7: Alternative Resolution III 2^{6-3} Fractional Factorial Design

Factorial effects contrasts						**Treatment combinations**
D	E	F	$A=DE$	$B=DF$	$C=EF$	
−	−	−	+	+	+	*abc*
+	−	−	−	−	+	*cd*
−	+	−	−	+	−	*be*
+	+	−	+	−	−	*ade*
−	−	+	+	−	−	*af*
+	−	+	−	+	−	*bdf*
−	+	+	−	−	+	*cef*
+	+	+	+	+	+	*abcdef*

all the design contrasts). The defining relation for this design, which includes the *generalized interactions* of these *generators*, becomes as follows:

$$I = ABD = ACE = BCF = BCDE = ACDF = ABEF = DEF$$

Other *words* in the defining relation (such as $ABEF$, $ACDF$, etc.) are obtained by equating to the generators their products, taking two at a time, then three at a time, without repetition. Each set of letters, which make up an effect in the defining relation, is called a word.

We could have obtained a 2_{III}^{6-3}, by writing down a full 2^3 design in any other three factors in the 2^6 factorial design, provided their contrasts are written in standard order. The other factors in the 2^6 factorial design are then associated with the first-order interactions of the factors chosen to obtain the generators; the defining relation is then obtained as previously described (by multiplying the generators). For instance, we can write a full 2^3 factorial design in D, E, and F and then obtain the generators as $A=DE$, $B=DF$, and $C=EF$, so that the full defining relation for this design becomes $I=ADE=BDF=CEF=ABEF=ACDF=BCDE=ABC$. In both configurations of the design, the alias structure is obtained for each main effect or interaction, by multiplying the defining relation by the factorial effect, whose aliases in the design are required as explained earlier. The alternative design with the defining relation is $I=ADE=BCDF=CEF=ABEF=ACDF=BCDE=ABC$, as given in Table 9.7.

Note that for each of the 2_{III}^{6-3} designs described earlier, there are eight possible designs that could be constructed, using arrangements of signs in the defining relations as follows:

$$I = \pm ABD \quad I = \pm ACE \quad I = \pm BCF$$
$$I = \pm ADE \quad I = \pm BDF \quad I = \pm CEF$$

In an exploratory experiment, 7 degrees of freedom available in this design could be used to estimate six main effects and one first-order interaction from the responses, after the experiment has been performed. Each of these estimated effects has eight factorial effects aliased with it. Other factorial designs of resolution III are 2_{III}^{5-2} and 2_{III}^{7-4}. The 2_{III}^{5-2} design could be obtained from 2_{III}^{6-3} in Table 9.6 by dropping F from it, as well as $F = BC$, while 2_{III}^{7-4} is obtained from the same design (Table 9.6) by adding G to the design and associating G to the interaction, ABC. For the 2_{III}^{5-2} design mentioned above, the defining relation would be $I = ABD = ACE = BCDE$, while for 2_{III}^{7-4} the defining relation would be $I = ABD = ACE = BCF = ABCG = BCDE = ACDF = CDG = ABEF = BEG = AFG = DEF = ADEG = CEFG = BDFG = ABCDEFG$. We can also use the same principle, employed in the construction of the last design, to construct a 2_{III}^{15-11} in A, B, C, D, E, F, G, H, J, K, L, M, N, P, Q. We write down the contrasts of a full 2_{III}^{15-11} design in D, E, F, G in standard order and associate the first-, second-, and third-order interactions in this design, with the rest of the factors in the full 2^{15} factorial design. The generators of this design are $A = DE$, $B = DF$, $C = DG$, $H = EF$, $J = EG$, $K = GF$, $L = DEF$, $M = DEG$, $N = DFG$, $P = EFG$, $Q = DEFG$. The calculation of the defining contrasts of this design is left as an exercise for the reader.

9.4.2 Resolution IV Designs

A fractional design is of resolution IV, if no main effect is aliased with first-order interactions and some first-order interactions are aliased with other first-order interactions and higher order interactions. Such designs are important in experiments involving sequential fractional factorial replication because if second-order and higher order interactions are found not to be significant, it is possible to obtain estimates of the main effects of the design. Examples of such designs can be constructed for four, six, or seven factors yielding the 2_{IV}^{4-1}, 2_{IV}^{6-2}, and 2_{IV}^{7-3} designs, respectively. We can construct 2_{IV}^{6-2} design by writing down a 2^4 design in A, B, C, D in the standard order, and then obtaining the generators by equating the rest of the factors as follows: $E = ABC$, $F = BCD$ (Table 9.8).

TABLE 9.8: A Typical 2^{6-2} Factorial Design of Resolution IV

Factorial effects contrasts						Treatment combinations
A	B	C	D	$E = ABC$	$F = BCD$	
−	−	−	−	−	−	(1)
+	−	−	−	+	−	*ae*
−	+	−	−	+	+	*aef*
+	+	−	−	−	+	*abf*
−	−	+	−	+	+	*cef*
+	−	+	−	−	+	*acf*
−	+	+	−	−	−	*bc*
+	+	+	−	+	−	*abce*
−	−	−	+	−	+	*df*
+	−	−	+	+	+	*adef*
−	+	−	+	+	−	*bde*
+	+	−	+	−	−	*abd*
−	−	+	+	+	−	*cde*
+	−	+	+	−	−	*acd*
−	+	+	+	−	+	*bcdf*
+	+	+	+	+	+	*abcdef*

$$\begin{aligned}
I &= ABCE = BCDF = ADEF \\
A &= BCE = ABCDF = DEF \\
B &= ACE = CDF = ABDEF \\
AB &= CE = ACDEF = BDEF \\
C &= ABE = BDF = ACDEF \\
AC &= BE = ABDF = CDEF \\
BC &= AE = DF = ABCDEF \\
D &= ABCDE = BCF = AEF \\
AD &= BCDE = ABCF = EF \\
BD &= ACDE = CF = ABEF \\
ABD &= CDE = ACF = BEF \\
CD &= ABDE = BF = ACEF \\
ACD &= BE = ABF = CEF \\
E &= ABC = BCDEF = ADF \\
DE &= ABCD = BCEF = AF \\
F &= ABCEF = BCD = ADE
\end{aligned}$$

From this alias structure, it can be clearly seen that the main effects are all aliased with second-order and higher order interactions. The first-order interactions also satisfy our definition of the resolution IV design.

9.4.3 Resolution V Designs

These are fractional factorial designs, in which the main effects have no main effects or first-order interactions aliased with them. The first-order interactions in this design are free of aliases with one another. Each factorial effect which appears in the defining relation of this design must have at least five letters in it. Typical examples of this design are 2_{V}^{5-1} factorial designs with defining relations, $I = \pm ABCDE$. We have already presented one of these designs and the resulting alias structure in Section 9.2.1.

9.5 Fractional Replication in 3^k Factorial Designs

When each of the factors in the designs we considered had two levels, the highest fractional factorial design we discussed is the half fraction, represented as the 2^{k-1} factorial designs. In the 3^k factorial designs, there is an equivalent fraction, the 1/3 fraction, which leads to the 3^{k-1} fractional factorial. The principle that was applied in the 2^k designs for obtaining their half fractions could be used here to obtain treatment combinations that make up the 1/3 fractional of the 3^k factorial design. We choose a high-order interaction and use its contrasts to obtain the defining relation for the 1/3 fractional factorial.

TABLE 9.9: The 1/3 Fractional Designs of the 3^3 Factorial with $I = AB^2C^2$

Fraction I ($x_1 + 2x_2 + 2x_3 = 0$)	Fraction II ($x_1 + 2x_2 + 2x_3 = 1$)	Fraction III ($x_1 + 2x_2 + 2x_3 = 2$)
0 0 0	0 0 2	0 0 2
0 1 2	0 1 1	0 1 0
0 2 1	0 2 0	0 2 1
1 0 1	1 0 0	1 0 1
1 1 0	1 1 2	1 1 2
1 2 2	1 2 2	1 2 0
2 0 2	2 0 2	2 0 0
2 1 1	2 1 0	2 1 1
2 2 0	2 2 1	2 2 2

It is advisable to choose a high-order interaction of factors whose effect is not likely to be significant as the defining contrast.

An interaction effect of a 3^k factorial design is of the general form

$$AB^{a_2}C^{a_3}\cdots K^{a_k}$$

The defining relation for the 3^{k-1} fractional factorial design is obtained by making a choice between the possible set of treatment combinations that satisfy the following relation:

$$C = x_1 + a_2x_2 + \cdots + a_kx_k$$

C takes the values 0, 1, 2 (modulo 3), respectively, while $a_i = 1$ or 2, for $i = 1, \ldots, k-1$. C is the contrast of the factorial effect we have chosen. We can use all the treatment combinations that fall under $C = 0$, 1, or 2 to define the set of treatment combinations, which should appear in 3^{k-1} fractional factorial design. We consider 3^{3-1} as fractional factorial since 3^{2-1} fractional factorial is trivial (containing only three treatment combinations). Recall that we discussed the 3^3 factorial in Chapter 7 (it may be helpful to refer to the design before continuing). Suppose we choose the factorial effect AB^2C^2 as the defining contrast, then there are three sets of treatment combinations which could define the 1/3 fraction of this design (recall that we stated that although the components of the ABC factorial effect, namely, ABC, AB^2C, ABC^2, and AB^2C^2, do not appear in ANOVA, they are useful for confounding and fractional replication). In this case, $C = x_1 + 2x_2 + 2x_3$. There are nine treatments in each set, which represents the 1/3 fractional factorial of this design. The three possible 1/3 fractionals are given in Table 9.9.

If we had chosen another factorial effect AB^2C for the definition of the 3^{3-1} design, our choice would have been between three other 1/3 fractions of the design given in Table 9.10.

In the first fractional design, with the defining relation, $I = AB^2C^2$, we can write down the alias structure for the 3^{3-1} factorial design, which occurs by the choice of one of the three possible fractions. Each factorial effect, which is

TABLE 9.10: The 1/3 Fractional Designs of the 3^3 Factorial with $I = AB^2C$

Fraction I ($x_1 + 2x_2 + x_3 = 0$)	Fraction II ($x_1 + 2x_2 + x_3 = 1$)	Fraction III ($x_1 + 2x_2 + x_3 = 2$)
0 0 0	0 0 1	0 0 2
0 1 1	0 1 2	0 1 0
0 2 2	0 2 0	0 2 1
1 0 2	1 0 0	1 0 1
1 1 0	1 1 1	1 1 2
1 2 1	1 2 2	1 2 0
2 0 1	2 0 2	2 0 0
2 1 2	2 1 0	2 1 1
2 2 0	2 2 1	2 2 2

not in the defining relation of this design, is aliased with two other effects. The generalized interaction of this design is obtained as the square of the defining factorial effect. So, the defining relation is as follows:

$$I = AB^2C^2 = I^2 = (AB^2C^2)^2$$

Under this design, we obtain the alias structure. It is important to note that the first letter in any factorial effect is not allowed to have an exponent of more than one, otherwise, the whole factorial effect will need to be squared and the exponents reduced to modulo 3.

$$\begin{aligned} A &= A(AB^2C^2) = A(AB^2C^2)^2 = ABC = BC \\ B &= B(AB^2C^2) = B(AB^2C^2)^2 = AC^2 = ABC^2 \end{aligned}$$

Others are as follows:

$$\begin{aligned} C &= AB^2 = AB^2C \\ AB &= AC = BC^2 \end{aligned}$$

On the other hand, if we consider the alternative fractional design which has the following defining contrasts:

$$I = ABC = (ABC)^2$$

The alias structure would be as follows:

$$\begin{aligned} A &= AB^2C^2 = BC \\ B &= AB^2C = AC \\ C &= AB = ABC^2 \\ AB^2 &= AC^2 = BC^2 \end{aligned}$$

With the following defining relation

$$I = ABC^2 = (ABC^2)^2$$

the alias structure is as follows:

$$\begin{aligned} A &= AB^2C = BC^2 \\ B &= AB^2C^2 = AC^2 \\ C &= AB = ABC \\ AB^2 &= AC = BC \end{aligned}$$

Using the same principles for identifying the resolutions of the 2^k fractional designs, it is not difficult to identify the resolutions of the 3^k fractional factorial designs we have just constructed. We note that for all those designs, the main effects are aliased with two-factor interactions and higher, while the two-factor interactions are aliased with one another. We therefore conclude that they are all 3_{III}^{3-1} fractional factorial designs (designs of resolution III). Suppose we choose the AB^2C interaction with $C = x_1 + 2x_2 + x_3 = 0$, we can arrange this 1/3 fractional of the 3^3 factorial design in the form of a Latin-square design. Incidentally, this design, like the rest described above, is a 3_{III}^{3-1} fractional factorial. We assume that the rows represent A and the columns B; the interaction effects are assumed negligible in the design so that the main effects A, B, C are estimated according to the theory of Latin squares discussed in Chapter 5. The arrangement of this design in the form of Latin squares is shown below:

000	011	022
102	110	121
201	212	220

Anderson and McLean (1974) elaborate more on these special Latin squares for the fractionals of the 3^k factorial design.

Example 9.2

A 1/3 fraction of the 3^3 factorial design was used to study the growth of black-eye beans under different fertilizer treatments. At this stage, the experimenter was not interested in the combinations of levels of these fertilizer treatments and was therefore willing to assume that they were not significant and should be pooled together to estimate variance. The fertilizers were nitrogen (N), phosphorus (P), and potassium (K). Only nine different plants were grown on the same plot of land and received treatments based on the assumption that $I = AB^2C$ with the 1/3 fraction based on the choice of the contrast to the effect, $C = 0$. Yields of the plants in ounces are given below. Are the effects of the fertilizer treatments significant?

Treatments	Responses
0 0 0	28.20
0 1 1	35.10
0 2 2	41.90
1 1 0	34.80
1 2 1	33.60
1 0 2	29.30
2 0 1	31.40
2 1 2	42.30
2 2 0	44.90

The alias structure of this design is as follows:

$$\begin{aligned} I &= AB^2C = (AB^2C)^2 \\ A &= ABC^2 = BC^2 \\ B &= AC = ABC \\ C &= AB^2C^2 = AB^2 \\ AB &= AC^2 = B^2C^2 \end{aligned}$$

To analyze the responses, we shall use essentially the same method that we employed in dealing with the full 3^3 design, with a few modifications. We now calculate the total sum of squares, which is incidentally equal to the treatment sum of squares in this case as follows:

$$SS_{\text{total}} = SS_{\text{treatment}} = \frac{[(28.20)^2 + (35.10)^2 + \cdots + (44.90)^2 - (321.5)^2]}{9} = 287.92$$

We now list the treatment combinations according to the levels of factorial effects to calculate their sums of squares.

Under A (with contrast x_1)	$x_1 = 0$	$x_1 = 1$	$x_1 = 2$
	0 0 0	1 0 2	2 0 1
	0 1 1	1 1 0	2 1 0
	0 2 2	1 2 1	2 2 0
Treatment total	105.2	97.7	118.6

Under B (with contrast x_2)	$x_2 = 0$	$x_2 = 1$	$x_2 = 2$
	0 0 0	0 1 1	0 2 2
	1 0 2	1 1 0	1 2 1
	2 0 1	2 1 2	2 2 0
Treatment total	88.9	112.2	120.4

Under AB (with contrast $x_1 + x_2$)	$x_1 + x_2 = 0$	$x_1 + x_2 = 1$	$x_1 + x_2 = 2$
	0 0 0	0 1 1	0 2 2
	1 2 1	1 0 2	1 1 0
	2 1 2	2 2 0	2 0 1
Treatment total	104.1	109.3	108.1

Under C (with contrast x_3)	$x_3=0$	$x_3=1$	$x_3=2$
	0 0 0	0 1 1	1 0 2
	1 1 0	1 2 1	0 2 2
	2 2 0	2 0 1	2 1 2
Treatment total	107.9	100.1	113.5

$$SS_B = [(88.9)^2 + (112.2)^2 + (120.4)^2]/3 - (321.5)^2/9 = 178.0422$$
$$SS_A = [(105.2)^2 + (97.7)^2 + (118.6)^2]/3 - (321.5)^2/9 = 74.7356$$
$$SS_{AB} = [(104.1)^2 + (109.3)^2 + (108.1)^2]/3 - (321.5)^2/9 = 4.9422$$
$$SS_C = [(107.9)^2 + (100.1)^2 + (113.5)^2]/3 - (321.5)^2/9 = 30.1956$$

Since we are to use interaction of effects to estimate error and because of the alias structure, we disregard higher order interactions in the design and use AB interaction to estimate error, then we present the ANOVA for this problem (Table 9.11).

Because $F(2,2,0.05)=19.0$, only B is significant. If $\alpha=0.10$, which is very high, A could be considered significant at that level since $F(2,2,0.10)=9.0$. Collapsing the data into a single replicate of the 3^2 factorial design (in A and B only) and using the AB interaction as a measure of the error in the design clearly show that only B is significant and A fails at this stage to be significant at $\alpha=0.10$. This check is left as an exercise for the reader. □

9.5.1 One-Third Fraction of the 3^4 Factorial Design

In the 3^4 factorial design, there are 8 four-factor interactions; any of them could be used to define the 1/3 fractional design in which the main effects are aliased with three-factor interactions, while some of the two-factor interactions are aliased with one another. Suppose we define such a 1/3 fractional design with the defining relation, $I = ABCD$, then we can write down some of the aliases of this design as follows:

$$I = ABCD = (ABCD)^2$$
$$A = AB^2C^2D^2 = BCD$$
$$B = AB^2CD = ACD$$
$$AB = ABC^2D^2 = ABD$$
$$C = ABC^2D = ABD$$
$$AC = AB^2CD^2 = BD$$

TABLE 9.11: ANOVA for 3^{3-1} Fractional Factorial (Example 9.2)

Source	**Sum of squares**	**Degrees of freedom**	**Mean square**	***F*-ratio**
A	74.7356	2	37.368	15.3
B	178.042	2	89.021	36.03
C	30.1956	2	15.098	6.11
AB (residual)	4.9422	2	2.4711	

It can be seen that for this design, the main effects are aliased with three- and four-factor interactions, whereas the two-factor interactions are aliased with two-factor interactions. This obviously fits our definition of resolution IV design, indicating that this is a 3_{IV}^{4-1} fractional factorial design.

9.6 General 3^{k-p} Factorial Design

In Section 9.5, we concentrated on the 1/3 fraction of the 3^k factorial design. There are other fractions of this design, which are of importance apart from the 1/3 fraction; notable among these are the 1/9 and 1/27 fractions. Normally they would be denoted as 3^{k-2} and 3^{k-3} factorial designs. These are part of the general 3^{k-p} designs that could be constructed. To construct any of these designs, we need to write down the entire treatment combinations in full 3^{k-p} designs (in a manner similar to the construction of the 2^{k-p} fractional factorial) and then associate with appropriate interactions, the missing p factors to provide generators for the design. Suppose we wish to construct the 3^{5-2} factorial design (in A, B, C, D, E), we first write down the factorial effects in the full 3^3 factorial design (in A, B, C) and then proceed to associate with two interactions in the design, D and E. Since we have presented the full 3^3 factorial design in Chapter 7, we refrain from doing so here. The reader is advised to refer to it before reading further to follow the next steps in this study. Out of all the interactions in the 3^3 factorial design, if we choose AB^2C and ABC^2, then the generators of the 3^{5-2} fractional factorial design would be $D = AB^2C$ and $E = ABC^2$.

The generalized interactions of these effects are $AB^2CD \cdot ABC^2E = AD^2E^2$ and $ABCD\ (ABC^2E) = ACD^2E$; so that the defining relation for this design is $I = AB^2CD = ABC^2E = AD^2E^2 = ACD^2E$. The alias of any effect in this model can be found by multiplying I by that effect and by reducing the resulting indices of factors (modulo 3). We can obtain the alias structure for the main effects of this design as follows:

$$\begin{aligned}
A &= ABC^2D^2 = AB^2CE = ADE = AC^2DE \\
B &= ACD = AB^2C^2E = ABD^2E^2 = ABCD^2E \\
C &= AB^2C^2D = ABE = ACD^2E^2 = AC^2D^2E \\
D &= AB^2CD^2 = ABC^2DE = AE^2 = ACE \\
E &= AB^2CDE = ABC^2E^2 = AD^2 = ACD^2E^2
\end{aligned}$$

This is a 3_{III}^{5-2} fractional factorial. There are good texts on fractionals of factorial designs at three levels. Cox (1958) and, particularly, Connor and Zelen (1959) present large numbers of fractional factorials from the 3^k factorial designs. Further readings on fractional factorial designs at two and three levels, Plackett and Burman, and other fractional factorial designs can be found in

Box et al. (2005) and Montgomery (2005). Other publications can be found in Box and Hunter (1961a,b).

9.7 Confounding in 2^k and 3^k Factorial Designs

Sometimes, due to constraints on resources, it is not feasible to perform all the experiments in a full replicate of a factorial design under homogeneous conditions. In agriculture, experiments to study the effects of fertilizer applications on the yield of a crop may be such that different combinations of the fertilizer treatments cannot be applied to it on a single homogeneous plot of land; another plot whose characteristics are different has to be used. In experiments in the laboratory, resources may be such that only part of the full replicate of the experiment could be performed in a single laboratory; another laboratory with varying conditions may be used for the rest of the experiments in the full replicate of the design. On the other hand, the full replicate of the experiment could be carried out in the same laboratory with different shifts of workers with varying skills. Whenever there is a factor, which places constraints on an investigator's ability to perform a full replicate of a factorial design under homogeneous conditions, it is advisable to use a design technique called confounding. This ensures that whatever conclusions are reached (on the effects of the factors) are obtained under as homogeneous a condition as possible. Confounding is a technique developed to improve the efficiency of factorial designs when we cannot run the full replicate under homogeneous conditions. Blocking (complete and incomplete) is generally used in nonfactorial experiments, dealing with nonhomogeneity of experimental conditions, as we have discussed in the previous chapters. What incomplete block experiments are to nonfactorial experiments is what confounding is to factorial experiments. Confounding is a special technique for dividing a full replicate of a factorial experiment into blocks, where these blocks are of sizes less than the full replicate of the experiment.

Once confounding is used this way, information on one of the factorial effects is sacrificed while this effect is used to define the blocks. The factorial effect so used is called the defining contrast. Confounding makes the block contrasts to be the same as the defining contrast so that they both are indistinguishable from each other. Information on the factorial effect, which is used in defining the blocks, is given up to increase precision in the experiment by using smaller blocks, which are likely to be more homogeneous than the whole aggregate of experimental material. It is also necessary that if we wish to sacrifice information on a factorial effect, it should be a factorial effect that is not likely to be significant in the design. The practice, born out of experience, is that the highest order interaction is used, as it is rarely significant in factorial experiments. When we are willing to fully sacrifice information on the effect whose contrasts we use in the definition of blocks, then the factorial effect is said to be fully confounded with blocks (recall our definition of "confounded" in Chapter 2). Sometimes, we do not wish to sacrifice the information on any factorial effect in the design. In such circumstances, we use *partial* confounding. This technique enables us to recover some information on factorial effects that is used to define blocks. In Section 9.10 we shall treat partial confounding in full.

It is possible to divide a full replicate of a design in such a way that we assign specific fractions of the design in each block. For the 2^k factorial design, popular fractions assigned to blocks include $(1/2)^p$ for $p=1,2,3,\ldots,k$ with $p<k$. For the 3^k factorial designs, the fractions which are frequently assigned to blocks are the $(1/3)^p$ factorial for $p=1$, 2, 3 $(p<k)$. We discuss these designs in the following sections.

9.8 Confounding in 2^k Factorial Designs

The 2^k factorial designs, by their nature, are amenable to divisions into 2^p blocks, each with 2^{k-p} treatment combinations in it, provided that $p<k$. The smallest factorial design in this group, 2^2 (with factors A and B), can easily be divided into two blocks. We can use the contrasts of this factorial design to assign treatments to blocks. The typical 2^2 design is given in Table 9.12.

If we choose AB as the defining contrast for the blocks or the factorial effect to be confounded with blocks, we can divide the treatments into two blocks by assigning those treatments with the − sign to one block and those with the + sign to another block. Each block in this design will contain two treatment combinations. The blocks obtained are as follows in Table 9.13:

If the same exercise is carried out for a 2^3 factorial design in A, B, C, while using ABC interaction as the defining contrast for the blocks (we have presented the full contrasts of 2^2 and 2^3 factorial designs here for convenience; they were previously presented in Chapter 7), we obtain the blocks by assigning those treatment combinations with − signs under the ABC contrasts to one block while those with + signs go to another block (Tables 9.14 and 9.15).

TABLE 9.12: The 2^2 Factorial Design in A and B

Treatment combinations	Factorial effects contrasts			
	I	A	B	AB
(1)	+	−	−	+
a	+	+	−	−
b	+	−	+	−
ab	+	+	+	+

TABLE 9.13: Typical 2^2 Factorial Design in Two Blocks

Block I	Block II
(1)	a
ab	b

TABLE 9.14: Typical 2^3 Factorial Design in A, B, and C

Treatment combinations	Factorial effects contrasts							
	I	A	B	AB	C	AC	BC	ABC
(1)	+	−	−	+	−	+	+	−
a	+	+	−	−	−	−	+	+
b	+	−	+	−	−	+	−	+
ab	+	+	+	+	−	−	−	−
c	+	−	−	+	+	−	−	+
ac	+	+	−	−	+	+	−	−
bc	+	−	+	−	+	−	+	−
abc	+	+	+	+	+	+	+	+

TABLE 9.15: Typical 2^3 Factorial Design in Two Blocks

Block I	Block II
(1)	a
ab	b
ac	c
bc	abc

From the foregoing assignments of treatments to blocks, we see that there are two treatment combinations in each block when the 2^2 factorial design is assigned into blocks, while there are four in each block for a similar exercise with the 2^3 factorial design. There are other ways of obtaining these blocks. One method, which relies on digital notation (which we employed in Chapter 7), was used by Kempthorne (1952), Hicks (1973), Anderson and McLean (1974), Montgomery (1991), and Petersen (1985), in representing treatment combinations in a factorial design. Essentially, this method uses the modulo arithmetic to identify the treatment combinations in a factorial design, which should be assigned to a particular block. In a 2^k factorial design, each factor can have levels of either zero (low) or one (high). We recall from Chapter 7 that treatment combinations in a typical 2^4 factorial design with factors A, B, C, and D could be represented digitally as follows:

Notations for treatment combinations

Treatment combination	Digital	Treatment combination	Digital
(1)	0 0 0 0	d	0 0 0 1
a	1 0 0 0	ad	1 0 0 1
b	0 1 0 0	bd	0 1 0 1
ab	1 1 0 0	abd	1 1 0 1
c	0 0 1 0	cd	0 0 1 1
ac	1 0 1 0	acd	1 0 1 1
bc	0 1 1 0	bcd	0 1 1 1
abc	1 1 1 0	$abcd$	1 1 1 1

Suppose we wish to divide the treatments in a full replicate of the 2^4 factorial into two blocks, let us represent the digital notation of the levels of the factors A, B, C, and D by x_1, x_2, x_3, and x_4, respectively. It is clear that x's can assume values 0 or 1 only. If we write the linear combinations of the levels of the factors for each treatment combination, appearing in the design, to obtain their values (modulo 2), it is clear that the treatments will be divided into two; those with $C' = 0$ or $C' = 1$ as follows:

$$C' = a_1x_1 + a_2x_2 + a_3x_3 + a_4x_4 \text{ (modulo 2)} \qquad (9.1)$$

Since the a's indicate the presence or absence of a particular factor in the defining contrast, they can only take values 0 or 1, according to whether the factor is absent or present. In the 2^4 factorial design, if we choose $ABCD$ as the defining contrast to obtain the two blocks, then every factor will be present in the defining contrast and a's will all be equal to 1. Under this defining contrast, the blocks are assigned to the treatment combinations in the 2^4 factorial design as follows (using Equation 9.1):

$$
\begin{aligned}
(1){:}C' &= 1(0) + 1(0) + 1(0) + 1(0) = 0 \text{ modulo } 2 = 0\\
a{:}C' &= 1(1) + 1(0) + 1(0) + 1(0) = 1 \text{ modulo } 2 = 1\\
b{:}C' &= 1(0) + 1(1) + 1(0) + 1(0) = 1 \text{ modulo } 2 = 1\\
ab{:}C' &= 1(1) + 1(1) + 1(0) + 1(0) = 2 \text{ modulo } 2 = 0\\
c{:}C' &= 1(0) + 1(0) + 1(1) + 1(0) = 1 \text{ modulo } 2 = 1\\
ac{:}C' &= 1(1) + 1(0) + 1(1) + 1(0) = 2 \text{ modulo } 2 = 0\\
bc{:}C' &= 1(0) + 1(1) + 1(1) + 1(0) = 2 \text{ modulo } 2 = 0\\
abc{:}C' &= 1(1) + 1(1) + 1(1) + 1(0) = 3 \text{ modulo } 2 = 1\\
d{:}C' &= 1(0) + 1(0) + 1(0) + 1(1) = 1 \text{ modulo } 2 = 1\\
ad{:}C' &= 1(1) + 1(0) + 1(0) + 1(1) = 2 \text{ modulo } 2 = 0\\
bd{:}C' &= 1(1) + 1(0) + 1(0) + 1(0) = 1 \text{ modulo } 2 = 0\\
abd{:}C' &= 1(1) + 1(1) + 1(0) + 1(1) = 3 \text{ modulo } 2 = 1\\
cd{:}C' &= 1(0) + 1(0) + 1(1) + 1(1) = 2 \text{ modulo } 2 = 0\\
acd{:}C' &= 1(1) + 1(0) + 1(1) + 1(1) = 3 \text{ modulo } 2 = 1\\
bcd{:}C' &= 1(0) + 1(1) + 1(1) + 1(1) = 3 \text{ modulo } 2 = 1\\
abcd{:}C' &= 1(1) + 1(1) + 1(1) + 1(1) = 4 \text{ modulo } 2 = 0
\end{aligned}
$$

The following treatments, *abcd*, *cd*, *bd*, *ad*, *ab*, (1), *ac*, *bc* have $C' = 0$; while *bcd*, *acd*, *abd*, *d*, *abc*, *c*, *b*, *a*, have $C' = 1$. Both sets form the two blocks of the 2^4 factorial design as given in Table 9.16.

There are eight treatment combinations in each block. As a check that this method yields the same result, we recall that we can calculate the contrasts

TABLE 9.16: Typical 2^4 Factorial Design in Two Blocks

Contrast levels (blocks)	
0/I	**1/II**
(1)	a
ab	b
ac	c
ad	abc
bc	d
bd	abd
cd	acd
$abcd$	bcd

of $ABCD$ in the 2^4 factorial design as

$$\begin{aligned} ABCD &= (a-1)(b-1)(c-1)(d-1) \\ &= abcd + cd + bd + ad + bc + ac + ab \\ &\quad + (1) - bcd - acd - abd - d - abc - c - b - a \end{aligned}$$

Under the first method of assignment of treatment combinations to blocks, we would have assigned all those with − signs to one block and those with + signs to another. Clearly, the contrasts under this method indicate that the two methods discussed above are the same because they yield the same results. By this design, $ABCD$ is confounded with blocks. It is difficult, therefore, to distinguish between the block effects and the factorial effects, $ABCD$. In analyzing the results of these experiments, especially when sequential fitting with computers is adopted, it is advisable to fit the block effect first, before examining it to see how effective the blocking has been. It is obvious that the experimenter would wish to perform the ANOVA to determine whether some effects are significant. There are 15 degrees of freedom; one is used to estimate the block effect while the other 14 are taken up in the estimation of the rest of the factorial effects, leaving no degrees of freedom for the estimation of experimental error. This means that the experiments will need to be replicated to provide additional degrees of freedom for the estimation of error. If the design were to be replicated three times, for instance, we would have six blocks as follows (Table 9.17).

To analyze the responses from the three replicates of this design, we realize that we have 47 degrees of freedom; 14 are used to estimate the factorial effects, A, B, AB, C, AC, BC, ABC, D, AD, BD, ABD, CD, ACD, BCD, each with 1 degree of freedom. Five degrees of freedom are assigned to the blocks. In Cochran and Cox (1957), it is suggested that the 5 block degrees of freedom could be subdivided into 2 degrees of freedom for the replicates, 1 for the block effect ($ABCD$) and 2 for error

TABLE 9.17: Assignment of Three Replicates of the 2^4 Factorial Design into Six Blocks

Replicate I		Replicate II		Replicate III	
Block I	**Block II**	**Block III**	**Block IV**	**Block V**	**Block VI**
(1)	a	(1)	a	(1)	a
ab	b	ab	b	ab	b
ac	c	ac	c	ac	c
bc	abc	bc	abc	bc	abc
ad	d	ad	d	ad	d
bd	abd	bd	abd	bd	abd
cd	acd	cd	acd	cd	acd
$abcd$	bcd	$abcd$	bcd	$abcd$	bcd

TABLE 9.18: Typical Factorial Effects in the ANOVA for the 2^4 Factorial Design with Three Replicates in Six Blocks

Source	Degrees of Freedom
Replicates	2
Blocks ($ABCD$)	1
Error($ABCD \times$ Replicates)	2
A	1
B	1
AB	1
C	1
AC	1
BC	1
ABC	1
D	1
AD	1
BD	1
ABD	1
CD	1
ACD	1
BCD	1
Residual (replicates $\times$ effects)	28

associated with $ABCD$. Further, the error for $ABCD$ represents interaction between blocks and $ABCD$. The rest of the 28 degrees of freedom are assigned to residual effects, which comprise of the interactions between replicates and factorial effects. With these discussions of the subdivisions of the total sum of squares, the ANOVA for this design is presented in Table 9.18.

We illustrate confounding in 2^k factorial designs with an application.

Example 9.3

Three factors A (temperature), B (pressure), and C (catalyst concentration) are considered to influence the yield of a chemical reaction. These factors were all set at two levels. Because of the limitations in the laboratory, the 2^3 factorial design that was set up had to be run in two blocks of size four; ABC was used to define the blocks. Three replicates of the experiments were performed. The responses are as follows:

Replicate I		Replicate II		Replicate III	
Block I	**Block II**	**Block III**	**Block IV**	**Block V**	**Block VI**
(1) 4.6	*a* 10.1	(1) 4.4	*a* 7.8	(1) 4.8	*a* 9.2
ab 6.6	*b* 4.6	*ab* 6.3	b 5.9	*ab* 6.3	*b* 6.2
ac 6.2	*c* 7.6	*ac* 7.0	*c* 6.9	*ac* 7.5	*c* 8.0
bc 8.1	*abc* 7.7	*bc* 8.9	*abc* 8.2	*bc* 9.0	*abc* 8.2

We can use any of the methods for calculating contrasts of an effect to obtain the estimates of the effects in the design. We arrange the treatment combinations in a standard order and estimate the effects as usual (using the sum of all responses), disregarding the blocks as given in Table 9.19.

TABLE 9.19: Yates' Algorithm for Example 9.3

Treatment combinations	Responses	I	II	III	Estimates	Sum of squares
(1)	13.8	40.9	76.8	170.1	7.087	
a	27.1	35.9	93.3	12.3	1.025	6.3037
b	16.7	43.2	15.8	1.9	0.158	0.1504
ab	19.2	50.1	−3.5	−10.7	−0.892	4.7704
c	22.5	13.3	−5.0	16.5	1.375	11.3437
ac	20.7	2.5	6.9	−19.3	−1.608	15.5204
bc	26.0	−1.8	−10.8	11.9	0.992	5.9000
abc	24.1	−1.7	0.1	10.9	0.908	4.9504

There are three replicates with sums of responses of 55.5, 55.4, and 59.2. There are also six blocks with sums of responses of 25.5, 30, 26.6, 28.8, 27.6, and 31.6.

$$SS_{\text{total}} = [(4.6)^2 + (6.6)^2 + \cdots + (8.0)^2 + (8.2)^2] - \frac{(170.1)^2}{24} = 55.4663$$

For the blocks, we calculate the following sums of squares as

$$\begin{aligned} SS_{\text{replicates}} &= (1/8)[(55.5)^2 + (55.4)^2 + (59.6)^2] - (170.1)^2/24 = 1.1725 \\ SS_{\text{block(total)}} &= (1/4)[(25.5)^2 + (30)^2 + (26.6)^2 + (28.8)^2 + (27.6)^2 + (31.6)^2] \\ &\quad - (170.1)^2/24 = 6.3087 \end{aligned}$$

From the tables above, $SS_{ABC} = 4.9504$, so that $SS_{\text{E(ABC)}} = 6.3087 - 1.1725 - 4.9504 = 0.1858$.

$$\begin{aligned} SS_{\text{residual}} &= SS_{\text{total}} - SS_{\text{block(total)}} - SS_A - SS_B - \cdots - SS_{BC} \\ &= 55.4663 - 6.3087 - 6.3037 - \cdots - 5.900 = 5.1690. \end{aligned}$$

TABLE 9.20: ANOVA for Example 9.3

Source	Sum of squares	Degrees of freedom	Mean square	F-ratio
Replicates	1.1725	2	0.5863	1.361
Blocks (ABC)	4.9504	1	4.9504	11.4925
Error (replicates × blocks)	0.1858	2	0.0929	0.2157
A	6.3087	1	6.3037	14.6342
B	0.1504	1	0.1504	0.3492
AB	4.7704	1	4.7704	11.0746
C	11.3437	1	11.3437	26.3348
AC	15.5204	1	15.5204	36.0311
BC	5.9000	1	5.9000	13.6970
Residuals (replicates × effects)	5.1690	12	0.43075	13.6970
Total	55.4663	23		

Using the responses, we set up the ANOVA table for Example 9.3 (see Table 9.20).

Because $F(2, 12, 0.01) = 6.93$ and $F(1, 12, 0.01) = 9.33$, A, C, AB, AC, BC are significant. Blocking has also been effective since the blocking effect is significant. □

9.8.1 Blocking a Single Replicate of the 2^k Factorial Design

In Chapter 7, we discussed the designs in which constraints make it impossible to obtain more than one replicate of the designed experiment. It is still possible that for such designs, an extra constraint would necessitate the running of this single replicate of the experiment in a number of blocks. When it happens, confounding is used to ensure precision. As an example, suppose that only half of one replicate of an industrial experiment could be performed by one shift of workers, the other shift completes the other half. The shifts represent blocks whose contrasts could be confounded with the highest order interaction in the design as we had earlier explained. Other examples could be found where it is impossible to run a single replicate of an experiment under homogeneous conditions. The experiments are performed with the highest order interaction, confounded with blocks. In addition, if we know any other interactions of factors whose effects may not be significant, we could combine their sums of squares and use them to estimate the error sum of squares to be used in ANOVA. We illustrate this with an example.

Example 9.4

Recall Example 7.3. A study was carried out on the stability of a vehicle in a tunnel, and the factors that were believed would have effect on the stability of the vehicle were A, vehicle speed; B, air speed; C, level of embankment of the track; and D, load of the vehicle. Only a single replicate of this experiment had been run; however, on serious thought, the experimenter felt that one factor not simulated in the experiment is the presence of ice. Presence of ice or snow on the tracks was introduced as the fifth factor in the design, and a single replicate of the resulting 2^5 factorial design was obtained in two runs using *ABCDE* to define the blocks. The blocks as well as the responses of the experiment (percentage instability) are given in the table as follows:

Block I	**Block II**
(1) 38	*a* 40
ab 30	*b* 27
ac 75	*c* 58
bc 30	*abc* 32
ad 62	*d* 59
bd 53	*bcd* 53
cd 79	*acd* 75
abcd 54	*abd* 50
ae 54	*e* 45
be 34	*abe* 42
ce 65	*ace* 70
abce 40	*bce* 42
de 64	*ade* 74
abde 58	*bde* 48
acde 80	*cde* 75
abcde 60	*bcde* 61

Again, Yates' algorithm was used to analyze the data, making sure that the data are arranged in standard order regardless of the block from which they come. The analysis is presented in Table 9.21.

The experimenters assumed that all the third-order interactions are unlikely to be significant; their sums of squares were, in accordance with this assumption, pooled together to estimate the error sum of squares with 5 degrees of freedom that is shown as follows:

$$SS_{ABCD} + SS_{ABDE} + SS_{ABCE} + SS_{ACDE} + SS_{BCDE}$$
$$= 22.78 + 3.78 + 1.53 + 7.03 + 1.53 = 36.65$$

From the tables for F-distribution, $F(1, 5, 0.01) = 16.26$ and $F(1, 5, 0.05) = 6.61$. Effects with $**$ are significant at $\alpha = 0.01$ while those with $*$ are significant at $\alpha = 0.05$. Incidentally, the block effect (*ABCDE*) turned out to have a low sum of squares. It does not necessarily mean that blocking should have been abandoned. It is usually advisable to block when there is the likelihood that the experiments are not being performed under homogeneous conditions. The block effect may or may not turn out to be large, but the experimenter has insured against error occurring in the design of the experiment and increased the precision of his or her conclusions. ∎

TABLE 9.21: Yates' Algorithm for the Single Replicate of the 2^5 Factorial Design in Two Blocks

Treatment combinations	Responses	I	II	III	IV	V	Effects	Sum of squares
(1)	38	78	135	330	815	1727		
a	40	57	195	485	912	65	*A*	132.03**
b	27	133	224	392	21	299	*B*	2793.78**
ab	30	62	261	520	44	−29	*AB*	26.28
c	58	121	175	24	157	171	*C*	913.78**
ac	75	103	217	−3	142	−19	*AC*	11.28
bc	30	154	244	20	−15	−111	*BC*	385.03**
abc	32	107	276	24	−14	−17	*ABC*	9.03
d	59	99	5	−92	97	283	*D*	2502.78**
ad	62	76	19	−65	74	−23	*AD*	16.53
abd	50	82	−3	−66	−30	15	*ABD*	7.03
cd	79	138	17	−14	−79	−33	*CD*	34.03
acd	75	106	3	−1	−32	−19	*ACD*	11.28
bcd	53	155	20	−8	−5	49	*BCD*	75.03*
abcd	54	121	4	−6	−12	27	*ABCD*	22.78*
e	45	2	−21	60	155	97	*E*	294.03**
ae	54	3	−71	37	128	23	*AE*	16.53
be	34	17	−18	42	−27	15	*BE*	7.03
abe	42	2	−47	32	4	−1	*ABE*	0.03
ce	65	3	−23	14	27	−23	*CE*	16.53
ace	70	−3	−53	−3	10	−41	*ACE*	52.53
bce	42	−4	−32	−14	13	47	*BCE*	69.03**
abce	40	1	−34	−16	2	−7	*ABCE*	1.53
de	64	9	1	−50	−23	−27	*DE*	22.78
ade	74	8	−15	−29	−10	31	*ADE*	30.03
bde	48	5	−6	−30	−17	−17	*BDE*	9.03
abde	58	−2	5	−2	−2	−11	*ABDE*	3.78
cde	75	10	−1	−16	21	13	*CDE*	5.28
acde	80	10	−7	11	28	15	*ACDE*	7.03
bcde	61	5	0	−6	27	7	*BCDE*	1.53
abcde (blocks)	60	−1	−6	−6	0	−27	*ABCDE*	22.78

9.8.2 Blocking Fractionals of the 2^k Factorial Design

Sometimes, the number of treatments in a fractional replication of a factorial design is so large that it is difficult to run all the treatment combinations under homogeneous conditions. Confounding is frequently used in such designs to increase precision. The two methods that we described earlier could be applied to fractional factorial designs to divide the treatment combinations into blocks. To obtain two blocks from a fractional replication of 2^k factorial design, we could use the linear combination of contrasts of the factorial effect, which is to be confounded with blocks, or we could write the chosen effect and the contrasts of the treatment combinations (in terms of their signs) under the factorial effect. In the former method, we assign the treatment with the linear combinations of the levels of the factors under the effect equal to zero (modulo 2), to one block, while those with the values of the linear combinations of

TABLE 9.22: Contrasts of ABD for Assigning 2^{6-2} Factorial Design into Two Blocks

Treatment combinations	Contrasts	
	ABD	C'
(1)	−	0
ae	+	1
aef	+	1
abf	−	0
cef	−	0
acf	+	1
bc	+	1
abce	−	0
df	+	1
adef	−	0
bde	−	0
abd	+	1
cd	+	1
acd	−	0
bcdf	−	0
abcdef	+	1

the levels of factors in the effect equal to one (also modulo 2) are assigned to a different block. In the latter method, we assign the treatment combinations with + signs to one block and those with − signs to a different block. Consider as an illustration, the 2^{6-2} fractional factorial design (Table 9.8). The defining relation for this design is $I = ABCE = BCDF = ADEF$.

We could choose the factorial contrast ABD to be confounded with blocks. In doing so, we have chosen all the aliases of this factorial effect in the design to be confounded with blocks, namely, $ABD = CDE = ACF = BEF$. If we represent the linear combinations of the levels of the factors in the factorial effect ABD as $C' = x_1 + x_2 + x_4$ (modulo 2), we can calculate this for all the treatment combinations, which appear in this design. Further, we use the second method to obtain the contrast of the treatment combinations under ABD. Under the two methods described, the design is given in Table 9.22.

The blocks resulting from dividing the 2^{6-2} fractional factorial design into blocks, using ABD to define the blocks, are presented as I and II as follows:

$$\text{I} \quad (1), abf, cef, abce, adef, bde, acd, bcdf$$
$$\text{II} \quad ae, aef, acf, bc, df, abd, cd, abcdef$$

9.8.3 2^k Factorial Design in 2^p Blocks

We have treated the blocking of 2^k factorial design in 2^p blocks for $p = 1$; naturally, we next consider the case when $p = 2$. When 2^k factorial designs are divided into 2^p blocks, each column contains 2^{k-p} treatment combinations. When $p = 2$, there are usually 2^{k-2} treatment combinations in the 2^2 or four blocks. Experimental materials do not often come in large homogeneous quantities. Blocking the 2^k factorial experiment for $k \geq 4$ is very useful since it reduces the sizes of blocks and enhances precision. Suppose we decide to assign treatments in 2^6 factorial design in A, B, C, D, E, and F into four blocks, then each block will contain 16 treatments. If we define the blocks using ABC and DEF for instance, then apart from ABC and DEF, $ABCDEF$ is also confounded with the blocks since it is the generalized interaction of the defining contrasts. Table 9.23 presents the assignment of all the treatment combinations in this design into blocks. We could use the linear combinations of contrasts of ABC and DEF, that is, C' and C'' [$C' = x_1 + x_2 + x_3$ (modulo 2); $C'' = x_4 + x_5 + x_6$ (modulo 2)] for assigning the treatment combinations to the blocks. There are four permutations of the contrasts of both factorial effects that arise as [(0 0), (1 0), (0 1), and (1 1)]. Treatments that fall under each of these permutations of contrasts are assigned to one of the blocks. These four permutations of contrasts of the treatments under ABC and DEF define the blocks. Similarly, if we were to use the sign contrasts, there are four permutations of the treatments under ABC and DEF, which define the blocks as [(−−), (+−), (−+), (++)]. Each of these contrast permutations for the treatments under these two factorial effects defines the blocks. The results are the same regardless of which method is used. Each of the methods will yield the blocks in Table 9.23:

TABLE 9.23: Assignment of 2^6 Factorial Design into Four Blocks

Contrasts				
ABC	***DEF***	***ABC***	***DEF***	**Blocks**
0	0	−	−	*(1), ab, ac, bc, de, abde, acde, bcde, df, abdf, acdf, bcdf, bcef, ef, abef, acef*
0	1	−	+	*d, abd, acd, bcd, e, abe, ace, bce, f, abf, acf, bcf, def, abdef, acdef, bcdef*
1	0	+	−	*a, b, c, abc, ade, bde, cde, abcde, adf, bdf, cdf, abcdf, aef, bef, cef, abcef*
1	1	+	+	*ad, bd, cd, abcd, ae, be, ce, af, bf, cf, abcf, adef, abce bdef, cdef, abcdef*

The principle used in assigning treatment combinations in the 2^k factorial design to two and four blocks can be used to divide the same treatment combinations into 2^p blocks as long as p does not exceed $k - 1$. The principle involves the choice of p factorial effects to be confounded with blocks. The p effects and their generalized interactions are all confounded with blocks. We then use any of the two methods we have previously described to assign the treatment combinations into blocks. For instance, the 2^5 factorial design in A, B, C, D, and E can be divided into eight blocks by choosing three factorial

TABLE 9.24: Assignment of 2^5 Factorial Design into Eight Blocks

	Contrasts			
S. No.	ABC	BCE	CDE	**Blocks**
(1)	−	−	−	(1), *bcd, ace, abde*
(2)	+	−	−	*a, abcd, ce, bde*
(3)	+	+	−	*b, cd, abce, ade*
(4)	−	+	−	*ab, acd, bce, de*
(5)	+	+	+	*c, bd, ae, abcde*
(6)	−	+	+	*ac, abd, e, bcde*
(7)	−	−	+	*bc, d, abe, acde*
(8)	+	−	+	*abc, ad, be, cde*

effects to be confounded with blocks. If we choose ABC, BCE, and CDE, then in addition to them, their generalized interactions, ACD, AE, BD, and $ABDE$ are all confounded with blocks. There are exactly $2^p - p - 1$ factorial effects in the generalized interaction of p factorial effects of the 2^k factorial design. Using the contrasts of these defining effects, the eight blocks are given in Table 9.24.

In all these designs that we have discussed, the block with treatment (1) (which is the same as all factors set at low level) is regarded as the *principal block*. This block has an important group theoretical property: it can be used to generate all the other blocks in the design. The treatment combinations in any block of the 2^5 factorial design, which we have just assigned to blocks could be obtained from the principal block:

[(1), *bcd, ace, abde*]

This is achieved by multiplying the principal block by one treatment in the desired block. Also, when any member of the principal block, other than (1), is multiplied by another, the result is another member of the principal block. All multiplications must be subject to the exponents of the letter in treatment combinations, being reduced by modulo arithmetic (modulo 2 in this case), so that $a^3 = a$ (modulo 2) and $a^2 = (1)$ (modulo 2). Applying this to the design, we generate block two of Table 9.24 as follows:

$$\begin{aligned} a(1) &= a \\ a(bcd) &= abcd \\ a(ace) &= a^2ce = ce \\ a(abde) &= a^2bde = bde \end{aligned}$$

Similarly, we generate block three as follows:

$$\begin{aligned} b(1) &= b \\ b(bcd) &= b^2cd = cd \\ b(ace) &= abce \\ b(abde) &= ab^2de = ade \end{aligned}$$

Further, within the principal block are $ace\,(bcd) = abde$, and $abde\,(ace) = bcd$ both of which are also members of the principal block confirming the group theoretical properties we stated earlier.

The blocks obtained by this method agree with those obtained by other methods, which are displayed in Table 9.24. This *group theoretical property* (exhibited here by the principal block) was first mentioned in the earlier parts of Chapter 7 and reappeared in this chapter (Section 9.3) in our discussion of fractional factorial designs. This property can be used for two purposes: (1) to check the accuracy of the blocks obtained by other methods and (2) to generate the blocks. Once the principal block has been identified, it is easy to generate the subsequent blocks by choosing any treatment combination that is not in the principal block and using it to multiply every member of the principal block to obtain the next block. Another treatment combination that did not occur in the two blocks so far obtained is chosen and used to multiply every member of the principal block to obtain the next block. This process is continued until all the blocks are obtained.

9.9 Confounding in 3^k Factorial Designs

Just as it is sometimes necessary to assign treatments in 2^k factorial designs into blocks containing less than one replicate of the design, it is similarly necessary at times to block the 3^k factorial design. The properties of the 3^k factorial design makes it easy for it to be divided into 3^p blocks, with 3^{k-p} treatment combinations in each block, provided that $p < k$. In the Sections 7.5–7.6, we mentioned in the 3^k factorial designs that two-, three-, ..., and k-factor interactions contain other components to which they could be decomposed. We noted that although these interactions of 3^k factorial designs could be decomposed in this way, they are rarely tested in ANOVA but find their uses in fractional replication and confounding. For instance, the ABC interaction contains the components ABC, AB^2C, ABC^2, and AB^2C^2. Each of them could be chosen and used to split a 3^k factorial design into three blocks (we have demonstrated the use of these in fractional replication of the 3^k factorial designs, which we discussed in the preceding sections of this chapter). Since each of them has 2 degrees of freedom, if one of these factorial effects is confounded with blocks, when each replicate of the 3^k factorial design is to be broken into three blocks, the blocks will have 2 degrees of freedom belonging to the factorial effect which was used to define the block.

9.9.1 Assignment of 3^k Factorial Design in 3^p Blocks ($p = 1$)

The lowest value of p for which the 3^k factorial design could be split into 3^p blocks is one. It is therefore possible to split a 3^k factorial design into three blocks. This is achieved by choosing a defining contrast, and constructing a

TABLE 9.25: Assignment of the 3^3 Factorial Design into Three Blocks with AB^2C Confounded with Blocks

Blocks		
I	**II**	**III**
0 0 0	0 0 1	0 0 2
0 1 1	0 1 2	0 1 0
0 2 2	0 2 0	0 2 1
1 0 2	1 0 0	1 0 1
1 1 0	1 1 1	1 1 2
1 2 1	1 2 2	1 2 0
2 0 1	2 0 2	2 0 0
2 1 2	2 1 0	2 1 1
2 2 0	2 2 1	2 2 2

linear combination of the exponents and levels of all the factors in the factorial effect that is chosen to be confounded with blocks. Under this design, the linear combination is

$$C' = a_1x_1 + a_2x_2 + \cdots + a_kx_k \text{ (modulo 3)} \tag{9.2}$$

where $a_i = 0$, 1, and 2; $x_i = 0$, 1, and 2; ($x_i = 0$, low level of factor; $x_i = 1$, intermediate level of factor; $x_i = 2$, high level of factor). C' has values 0, 1, and 2 (modulo 3), enabling these levels of C' to be used to define the blocks.

The 3^3 factorial design can be split into three blocks by choosing the factorial effect AB^2C to be confounded with blocks. The appropriate linear combination of levels of this factorial effect is

$$C' = x_1 + 2x_2 + x_3 \text{ (modulo 3)} \tag{9.3}$$

Recall that the treatment combinations that fall under the different levels of this effect are given in Section 9.5. For the treatment combinations, we calculate their contrasts under AB^2C as follows:

$$000 \; C' = 1(0) + 2(0) + 1(0) = 0 \text{ modulo } 3 = 0$$

$$011 \; C' = 1(0) + 2(1) + 1(1) = 3 \text{ modulo } 3 = 0$$

$$012 \; C' = 1(0) + 2(1) + 1(2) = 4 \text{ modulo } 3 = 1$$

$$001 \; C' = 1(0) + 2(0) + 1(1) = 1 \text{ modulo } 3 = 1$$

$$022 \; C' = 1(0) + 2(2) + 1(2) = 6 \text{ modulo } 3 = 0$$

$$002 \; C' = 1(0) + 2(0) + 1(2) = 2 \text{ modulo } 3 = 2$$

$$010 \; C' = 1(0) + 2(1) + 1(0) = 2 \text{ modulo } 3 = 2$$

Calculations carried out in this way have led to the splitting of this design into three blocks as given in Table 9.25.

There are 27 treatment combinations in the 3^3 factorial design, and each block contains $3^{3-1} = 9$ treatment combinations. We illustrate this design and analysis of its responses. Let us treat the data of Example 7.5 as if the responses were obtained from three blocks.

Example 9.5

Analyze the responses of Example 7.5, as if the treatments were carried out in blocks of size nine using AB^2 to define the blocks. Recall that in Example 7.5, there were two replicates of the design. Assignment of each replicate of the design to three blocks means that we have six blocks each with nine treatments (actually there are three blocks and two replications for each block). In the analysis of the responses, we proceed as was the case when the treatments in the design were not assigned to blocks, so that sums of squares of the factorial effects in the design are obtained as in Example 7.5. In addition, there is the block total sum of squares, which is calculated directly using the block totals.

Since we are using AB^2 to define blocks, we note that the linear combination for the levels of this factorial effect is $C' = x_1 + 2x_2$ (modulo 3). The blocks obtained are as follows:

Replicate I					
Blocks					
I	**Responses**	**II**	**Responses**	**III**	**Responses**
0 0 0	4.20	0 2 0	6.70	0 1 0	5.11
0 0 1	9.80	0 2 1	12.40	0 1 1	11.90
0 0 2	14.60	0 2 2	16.10	0 1 2	13.30
1 1 0	14.30	1 0 0	12.50	1 2 0	15.38
1 1 1	20.96	1 0 1	15.51	1 2 1	19.20
1 1 2	21.60	1 0 2	18.70	1 2 2	22.40
2 2 0	21.30	2 1 0	19.86	2 0 0	16.80
2 2 1	27.30	2 1 1	26.37	2 0 1	19.00
2 2 2	31.40	2 1 2	29.20	2 0 2	24.20
Total	163.3		157.96		147.29
IV	**Responses**	**V**	**Responses**	**VI**	**Responses**
0 0 0	4.93	0 2 0	7.00	0 1 0	5.31
0 0 1	10.10	0 2 1	13.15	0 1 1	11.34
0 0 2	13.79	0 2 2	17.10	0 1 2	15.64
1 1 0	15.35	1 0 0	13.00	1 2 0	14.90
1 1 1	20.96	1 0 1	15.21	1 2 1	20.10
1 1 2	24.27	1 0 2	17.98	1 2 2	23.40
2 2 0	24.00	2 1 0	20.88	2 0 0	15.96
2 2 1	27.20	2 1 1	24.38	2 0 1	18.60
2 2 2	29.64	2 1 2	29.64	2 0 2	22.58
Total	170.24		158.35		147.83

Recall the totals for the factorial effects which were as follows (Table 9.26).

These totals were used as in Example 7.5 to calculate the sums of squares for the factorial effects. Since these sums of squares have been calculated in Chapter 7, we shall merely use their values here. In addition, the total sums

TABLE 9.26: Response Totals for Factorial Effects in Example 7.5

Factorial effects	0	1	2	Sum of squares
A	192.57	323.57	428.83	1556.66
B	267.47	328.73	348.77	199.33
AB	323.65	316.35	304.97	9.85
AB^2	333.54	316.31	295.12	41.15
C	237.48	321.33	386.16	617.40
AC	304.45	321.30	319.22	9.38
AC^2	310.22	318.91	315.84	2.16
BC	320.91	309.18	314.88	3.82
BC^2	321.28	312.01	311.68	3.30
ABC	309.72	317.12	318.13	2.34
AB^2C	313.56	315.62	315.79	0.17
ABC^2	314.57	317.68	312.72	0.70
ABC^2	312.63	315.38	316.96	0.53

of squares for the blocks (total) and their partitions, that is, replicates and error × blocks are obtained as follows:

$$SS_{\text{total}} = [(4.2)^2+(9.8)^2+\cdots+(18.6)^2+(22.58)^2]-\frac{(944.97)^2}{54} = 2468.46$$

$$SS_{\text{block(total)}} = \frac{[(163.3)^2+(157.96)^2+\cdots+(147.83)^2]}{9}-\frac{(944.97)^2}{54} = 43.8483$$

$$SS_{\text{replicates}} = \frac{1}{27}(476.42)^2+(468.55)^2-\frac{(944.97)^2}{54} = 1.1470$$

$$\begin{aligned} SS_{\text{E}(AB^2)} &= SS_{\text{block(total)}} - SS_{AB^2} - SS_{\text{replicates}} \\ &= 43.8483 - 41.15 - 1.1470 = 1.5513 \end{aligned}$$

These three sums of squares, $SS_{\text{E}(AB^2)}$, SS_{AB^2}, and $SS_{\text{replicates}}$ with degrees of freedom, 2, 2, and 1, respectively, represent a partition of the block sum of squares (Cochran and Cox, 1957) which is similar to the same partition in the 2^k factorial designs discussed in the previous sections on confounding in 2^k factorial designs (Section 9.8). We present the ANOVA table for this design in Table 9.27.

TABLE 9.27: ANOVA for the 3^3 Factorial Design in Six Blocks ($I = AB^2$)

Source	Sum of squares	Degrees of freedom	Mean square	*F*-ratio
Replicates	1.1470	1	1.1470	—
Blocks (AB^2)	41.1500	2	20.5750	—
Error (AB^2)	1.5513	2	0.7756	—
A	1556.6600	2	778.3300	984.6202
B	199.3300	2	99.6650	126.0804
AB	9.8500	2	4.9250	6.2303

Continued

TABLE 9.27: Continued

Source	Sum of squares	Degrees of freedom	Mean square	F-ratio
C	617.4000	2	308.7000	390.5185
AC interaction	11.5400	4	2.8850	3.6496
BC interaction	7.1200	4	1.7800	2.5177
ABC interaction	3.7400	8	0.4675	0.5914
Residual	18.9717	24	0.7905	
Total	2468.4600	53		

From the tables, $F(1, 24, 0.01) = 7.82$, $F(2, 24, 0.01) = 5.61$, $F(4, 24, 0.01) = 4.22$, and $F(8, 24, 0.01) = 3.36$. A, B, AB, and C are significant. The blocking was also effective since there is a large mean square for the block effect. □

9.9.2 Assignment of the 3^k Factorial Design into 3^p Blocks ($p = 2$)

The next value of p for which we could divide the 3^k factorial design into 3^p blocks is 2. We can split the 3^k factorial design in $3^2 = 9$ blocks, each with 3^{k-2} units. As was the case in the 2^k factorial design, we shall choose two factorial effects to define the blocks. With the choice of the factorial effects, we shall determine the generalized interaction of these effects, which is automatically confounded with blocks. The generalized interactions for two factorial effects X, Y are XY and X^2Y (or XY^2), with the exponents reduced by modulo arithmetic (modulo 3). This ensures that the first letter in the factorial effect obtained by the multiplications conforms with the convention of not having an exponent of more than one; otherwise, the effect will need to be squared, after which the rules governing exponents of the letters are applied again until we obtain an effect that satisfies the convention. As an illustration of this design, we use the factorial effects AB^2 and AC^2 for the definition of blocks of the 3^{3-2} factorial design. In this design, we shall have $3^2 = 9$ blocks and each containing $3^{3-2} = 3$ treatment combinations. We can obtain the generalized interaction of these effects as

$$AB^2(AC^2) = A^2B^2C^2 = (A^2B^2C^2)^2 = ABC$$
$$(AB^2)^2(AC^2) = A^3B^4C^2 = BC^2$$

From this, it can be seen that apart from AB^2 and AC^2, BC^2 and ABC are all confounded with blocks. However, to define the blocks, we need only the two original factorial effects. As usual, we shall use the linear combinations of the levels of AB^2 and AC^2, which are $C_1 = x_1 + 2x_2$ (modulo 3) and $C_2 = x_1 + 2x_3$ (modulo 3) to identify which treatment combinations belong to any block. Recall that C can take values, 0, 1, and 2; there are nine different arrangements of these contrasts which give the nine blocks of the design. These arrangements of contrasts are 0 0; 0 1; 0 2; 1 0; 1 1; 1 2; 2 0; 2 1; 2 2. The blocks for this design and their members are listed in Table 9.28.

TABLE 9.28: Assignment of 3^3 Factorial Design into Nine Blocks Defined by AB^2 and AC^2

Contrasts of AB^2 and AC^2		Block	Responses		
0	0	1	000	111	222
0	1	2	002	110	221
0	2	3	001	112	220
1	0	4	020	101	212
1	1	5	022	100	211
1	2	6	021	102	210
2	0	7	010	121	202
2	1	8	012	120	201
2	2	9	011	122	200

TABLE 9.29: The 3^3 Factorial Design with Nine Blocks/Replicate

Blocks	Responses			Total
1	4.20	18.80	31.4	54.4
2	4.93	20.96	29.64	55.53
3	14.60	14.30	27.3	56.20
4	13.79	15.35	27.2	56.34
5	9.80	21.60	21.3	52.70
6	10.10	24.27	24.0	58.37
7	6.70	15.51	29.72	51.93
8	7.0	15.22	29.64	51.86
9	16.20	12.50	26.37	55.07
10	17.10	13.0	24.38	54.48
11	12.40	18.7	19.86	50.96
12	13.15	17.98	20.88	52.01
13	5.11	19.20	24.20	48.51
14	5.31	20.10	22.58	47.99
15	13.30	15.38	19.00	47.68
16	15.64	14.90	18.60	49.14
17	11.90	22.40	16.80	51.10
18	11.34	23.40	15.96	50.70

As we did earlier, we can carry out the analysis to show how ANOVA could be performed for this design.

Example 9.6

Suppose that the 3^3 factorial design of Example 7.5 was blocked, and the blocks were defined by confounding AC^2 and AB^2 with blocks. Reanalyze the responses of the experiment. As we have seen, there are nine blocks in each replicate of the 3^3 factorial design, but in Example 7.5, we have two replicates of the design, meaning that each block of the design is replicated twice. Therefore there are 18 blocks altogether in the entire experiment. AC^2, AB^2, BC^2, and ABC are all confounded with the blocks. The responses for the blocks are listed in Table 9.29.

TABLE 9.30: ANOVA for the 3^3 Factorial Design in Six Blocks ($I = AB^2$)

Source	Sum of squares	Degrees of freedom	Mean square	*F*-ratio
Replicates	1.1470	1	1.1470	—
Blocks ($AB \ldots ABC$)	48.9500	8	6.1187	—
Error ($AB \ldots ABC$)	5.0953	8	0.6369	—
A	1556.6600	2	778.3300	908.0971
B	199.3300	2	99.6650	116.2816
AB	9.8500	2	4.9250	5.7461
C	617.4000	2	308.7000	360.1680
AC	9.3800	2	4.6900	5.4719
BC	3.8200	2	1.9100	2.2284
ABC interaction	1.400	6	0.2333	0.2722
Residual	15.4277	18	0.8571	
Total	2468.4600	53		

Except for the block (total) sum of squares and their partitions, the rest of the factorial effects had their sums of squares calculated as in Example 7.5. We calculate the sums of squares (blocks) and its partitions as follows:

$$SS_{\text{block(total)}} = \frac{[(54.4)^2 + (55.53)^2 + \cdots + (50.70)^2]}{3} - \frac{(944.97)^2}{54} = 55.19235$$

$$SS_{\text{replicates}} = \frac{[(476.42)^2 + (468.55)^2]}{27} - \frac{(944.97)^2}{54} = 1.1470$$

$$\begin{aligned} SS_{\text{E}(AB^2+AC^2+BC^2+ABC)} &= SS_{\text{block(total)}} - SS_{AB^2} - SS_{AC^2} \\ &\quad - SS_{BC^2} - SS_{ABC} - SS_{\text{replicates}} \\ &= 55.1923 - 41.15 - 2.16 - 3.30 \\ &\quad - 2.34 - 1.147 = 5.0953 \end{aligned}$$

The ANOVA for this design is given in Table 9.30.

From the tables, $F(1, 18, 0.01) = 8.29$, $F(2, 18, 0.01) = 6.01$, $F(6, 18, 0.01) = 4.01$, and $F(8, 18, 0.01) = 3.71$. Under this design, only A, B, and C are significant at $\alpha = 1\%$. Blocking is also effective since the block mean square is relatively large. □

9.9.3 Assignment of the 3^k Factorial Design into 3^p Blocks ($p > 3$)

The methods used in the previous examples could be extended and used when the number of blocks to which the 3^k factorial design is to be assigned exceeds nine. Usually, to assign treatment combinations into 3^p blocks, $p(p < k)$ factorial effects would have to be chosen. These p factorial effects, on the other hand, have $(3^p - 2p - 1)/2$ other factorial effects in their generalized interactions, which are automatically confounded with blocks, as are the chosen p factorial effects. Methods of analysis follow the same patterns as those obtained in the solutions of Examples 9.4 and 9.5. For instance, to assign a 3^5 factorial design into $3^3 = 27$ blocks, we need to choose three factorial effects to define the blocks. There will be $3^2 = 9$ units in each block. Furthermore, there will be $(27 - 6 - 1)/2 = 10$ other factorial effects confounded with blocks, as well as the chosen three effects, making a total of 13 confounded effects.

Suppose we choose *ABCD*, *BCDE*, *ACDE* to define the blocks for the 3^5 factorial designs in A, B, C, D, E, the generalized interactions of these effects are as follows:

$$
\begin{aligned}
(ABCD)(BCDE) &= AB^2C^2D^2E \\
(ABCD)(ACDE) &= A^2BC^2D^2E = AB^2CDE^2 \\
(BCDE)(ACDE) &= ABC^2D^2E^2 \\
(ABCD)(BCDE)(ACDE) &= ABE \\
(ABCD)(BCDE)^2 &= AE^2 \\
(ABCD)(ACDE)^2 &= BE^2 \\
(ABCD)^2(BCDE)(ACDE) &= DE^2 \\
(ACDE)^2(BCDE) &= AB^2 \\
(ABCD)(BCDE)^2(ACDE) &= AC^2D^2 \\
(ABCD)(BCDE)(ACDE)^2 &= BC^2D^2
\end{aligned}
$$

It is also possible to assign the 3^5 factorial design into 81 blocks of three units each. Four factorial effects would need to be chosen to define the blocks; altogether, there will be 40 factorial effects that are confounded with blocks.

9.10 Partial Confounding in Factorial Design

At the beginning of Section 9.7 while discussing confounding, we mentioned that when a factorial effect is fully confounded with blocks, information on that factorial effect is lost. However, it is possible to use a technique called *partial confounding* to recover some information on that factorial effect. This can only happen when there is replication. In this way, information is not totally lost on any effect in the factorial design. Partial confounding is a technique for confounding different factorial effects in different replicates (and hence, blocks) of a factorial design. This enables information on all the factorial effects in the design to be recovered, for some of the confounded effects, from particular replicates of the design.

9.10.1 Partial Confounding in 2^k Factorial Design

The technique of partial confounding simply uses the methods of confounding which have been described in the Sections 9.7 through 9.9, but the techniques are applied in such a way that only one factorial effect is confounded in each replicate of the design. We can illustrate this by using the replicates of 2^3 factorial design. Recall Example 9.3 in which we confounded the blocks with

TABLE 9.31: Partial Confounding in a 2^3 Factorial Design

Replicate I		Replicate II		Replicate III	
Block I	**Block II**	**Block III**	**Block IV**	**Block V**	**Block VI**
(1)	a	(1)	b	(1)	a
ab	b	a	ab	b	ab
ac	c	bc	c	ac	c
bc	abc	abc	ac	abc	bc

ABC and had six blocks. Suppose we change this design so that we confound ABC in the first replicate, confound BC in the second replicate, and confound the third replicate with AC. The resulting blocks are as given in Table 9.31.

In this design, each of the interactions, AC, BC, ABC, is confounded in only two of the six blocks. We can therefore recover information on each effect in four blocks or two of the replicates of the experiment.

Example 9.7

Assume that the responses of Example 9.3 were obtained from a 2^3 factorial design that had the factorial effect ABC confounded with blocks in the first replicate, BC confounded with blocks in the second replicate, and AC confounded with blocks in the third replicate. The new blocks are as follows:

Replicate I		Replicate II		Replicate III	
Block I	**Block II**	**Block III**	**Block IV**	**Block V**	**Block VI**
(1) 4.6	a 10.1	(1) 4.4	b 5.9	(1) 4.8	a 9.2
ab 6.6	b 4.6	a 7.8	ab 6.3	b 6.2	ab 7.0
ac 6.2	c 7.6	bc 8.9	c 6.9	ac 7.5	c 8.0
bc 8.2	abc 7.7	abc 8.4	ac 7.0	abc 8.2	bc 9.0

The sum of squares of the factorial effects are calculated as in Example 9.3. However, we obtain the block sum of squares using the blocks we have just created as follows:

$$SS_{\text{block}} = [(25.6)^2 + (30)^2 + \cdots + (33.2)^2]/4 - \frac{171^2}{24} = 10.687$$

$$SS_{\text{replicates}} = \left(\frac{1}{8}\right)[(55.5)^2 + (55.6)^2 + (59.6)^2] - \frac{171^2}{24} = 1.1725$$

From replicates I and II, we calculate the sum of squares for AC as follows:

$$SS_{AC} = \frac{(9 + 10.5 + 16.1 + 13.2 - 17.9 - 12.9 - 17.1 - 14.5)^2}{16} = 11.56$$

Similarly, we obtain the sum of squares for BC from replicates I and III as follows:

$$SS_{BC} = \frac{(9.4 + 15.9 + 19.3 + 17.2 - 15.6 - 13.7 - 13.6 - 10.8)^2}{16} = 4.101$$

TABLE 9.32: ANOVA for Example 9.7 on Partial Confounding

Source	Sum of squares	Degrees of freedom	Mean square	F-ratio
Replicates	1.1725	2	0.5868	—
Blocks (*ABC*,*BC*,*AC*)	9.5145	3	3.1715	—
A	6.3087	1	6.3087	15.442
B	0.1504	1	0.1504	0.368
AB	4.7704	1	4.7704	11.676
C	11.3437	1	11.3437	27.765
AC from replicates I and II	11.5600	1	11.5600	28.295
BC from replicates I and III	4.101	1	4.1010	10.038
ABC from replicates II and III	2.031	1	2.031	4.971
Residual	4.4941	11	0.40855	
Total	55.4463	23		

We finally obtain the sum of squares for *ABC* from replicates II and III as follows:

$$SS_{ABC} = \frac{(17.0 + 12.1 + 14.9 + 16.6 - 9.2 - 13.3 - 14.5 - 17.9)^2}{16} = 2.031$$

The ANOVA for this design is given in Table 9.32.

From the Tables, $F(1, 11, 0.01) = 9.65$, $F(2, 11, 0.01) = 7.21$, and $F(3, 11, 0.01) = 6.22$. Only A, C, AB, AC, and BC are significant. The blocking was also effective since the mean square for block is relatively large. ☐

9.10.2 Partial Confounding in 3^k Factorial Design

For the same reasons why partial confounding is used in 2^k factorial designs, the technique can also be used in 3^k factorial designs. Suppose we decide to divide the 3^3 factorial design, which we dealt with in the previous sections in blocks (see Examples 9.3 and 9.7). If we do not wish to forgo the entire information on any of the factorial effects confounded with the blocks, we could use partial confounding to assign the treatments in the design into blocks. In particular, if we have two replicates of the 3^3 factorial design, we could assign treatments in each replicate into three blocks. This is achieved by choosing two factorial effects to be confounded with the blocks and confounding only one factorial effect with blocks in each replicate of the design. If we choose the factorial effects, AB^2 and AC^2, to be confounded with blocks, we confound AB^2 with blocks in replicate I and AC^2 with blocks in replicate II. There are of course many other factorial effects, which could be chosen and confounded with blocks. Each choice of effects produces a different design, and therefore different results, when the responses are analyzed.

Example 9.8

Consider the design of the 3^3 factorial design (Example 7.3) to have been arranged according to the design described above. The first three blocks belong to the first replicate (and are confounded with AB^2), while the other three come from the second replicate and are confounded with AC^2. Analyze the responses.

Replicate I							
	0 0 0	4.2	0 2 0	6.7	0 1 0	5.11	
	0 0 1	9.8	0 2 1	12.4	0 1 1	11.90	
	0 0 2	14.6	0 2 2	16.2	0 1 2	13.30	
	1 1 0	14.3	1 0 0	12.5	1 2 0	15.38	
	1 1 1	18.8	1 0 1	15.51	1 2 1	19.20	
	1 1 2	21.6	1 0 2	18.70	1 2 2	22.40	
	2 2 0	21.3	2 1 0	19.86	2 0 0	16.80	
	2 2 1	27.3	2 1 1	26.37	2 0 1	19.00	
	2 2 2	31.4	2 1 2	29.72	2 0 2	24.20	
Total		163.3		157.96		147.29	468.55

Replicate II							
	0 0 0	4.93	0 0 2	13.79	0 0 1	10.10	
	0 1 0	5.31	0 1 2	15.64	0 1 1	11.34	
	0 2 0	7.0	0 2 2	17.10	0 2 1	13.15	
	1 0 1	15.22	1 0 0	13.0	1 0 2	17.98	
	1 1 1	20.96	1 1 0	15.35	1 1 2	24.27	
	1 2 1	20.10	1 2 0	14.90	1 2 2	23.40	
	2 0 2	22.58	2 0 1	18.60	2 0 0	15.96	
	2 1 2	29.64	2 1 1	24.38	2 1 0	20.88	
	2 2 2	29.64	2 2 1	27.20	2 2 0	24.00	
Total		155.38		159.96		161.08	476.42

$$SS_{\text{block}} = \frac{[(163.3)^2 + (157.96)^2 + \cdots + (161.08)^2]}{9} - \frac{(944.97)^2}{54} = 17.9398$$

$$SS_{\text{replicates}} = \frac{[(468.55)^2 + (476.42)^2]}{27} - \frac{(944.97)^2}{54} = 1.1470$$

$$SS_{AB^2} = \frac{[(170.24)^2 + (158.35)^2 + (147.83)^2]}{9} - \frac{(476.42)^2}{27} = 27.9352$$

$$SS_{AC^2} = \frac{[(154.84)^2 + (158.95)^2 + (154.76)^2]}{9} - \frac{(468.55)^2}{27} = 1.2761$$

Other sum of squares remain as obtained in Example 7.3. The ANOVA for this design is presented in Table 9.33.

From the tables, $F(1,22,0.01)=7.95$, $F(2,22,0.01)=5.72$, $F(4,22,0.01)=3.99$, and $F(8,22,0.01)=3.45$. The effects marked with $*$ are significant at $\alpha=0.01$. Blocking was effective as indicated by the fact that the block mean square is large. The half information obtained on AB^2 was also significant. ▯

TABLE 9.33: ANOVA for the Partial Confounding of 3^3 Factorial Design with AB^2 Confounded in Replicate I and with AC^2 Confounded in Replicate II

Source	Sum of squares	Degrees of freedom	Mean square	F-ratio
Replicates	1.1470	1	1.1470	—
Blocks(AB, AC)	17.9398	4	4.4849	—
A	1566.6600	2	778.3300	2562.6333*
B	199.3300	2	99.6650	328.1447*
AB	9.8500	2	4.9250	16.2154*
AB^2 (from replicate II only)	27.9352	2	13.7600	45.3045*
C	617.4000	2	308.7000	1016.3876*
AC	9.3800	2	4.6900	15.4417*
AC^2 (from replicate I only)	1.2761	2	0.6385	2.1022
BC interaction	7.1200	4	1.7800	5.8606*
ABC interaction	3.7400	8	0.4675	1.5392
Residual	6.6819	22	0.3037	
Total	2468.4600	53		

9.10.3 Other Confounding Systems

With these few examples drawn from the 2^k and 3^k factorial designs, we have sought to describe the techniques of partial confounding in replicated factorial designs. The principles used here for the 2^k and 3^k designs can be extended to p^k designs provided that to avoid confusion we use contrasts of the form 000, 001, and 002. Discussion of these systems can be found in Andersen and McLean (1974) and Montgomery (2005).

9.11 Exercises

Exercise 9.1

A 2^3 factorial experiment was performed to study the yield of a chemical plant under two levels of A (pressure) B (temperature), and C (catalyst concentration). However, because it was not possible to run a full replicate each, using the same shift of workers, it was decided to run four replicates of the experiment. The four replicates were fully confounded with the ABC effect as shown below.

Blocks			
I	II	III	IV
(1) 21.1	a 19.7	(1) 22.8	a 27.9
ab 21.2	b 19.3	ab 24.1	b 25.0
ac 20.8	c 21.0	ac 23.2	c 26.9
bc 21.4	abc 22.5	bc 21.1	abc 25.4

Blocks			
V	VI	VII	VIII
(1) 25.4	*a* 24.2	(1) 21.2	*a* 25.6
ab 22.5	*b* 23.2	*ab* 23.9	*b* 25.7
ac 25.3	*c* 25.6	*ac* 20.6	*c* 23.0
bc 22.4	*abc* 24.7	*bc* 23.5	*abc* 24.8

Analyze fully the responses of the experiment and comment on the effectiveness of confounding in the design. □

Exercise 9.2

If the experiment in Exercise 9.1 is to be redesigned in such a way that a different factorial effect is to be confounded with blocks in each replicate, write down the design so that AB is confounded with blocks in replicate I, AC in replicate II, BC in replicate III, and ABC in replicate IV. Carry out the analysis of the redesigned experiment with responses in Exercise 9.1, as if they were obtained under this new design, and comment on the information recovered for the effects that were partially confounded with blocks. □

Exercise 9.3

Suppose that in Example 7.5, it was not possible to carry out all the experiments in each replicate of the design under homogeneous conditions. To introduce confounding, let the factorial effects ABC be confounded in replicate I, and ABC^2 be confounded in replicate II. Write down the design using the responses observed in Example 7.5. □

Exercise 9.4

Redesign the experiment in Exercise 9.3, assigning the three replicates into nine blocks, using ABC^2 effect for the definition of blocks. Using responses from Exercise 9.3, carry out the analysis as if the responses were obtained from the redesigned experiment and observe the differences between the result and those of Exercise 9.3. □

Exercise 9.5

A 2^5 factorial design is to be run in blocks of size four. How many factorial effects are needed to define the blocks? For a 2^5 factorial design in A, B, C, D, E, choose second- or higher order interaction effects to obtain the design described above. How many other effects are aliased with blocks? □

Exercise 9.6
In a 2^4 factorial design in A, B,C, and D, the blocks are defined as follows:

Block I	(1)	ab	cd	$abcd$
Block II	a	b	acd	bcd
Block III	c	abc	d	abd
Block IV	ac	bc	ad	bd

Identify all the factorial effects that are aliased with blocks.
Analyze the data and comment on the effectiveness of blocking. □

Exercise 9.7
Generate a 2^{6-3} factorial design in A, B, C, D, E, and F using $I = ABD = BCE = ACF$ and products. List the alias structure. What is the resolution of the design? □

Exercise 9.8
Generate a 3^4 factorial design and assign them into blocks of size nine using $ABCD$ and BCD. □

Exercise 9.9
A 3^5 factorial design is to be run in blocks of size nine. If $ABDE$, ACD, and BCE are chosen for the definition of the blocks, list all the effects that are confounded with blocks. □

Exercise 9.10
The terms *defining contrasts* and *generalized interactions* are used in the theory of fractional replication of 2^k designs. Explain what they mean. In an exploratory experiment, six factors, A, B, C, D, E, and F, are to be studied at two levels. However, only eight experimental units are available. Information is required on all main effects, and the factors are written in order of importance. Write down a list of eight treatments that form the fractional replicate of the 2^6 design. List the quantities that can be estimated if this design is used. What further assumption should be imposed to estimate the main effects? □

Exercise 9.11
Design an experiment in which a 2^4 factorial design is partially confounded in two replicates, with $ABCD$ confounded in the first replicate and ACD confounded in the second. □

Exercise 9.12
From a 2^6 factorial design in A, B, C, D, E, and F, it is possible to obtain 2_{III}^{6-3} design as suggested in this chapter. Obtain this design using ACD, ABF, and BCE in the defining relation and write down its alias structures. ▯

Exercise 9.13
From a 2^7 factorial design in A, B, C, D, E, F, and G, it is possible to obtain 2_{III}^{7-4} design as suggested in this chapter. Obtain this design and write down its alias structures. ▯

Exercise 9.14
By equating $E = ABD$ and $F = BCD$, construct a 2^{6-2} factorial design. By examining its alias structure, show that this design is of resolution IV. ▯

Exercise 9.15
Write a SAS program to analyze the responses of the partially confounded 3^3 factorial design in Exercise 9.3. ▯

Exercise 9.16
The experiment designed in Exercise 9.6 was run in two replicates, and the following responses were obtained:

	Replicate I			
Block I	(1) 74	*ab* 100.8	*cd* 100.0	*abcd* 129.2
Block II	*a* 110.8	*b* 88.8	*acd* 137.6	*bcd* 86.0
Block III	*c* 78.4	*abc* 98.4	*d* 78.0	*abd* 131.2
Block IV	*ac* 115.6	*bc* 69.2	*ad* 118.4	*bd 86.4*

	Replicate II			
Block I	(1) 124.8	*ab* 128.4	*cd* 100.8	*abcd* 150.8
Block II	*a* 117.6	*b* 100.8	*acd* 146.8	*bcd* 110.0
Block III	*c* 89.6	*abc*135.6	*d* 110.4	*abd* 132.4
Block IV	*ac* 132.8	*bc* 84.4	*Ad* 137.6	*bd* 99.6

Write a SAS program to analyze the responses and reach conclusions. Was blocking effective? ▯

Exercise 9.17
Write a SAS program to carry out the analysis of Exercise 9.1. ▯

References

Anderson, V.L. and McLean, R.A. (1974) *Design of Experiments: A Realistic Approach*, Marcel Dekker, New York.

Box, G.E.P. and Hunter, J.S. (1961a) The 2k-p fractional factorial designs, Part I, *Technometrics*, 3, 311–352.

Box, G.E.P. and Hunter, J.S. (1961b) The 2k-p fractional factorial designs, Part II, *Technometrics*, 3, 449–458.

Box, G.E., Hunter, J.S., and Hunter, W.G. (2005) *Statistics for Experimenters: Design, Innovation, and Discovery.* 2nd ed., Wiley, New York.

Cochran, W.G. and Cox, G.M. (1957) *Experimental Designs.* 2nd ed., Wiley, New York.

Connor, W.S. and Zelen, M. (1959) *Fractional Factorial Experimental Designs for Factors at Three Levels Applied Mathematics Series No. 54*, National Bureau of Standards, Washington, D.C.

Cox, D.R. (1958) *Planning Experiments*, Wiley, New York.

Hicks, R. (1973) *Fundamental Concepts in the Design of Experiments*, 2nd ed., Holt Rinehart and Winston, New York.

Kempthorne, O. (1952) *The Design and Analysis of Experiments*, Wiley, New York.

Montgomery, D.C. (2005) *Design and Analysis of Experiments*, 6th ed., Wiley, New York.

Montgomery, D.C. (1991) *Design and Analysis of Experiments*, 3rd ed., Wiley, New York.

Petersen, R. (1985) *Design and Analysis of Experiments*, Marcel Dekker, New York.

Chapter 10

Balanced Incomplete Blocks, Lattices, and Nested Designs

In the previous chapters, we came across a number of designs that were constructed to eliminate the effects of extraneous factors, while studying the effects of treatments or factors on some responses. One of the chief instruments used to achieve precision in these designs was blocking. A block is represented by the row or column when Latin and Graeco-Latin square designs are employed. In the randomized complete block designs, we assume that blocks always contained enough experimental units to ensure that all treatments or treatment combinations (if factors set at different levels are used) occur at least once in each block. However, in practice, performing the complete block experiment is impossible under certain circumstances. The early part of this chapter aims to explore some of the techniques used in dealing with such situations. The full and partially balanced lattices as well as designs in which nesting of factors occur are studied in the latter part of this chapter.

Suppose that an experiment involving seven separate treatments is being performed in an industrial laboratory. A shift of workers can only perform four of these experiments in a shift; the other three treatments needed to complete a full replicate of the design would have to be performed by another shift of workers. In a study like this, we would generally need to eliminate the differences between the shifts of workers since each shift constitutes a block of varying expertise and skill. However, if all the treatments are considered to be equally important, with each shift being able to perform only four of the seven experiments at a time, we need to balance the assignments of these treatments among shifts, so as not to give any treatment undue advantage over the other. To satisfy the requirements of the study, seven shifts of workers will be needed, each performing four experiments at a time, completing four replicates of the design. The experiment we have just described is a typical balanced incomplete block design (BIBD), which is the subject of the next few sections of this chapter.

10.1 Balanced Incomplete Block Design

BIBD is a design in which all the treatments are regarded as being of equal importance in all the blocks, regardless of the fact that the experiments are carried out in incomplete blocks. The properties of such designs include the following:

- Each block contains the same number of treatments.
- Each treatment occurs the same number of times in the entire experiment.
- Each pair of treatments occurs the same number of times in each block, and appears as many times as any other pair of treatments in the entire design.

Other types of designs that are arranged in incomplete blocks (e.g., partially balanced designs) do not satisfy these criteria. BIBD is complex because of the requirement that all the treatments in all the blocks that appear in the design should be balanced. Five quantities are used to define the design as well as to achieve the balance required: the number of treatments, the number of replications of the design (number of times each treatment occurs in the entire experiment), the number of blocks, the number of experimental units in each block, and the number of times each pair of treatments is compared in the same block. Obtaining this balanced design is not an easy task. In fact, a BIBD may not exist for some combinations of the above five quantities. Tables of BIBDs have been presented in Fisher and Yates (1953), Davies (1956), Cochran and Cox (1957), and Cox (1958).

10.1.1 Balanced Incomplete Block Design—The Model and Its Analysis

We define the BIBD design in terms of a linear model. The model that represents this design is

$$y_{ij} = \mu + \beta_i + \tau_j + \varepsilon_{ij} \quad i = 1, \ldots, b \quad j = 1, \ldots, t \tag{10.1}$$

where μ is the overall mean, β_i is the effect of block i, τ_j is the effect of the jth treatment, $\epsilon_{ij} \sim \text{NID}(0, \sigma^2)$. This model is based on the assumptions that there are altogether t treatments and the experiment is replicated r times; there are b blocks each of which contains k units; the comparison of a given pair of treatments in the same block occurs only λ times. For the fixed model, $\sum_{i=1}^{b} \beta_i = \sum_{j=1}^{t} \tau_j = 0$.

Most analyses of this design are confined to fixed effects models. However, Yates (1940) noted that when the block effects are considered as uncorrelated random variables, additional information could be obtained on the treatment effects, τ_j. Subsequently, using a model that incorporates this assumption of uncorrelated block effects, it is possible to obtain combined linear estimators for treatment effects [see John (1971)]. We shall confine the analysis to the fixed effects model. Examples involving random effects can be found in the above text.

Since there are k units in each block, the number of pairs of treatments in each block is ${}^kC_2$ intra-block pairs and, therefore, there are altogether $bk(k-1)/2$ intra-block pairs. Furthermore, since there are t treatments in the design, there are ${}^tC_2$ pairs of treatments and $\lambda t(t-1)/2$ inter-block pairs. Hence,

$$\frac{bk(k-1)}{2} = \frac{\lambda t(t-1)}{2} \tag{10.2}$$

But $bk = tr =$ total number of units in the design, so that

$$tr(k-1) = \lambda t(t-1) \Rightarrow \lambda = \frac{r(k-1)}{t-1} \tag{10.3}$$

Since λ is the number of comparisons of a pair of treatments in the same block, it must be an integer. Hence, the balancing of the design involves ensuring that all pairs of treatments occur the same number of times in each block. This emphasizes that there is a combination of parameters for which this design would not exist. If $t = b$, the design is said to be symmetric. Incidentally, the example we used to illustrate this design at the beginning of this chapter is symmetric with $t = b = 7$. As we mentioned earlier, this design needs to be balanced, taking into consideration all the constraints related to treating all treatments equally in the experiment. In performing this experiment all other requirements of a good design such as randomization should be taken into account. Thereafter, in analyzing the responses, we can obtain the total sum of squares as in other designs, using all the observations:

$$SS_{\text{total}} = \sum_{i=1}^{b}\sum_{j=1}^{t} y_{ij}^2 - \frac{y_{..}^2}{tr} \tag{10.4}$$

This sum of squares can be partitioned as follows:

$$SS_{\text{total}} = SS_{\text{treatment}} + SS_{\text{block}} + SS_{\text{residual}} \tag{10.5}$$

The sum of squares for treatment needs to be adjusted to allow for the effects of different blocks since all the treatments do not appear in all the blocks. This is done because the treatment totals, $y_{.1}, y_{.2}, \ldots, y_{.t}$, are affected by the differences between the blocks from which they are obtained and the block effects need to be removed. We obtain the sum of squares for the blocks as

$$SS_{\text{block}} = \sum_{i=1}^{b} \frac{y_{i.}^2}{k} - \frac{y_{..}^2}{tr} \tag{10.6}$$

Since there are b blocks, the block sum of squares has $b-1$ degrees of freedom. Similarly, since there are t treatments in the design, treatment has $t-1$ degrees of freedom.

10.1.2 Estimation of Treatment Effect and Calculation of Treatment Sum of Squares for Balanced Incomplete Block Design

Consider the naive estimator (connected with treatment j):

$T_j = \sum$(responses on treatment j) (T_j is the sum of r units).

The block sum (for all the blocks in which treatment j appear) is

$B_j = \sum$(all responses in all blocks containing treatment j).

Let

$$Q_j = kT_j - B_j \tag{10.7}$$

We can estimate the treatment effect of the jth treatment as

$$\hat{\tau}_j = \frac{Q_j}{kt} \tag{10.8}$$

We obtain the treatment sum of squares as

$$SS_{\text{treatment}} = \frac{1}{\lambda kt}\sum_j Q_j^2 \tag{10.9}$$

With all sums of squares calculated and the residual sum of squares obtained by subtraction from the total sum of squares, the ANOVA for this design is carried out as below (Table 10.1).

TABLE 10.1: Typical ANOVA for BIBD Experiment

Source	Sum of squares	Degrees of freedom	Mean square	F-ratio
Block	SS_{block}	$b-1$	$\frac{SS_{\text{block}}}{b-1}$	$\frac{MS_{\text{block}}}{MS_{\text{residual}}}$
Treatment (adjusted)	$SS_{\text{treatment(adjusted)}}$	$t-1$	$\frac{SS_{\text{treatment}}}{t-1}$	$\frac{MS_{\text{treatment}}}{MS_{\text{residual}}}$
Residual	SS_{residual}	$bk-t-b+1$	$\frac{SS_{\text{residual}}}{bk-t-b+1}$	
Total	SS_{total}	$bk-1$		

Example 10.1

A species of grass used for making hay was subjected to five fertilizer treatments. The experiment was carried out in ten blocks each of which contained only three out of the five treatments leading to a BIBD. The responses obtained from the experiments (representing weights of hay harvested from a specified area for each block) are presented below. Are the treatments and block effects significant?

	Blocks									
Treatment	**1**	**2**	**3**	**4**	**5**	**6**	**7**	**8**	**9**	**10**
1	2.41		4.33	4.57		2.32	4.87			5.47
2		2.82	4.92			2.61		3.22	4.32	4.95
3				4.1	4.67	3.14	5.27	5.81	6.11	
4	5.44	6.1	7.03		3.82		6.88		6.43	
5	4.17	5.55		5.67	5.67			4.07		6.54

We can calculate the constant λ for this design. Since $r = 4$; $t = 9$; $b = 12$; and $k = 3$; then

$$\lambda = \frac{6(3-1)}{5-1} = 3$$

We can list the treatment and block totals as in the table below:

	Blocks										
Treatment	**1**	**2**	**3**	**4**	**5**	**6**	**7**	**8**	**9**	**10**	**Total**
1	2.41		4.33	4.57		2.32	4.87			5.47	23.97
2		2.82	4.92			2.61		3.22	4.32	4.95	22.84
3				4.1	4.67	3.14	5.27	5.81	6.11		29.1
4	5.44	6.1	7.03		3.82		6.88		6.43		35.7
5	4.17	5.55		5.67	5.67			4.07		6.54	31.67
Total	12.02	14.47	16.28	14.34	14.16	8.07	17.02	13.1	16.86	16.96	143.28

$$SS_{\text{total}} = [(2.41)^2 + (4.33)^2 + \cdots + (5.47)^2] - \frac{(143.28)^2}{30} = 51.32252$$

$$SS_{\text{block}} = [(21.78)^2 + (22.09)^2 + \cdots + (21.51)^2] - \frac{142^2}{30} = 23.48052$$

To calculate the $SS_{\text{treatment}}$ we obtain the Q_j for each treatment j as follows:

$$Q_1 = 3(23.97) - (12.02 + 16.28 + 14.34 + 8.07 + 17.02 + 16.96) = -12.78$$

$$Q_2 = 3(22.84) - (14.47 + 16.28 + 13.1 + 8.07 + 16.86 + 16.96) = -17.22$$

$$Q_3 = 3(29.1) - (14.34 + 14.16 + 8.07 + 17.02 + 13.1 + 16.86) = 3.75$$

$$Q_4 = 3(35.7) - (12.02 + 16.28 + 14.47 + 16.28 + 14.16 + 17.02 + 16.86) = 16.29$$

$$Q_5 = 3(31.67) - (12.02 + 14.47 + 14.34 + 14.16 + 13.1 + 16.96) = 9.96$$

$$SS_{\text{treatment}} = \frac{[(-12.78)^2 + (-17.22)^2 + (3.75)^2 + (16.29)^2 + (9.96)^2]}{1 \times 3 \times 9} = 18.633$$

$$SS_{\text{residual}} = 51.32252 - 23.48052 - 18.633 = 9.209$$

TABLE 10.2: ANOVA for Example 10.1

Source	Sum of squares	Degrees of freedom	Mean square	*F*-ratio	*p*-Value
Block	23.48052	9	2.608947	4.532864	0.004247052
Treatment	18.633	4	4.65825	8.093387	0.000164431
Error	9.209	16	0.575562		
Total	51.32252	29			

From the tables, $F(9,16,0.01)=3.7804$ and $F(4,16,0.01)=4.7726$; consequently, both the treatment and the blocks are significant at $\alpha=1\%$. □

10.1.3 SAS Analysis of Responses of Experiment in Example 10.1

Here we write a program to carry out the analysis of the responses of the experiment in Example 10.1:

SAS PROGRAM

```
data bibd1;
input treat block y@@;
datalines;
1 1 2.41 1 3 4.33 1 4 4.57 1 6 2.32 1 7 4.87 1 10 5.47
2 2 2.82 2 3 4.92 2 6 2.61 2 8 3.22 2 9 4.32 2 10 4.95
3 4 4.1 3 5 4.67 3 6 3.14 3 7 5.27 3 8 5.81 3 9 6.11
4 1 5.44 4 2 6.1 4 3 7.03 4 5 3.82 4 7 6.88 4 9 6.43
5 1 4.17 5 2 5.55 5 4 5.67 5 5 5.67 5 8 4.07 5 10 6.54
;
proc glm data=bibd1; class block treat; model y=block treat/ss1;
random block/test; run;
```

SAS OUTPUT

```
                         The SAS System
                        The GLM Procedure

                     Class Level Information

           Class          Levels    Values

           block              10    1 2 3 4 5 6 7 8 9 10

           treat               5    1 2 3 4 5

             Number of Observations Read          30
             Number of Observations Used          30
```

```
                              The SAS

                          The GLM Procedure

Dependent Variable: y
                                        Sum of
      Source                  DF       Squares     Mean Square    F Value    Pr > F

      Model                   13    42.11352000     3.23950154       5.63    0.0008
      Error                   16     9.20900000     0.57556250
      Corrected Total         29    51.32252000

                    R-Square     Coeff Var      Root MSE      y Mean

                    0.820566      15.88481      0.758658    4.776000

      Source              DF     Type I SS     Mean Square    F Value    Pr > F

      block                9   23.48052000      2.60894667       4.53    0.0042
      treat                4   18.63300000      4.65825000       8.09    0.0009
```

We can compare the treatment effects using a number of methods we have previously studied. If we decided to use the Duncan's multiple range test (discussed in Chapter 2) for the comparison, we could carry out the comparisons of the treatment effects using the value of the MS_{residual} in the above ANOVA (Table 10.2). We shall carry out this analysis when we have considered the precision of the estimates of treatment effects obtained under the model we have assumed (Equation 10.1). We refer to the estimator $\hat{\tau}_j$ of τ_j (see Equation 10.8) and state that it is unbiased for τ_j and by the theory of sufficient statistics is MLE, MVUE, and BLUE. However, we shall only show here that $\hat{\tau}_j$ is unbiased. Interested readers are referred to Silvey (1975), Mood et al. (1974), and Cox and Hinkley (1974) for the treatment of basic theories of maximum likelihood estimation, sufficiency of a statistic, and completeness of an estimator. Thereafter, it should be easy to verify our assertions about $\hat{\tau}_j$. We show that $\hat{\tau}_j$ is unbiased for τ_j (Recall Equation 10.1, we discuss EQ_j). From Equation 10.1, we know that

$$EY_{ij} = \mu + \beta_i + \tau_j$$

$$ET_j = r\mu + r\tau_j + \sum_{i=1}^{b} \beta_i$$

($\sum \beta_i$ is the sum of the block effects for all blocks containing the treatment j)

$$EB_j = rk\mu + r\tau_j + \sum_{j' \neq j} (\tau_{j'}) + k \sum \beta_i$$

($\sum \beta_i$ is same as above)

$$EQ_j = E(kT_j - B_j) = r(k-1)\tau_j - \lambda \sum_{j' \neq j} (\tau_{j'}) \tag{10.10}$$

We normally assume (for the fixed effects model) that $\sum_j \tau_j = 0$, which means that $\tau_j = -\sum_{j' \neq j} (\tau_{j'})$ and so

$$EQ_j = r(k-1)\tau_j + \lambda \tau_j \tag{10.11}$$

But recall that $\lambda = r(k-1)/(t-1)$ (see Equation 10.3); substituting this in Equation 10.10, we obtain

$$EQ_j = (t-1)\lambda\tau_j + \lambda\tau_j = t\lambda\tau_j$$

$$\Rightarrow E\frac{Q_j}{t\lambda} = \tau_j$$

$$\Rightarrow \hat{\tau} = \frac{Q_j}{t\lambda}$$

Thus, we have shown that $\hat{\tau}_j$ is unbiased for τ_j.

10.1.4 Precision of the Estimates and Confidence Intervals

We need to know the precision of our estimates to use them in the construction of confidence intervals (CIs) for the unknown treatment effects τ_j. Apart from this, we can test the hypothesis for the differences between two treatment effects if we know or can estimate the variance of the difference between them.

$$\text{Consider } \hat{\tau}_j = \frac{Q_j}{t\lambda}$$

$$\Rightarrow \text{Var}(\hat{\tau}_j) = \left[\frac{1}{t\lambda}\right]^2 \text{Var}(Q_j)$$

$$\begin{aligned} \text{But Var}(Q_j) &= \text{Var}(kT_j - B_j) \\ &= \text{Var}[(k-1)T_j + \text{Var}(B_{j'})] \ \ (\text{where } B_{j'} = B_j - T_j) \end{aligned}$$

10.1.4.1 An Aside

We take a brief aside to understand some concepts we need for finding out the precision for the estimate of the treatment effect. Consider the BIBD in Table 10.3.

To illustrate B_j we use Table 10.3. In the typical BIBD shown in that table, let A be treatment 1, B be treatment 2, and C be treatment 3. The subscripts in Table 10.3 are consistent with the definitions and the subscripts in Equation 10.1. If we consider A as the treatment of interest, then $T_1 = A_{11} + A_{21}$; $B_1 = A_{11} + A_{21} + B_{12} + C_{23}$; while $B_{1'} = B_{32} + C_{23}$. Usually, for each treatment j, $B_{j'}$ comprises $kr - r$ units. In the design shown in Table 10.3, $r = 2$, $k = 2$, $t = 3$, and $b = 3$; $B_{1'}$ has $2(2) - 2 = 2$ units.

TABLE 10.3: Typical Balanced Incomplete Block Design

	Blocks	
A_{11}	B_{12}	*
A_{21}	*	C_{23}
*	B_{32}	C_{33}

10.1.4.2 End of Aside

Having understood $B_{j'}$ we continue with our attempt to find the variance of Q_j.

Now $\text{Var}(B_{j'}) = r(k-1)\sigma^2$; ($\sigma^2$ for each unit). Likewise

$$\begin{aligned}&\text{Var}[(k-1)T_j] = (k-1)^2\text{Var}(T_j) = (k-1)r\sigma^2 \text{ (there are } r \text{ units in } T_j)\\&\Rightarrow \text{Var}(Q_j) = r(k-1)^2\sigma^2 + r(k-1)\sigma^2 = rk(k-1)\sigma^2 = \lambda(t-1)k\sigma^2\end{aligned}$$

so that

$$\text{Var}(\hat{\tau}_j) = \frac{k(k-1)r\sigma^2}{\lambda t} = \frac{(t-1)k\sigma^2}{\lambda t^2} \tag{10.12}$$

Having obtained $\text{Var}(\hat{\tau}_j)$, we are now in a position to construct CIs for τ_j, if we know σ^2. However, σ^2 is often unknown; we can construct CIs for τ_j with an estimate of the common variance, $\hat{\sigma}^2 = MS_{\text{residual}} = SS_{\text{residual}}/(bk-t-b+1)$. To obtain the CI for τ_j, note that

$$\frac{(\hat{\tau}_j - \tau_j)}{\sqrt{\dfrac{kMS_{\text{residual}}}{\lambda t}}} \sim t(bk-t-b+1) \tag{10.13}$$

so that the standard error for $\hat{\tau}_j$ becomes

$$s = \sqrt{\frac{kMS_{\text{residual}}}{\lambda t}} \tag{10.14}$$

The $100(1-\alpha)\%$ CI for the treatment effect τ_j is

$$\hat{\tau}_j \pm t(\alpha/2, bk-t-b+1)\sqrt{\frac{kMS_{\text{residual}}}{\lambda t}} \tag{10.15}$$

10.2 Comparison of Two Treatments

When we wish to compare two treatments, τ_j and τ_s, we usually employ the difference $d = \hat{\tau}_j - \hat{\tau}_s$; we, therefore, need to find $\text{Var}(Q_j - Q_s)$. Since treatments s and j may occur in the same block, the quantities, Q_j and Q_s, are obviously correlated. Consequently,

$$\text{Var}(Q_j - Q_s) = \text{Var}(Q_j) - 2\text{Cov}(Q_j, Q_s) + \text{Var}(Q_s) \tag{10.16a}$$

$$\begin{aligned}\text{Cov}(Q_j, Q_s) &= \text{Cov}(kT_j - B_j, kT_s - B_s)\\&= \text{Cov}(kT_j, kT_s) - k\text{Cov}(T_j, B_s) - k\text{Cov}(T_s, B_j) + \text{Cov}(B_j, B_s)\end{aligned}$$

Obviously, $\text{Cov}(T_j, T_s) = 0$ (since they are disjoint sets)

$$\begin{aligned}\text{Cov}(T_j,\, B_s) &= \sigma^2 \times (\text{number of occurrences of treatment, } s \text{ in blocks containing treatment } j)\\ &= \sigma^2\lambda\ (\lambda = \text{number of comparisons of a pair of treatments in the same block})\\ \text{Cov}(B_j, B_s) &= k\sigma^2 \times (\text{number of blocks containing both } j \text{ and } s \text{ treatments})\\ &= \lambda k\sigma^2\end{aligned}$$

Hence

$$\text{Cov}(Q_j, Q_s) = \lambda k\sigma^2 - 2\lambda k\sigma^2 = -\lambda k\sigma^2$$

Substituting these results in Equation 10.16, we obtain

$$\text{Var}(Q_j - Q_s) = 2\lambda(t-1)k\sigma^2 + 2\lambda k\sigma^2 = 2\lambda tk\sigma^2$$

Now $\text{Var}(\hat{\tau}_j - \hat{\tau}_s) = \dfrac{2\lambda tk\sigma^2}{\lambda^2 t^2} = \dfrac{2k\sigma^2}{\lambda t}$. CI for the difference $\hat{\tau}_j - \hat{\tau}_s = d$ is obtained by using the fact that

$$\frac{d - (\hat{\tau}_j - \tau_s)\sqrt{\lambda t}}{\sqrt{(2k\sigma^2)}} \sim t(bk - t - b + 1) \tag{10.16b}$$

as

$$d \pm t(\alpha/2, bk - t - b + 1) \times \sqrt{\frac{2k\sigma^2}{\lambda t}} \tag{10.16c}$$

Consider Example 10.1, we now use Duncan's multiple range test to compare the treatments in the design. From Equation 10.8, we know that

$$\hat{\tau}_j = Q_j/t\lambda$$

For Example 10.1, $\lambda = 1$, $t = 9$

$\Rightarrow \hat{\tau}_1 = -57.75/15 = -3.85 \quad \hat{\tau}_2 = -38.46/15 = -2.564$

$\hat{\tau}_3 = -7.91/15 = -0.52733 \quad \hat{\tau}_4 = 97.12/15 = 6.47466 \quad \hat{\tau}_5 = 7/15 = 0.46666$

We now arrange these estimates of treatment effects in ascending order of their magnitude in Table 10.4.

TABLE 10.4: Ranked Treatment Effects

Rank	**Treatment effect**	**Value**
1	$\hat{\tau}_1$	−3.85
2	$\hat{\tau}_2$	−2.564
3	$\hat{\tau}_3$	−0.527333333
4	$\hat{\tau}_5$	0.466666667
5	$\hat{\tau}_4$	6.474666667

From Table 10.2, we note that $MS_{\text{residual}} = 0.575562$; consequently, the standard error for each estimate of treatment effect is

$$s = \sqrt{\frac{kMS_{\text{residual}}}{\lambda t}} = \sqrt{\frac{3(0.7587658)}{5 \times 3}} = 0.389527$$

Testing for differences between treatment effects at $\alpha = 0.05$, we obtain the values $r_{0.05}(p, f)$ from the least significant ranges (LSR_i), noting that $f = 16$ as follows:

$$\text{LSR}_2 = r_{0.05}(2, 16) \times s = 3 \times 0.389527 = 1.168582$$

$$\text{LSR}_3 = r_{0.05}(3, 16) \times s = 3.15 \times 0.389527 = 1.227011$$

$$\text{LSR}_4 = r_{0.05}(4, 16) \times s = 3.23 \times 0.389527 = 1.258174$$

$$\text{LSR}_5 = r_{0.05}(5, 16) \times s = 3.30 \times 0.389527 = 1.28544$$

We compare the ranked estimates of treatment effects:

$$4\,\text{vs}\,1 => 6.47466 - (-3.85) = 10.32467 > 1.28544\ (\text{LSR}_5)$$

$$4\,\text{vs}\,2 => 6.47466 - (-2.564) = 9.038667 > 1.258174\ (\text{LSR}_4)$$

$$4\,\text{vs}\,3 => 6.47466 - (-0.52733) = 7.002 > 1.227011\ (\text{LSR}_3)$$

$$4\,\text{vs}\,5 => 6.47466 - (0.46666) = 6.008 > 1.168582\ (\text{LSR}_2)$$

$$5\,\text{vs}\,1 => 0.46666 - (-3.85) = 4.316667 > 1.258174\ (\text{LSR}_4)$$

$$5\,\text{vs}\,2 => 0.46666 - (-2.564) = 3.030667 > 1.227011\ (\text{LSR}_3)$$

$$5\,\text{vs}\,3 => 0.46666 - (-0.52733) = 0.994 < 1.168582\ (\text{LSR}_2)$$

$$3\,\text{vs}\,2 => -0.52733 - (-2.564) = 2.036667 > 1.227011\ (\text{LSR}_3)$$

$$3\,\text{vs}\,1 => -0.52733 - (-3.85) = 3.322667 > 1.168582\ (\text{LSR}_2)$$

$$2\,\text{vs}\,1 => -2.564 - (-3.85) = 1.286 > 1.168582\ (\text{LSR}_2)$$

From these comparisons, we could assign the estimates of treatment effects into groups that are not significantly different from each other:

$$(\hat{\tau}_4); \quad (\hat{\tau}_5); \quad (\hat{\tau}_3); \quad (\hat{\tau}_2, \hat{\tau}_1)$$

10.3 Orthogonal Contrasts in Balanced Incomplete Block Designs

Sometimes, a good reason exists for supposing that there is a specific linear dependence between levels of treatments employed in the BIBD. This means that the $SS_{\text{treatment}}$ could be broken up and part of it could be represented as a linear combination of adjusted treatment totals, $\sum C_j Q_j$. The contrasts C_j must be applied to the adjusted treatment sum of squares and not on the treatment totals $y_{.}$ as was the case in Chapter 2 in the use of orthogonal polynomials to fit a response curve to treatments. A contrast $\sum C_j Q_j$ is such that $\sum C_j = 0$; sometimes, though rarely, we could have $\sum C_j^2 = 1.0$. In such situations, the variance of $\sum C_jQ_j$ is of great importance and is required. We derive the formula for the variance when this happens

$$\begin{aligned}\mathrm{Var}\left(\sum C_jQ_j\right) &= \sum C_j^2\mathrm{Var}Q_j + \sum_{i\neq j} C_iC_j\mathrm{Cov}(Q_i,Q_j)\\ &= \sum C_1^2\mathrm{Var}Q_1 + \sum_{i\neq j} C_iC_j\mathrm{Cov}(Q_1,Q_2)\end{aligned}$$

(we have replaced Q_j and Q_i by Q_1 and Q_2 as this does not affect the variance or covariance in any way)

$$\mathrm{Var}\left(\sum C_jQ_j\right) = \sum C_j^2[\mathrm{Var}Q_1 - \mathrm{Cov}(Q_1,Q_2)] + \sum_{i\neq j'} C_iC_j\mathrm{Cov}(Q_1,Q_2)$$

Recall that $\mathrm{Var}(Q_j)=\lambda(t-1)k\sigma^2$ and $\mathrm{Cov}(Q_1,Q_2)=-\lambda k\sigma^2$; so that

$$\mathrm{Var}\sum C_jQ_j = \sum C_j^2[\lambda(t-1)k\sigma^2 + \lambda k\sigma^2] = \lambda kt\sigma^2\left[\sum C_j^2\right]$$

Further, if $\sum b_jQ_j$ is another contrast that is orthogonal to $\sum C_jQ_j$, that is, $\sum C_jb_j=0$, then

$$\begin{aligned}\mathrm{Cov}(C_jQ_j,b_jQ_j) &= \sum C_jb_j\mathrm{Var}Q_1 + \sum_{i\neq j}\sum C_ib_j\mathrm{Cov}(Q_1,Q_2)\\ &= \sum C_jb_j[\mathrm{Var}Q_1 - \mathrm{Cov}(Q_1,Q_2)]\\ &\quad + \left(\sum C_i\sum b_j\right)\mathrm{Cov}(Q_1,Q_2)\\ &= 0+0=0\end{aligned}$$

Orthogonal contrasts are (by normal distribution theory) independent. If we choose *a priori* contrasts $\sum C_jQ_j$, $\sum b_jQ_j$, that are orthogonal to each other, then

$$SS_{\text{treatment}} = SS_{\sum C_jQ_j} + SS_{\sum b_jQ_j} + \cdots + (\text{possibly})SS_{\text{treatment}'}$$

We can carry out the usual ANOVA involving the sums of squares obtained by splitting the treatment sum of squares and the sum of squares (block) as indicated below: $c', b', \ldots$ are the degrees of freedom corresponding to the portions of the different parts of the treatment sum of squares arising from these orthogonal contrasts. The model that leads to the ANOVA with the splitting of the treatment sum of squares is

$$EY_{ij} = \mu + \beta_i + \gamma_c(C_j) + \gamma_b(b_j) + \cdots + \tau_j$$

and the typical ANOVA that applies when this model is adopted, is presented in Table 10.5.

TABLE 10.5: ANOVA When Treatment Sum of Squares is Split into Orthogonal Contrasts

Source	Sum of squares	Degrees of freedom
Blocks	SS_{block}	$b-1$
Contrast C_j	$SS_{\sum C_j Q_j}$	c'
Contrast b_j	$SS_{\sum b_j Q_j}$	b'
⋮	⋮	⋮
Treatment	$SS_{\text{treatment}'}$	$t-c'-b'-\cdots-1$
Residual	SS_{residual}	$bk-b-t+1$
Total	SS_{total}	$bk-1$

10.4 Lattice Designs

Sometimes, a design is constructed in such a way that there are exactly k^2 treatments in a single replicate of the design; these k^2 treatments are arranged in blocks of size k. The resulting incomplete blocks are arranged in groups, which form separate complete replicates. These designs are called lattice designs. Lattices are composed of two or more replicates. It is possible, therefore, to use techniques for analyzing randomized complete block experiments to analyze the responses of this design.

10.4.1 Balanced Lattice Designs

Sometimes, a replicate of an experiment is performed to contain k^2 units, with $1/k$ of all the units in the design in each block. The number of blocks required for this design is k. Although lattices are designed so that $1/k$ of all the treatments which are in a single replicate appears in a block, this design could be balanced. When lattices are balanced, the treatments are arranged in such a way that each treatment occurs together with every other treatment once in the same block. This means that each pair of treatments is compared with the same precision in the entire design. To obtain a balanced design in which the blocks must each be of size k units (with k^2 treatments in a single replicate of the design), we need to replicate the design $k+1$ times to achieve balance. All these constraints imply that there has to be $k(k+1)$ blocks in the balanced lattice design since the total number of units in a balanced design is $k^2(k+1)$. If we consider this design as a particular form of BIBD, we note that it satisfies the requirement that λ be an integer. λ for the design is obtained as

$$\lambda = (k+1)(k-1)(k^2-1) = 1$$

Due to the restrictions on this design, it is not always possible, in practice, to construct the design for all values of k. It is known that this design does not exist for $k^2 = 36$, 100, and 144. General methods for constructing balanced lattices could be found in classical texts such as Kempthorne (1952). Once the design has been constructed, the randomization of the design could be achieved by permuting rows and columns (blocks) for each square (or full

TABLE 10.6: Typical Summary of Responses from a Balanced Lattice

Replicate	Blocks	Responses				Total
1	1	$y_{11(1)}$	$y_{12(1)}$	$\cdots$	$y_{1k(1)}$	B_{11}
	2	$y_{1k+1(2)}$	$y_{1k+2(2)}$	$\cdots$	$y_{12k(2)}$	B_{12}
	$\vdots$				$\vdots$	$\vdots$
	k	$y_{1k^2-k(k)}$	$y_{1k^2-k+1(k)}$	$\cdots$	$y_{1k^2(k)}$	B_{1k}
						R_1
2	1	$y_{21(1)}$	$y_{22(1)}$	$\cdots$	$y_{2k(1)}$	B_{21}
	2	$y_{2k+1(2)}$	$y_{2k+2(2)}$	$\cdots$	$y_{22k(2)}$	B_{22}
	$\vdots$				$\vdots$	$\vdots$
	k	$y_{2k^2-k(k)}$	$y_{2k^2-k+1(k)}$	$\cdots$	$y_{2k^2(k)}$	B_{2k}
						R_2
$\vdots$	$\vdots$	$\vdots$	$\vdots$	$\vdots$	$\vdots$	$\vdots$
$k+1$	1	$y_{k+11(1)}$	$y_{k+12(1)}$	$\cdots$	$y_{k+1k(1)}$	B_{k+11}
	2	$y_{k+1k+1(2)}$	$y_{k+1k+2(2)}$	$\cdots$	$y_{k+12k(2)}$	B_{k+12}
	$\vdots$				$\vdots$	$\vdots$
	k	$y_{k+1k^2-k(k)}$	$y_{k+1k^2-k+1(k)}$	$\cdots$	$y_{k+1k^2(k)}$	B_{k+1k}
						R_{k+1}

replicate) of the design. Randomization is, therefore, carried out as should have been the case in a typical normal Latin square design (each replicate is treated as a Latin square for randomization). Cox (1958) discusses the relationship between Latin squares, Graeco-Latin squares, extended Graeco-Latin square design, and the existence of balanced lattices. In addition to employing methods of analysis used for randomized block experiments for this design, it should also be possible for the analysis to be carried out by using the techniques unemployed for the analysis of the usual BIBDs. Further discussion of this design appears in Kempthorne (1952) and Cochran and Cox (1957). Tables showing possible balanced lattices are presented in Cox (1958) and Peterson (1985) for $k = 3$, 4, 5, 7, 8, and 9. Since there are k^2 treatments replicated $(k+1)$ times, let $y_{ij(l)}$ be the response to the jth treatment applied to a unit in the lth block of the ith replicate of the balanced lattice design. The data obtained for such an experiment could be arranged as shown in Table 10.6.

B_{il} = the sum for the k units in the lth block of the ith replicate

$R_i = \sum B_{il}$ = sum of all the responses on the ith replicate.

We could represent the sum of all responses on the jth treatment as

$$T_j = \sum y_{ij(l)} \tag{10.17}$$

Let B_j = sum of all the responses on all treatments in all the blocks in which treatment j appears; then

$$Q_j = kT_j - (k+1)B_j + \sum_i R_i$$

or (10.18)

$$Q_j = kT_j - (k+1)B_j + \sum_j T_j$$

The sums of square of this design can be computed as follows:

$$SS_{\text{total}} = \sum_i \sum_j y_{ij(l)}^2 - \frac{y_{..(.)}^2}{k^2(k+1)} \tag{10.19a}$$

$$SS_{\text{treatment}} = SS_{\text{treat}} = \frac{1}{(k+1)}\sum_j T_j - \frac{y_{..(.)}^2}{k^2(k+1)} \tag{10.19b}$$

$$SS_{\text{replicate}} = SS_{\text{rep}} = \frac{1}{k^2}\sum_i R_i^2 - \frac{y_{..(.)}^2}{k^2(k+1)} \tag{10.19c}$$

$$SS_{\text{block(adjusted)}} = SS_{bl(a)} = \frac{k+1}{k^3}\sum_j Q_j^2 \tag{10.19d}$$

$$SS_{\text{intra-block error}} = SS_{\text{ibe}} = SS_{\text{total}} - SS_{\text{treat}} - SS_{\text{rep}} - SS_{bl(a)} \tag{10.19e}$$

The typical ANOVA for this design is given in Table 10.7.

TABLE 10.7: ANOVA for a Typical Balanced Lattice Design

Source	**Sum of squares**	**Degrees of freedom**	**Mean square**	***F*-ratio**
Replication	$SS_{\text{replicate}}$	k	$\frac{SS_{\text{replicate}}}{k}$	$\frac{MS_{\text{replicate}}}{MS_{\text{ibe}}}$
Treatment	$SS_{\text{treatment}}$	k^2-1	$\frac{SS_{\text{treatment}}}{k}$	$\frac{MS_{\text{treatment}}}{MS_{\text{ibe}}}$
Block (adjusted)	$SS_{\text{block(adjusted)}}$	k^2-1	$\frac{SS_{\text{block(adjusted)}}}{(k-1)}$	
Intra-block error	SS_{ibe}	$(k-1)(k^2-1)$	$\frac{MS_{\text{ibe}}}{(k-1)(k^2-1)}$	
Total	SS_{total}	$k^2(k+1)-1$		

Example 10.2

At Umudike Agricultural Research Institute, a study was being carried out on the yield of yams under nine different fertilizer dressings. To ensure that the result of the study is widely valid, four different types of soil, whose characteristics are representative of the kinds of soils on which yam is normally grown, were chosen for the study. These different soils were located in patches, and because of this, they used a balanced lattice design to study the yield of yam under the different fertilizer dressings. Yam seeds were planted instead of tubers to eliminate any yam effect. The responses obtained (weight of the yams harvested in kg) are given below. Treatments are shown in brackets and blocks are shown as well:

Replicate	Blocks				Total
1	1	(1) 5.2	(2) 6.8	(3) 6.6	18.60
	2	(4) 7.8	(5) 7.2	(6) 4.2	19.20
	3	(7) 8.6	(8) 7.0	(9) 6.8	22.40
					60.2
2	1	(1) 4.4	(4) 6.9	(7) 6.7	18.0
	2	(2) 8.4	(5) 6.7	(8) 6.0	21.1
	3	(3) 4.8	(6) 4.5	(9) 6.3	15.6
					54.7
3	1	(1) 5.3	(5) 6.8	(9) 6.7	18.80
	2	(7) 6.7	(2) 8.1	(6) 5.2	20.00
	3	(4) 8.3	(8) 6.5	(3) 5.3	20.10
					58.9
4	1	(1) 6.3	(8) 5.2	(6) 4.6	16.10
	2	(4) 8.7	(2) 7.0	(9) 5.2	20.90
	3	(7) 7.1	(5) 6.3	(3) 4.4	17.80
					54.8

□

For this design, $k = 3$. To analyze the responses, we calculate T_j, B_j, and Q_j, and the sums of squares:

j	T_j	B_j	Q_j
1	21.20	71.50	6.2
2	30.30	80.60	−2.90
3	21.10	72.10	3.50
4	31.70	78.20	10.90
5	27.00	76.90	2.00
6	18.50	70.90	0.50
7	29.10	78.20	3.10
8	24.70	79.70	−16.10
9	25.00	77.70	−7.20
Total	228.60		0.00

$$\sum_i R_i = 228.60 = \sum_j T_j. \text{ Note also that } \sum_j Q_j = 0.0$$

$$\begin{aligned} SS_{\text{total}} &= [(5.2)^2 + (6.8)^2 + \cdots + (5.2)^2 + (4.6)^2] - \frac{228^2}{36} \\ &= 56.11 \\ SS_{\text{replicate}} &= \frac{[(60.2)^2 + (54.7)^2 + (58.9)^2 + (54.8)^2]}{9} - \frac{228^2}{36} \\ &= 2.6544 \\ SS_{\text{treatment}} &= \frac{[(21.2)^2 + (30.3)^2 + \cdots + (24.7)^2 + 25^2]}{4} - \frac{228^2}{36} \\ &= 41.085 \end{aligned}$$

$$SS_{\text{block(adjusted)}} = \frac{[(6.2)^2 + (-2.9)^2 + \cdots + (-16.1)^2 + (-7.2)^2]}{27 \times 4}$$
$$= 4.6557$$

$$SS_{\text{ibe}} = 56.11 - 2.6544 - 41.085 - 4.6557 = 7.7149$$

The ANOVA for this problem is presented in Table 10.8.

TABLE 10.8: ANOVA for Example 10.2

Source	Sum of squares	Degrees of freedom	Mean square	F-ratio
Replicate	2.6544	3	0.8848	1.8349
Treatment (unadjusted)	41.0850	8	5.1356	10.6504
Block (unadjusted)	4.6557	8	0.5820	1.2069
Intra-block error	7.7149	16	0.4822	
Total	56.11	35		

Since $F(3, 16, 0.01) = 5.29$ and $F(8, 16, 0.01) = 3.89$, we see that the adjusted treatment means are significant. We performed the ANOVA by using the adjusted block effects to estimate the block sum of squares. The treatment totals were not adjusted for the ANOVA.

We present a SAS program for the analysis of the responses of the Umudike experiment and the ensuing output.

SAS PROGRAM

```
data Umudike_BLat;
input group block treatment y@@;
datalines;
 1 1 1 5.2 1 1 2 6.8 1 1 3 6.6 1 2 4 7.8 1 2 5 7.2 1 2 6 4.2
 1 3 7 8.6 1 3 8 7 1 3 9 6.8 2 1 1 4.4 2 1 4 6.9 2 1 7 6.7
 2 2 2 8.4 2 2 5 6.7 2 2 8 6 2 3 3 4.8 2 3 6 4.5 2 3 9 6.3
 3 1 1 5.3 3 1 5 6.8 3 1 9 6.7 3 2 7 6.7 3 2 2 8.1 3 2 6 5.2
 3 3 4 8.3 3 3 8 6.5 3 3 3 5.3 4 1 1 6.3 4 1 8 5.2 4 1 6 4.6
 4 2 4 8.7 4 2 2 7 4 2 9 5.2 4 3 7 7.1 4 3 5 6.3 4 3 3 4.4
;
proc lattice data=Umudike_BLat; run;
```

SAS OUTPUT

```
                          The SAS System

                      The Lattice Procedure

                    Analysis of Variance for y

                                                  Sum of      Mean
     Source                                DF    Squares    Square

     Replications                           3     2.6544    0.8848
     Blocks within Replications (Adj.)      8     4.6557    0.5820
       Component B                          8     4.6557    0.5820
```

```
Treatments (Unadj.)                  8   41.0850   5.1356
Intra Block Error                   16    7.7148   0.4822
Randomized Complete Block Error     24   12.3706   0.5154
Total                               35   56.1100   1.6031

                 Additional Statistics for y

         Variance of Difference           0.2411
         LSD at .01 Level                 1.3733
         LSD at .05 Level                 1.0134
         Efficiency Relative to RCBD      101.12

                   Treatment Means for y

                    Treatment     Mean

                        1        5.3000
                        2        7.5750
                        3        5.2750
                        4        7.9250
                        5        6.7500
                        6        4.6250
                        7        7.2750
                        8        6.1750
                        9        6.2500
```

10.4.2 Adjustment of Treatment Totals and Further Analysis

At this stage, we need to compare the $MS_{\text{block(adjusted)}}$ and MS_{ibe}. It is suggested (see Peterson, 1985) that if $MS_{\text{block(adjusted)}} \leq MS_{\text{ibe}}$, then blocking is ineffective. It will be better to analyze the responses as a completely randomized block experiment using the replicates as blocks. On the other hand, if $MS_{\text{block(adjusted)}} > MS_{\text{ibe}}$, then blocking has been effective. In such situations, further analysis of the responses could be carried out—the treatment means are then adjusted, and a further ANOVA is performed that is based on an approximation of the F-distribution. If we define a quantity d as

$$d = \frac{[MS_{\text{block(adjusted)}} - MS_{\text{ibe}}]}{k^2 MS_{\text{block(adjusted)}}} \tag{10.20}$$

then the adjusted treatment totals are obtained as

$$T_{j(\text{adjusted})} = T_j - dQ_j \tag{10.21}$$

Since our aim is to conduct tests of significance and construct CIs for adjusted treatment means, we need to find an effective mean square residuals (MS_{residual}), which corresponds to the adjusted treatment means. This quantity is obtained as

$$MS_{\text{residual}} = MS_{\text{ibe}}[1 + kd] \tag{10.22}$$

MS_{residual} has $(k-1)(k-1)$ degrees of freedom. With this we could carry out the test for significant difference between treatments (this time, for adjusted treatment means). The sum of squares for the adjusted treatment means is obtained as

$$SS_{\text{treatment(adjusted)}} = \frac{1}{(k+1)}\sum_j T_j^2 - \frac{y^2_{..(.)}}{k^2(k+1)} \tag{10.23}$$

and

$$MS_{\text{treatment(adjusted)}} = \frac{SS_{\text{treat(adjusted)}}}{k^2-1}, \tag{10.24}$$

The statistic for this test, whose distribution is approximately F, is

$$F_{\text{cal}} = \frac{MS_{\text{treat(adjusted)},}}{MS_{\text{residual}}}$$

with k^2-1, $(k-1)(k^2-1)$ degrees of freedom. Since there are $k+1$ replicates in which each treatment appears once, we can obtain the adjusted treatment means from the adjusted treatment totals as

$$\hat{\bar{T}}_{j(\text{adjusted})} = \frac{\hat{T}_{j(\text{adjusted})}}{(k+1)} \tag{10.25}$$

The variance of $\hat{\bar{T}}_{j(\text{adjusted})}$ by analogy with that of $\text{Var}(\hat{\tau}_j)$ of BIBD that we discussed in the previous section is

$$\text{Var}[\hat{\bar{T}}_{j(\text{adjusted})}] = \frac{MS_{\text{residual}}}{k+1} \tag{10.26}$$

Often, when ANOVA indicates significant difference between treatments, we compare two treatment means to know whether they are significantly different from each other; the variance for the difference between two treatments $\hat{T}_i$ and $\hat{\bar{T}}_j$, $d_{ij} = \hat{\bar{T}}_i - \hat{\bar{T}}_j$ is given as

$$\text{Var}(d_{ij}) = \frac{2MS_{\text{residual}}}{(k+1)} \tag{10.27}$$

As usual, under $H_0 : \mu_{d_{ij}} = 0$;

$$\frac{\hat{\bar{T}}_i - \hat{\bar{T}}_j}{\sqrt{\left[\dfrac{2MS_{\text{residual}}}{(k+1)}\right]}} \sim t[(k-1)(k^2-1)] \tag{10.28}$$

TABLE 10.9: Adjusted Treatment Totals and Means

T_j	$T_{j(\text{adjusted})}$	$\hat{\overline{T}}_{j(\text{adjusted})}$
21.20	21.3426	5.3357
30.30	30.2333	7.5583
21.10	21.1805	5.2951
31.70	31.9507	7.9877
27.00	27.0460	6.7615
18.50	18.5115	4.6279
29.10	29.1713	7.2928
24.70	24.3298	6.0824
25.00	24.8344	6.2086

As an illustration, it is clear from analysis of the responses of Example 10.3 that $MS_{\text{block(adjusted)}} > MS_{\text{ibe}}$ (but only just) notwithstanding, we can demonstrate the application of the theories we have described in further analysis of the responses of the experiment. To obtain $\overline{T}_{j(\text{adjusted})}$ and its mean $\hat{\overline{T}}_{j(\text{adjusted})}$, we calculate the quantity d using Equation 10.20:

$$d = \frac{(0.5820 - 0.4822)}{0.4822(9)} = 0.022996$$

Having obtained the value of d as indicated, we can calculate the adjusted treatment totals and means using Equations 10.21 and 10.25. The values obtained are shown in Table 10.9.

$$MS_{\text{residual}} = 0.8422[1 + 3(0.23)] = 0.5155$$

$$SS_{\text{treatment(adjusted)}} = \frac{(21.3426^2 + \cdots + 24.8344^2)}{4} - \frac{(228.60)^2}{36} = 41.5982$$

$$MS_{\text{treatment(adjusted)}} = \frac{41.5982}{4} = 5.1998$$

We compute the statistic as:

$$F_{\text{cal}} = \frac{5.1998}{0.5155} = 10.0869$$

The test indicates that there are significant differences between adjusted treatment means at $\alpha = 1\%$ since $F(8, 16, 0.01) = 3.89$.

We can compare a pair of adjusted treatment means, $\overline{T}_1$ and $\overline{T}_2$, to see if they are significantly different from each other. To do so, we obtain d_{21} as

$$d_{21} = \hat{\overline{T}}_2 - \hat{\overline{T}}_1 = 7.5583 - 5.3357 = 2.2226$$

$$\text{Var}(d_{21}) = \frac{2(0.5155)}{4} = 0.2578 \Rightarrow s = \sqrt{0.2578}$$

To compare $\hat{\overline{T}}_1$ and $\hat{\overline{T}}_2$, we have

$$t_{\text{cal}} = \frac{2.2226}{\sqrt{0.2578}} = 4.3774$$

But $t(16,\ 0.01) = 2.5839$, indicating that $\overline{T}_1$ and $\overline{T}_2$ are significantly different.

We can also compare $\overline{T}_8$ and $\overline{T}_7$. $D_{87} = -1.2104$; since $\text{Var}(d_{87}) = \text{Var}(d_{21})$, we obtain the test statistic as

$$t_{\text{cal}} = \frac{-1.2104}{\sqrt{0.2578}} = -2.3839$$

We conclude that $\overline{T}_8$ and $\overline{T}_7$ are not significantly different from each other. All the pairs of treatments could be similarly compared. We note that any of the methods for multiple comparison of means (LSD, Duncan's multiple range test, discussed in Chapter 2) could be used for the comparison of these treatments.

10.5 Partially Balanced Lattices

Partially balanced lattices arise when the number of replicates required to achieve balance in a lattice is large. Limited resources lead the investigator to accept that balancing the design would not be achieved and that a partially balanced design would be the only resort. Peterson (1985) suggests that when designs are such that the number of units required in a block to balance the design is more than seven, partially balanced lattices should rather be used. The partially balanced lattices have an advantage because they exist for $k = 6$, 10, and 12 for which the balanced lattices may not exist. The partially balanced lattices are of different types; each must, however, contain treatments whose number is an exact square. The differences between the partially balanced lattices lie in the number of replicates in each of them. When only two replicates of a lattice design with k^2 treatments are run in such a way that the entire $2k^2$ units in the two replicates are assigned to $2k$ blocks, each of size k, the design is called a simple lattice. For $k = 2$, 3, this design is not recommended since the degrees of freedom for the residual are unacceptably low and would, therefore, affect the efficiency of the design relative to the randomized complete block design, which could also be used for the analysis of the responses. A lattice design with k treatments, replicated three times with $3k^2$ units in this design assigned to $3k$ blocks, each of size k, is called a triple lattice. A quadruple lattice is obtained when four replicates of a lattice design with $4k^2$ treatments are assigned into $4k$ blocks of k units each.

10.5.1 Analysis of Partially Balanced Lattices

The responses of the simple, triple, or quadruple lattices could be considered to have come from Table 10.5, by taking the first $r = 2, 3, 4$ replicates to obtain the simple, triple, and quadruple lattices. The same method of analysis can be used for all these partial lattices; we, therefore, outline a general procedure for all of them.

Let $y_{ij(1)}$ = response to the jth treatment applied to a unit in the lth block of the ith replicate

B_{il} = sum of all the k units in the lth block of the ith replicate

$R_i = \sum_l B_{il}$ = sum of all the responses in the ith replicate

A_{il} = the difference between the sum (for all replicates) of all responses for all treatments in the lth block and the quantity rB_{il} (note that r is the number of replicates)

$$A_{il} = \sum_l A_{il}$$

bA_{il} = adjusted A_{il} (the sum of all bA_{il} for the entire design is 0). The constant b is given as

$$b = \frac{[MS_{bl(\text{adjusted})} - MS_{\text{ibe}}]}{k(r-1)MS_{bl(\text{adjusted})}} \tag{10.29}$$

The quantities to be used in the calculation of b are obtained after ANOVA for the partially balanced lattice (just as we obtained the quantity d for the balanced lattices of Equation 10.20).

$$T_j = \sum_i \sum_j y_{ij(l)}$$

$$y_{..(.)} = \sum_i R_i = \sum_j T_j$$

T_j and $y_{..(.)}$ are the treatment totals and overall totals for all treatments, respectively.

$\hat{T}_{(\text{adjusted})} = T_j$ + sum of all bA_{il} for all blocks containing the treatment j.

$$\hat{\bar{T}}_{j(\text{adjusted})} = \frac{\hat{T}_{j(\text{adjusted})}}{r}$$

$\hat{\bar{T}}_{j(\text{adjusted})}$ and $\hat{T}_{j(\text{adjusted})}$ are the adjusted treatment mean and total for the jth treatment. We can calculate SS_{total}, $SS_{\text{replicate}}$, and $SS_{\text{treatment}}$ as in Equations 10.19a through 10.19c; however, these sums of squares now have $rk^2 - 1$, $r - 1$, and $k^2 - 1$ degrees of freedom, respectively. To obtain the sum of squares for the adjusted block, $SS_{\text{blocks(adjusted)}}$, we use the following formula:

$$SS_{bl(\text{adjusted})} = \frac{1}{kr(r-1)}\sum_i \sum_j A_{il}^2 - \frac{1}{rk^2(r-1)}\sum_i A_i^2 \tag{10.30}$$

$$MS_{bl(\text{adjusted})} = \frac{SS_{\text{block(adjusted)}}}{r(k-1)} \tag{10.31}$$

$$MS_{\text{ibe}} = \frac{1}{[(k-1)(rk-k-1)]}[SS_{\text{total}} - SS_{\text{replicate}} - SS_{\text{treatment}} - SS_{bl(\text{adjusted})}]$$

The typical ANOVA that would apply if either a simple, a triple, or a quadruple lattice is being considered ($r = 2, 3, 4$) is shown in Table 10.10.

TABLE 10.10: Typical ANOVA for Simple, Triple, and Quadruple Lattices

Source	Sum of squares	Degrees of freedom	Mean square	F-ratio
Replication	$SS_{\text{replicate}}$	$r-1$	$\frac{SS_{\text{replicate}}}{r-1}$	$\frac{MS_{\text{replicate}}}{MS_{\text{ibe}}}$
Treatment	$SS_{\text{treatment}}$ (unadjusted)	k^2-1	$\frac{SS_{\text{treatment(unadjusted)}}}{k^2-1}$	$\frac{MS_{\text{treatment(unadjusted)}}}{MS_{\text{ibe}}}$
Block (adjusted)	SS_{block} (adjusted)	$r(k-1)$	$\frac{SS_{\text{block(adjusted)}}}{r(k-1)}$	$\frac{MS_{\text{block(adjusted)}}}{MS_{\text{ibe}}}$
Intra-block error	SS_{ibe}	$(k-1)$ $(rk-k-1)$	$\frac{MS_{\text{ibe}}}{(k-1)(rk-k-1)}$	
Total	SS_{total}	rk^2-1		

After ANOVA, if $MS_{\text{block(adjusted)}} > MS_{\text{ibe}}$, we can adjust the treatment totals as we did for lattice designs. We can then obtain the effective error sum of squares that reflects the adjusted treatment totals as follows:

$$SS_{\text{residual}} = \left[1 + \frac{rkb}{(k+1)}\right] MS_{ibe} \tag{10.32}$$

An adjusted treatment sum of squares is then calculated, and the test of significance performed is based on an approximate F-distribution. To obtain the adjusted treatment sum of squares, we first calculate the unadjusted block sum of squares as

$$SS_{\text{block(unadjusted)}} = \frac{1}{k}\sum_i \sum_l B_{il}^2 - \frac{y_{..(.)}^2}{rk^2} \quad i = 1, 2, \ldots, k \tag{10.33}$$

Then, we obtain the adjusted treatment sum of squares as

$$SS_{\text{treatment(adjusted)}} = SS_{\text{treatment(unadjusted)}} - bk(r-1) \times \left[\frac{rSS_{bl(\text{unadjusted})}}{(r-1)(1+kb)} - SS_{bl(\text{adjusted})}\right] \tag{10.34}$$

The approximate F for this test is

$$F_{\text{app.}} = \frac{SS_{\text{treatment(adjusted)}}}{(k^2-1)MS_{\text{residual}}} \tag{10.35}$$

with $k^2 - 1$ and $(k-1)(rk-k-1)$ degrees of freedom. It is possible to construct CIs for the adjusted treatment means. Also, adjusted treatments could be compared with each other. For these purposes, we note that

$$\text{Var}(T) = \frac{MS_{\text{residual}}}{r} \tag{10.36}$$

If treatments i and j are in different blocks, the variance of the difference between them is

$$\text{Var}(d_{ij}) = \frac{[1+(r-1)b](2MS_{\text{ibe}})}{r} \tag{10.37}$$

If $k > 4$, then as Peterson (1985) suggests, it is sufficient to use the effective error mean square for the variance of any difference d_{ij}, which is

$$\text{Var}(d_{ij}) = \frac{2MS_{\text{residual}}}{r} \tag{10.38}$$

Example 10.3

Suppose that in Example 10.2, only the first three replicates had been run, leading to a triple lattice design. Analyze the responses. The data is reproduced here for convenience:

		Responses				
Replicate	**Blocks**				**Total**	A_{il}
1	1	(1) 5.2	(2) 6.8	(3) 6.6	18.60	−0.90
	2	(4) 7.8	(5) 7.2	(6) 4.2	19.20	0
	3	(7) 8.6	(8) 7.0	(9) 6.8	22.40	4.9
					60.2	−6.8
2	1	(1) 4.4	(4) 6.9	(7) 6.7	18.0	5.9
	2	(2) 8.4	(5) 6.7	(8) 6.0	21.1	0.2
	3	(3) 4.8	(6) 4.5	(9) 6.3	15.6	3.6
					54.7	9.7
3	1	(1) 5.3	(5) 6.8	(9) 6.7	18.80	−1.0
	2	(7) 6.7	(2) 8.1	(6) 5.2	20.00	−0.8
	3	(4) 8.3	(8) 6.5	(3) 5.3	20.10	−1.1
					58.9	−2.9

□

Using the results stated in the previous section, we obtain the sums of squares as follows:

$$\begin{aligned} SS_{\text{block(adjusted)}} &= \frac{[(-0.9)^2 + 0^2 + \cdots + (-1.1)^2]}{3^2(2)} \\ &\quad - \frac{[(-6.8)^2 + (10.7)^2 + (2.9)^2]}{3^3(2)} \\ &= \frac{87.68}{18} - \frac{169.14}{54} = 2.038889 \end{aligned}$$

To obtain the treatment sum of squares, we calculate the treatment totals.

Treatment	I	II	III	IV	V	VI	VII	VIII	IX
Total	14.9	23.3	16.7	23.0	20.7	13.7	22.0	19.5	19.80

$$SS_{\text{treatment(unadjusted)}} = \frac{[(14.9)^2 + (23.2)^2 + \cdots + (19.8)^2]}{3} - \frac{(173.8)^2}{27} = 31.503$$

$$SS_{\text{replicate}} = \frac{[(60.2)^2 + (54.7)^2 + (58.9)^2]}{9} - \frac{(173.8)^2}{27} = 1.8363$$

$$SS_{\text{total}} = [(5.2)^2 + (6.8)^2 + \cdots + (6.5)^2 + (5.3)^2] - \frac{173.8}{27} = 39.883$$

$$SS_{\text{ibe}} = 39.883 - 1.7389 - 2.038889 - 31.503 = 4.504811$$

The ANOVA for these responses is given in Table 10.11.

TABLE 10.11: ANOVA for Triple Lattice in Example 10.3

Source	**Sum of squares**	**Degrees of freedom**	**Mean square**	***F*-ratio**
Replicate	1.8363	2	0.9182	1.9109
Treatment (unadjusted)	31.503	8	3.9379	8.1954
Block (adjusted)	1.7389	6	0.2898	0.6031
Intra-block error (ibe)	4.8048	10	0.4805	
Total	39.8830	26		

$F(2,\ 10,\ 0.01) = 7.56$; $F(8, 10, 0.01) = 5.06$; and $F(6, 10, 0.01) = 5.38$; from which we conclude that only the treatment sum of squares is significant. It is also clear that $MS_{\text{blocks(adj)}} < MS_{\text{ibe}}$. Blocking in this design is, therefore, ineffective and as was the case for balanced lattices, we analyze the data as if they came from a randomized block design with the replicates regarded as blocks. ANOVA performed according to our results is shown in Table 10.12 below.

TABLE 10.12: ANOVA for Example 10.3 Treated as a Randomized Block Design

Source	**Sum of squares**	**Degrees of freedom**	**Mean square**	***F*-ratio**
Replicate (block)	1.8363	2	0.9182	2.2451
Treatment	31.5030	8	3.7379	9.1395
Residual	6.5437	16	0.4090	
Total	39.8830	26		

Using this method of analysis we see that only the treatment effects are significant since $F(8, 16, 0.01) = 3.89$ and $F(2, 16, 0.01) = 6.23$. We write a SAS program to analyze the responses in the above triple lattice design and present the SAS output.

SAS PROGRAM

```
data tripleLat;
input group block treatment y@@;
datalines;
1 1 1 5.2 1 1 2 6.8 1 1 3 6.6 1 2 4 7.8 1 2 5 7.2 1 2 6 4.2
1 3 7 8.6 1 3 8 7 1 3 9 6.8 2 1 1 4.4 2 1 4 6.9 2 1 7 6.7
2 2 2 8.4 2 2 5 6.7 2 2 8 6 2 3 3 4.8 2 3 6 4.5 2 3 9 6.3
3 1 1 5.3 3 1 5 6.8 3 1 9 6.7 3 2 7 6.7 3 2 2 8.1 3 2 6 5.2
3 3 4 8.3 3 3 8 6.5 3 3 3 5.3
;
proc lattice data=tripleLat; run;
```

SAS OUTPUT

```
                          The SAS System

                      The Lattice Procedure

                     Analysis of Variance for y

                                                  Sum of        Mean
Source                                  DF       Squares      Square

Replications                             2        1.8363      0.9181
Blocks within Replications (Adj.)        6        2.0389      0.3398
  Component B                            6        2.0389      0.3398
Treatments (Unadj.)                      8       31.5030      3.9379
Intra Block Error                       10        4.5048      0.4505
Randomized Complete Block Error         16        6.5437      0.4090
Total                                   26       39.8830      1.5340


                    Additional Statistics for y

              Variance of Difference          0.3003
              LSD at .01 Level                1.6006
              LSD at .05 Level                1.1617
              Efficiency Relative to RCBD    90.7876


                    Treatment Means for y

                    Treatment          Mean

                        1            4.9667
                        2            7.7667
                        3            5.5667
                        4            7.6667
                        5            6.9000
                        6            4.6333
                        7            7.3333
                        8            6.5000
                        9            6.6000
```

10.6 Nested or Hierarchical Designs

Some experiments involving two factors or more are performed for which the levels of one factor are different for the different levels of the other factors. These classes of experiments are called nested or hierarchical designs. The designs are similar to the multistage sampling in the theory of sampling. This design is, however, different from the completely randomized design. In a typical completely randomized design with two factors A and B (such as the two-way layout discussed in Chapter 3), every level of B occurs for every level of A and the two factors are said to be completely crossed. There are situations where the factors or treatments cannot be completely crossed. One of the classes of designs used in such situations is the nested design. A factor B is said to be nested within the levels of A if some levels of the factor B occur only for a single level of the factor A. $B(A)$ indicates that B is nested within A. B is the nested factor while A is referred to as the nest factor. The design described above is referred to as a two-stage nested or hierarchical design. For example, a company has a number of technicians in each of the four laboratories (based in Glasgow [G], London [L], Melbourne [M], and New York [N]). Each of these technicians synthesizes small quantities of a rare drug. The company suspects that there are differences in the purities of drugs from each laboratory and decides to compare the purities of the drugs. We can illustrate the major difference between the completely crossed design and the nested design by describing two possible methods for comparing the purities of these drugs. One method is to choose three technicians at random from each of the laboratories to determine the purity of three out of the four drugs: this is the typical nested or hierarchical design. In this design, 12 technicians are required, three working in each of the four laboratories, but each technician working only in three out of the four laboratories. The technicians are, therefore, not comparable over all the laboratories and, therefore, there cannot be any interaction between technicians and laboratories. This is also an example of a balanced nested design. The second method is the completely crossed design. In this design, only three technicians are used and each goes to each of the four laboratories to perform three tests on all the four small quantities. If this experiment is replicated, it may be possible to calculate the interaction between the technicians and the laboratories.

We show the two designs below:

Completely crossed design

Factor B (Technicians)	**Factor A (Laboratories)**			
	G	**L**	**M**	**N**
1	T_1	T_1	T_1	T_1
2	T_2	T_2	T_2	T_2
3	T_3	T_3	T_3	T_3

Nested or hierarchical design

Factor B (Technicians)	Factor A (Laboratories)			
	G	**L**	**M**	**N**
1	T_1	×	×	×
2	T_2	×	×	×
3	T_3	×	×	×
4	×	T_4	×	×
5	×	T_5	×	×
6	×	T_6	×	×
7	×	×	T_7	×
8	×	×	T_8	×
9	×	×	T_9	×
10	×	×	×	T_{10}
11	×	×	×	T_{11}
12	×	×	×	T_{12}

In the above designs, we show only one replicate of the experiment. One consequence of nesting in the nested design is that nested effects are not comparable across the main effect. This is because some of the levels of the nested factor do not occur for all the levels of the main effect.

10.6.1 The Model, Assumptions, and Statistical Analysis

For the nested design, with two factors A at a levels, and B at b levels, and n replicates, the underlying design is

$$y_{ijk} = \mu + \tau_i + \beta_{j(i)} + \varepsilon_{(ij)k} \quad i = 1, \ldots, a \quad j = 1, \ldots, b \quad k = 1, \ldots, n \tag{10.39}$$

Analysis of the responses is based on least squares. The corrected total sum of squares can be subdivided into its component parts as follows:

$$SS_{\text{total}} = \sum_{i=1}^{a}\sum_{j=1}^{b}\sum_{k=1}^{n}(y_{ijk} - \overline{y}_{...})^2 = bn\sum_{i=1}^{a}(\overline{y}_{i..} - \overline{y}_{...})^2 + n\sum_{i=1}^{a}\sum_{j=1}^{b}(\overline{y}_{ij} - \overline{y}_{i..})^2 + \sum_{i=1}^{a}\sum_{j=1}^{b}\sum_{k=1}^{n}(y_{ijk} - \overline{y}_{ij.})^2 + CPT \tag{10.40}$$

(the cross product terms (CPT) are all zero). So that

$$SS_{\text{total}} = SS_A + SS_{B(A)} + SS_{\text{residual}} \tag{10.41}$$

The total degrees of freedom for the design is $abn-1$. The main effect A has $a-1$ degrees of freedom; the nested factor $B(A)$ has $a(b-1)$ degrees of freedom while the residual has $ab(n-1)$ degrees of freedom. The ANOVA for a two-stage nested design is presented in Table 10.13.

To simplify the manual calculation of the sums of squares required in the analysis of variance for this design, the following forms of the formulae are adopted:

$$SS_A = \frac{1}{bn}\sum_{i=1}^{a} y_{i..}^2 - \frac{y_{...}^2}{abn} \tag{10.42}$$

$$SS_{B(A)} = \frac{1}{n}\sum_{i=1}^{a} y_{ij.}^2 - \frac{1}{bn}\sum_{i=1}^{a} y_{i..}^2 \tag{10.43}$$

$$SS_{\text{residual}} = \sum_{i=1}^{a}\sum_{j=1}^{b}\sum_{k=1}^{n} y_{ijk}^2 - \frac{1}{n}\sum_{i=1}^{a}\sum_{j=1}^{b} y_{ij.}^2 \tag{10.44}$$

$$SS_{\text{total}} = \sum_{i=1}^{a}\sum_{j=1}^{b}\sum_{k=1}^{n} y_{ijk}^2 - \frac{y_{...}^2}{abn} \tag{10.45}$$

TABLE 10.13: Typical ANOVA for the Two-Stage Nested Design

Source	Sum of squares	Degrees of freedom	Mean square	*F*-ratio
A	SS_A	$a-1$	$SS_A/(a-1)$	$\text{MS}_A/MS_{\text{residual}}$
$B(A)$	$SS_{B(A)}$	$a(b-1)$	$SS_{B(A)}/a(b-1)$	$\text{MS}_{B(A)}/MS_{\text{residual}}$
Residual	SS_{residual}	$ab(n-1)$	$SS_{\text{residual}}/ab(n-1)$	
Total	SS_{total}	$abn-1$		

In the model stated in Equation 10.39, the effects can either be fixed or random. A factor that is considered to be random is better understood if we regard the levels of that factor as a random sample from a population of similar potential levels. The treatment effect is, therefore, considered random with an unknown variance σ^2; the nested effect is similarly random with unknown variance σ^2, if both nested and nesting factors are regarded as random. In the two-stage nested design, we could have both factors A and B fixed (or nonrandom) or we can have A fixed and B random. If A and B are fixed, we could carry out the analysis of variance described in Table 10.13. On the other hand, if one or the two effects are random, the analysis is different and is referred to as analysis of components of variance. The assumptions about the model determine the best method for testing the significance of the effects of the factors A and B. If A and B have fixed effects, then $\sum_{i=1}^{a}\tau_i=0$ while

$\sum_{j=1}^{b}\beta_{j(i)}=0$ (for $i=1,\ldots,a$). This means that the sum of all the A treatment effects is zero while the sum of all the B treatment effects within each level of A is zero. In this case, we can carry out the test of the significance of the effects of the factors A and B (i.e., $H:\tau_i=0$ and $H:\beta_j(i)=0$) by dividing the MS_A and $MS_{B(A)}$ by MS_{residual}, which is the usual analysis of variance. When A is fixed and B is random, then $\sum_{i=1}^{a}\tau_i=0$ and $\beta_j(i)\sim \text{NID}(0,\,\sigma_\beta^2)$, which means that the sum of all effects of A is zero while the effect of the nested factor B within each level of A is a normally independent and identically distributed random variable with mean zero and variance σ_β^2. In this case, we test the hypothesis $H_0:\tau_i=0$ using the statistic $MS_A/MS_{\text{residual}}$ while we test the hypothesis, $H_0:\sigma_\beta^2=0$ by employing the statistic, $MS_{B(A)}/MS_{\text{Error}}$. On the other hand, if both nest and nested factors, A and B are random, then $\tau_i\sim \text{NID}(0,\,\sigma_\tau^2)$ and $\beta_j(i)\sim \text{NID}(0,\,\sigma_\beta^2)$. In such cases, the hypotheses, $H_0:\sigma_\tau^2$ and $H_0:\sigma_\beta^2=0$, are, respectively, tested using the statistics, $MS_A/MS_{B(A)}$ and $MS_{B(A)}/MS_{\text{Error}}$. In Table 10.14, we present the different model assumptions and the expected mean squares for each of them.

TABLE 10.14: Expected Mean Squares in the Two-Stage Nested Design

E **(Mean square)**	***A*, *B* Fixed**	***A* Fixed, *B* Random**	***A*, *B* Random**
$E\ (MS_A)$	$\sigma^2+\dfrac{bn\sum_{i=1}^{a}\tau_i^2}{a-1}$	$\sigma^2+n\sigma_\beta^2+\dfrac{bn\sum_{i=1}^{a}\tau_i^2}{a-1}$	$\sigma^2+n\sigma_\beta^2+bn\sigma_\tau^2$
$E\ [MS_{B(A)}]$	$\sigma^2+\dfrac{bn\sum_{i=1}^{a}\sum_{j=1}^{b}\beta_{j(i)}^2}{a(b-1)}$	$\sigma^2+n\sigma_\beta^2$	$\sigma^2+n\sigma_\beta^2$
$E\ (MS_{\text{residual}})$	σ^2	σ^2	σ^2

When B is random, the variance of its effect is estimated by the quantity

$$\hat{\sigma}_\beta^2=\frac{1}{n}\left[\frac{SS_{B(A)}}{a(b-1)}-\frac{SS_{\text{residual}}}{ab(n-1)}\right]$$

Also when A is random, the variance of its effect is estimated as

$$\hat{\sigma}_\tau^2=\frac{1}{bn}\left[\frac{SS_A}{(a-1)}-\frac{SS_{B(A)}}{a(b-1)}\right]$$

Example 10.4

Consider the example discussed in Section 10.6. The purity of the drug was determined, three times by each of the three technicians in each of the four laboratories. The responses from the experiment are as follows:

Laboratories											
I			II			III			IV		
Technicians											
I	**II**	**III**	**IV**	**V**	**VI**	**VII**	**VIII**	**IX**	**X**	**XI**	**XII**
88	82	83	81	89	91	94	90	90	87	80	94
89	83	80	78	84	81	90	89	85	80	81	89
91	86	84	79	80	87	88	87	87	82	84	88

We analyze the responses to compare the drugs from the laboratories and study the differences between technicians.

By analogy with our discussions above, let laboratories be the factor A and technicians be the factor B. Then the factor B is nested within factor A. We note that $y_{..} = 3081$, $y_1 = 766$, $y_2 = 750$, $y_{3.} = 800$, $y_{4.} = 765$.

$$y_{11.} = 268 \quad y_{12.} = 251 \quad y_{13.} = 247 \quad y_{24.} = 238 \quad y_{25.} = 253 \quad y_{26.} = 259$$
$$y_{37.} = 272 \quad y_{38.} = 266 \quad y_{39.} = 262 \quad y_{410.} = 249 \quad y_{411.} = 245 \quad y_{412.} = 271$$

The sums of squares required for ANOVA are (Table 10.15)

$$SS_A = \frac{(766^2 + 750^2 + 800^2 + 765^2)}{9} - \frac{3081^2}{36} = 148.972$$

$$SS_{B(A)} = \frac{(268^2 + 251^2 + \cdots + 245^2 + 271^2)}{3} - \frac{(766^2 + 750^2 + 800^2 + 765^2)}{9}$$
$$= 308.4445$$

$$SS_{\text{total}} = (88^2 + 82^2 + 83^2 + \cdots + 84^2 + 88^2) - \frac{3081^2}{36} = 666.75$$

$$SS_{\text{residual}} = 666.75 - 148.972 - 308.4445 = 209.3335$$

TABLE 10.15: ANOVA for the Drug Purity Problem

Source	**Sum of squares**	**Degrees of freedom**	**Mean square**	***F*-ratio**
Laboratories (A)	148.972	3	49.657	5.693
Technicians [$B(A)$]	308.4445	8	8.54	4.419
Residual	209.3335	24	8.722	
Total	666.75	35		

From the tables, $F(3, 24, 0.01) = 4.72$ and $F(8, 24, 0.01) = 3.36$. Clearly, both A and $B(A)$ are significant. We can carry out further analysis of the residuals using principles described in Chapter 2 to check that design assumptions are satisfied.

We present a SAS program to carry out the analysis of the responses of the designed experiment in Example 10.4.

SAS PROGRAM

```
data purity1;

input lab tech y@@;

datalines;

1 1 88 1 1 89 1 1 91 1 2 82 1 2 83 1 2 86 1 3 83 1 3 80 1 3 84 2 4 81
2 4 78 2 4 79 2 5 89 2 5 84 2 5 80 2 6 91 2 6 81 2 6 87 3 7 94 3 7 90
3 7 88 3 8 90 3 8 89 3 8 87 3 9 90 3 9 85 3 9 87 4 10 87 4 10 80 4 10 82
4 11 80 4 11 81 4 11 84 4 12 94 4 12 89 4 12 88
;

proc glm data=purity1;
class lab tech;
model y=lab tech(lab); run;
```

SAS OUTPUT

```
                         The SAS System

                        The GLM Procedure

                     Class Level Information

              Class     Levels  Values

              lab            4  1 2 3 4

              tech          12  1 2 3 4 5 6 7 8 9 10 11 12

                  Number of Observations Read      36
                  Number of Observations Used      36

                          The SAS System

                         The GLM Procedure

Dependent Variable: y

                                  Sum of
    Source          DF           Squares    Mean Square    F Value    Pr > F

    Model           11       457.4166667     41.5833333       4.77    0.0007

    Error           24       209.3333333      8.7222222
```

```
Corrected Total  35      666.7500000

                R-Square  Coeff Var   Root MSE    y Mean

                0.686039   3.450836   2.953341   85.58333


  Source            DF       Type I SS     Mean Square    F Value     Pr > F

  lab                3     148.9722222      49.6574074       5.69     0.0043
  tech(lab)          8     308.4444444      38.5555556       4.42     0.0022


  Source            DF      Type III SS    Mean Square   F Value      Pr > F

  lab                3       148.9722222    49.6574074       5.69     0.0043
  tech(lab)          8       308.4444444    38.5555556       4.42     0.0022
```

10.6.2 Unbalanced Two-Stage Nested Design

The two-stage nested design described in the previous section can strictly be regarded as a balanced two-stage nested design. The design is balanced because, for each level of B that is nested in A, the observations are replicated the same number of times. In the more general case of this design, the number of observations made at each level of B is different. Hence, the design model could best be written as

$$Y_{ijk} = \mu + \tau_i + \beta_{j(i)} + \varepsilon_i \tag{10.46}$$

where $i = 1, \ldots, a; j = 1, \ldots, b$; and $k = 1, \ldots, N_{ij}$; so that

$$N = \sum_{i=1}^{a} N_i \quad \text{or} \quad N = \sum_{i=1}^{a}\sum_{j=1}^{b} N_{ij}$$

As usual, $\beta_{j(i)}$ is the effect of the jth level of B that is nested in A; τ_i is the effect of the ith level of A, while μ is the overall average. N_{ij} is the number of observations under the jth level of B that occurs under the ith level of A.

We obtain the following:

$$\overline{y}_{ij.} = \frac{\sum_{k=1}^{N_{ij}} y_{ijk}}{N_{ij}} \tag{10.47}$$

$$\overline{y}_{i..} = \frac{\sum_{j=1}^{b_i} \sum_{k=1}^{N_{ij}} y_{ijk}}{N_{i.}} = \frac{\sum_{j=1}^{b_i} N_{ij} y_{ij.}}{N_i} \tag{10.48}$$

$$\overline{y}_{...} = \frac{\sum_{i=1}^{a}\sum_{j=1}^{b_i}\sum_{k=1}^{N_{ij}} y_{ijk}}{N} = \frac{\sum_{j=1}^{b_i} N_i y_i}{N} \tag{10.49}$$

The sums of squares for this design are given as follows:

$$SS_{\text{total}} = \sum_{i=1}^{a}\sum_{j=1}^{b_i}\sum_{k=1}^{N_{ij}} (y_{ijk} - \overline{y}_{...})^2$$

with $N-1$ degrees of freedom;

$$SS_A = \sum_{i=1}^{a} N_i(\overline{y}_i - \overline{y}_{...})^2$$

with $a-1$ degrees of freedom;

$$SS_{B(A)} = \sum_{i=1}^{a}\sum_{j=1}^{b_i} N_{ij}(\overline{y}_{ij.} - \overline{y}_{i..})^2$$

with $d_{b(a)} = \sum_{i=1}^{a}(b_i - 1)$ degrees of freedom, while

$$SS_{\text{residual}} = \sum_{i=1}^{a}\sum_{j=1}^{b_i}\sum_{k=1}^{N_{ij}} (y_{ijk} - \overline{y}_{ij.})^2$$

with $d_{\text{res}} = \sum_{i=1}^{a}\sum_{j=1}^{b_i}(N_{ij} - 1)$ degrees of freedom.

The following hypotheses may be of interest under this model:

$$H_a\colon \tau_1 = \tau_2 = \cdots = \tau_a \quad \text{vs} \quad H_1\colon \text{at least one } \tau_i \text{ is different}$$

$$H_{b(a)}\colon \beta_1(i) = \beta_2(i) = \cdots = \beta_{bi}(i) \quad \text{vs} \quad H_1\colon \text{at least one } \beta_j(i) \text{ is different}$$

These hypotheses are tested by assuming the side conditions that

$$\sum_{i=1}^{a} N_i\tau_i = 0; \quad \sum_{i=1}^{a} \frac{N_{ij}\beta_{j(i)}}{N_i} = 0$$

H_a is tested in ANOVA by computing the quantity

$$F_{\text{cal}} = \left(\frac{SS_A}{a-1}\right) \Bigg/ \left(\frac{SS_{\text{residual}}}{\sum_{i=1}^{a}\sum_{j=1}^{b_i}(N_{ij}-1)}\right)$$

The hypothesis is rejected at $\alpha\%$ level of significance if $F_{\text{cal}} > F(a-1, d_{\text{res}}, \alpha)$. Similarly, to test $H_{b(a)}$, we compute the quantity:

$$F_{\text{cal}} = \left(\frac{SS_{B(A)}}{\sum_{i=1}^{a} (b_i - 1)} \right) \Bigg/ \left(\frac{SS_{\text{residual}}}{\sum_{i=1}^{a} \sum_{j=1}^{b_i} (N_{ij} - 1)} \right)$$

Again, the hypothesis is rejected at $\alpha\%$ level of significance if $F > F(d_{b(a)}, d_{\text{res}}, \alpha)$. As usual, we could use the diagnostic techniques of Chapter 2 to check that the basic assumptions of the model are satisfied.

Example 10.5

A company that produces seasoning oils from ginger roots at three production plants uses four different grades of the roots. The management suspects that the yields from the four grades of roots are different. Since the skills at the plants vary, a study was carried out with each production plant processing the four grades of roots in unequal replicates. Three different batches of the four grades of root were used at the three plants. The data show percentage yield:

Production plants	**I**				**II**				**III**			
Root grades	**1**	**2**	**3**	**4**	**1**	**2**	**3**	**4**	**1**	**2**	**3**	**4**
	20	20	33	22	22	40	21	41	28	19	30	29
	21	27	41	24	23	38	36	43	26	32	35	38
	30	32	36	31	28	33	34	44	24	25	37	40
	47			28	24			41	21			27

□

Analyze the data.

Let A represent the production plants while B represents the root grades that are nested in A.

$\overline{y}_{1..} = 412/14 = 29.4286$ $\overline{y}_{2..} = 468/14 = 33.4286$ $\overline{y}_{3..} = 411/14 = 29.35714$

$\overline{y}_{11} = 118/4 = 29.5$ $\overline{y}_{12} = 79/3 = 26.333$ $\overline{y}_{13.} = 110/3 = 36.667$

$\overline{y}_{14.} = 105/4 = 26.25$ $\overline{y}_{21.} = 97/4 = 24.25$ $\overline{y}_{22.} = 111/3 = 37.0$

$\overline{y}_{23.} = 91/3 = 30.333$ $\overline{y}_{24.} = 169/4 = 42.25$ $\overline{y}_{31.} = 99/4 = 24.75$

$\overline{y}_{32.} = 70/3 = 23.333$ $\overline{y}_{33.} = 102/3 = 34.0$ $\overline{y}_{34.} = 134/4 = 33.5.$

For ANOVA, we calculate the sums of squares for the different factors in this design (Table 10.16):

$$
\begin{aligned}
SS_{\text{total}} &= (20^2 + 21^2 + 30^2 + \cdots + 40^2 + 27^2) - \frac{1291^2}{42} = 2432.1190 \\
SS_A &= 14(29.4286 - 30.7380)^2 + 14(33.4286 - 30.7380)^2 \\
&\quad + 14(29.35714 - 30.7380)^2 = 152.0476 \\
SS_{B(A)} &= 4(29.5 - 29.4286)^2 + 3(26.333 - 29.4286)^2 \\
&\quad + 3(36.667 - 29.4286)^2 + 4(33.5 - 29.4286)^2 \\
&\quad + 4(24.25 - 33.4286)^2 + 3(37 - 33.4286)^2 \\
&\quad + (30.333 - 33.4286)^2 + 4(42.25 - 33.4286)^2 \\
&\quad + 4(24.75 - 29.35714)^2 + (23.333 - 29.35714)^2 \\
&\quad + 3(34 - 29.35714)^2 + 4(33.5 - 29.35714)^2 = 1208.405 \\
SS_{\text{residual}} &= 2432.1190 - 152.0476 - 1208.405 = 1071.667
\end{aligned}
$$

TABLE 10.16: ANOVA for the Root-Oil Problem

Source	Sum of squares	Degrees of freedom	Mean square	F-ratio	p-Value
A	152.0476	2	76.0238	2.12819281	0.136677
$B(A)$	1208.405	9	134.26733	3.75864891	0.002911
Residual	1071.667	30	35.722233		
Total	2432.119	41			

From the tables, we find that $F(2, 30, 0.05) = 3.32$ and $F(9, 30, 0.05) = 2.21$. Therefore, only $B(A)$ is significant.

We present a SAS program that analyzes the responses of root-oil experiment of Example 10.5.

SAS PROGRAM

```
data rootgrades;
input prodP rootG y@@;
datalines ;
1 1 20 1 1 21 1 1 30 1 1 47 1 2 20 1 2 27 1 2 32 1 3 33 1 3 41 1 3 36
1 4 22 1 4 24 1 4 31 1 4 28 2 1 22 2 1 23 2 1 28 2 1 24 2 2 40 2 2 38
2 2 33 2 3 21 2 3 36 2 3 34 2 4 41 2 4 43 2 4 44 2 4 41 3 1 28 3 1 26
3 1 24 3 1 21 3 2 19 3 2 32 3 2 25 3 3 30 3 3 35 3 3 37 3 4 29 3 4 38
3 4 40 3 4 27
;
proc glm data=rootgrades;
classf prodP rootG;
model y=prodP prodP(rootG); run;
```

SAS OUTPUT

```
                          The SAS System

                        The GLM Procedure

                     Class Level Information

                    Class     Levels    Values

                    prodP          3     1 2 3

                    rootG          4   1 2 3 4

               Number of Observations Read      42
               Number of Observations Used

                          The SAS System
                        The GLM Procedure

Dependent Variable: y

                                      Sum of
  Source                  DF         Squares   Mean Square   F Value   Pr > F

  Model                   11     1360.452381    123.677489      3.46   0.0033

  Error                   30     1071.666667     35.722222

  Corrected Total         41     2432.119048


                R-Square   Coeff Var   Root MSE    y Mean

                0.559369   19.44430    5.976807   30.73810

   Source                DF      Type I SS    Mean Square   F Value    Pr > F

   prodP                  2     152.047619      76.023810      2.13    0.1367
   prodP(rootG)           9    1208.404762     134.267196      3.76    0.0029


   Source                DF    Type III SS    Mean Square   F Value    Pr > F

   prodP                  2     140.837302     70.418651       1.97    0.1569
   prodP(rootG)           9    1208.404762    134.267196       3.76    0.0029
```

10.6.3 Higher Order Nested Designs

We saw in our study of factorial designs that it is possible to have higher order designs. We noticed that as the order of the factorial design increased, the complexity of the design and the cumbersomeness of the analysis increased. It is also possible to have higher order nested designs. We have just discussed the two-stage nested design. The next possible order of nested designs is the three-stage nested design. For such designs the appropriate model is

$$y_{ijkl} = \mu + \beta_{j(i)} + \gamma_{k(ij)} + \varepsilon_{ijkl} \tag{10.50}$$

where $i = 1, \ldots, a$; $j = 1, \ldots, b$; $k = 1, \ldots, c$; and $l = 1, \ldots, N_{ijk}$.

Just as it is possible to have an m-stage multistage sampling, it is possible to have an m-stage nested design. However, the design and its analysis get more complicated as the stages grow.

A common variation of the nested design is that in which some factors are crossed while others are nested. These designs are often called partial hierarchical designs. We discuss these designs in the next section.

10.7 Designs with Nested and Crossed Factors

Sometimes in experiments involving more than two factors, it is possible for some of the factors to be crossed with each other while some factors are nested in others. Such designs are said to be partially nested or hierarchical. Consider a design with factors A, B, and C with C nested in B. For this design, all the levels of C are different for all the levels of B. The suitable model for this design is

$$y_{ijkl} = \mu + \tau_i + \gamma_j + \beta_{k(j)} + (\tau\gamma)_{(ij)} + (\tau\beta)_{ik(j)} + \varepsilon_{ijkl} \tag{10.51}$$

where $i = 1, 2, \ldots, a$; $j = 1, \ldots, b$; $k = 1, \ldots, c$; and $l = 1, \ldots, n$.

The symbol τ_i is the effect of A, γ_j is the effect of B while $\beta_{k(j)}$ is the effect of kth level of C within B, $(\tau\gamma)_{ij}$ is the effect of AB interaction while $(\tau\beta)_{ik(j)}$ is the $A \times C$ within B interaction effect while ε_{ijkl} is the error. In this design, there is no BC interaction because all the levels of C do not occur for all the levels of B. Similarly, there is no second-order interaction ABC for the same reasons.

The sums of squares and degrees of freedom required for the test of hypotheses in this design are as follows:

$$SS_{\text{total}} = \sum_{i=1}^{a}\sum_{j=1}^{b}\sum_{k=1}^{c}\sum_{l=1}^{n} y_{ijkl}^2 - \frac{y_{....}^2}{abcn} \tag{10.52}$$

$$SS_A = \sum_{i=1}^{a} \frac{y_{i...}^2}{bcn} - \frac{y_{....}^2}{abcn} \tag{10.53}$$

$$SS_B = \sum_{j=1}^{b} \frac{y_{.j..}^2}{acn} - \frac{y_{....}^2}{abcn} \tag{10.54}$$

$$SS_{AB} = \sum_{i=1}^{a}\sum_{j=1}^{b} \frac{y_{ij..}^2}{cn} - \frac{y_{....}^2}{abcn} - SS_A - SS_B \tag{10.55}$$

$$SS_{C(B)} = \sum_{j=1}^{b}\sum_{k=1}^{c} \frac{y_{.jk.}^2}{an} - \sum_{j=1}^{b} \frac{y_{.j..}^2}{acn} \tag{10.56}$$

$$SS_{AC(B)} = \sum_{i=1}^{a}\sum_{j=1}^{b}\sum_{k=1}^{c} \frac{y_{ijk.}^2}{n} - \sum_{j=1}^{b}\sum_{k=1}^{c} \frac{y_{.jk.}^2}{an} - \sum_{i=1}^{a}\sum_{j=1}^{c} \frac{y_{ij..}^2}{cn} + \sum_{k=1}^{c} \frac{y_{.j..}^2}{acn} \tag{10.57}$$

$$SS_{\text{residual}} = SS_{\text{total}} - SS_A - SS_{AB} - SS_{C(B)} - SS_{AC(B)} \tag{10.58}$$

SS_{total} has $abcn-1$ degrees of freedom. Of these, SS_A has $a-1$, SS_B has $b-1$, SS_{AB} has $(a-1)(b-1)$, $SS_{C(B)}$ has $b(c-1)$, $SS_{B(AC)}$ has $b(a-1)(c-1)$ while the SS_{residual} has $abc(n-1)$ degrees of freedom.

Example 10.6

An experiment was performed to compare three methods for teaching statistics to higher school students. It was decided to study any possible differences between performances of boys and girls. The three teaching methods were student-centered learning (SCL), lecture and tutorials (LAT), and mixture of student-centered and lecture (MSL). Since the schools were not co-educational, three different teams from the three who taught at the boys' schools were chosen to handle the teaching of students at each of the three girls' schools. Each class involved in the experiment was divided at random into three groups, each of which was taught with one of the three different methods. At the end of the syllabus, three students from each group were chosen at random to represent each school at the examination that followed. Their scores out of 50 are recorded below:

	Schools					
	Boys			Girls		
Teaching method	**I**	**II**	**III**	**I**	**II**	**III**
SCL	19	18	21	27	11	17
	39	10	14	25	15	16
	25	33	14	23	26	10
LAT	38	23	28	36	35	32
	20	24	30	34	31	28
	41	22	26	30	26	31
MSL	36	38	35	33	29	28
	32	30	31	24	30	30
	27	28	26	22	31	28

▯

Analyze the data and state your conclusions. For analyses, let the factor teaching method with the levels SCL, LAT, and MSL be A; let gender (with levels of boys and girls) be B; and let C be the factor schools that has three levels, which are clearly nested in the gender B. The model of Equation 10.51, therefore, applies to

$$SS_{\text{total}} = (19^2 + 39^2 + 25^2 + \cdots + 28^2 + 30^2 + 28^2) - \frac{1436^2}{54} = 3031.037$$

$$SS_A = \frac{(363^2 + 535^2 + 538^2)}{18} - \frac{1436^2}{54} = 1115.148$$

$$SS_B = \frac{(728^2 + 708^2)}{27} - \frac{1436^2}{54} = 7.407$$

$$\begin{aligned} SS_{AB} &= \frac{(193^2 + 170^2 + 252^2 + 283^2 + 283^2 + 255^2)}{9} - 1115.148 - 7.407 \\ &= 118.9265 \end{aligned}$$

$$\begin{aligned} SS_{C(B)} &= \frac{(277^2 + 226^2 + 225^2 + 254^2 + 234^2 + 220^2)}{9} - \frac{(728^2 + 708^2)}{27} \\ &= 261.41 \end{aligned}$$

$$\begin{aligned} SS_{AC(B)} &= \frac{(83^2 + 61^2 + 49^2 + 99^2 + \cdots + 91^2 + 86^2)}{3} - \frac{346102}{9} - \frac{354856}{9} \\ &\quad + 38194.37037 = 308.1482 \end{aligned}$$

$$\begin{aligned} SS_{\text{residual}} &= 3031.037 - 1115.148 - 7.407 - 118.9265 - 261.41 - 308.1482 \\ &= 1219.9972 \end{aligned}$$

The analysis is presented in Table 10.17.

TABLE 10.17: ANOVA for the Teaching Method Problem

Source	**Sum of squares**	**Degrees of freedom**	**Mean square**	***F*-ratio**
Method	1115.148	2	557.574	16.4529
Gender	7.407	1	7.407	0.2186
Method*gender	118.9265	2	59.463	1.7546
School (gender)	261.410	4	65.353	1.9284
Method*school (gender)	308.1482	8	38.518	1.1365
Residual	1219.9973	36	33.889	
Total	3031.0370	53		

From the tables, $F(1, 36, 0.05) = 4.116$, $F(2, 36, 0.05) = 3.182$, $F(4, 36, 0.05) = 2.642$, and $F(8, 36, 0.05) = 2.216$. Clearly, only the teaching method factor is significant. As usual, all the basic assumptions about the design could be checked, using the methods outlined in Chapter 2. It is also possible to go further and compare the teaching methods to each other using one of the methods described in Chapter 2 and used in other chapters.

10.7.1 SAS Program for Example 10.6

Next, we write a program to analyze the responses of the designed experiment in Example 10.6 and present the output after analysis.

SAS PROGRAM

```
data learning;
input TeachM$ Gender$ School Y@@;
datalines;
SCL Boys 1 19 SCL Boys 1 39 SCL Boys 1 25
SCL Boys 2 18 SCL Boys 2 10 SCL Boys 2 33
SCL Boys 3 21 SCL Boys 3 14 SCL Boys 3 14
SCL girls 1 27 SCL girls 1 25 SCL girls 1 23
SCL girls 2 11 SCL girls 2 15 SCL girls 2 26
SCL girls 3 17 SCL girls 3 16 SCL girls 3 10
LAT Boys 1 38 LAT Boys 1 20 LAT Boys 1 41
LAT Boys 2 23 LAT Boys 2 24 LAT Boys 2 22
LAT boys 3 28 LAT Boys 3 30 LAT Boys 3 26
LAT girls 1 36 LAT girls 1 34 LAT girls 1 30
LAT girls 2 35 LAT girls 2 31 LAT girls 2 26
LAT girls 3 32 LAT girls 3 28 LAT girls 3 31
MSL Boys 1 36 MSL Boys 1 32 MSL Boys 1 27
MSL Boys 2 38 MSL Boys 2 30 MSL Boys 2 28
MSL boys 3 35 MSL Boys 3 31 MSL Boys 3 26
MSL girls 1 33 MSL girls 1 24 MSL girls 1 22
MSL girls 2 29 MSL girls 2 30 MSL girls 2 31
MSL girls 3 28 MSL girls 3 30 MSL girls 3 28
;
proc print data=learning; run;
proc glm data=learning;
class TeachM Gender School;
model y=TeachM Gender TeachM*Gender School(Gender) TeachM*School(Gender); run;
```

SAS OUTPUT

```
                    The GLM Procedure

                 Class Level Information

               Class     Levels  Values

               TeachM         3  LAT MSL SCL

               Gender         3  Boys girls
               School         3  1 2 3
```

```
                    Number of Observations Read      54
                    Number of Observations Used      54
                           The GLM Procedure

Dependent Variable: Y
                                        Sum of
  Source                      DF       Squares    Mean Square    F Value    Pr > F

  Model                       17   1811.037037     106.531590       3.14    0.0019
  Error                       36   1220.000000      33.888889
  Corrected Total             53   3031.037037


                  R-Square   Coeff Var    Root MSE     Y Mean

                  0.597497    21.89112    5.821416   26.59259


  Source                        DF     Type I SS    Mean Square    F Value    Pr > F


  TeachM                         2   1115.148148     557.574074      16.45    <.0001
  Gender                         1      7.407407       7.407407       0.22    0.6429
  TeachM*Gender                  4    118.925926      59.462963       0.75    0.1874
  School(Gender)                 4    261.407407      65.351852       1.93    0.1268
  Teach*School(Gender)           8    308.148148      38.518519       1.14    0.3633

  Source                        DF   Type III SS    Mean Square    F Value    Pr > F


  TeachM                         2   1115.148148    557.574074      16.45    <.0001
  Gender                         1      7.407407      7.407407       0.22    0.6429
  TeachM*Gender                  2    118.925926     59.462963       1.75    0.1874
  School(Gender)                 4    261.407407     65.351852       1.93    0.1268
  Teach*School(Gender)           8    308.148148     38.518519       1.14    0.3633
```

10.8 Exercises

Exercise 10.1

In an industrial experiment, the yield of the process was studied under seven different levels of reactant concentrations. However, only four of the processes can be run in a day so a symmetric BIBD with four replicates was used. The table below shows the responses in percentage yields of the process. Write down the model for this design, analyze the responses, and decide whether the processes are significantly different. Use $\alpha = 5\%$.

	Process						
Shifts	**I**	**II**	**III**	**IV**	**V**	**VI**	**VII**
1			70	79	77	79	
2		72		53	44		61
3	65	55	64			64	
4	51	55	40	53			
5			73	54	43		75
6	43				44	40	63
7	44	52				44	69

□

Exercise 10.2

An experiment was performed to compare five treatments. Since only four of these could be accomplished in a day, 5 days were used, leading to a BIBD. The responses of the experiments are as follows:

	Days (blocks)				
Treatments	**I**	**II**	**III**	**IV**	**V**
1	4.4		4.33	4.57	11.8
2	6	6.8	7.1	7.9	
3		8.9	10.5	9.1	10.67
4	15.44	16.1	17.03		17.8
5	10.17	10.65		10.67	10.67

Analyze the results of this experiment and compare the treatment effects. What is the value of λ for this design? Examine the residuals of this design and report on the satisfaction of the basic assumptions. □

Exercise 10.3

Which of the following proposed designs can be constructed as BIBDs? For those identified as BIBDs, write down the assignment of units into blocks.

(1) $t=8,\ k=4,\ b=14,\ r=8$

(2) $t=14,\ k=3,\ b=28,\ r=6$

(3) $t=16,\ k=6,\ b=24,\ r=9$

(4) $t=8,\ k=4,\ b=14,\ r=7$

(5) $t=9,\ k=3,\ b=16,\ r=4$

□

Exercise 10.4

Construct all the symmetric BIBDs for $t=3,\ldots 6$. □

Exercise 10.5

An industrial experiment was performed that involved 10 treatments for which a BIBD could only be accomplished in 15 shifts. Write down the model for this design. The responses for the experiment are as follows:

	Treatments									
Blocks	**I**	**II**	**III**	**IV**	**V**	**VI**	**VII**	**VIII**	**IX**	**X**
1		19.6	16.8			10.6				6.6
2	18.7	16.7		10.4	10.0					
3		18		14.3		6.8	8.3			
4		18.8			14.4			8.4		6.8
5		18.6					10.1	8.6	6.6	
6	20.0			14.5				8.2		4.8
7	18.3		16.7		12.8		6.8			
8	18.3	18.3	16.2						6.3	
9					18.4		12.3		10.1	4.0
10				16.7	18.0	12.0			6.3	
11			18.3	16.1					6.7	4.6
12	18.7					12.7	12.6			4.8
13	18.8					14.3		10.4	8.0	
14			18	16.3			8.8	6.8		
15			18.3		16.3	12.3		6.8		

Analyze the data and draw conclusions. Take the level of significance to be 5%. Was it necessary to separate data based on shifts? □

Exercise 10.6

Two laboratories were used to determine the purity of a chemical compound synthesized from three sources. Within each of these laboratories, three technicians were used to carry out the analysis.

	Lab I			Lab II		
Technicians	**I**	**II**	**III**	**IV**	**V**	**VI**
Source 1	10	8	11	17	11	17
	9	10	14	15	15	16
	15	13	14	13	11	10
Source 2	8	13	18	26	25	22
	20	24	30	24	21	28
	21	22	16	20	26	21
Source 3	26	18	25	23	29	28
	22	20	21	14	20	20
	17	18	16	12	18	18

Analyze the data and state your conclusions. □

Exercise 10.7

Suppose that only a simple lattice was studied for Example 10.2 leading to the following data. Analyze the responses of the experiment and reach conclusion.

Replicate	Blocks			
1	1	(1) 5.2	(2) 6.8	(3) 6.6
	2	(4) 7.8	(5) 7.2	(6) 4.2
	3	(7) 8.6	(8) 7.0	(9) 6.8
2	1	(1) 4.4	(4) 6.9	(7) 6.7
	2	(2) 8.4	(5) 6.7	(8) 6.0
	3	(3) 4.8	(6) 4.5	(9) 6.3

□

Exercise 10.8

Write a SAS program to analyze the responses of Exercise 10.7. □

Exercise 10.9

The data below present the yield of alfalfa under nine different fertilizer treatments, studied in 12 incomplete blocks leading to a BIBD.

	Treatments								
Blocks	**1**	**2**	**3**	**4**	**5**	**6**	**7**	**8**	**9**
1	59	26	38						
2				85	92	69			
3	63			70			68		
4		26			98			59	
5							74	52	27
6			31			60			60
7	62				85				85
8		23				73	75		
9			49	74				51	
10	52					76		43	
11		18		79					79
12			42		84		81		

Analyze the data and determine whether treatment and block are significant at 5%. Obtain estimates of treatment effects. □

Exercise 10.10

A gage repeatability experiment was performed by Lexon Corporation using three instruments to measure parts set to diameter of 20 mm. It was thought that operator differences might affect the outcomes. Each of three operators used three of the same type of instrument to make four determinations of the diameters of a part, leading to a nested design. The responses are as follows:

Operator	I	II	III
	I1: 20.5 20.2 20.3 20.4	I4: 20.7 20.8 20.8 20.4	I7: 20.8 20.9 21.0 20.9
	I2: 21.0 20.7 20.8 21.1	I5: 21.2 21.3 20.8 20.6	I8: 20.6 20.5 20.2 20.3
	I3: 20.1 20.9 20.6 20.7	I6: 20.8 21.3 21.4 21.4	I9: 21.4 21.5 20.7 21.3

Analyze the data and test for significance of operator and instrument at 5% level of significance. □

Exercise 10.11
Write a SAS program to carry out the analysis of the responses for the experiment described in Exercise 10.10. □

Exercise 10.12
Write a SAS program to analyze the responses of the experiment described in Exercise 10.6. □

Exercise 10.13
Write a SAS program to analyze the responses of the experiment described in Exercise 10.9. □

Exercise 10.14
Obtain the estimates of treatment effects in the design of Exercise 10.9 and use LSD method to compare the treatment effects at 1% level of significance. □

Exercise 10.15
Write a SAS program to analyze the data of Exercise 10.5, extract residuals, and fitted values, and carry out a normal probability plot of the residuals. □

Exercise 10.16
Write a SAS program to analyze the responses of the experiment described in Exercise 10.2. □

References

Cochran, W.G. and Cox, G.M. (1957) *Experimental Designs*, 2nd ed., Wiley, New York.

Cox, D.R. (1958) *Planning Experiments,* Wiley, New York.

Cox, D.R. and Hinkley, D.V. (1974) *Theoretical Statistics*, Chapman and Hall, London.

Davies, O.L. (1956) *Design and Analysis of Industrial Experiments*, 2nd ed., Oliver and Boyd, Edinburgh.

Fisher, R.A. and Yates, F. (1953) *Statistical Tables for Biological, Agricultural, and Medical Research*, 4th ed., Oliver and Boyd, Edinburgh.

John, P.W.M. (1971) *Statistical Design and Analysis of Experiments*, Macmillan, New York.

Kempthorne, O. (1952) *The Design and Analysis of Experiments*, John Wiley, NY.

Mood, A.M., Graybill, F.A., and Boes, D.C. (1974) *Introduction to Theory of Statistics*, 3rd ed., McGraw-Hill, New York.

Peterson, R.G. (1985) *Design and Analysis of Experiments*, Marcel Dekker, NY.

Silvey, S.D. (1975) *Statistical Inference*, Chapman and Hall, London.

Yates, F. (1940) The recovery of inter-block information in balanced incomplete block designs. *Annals of Eugenics*, 10, 317–325.

Chapter 11

Methods for Fitting Response Surfaces and Analysis of Covariance

In the previous chapters on factorial experiments, we came across factors that were either quantitative or qualitative. We know that when a factor in a system is quantitative, it is often possible to represent the responses in terms of a function of the factor plus random error. We illustrated this method when we used orthogonal polynomials and regression analysis to represent the responses in terms of the levels of the explanatory variables (see Example 2.3). The same was true of the regression approaches to ANOVA we adopted elsewhere in which we employed orthogonal contrasts. The aim in using the orthogonal polynomials or contrasts in the above examples was to enable us find a suitable function, which could approximate the responses in terms of the levels of the factors in the designed experiment.

In Example 2.3, we fitted the polynomial model to the responses when a single quantitative factor set at different levels was deemed to give rise to the observed responses. The usual aim of the study is to fit a functional relationship between the responses and the quantitative factors in the design. In some studies, interest could lie in the optimization of the responses obtained when a number of quantitative factors are believed to be responsible for the observed phenomenon; this is more so in industrial experiments, where the optimum yield might be of interest. To achieve optimization of responses, the techniques of response surface methodology (RSM) are used.

The RSM uses a collection of mathematical and statistical techniques to explore an unknown response function that describes the responses of a system over an experimental region represented by k variables $x_1, x_2, \ldots, x_k$. RSM assumes that there is a functional relationship between the independent variables $x_1, x_2, \ldots, x_k$ and the response function $f(x_1, x_2, \ldots, x_k)$, which itself is unknown. The variables $x_1, x_2, \ldots, x_k$ are assumed to be continuous and quantitative; they are assumed to change the responses as their values change in the experimental region. It is also assumed that these variables can be set at different levels with very negligible error. It is further assumed that the response function $f(x_1, x_2, \ldots, x_k)$ could be represented within a small region of the entire space of independent variables by a low-order polynomial. There are two basic aims of RSM. One is to identify this functional relationship between the response and the explanatory variables, which is usually very difficult to achieve. The other is to achieve an optimum response or (what is more easily achievable) to identify the values of $x_1, x_2, \ldots, x_k$ at which some optimum-operating conditions are satisfied.

The RSM uses multiple regression techniques to sequentially fit low-order polynomial to sample data and carry out tests which should indicate the next appropriate action. Since the objective of this method is to attain an optimum, sequential fitting of the functional forms in Equations 11.1 and 11.2 is adopted for fitting first and second-order models. The procedure involves the fitting of first-order models that contain all the independent variables. These first-order models are of the form:

$$y = f(x) + \varepsilon = b_0 + b_1 x_1 + \cdots + b_k x_k + \varepsilon \tag{11.1}$$

As the investigation continues, if there is any reason to believe that there is some curvature in the system, a polynomial of higher degree is fitted to approximate the responses in terms of the individual variables. The next such polynomial function that could be fitted is of the form:

$$y = f(x) + \varepsilon = b_0 + \sum b_i x_i + \sum b_{ii} x_i^2 + \sum_{i<j}\sum b_{ij} x_i x_j + \varepsilon \tag{11.2}$$

This is a quadratic or second-order model.

Before using RSM to study a process with an objective to map out its route through a response region or determine its optimum, it is usually advisable to carry out a screening experiment to identify factors, whose levels exert significant effects on the responses of the process. Other factors, which do not bring significant effect to bear on the responses, are not designed into the study. Recall that previously we stated that factorial experiments are used for exploratory studies; the full, single replicate, and fractionals of the 2^k and 3^k factorial designs are useful for screening purposes. For other useful screening designs, see Tabachnick and Fidell (2007).

Part of the RSM requires that a simple linear model to be first investigated as the possible functional relationship between the levels of the factors and the responses. Thereafter, the next functional relationship that is examined in the response region is quadratic. It is always desirable to keep the investigation as simple and manageable as possible; very few studies ever investigate higher-order interactions or higher-order polynomial relationships. For most studies, a second-order model is found to adequately represent the functional relationship between the quantitative factor and the response.

11.1 Method of Steepest Ascent

The general theory of using regression to study response surfaces have in recent years been given prominence by Hotelling (1941) and Box and

Wilson (1951). They generally discussed the use of regression techniques as tools for studying unknown response surfaces, which could arise within an experimental region. Among other publications that discussed these techniques are Myers (1971), John (1971), Anderson and McLean (1974), Montgomery (2005), Myers and Montgomery (1995), Petersen (1985), Box and Draper (1987), and Tabachnick and Fidell (2007).

One of the methods to emerge from this general use of regression techniques to study response surfaces, which arise due to changes in the levels of the quantitative factors employed in an experiment, is the method of steepest ascent. It has often been described as being equivalent to climbing a mountain in fog. This is because the response surface is unknown. If the purpose of an experiment is to arrive at an operating condition that yields maximum response, the method of steepest ascent is used. On the other hand, if interest is in the minimization of response, we speak of the method of steepest descent.

In the method of steepest ascent, for instance, a 2^k factorial or fractional factorial experiment is first designed (which includes screening) and the k factors involved determined (all the factors must be quantitative). An assumption is made as to what is the maximum point in the factor space; the k factors are then set at values, which correspond to this maximum, which is now regarded as the center point. The 2^k factorial experiment is now performed using the values of these k factors corresponding to the center point. Thereafter, the k factors have their levels set a little bit further from the center point in either direction of their axes, all of which pass through the center point. It is often useful if the numbers of steps before and after the center point are of the same magnitude for all the k variables. This is achieved by taking equal steps before and after the center point for each variable. A first-order regression analysis employing the model in Equation 11.1 is used to study the result of the initial experiment with a view to ascertaining in which direction the next exploration of the response surface should proceed. If the objective of the experiment is the maximization of the responses, it is normal to follow the path of steepest ascent once it is established, making observations in equally spaced intervals, until there is a leveling-off in the observed response. This leveling-off is equivalent to arriving at the vicinity of an optimum. This is usually indicated by the lack of fit of the first-order model in Equation 11.1 at this stage. The values of the responses as well as the variables at this point are then used as a start for choosing another path of steepest ascent. Another alternative at this point is to fit a second-order model of the type in Equation 11.2 by making more observations for this purpose and analyzing them by regression techniques. The optimum-operating levels for the factors are determined using the data. We speak more of optimum-operating levels because the response surface is not known and whatever optimum is observed could well be a local optimum. Most experiments are, therefore, confined to finding optimum-operating conditions. This idea is useful in industrial experimentation where the objective is usually set to achieve target optimum operating conditions.

11.2 Designs for Fitting Response Surfaces

A proper choice of experimental design for a response surface study helps ensure that we succeed in fitting and analyzing the response surface. A good choice of design will ensure that data points are reasonably distributed within the region of interest. It will ensure that we are able to fit models and test for adequacy at each stage of the investigation. The design should be efficient, allowing us to elicit enough information about the region of interest at each stage of the investigation with as few runs as possible. It should allow us to investigate the surface without setting the factors in the model at many, and therefore, unwieldy number of levels. We should be able to fit the model parameters with relative ease without recourse to tedious mathematics. We should be able to estimate errors without having to depend on information that is extraneous to the study process. Next, we will discuss only designs for fitting first-order models; we will discuss the designs for fitting second-order models later.

11.2.1 Designs for Fitting First-Order Models

To fit the first-order model described in Equation 11.1, we need a design that would minimize the estimates of the parameters $\hat{b}_i$. Designs that achieve this aim are described as orthogonal first-order designs. Consider the following solution for parameters of a first-order model:

$$\hat{\mathbf{b}} = (\mathbf{X}'\mathbf{X})^{-1}\mathbf{X}'\mathbf{Y} \tag{11.3a}$$

$$\begin{bmatrix} \hat{b}_0 \\ \hat{b}_1 \\ \vdots \\ \hat{b}_k \end{bmatrix} = \begin{bmatrix} n & \sum_{j=1}^{n} x_{1j} & \cdots & \sum_{j=1}^{n} x_{kj} \\ \sum_{j=1}^{n} x_{1j} & \sum_{j=1}^{n} x_{1j}^2 & \cdots & \sum_{j=1}^{n} x_{1j}x_{kj} \\ \vdots & \vdots & \vdots\vdots\vdots & \vdots \\ \sum_{j=1}^{n} x_{kj} & \sum_{j=1}^{n} x_{1j}x_{kj} & \cdots & \sum_{j=1}^{n} x_{kj}^2 \end{bmatrix}^{-1} \begin{bmatrix} \sum_{j=1}^{n} y_j \\ \sum_{j=1}^{n} x_{1j}y_j \\ \vdots \\ \sum_{j=1}^{n} x_{mj}y_j \end{bmatrix} \tag{11.3b}$$

A first-order design such that in Equation 11.3a (which is the same as Equation 11.3b), all the off-diagonal elements of the $\mathbf{X}'\mathbf{X}$ matrix are made up of positive diagonal elements while the off-diagonals are all zero, is said to be orthogonal. Thus, for orthogonal first-order designs, the following should be true:

$$\begin{bmatrix} \hat{b}_0 \\ \hat{b}_1 \\ \vdots \\ \hat{b}_k \end{bmatrix} = \begin{bmatrix} n & 0 & \dots & 0 \\ 0 & \sum_{j=1}^{n} x_{1j}^2 & \dots & 0 \\ \vdots & \vdots & \vdots\vdots\vdots & \vdots \\ 0 & 0 & \dots & \sum_{j=1}^{n} x_{kj}^2 \end{bmatrix}^{-1} \begin{bmatrix} \sum_{j=1}^{n} y_j \\ \sum_{j=1}^{n} x_{1j} y_j \\ \vdots \\ \sum_{j=1}^{n} x_{mj} y_j \end{bmatrix} \tag{11.3c}$$

The class of first-order designs that satisfy the orthogonality property (Equation 11.3c) includes the full 2^k factorial designs and fractionals of 2^k factorial designs, which do not have their main effects aliased with each other.

Use of unreplicated 2^k factorial designs or fractionals of 2^k factorial designs does not provide us with enough responses to estimate error. In practice, in RSM, measurements are made at the center of the design to ensure that error can be estimated. At the center of the design, the dummy variable representation of each of the predictor variables in the design takes the value zero; that is *dummy* $(x_1, x_2, \dots, x_k) = (0, 0, \dots, 0)$.

We can also fit first-order models by using the class of orthogonal designs called the simplex designs. A regular sided figure with $k+1$ vertices is called a simplex. For instance, if we are investigating only two factors, we can use an equilateral triangle; for three factors, the corresponding figure is the regular tetrahedron. If the number of factors to be studied was four, then a simplex required would be a polytope (see Figures 11.1 and 11.2).

We will discuss designs for fitting second-order models when we reach that stage in our application of the RSM. Meanwhile, let us fit a first-order model to a process under investigation.

11.3 Fitting a First-Order Model to the Response Surface

To fit a first-order model of the form in Equation 11.1, we simply use the principles of least squares regression analysis (see regression theory in Chapter 1). Least squares regression can also be found in many standard texts in statistics. We think it is best to illustrate how this is done with a problem, and we intend to use the same problem to illustrate RSM or more precisely, the method of steepest ascent described in Subsections 11.1 and 11.2.

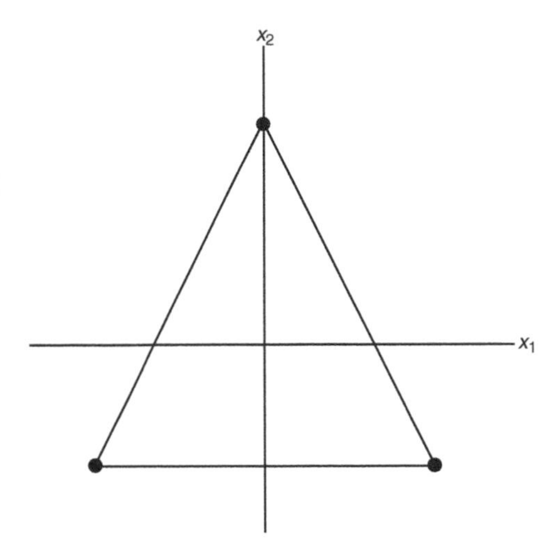

FIGURE 11.1: Simplex design for fitting first-order models for two factors.

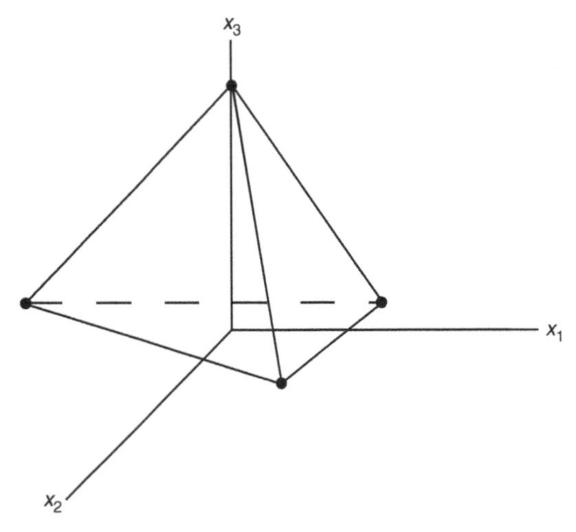

FIGURE 11.2: Simplex design for fitting first-order models for three factors.

Example 11.1

Three factors, temperature (A), time (B), and reactant concentration (C), are believed to affect the yield of a chemical process. To explore the response surface of the system, it was determined following the initial screening that the temperature 44°C, time 24 min, and reactant concentration of 14% would represent the center of the design. The yield was measured at this point and it was 12.0 kg. The regions of exploration chosen were $A \subseteq [40, 48]$ °C, $B \subseteq [20, 28]$,

TABLE 11.1: Results of the Chemical Process Experiment

Treatment combinations	A	B	C	x_1	x_2	x_3	Responses
(1)	40	20	10	−1	−1	−1	12
a	48	20	10	1	−1	−1	14.4
b	40	28	10	−1	1	−1	10.8
ab	48	28	10	1	1	−1	13.6
c	40	20	18	−1	−1	1	10.4
ac	48	20	18	1	−1	1	16.6
bc	40	28	18	−1	1	1	9
abc	48	28	18	1	1	1	16
y_{01}	44	24	14	0	0	0	13
y_{02}	44	24	14	0	0	0	11.8
y_{03}	44	24	14	0	0	0	13.6
y_{04}	44	24	14	0	0	0	12
y_{05}	44	24	14	0	0	0	13.2

$C \subseteq [10, 18]$ % and the surface was studied with a 2^k factorial design. Observations were, therefore, made for each combination of treatments. The aim was to explore this response surface formed by the $k = 3$ variables with a view to arriving at a function, which describes the maximum operating condition for the process. The data obtained for these set of initial experiments are given in Table 11.1.

To analyze the results, the natural variables (A, B, C) are replaced by dummy variables x_1, x_2, and x_3. To obtain these dummy variables, we note that at the center of the design, $A = 44$, $B = 24$, and $C = 14$, so that

$$x_1 = \frac{[A - 44]}{4}$$

$$x_2 = \frac{[B - 24]}{4}$$

$$x_3 = \frac{[C - 14]}{4}$$

These dummy variables x_1, x_2, and x_3 take the values −1, 0, and 1 corresponding to the values of the factors A, B, and C before the center, at the center, and after the center, respectively. As stated earlier, regression was used to analyze the responses of the experiment. One of the techniques employed in this type of design is to take a number of extra observations at the center of the design. Apart from the first observation taken at the center, five other observations were also taken at the center of the design (these observations at the center were added to facilitate the estimation of experimental error). It is suggested that for more efficient designs, the number of center points, n_c, should be chosen so that $3 \leq n_c \leq 5$.

The next stage is to fit the first-order model of the form

$$y = b_0 + b_1 x_1 + b_2 x_2 + b_3 x_3 + \varepsilon \tag{11.4}$$

(the x's are all dummy variables). The appropriate model set in multiple regression terms can be written in matrix form as

$$\mathbf{y} = \mathbf{bX} + \boldsymbol{\varepsilon} \tag{11.5a}$$

$$\hat{\mathbf{b}} = (\mathbf{X}'\mathbf{X})^{-1}\mathbf{X}'\mathbf{y} \tag{11.5b}$$

For ANOVA, we note that the sums of squares regression (i.e., due to b) is obtained as

$$SS_{\text{regression}} = \hat{\mathbf{b}}\mathbf{X}'\mathbf{y} - \frac{\left(\sum y\right)^2}{n} \tag{11.6}$$

The rest are obtained using the following:

$$SS_{\text{total}} = \sum y_i^2 - \frac{1}{n}\left(\sum y_i\right)^2 \tag{11.7}$$

$$SS_{\text{residual}} = SS_{\text{total}} - SS_{\text{regression}} \tag{11.8}$$

Using first-order model, regression equation is fitted overlooking the interaction terms and the pure quadratic (x^2's). Since we are exploring an unknown response surface, we have to reckon with the fact that the fitted model may not be adequate. Consequently, when we fit the first-order model, we test for lack of fit. The way to carry out the lack of fit analysis is to recognize that the residual sum of squares is calculated as in Equation 11.8 and contains three sums of squares: $SS_{\text{pure quadratic}}, SS_{\text{interaction term}}, SS_{\text{pure error}}$. Each of these components can be calculated and tabulated with the rest of the components in ANOVA to test whether they are significant and, thereby, decide whether the model fitted is adequate; this enables the experimenter to feel his/her way toward the next stage in the path of steepest ascent.

The sums of squares interaction which are contained in the $SS_{\text{residuals}}$ are obtained by calculating the equivalent dummy variables corresponding to the interactions of x's. They are simply obtained as the products of dummy variables corresponding to the individual x's, which appear in the interaction.

For instance, Table 11.2 shows these dummy variables as they would appear in a typical 2^3 factorial design replicated five times at the center of the design as this corresponds to the requirements for further analysis pursued in Example 11.2.

TABLE 11.2: Dummy Variables for Interaction Terms in a 2^3 Factorial Design

x_1x_2	x_1x_3	x_2x_3	$x_1x_2x_3$	x_1	x_2	x_3	**Responses**
1	1	1	−1	−1	−1	−1	12
−1	−1	1	1	1	−1	−1	14.4
−1	1	−1	1	−1	1	−1	10.8
1	−1	−1	−1	1	1	−1	13.6
1	−1	−1	1	−1	−1	1	10.4
−1	1	−1	−1	1	−1	1	16.6
−1	−1	1	−1	−1	1	1	9
1	1	1	1	1	1	1	16
0	0	0	0	0	0	0	13
0	0	0	0	0	0	0	11.8
0	0	0	0	0	0	0	13.6
0	0	0	0	0	0	0	12
0	0	0	0	0	0	0	13.2

With manual calculations, we could use the table below to obtain the interaction sums of squares or use the matrix form of Equation 11.9:

	x_1y	x_2y	x_3y	x_1x_2y	x_1x_3y	x_2x_3y	$x_1x_2yx_3y$
	−12	−12	−12	12	12	12	−12
	14.4	−14.4	−14.4	−14.4	−14.4	14.4	14.4
	−10.8	10.8	−10.8	−10.8	10.8	−10.8	10.8
	13.6	13.6	−13.6	13.6	−13.6	−13.6	−13.6
	−10.4	−10.4	10.4	10.4	−10.4	−10.4	10.4
	16.6	−16.6	16.6	−16.6	16.6	−16.6	−16.6
	−9	9	9	−9	−9	9	−9
	16	16	16	16	16	16	16
	0	0	0	0	0	0	0
	0	0	0	0	0	0	0
	0	0	0	0	0	0	0
	0	0	0	0	0	0	0
	0	0	0	0	0	0	0
Total	18.4	−4	1.2	1.2	8	0	0.4

If we denote each column of contrasts listed in Table 11.2 by its heading and consider that column to be a vector, then the sums of squares for the interaction are obtained as follows:

$$\begin{aligned} SS_{12} &= \frac{[(x_1x_2)'y]^2}{2^3} \\ SS_{13} &= \frac{[(x_1x_3)'y]^2}{2^3} \\ SS_{23} &= \frac{[(x_2x_3)'y]^2}{2^3} \\ SS_{123} &= \frac{[(x_1x_2x_3)'y]^2}{2^3} \end{aligned} \tag{11.9}$$

The sums of squares are for the interactions written in brackets in Equation 11.9; the sum of all of them

$$SS_{\text{interaction}} = SS_{12} + SS_{13} + SS_{23} + SS_{123}$$

is used in ANOVA and for this particular problem is considered to have 3 degrees of freedom. The sum of squares for the pure quadratic terms is calculated by using two means: (i) the mean of all the observations belonging to 2^3 factorial part of this design, $\bar{y}_{\text{f}}$ (which excludes the center) and (ii) the mean of the observations obtained at the center of the design, denoted by $\bar{y}_{\text{c}}$. The sum of squares pure quadratic is obtained as

$$SS_{\text{pure quadratic}} = \frac{n_{\text{f}} n_{\text{c}} (\bar{y}_{\text{f}} - \bar{y}_{\text{c}})^2}{(n_{\text{f}} + n_{\text{c}})} \tag{11.10}$$

where n_{f} and n_{c} are the numbers of observations in the two sets of observations described in (i) and (ii). For this particular problem, the pure quadratic term in Example 11.1 has 1 degree of freedom. It should be noted that if $\bar{y}_{\text{f}} - \bar{y}_{\text{c}}$ is large, this would indicate the presence of a quadratic curvature; instead of the linear relationship assumed in Equation 11.4 or Equation 11.5a, the model would need to be augmented with quadratic terms.

The sum of squares pure error is obtained by using only data determined at the center of the design which means that

$$SS_{\text{pure error}} = \sum y_{0i}^2 - \frac{1}{n_{\text{c}}} \left(\sum y_{0i} \right)^2 \tag{11.11a}$$

Of course, it is possible to obtain this by subtraction from SS_{residual} by removing all other sums of squares from it. To estimate regression parameters and perform ANOVA, we note that from Table 11.1 we can obtain the matrix $\mathbf{X}$ that appears in Equations 11.5a and 11.5b and rewrite Equation 11.5a, for example, as shown in the following equations, that is,

$$\mathbf{y} = \mathbf{X}'\mathbf{b} + \boldsymbol{\varepsilon} \tag{11.11b}$$

$$\begin{bmatrix} 12 \\ 14.4 \\ 10.8 \\ 13.6 \\ 10.4 \\ 16.6 \\ 9 \\ 16 \\ 13 \\ 11.8 \\ 13.6 \\ 12 \\ 13.2 \end{bmatrix} = \begin{bmatrix} 1 & -1 & -1 & -1 \\ 1 & 1 & -1 & -1 \\ 1 & 1 & 1 & -1 \\ 1 & -1 & -1 & 1 \\ 1 & -1 & -1 & 1 \\ 1 & 1 & -1 & 1 \\ 1 & -1 & 1 & 1 \\ 1 & 1 & 1 & 1 \\ 1 & 0 & 0 & 0 \\ 1 & 0 & 0 & 0 \\ 1 & 0 & 0 & 0 \\ 1 & 0 & 0 & 0 \\ 1 & 0 & 0 & 0 \end{bmatrix} \begin{bmatrix} b_0 \\ b_1 \\ b_2 \\ b_3 \end{bmatrix} + \begin{bmatrix} \varepsilon_1 \\ \varepsilon_2 \\ \varepsilon_3 \\ \varepsilon_4 \\ \varepsilon_5 \\ \varepsilon_6 \\ \varepsilon_7 \\ \varepsilon_8 \\ \varepsilon_9 \\ \varepsilon_{10} \\ \varepsilon_{11} \\ \varepsilon_{12} \\ \varepsilon_{13} \end{bmatrix} \tag{11.11c}$$

From the equation above, we can show that

$$\mathbf{X'y} = \begin{bmatrix} 166.4 \\ 18.5 \\ -4 \\ 1.2 \end{bmatrix} \quad \mathbf{X'X} = \begin{bmatrix} 13 & 0 & 0 & 0 \\ 0 & 8 & 0 & 0 \\ 0 & 0 & 8 & 0 \\ 0 & 0 & 0 & 8 \end{bmatrix} \Rightarrow (\mathbf{X'X})^{-1} = \begin{bmatrix} \frac{1}{13} & 0 & 0 & 0 \\ 0 & \frac{1}{8} & 0 & 0 \\ 0 & 0 & \frac{1}{8} & 0 \\ 0 & 0 & 0 & \frac{1}{8} \end{bmatrix}$$

$$\Rightarrow \hat{\mathbf{b}} = (\mathbf{X'X})^{-1}\mathbf{X'y} = \begin{bmatrix} 12.8 & 2.3 & -0.5 & -0.15 \end{bmatrix}$$

The fitted first-order model is

$$y = 12.8 + 2.3x_1 - 0.5x_2 - 0.15x_3 \tag{11.12}$$

The sums of squares for this problem are

$$SS_{\text{total}} = 2185.12 - \frac{(166.4)^2}{13} = 55.2$$

$$SS_{\text{regression}} = [12.8 \quad 2.3 \quad -0.5 \quad 0.15] \begin{bmatrix} 166.4 \\ 18.5 \\ -4 \\ 1.2 \end{bmatrix} - \frac{(166.4)^2}{13} = 44.5$$

$$SS_{\text{residual}} = 55.2 - 44.5 = 10.7$$

$$SS_{\text{pure quadratic}} = 8 \times 5 \times (12.85 - 12.72)^2 \times \frac{1}{13} = 0.052$$

$$SS_{12} = \frac{(1.2)^2}{2^3} = 0.18$$

$$SS_{13} = \frac{8^2}{2^3} = 8$$

$$SS_{23} = \frac{0^2}{2^3} = 0$$

$$SS_{123} = \frac{(0.4)^2}{2^3} = 0.02$$

$$SS_{\text{interaction}} = 0.18 + 8 + 0 + 0.02 = 8.2$$

$$SS_{\text{pure error}} = [(13)^2 + (11.8)^2 + (13.6)^2 + (12)^2 + (13.2)^2] - \frac{(63.6)^2}{5} = 2.448$$

TABLE 11.3: ANOVA for First-Order Model Fitted to Data of Example 11.1

Source	Sum of squares	Degrees of freedom	Mean square	F-ratio	p-Value
Regression (β_1, β_2, β_3)	44.5	3	14.833333	12.47663	0.0014747*
Residual	10.7	9	1.1888889		
Interaction	8.2	4	2.05	3.3496732	0.13426223
Pure quadratic	0.052	1	0.052	0.0849673	0.78516683
Pure error	2.448	4	0.612		
Total	55.2	12			

Only those with * are significant at 1% level. Clearly, only the first-order model is indicated by the data at this stage of the study (see Table 11.3).

The conclusions reached in the analysis of the responses of the first-order model of Example 11.1 indicate that to move away from the center of the design ($x_1 = x_2 = x_3 = 0$) in the path of steepest ascent, we need to move 2.3 units in the direction x_1, -0.5 units in the direction x_2, and 0.15 units in the x_3 direction. It is required at this stage that a choice be made as to what should be considered as a unit step for the experiment; that is, the change in the value of one of the factors (or controlled variables), which could be regarded as a unit step. This step can be chosen conveniently by the investigator so as to lend its use to easy mathematical manipulation. Suppose that the investigator settled for a change of 4℃ in temperature as the unit step, then Δx_1 (the change in the dummy variable x_1) is equal to 1.0; the equivalent change in the x_2 direction is $(-0.5/2.3)\Delta x_1$, which is -0.2174. Similarly, the corresponding change in x_3 direction is $\Delta x_3 = (0.15/2.3)\Delta x_1 = 0.0652$. The values for x_1, x_2, and x_3 are translated into natural variables and used to run the experiment. The results of the preceding analysis indicate that the next set of experiments should be carried out at 48℃ for 22.087 min, and 14.26% reactant concentration. This is the next step in the path of steepest ascent.

The usual procedure is to keep on moving in equal steps in the path of steepest ascent and keep on observing the yield or response of these steps until a decrease in response is observed for the first time. The values of the natural variables required to run the experiment at each step are obtained by transforming the dummy into natural variables. Suppose the investigator continues his/her observations and after recording increases in response through seven steps, observes that step eight shows a decrease; he/she could take one of the two possible actions, fit a first-order model in the general vicinity of this point, and test whether the quadratic and interaction terms are significant or else he/she could fit a second-order model in the same surroundings. If the first action is taken, he/she would need to fit a second-order model if the test shows that the quadratic terms are significant. Otherwise, the point could become the focus for the choice of another path of steepest ascent. If he/she decides to fit the first-order model, then the data obtained are given in Table 11.4

$$\mathbf{X'y} = \begin{bmatrix} 257.2 \\ 3.3 \\ 7.1 \\ 8.7 \end{bmatrix}; \quad \mathbf{X'X} = \begin{bmatrix} 13 & 0 & 0 & 0 \\ 0 & 8 & 0 & 0 \\ 0 & 0 & 8 & 0 \\ 0 & 0 & 0 & 8 \end{bmatrix} \Rightarrow (\mathbf{X'X})^{-1} = \begin{bmatrix} \frac{1}{13} & 0 & 0 & 0 \\ 0 & \frac{1}{8} & 0 & 0 \\ 0 & 0 & \frac{1}{8} & 0 \\ 0 & 0 & 0 & \frac{1}{8} \end{bmatrix}$$

$$\Rightarrow \hat{\mathbf{b}} = (\mathbf{X'X})^{-1}\mathbf{X'y} = \begin{bmatrix} 19.78462 & 0.425 & 0.8875 & 1.0875 \end{bmatrix}$$

TABLE 11.4: Data Obtained for First-Order Model at the Eighth Step

Treatment combinations	Variables						Responses
	Natural			Dummy			
	A	B	C	x_1	x_2	x_3	
(1)	68	12	12	−1	−1	−1	17
a	76	12	12	1	−1	−1	17.5
b	68	20	12	1	1	−1	18.5
ab	76	20	12	1	1	−1	20
c	68	12	20	−1	−1	1	19.3
ac	76	12	20	1	−1	1	20
bc	68	20	20	−1	1	1	20.9
abc	76	20	20	1	1	1	21.5
y_{01}	72	16	16	0	0	0	20.4
y_{02}	72	16	16	0	0	0	20.8
y_{03}	72	16	16	0	0	0	20.5
y_{04}	72	16	16	0	0	0	20.3
y_{05}	72	16	16	0	0	0	20.4

The fitted model is

$$y = 19.78462 + 0.4125x_1 + 0.8875x_2 + 1.0875x_3 \tag{11.13}$$

$$SS_{\text{total}} = \sum_{i=1}^{13} y_i^2 - \frac{\left(\sum_{i=1}^{13} y_i\right)}{13} = 5110.44 - \frac{(257.2)^2}{13} = 21.83692$$

$$\begin{aligned} SS_{\text{regression}} &= \hat{\mathbf{b}}\mathbf{X}'\mathbf{y} - \frac{\left(\sum_{i=1}^{13} y_i\right)}{13} \\ &= \begin{bmatrix}19.78462 & 0.4125 & 0.8875 & 1.0875\end{bmatrix}\begin{bmatrix}257.2\\3.3\\7.1\\8.7\end{bmatrix} = 17.12375 \end{aligned}$$

$$SS_{\text{pure quadratic}} = \frac{n_1 n_2(\bar{y}_1 - \bar{y}_2)^2}{n_1 + n_2} = \frac{8 \times 5(20.5 - 19.3375)^2}{8+5} = 4.158173$$

$$\begin{aligned} SS_{\text{interaction}} &= \frac{[(x_1x_2)'y]}{2^3} + \frac{[(x_1x_3)'y]}{2^3} + \frac{[(x_2x_3)'y]}{2^3} + \frac{[(x_1x_2x_3)'y]}{2^3} \\ &= \frac{(0.9)^2}{8} + \frac{(-0.7)^2}{8} + \frac{(-0.9)^2}{8} + \frac{(-1.1)^2}{8} \\ &= 0.415 \end{aligned}$$

$$SS_{\text{pure error}} = \sum y_{0i}^2 - \frac{1}{n_c}\left(\sum y_{0i}\right)^2$$
$$= [(20.4)^2 + (20.8)^2 + (20.5)^2 + (20.3)^2 + (20.5)^2] - \frac{(102.5)^2}{5} = 0.14$$

The ANOVA for the model is given in Table 11.5.

TABLE 11.5: ANOVA for Example 11.1: Fitted First-Order Model at the Eighth Step

Source	Sum of squares	Degrees of freedom	Mean square	F-ratio	p-Value
Regression (β_1, β_2, β_3)	17.12375	3	5.707917	10.8995	0.002366**
Residual	4.71317	9	0.523686		
Interaction	0.415	4	0.059286	2.964286	0.099754
Pure quadratic	4.157023	1	4.157023	207.8512	1.84E−06**
Pure error	0.14	7	0.02		
Total	21.83692	12			

In Table 11.5, those with ** are significant at 1% level. The analysis of the data obtained at this stage indicates that a quadratic model should be fitted to the data. Before we fit the second-order model, we shall first discuss the general procedure for fitting second-order models.

Once the change (decrease or increase as the case may be) in the response and fitting of first-order model along with testing indicate that the quadratic terms are significant, the next step is to fit a quadratic model to the response surface at this point. The second-order model is used to approximate the response surface at this stage. Almost always, it is found that second-order model is an adequate approximation to the response surface at this stage. Rarely are higher polynomials and interactions indicated. The model to be fitted is obtained from Equation 11.2 as

$$y = b_0 + \sum_{i=1}^{k} b_i x_i + \sum_{i=1}^{k} b_{ii} x_i^2 + \sum_{i<j}^{k}\sum^{k} b_{ij} x_i x_j \tag{11.14}$$

Notice that the model does include interaction terms, which may or may not form part of the final fitted equation; however, they are examined. □

11.3.1 Designs for Fitting Second-Order Surfaces

Some designs are known to have desirable qualities for fitting second-order models. Among the desirable qualities are rotatability, orthogonality, and sphericity. We will discuss these qualities in the following sections (Sections 11.3.1.1 and 11.3.1.2). Besides, if we intend to use a design for fitting second-order models, it is important that such a design gives good predictions of the surface in the region of interest. Models that have such desirable qualities should be chosen for fitting second-order surfaces.

11.3.1.1 Use of 3^k Factorial Designs

When the procedure in RSM indicates that a second-order surface should be fitted, the general form of the surface to be fitted is shown in Equation 11.14. From Equation 11.14, we know that
when $k=2$

$$y = b_0 + b_1x_1 + b_2x_2 + b_{11}x_1^2 + b_{22}x_2^2 + b_{12}x_1x_2$$

when $k=3$

$$\begin{aligned} y = b_0 + b_1x_1 + b_2x_2 + b_3x_3 + b_{11}x_1^2 + b_{22}x_2^2 + b_{33}x_3^2 \\ + b_{12}x_1x_2 + b_{13}x_1x_3 + b_{23}x_2x_3 \end{aligned}$$

Clearly, it is impossible to estimate all the parameters of the model unless each variable in the design has at least three levels. If the 2^3 factorial design were used to study the system, there are only eight units in this design, which yield 7 degrees of freedom. As all of these will be used in estimating b_0, b_1, b_2, b_3, b_{11}, b_{22}, b_{33}, there will be no other units available for the estimation of b_{12}, b_{13}, and b_{23}. This difficulty would be overcome if a 3^k factorial design were used. When a 3^k factorial design is used, there will be 27 units, and therefore, 26 degrees of freedom. The parameters of the second-order response surface with three variables can all be estimated, leaving enough degrees of freedom for the estimation of error.

The use of 3^k factorial designs for fitting second-order response surfaces, while satisfactory with respect to its ability to provide enough degrees of freedom for the estimation of parameters, is inefficient when $k>3$. Usually, we estimate only ${}^kC_2+2k+1$ parameters; it is therefore a waste of resources to perform 3^k separate experiments for this purpose. For instance, when four variables are involved (i.e., $k=4$) using a 3^4 factorial design would require us to perform 81 experiments to estimate 15 parameters. It is better to use designs, which require fewer resources for the purpose.

11.3.1.2 Central Composite Designs

The search for more efficient designs for fitting second-order surfaces led to the development of the central composite designs (CCD) (Box and Wilson, 1951; Myers, 1971; and others). These are the most widely used designs for fitting second-order response surfaces. The design, in general, has three parts: (i) a basic 2^k factorial or fractional factorial design, (ii) n_c extra points ($n_c \geq 1$) at the center of the design, and (iii) $2k$ extra points each of them being obtained at an extreme of the axis for each factor while all other factorials are restricted at the center. It means that CCD has got altogether, 2^k+2k+n_c units. The values of the factors at the extremes of the axes of the factors are restricted to the values $\pm\alpha$ while other combinations of treatments in the 2^k factorial section of the design are represented by $+1$ and -1 for high and low levels of each of the factors represented in that treatment combination.

n_c and α are chosen with a view to satisfying some requirements of the design. Some of these requirements are rotatability, sphericity, and orthogonality.

(a) *Rotatability*

Sometimes, it is discovered that the responses of a design are estimated with a higher precision at some points more than others. This problem usually arises as a result of the relationship of the true response surface to the design point in the space of the factors. One of the ways of obviating this problem is to make use of one of the class of rotatable designs. To obtain a rotatable CCD, the constant, α is chosen such that $\alpha = F^{1/4}$ ($F = 2^k$, the number of units in the factorial portion of the composite design). In a design with $k = 3$, then $\alpha = (2^3)^{1/4} = 1.6818$. If a fractional factorial design is used, then $\alpha = (2^{k-p})^{1/4}$ (see Figures 11.3 and 11.4).

All points in a rotatable design are equidistant from the center point. Once a rotatable design is obtained as described above, it loses its orthogonality properties. This is because the 2^k extra axial points (often called star points) added to the design do not satisfy conditions for orthogonality.

Since the rest of the design excluding the axial points form an orthogonal design, it is possible to run the factorial experiment to study the 2^k factorial or

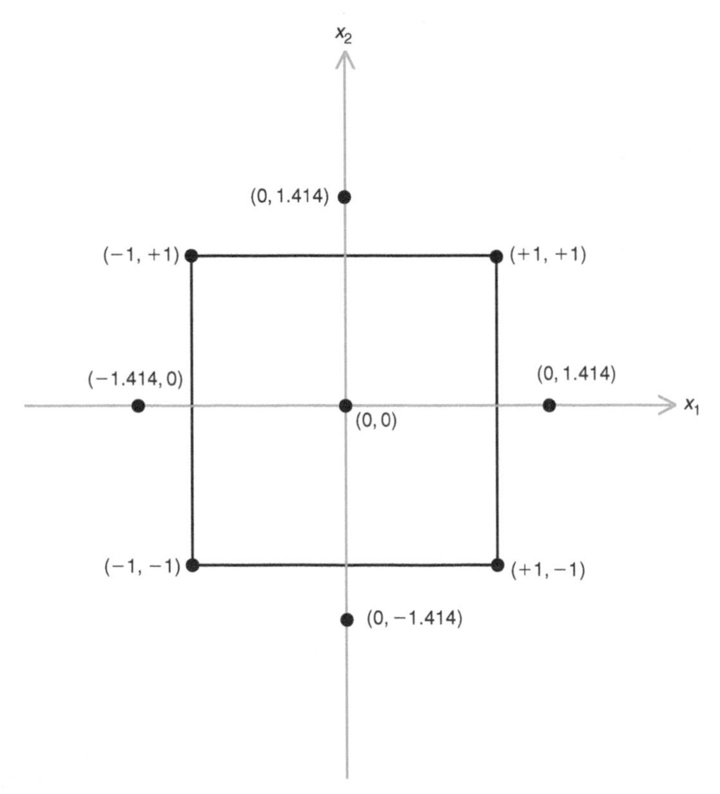

FIGURE 11.3: Central composite design for two factors when $\alpha = 1.414$.

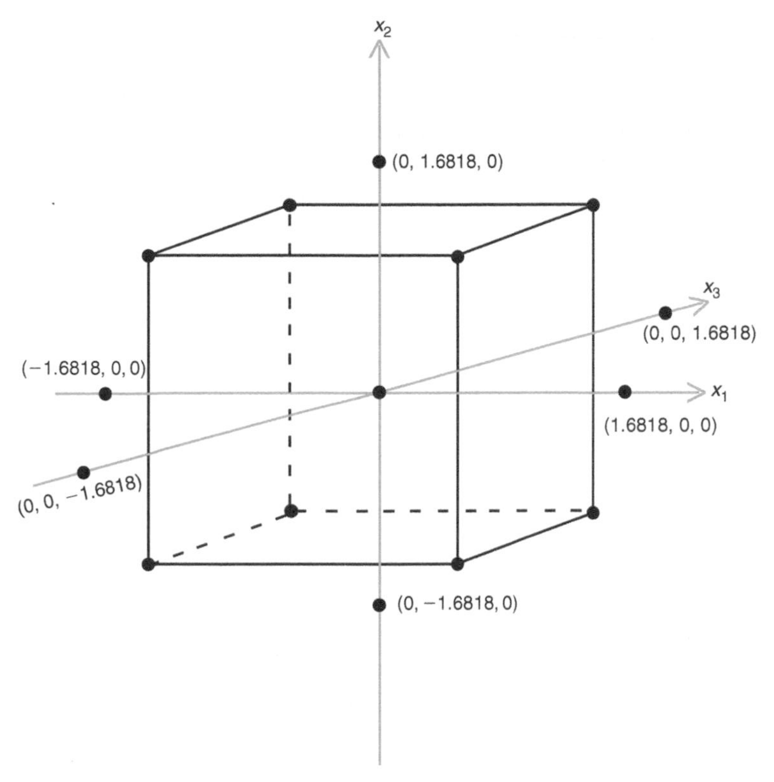

FIGURE 11.4: Central composite design for three factors when $\alpha = 1.6818$.

fractional factorial part of the design before running the rest. By doing so, it is possible to study the main effects as well as the linear by linear interactions before examining the entire model.

The major thing achieved by using rotatable designs is that variances of estimates are equalized. Rotatable designs are, therefore, used if there is a tendency for the variances of estimates obtained from one axis of the design to be different from those obtained from other axes.

(b) *Sphericity*

In exploring response surfaces, it is generally accepted that the nature of the response surface being explored is not known in advance. It is considered appropriate to adopt designs that generate information so that the estimated responses have constant variance at all points, which are equidistant from the origin of the design (Box and Hunter, 1957). The number of center points generated in a CCD, n_c is related to rotatability and helps ensure a spherically uniform variance of the predicted response. The rotatability property helps assure sphericity that is uniform variance for the predicted responses. This quality is intuitively appealing. In practice, we do not need perfect conformance to this requirement; approximate sphericity will be sufficient.

(c) *Orthogonality*

To obtain an orthogonal CCD, α is chosen in such a way that all the main effects and interactions estimated in the design are orthogonal to each other. Moreover, estimates of the parameters of this design are required to be uncorrelated. To achieve this, we note that $\mathbf{X}'\mathbf{X}$, obtained when fitting the second-order design, must be diagonalized and this is achieved when

$$FT - 4F\alpha^2 - 4\alpha^4 = 0 \tag{11.15}$$

or

$$\alpha = \left\{\left(\frac{1}{4n^2}\right)\left[(F+T)^{1/2} - F^{1/2}\right]\cdot F\right\}^{1/4} \tag{11.16}$$

where F = number of observations in the factorial part of the design; $T = 2k + n_c$; n = number of replicates of the design.

If $n_c = 5$; $k = 3$, $F = 2^3$, $n = 1$, then

$$\alpha = \left\{\left(\frac{1}{4(1)^2}\right)\left[\sqrt{(8+13)} - \sqrt{8}\right]\cdot 8\right\}^{1/4} = 1.36859$$

We shall use the above result when fitting a second-order model to the responses we have obtained for the method of steepest ascent for Example 11.1.

In general, it should be pointed out that the n_c in CCD must be, at least, equal to 1 if error is to be estimated.

11.4 Fitting and Analysis of the Second-Order Model

The analysis of second-order models is usually referred to as the canonical analysis. Interest in a study being conducted may lie in the maximization of the response. In that case, the maximum point is identified when partial differentials with respect to the variables are equated to zero to solve for the estimates of the x_i's, $\hat{x}_i$. This means that

$$\frac{\partial y}{\partial x_1} = \frac{\partial y}{\partial x_2} = \cdots = \frac{\partial y}{\partial x_k} = 0 \tag{11.17}$$

The estimated values of x_i obtained by solving these equations could belong to a maximum point, a minimum point, or a point of inflexion. The general matrix notation for the second-order model can be written as

$$\mathbf{y} = b_0 + x'b + x'\mathbf{B}x \tag{11.18}$$

where

$$\mathbf{x} = [x_1 \ x_2 \ \dots \ x_k]'; \quad \mathbf{b} = [b_1 \ b_2 \ \dots \ b_k]'$$

$$\frac{\partial y}{\partial x} = b + 2\mathbf{B}x \tag{11.19}$$

indicating that

$$\mathbf{x}_0 = -(1/2)\mathbf{bB}^{-1} \tag{11.20}$$

As usual, we substitute back in Equation 11.19 to obtain the estimated value of the response at the turning point. This value is

$$\hat{y}_0 = b_0 + (1/2)x_0'b \tag{11.21}$$

To perform the canonical analysis and, therefore, be able to characterize the turning point, it is usual to transform the fitted model into one with new coordinate system with origin at the stationary point x_0. The axes of the system are then rotated until they are parallel to the fitted response surface. The resulting model is of the form:

$$y_0 = \lambda_0 + \lambda_1 t_1^2 + \lambda_2 t_2^2 + \cdots + \lambda_k t_k^2 \tag{11.22}$$

where λ_i's are linear combinations of the coefficients of the fitted model. Incidentally, λ_i's are the characteristic roots or eigenvalues of the matrix $\mathbf{B}$, which appear in Equations 11.19 and 11.20.

We can now characterize the nature of the response surface at the stationary points by looking at the signs and magnitudes of the eigenvalues, λ_i. If the λ_i's are all negative, then we shall consider the stationary point x_0 as a maximum point; x_0 is a minimum if all the λ_i are positive. On the other hand, if λ_i have different signs, then x_0 represents a point of inflexion. The response surface is steepest in the direction of the t_i for which $|B|$ has the largest value. If one or more of λ_i are small (approximately zero), the system is considered insensitive to the variables, t_i, then the type of surface is called a stationary ridge. On the other hand, if the stationary point has some $\lambda_i \approx 0$; furthermore, if the stationary point is far outside the region of exploration for the fitting of the second-order model, then the surface could be a rising ridge.

Example 11.1 Continued

To proceed further in the method of steepest ascent described in the previous sections (Section 11.3), after the quadratic terms were found significant, it was decided to fit a second-order model using an orthogonal design. The axial points were located in accordance with the results obtained in Section 11.3.1.2. Observations were carried out at these axial points, and the responses were used to further augment the responses earlier given in Table 11.4. This table is now reproduced in Table 11.6 showing the response at the axial points.

TABLE 11.6: Data Obtained for Second-Order Model at the Eighth Step

	Variables						
	Natural			Dummy			
Treatment combinations	A	B	C	x_1	x_2	x_3	Responses
−1	68	12	12	−1	−1	−1	17
a	76	12	12	1	−1	−1	17.5
b	68	20	12	−1	1	−1	18.5
ab	76	20	12	1	1	−1	20
c	68	12	20	−1	−1	1	19.3
ac	76	12	20	1	−1	1	20
bc	68	20	20	−1	1	1	20.9
abc	76	20	20	1	1	1	21.5
$y01$	72	16	16	0	0	0	20.4
$y02$	72	16	16	0	0	0	20.8
$y03$	72	16	16	0	0	0	20.5
$y04$	72	16	16	0	0	0	20.3
$y05$	72	16	16	0	0	0	20.5
$ya1$	77.474	16	16	1.36859	0	0	22.4
$ya2$	66.526	16	16	−1.36859	0	0	16.7
$ya3$	72	21.474	16	0	1.36859	0	21.5
$ya4$	72	10.526	16	0	−1.3686	0	18
$ya5$	72	16	21.474	0	0	1.36859	21.5
$ya6$	72	16	10.526	0	0	−1.36859	18.5

The fitted second-order model using methods described in Section 11.4 above is

$$y = 20.4977 + 0.95408x_1 + 1.01226x_2 + 1.09022x_3 + 0.5032x_1^2 + 0.3964x_2^2 - 0.263x_3^2 + 0.1125x_1x_2 - 0.0875x_1x_3 - 0.1125x_2x_3 \quad (11.23a)$$

From the results we see that

$$\mathbf{B} = \begin{bmatrix} -0.50320 & 0.05625 & -0.04375 \\ 0.05625 & -0.39640 & 0.05625 \\ -0.04375 & -0.05625 & -0.26300 \end{bmatrix} \quad \mathbf{b}' = \begin{bmatrix} 0.94508 \\ 1.01226 \\ 1.09022 \end{bmatrix}$$

With $\mathbf{x_0} = \mathbf{B}^{-1}\mathbf{b}'$, then

$$\mathbf{x_0} = -\frac{1}{2}\begin{bmatrix} -2.06303 & -0.35213 & 0.41850 \\ -0.35213 & -2.66177 & 0.62787 \\ 0.41850 & 0.62787 & -4.00619 \end{bmatrix} \begin{bmatrix} 0.94508 \\ 1.01226 \\ 1.09022 \end{bmatrix}$$

$$\mathbf{x_0} = \begin{bmatrix} 0.92496 \\ 1.17134 \\ 1.66827 \end{bmatrix} \quad (11.23b)$$

We can obtain the values of the natural variables at this stationary point by substituting back into the equation with which we defined the dummy variables. The stationary point occurs at

$$\frac{A-72}{4}=0.92496 \quad \frac{B-16}{4}=1.17134 \quad \frac{C-16}{4}=1.66827$$

which gives the values $A=75.70$, $B=20.685$, and $C=22.651$. The stationary point yields the predicted response $\hat{y}=22.65686$. To perform the canonical analysis of the model, we obtain the eigenvalues of the matrix $\mathbf{B}$. Press et al. (1986) provide excellent numerical recipes that can help us compute eigenvalues. By using such recipes or otherwise to solve the following equation in λ (λ_i's are the eigenvalues), we obtain the eigenvalues for the canonical analysis:

$$|\mathbf{B}-\lambda\mathbf{I}| = \begin{vmatrix} -0.50320-\lambda & 0.05625 & -0.04375 \\ 0.05625 & -0.39640-\lambda & -0.05625 \\ -0.04375 & -0.05625 & -0.26300-\lambda \end{vmatrix} = \mathbf{0}$$

The eigenvalues are $\lambda_1=-0.528737$, $\lambda_2=-0.405087$, and $\lambda_3=-0.228776$. The canonical form of the fitted model (using Equation 11.22) is

$$\hat{y}=22.65686-0.528737t_1^2-0.405087t_2^2-0.228776t_3^2 \tag{11.24}$$

Since all the eigenvalues are negative, and the stationary point is within the region of exploration, the point is a maximum. The surface is steepest in the direction of t_1 since $|\lambda_1|>|\lambda_2|>|\lambda_3|$. It may be of interest at this stage to find out the relationship between the canonical variable and the design variable. Sometimes, if for some reasons it is impossible to run the system at the stationary point, the investigator may wish to "back away" from this point; one useful method for doing this is to convert points in the design variables x_i space into the t_i space.

In general, the relationship between x_i and t_i which can be used for the conversion is obtained by finding an orthogonal $p\times p$ matrix (if there are p factors in the design), say $\mathbf{W}$ so that:

$$t=\mathbf{W}'(x-x_0) \tag{11.25}$$

The columns of $\mathbf{W}$ are made up of the normalized eigenvectors associated with the eigenvalues, λ_i's, which are found in accordance with Equation 11.22. Thus, using the relationship that exists between w_i, the ith column of $\mathbf{W}$, the matrix $\mathbf{B}$, and the eigenvalues λ_i:

$$(\mathbf{B}-\lambda\mathbf{I})w_i=0 \tag{11.26}$$

so that $\sum_{j=1}^{p} w_{ji}^2=1$, we can find the relationship between t and x.

Applying the foregoing to the analysis of the second-order model above, which yielded the λ_i's, we see that when $\lambda = -0.528737$, we obtain

$$\begin{pmatrix} -0.5032 + 0.528737 & 0.05625 & -0.04375 \\ 0.05625 & -0.3964 + 0.528737 & -0.05625 \\ -0.04375 & -0.05625 & -0.263 + 0.528737 \end{pmatrix} \times \begin{pmatrix} w_{11} \\ w_{21} \\ w_{31} \end{pmatrix} = \begin{pmatrix} 0 \\ 0 \\ 0 \end{pmatrix} \tag{11.27a}$$

$$\Rightarrow$$

$$\begin{aligned} 0.025537w_{11} + 0.05625w_{21} - 0.04375w_{31} &= 0 \\ 0.05625w_{11} + 0.132337w_{21} - 0.05625w_{31} &= 0 \\ -0.04375w_{11} - 0.05625w_{21} + 0.265737w_{31} &= 0 \end{aligned} \tag{11.27b}$$

It is easy to see that the solution to either Equation 11.27a or Equation 11.27b which will satisfy the constraint that $w_{11}^2 + w_{21}^2 + w_{31}^2 = 1$ is not unique. We, therefore, assign arbitrary solutions to one of the variables, solve for the two other variables, and normalize the solutions to satisfy the constraint. Thus, if we set the unnormalized value of $w_{31} = 1$ then we obtain

$$\begin{aligned} 0.025537w_{11} + 0.05625w_{21} &= 0.04375 \\ 0.05625w_{11} + 0.132337w_{21} &= 0.05625 \\ -0.04375w_{11} - 0.05625w_{21} &= -0.265737 \end{aligned} \tag{11.27c}$$

By subtracting the third derivation of Equation 11.27c from the first and second, we are left with the following two equations in two unknowns

$$\begin{aligned} 0.069287w_{11} + 0.1125w_{21} &= 0.30949 \\ 0.10w_{11} + 0.188587w_{21} &= 0.32199 \end{aligned} \tag{11.27d}$$

Solving Equation 11.27d, we obtain the unnormalized solutions as

$$w_{11} = 12.188473 \quad w_{21} = -4.755669 \quad w_{31} = 1$$

Next, we normalize the solutions and obtain the final values of w_{11}, w_{21}, and w_{31} as follows:

$$w_{11} = \frac{12.188473}{\sqrt{12.188473^2 + (-4.755669)^2 + 1^2}} = 0.92888922$$

$$w_{21} = \frac{-4.755669}{\sqrt{12.188473^2 + (-4.755669)^2 + 1^2}} = -0.362431756$$

$$w_{31} = \frac{1}{\sqrt{12.188473^2 + (-4.755669)^2 + 1^2}} = 0.07621$$

The above values for w_{11}, w_{21}, and w_{31}, represent the first column of the matrix $\mathbf{W}$. By similar method, when $\lambda = -0.405087$ and when we set $w_{32} = 1$, then the unnormalized values of w_{11}, w_{22}, and w_{32} are obtained as 0.693192, 1.986861, and 1.0 leading to the normalized values of 0.29752835, 0.852790382, and 0.429215. These normalized values make up the second column of the matrix $\mathbf{W}$. By the same token, when $\lambda = -0.228776$, if we set $w_{33} = 1$, then the unnormalized values for w_{13}, w_{23}, and w_{33} would be −0.245056, −0.417796, and 1.0. Their normalized equivalents are −0.220547, −0.376010607, and 0.899986. These values make up the third column of the $\mathbf{W}$ matrix, leading to

$$\mathbf{W} = \begin{pmatrix} 0.92888922 & 0.29752835 & -0.220547 \\ -0.362431756 & 0.852790382 & -0.376010607 \\ 0.07621 & 0.429215 & 0.899986 \end{pmatrix} \tag{11.28}$$

It is easily verified that $\mathbf{W}^{\mathrm{T}}\mathbf{W} = \mathbf{I}$, satisfying the orthogonality of the matrix $\mathbf{W}$.

Therefore, with reference to Equations 11.23a and 11.25, the relationship between t and x is obtained as

$$\begin{pmatrix} t_1 \\ t_2 \\ t_3 \end{pmatrix} = \begin{pmatrix} 0.92888922 & -0.362431756 & 0.07621 \\ 0.29752835 & 0.852790382 & 0.429215 \\ -0.220547 & -0.376010607 & 0.899986 \end{pmatrix} \begin{pmatrix} x_1 - 0.92496 \\ x_2 - 1.17134 \\ x_3 - 1.66827 \end{pmatrix} \tag{11.29}$$

For the purpose of exploring the surface in the vicinity of the stationary point, we can use Equation 11.29 to determine the next point in the (t_1, t_2, t_3) space where we are supposed to make observations. We convert the values to (x_1, x_2, x_3) and carry out the measurements. □

11.5 Analysis of Covariance

In the behavioral sciences, large error terms are often common in the analysis of results of experiments. For these designs steps are taken that should reduce the sizes of the error terms and improve the sensitivity of the analysis. Often, attempts are made to eliminate these errors by choosing as homogeneous a set of subjects as possible for experiments. The shortcoming of this kind of solution is that the outcome of such an experiment cannot be generalized but can be applied only to the narrow set of subjects represented by those used in the experiment. Another way in which this type of problem is dealt with is while using the repeated measures design, which deals with variations in individuals by using the same subjects for all treatments.

In Chapter 2, we have treated the randomized block design. In Chapter 3 also, we have treated the Latin and Graeco-Latin squares designs. In all these designs, blocking was used as a way of isolating and eliminating from the error effect, the effects of some extraneous factors, whose contributions could have affected the inferences we drew on the responses and could have made our design and analysis less precise than desirable. The same result was achieved when we used the repeated measures design on one factor, which is simply a peculiar form of the randomized complete block experiment in which extra assumptions are made about adjacent sets of measurements made on the same individuals. The methods mentioned in the past sections represent different methods of dealing with large error terms and improving the precisions of different designs.

The same principle used in the above designs also applies to the analysis of covariance. Suppose that an experiment is being performed, whose response is y, but there is another variable, x, which is linearly related to y, which cannot be controlled in the actual experiment, but can be measured along with the response variable, then x is called a covariate—a concomitant or control variable. ANCOVA is a technique that enables us to adjust the response variable for the effects of the covariate and subsequently adjust the experimental error in the analysis of the responses with an objective of improving the sensitivity of the design. Obviously, the effectiveness of ANCOVA as a tool for dealing with error depends on the extent of the linearity of the relationship between y and x. The adjustment removes the covariate effect on the responses, which if left, could have increased the size of the error mean-square and made it difficult to detect actual differences between treatments and factor levels. The magnitude of the adjustment will depend on the extent to which the covariate x is linearly correlated to the response variable y and on the differences actually observed on the covariates.

11.6 One-Way Analysis of Covariance

The completely randomized design with one covariate is often referred to as the one-way analysis of covariance design. In this design, the response variable is y, with a single covariate x. The model is

$$y_{ij} = \mu + \alpha_i + \gamma x_{ij} + \varepsilon_{ij} \quad i = 1, \ldots, k \quad j = 1, \ldots, N_i \tag{11.30a}$$

Expressed in this way, the model emphasizes that ANCOVA is the simultaneous study of several regressions. The error is $\varepsilon_{ij} \sim \text{NID}(0, \sigma^2)$. This model contains the term γx_{ij} that accounts for the influence of the covariate, x. The term μ is not the overall mean as was the case in the completely randomized design. In this case, the overall mean is $\mu + \gamma\bar{x}$. It is also possible to state this model in a way that emphasizes the analysis of variance aspect of ANCOVA. The model that shows this in a clearer manner is

$$y_{ij} = \mu + \alpha_i + \gamma(x_{ij} - \bar{x}) + \varepsilon_{ij} \quad i = 1, \ldots, k \quad j = 1, \ldots, N_i \tag{11.30b}$$

Under these models, apart from the errors that are normally and independently distributed with zero mean and variance σ^2, y_{ij} is the jth observation of the response taken at the ith level of the treatment or ith level of the factor. x_{ij} is the covariate corresponding to y_{ij}; $\gamma \neq 0$ is the linear regression coefficient, indicating that the relationship between x_{ij} and y_{ij} is strictly linear; the effect of the ith treatment or ith level of the factor is α_i in Equations 11.30a and 11.30b; $\bar{x}$ is the mean of the covariate, and the overall mean is μ. It is assumed that $\sum_{i=1}^{k} \alpha_i = 0$. To test hypothesis, one of the things we need to do is estimate σ^2, the variance. The error sum of squares, SS_{E}, is obtained by using the SS_{residual} in the completely randomized design and by adjusting it for the effect of regression, where

$$SS_{\text{residual}} = \sum_{i=1}^{k} \sum_{j=1}^{N_i} y_{ij}^2 - \sum_{i=1}^{k} N_i \bar{y}_{i.}^2 \tag{11.31}$$

We adjust this usual residual sum of squares for the effect of the regression or the concomitant variable, x, on the response variable y, and obtain SS_{E} as

$$SS_{\mathrm{E}} = \sum_{i=1}^{k} \sum_{j=1}^{N_i} y_{ij}^2 - \sum_{i=1}^{k} N_i \bar{y}_{i.}^2 - \left(\sum_{i=1}^{k} \sum_{j=1}^{N_i} y_{ij} x_{ij} - \sum_{i=1}^{k} N_i \bar{y}_{i.} \bar{x}_{i.} \right) \hat{\gamma} \tag{11.32}$$

where

$$\hat{\gamma} = \frac{\left(\sum_{i=1}^{k} \sum_{j=1}^{N_i} y_{ij} x_{ij} - \sum_{i=1}^{k} N_i \bar{y}_{i.} \bar{x}_{i.} \right)}{\sum_{i=1}^{k} \sum_{j=1}^{N_i} x_{ij}^2 - \sum_{i=1}^{k} N_i \bar{x}_{i.}^2} \tag{11.33}$$

We can define a number of quantities that reduce the formulae in Equations 11.31 through 11.33 into more manageable forms. Using the following:

$$E_{yy} = \sum_{i=1}^{k} \sum_{j=1}^{N_i} y_{ij}^2 - \sum_{i=1}^{k} N_i \bar{y}_{i.}^2 \tag{11.34}$$

$$E_{xx} = \sum_{i=1}^{k} \sum_{j=1}^{N_i} x_{ij}^2 - \sum_{i=1}^{k} N_i \bar{x}_{i.}^2 \tag{11.35}$$

$$E_{xy} = \sum_{i=1}^{k} \sum_{j=1}^{N_i} y_{ij} x_{ij} - \sum_{i=1}^{k} N_i \bar{y}_{i.} \bar{x}_{i.} \tag{11.36}$$

It is clear that

$$SS_{\mathrm{E}} = E_{yy} - \frac{E_{xy}^2}{E_{xx}} \tag{11.37}$$

and

$$\hat{\gamma} = \frac{E_{xy}}{E_{xx}} \tag{11.38}$$

Since SS_{E} has $n = N - k - 1$ degrees of freedom, we can estimate the variance σ^2 as

$$\hat{\sigma}^2_{y|x} = \frac{\frac{E_{yy} - E^2_{xy}}{E_{xx}}}{N - k - 1} \tag{11.39}$$

We can now test the hypothesis that the regression coefficient in the design γ is zero, which if accepted would indicate that there is no linear relationship between y and x, which is a violation of one of the basic assumptions in ANCOVA.

$$H_0 : \gamma = 0 \tag{11.40}$$

using the hypothesis sum of squares, SS_γ, where

$$SS_\gamma = \frac{E^2_{xy}}{E_{xx}} \tag{11.41}$$

so that the test statistic for the hypothesis of Equation 11.40 is

$$F_{\mathrm{cal}} = \frac{SS_\gamma}{\sigma^2_{y|x}} \tag{11.42}$$

The hypothesis is rejected if $F_{\mathrm{cal}} > F(1, N-k-1, \alpha)$. To test the hypothesis of "no treatment effect," we refer back to some of the quantities from the regression model and define a few more

$$S_{yy} = \sum_{i=1}^{k}\sum_{j=1}^{N_i} (y_{ij} - \bar{y}_{..})^2 \tag{11.43}$$

$$S_{xy} = \sum_{i=1}^{k}\sum_{j=1}^{N_i} (y_{ij} - \bar{y}_{..})(x_{ij} - \bar{x}_{..}) \tag{11.44}$$

$$S_{xx} = \sum_{i=1}^{k}\sum_{j=1}^{N_i} (x_{ij} - \bar{x}_{..})^2 \tag{11.45}$$

$$T_{yy} = \sum_{i=1}^{k}\sum_{j=1}^{N_i} (y_{i.} - \bar{y}_{..})^2 \tag{11.46}$$

$$T_{xx} = \sum_{i=1}^{k}\sum_{j=1}^{N_i} (x_{i.} - \bar{x}_{..})^2 \tag{11.47}$$

$$T_{xy} = \sum_{i=1}^{k}\sum_{j=1}^{N_i} (x_{i.} - \bar{x}_{..})(y_{i.} - \bar{y}_{..}) \tag{11.48}$$

The quantities, S_{yy}, S_{xy}, and S_{xx} of Equations 11.43 through 11.45, show the total sums of squares and cross products. T_{xx}, T_{yy}, and T_{xy} of Equations 11.46 through 11.48 show the sums of squares and cross products for treatments, while E_{xx}, E_{yy}, and E_{xy} of Equations 11.34 through 11.36 are the sums of squares and cross products related to error.

There are two ways of presenting ANCOVA. The first shows ANCOVA as an adjusted ANOVA. For such tests, the table required is of the form shown in Table 11.7.

The second is displayed in Table 11.8. The analysis is in a format that shows all the sums of squares and cross products required in ANCOVA as well as those required for testing hypothesis about treatment effects (Table 11.8).

We note that if we adjust the model of Equation 11.30a by assuming that there is no treatment effect, then it simply changes to

$$y_{ij} = \mu + \gamma x_{ij} + \varepsilon_{ij} \tag{11.49}$$

TABLE 11.7: Analysis of Covariance as an Adjusted ANOVA

Source	Sum of squares	Degrees of freedom	Mean square	F-ratio
Regression	$\frac{S_{xy}^2}{S_{xx}}$	1		
Treatment	$SS_e^* - SS_e = S_{yy} - \frac{S_{xy}^2}{S_{xx}} - \left[E_{yy} - \frac{E_{xy}^2}{E_{xx}}\right]$	$k-1$	$\frac{SS_e^* - SS_e}{k-1}$	$\frac{SS_e^* - SS_e}{(k-1)\,MS_E}$
Error	$SS_e = \left[E_{yy} - \frac{E_{xy}^2}{E_{xx}}\right]$	$N-k-1$	$MS_E = \frac{SS_e}{(N-k-1)}$	
Total	S_{yy}	$N-1$		

TABLE 11.8: Analysis of Variance for a Single-Factor Experiment with One Concomitant Variable

		Sum of squares				Adjusted for regression	
Source of variation	**Degrees of freedom**	x	xy	y	y	**Degrees of freedom**	**Mean square**
Treatment	$K-1$	T_{xx}	T_{xy}	T_{yy}			
Error	$N-k$	E_{xx}	E_{xy}	E_{yy}	$SS_e = \left[E_{yy} - \frac{E_{xy}^2}{E_{xx}}\right]$	$N-k-1$	$\frac{SS_e}{(N-k-1)}$
Total	$N-1$	S_{xx}	S_{xy}	S_{yy}	$SS_e^* = \left[S_{yy} - \frac{S_{xy}^2}{S_{xx}}\right]$	$N-2$	
Adjusted treatment					$SS_e^* - SS_e$	$k-1$	$\frac{SS_e^* - SS_e}{k-1}$

so that the least squares estimators of μ and γ are $\hat{\mu}=\bar{y}+\gamma\bar{x}$ and $\hat{\gamma}=S_{xy}/S_{xx}$. Under this model, we obtain the error sum of squares as

$$SS_{\text{e}}^{*}=S_{yy}-\frac{S_{xy}^{2}}{S_{xx}} \tag{11.50}$$

with a total of $N-2$ degrees of freedom. The quantity S_{xy}^{2}/S_{xx} in Equation 11.50 is an adjustment due to regression of y on x. However, we note that the actual model in Equation 11.30a contains the treatment effects, α_i; the adjusted sum of squares for the treatment effects is obtained as the difference between SS_{e}^{*} and SS_{e}. Consequently, the sum of squares due to the treatment effects is

$$SS_{\text{treatment(adjusted)}}=SS_{e}^{*}-SS_{e} \tag{11.51}$$

with $N-2-(N-k-1)=k-1$ degrees of freedom.

To test the hypothesis of "no treatment effect," that is,

$$H_0:\alpha_1=\cdots=\alpha_k=0 \tag{11.52}$$

we obtain the statistic

$$F_{\text{cal}}=\frac{\left[\dfrac{SS_{\text{e}}^{*}-SS_{\text{e}}}{k-1}\right]}{\left[\dfrac{SS_{\text{e}}}{N-k-1}\right]} \tag{11.53}$$

which is compared with $F(k-1,\ N-k-1,\ \alpha)$ and if $F_{\text{cal}}>F(k-1,\ N-k-1,\ \alpha)$ we reject the null hypothesis.

Example 11.2

In an experiment in language teaching to young children, pupils of different ages were divided into three groups and passed through three teaching methods and tested to see whether there is any difference between the teaching methods. In this case, the concomitant variable is age (x). The data obtained are

	T_1		T_2		T_3	
Individuals	X	Y	X	Y	X	Y
1	4	43	5	58	8	62
2	7	58	6	60	9	66
3	4	34	5	52	8	65
4	4	48	4	52	7	57
5	5	52	4	49	7	53
6	5	55	5	54	6	59
7	5	53	4	59	7	54
8	5	59	6	65	6	55
Total	39	402	39	449	58	471

Assuming the model of Equation 11.30, we calculate the quantities required for ANCOVA:

$$N_1 = 8 \quad N_2 = 8 \quad N_3 = 8 \quad N = 24 \quad \bar{y} = 1322/24 = 55.083$$
$$\bar{x} = 136/24 = 5.6667 \quad \bar{y}_1 = 402/8 = 50.25 \quad \bar{y}_2 = 449/8 = 56.125$$
$$\bar{y}_3 = 471/8 = 58.875 \quad \bar{x}_1 = 39/8 = 4.875 \quad \bar{x}^2 = 39/8 = 4.875$$
$$\bar{x}^3 = 58/8 = 7.25$$

$$\begin{aligned} E_{yy} &= [43^2 + 58^2 + \cdots + 54^2 + 55^2 - 8[50.25^2 + 56.125^2 + 58.875^2] \\ &= 73992 - 8(9141.34375) = 861.25 \end{aligned}$$

$$E_{xx} = [4^2 + 7^2 + \cdots + 7^2 + 6^2] - [2(4.875)^2 + 7.25^2] = 820 - 800.75 = 19.25$$

$$\begin{aligned} E_{xy} &= [4 \times 43 + 58 \times 7 + \cdots + 6 \times 55] - 8[4.875(50.25 + 58.875) \\ &\qquad + 58.875 \times 7.25] \\ &= 7653 - 7563.375 = 89.625 \end{aligned}$$

$$S_{yy} = [(43 - 55.083)^2 + (58 - 55.083)^2 + \cdots + (55 - 55.083)^2] = 1171.8$$
$$S_{xx} = [(4 - 5.6667)^2 + (7 - 5.6667)^2 + \cdots + (6 - 5.6667)^2] = 49.33333$$
$$S_{xy} = [(4 \times 43 - 5.6667 \times 55.083) + \cdots + (6 \times 55 - 5.6667 \times 55.083)] = 161.66667$$
$$T_{yy} = [402^2 + 449^2 + 471^2]/8 - [1322^2/24] = 310.5833$$
$$T_{xy} = [39 \times 402 + 39 \times 449 + 58 \times 471]/8 - [1322 \times 136/24] = 72.0417$$
$$T_{xx} = [39^2 + 39^2 + 58^2]/8 - [136^2/24] = 30.0833$$

By regression theory

$$\hat{\gamma} = \frac{S_{xy}}{S_{xx}} = \frac{161.66667}{49.33333} = 3.277027$$

Thus, the uncorrected sum of squares for testing for the slope γ is

$$SS_{\hat{\gamma}} = \hat{\gamma} S_{xy} = 3.277027 \times 161.66667 = 529.786$$

For testing for slope γ, the sum of square is

$$SS_x = \frac{E_{xy}^2}{E_{xx}} = \frac{(89.625)^2}{19.25} = 417.28$$
$$SS_e = E_{yy} - \frac{E_{xy}^2}{E_{xx}} = 861.25 - \frac{(89.625)^2}{19.25} = 443.9699$$
$$SS_e^* = S_{yy} - \frac{S_{xy}^2}{S_{xx}} = 1171.8 - \frac{(161.67)^2}{49.333} = 641.9885$$

To test the hypothesis, $H_0 : \gamma = 0$, we calculate the test statistic

$$F_{\text{cal}} = \frac{\left(\dfrac{E_{xy}^2}{E_{xx}}\right)}{(N-k-1)\left(E_{yy} - \dfrac{E_{xy}^2}{E_{xx}}\right)}$$

$$= \frac{\left(\dfrac{89.625^2}{19.25}\right)}{(24-3-1)\left(861.625 - \dfrac{89.625^2}{19.25}\right)} = 18.7977$$

From the tables, $F(1, 20, 0.01) = 8.10$, therefore, we reject the null hypothesis. The regression effect is therefore significant.

To test the null hypothesis that "there is no treatment effect," that is,

$$H_0 : \alpha_1 = \alpha_2 = \cdots = \alpha_k = 0$$

we obtain the test statistic,

$$F_{\text{cal}} = \frac{N-k-1}{k-1} \times \frac{SS_{\text{e}}^* - SS_{\text{e}}}{SS_{\text{e}}}$$

$$SS_{\text{e}}^* - SS_{\text{e}} = 641.9885 - 443.9699 = 198.0186$$

so that

$$F_{\text{cal}} = \frac{20}{2} \times \frac{198.0186}{443.9699} = 4.4602$$

From the table, we obtain $F(2, 20, 0.05) = 3.49 < F_{\text{cal}}$, so that we can reject the null hypothesis, H_0 at 5% level of significance. We present the above analysis in an ANCOVA table (see Table 11.9). □

TABLE 11.9: Analysis of Covariance for the Language Teaching Problem

Source	yy	xx	xy	**Degrees of freedom**	**Mean square**	***F*-ratio**
Unadjusted treatment (T)	310.58	30.08	72.04	2		
Error (components)(E)	861.25	19.25	89.625			
Error (unadjusted)(SS_{e}^*)	641.99			20		
Total	1171.8	49.33	161.67	23		
Error (adjusted)(SS_{e})	443.97			20	22.198	
Treatment (adjusted)	198.018			2	99.009	4.4603

11.6.1 Using SAS to Carry Out Covariance Analysis

In covariance analysis, SAS usually outputs the Type I SS and Type III SS. The Type III SS represents the sums of squares for factors in the model adjusted for the covariate. The Type I SS gives the between-factor sums of squares that are obtained for the analysis-of-variance model. This measures the difference between arithmetic means of the responses for different levels of the factor without regarding the covariate. The Type III SS gives the sum of squares of factors adjusted for the covariate. This measures the differences between the least squares means of the different levels of the factor while controlling for the covariate. For ANCOVA, the theory discussed above and results of manual analysis agree with SAS in every particular detail. In general, the Type I SS is the unadjusted sum of squares for the factors while the Type III SS presents the adjustments due to the covariate.

SAS PROGRAM

```
data examp13_1;
do i=1 to 8;
do teach='t1', 't2', 't3';
input x y @@; output;
end; end;
cards;
4     43     5     58     8     62
7     58     6     60     9     66
4     34     5     52     8     65
4     48     4     52     7     57
5     52     4     49     7     53
5     55     5     54     6     59
5     53     4     59     7     54
5     59     6     65     6     55
;
proc print data=examp13_1;
sum x y;
proc glm data=examp13_1;
class teach;
model y=x teach; run;
```

SAS OUTPUT

Obs	i	teach	x	y
1	1	t1	4	43
2	1	t2	5	58
3	1	t3	8	62
4	2	t1	7	58
5	2	t2	6	60
6	2	t3	9	66
7	3	t1	4	34
8	3	t2	5	52
9	3	t3	8	65
10	4	t1	4	48
11	4	t2	4	52

```
          Obs      i      teach      x       y

           12      4      t3         7      57
           13      5      t1         5      52
           14      5      t2         4      49
           15      5      t3         7      53
           16      6      t1         5      55
           17      6      t2         5      54
           18      6      t3         6      59
           19      7      t1         5      53
           20      7      t2         4      59
           21      7      t3         7      54
           22      8      t1         5      59
           23      8      t2         6      65
           24      8      t3         6      55
                                   ===    ====
                                   136    1322

                        The GLM Procedure

                     Class Level Information

                  Class    Levels        Values

                  teach         3    t1  t2  t3

                Number of Observations Read     24
                Number of Observations Used     24

                        The GLM Procedure

Dependent Variable: y

                                      Sum of
     Source               DF         Squares     Mean Square    F Value    Pr > F
     Model                 3      727.863366      242.621122      10.93    0.0002
     Error                20      443.969968       22.198498
     Corrected Total      23     1171.833333

              R-Square      Coeff Var      Root MSE       y Mean
              0.621132      8.553455       4.711528      55.08333

     Source       DF      Type I SS       Mean Square     F Value      Pr > F

     x             1      529.7860360     529.7860360       23.87      <.0001
     teach         2      198.0773298      99.0386649        4.46      0.0250

     Source       DF      Type III SS     Mean Square     F Value      Pr > F

     x             1      417.2800325     417.2800325       18.80      0.0003
     teach         2      198.0773298      99.0386649        4.46      0.0250
```

11.7 Other Covariance Models

The principle of analysis of covariance can be embedded into virtually every experimental design. Apart from the completely randomized design, this test is also popularly applied to the randomized complete block design. Montgomery

(2005), Dunn and Clark (1987), Keppel (1983) and others have presented this design. Since we can apply the principle to virtually every design, we present the analysis of covariance for the two-way experimental layout (for two factors, A and B), with replication. The model for this design with a single concomitant variable x is

$$y_{ijk} = \mu + \alpha_i + \beta_j + \alpha\beta_{(ij)} + \gamma x_{ij} + \varepsilon_{ijk} \tag{11.54}$$

This model can be rewritten in a version that emphasizes the analysis of variance aspect of ANCOVA as

$$y_{ijk} = \mu + \alpha_i + \beta_j + \alpha\beta_{(ij)} + \gamma(x_{ij} - \bar{x}) + \varepsilon_{ijk} \tag{11.55}$$

where $i=1,\ldots,a$, $j=1,\ldots,b$, and $k=1,\ldots,n$ (for the models in Equations 11.54 and 11.55). In both forms of the model, $\varepsilon_{ijk} \sim \text{NID}(0,\sigma^2)$, α_i and β_j are the effects of ith and jth levels of the factors A and B, respectively, while γ represents the regression coefficient, indicating a strict linear relationship between y and x, and it is assumed that $\gamma \neq 0$. We also assume that $\sum_{i=1}^{a} \alpha_i = \sum_{j=1}^{b} \beta_j = 0$. $\alpha\beta_{(ij)}$ is the effect of interaction of ith level of A and jth level of B. Also, $\sum_{i=1}^{a} \alpha\beta_{(ij)} = \sum_{j=1}^{b} \alpha\beta_{(ij)} = 0$.

To carry out an analysis of covariance for the two-way experimental design with replication, we define a number of quantities that we need. They are mainly sums of squares and cross products for all the factorial effects in the design. We use essentially the same formulae employed in the study of the two-way layout in Chapter 3 for calculating these sums of squares both for the response and for the covariate. Further, we calculate the sums of cross-products of the two.

For all responses and covariate (y and x)

$$S_{yy} = \sum_{i=1}^{a}\sum_{j=1}^{b}\sum_{k=1}^{n} y_{ijk}^2 - \frac{y_{\ldots}^2}{abn}$$

$$S_{xx} = \sum_{i=1}^{a}\sum_{j=1}^{b}\sum_{k=1}^{n} x_{ijk}^2 - \frac{x_{\ldots}^2}{abn}$$

$$S_{xy} = \sum_{i=1}^{a}\sum_{j=1}^{b}\sum_{k=1}^{n} x_{ijk}^2 y_{ijk}^2 - \frac{(x_{\ldots} y_{\ldots})}{abn}$$

$$A_{yy} = \frac{1}{bn}\sum_{i=1}^{a} y_{i..}^2 - \frac{y_{\ldots}^2}{abn}$$

$$A_{xx} = \frac{1}{bn}\sum_{i=1}^{a} x_{i..}^2 - \frac{x_{\ldots}^2}{abn}$$

$$A_{xy} = \frac{1}{bn}\sum_{i=1}^{a} x_{i..} \times y_{i..} - \frac{x_{\ldots} y_{\ldots}}{abn}$$

For factor B and the interaction AB

$$B_{yy} = \frac{1}{an}\sum_{j=1}^{b} y_{.j.}^2 - \frac{y_{...}^2}{abn}$$

$$B_{xx} = \frac{1}{an}\sum_{j=1}^{b} x_{.j.}^2 - \frac{x_{...}^2}{abn}$$

$$B_{xy} = \frac{1}{an}\sum_{j=1}^{b} x_{.j.} \times y_{.j.} - \frac{x_{...}y_{...}}{abn}$$

$$AB_{yy} = \frac{1}{n}\sum_{i=1}^{a}\sum_{j=1}^{b} y_{ij.}^2 - \frac{y_{...}^2}{abn} - A_{yy} - B_{yy}$$

$$AB_{xx} = \frac{1}{n}\sum_{i=1}^{a}\sum_{j=1}^{b} x_{ij.}^2 - \frac{x_{...}^2}{abn} - A_{xx} - B_{yy}$$

$$AB_{xy} = \frac{1}{n}\sum_{i=1}^{a}\sum_{j=1}^{b} x_{ij.} \cdot y_{ij.} - \frac{x_{...}y_{...}}{abn} - A_{xy} - B_{xy}$$

To obtain the adjusted effects to be used in the test of hypothesis, we define the following quantities related to error

$$E_{yy} = S_{yy} - A_{yy} - B_{yy} - AB_{yy}$$
$$E_{xx} = S_{xx} - A_{xx} - B_{xx} - AB_{xx}$$
$$E_{xy} = S_{xy} - A_{xy} - B_{xy} - AB_{xy}$$

We obtain the adjustment factors for error and total sums of squares as

$$AF_{\mathrm{E}} = \frac{E_{xy}^2}{E_{xx}}$$
$$AF_{\mathrm{total}} = \frac{S_{xy}^2}{S_{xx}}$$

The adjustment for the factorial effects, which appear in the design are accomplished as follows:

$$AFX_{\mathrm{effect}} = SSX_{\mathrm{effect}} + SSX_{\mathrm{E}}$$
$$AFCP_{\mathrm{effect}} = SSCP_{\mathrm{effect}} + SSCP_{\mathrm{E}}$$
$$AF_{\mathrm{effect}} = \frac{(AFCF_{\mathrm{effect}})^2}{AFX_{\mathrm{effect}}}$$

(where AF = adjustment factor; AFX = adjustment factor for the X sum of squares; $SSX = X$ sum of squares; $AFCP$ = adjustment factor for cross product; $SSCP$ = sum of cross-products). The adjusted error and total sums of squares are

$$E_{yy(\text{adjusted})} = E_{yy} - AF_{\text{E}}$$

$$S_{yy(\text{adjusted})} = S_{yy} - \frac{S_{xy}^2}{S_{xx}}$$

For the factorial effects A, B, AB

$$AFX_A = A_{xx} + E_{xx}$$

$$AFCP_A = A_{xy} + E_{xy}$$

$$AFX_B = B_{xx} + E_{xx}$$

$$AFCP_B = B_{xy} + E_{xy}$$

$$AFX_{AB} = AB_{xx} + E_{xx}$$

$$AFCP_{AB} = AB_{xy} + E_{xy}$$

$$AF_A = \frac{A_{xy} + E_{xy}}{A_{xx} + E_{xx}}$$

$$AF_B = \frac{B_{xy} + E_{xy}}{B_{xx} + E_{xx}}$$

$$AF_{AB} = \frac{AB_{xy} + E_{xy}}{AB_{xx} + E_{xx}}$$

Applying these adjustments to the effects in the two-way layout with replication, we obtain

$$SS_{A(\text{adjusted})} = A_{yy} - \frac{(A_{xy} + E_{xy})^2}{A_{xx} + E_{xx}} + \frac{E_{xy}^2}{E_{xx}}$$

$$SS_{B(\text{adjusted})} = B_{yy} - \frac{(B_{xy} + E_{xy})^2}{B_{xx} + E_{xx}} + \frac{E_{xy}^2}{E_{xx}}$$

$$SS_{AB(\text{adjusted})} = AB_{yy} - \frac{(AB_{xy} + E_{xy})^2}{AB_{xx} + E_{xx}} + \frac{E_{xy}^2}{E_{xx}}$$

With these adjusted sums of squares for the error and factorial effects, we can perform the ANCOVA for the two-way replicated design with one covariate. The general ANCOVA table that is applicable to this design is presented below (Table 11.10).

TABLE 11.10: ANCOVA for a Typical Full Two-Factor Experiment with One Covariate

	Unadjusted sum of squares and cross products			Components of adjustment factors			Adjusted for regression			
Source	y	x	xy	x	xy	**Adjustment factors**	**Sum of squares (y)**	**Degrees of freedom**	**Mean square**	***F*-ratio**
A	A_{yy}	A_{xx}	A_{xy}	AFX_A	$AFCP_A$	AF_A	$SS_{A(\text{adjusted})}$	$a-1$	$\frac{SS_{A(\text{adjusted})}}{a-1}$	$\frac{MS_{A(\text{adjusted})}}{MS_{E(\text{adjusted})}}$
B	B_{yy}	B_{xy}	B_{xy}	AFX_B	$AFCP_B$	AF_B	$SS_{B(\text{adjusted})}$	$b-1$	$\frac{SS_{B(\text{adjusted})}}{b-1}$	$\frac{MS_{B(\text{adjusted})}}{MS_{E(\text{adjusted})}}$
AB	AB_{yy}	AB_{xy}	AB_{xy}	AFX_{AB}	$AFCP_{AB}$	AF_{AB}	$SS_{AB(\text{adjusted})}$	$(a-1)(b-1)$	$\frac{SS_{AB(\text{adjusted})}}{(a-1)(b-1)}$	$\frac{MS_{AB(\text{adjusted})}}{MS_{E(\text{adjusted})}}$
Error	E_{yy}	E_{xx}	E_{xy}	E_{xx}	E_{xy}	AF_E	$E_{yy(\text{adjusted})}$	$ab(n-1)-1$	$\frac{E_{yy(\text{adjusted})}}{ab(n-1)-1}$	
Total	S_{yy}	S_{xx}	S_{xy}	S_{xx}	S_{xy}	AF	$S_{yy(\text{adjusted})}$	$abn-2$		

Example 11.3

Three new methods (Factor A) for instructing students (chosen from three different schools—Factor B) in a course are being compared to see whether the students' performances under the teaching conditions are the same. The students had just finished another course taught by the same methods, which is a prerequisite for this new course being studied, and it was believed that their scores in the prerequisite course could affect the outcome of the test and distort the findings. It was decided to use ANCOVA to analyze the results of the study, treating the scores in the prerequisite as the covariate. Nine students were chosen from each school, and three were assigned to each of the three teaching methods, and the responses are presented in the table below.

	Factor $B\rightarrow$					
Factor $A\downarrow$	X	Y	X	Y	X	Y
I	11	16	8	10	5	11
	7	2	5	7	8	14
	6	5	9	14	8	17
II	9	7	9	6	4	8
	10	11	9	19	7	14
	5	1	7	10	9	19
III	10	8	12	8	7	14
	13	14	11	16	9	7
	8	9	10	16	9	10
Totals	79	73	80	106	66	114

$$x_{...} = 225 \quad y_{...} = 293 \quad y_{.1.} = 73 \quad y_{.2.} = 106 \quad y_{.3.} = 114 \quad x_{.1.} = 79$$

$$x_{.2.} = 80 \quad x_{.3.} = 66 \quad y_{1..} = 96 \quad y_{2..} = 95 \quad y_{3..} = 102 \quad x_{1..} = 67$$

$$x_{2..} = 69 \quad x_{3..} = 89$$

$$S_{yy} = [16^2 + 2^2 + \cdots + 7^2 + 10^2] - \frac{293^2}{27} = 3787 - \frac{293^2}{27} = 607.4074$$

$$S_{xx} = [11^2 + 7^2 + \cdots + 9^2 + 9^2] - \frac{225^2}{27} = 2001 - \frac{225^2}{27} = 126$$

$$S_{xy} = [11 \times 16 + \cdots + 9 \times 10] - \frac{225(293)}{27} = 2555 - \frac{225(293)}{27} = 113.333$$

$$A_{yy} = \frac{(96^2 + 95^2 + 102^2)}{9} - \frac{293^2}{27} = \frac{28645}{9} - \frac{293^2}{27} = 3.1852$$

$$A_{xx} = \frac{(67^2 + 69^2 + 89^2)}{9} - \frac{225^2}{27} = \frac{17171}{9} - \frac{225^2}{27} = 32.8889$$

$$A_{xy} = \frac{(96 \times 67 + 95 \times 69 + 102 \times 89)}{9} - \frac{293(225)}{27} = 10.0$$

$$B_{yy} = \frac{(73^2 + 106^2 + 114^2)}{9} - \frac{293^2}{27} = \frac{29561}{9} - \frac{293^2}{27} = 104.9629$$

$$B_{xx} = \frac{(79^2 + 80^2 + 66^2)}{9} - \frac{225^2}{27} = \frac{16997}{9} - \frac{225^2}{27} = 13.5556$$

$$B_{xy} = \frac{(73 \times 79 + 106 \times 80 + 114 \times 66)}{9} - \frac{293(225)}{27} = -22.6667$$

$$AB_{yy} = \frac{(23^2 + 19^2 + 31^2 + 31^2 + 35^2 + 40^2 + 42^2 + 4^2 + 31^2)}{3} - \frac{293^2}{27} - 104.9629 - 3.1852 = 59.93$$

$$AB_{xx} = \frac{(24^2 + 24^2 + 31^2 + 22^2 + 23^2 + 28^2 + 21^2 + 20^2 + 25^2)}{3} - \frac{225^2}{27} - 13.5556 - 32.8889 = 4.2215$$

$$AB_{xy} = \frac{(23 \times 24 + 24 \times 19 + 31 \times 31 + \cdots + 31 \times 25)}{3} - \frac{293(225)}{27} - 10 + 22.6667 = 12$$

$$E_{yy} = 607.4074 - 104.9629 - 3.1852 - 59.93 = 439.35$$

$$E_{xx} = 126 - 13.5556 - 32.8889 - 4.2215 = 75.33$$

$$E_{xy} = 113.33 + 22.6667 - 10 - 12 = 114$$

$$AF_{\text{total}} = \frac{(113.33)^2}{126} = 101.93$$

$$AF_B = \frac{(114 - 22.6667)^2}{13.5556 + 75.33} = 93.848$$

$$AF_A = \frac{(10 + 114)^2}{32.8889 + 75.33} = 142.08$$

$$AF_{AB} = \frac{(12 + 114)^2}{4.2215 + 75.33} = 199.57$$

$$AF_{\text{E}} = \frac{114^2}{75.33} = 172.52$$

$$E_{yy(\text{adjusted})} = 439.35 - 172.52 = 266.83$$

$$S_{yy(\text{adjusted})} = 607.41 - 101.93 = 505.48$$

$$SS_{A(\text{adjusted})} = 3.18 - 142.08 + 172.52 = 33.62$$

$$SS_{B(\text{adjusted})} = 104.96 - 93.84 + 172.52 = 183.64$$

$$SS_{AB(\text{adjusted})} = 59.93 - 199.57 + 172.52 = 32.87$$

TABLE 11.11: Unadjusted Sums of Squares and Cross Products and Components of Adjustment for ANCOVA for the Two-Factor Experiment with One Covariate for Example 11.3

	Unadjusted sums of squares and cross products			Components of adjustment	
Source	y	x	xy	X	xy
A	3.18	32.89	10	108.22	124
B	104.96	13.56	−22.67	88.89	91.33
AB	59.92	4.22	12	79.55	126
Error	439.35	75.33	114	75.33	114
Total	607.41	126	113.33	126	113.33

TABLE 11.12: Adjustment Factors and Adjustment for Regression (ANCOVA for the Two-Factor Experiment with One Covariate for Example 11.3)

		Adjusted for regression			
Source	**Adjustment factors**	**Sum of squares**	**Degrees of freedom**	**Mean square**	***F*-ratio**
A	142.08	33.63	2	16.81	1.07
B	93.84	183.64	2	91.82	5.85
AB	199.57	32.87	4	8.22	0.52
Error	172.52	266.83	17	15.69	
Total	505.48	505.48	25		

The analysis of covariance is presented in Tables 11.11 and 11.12:

From the tables, $F(0.05, 2, 17) = 3.59$ and $F(0.05, 4, 17) = 2.96$ so that only the teaching methods A is significant. In Table 11.12, it is clear that the adjusted sums of squares of effects and error do not add up to the adjusted total sum of squares. This is because in analysis of covariance the adjusted sums of squares depend on the estimate of γ, and, therefore, are not orthogonal as is the case in the analysis of variance. Therefore, it is meaningless to try to estimate the variances of effects or proportions of variance accounted for by each effect. ▯

11.7.1 Tests about Significance of Regression

We can test the hypothesis about the effect, γ, that is, in the test of the hypothesis, there is no regression effect. We state the hypothesis as

$$H_0 : \gamma = 0 \quad \text{vs} \quad H_1 : \gamma \neq 0$$

The statistic to use for this test is

$$F_{\text{cal}} = \frac{AF_{\text{E}}}{MSE_{\text{adjusted}}}$$

The hypothesis is rejected for the two-way layout with replication if $F > F[\alpha, 1, ab(n-1) - 1]$. In Example 13.3, we test the hypothesis about the regression effect, γ, using the statistic

$$F_{\text{cal}} = \frac{172.52}{15.69} = 10.99$$

From the tables, $F(0.05, 1, 17) = 8.40$ so that the regression effect is significant.

SAS PROGRAM

```
DATA NEWEXAMP11_2;
dO I=1 TO 3;
a=1;
DO b=1 TO 3;
INPUT X Y@@; OUTPUT;
END; END;
DO I=1 TO 3;
a=2;
DO b=1 TO 3;
INPUT X Y@@; OUTPUT;
END; END;
DO I=1 TO 3;
a=3;
DO b=1 TO 3;
INPUT X Y@@; OUTPUT;
END; END;
CARDS;
11   16    8    10   5    11
 7    2    5     7   8    14
 6    5    9    14   8    17
 9    7    9     6   4     8
10   11    9    19   7    14
 5    1    7    10   9    19
10    8   12     8   7    14
13   14   11    16   9     7
 8    9   10    16   9    10
;
proc print data=NEWexamp11_2;
sum x y;
proc glm data=NEWexamp11_2;
class a b;
model y = X a|b; run;
```

SAS OUTPUT

Obs	I	a	b	X	Y
1	1	1	1	11	16
2	1	1	2	8	10
3	1	1	3	5	11
4	2	1	1	7	2
5	2	1	2	5	7
6	2	1	3	8	14
7	3	1	1	6	5
8	3	1	2	9	14
9	3	1	3	8	17
10	1	2	1	9	7
11	1	2	2	9	6
12	1	2	3	4	8
13	2	2	1	10	11
14	2	2	2	9	19
15	2	2	3	7	14
16	3	2	1	5	1
17	3	2	2	7	10
18	3	2	3	9	19
19	1	3	1	10	8
20	1	3	2	12	8
21	1	3	3	7	14
22	2	3	1	13	14
23	2	3	2	11	16
24	2	3	3	9	7
25	3	3	1	8	9
26	3	3	2	10	16
27	3	3	3	9	10
				===	===
				225	293

```
              The GLM Procedure

           Class Level Information

          Class     Levels     Values
            a         3         1  2  3
            b         3         1  2  3

       Number of Observations Read    27
       Number of Observations Used    27

              The GLM Procedure

Dependent Variable: Y

                                 Sum of
    Source              DF      Squares    Mean Square    F Value    Pr > F

    Model                9  340.5873484     37.8430387       2.41    0.0565
    Error               17  266.8200590     15.6952976
    Corrected Total     26  607.4074074

         R-Square     Coeff Var     Root MSE     Y Mean
         0.560723     36.50740      3.961729     10.85185
```

Source	DF	Type I SS	Mean Square	F Value	Pr > F
X	1	101.9400353	101.9400353	6.49	0.0208
a	2	15.9229542	7.9614771	0.51	0.6110
b	2	189.8438179	94.9219089	6.05	0.0104
a*b	4	32.8805410	8.2201353	0.52	0.7197

Source	DF	Type III SS	Mean Square	F Value	Pr > F
X	1	172.5132743	172.5132743	10.99	0.0041
a	2	33.6204308	16.8102154	1.07	0.3647
b	2	183.6312373	91.8156186	5.85	0.0117
a*b	4	32.8805410	8.2201353	0.52	0.7197

11.7.2 Testing for Homogeneity of Slopes across Levels of Factors and Cells

The analysis of the responses of the two-factor factorial design with one concomitant variable was illustrated in Example 11.3. The testing for homogeneity of slopes for such a design can be accomplished in a similar manner as we did in Section 11.6.2 for the one-way ANCOVA with one covariate; however, the process is more complicated than was presented in Section 11.6.2. We present no methods of manual analysis here. It will suffice for us to describe what is needed and to carry out the test using the SAS software.

To test for homogeneity of slopes, we need to test for interaction between each of the main factors and the covariate. We also need to test for the interaction of the generalized interaction of the factors in the design and the concomitant variable. Thus, we need to test for significance of AX, BX, and ABX in the two-factor factorial design if A and B are the factors in the design and X is the covariate. If none of these is significant, then homogeneity of slopes is established.

We present a SAS program that carries out the test for Example 11.3 as follows:

SAS PROGRAM: TESTING HOMOGENEITY OF SLOPE FOR EXAMPLE 11.3

```
DATA NEWEXAMP11_2;
dO I=1 TO 3;
a=1;
DO b=1 TO 3;
INPUT X Y@@; OUTPUT;
END; END;
DO I=1 TO 3;
a=2;
DO b=1 TO 3;
INPUT X Y@@; OUTPUT;
END; END;
DO I=1 TO 3;
a=3;
```

```
DO b=1 TO 3;
INPUT X Y@@; OUTPUT;
END; END;
CARDS;
11    16    8     10    5     11
7     2     5     7     8     14
6     5     9     14    8     17
9     7     9     6     4     8
10    11    9     19    7     14
5     1     7     10    9     19
10    8     12    8     7     14
13    14    11    16    9     7
8     9     10    16    9     10
;
proc print data=NEWexamp11_2;
sum x y;
proc glm data=NEWexamp11_2;
class a b;
model y = X a|b a*x b*x a*b*x; run;
```

SAS OUTPUT

```
SAS OUTPUT:

                              The GLM Procedure

                           Class Level Information

                        Class         Levels    Values

                        a                  3    1 2 3

                        b                  3    1 2 3

                   Number of Observations Read          27
                   Number of Observations Used          27

                              The GLM Procedure

Dependent Variable: Y

                                       Sum of
  Source                    DF        Squares     Mean Square    F Value    Pr > F

  Model                     17    475.4912709      27.9700748       1.91    0.1625

  Error                      9    131.9161365      14.6573485

  Corrected Total           26    607.4074074

                 R-Square     Coeff Var      Root MSE        Y Mean

                 0.782821      35.27962      3.828492      10.85185
```

Source	DF	Type I SS	Mean Square	F Value	Pr > F
X	1	101.9400353	101.9400353	6.95	0.0270
a	2	15.9229542	7.9614771	0.54	0.5988
b	2	189.8438179	94.9219089	6.48	0.0181
a*b	4	32.8805410	8.2201353	0.56	0.6972
X*a	2	58.4703612	29.2351806	1.99	0.1919
X*b	2	28.7903749	14.3951875	0.98	0.4113
X*a*b	4	47.6431863	11.9107966	0.81	0.5478

Source	DF	Type III SS	Mean Square	F Value	Pr > F
X	1	15.1454879	15.1454879	1.03	0.3359
a	2	91.2136437	45.6068219	3.11	0.0939
b	2	93.6531682	46.8265841	3.19	0.0895
a*b	4	39.9631338	9.9907835	0.68	0.6220
X*a	2	107.3308245	53.6654123	3.66	0.0686
X*b	2	50.8215854	25.4107927	1.73	0.2307
X*a*b	4	47.6431863	11.9107966	0.81	0.5478

None of the interactions AX, BX, and ABX is significant. Therefore, we have not got enough evidence to lead us to reject the null hypothesis that the slopes are homogeneous. We can therefore conclude that the findings in the ANCOVA carried out previously on the designed experiment stands since this important premise of the model and analysis is not violated.

11.7.3 Using Regression for ANCOVA (Application to Two-Way ANCOVA)

Next, we use the effects coding method to define the dummy variables for the regression approach to the two-way ANCOVA applied to Example 11.2 as:

$$V_i = \begin{cases} -1 & \text{if } y_{ij} \text{ is under level 3 of } A \\ 1 & \text{if } y_{ij} \text{ is under level } i \text{ of } A \\ 0 & \text{otherwise} \end{cases}$$
$$\text{for } i = 1, 2$$

Also,

$$Z_j = \begin{cases} -1 & \text{if } y_{ij} \text{ is under level 3 of } B \\ 1 & \text{if } y_{ij} \text{ is under level } j \text{ of } B \\ 0 & \text{otherwise} \end{cases}$$
$$\text{for } j = 1, 2$$

The product of V_i and Z_j will test the interaction between the factors A and B. In order to test for homogeneity of slopes, we again obtain V_iX, Z_jX, V_iZ_jX and test for their significance. The SAS analysis for data of Example 11.3 was recoded and regression was used to carry out ANCOVA as follows:

SAS PROGRAM

```
data examp11_3effect;
input v1 v2 z1 z2 x y @@;
s1=v1*z1; s2=v1*z2; s3=v2*z1; s4=v2*z2;
h1=v1*x; h2=v2*x; h3=z1*x; h4=z2*x;
h5=v1*z1*x; h6=v1*z2*x; h7=v2*z1*x; h8=v2*z2*x;
cards;
1 0 1 0 11 16 1 0 1 0 7 2 1 0 1 0 6 5
1 0 0 1 8 10 1 0 0 1 5 7 1 0 0 1 9 14
1 0 -1 -1 5 11 1 0 -1 -1 8 14 1 0 -1 -1 8 17
0 1 1 0 9 7 0 1 1 0 10 11 0 1 1 0 5 1
0 1 0 1 9 6 0 1 0 1 9 19 0 1 0 1 7 10
0 1 -1 -1 4 8 0 1 -1 -1 7 14 0 1 -1 -1 9 19
-1 -1 1 0 10 8 -1 -1 1 0 13 14 -1 -1 1 0 8 9
-1 -1 0 1 12 8 -1 -1 0 1 11 16 -1 -1 0 1 10 16
-1 -1 -1 -1 7 14 -1 -1 -1 -1 9 7 -1 -1 -1 -1 9 10
;
proc print data=examp11_3effect;
sum y v1-v2 z1-z2 s1-s4 x h1-h8;
proc glm data=examp11_3effect;
model y=x v1-v2 z1-z2 s1-s4 h1-h8; run;
```

SAS OUTPUT

Obs	v1	v2	z1	z2	x	y	s1	s2	s3	s4	h1	h2	h3	h4	h5	h6	h7	h8
1	1	0	1	0	11	16	1	0	0	0	11	0	11	0	11	0	0	0
2	1	0	1	0	7	2	1	0	0	0	7	0	7	0	7	0	0	0
3	1	0	1	0	6	5	1	0	0	0	6	0	6	0	6	0	0	0
4	1	0	0	1	8	10	0	1	0	0	8	0	0	8	0	8	0	0
5	1	0	0	1	5	7	0	1	0	0	5	0	0	5	0	5	0	0
6	1	0	0	1	9	14	0	1	0	0	9	0	0	9	0	9	0	0
7	1	0	-1	-1	5	11	-1	-1	0	0	5	0	-5	-5	-5	-5	0	0
8	1	0	-1	-1	8	14	-1	-1	0	0	8	0	-8	-8	-8	-8	0	0
9	1	0	-1	-1	8	17	-1	-1	0	0	8	0	-8	-8	-8	-8	0	0
10	0	1	1	0	9	7	0	0	1	0	0	9	9	0	0	0	9	0
11	0	1	1	0	10	11	0	0	1	0	0	10	10	0	0	0	10	0
12	0	1	1	0	5	1	0	0	1	0	0	5	5	0	0	0	5	0
13	0	1	0	1	9	6	0	0	0	1	0	9	0	9	0	0	0	9
14	0	1	0	1	9	19	0	0	0	1	0	9	0	9	0	0	0	9
15	0	1	0	1	7	10	0	0	0	1	0	7	0	7	0	0	0	7
16	0	1	-1	-1	4	8	0	0	-1	-1	0	4	-4	-4	0	0	-4	-4
17	0	1	-1	-1	7	14	0	0	-1	-1	0	7	-7	-7	0	0	-7	-7
18	0	1	-1	-1	9	19	0	0	-1	-1	0	9	-9	-9	0	0	-9	-9
19	-1	-1	1	0	10	8	-1	0	-1	0	-10	-10	10	0	-10	0	-10	0
20	-1	-1	1	0	13	14	-1	0	-1	0	-13	-13	13	0	-13	0	-13	0
21	-1	-1	1	0	8	9	-1	0	-1	0	-8	-8	8	0	-8	0	-8	0
22	-1	-1	0	1	12	8	0	-1	0	-1	-12	-12	0	12	0	-12	0	-12

Obs	v1	v2	z1	z2	x	y	s1	s2	s3	s4	h1	h2	h3	h4	h5	h6	h7	h8
23	-1	-1	0	1	11	16	0	-1	0	-1	-11	-11	0	11	0	-11	0	-11
24	-1	-1	0	1	10	16	0	-1	0	-1	-10	-10	0	10	0	-10	0	-10
25	-1	-1	-1	-1	7	14	1	1	1	1	-7	-7	-7	-7	7	7	7	7
26	-1	-1	-1	-1	9	7	1	1	1	1	-9	-9	-9	-9	9	9	9	9
27	-1	-1	-1	-1	9	10	1	1	1	1	-9	-9	-9	-9	9	9	9	9
	==	==	==	==	===	===	==	==	==	==	===	===	===	===	===	===	===	===
	0	0	0	0	225	293	0	0	0	0	-22	-20	13	14	-3	-7	-2	-3

The GLM Procedure

Number of Observations Read	27
Number of Observations Used	27

The GLM Procedure

Dependent Variable: y

Source	DF	Sum of Squares	Mean Square	F Value	Pr > F
Model	17	475.4912709	27.9700748	1.91	0.1625
Error	9	131.9161365	14.6573485		
Corrected Total	26	607.4074074			

R-Square	Coeff Var	Root MSE	y Mean
0.782821	35.27962	3.828492	10.85185

Source	DF	Type I SS	Mean Square	F Value	Pr > F
x	1	101.9400353	101.9400353	6.95	0.0270
v1	1	13.4276777	13.4276777	0.92	0.3635
v2	1	2.4952765	2.4952765	0.17	0.6896
z1	1	189.8166291	189.8166291	12.95	0.0058
z2	1	0.0271888	0.0271888	0.00	0.9666
s1	1	17.0795737	17.0795737	1.17	0.3085
s2	1	0.4819193	0.4819193	0.03	0.8601
s3	1	15.3170895	15.3170895	1.05	0.3333
s4	1	0.0019586	0.0019586	0.00	0.9910
h1	1	42.0987173	42.0987173	2.87	0.1244
h2	1	16.3716439	16.3716439	1.12	0.3181
h3	1	14.8819147	14.8819147	1.02	0.3400
h4	1	13.9084602	13.9084602	0.95	0.3555
h5	1	10.4221194	10.4221194	0.71	0.4209
h6	1	9.6602401	9.6602401	0.66	0.4378
h7	1	25.7835566	25.7835566	1.76	0.2174
h8	1	1.7772703	1.7772703	0.12	0.7357

Source	DF	Type III SS	Mean Square	F Value	Pr > F
x	1	15.14548790	15.14548790	1.03	0.3359
v1	1	50.18465064	50.18465064	3.42	0.0973
v2	1	35.54708458	35.54708458	2.43	0.1538
z1	1	91.70454806	91.70454806	6.26	0.0338
z2	1	23.64234811	23.64234811	1.61	0.2359
s1	1	9.19778676	9.19778676	0.63	0.4486
s2	1	11.69319586	11.69319586	0.80	0.3950

Source	DF	Type III SS	Mean Square	F Value	Pr > F
s3	1	20.51344010	20.51344010	1.40	0.2671
s4	1	5.50843117	5.50843117	0.38	0.5550
h1	1	47.12303442	47.12303442	3.21	0.1066
h2	1	32.99411873	32.99411873	2.25	0.1678
h3	1	50.69630479	50.69630479	3.46	0.0958
h4	1	16.07568312	16.07568312	1.10	0.3223
h5	1	6.08451358	6.08451358	0.42	0.5355
h6	1	5.27649054	5.27649054	0.36	0.5633
h7	1	22.98920128	22.98920128	1.57	0.2420
h8	1	1.77727026	1.77727026	0.12	0.7357

Parameter	Estimate	Standard Error	t Value	Pr > |t|
Intercept	7.88482960	5.22788796	1.51	0.1658
x	0.58540582	0.57589474	1.02	0.3359
v1	-11.43000664	6.17715839	-1.85	0.0973
v2	-10.60767838	6.81155170	-1.56	0.1538
z1	-15.29961656	6.11663853	-2.50	0.0338
z2	11.23269177	8.84435661	1.27	0.2359
s1	5.94003170	7.49850700	0.79	0.4486
s2	-8.91828396	9.98486945	-0.89	0.3950
s3	9.49865583	8.02916920	1.18	0.2671
s4	-7.25984298	11.84243395	-0.61	0.5550
h1	1.29737806	0.72356514	1.79	0.1066
h2	1.17837864	0.78540616	1.50	0.1678
h3	1.25043378	0.67235750	1.86	0.0958
h4	-0.97643146	0.93236241	-1.05	0.3223
h5	-0.56178909	0.87194296	-0.64	0.5355
h6	0.67057066	1.11763417	0.60	0.5633
h7	-1.15707538	0.92390522	-1.25	0.2420
h8	0.46264700	1.32861981	0.35	0.7357

The analysis agrees with the results obtained previously by manual and other methods.

11.7.4 Multiple Covariates in ANCOVA

In most of the designed experiments, it is a practice to avoid too many covariates, so that often only one covariate is measured. However, sometimes, more than one covariate could be needed in order to explain the observed responses. Moreover, studies in which more than one covariate is involved are routinely encountered in behavioral research. If we assume homogeneity of slopes for the design, the model for such a design is

$$y_{ij} = \mu + \alpha_i + \beta(x_{ij} - \bar{x}_{i.}) + \gamma(z_{ij} - \bar{z}_{i.}) + \varepsilon_{ij} \quad i = 1, 2, \ldots, a \;\; j = 1, 2, \ldots, n$$

Further details about computation needed for this type of design can be found in Kirk (1982). Although here we treat only an application to one-way ANCOVA, the principle can easily be extended to cover other designs such as randomized complete block and two-factor factorial experiments.

We present an example of application to one-way ANCOVA and the ensuing SAS analysis.

Example 11.4
An experiment was performed to test three new methods of teaching spelling in an elementary school. Twenty four students of approximate abilities were randomly assigned to each teaching method. The students had taken spelling in the preceding couple of school terms and it was thought that their performances in these previous classes would affect their responses to the new teaching method. The two scores were treated as concomitant variables X, Z and the responses (scores in spelling) Y is to be analyzed as ANCOVA with two covariates. The responses are presented as follows:

A								
L1			**L2**			**L3**		
Y	*X*	*Z*	*Y*	*X*	*Z*	*Y*	*X*	*Z*
4	37	10	4	42	9	7	56	16
7	52	21	5	44	16	8	60	20
4	28	19	4	37	13	7	59	19
4	42	15	3	36	10	6	51	15
2	27	10	2	33	18	5	47	17
3	30	14	3	38	10	6	53	13
3	28	5	4	43	14	5	48	8
3	34	6	3	40	12	6	49	9

We write a SAS program to analyze the responses of the experiment. We include a test for the homogeneity of slopes for X and Z in the program across the levels of teaching method A. □

SAS PROGRAM

```
data ancov3(drop=j);
do j=1 to 8;
do A=1 to 3;
input x y z @@;
output; end; end;
cards;
4     37    10    4     42    9     7     56    16
7     52    21    5     44    16    8     60    20
4     28    19    4     37    13    7     59    19
4     42    15    3     36    10    6     51    15
2     27    10    2     33    18    5     47    17
3     30    14    3     38    10    6     53    13
3     28    5     4     43    14    5     48    8
3     34    6     3     40    12    6     49    9
;
proc print data=ancov3;
proc glm data=ancov3;
class A;
model y=x z a a*x a*z; run;
```

SAS OUTPUT

Obs	A	x	y	z
1	1	4	37	10
2	2	4	42	9
3	3	7	56	16
4	1	7	52	21
5	2	5	44	16
6	3	8	60	20
7	1	4	28	19
8	2	4	37	13
9	3	7	59	19
10	1	4	42	15
11	2	3	36	10
12	3	6	51	15
13	1	2	27	10
14	2	2	33	18
15	3	5	47	17
16	1	3	30	14
17	2	3	38	10
18	3	6	53	13
19	1	3	28	5
20	2	4	43	14
21	3	5	48	8
22	1	3	34	6
23	2	3	40	12
24	3	6	49	9

```
                    The GLM Procedure

                 Class Level Information

              Class          Levels    Values

              A                   3    1 2 3

         Number of Observations Read          24
         Number of Observations Used          24
                    The GLM Procedure

Dependent Variable: y
```

Source	DF	Sum of Squares	Mean Square	F Value	Pr > F
Model	8	2081.265152	260.158144	25.14	<.0001
Error	15	155.234848	10.348990		
Corrected Total	23	2236.500000			

R-Square	Coeff Var	Root MSE	y Mean
0.930590	7.614165	3.216985	42.25000

Source	DF	Type I SS	Mean Square	F Value	Pr > F
x	1	1888.015152	1888.015152	182.43	<.0001
z	1	2.649788	2.649788	0.26	0.6202
A	2	162.443525	81.221762	7.85	0.0047
x*A	2	16.284055	8.142028	0.79	0.4732
z*A	2	11.872632	5.936316	0.57	0.5754

Source	DF	Type III SS	Mean Square	F Value	Pr > F
x	1	342.4381830	342.4381830	33.09	<.0001
z	1	1.9562965	1.9562965	0.19	0.6699
A	2	33.5131905	16.7565952	1.62	0.2308
x*A	2	25.4733182	12.7366591	1.23	0.3200
z*A	2	11.8726320	5.9363160	0.57	0.5754

11.7.5 Dealing with the Failure of the Homogeneity of Slopes Assumption in ANCOVA

There are often difficulties in interpreting the results of ANCOVA. The use of ANCOVA in experimental designs is to serve the purpose of eliminating extraneous variation and increasing the precision of the analysis. If treatment affects the covariates, then removal of extraneous variation may wind up removing some variance due to treatment. This should be avoided by measuring the covariate before the inception of the experiment.

Not infrequently, the assumptions of ANCOVA, and in particular the assumption of homogeneity of variance fail. A number of methods are being suggested for dealing with an experiment in which the slope homogeneity is in doubt.

11.7.5.1 Blocking the Concomitant Variable

One of the alternatives to ANCOVA, which deals with heterogeneity of slopes, is the use of blocking. The concomitant variable is grouped into blocks by subdividing the range of values of the covariate to create levels of block. For instance, values of the covariate can be set into levels of the block such as low, medium, and high. Thus, the levels of block become the levels of another factor which is deemed to be crossed with the factor of interest in a factorial design. Here, after analysis, we can easily interpret the main effect of the factor of interest. Moreover, with this arrangement, we have removed the variation due to the concomitant variable from the design and assessed it as a separate main effect. The interaction between the concomitant variable treated as block with the main factor captures the aspects of the design that has to do with the assumption of homogeneity of regression and can be interpreted. The advantages of such a design are that it does not require the special assumptions needed for ANCOVA. Also, the relationship between the concomitant variable and the response variable need not be linear. This blocking is more powerful than ANCOVA if the relationship between the response variable and the concomitant variable is not linear. However, if there is a linear relationship between them, blocking would be less powerful than ANCOVA. Blocking would still be preferable for many experimental designs but not for nonexperimental research.

11.7.5.2 The Johnson–Neyman Method

Another method for dealing with the failure of the assumption of homogeneous slope were developed by Johnson and Neyman (1936). ANCOVA with heterogeneous slopes is analogous to a two-factor factorial design in factors A and B in which the AB interaction is significant. When this happens, then as we illustrated earlier, the analysis of the interaction effect usually seeks to indentify the levels of B at which simple effects of A are present and vice versa. In one way ANCOVA in which there is heterogeneity of slopes, the Johnson–Neyman method aids the researcher in identifying the values of the concomitant variable X which are associated with significant group differences in Y, the response variable. The method helps in identifying the regions of significance and nonsignificance. The computations required for this method were treated in Section 13.2 of Huitema (1980) along with an application.

Other methods for dealing with heterogeneity of slopes are presented in Chapter 9 of Maxwell and Delaney (1990).

11.8 Exercises

Exercise 11.1

After some steps in the path of steepest ascent, a chemical process represented by 2^3 factorial design was augmented with five measurements at the center.

x_1	x_2	x_3	y	x_1	x_2	x_3	y
−1	−1	−1	17	1	1	1	25.6
1	−1	−1	22.1	0	0	0	19.5
−1	1	−1	17.4	0	0	0	19.2
1	1	−1	20.4	0	0	0	20
−1	−1	1	21.2	0	0	0	19.1
1	−1	1	22	0	0	0	17.9
−1	1	1	26				

Fit a first-order model to the data and determine whether the model fits. □

Exercise 11.2

In Example 11.1, suppose that a second system manager instead of using an orthogonal design choose a rotatable design and obtain the α for his/her design. The responses of the process after the rotatable design was chosen are presented below.

x_1	x_2	x_3	y	x_1	x_2	x_3	y
−1	−1	−1	17.0	0	0	0	19.5
1	−1	−1	22.1	0	0	0	19.2
−1	1	−1	19.8	α	0	0	23.4
1	1	−1	20.4	$-\alpha$	0	0	21.1
−1	−1	1	21.2	0	α	0	20.6
1	−1	1	22.0	0	$-\alpha$	0	16.4
−1	1	1	25.6	0	0	α	23.4
1	1	1	26.0	0	0	$-\alpha$	19.5

Fit a second-order model to the design and perform the canonical analysis. How does your conclusion vary from that reached in Exercise 11.1? □

Exercise 11.3

In exploring the response surface of a process involving three factors, it was decided to use a rotatable 2^3 CCD to fit a second order model around the point where lack of fit was significant for a first order model. The responses obtained at the points chosen are as follows:

x_1	x_2	x_3	y
−1	−1	−1	12.99
1	−1	−1	1.53
−1	1	−1	10.64
1	1	−1	14.47
−1	−1	1	7.72
1	−1	1	2.75
−1	1	1	5.39
1	1	1	18.95
0	0	0	15.23
0	0	0	15.65
0	0	0	16.23
0	0	0	14.30
0	0	0	14.78
0	0	0	17.13
−1.6818	0	0	5.91
1.6818	0	0	2.86
0	−1.6818	0	14.03
0	1.6818	0	2.83
0	0	−1.6818	18.53
0	0	1.6818	19.66

Fit a second-order model to the data and perform canonical analysis and determine the nature of the stationary point. □

Exercise 11.4

Suppose that in Exercise 11.3, instead of a rotatable CCD an orthogonal design was used so that the following data were obtained for the star points:

x_1	x_2	x_3	y
$-\alpha$	0	0	4.71
α	0	0	1.76
0	$-\alpha$	0	1.57
0	α	0	12.23
0	0	$-\alpha$	16.75
0	0	α	17.22

Determine the value of α and hence fit the second-order model and perform the canonical analysis. Compare your conclusions about the nature of the stationary point arising from both designs. □

Exercise 11.5

In a study to explore the effect of small increases in the sizes of the three factors that are believed to influence the response of a process, the following data were obtained.

x_1	x_2	x_3	y	x_1	x_2	x_3	y
−1	−1	−1	11.45	1	1	1	16.85
1	−1	−1	20.2	0	0	0	14.1
−1	1	−1	8.75	0	0	0	14.8
1	1	−1	11.6	0	0	0	14.8
−1	−1	1	10.1	0	0	0	14.3
1	−1	1	18.85	0	0	0	14.3
−1	1	1	7.55				

Fit a first-order model to the data and determine the next action to be taken. □

Exercise 11.6

A group of pupils were divided into three groups and subjected to three new methods of learning basic mathematics, however, it is thought that their previous performances in arithmetic (X) in the previous year affected their performances after the course (Y). To see whether there is a differential performance according to gender, the group was also divided accordingly. Analyze the data to compare the teaching methods as well as performances according to sex.

	A_1		A_2		A_3	
Gender	X	Y	X	Y	X	Y
Boys	27	9	32	28	30	33
	27	8	21	23	19	22
	25	9	23	23	18	24
	25	13	30	22	22	22
	24	7	23	27	17	19
	22	12	24	30	23	19
Girls	27	20	23	20	22	27
	28	18	25	28	24	17
	35	17	21	18	19	20
	28	31	14	14	15	18
	26	21	19	19	22	20
	28	30	32	35	14	16

□

Exercise 11.7
In a study conducted in a textile training establishment, it was assumed that the number of months of training of the apprentices would affect their performances in the three tests, T_1, T_2, and T_3, which were administered. In instructing the apprentices, two methods, A and B, were used. Their scores in the tests (Y) as well as months of training (X) are as follows:

	Tests					
	T_1		T_2		T_3	
Method	X	Y	X	Y	X	Y
I	5	27	3	23	3	28
	3	21	4	22	2	26
	5	26	4	20	5	26
	6	31	2	18	3	24
	1	23	5	22	1	20
II	5	24	1	13	3	32
	3	18	2	12	2	30
	5	23	2	10	5	30
	6	28	1	8	3	28
	1	20	3	12	1	24

Using regression analysis, show that there is a linear relationship between the months of experience and the scores in the tests. Perform the analysis of covariance to reach the conclusions about the performances of the apprentices in the different tests. Is there any significant difference in the performances due to method of instruction? □

Exercise 11.8
Three groups of children subjected to the same instruction (grouped by social strata) were studied for keyboard skills. The ages of the children were taught to affect the outcomes (number of words typed per minute), so a one-way ANCOVA was used for the study, and the responses are as follows:

Group I		**Group II**		**Group III**		**Group I**		**Group II**		**Group III**	
X	Y	X	Y	X	Y	X	Y	X	Y	X	Y
11	19	10	25	5	13	4	12	10	28	9	16
6	17	7	18	8	19	9	24	7	20	6	12
5	15	11	27	7	18	12	26	4	14	6	15
8	23	9	25	12	22	5	14	6	17	9	17
9	19	5	16	5	14	4	17	8	21	10	20

Using regression analysis, show that there is a linear relationship between the ages of the children and the scores in the tests. Perform the analysis of covariance to reach conclusions about the performances of the children in the different groups. Is there any significant difference in the performances due to group? □

Exercise 11.9
Write a SAS program to analyze the data of Exercise 11.6. □

Exercise 11.10
Write a SAS program to analyze the data of Exercise 11.7. □

Exercise 11.11
Write a SAS program to analyze the data of Exercise 11.8. □

References

Anderson, V.L. and McLean, R.A. (1974) *Design of Experiments: A Realistic Approach*, Marcel Dekker, New York.

Box, G.E.P. and Draper, N.R. (1987) *Empirical Model Building and Response Surfaces*, John Wiley & Sons, New York.

Box G.E.P and Hunter J.S. (1957) Multifactor designs for exploring response surfaces, *Ann. Math. Stat.*, 28, 195–241.

Box, G.E.P. and Wilson, K.B. (1951) On the experimental attainment of optimum conditions, *Journal of the Royal Statisticsl Society, Series B*, 13, 1–45.

Dunn, O.J. and Clark, V.A. (1987) *Applied Statistics: Analysis of Variance and Regression*, 2nd ed., Wiley, New York.

Hotelling, H. (1941) Experimental determination of maximum of a function, *Ann. Math. Stat.*, 12, 20.

Huitema, B.E. (1980) *The Analysis of Covariance and Alternatives*, Wiley, New York.

John, P.W.M. (1971) *Statistical Design and Analysis of Experiments*, Macmillan, New York.

Johnson, P.O. and Neyman, J. (1936) Tests of certain linear hypotheses and their application to some educational problems. In J. Neyman and E.S. Pearson (Eds.), *Statistical Research Memoris*, 1936, 1, 57–93.

Keppel, G. (1983) *Design and Analysis: A Researcher's Handbook*, 2nd ed., Prentice Hall, Englewood Cliffs, NJ.

Kirk, R.E. (1982) *Experimental Design*, 2nd ed., Brooks/Cole, Belmont, CA.

Maxwell, S.E. and Delaney, H.D. (1990) *Designing Experiments and Analyzing Data: A Model Comparison Perspective.* Wadsworth Publishing Company, Belmont, CA.

Montgomery, D.C. (2005) *Design and Analysis of Experiments*, 6th ed., Wiley, New York.

Myers, R.H. and Montgomery, D.C. (1995) *Response Surface Methodology: Process and Product Optimization Using Designed Experiments*, Wiley, New York.

Myers, R.H. (1971) *Response Surface Methodology*, Allen and Bacon Inc, Boston.

Petersen, R. (1985) *Design and Analysis of Experiments*, Marcel Dekker, New York.

Press, W.H., Flannery, B.P. Teukolsky, S.A. and Vetterling, W.T. (1986) *Numerical Recipes: The Art of Scientific Computing*, Cambridge University Press, Cambridge MA.

Tabachnick, B.G. and Fidell, L.S. (2007) *Experimental Designs Using ANOVA*. Duxbury Press, Belmont, CA.

Chapter 12

Multivariate Analysis of Variance (MANOVA)

12.1 Link between ANOVA and MANOVA

In the previous chapters, we discussed the univariate analysis of variance ANOVA. In those sections, we studied the one-way and two-way classifications as well as full two-factor factorial experiments—the two-factor factorial experiment being a two-way classification with replication. In those studies, only one response or independent variable was elicited for each of the units in the designed experiments. However, with recent advances in sciences and the advent of more sophisticated measurement equipment, it is now possible to measure a number of responses of a chemical system in a designed experiment simultaneously. Some industrial processes have more than one yield for the settings of the system control factors. In chemical studies, some spectrophotometers can simultaneously measure several spectra. In distillations from raw material, more than one response can be simultaneously elicited. In educational research, students can be subjected to a number of tests under different teaching methods.

In the univariate one-way analysis of variance which we discussed earlier, we had a set of samples that were suspected to have arisen from the same population. The ANOVA is the process of testing to ensure that such samples do not actually arise from the same population. In the process, we assume that the samples were normally distributed with the same variance in each population. The research situation in the one-way ANOVA is to determine whether there is any significant difference between the means of the samples which should, in turn, indicate to us that they come from different populations. On the other hand, MANOVA represents a generalization of the principle underlying the univariate ANOVA to cover cases in which multiple responses are obtained from each individual in the designed experiment.

The MANOVA is distinguished from the ANOVA by the fact that the dependent variable in the design is a vector variable corresponding to different attributes (responses) being investigated for each subject. For instance, if in educational psychology experiments some pupils have been subjected to instructions, using some levels of a particular early learning technique, the response vector may include their spelling scores, reading ability, and some other cognitive abilities. As indicated earlier, in a chemical experiment, it is possible for different types of spectra to arise simultaneously from an experiment; these spectra form the elements of the response vectors that are to be analyzed. Each of the response vectors is assumed to have a multivariate normal distribution with the same variance–covariance matrix. The issue investigated in one-way MANOVA is whether the different populations have the same centroids or mean vectors, that is, to discover whether the populations have mean vectors, which are located at different points in the parameter space.

12.2 One-Way MANOVA

The design of the experiments for the univariate CRDs is the same as the design for their multivariate equivalents. The major difference is in the information elicited from both experiments. Although a single response variable is measured in the univariate case, the models used for multivariate analysis of variance designs assume that the dependent (response) variable is vector-valued. This variable is also assumed to be distributed as multivariate normal, with the same dispersion matrix for all the k groups (just as in the univariate, there are k groups in the one-way version of the CRD design for MANOVA). The research being carried out under MANOVA is to investigate whether the different groups that give rise to this dependent multivariate response variable could be said to have the same mean vector.

As was the case for the univariate ANOVA, the MANOVA is based on an assumption of a linear model which is of the form

$$\boldsymbol{Y}_{ij} = \overline{\boldsymbol{Y}}_{..} + (\overline{\boldsymbol{Y}}_{i.} - \overline{\boldsymbol{Y}}_{..}) + (\boldsymbol{Y}_{ij} - \overline{\boldsymbol{Y}}_{i.}) \tag{12.1}$$

where $\boldsymbol{Y}_{ij}$ is the dependent vector variable for the jth subject in the ith sample for $i = 1, 2, \ldots, k$ and k is the number of populations under study. $\overline{\boldsymbol{Y}}_{..}$ is the MANOVA equivalent of the grand mean in the ANOVA, and it is referred to as the grand centroid; that is, the vector of total sample means. $\overline{\boldsymbol{Y}}_{i.}$ is the centroid for the ith sample. As usual, the grand centroid does not feature in the MANOVA. So we can deduct this vector from both sides of Equation 12.1, and the resultant equation is

$$\boldsymbol{y}_{ij} = (\overline{\boldsymbol{Y}}_{i.} - \overline{\boldsymbol{Y}}_{..}) + (\boldsymbol{Y}_{ij} - \overline{\boldsymbol{Y}}_{i.}) \tag{12.2}$$

where

$$\boldsymbol{y}_{ij} = \boldsymbol{Y}_{ij} - \overline{\boldsymbol{Y}}_{..}$$

By splitting $\boldsymbol{y}_{ij}$ (which is the deviation of the response for the jth subject from the grand centroid) as indicated in Equation 12.2, we have partitioned this quantity into two, the two partitions being vectors. The first partition has to do with the hypothesis we are testing, that is, there are no differences in locations of the means of the groups. The second has to do with the error or residual effect (i.e., the deviations of the responses from the centroids of the samples).

We can obtain the total sum of squares (bearing in mind that the partitions mentioned above are column vectors) using vector multiplication as follows:

$$\sum_{i=1}^{k}\sum_{j=1}^{N_i} \boldsymbol{y}_{ij}\boldsymbol{y}'_{ij} = \sum_{i=1}^{k}\sum_{j=1}^{N_i} (\overline{\boldsymbol{Y}}_{i.} - \overline{\boldsymbol{Y}}_{..})(\overline{\boldsymbol{Y}}_{i.} - \overline{\boldsymbol{Y}}_{..})' + \sum_{i=1}^{k}\sum_{j=1}^{N_i} (\boldsymbol{Y}_{ij} - \overline{\boldsymbol{Y}}_{i.})(\boldsymbol{Y}_{ij} - \overline{\boldsymbol{Y}}_{i.})' \tag{12.3a}$$

By analogy with terms used in the one-way ANOVA, we can see that the term on the left-hand side of Equation 12.3a is the total sum of squares. If we call this term T, then

$$\boldsymbol{T} = \sum_{i=1}^{k}\sum_{j=1}^{N_i} \boldsymbol{y}_{ij}\boldsymbol{y}'_{ij} \tag{12.3b}$$

The first term on the right-hand side is the "between groups" sum of squares, which we can represent by B, so that

$$\boldsymbol{B} = \sum_{i=1}^{k}\sum_{j=1}^{N_i} (\overline{\boldsymbol{Y}}_{i.} - \overline{\boldsymbol{Y}}_{..})(\overline{\boldsymbol{Y}}_{i.} - \overline{\boldsymbol{Y}}_{..})' = \sum_{i=1}^{k} \boldsymbol{N}_i(\overline{\boldsymbol{Y}}_{i.} - \overline{\boldsymbol{Y}}_{..})(\overline{\boldsymbol{Y}}_{i.} - \overline{\boldsymbol{Y}}_{..})' \tag{12.3c}$$

The other partition of the right-hand side of Equation 12.3a has to do with the "within group" variation; we could represent it by the quantity W, so that

$$\boldsymbol{W} = \sum_{i=1}^{k}\sum_{j=1}^{N_i} (\boldsymbol{Y}_{ij} - \overline{\boldsymbol{Y}}_{i.})(\boldsymbol{Y}_{ij} - \overline{\boldsymbol{Y}}_{i.})' \tag{12.3d}$$

we can rewrite Equation 12.3a as

$$\boldsymbol{T} = \boldsymbol{B} + \boldsymbol{W} \tag{12.4}$$

By analogy with the one-way ANOVA, we note that if we divide the partitions of the total sum of squares and cross products for the MANOVA by their degrees of freedom, we obtain the estimators of the common population dispersion matrix, Δ, if the null hypothesis of equality of group centroids holds. On the other hand, if the null hypothesis does not hold, the matrix Δ arising due to differences between groups would be significantly different from the population variance–covariance matrix (whose estimate is provided by the within group error matrix). This provides a basis for making decision regarding the locations of the population centroids. Dividing these partitions by their degrees of freedom we obtain

$$\boldsymbol{\Delta}_B = \left[\frac{1}{k-1}\right]\boldsymbol{B} \tag{12.5}$$

$$\boldsymbol{\Delta}_W = \left[\frac{1}{N-k}\right]\boldsymbol{W} \tag{12.6}$$

where

$$N = \sum_{i=1}^{k} N_i$$

It is the comparison of Δ_B and Δ_W that provides us with the basis for making decision regarding the equality of centroids, which is the null hypothesis tested in the case of one-way MANOVA.

The model for the one-way MANOVA can be rewritten in terms of grand centroid, treatment effects, and error as

$$\boldsymbol{Y}_{ij} = \boldsymbol{\mu} + \boldsymbol{\gamma}_i + \boldsymbol{\varepsilon}_{ij} \tag{12.7}$$

where $\boldsymbol{\varepsilon}_{ij} \sim IN_p(\mathbf{0}, \boldsymbol{\Sigma})$; that $\boldsymbol{\varepsilon}_{ij}$ is an independent normal p-vector variable so that

$$\begin{aligned}
\mu &= (\mu_{11}, \mu_{12}, \ldots, \mu_{1p})' \\
\gamma_1' &= (\gamma_{11}, \gamma_{12}, \ldots, \gamma_{1p})' \\
\gamma_2' &= (\gamma_{21}, \gamma_{22}, \ldots, \gamma_{2p})' \\
&\vdots \\
\gamma_k' &= (\gamma_{k1}, \gamma_{k2}, \ldots, \gamma_{kp})'
\end{aligned}$$

Thus, the hypothesis to be tested is

$$H_0 : \gamma_i = \gamma \quad \text{vs} \quad H_1 : \gamma_i \neq \gamma \quad \text{for some } i = 1, 2, \ldots, k \tag{12.8}$$

The constant γ in Equation 12.8 could be set to the null vector, $\mathbf{0}$. This means that under this hypothesis, the best estimate for the common centroid for all the populations involved in the model is the grand centroid. That is,

$$\overline{\boldsymbol{Y}} = \frac{1}{N} \sum_{i=1}^{k} \sum_{j=1}^{N_i} \boldsymbol{Y}_{ij} \tag{12.9}$$

To recapitulate, we note that CRD experiments were usually designed under the univariate ANOVA for testing the differences between the means of k different groups and the means of responses at k different levels of a factor. Only one response variable was studied under the ANOVA model. In MANOVA, the basic principles and designs of the CRD is maintained, however, more than one response variable is obtained at the same time from the same experimental unit. The randomization of the order of experimentation used for the CRD is also employed for the one-way MANOVA. Although the experimental principles are the same, analyses are different, as one model is univariate while the other is multivariate.

12.2.1 Tests for Equality of Vectors of Treatment Effects

In the investigation of any differences between groups under the univariate ANOVA, there is only one test that can be used; this is the ratio of the mean squares between groups and of within groups, which is compared with the appropriate critical values under the applicable F-distribution. Unlike the univariate case where there is only one test, there is a number of tests that could be used for the multivariate analysis of variance. All the tests are based on the roots $\lambda_1, \lambda_2, \ldots, \lambda_s$ of the following determinantal equation:

$$|\boldsymbol{B} - \lambda \boldsymbol{W}| = \boldsymbol{0} \tag{12.10}$$

These roots are often called the eigenvalues. One of these tests was devised by Wilks in 1932. It uses the ratio of the determinants of the within groups and the total sum of squares, usually denoted by Λ (called the Wilks' Lambda) and is of the form

$$\boldsymbol{\Lambda} = \frac{|\boldsymbol{W}|}{|\boldsymbol{T}|} \tag{12.11a}$$

or

$$\boldsymbol{\Lambda} = \frac{|\boldsymbol{W}|}{|\boldsymbol{B} + \boldsymbol{W}|} \tag{12.11b}$$

Λ is the ratio of determinants of $\boldsymbol{W}$ and $\boldsymbol{T}$, that is, the ratio of the within group sum of squares and the total sum of squares in MANOVA CRD. For other designs, this formula will be adjusted along the same principles. The Wilks' Lambda is the statistic used for testing the equality of mean vectors. The formula for Wilks' Lambda is given in Equation 12.11b. It can be shown that

$$\Lambda = \prod_{i=1}^{s} \left(\frac{1}{1 + \lambda_i} \right) \tag{12.12}$$

where λ_i is as described in Equation 12.10.

Using Equation 12.12, the null hypothesis in Equation 12.8 is tested and rejected at $\alpha\%$ level of significance if

$$\Lambda = \prod_{i=1}^{s} \left(\frac{1}{1 + \lambda_i} \right) < U^{\alpha}(p, k-1, N-k) \tag{12.13}$$

where s is $\min(k-1,\ p)$, k the number of treatments, and p the number of response measures simultaneously obtained from each subject in the group, N the total number of observations in all the groups, and $U^{\alpha}(p, k-1, N-k)$ represents the lower percentage point of the distribution of Wilks' Lambda criterion.

The second method that could be used for the purpose of this test is the Roy's largest root criterion (Roy, 1957), which arose from the generalization of Scheffe's method by Roy and Bose (1953). Roy had proposed that the largest root of Equation 12.10, that is, λ_1 be used as a multivariate test statistic. He derived the distribution of θ_s under H_0. Pillai (1960, 1965, 1967) obtained tables of θ_s. Essentially, the null hypothesis H_0 is rejected if

$$\theta = \frac{\lambda_1}{1 + \lambda_1} > \theta^{\alpha}(s, m, n) \tag{12.14}$$

where s is $\min(v_{\text{h}}, p)$ and v_{h} the degree of freedom of effect being considered under the null hypothesis, which in this particular case is $\boldsymbol{B}$.

$$m = \frac{|k - p - 1| - 1}{2} \quad \text{and} \quad n = \frac{N - k - p - 1}{2} \tag{12.15}$$

$\theta^{\alpha}(s, m, n)$ in Equation 12.14 is the upper percentage point of the distribution of Roy's largest root criterion.

Another method which could be employed for this test is that proposed by Lawley (1938) and Hotelling (1951). The method which we shall refer to as Lawley–Hotelling trace criterion is otherwise known as Hotelling's generalized statistic, T_0^2. Under this proposal for testing the above hypothesis, the test statistic to be used is $\boldsymbol{U}^{(s)}$, where

$$U^{(s)} = \frac{T_0^2}{N - k} = \sum_{i=1}^{k} \lambda_i \tag{12.16}$$

the hypothesis H_0 is rejected if

$$U^{(s)} = \frac{T_0^2}{N - k} = \sum_{i=1}^{k} \lambda_i > U_0^{\alpha}(s, m, n) \tag{12.17}$$

Pillai (1960) obtained tables of critical points of the sum of all the roots of Equation 12.10, that is, $\sum_{i=1}^{k} \lambda_i > U_0^{\alpha}(s, m, n)$ tables of the Hotelling's criterion, which is a slight amendment to Equation 12.17, was also obtained by Davis (1970).

$U_0^{\alpha}(s, m, n)$ is the upper percentage point of the distribution of the Lawley–Hotelling trace criterion.

In addition, Pillai (1960) suggested another test criterion that enables us to make decisions about H_0: the equality of vector of effects. This is referred to as the Pillai's trace criterion, which involves the sum of all the roots of

$$|\boldsymbol{B} - \boldsymbol{\theta}(\boldsymbol{B} + \boldsymbol{W})| = 0 \tag{12.18}$$

The statistic which is used for this test is $V^{(s)}$; the null hypothesis H_0 is rejected if

$$U^{(s)} = \sum_{i=1}^{s} \theta_i = \sum_{i=1}^{s} \left[\frac{\lambda_i}{1 + \lambda_i} \right] > V^{\alpha}(s, m, n) \tag{12.19}$$

θ_i are the roots of the determinantal Equation 12.18. $V^{\alpha}(s, m, n)$ in Equation 12.19 represents the upper percentage point of the distribution of Pillai's trace criterion.

From Equation 12.19, we can see that although Pillai's trace statistics was obtained as sum of roots of Equation 12.18, its roots are related to the roots of Equation 12.10. This confirms that all the statistics stated above are functions of the roots of Equation 12.10.

Extensive tables of the critical regions for the tests under all the above criteria appear in many publications (Kres, 1983; Pillai, 1960; Timm, 1975) among others. In particular, Kres (1983) contains more tables for multivariate analysis than those discussed here. Also, a number of statistical packages (SAS, SPSS, MINITAB among others) now provide the critical values of the test statistics for each of the above criteria as part of their solutions to MANOVA problems. In this chapter, we shall use the critical values of the test statistics obtained under the criteria without including the tables in the chapter. However, in addition to these, we include equivalent tests based on the χ^2- as well as F-approximations for these tests, which are discussed in the following sections. In this chapter, the inclusion of the criteria helps to show the evolution of testing in MANOVA to the present day, when tests based on the F-distribution has gained accuracy and acceptability.

12.2.2 Alternative Testing Methods

Although tables of those statistics discussed in Section 12.2.1 were and are still widely available, attempts have been made to obtain good approximate tests based on χ^2- and F-distributions of functions of the test statistics described. For instance, some transformations of the Wilks' Lambda are known to be approximately distributed as F or χ^2. One of such approximations was derived by Bartlett (1947); however, Rao (1952) later obtained an approximation of the Lambda, which is F-distributed. An evaluation of the Rao's approximation indicates that it gives closer approximations to the cumulants of Lambda even for small degrees of freedom. Comparative Monte-Carlo studies of Rao's approximation and two other approximations were carried out by Lohnes (1961).

The χ^2 approximation obtained by Bartlett (1947) depends on Wilks' Lambda, which we discussed in the previous section (Sections 12.2.1 and 12.2.2) and is of the form

$$\begin{aligned}\chi_f^2 &\sim -\left[v_{\mathrm{e}} + v_{\mathrm{h}} - \left(\frac{p + v_{\mathrm{h}} + 1}{2}\right)\right] \ln(\Lambda) \\ &= \left[v_{\mathrm{e}} + v_{\mathrm{h}} - \left(\frac{p + v_{\mathrm{h}} + 1}{2}\right)\right] \ln\left\{\sum_{i=1}^{s}\left[\frac{1}{1+\lambda_i}\right]\right\} \qquad (12.20)\end{aligned}$$

where v_{e} is the error degree of freedom, v_{h} the hypothesis degree of freedom, and p the number of responses obtained from each individual. The degree of freedom of the χ^2 statistic is

$$f = pv_{\mathrm{h}}$$

The approximation obtained by Rao (1952) also depends on Wilks' Lambda and has the form

$$F_c = \left(\frac{1-x}{x}\right)\left(\frac{n_2}{n_1}\right) \qquad (12.21a)$$

where

$$x = \Lambda^{1/s}$$
$$n_1 = pv_{\text{h}} \tag{12.21b}$$

$$n_2 = s\left[v_{\text{e}} + v_{\text{h}} - \left(\frac{p + v_{\text{h}} + 1}{2}\right)\right] - \frac{pv_{\text{h}} - 2}{2} \tag{12.21c}$$

$$s = \sqrt{\frac{(pv_{\text{h}})^2 - 4}{p^2 + v_{\text{h}}^2 - 5}} \tag{12.21d}$$

where F_c is F distributed with n_1 and n_2 degrees of freedom for numerator and denominator, respectively. The above formula for Rao's approximation applies in general to MANOVA for all designed experiments in which $p > 1$. In general, any of the two approximations could be used for our tests in place of other criteria discussed in the previous section (Sections 12.2.1 and 12.2.2). However, studies have shown that when v_{e} is small compared with v_{h} and p, it is better to use Rao's approximation as it provides more exact values. To use Rao's approximation for one-way MANOVA in particular, we rewrite the formulas in terms of this design. In practice, to use the approximation, we need to know a number of functions of the parameters of the design under study. The parameters are g, the number of groups; p, the number of measurements; and N, the total number of subjects in all the groups. From these parameters, we obtain the following functions:

$$s = \sqrt{\left[\frac{p^2(g-1)^2 - 4}{p^2 + (g-1)^2 - 5}\right]} \tag{12.22a}$$

$$n_1 = p(g-1) \tag{12.22b}$$

$$n_2 = s\left[(N-1) - \frac{p + (g-1) + 1}{2}\right] - \frac{p(g-1) - 2}{2} \tag{12.22c}$$

If we define $x = \Lambda^{1/s}$ then

$$F_c = \left(\frac{1-x}{x}\right)\left(\frac{n_2}{n_1}\right)$$

F_c is F-distributed with n_1 and n_2 degrees of freedom. In Table 12.1, we summarize the variance ratio test for special cases in terms of p and q, with $q = g - 1$.

Once we have identified which of the methods we wish to use for testing, we can obtain the MANOVA table just like we did for ANOVA in the univariate case (Table 12.2).

We can then compare the test statistic obtained on the basis of these quantities with tabulated critical regions for the distribution of the statistic at the

TABLE 12.1: Test Using Wilks' Lambda—Some Special Cases

Parameter p	Parameter q	n_1	n_2	F-ratio
Any	1	p	$N-p-1$	$\left(\frac{1-\Lambda}{\Lambda}\right)\left(\frac{N-p-1}{p}\right)$
Any	2	$2p$	$2(N-p-2)$	$\left(\frac{1-\Lambda^{1/2}}{\Lambda^{1/2}}\right)\left(\frac{N-p-1}{p}\right)$
1	Any	q	$N-q-1$	$\left(\frac{1-\Lambda}{\Lambda}\right)\left(\frac{N-q-1}{q}\right)$
2	Any	$2q$	$2(N-q-2)$	$\left(\frac{1-\Lambda^{1/2}}{\Lambda^{1/2}}\right)\left(\frac{N-q-1}{q}\right)$

TABLE 12.2: MANOVA Table for the One-Way Design with k Treatments

Source	Degrees of freedom	Sum of squares and cross products
Treatments ($\boldsymbol{B}$)	$k-1$	$\boldsymbol{B}=\sum_{i=1}^{k}\sum_{j=1}^{N_i}(\overline{\boldsymbol{Y}}_{i.}-\overline{\boldsymbol{Y}}_{..})(\overline{\boldsymbol{Y}}_{i.}-\overline{\boldsymbol{Y}}_{..})'$
Within treatments (Error) ($\boldsymbol{W}$)	$N-k$	$\boldsymbol{W}=\sum_{i=1}^{k}\sum_{j=1}^{N_i}(\boldsymbol{Y}_{ij}-\overline{\boldsymbol{Y}}_{i.})(\boldsymbol{Y}_{ij}-\overline{\boldsymbol{Y}}_{i.})'$
Total ($\boldsymbol{T}$)	$N-1$	

level of significance α that has been chosen for our test; this is done by comparing it with the appropriate percentage point of the chosen test criterion. We consider a designed experiment and the analysis of its responses in the following example.

Example 12.1

An experiment was performed to test four methods of teaching mathematics to first year high school students. Four samples of first year high school students were subjected to four different teaching methods for mathematics. In the tests that ensued, the students obtained the following scores in algebra (Al), arithmetic (Ar), and geometry (Ge):

Method I (11 subjects)			Method II (11 subjects)			Method III (11 subjects)			Method IV (11 subjects)		
Al	Ar	Ge	Al	Ar	Ge	Al	Ar	Ge	Al	Ar	Ge
60	64	70	71	74	77	64	45	43	43	57	59
59	72	54	76	69	58	51	64	55	61	56	46
64	59	61	79	71	69	49	44	59	61	62	59

(Continued)

Method I (11 subjects)			Method II (11 subjects)			Method III (11 subjects)			Method IV (11 subjects)		
Al	**Ar**	**Ge**	**Al**	**Ar**	**Ge**	**Al**	**Ar**	**Ge**	**Al**	**Ar**	**Ge**
64	59	53	58	75	64	59	55	56	57	58	59
59	61	59	90	73	67	43	61	45	43	33	35
49	44	45	88	81	57	45	48	30	55	45	31
45	58	62	91	78	71	35	37	54	31	66	40
48	45	44	62	71	82	55	58	47	49	49	66
67	55	47	56	61	50	50	55	55	57	51	62
62	61	58	62	56	69	52	52	40	58	53	49
66	57	52	62	58	64	46	46	47	64	63	63

Assuming multivariate normality for the test scores, carry out a test to compare the four methods of teaching mathematics to first year high school students using 5% level of significance.

Analysis:
If we are to carry out the analysis manually, we could reduce the sizes of the numbers we deal with and complexity of our calculations by subtraction of a constant from all the scores across the teaching methods. This subtraction is not going to affect the MANOVA as we have often proven in univariate ANOVA.

Al	Ar	Ge	Al	Ar	Ge	Al	Ar	Ge	Al	Ar	Ge
0	4	10	11	14	17	4	−15	−17	−17	−3	−1
−1	12	−6	16	9	−2	−9	4	−5	1	−4	−14
4	−1	1	19	11	9	−11	−16	−1	1	2	−1
4	−1	−7	−2	15	4	−1	−5	−4	−3	−2	−1
−1	1	−1	30	13	7	−17	1	−15	−17	−27	−25
−11	−16	−15	28	21	−3	−15	−12	−30	−5	−15	−29
−15	−2	2	31	18	11	−25	−23	−6	−29	6	−20
−12	−15	−16	2	11	22	−5	−2	−13	−11	−11	6
7	−5	−13	−4	1	−10	−10	−5	−5	−3	−9	2
2	1	−2	2	−4	9	−8	−8	−20	−2	−7	−11
6	−3	−8	2	−2	4	−14	−14	−13	4	3	3

In calculating the elements of matrices required for MANOVA, we shall adopt systematic calculation without any rigorous statements. As the first step in this process, for Methods I, II, III, and IV, we obtain the sums of the observations under each of the three tests Al, Ar, and Ge.

Method I:

$$\sum(\text{Al}) = 0 - 1 + 4 + 4 - 1 - 11 - 15 - 12 + 7 + 2 + 6 = -17$$

$$\sum(\text{Ar}) = 4 + 12 - 1 - \cdots - 3 = -25$$

$$\sum(\text{Ge}) = 10 - 6 + \cdots - 8 = -55$$

$$\sum(\text{Al})^2 = 613 \qquad \sum(\text{Ar})^2 = 683 \qquad \sum(\text{Ge})^2 = 909$$
$$\sum(\text{Al})(\text{Ar}) = 314 \qquad \sum(\text{Al})(\text{Ge}) = 167 \qquad \sum(\text{Ar})(\text{Ge}) = 536$$

Method II:

$$\sum(\text{Al}) = 135 \qquad \sum(\text{Ar}) = 107 \qquad \sum(\text{Ge}) = 68$$
$$\sum(\text{Al})^2 = 3415 \qquad \sum(\text{Ar})^2 = 1699 \qquad \sum(\text{Ge})^2 = 1250$$
$$\sum(\text{Al})(\text{Ar}) = 2019 \qquad \sum(\text{Al})(\text{Ge}) = 895 \qquad \sum(\text{Ar})(\text{Ge}) = 793$$

Method III:

$$\sum(\text{Al}) = -111 \qquad \sum(\text{Ar}) = -95 \qquad \sum(\text{Ge}) = -129$$
$$\sum(\text{Al})^2 = 1743 \qquad \sum(\text{Ar})^2 = 1485 \qquad \sum(\text{Ge})^2 = 2255$$
$$\sum(\text{Al})(\text{Ar}) = 1143 \qquad \sum(\text{Al})(\text{Ge}) = 1304 \qquad \sum(\text{Ar})(\text{Ge}) = 1147$$

Method IV:

$$\sum(\text{Al}) = -81 \qquad \sum(\text{Ar}) = -67 \qquad \sum(\text{Ge}) = -91$$
$$\sum(\text{Al})^2 = 1605 \qquad \sum(\text{Ar})^2 = 1283 \qquad \sum(\text{Ge})^2 = 2235$$
$$\sum(\text{Al})(\text{Ar}) = 589 \qquad \sum(\text{Al})(\text{Ge}) = 1117 \qquad \sum(\text{Ar})(\text{Ge}) = 1051$$

Next, for all the methods (I, II, III, and IV), we obtain the overall sum of the squares and cross products of all observations under each of the tests, Al, Ar, Ge.

$$\sum(\text{Al})^2 = 613 + 3415 + 1743 + 1605 = 7376$$
$$\sum(\text{Ar})^2 = 683 + 1699 + 1485 + 1283 = 5150$$
$$\sum(\text{Ge})^2 = 909 + 1250 + 2255 + 2235 = 6649$$
$$\sum(\text{AlAr}) = 314 + 2019 + 1143 + 589 = 4065$$
$$\sum(\text{AlGe}) = 167 + 895 + 1304 + 1117 = 3483$$
$$\sum(\text{ArGe}) = 536 + 793 + 1147 + 1051 = 3527$$

We calculate the overall sums of all observations under each of the tests, Al, Ar, Ge as

$$\sum(\text{Al}) = -74 \quad \sum(\text{Ar}) = -80 \quad \sum(\text{Ge}) = -207$$

Now, we calculate the elements of $\boldsymbol{T}$ (the matrix of total sum of squares and cross products) for the tests. First we obtain the diagonal elements.

For Al, we obtain

$$T_{11} = 7376 - (-74)^2/44 = 7251.5454$$

For Ar, we obtain

$$T_{22} = 5150 - (-80)^2/44 = 5004.545$$

For Ge, we obtain

$$T_{33} = 6649 - (-207)^2/44 = 5675.1591$$

Similarly, we obtain the off-diagonal elements, so that
For AlAr

$$T_{12} = T_{21} = 4065 - (-74)(-80)/44 = 4081.7727$$

For AlGe

$$T_{13} = T_{31} = 3483 - (-74)(-207)/44 = 3134.8636$$

For ArGe

$$T_{23} = T_{32} = 3527 - (-80)(-207)/44 = 3150.6364$$

The matrix $\boldsymbol{T}$ is

$$\boldsymbol{T} = \begin{bmatrix} 7251.5454 & 4081.7727 & 3134.8636 \\ 4081.7727 & 5004.545 & 3150.6364 \\ 3134.8636 & 3150.6364 & 5675.1591 \end{bmatrix}$$

To obtain the $\boldsymbol{W}$ matrix, we first calculate its diagonal elements as follows:

$$W_{11} = 613 - \frac{(-17)^2}{11} + 3415 - \frac{(135)^2}{11} + 1743 - \frac{(-111)^2}{11} + 1605 - \frac{(-81)^2}{11} = 3976.3636$$

$$W_{22} = 683 - \frac{(-25)^2}{11} + 1699 - \frac{(107)^2}{11} + 1485 - \frac{(-95)^2}{11} + 1283 - \frac{(-67)^2}{11} = 2823.8282$$

$$W_{33} = 909 - \frac{(-55)^2}{11} + 2255 - \frac{(-129)^2}{11} + 1250 - \frac{(68)^2}{11} + 2235 - \frac{(-91)^2}{11} = 3688.00$$

We obtain the subdiagonal elements as

$$W_{12} = W_{21} = 314 - \frac{(-17)(-25)}{11} + 2019 - \frac{(135)(107)}{11} + 1143 - \frac{(-111)(-95)}{11}$$
$$+ 589 - \frac{(-81)(-67)}{11} = 1261.1818$$
$$W_{13} = W_{31} = 167 - \frac{(-17)(-55)}{11} + 895 - \frac{(135)(68)}{11} + 1304 - \frac{(-111)(-129)}{11}$$
$$+ 1117 - \frac{(-81)(-91)}{11} = 591.6364$$
$$W_{23} = W_{32} = 536 - \frac{(-25)(-55)}{11} + 793 - \frac{(107)(68)}{11} + 1147 - \frac{(-95)(-129)}{11}$$
$$+ 1051 - \frac{(-67)(-91)}{11} = 1072.1818$$

$$\boldsymbol{W} = \begin{bmatrix} 3976.3636 & 1261.1818 & 591.6364 \\ 1261.1818 & 2823.8282 & 1072.1818 \\ 591.6364 & 1072.1818 & 3688 \end{bmatrix}$$

We obtain the matrix $\boldsymbol{B}$ using the relation

$$\boldsymbol{B} = \boldsymbol{T} - \boldsymbol{W} \tag{12.23}$$

so that

$$\boldsymbol{B} = \begin{bmatrix} 3275.181818 & 2669.272727 & 2543.2273 \\ 2669.272727 & 2180.727273 & 2078.454545 \\ 2543.2273 & 2078.454545 & 1987.159091 \end{bmatrix}$$

To utilize the different criteria described in the previous sections for testing the hypothesis of equality of means of all groups to the centroid, we need to solve the characteristic equation in Equation 12.10. To do this, we rewrite the equation as

$$|\boldsymbol{B}\boldsymbol{W}^{-1} - \lambda \boldsymbol{I}| = 0 \tag{12.24}$$

We use a suitable computer program to invert the matrix $\boldsymbol{W}$ and hence obtain $\boldsymbol{B}\boldsymbol{W}^{-1}$. Press et al. (1988) have written subroutines and procedures in Fortran, Pascal, and C languages, which we can modify slightly to use them for inversion of matrices as well as computation of eigenvalues or roots for Equation 12.24. There are several other applications that can be used to compute eigenvalues, such as MAPLE. Thus, we obtain $\boldsymbol{BW}^{-1}$ as

$$\boldsymbol{BW}^{-1} = \begin{bmatrix} 3275.181818 & 2669.272727 & 2543.2273 \\ 2669.272727 & 2180.727273 & 2078.454545 \\ 2543.2273 & 2078.454545 & 1987.159091 \end{bmatrix}$$

$$\times \begin{bmatrix} 3976.363636 & 1261.181818 & 591.6364 \\ 1261.181818 & 2823.818182 & 1072.181818 \\ 591.6364 & 1072.181818 & 3688 \end{bmatrix}^{-1}$$

$$= \begin{bmatrix} 0.59557 & 0.51001 & 0.44578 \\ 0.48467 & 0.41741 & 0.36447 \\ 0.46162 & 0.39726 & 0.34927 \end{bmatrix}$$

Solving Equation 12.24 with a computer program shows that the eigenvalues for $\boldsymbol{BW}^{-1}$ are $\lambda_1 = 1.3588$, $\lambda_2 = 2.4004 \times 10^{-3}$, and $\lambda_3 = 1.0151 \times 10^{-3}$

$$\text{Again, } |\boldsymbol{W}| = 3.1585 \times 10^{10} \quad |\boldsymbol{T}| = 7.4758 \times 10^{10}$$

$$\text{Wilks' criterion, } \Lambda = \frac{|\boldsymbol{W}|}{|\boldsymbol{T}|} = 0.422496589$$

$$\text{Roy's criterion: } \theta_s = \frac{\lambda_1}{1+\lambda_1} = \frac{1.3588}{2.3588} = 0.576055$$

$$\text{Pillai's criterion: } V^{(s)} = \sum_{i=1}^{s} \frac{\lambda_i}{1+\lambda_i} = 0.579464$$

$$\text{Lawley–Hotelling criterion: } U^{(s)} = \sum_{i=1}^{s} \lambda_i = 1.3622155$$

It should be noted that Roy's criterion used here for manual analysis (although it is based on the largest root) is different from Roy's largest root criterion implemented in SAS. In SAS, the test statistic is the largest root λ_1. The Roy's criterion that we use here for manual analysis is

$$\theta_s = \frac{\lambda_1}{1+\lambda_1}$$

To make decisions using the tables (Kres, 1983; Timm, 1975), we need to know a number of parameters that are defined in Equation 12.15. We can see that $p=3$, $k=4$, $N=44$, $s=3$, $m=-0.5$, and $n=18.00$. From the tables and from our calculations, we conclude that

$$\Lambda < U^{0.01}(3, 3, 40) = 0.577483$$
$$V^{(s)} > V^{0.01}(3, -0.5, 18.0) = 0.3475$$
$$V^{(s)} > V^{0.01}(3, -0.5, 18.0) = 0.4714$$

Clearly, all the criteria agree that we should reject the hypothesis of equality of centroids. We therefore conclude that there is probably enough evidence to believe that the four teaching methods have significant effects.

Using the χ^2 approximation in Equation 12.20, we note that $v_e = 40$, $v_h = 3$, $p = 3$, $f = 9$, and $\Lambda = 0.422496589$, so that

$$\chi_f^2 = -(40 + 3 - ((3 + 3c + 1)/2)) \ln(0.422496589) = 34.03216$$

From the tables (Appendix A4), $\chi^2_{(0.01,9)} = 21.67$. Because $\chi_f^2 > \chi^2_{(0.01,9)}$, we reject the hypothesis of equality of centroids. The teaching methods are therefore significantly different from each other at $\alpha = 1\%$.

Now, we illustrate the application of the Rao approximation; since this is a one-way MANOVA, we use the form shown in Equation 12.22.

$$p = 3 \quad N = 44 \quad g = 4 \quad s = \sqrt{77/13} = 2.4337 \quad n_1 = 3(4-1) = 9$$

$$n_2 = 2.4337[43 - (3 + 3 + 1)/2] - [3(4-1) - 2]/2 = 92.6312$$

$$\Lambda^{1/s} = 0.701154557$$

$$F_c = [(1 - 0.701154557)/0.701154557](92.6312/9) = 4.386798$$

By interpolation in table of the F-distribution (Appendix A5 through A7), we obtain $F(0.01, 9, 92.63) = 2.3067$, from which it is clear that $F_c > F(0.01, 9, 92.63)$ and that the hypothesis of equality of centroids should be rejected. □

12.2.3 SAS Analysis of Responses of Example 12.1

Next, we present a SAS program to carry out the analysis of the responses of Example 12.1.

SAS PROGRAM

```
data examp12_1;
do j=1 to 11;
do group=1 to 4;
input y1 y2 y3@@;y1=y1-60; y2=y2-60; y3=y3-60;
output;
end; end;
datalines;
60 64 70 71 74 77 64 45 43 43 57 59 59 72 54 76 69 58 51 64 55 61 56 46
64 59 61 79 71 69 49 44 59 61 62 59 64 59 53 58 75 64 59 55 56 57 58 59
59 61 59 90 73 67 43 61 45 43 33 35 49 44 45 88 81 57 45 48 30 55 45 31
45 58 62 91 78 71 35 37 54 31 66 40 48 45 44 62 71 82 55 58 47 49 49 66
67 55 47 56 61 50 50 55 55 57 51 62 62 61 58 62 56 69 52 52 40 58 53 49
66 57 52 62 58 64 46 46 47 64 63 63
;
proc print data=examp12_1;
proc glm data=examp12_1;
class group;
model y1-y3=group;
manova h=_all_ /printe printh;
run;
```

SAS OUTPUT

Obs	j	group	y1	y2	y3
1	1	1	0	4	10
2	1	2	11	14	17
3	1	3	4	-15	-17
4	1	4	-17	-3	-1
5	2	1	-1	12	-6
6	2	2	16	9	-2
7	2	3	-9	4	-5
8	2	4	1	-4	-14
9	3	1	4	-1	1
10	3	2	19	11	9
11	3	3	-11	-16	-1
12	3	4	1	2	-1
13	4	1	4	-1	-7
14	4	2	-2	15	4
15	4	3	-1	-5	-4
16	4	4	-3	-2	-1
17	5	1	-1	1	-1
18	5	2	30	13	7
19	5	3	-17	1	-15
20	5	4	-17	-27	-25
21	6	1	-11	-16	-15
22	6	2	28	21	-3
23	6	3	-15	-12	-30
24	6	4	-5	-15	-29
25	7	1	-15	-2	2
26	7	2	31	18	11
27	7	3	-25	-23	-6
28	7	4	-29	6	-20
29	8	1	-12	-15	-16
30	8	2	2	11	22
31	8	3	-5	-2	-13
32	8	4	-11	-11	6
33	9	1	7	-5	-13
34	9	2	-4	1	-10
35	9	3	-10	-5	-5
36	9	4	-3	-9	2
37	10	1	2	1	-2
38	10	2	2	-4	9
39	10	3	-8	-8	-20
40	10	4	-2	-7	-11
41	11	1	6	-3	-8
42	11	2	2	-2	4
43	11	3	-14	-14	-13
44	11	4	4	3	3

The GLM Procedure

Class Level Information

Class	Levels	Values
group	4	1 2 3 4

```
                    Number of Observations Read    44
                    Number of Observations Used    44

                            The GLM Procedure

Dependent Variable: y1
                                      Sum of
      Source                DF       Squares     Mean Square    F Value    Pr > F

      Model                  3   3275.181818     1091.727273      10.98    <.0001

      Error                 40   3976.363636       99.409091

      Corrected Total       43   7251.545455

               R-Square     Coeff Var      Root MSE       y1 Mean

               0.451653     -592.8352      9.970411     -1.681818

      Source                DF     Type I SS     Mean Square    F Value    Pr > F

      group                  3   3275.181818     1091.727273      10.98    <.0001

      Source                DF   Type III SS     Mean Square    F Value    Pr > F

      group                  3   3275.181818     1091.727273      10.98    <.0001

                            The GLM Procedure

Dependent Variable: y2

                                      Sum of
      Source                DF       Squares     Mean Square    F Value    Pr > F

      Model                  3   2180.727273      726.909091      10.30    <.0001

      Error                 40   2823.818182       70.595455

      Corrected Total       43   5004.545455

               R-Square     Coeff Var      Root MSE       y2 Mean

               0.435749     -462.1161      8.402110     -1.818182

      Source                DF     Type I SS     Mean Square    F Value    Pr > F

      group                  3   2180.727273      726.909091      10.30    <.0001

      Source                DF   Type III SS     Mean Square    F Value    Pr > F

      group                  3   2180.727273      726.909091      10.30    <.0001

                            The GLM Procedure

Dependent Variable: y3

                                      Sum of
      Source                DF       Squares     Mean Square    F Value    Pr > F

      Model                  3   1987.159091      662.386364       7.18    0.0006

      Error                 40   3688.000000       92.200000

      Corrected Total       43   5675.159091
```

R-Square	Coeff Var	Root MSE	y3 Mean
0.350150	-204.1022	9.602083	-4.704545

Source	DF	Type I SS	Mean Square	F Value	Pr > F
group	3	1987.159091	662.386364	7.18	0.0006

Source	DF	Type III SS	Mean Square	F Value	Pr > F
group	3	1987.159091	662.386364	7.18	0.0006

The GLM Procedure
Multivariate Analysis of Variance

E = Error SSCP Matrix

	y1	y2	y3
y1	3976.3636364	1261.1818182	591.63636364
y2	1261.1818182	2823.8181818	1072.1818182
y3	591.63636364	1072.1818182	3688

Partial Correlation Coefficients from the Error SSCP Matrix / Prob > |r|

DF = 40	y1	y2	y3
y1	1.000000	0.376371	0.154496
	0.0153		0.3348
y2	0.376371	1.000000	0.332242
	0.0153		0.0338
y3	0.154496	0.332242	1.000000
	0.3348	0.0338	

The GLM Procedure
Multivariate Analysis of Variance

H = Type III SSCP Matrix for group

	y1	y2	y3
y1	3275.1818182	2669.2727273	2543.2272727
y2	2669.2727273	2180.7272727	2078.4545455
y3	2543.2272727	2078.4545455	1987.1590909

Characteristic Roots and Vectors of: E Inverse * H, where
H = Type III SSCP Matrix for group
E = Error SSCP Matrix

Characteristic Root	Percent	Characteristic Vector V'EV=1 y1	y2	y3
1.35883727	99.75	0.00891454	0.00765689	0.00670054
0.00240352	0.18	-0.01202001	0.00261429	0.01270062
0.00101859	0.07	-0.00832905	0.01969135	-0.00994675

```
MANOVA Test Criteria and F Approximations for the Hypothesis of No Overall
                              group Effect
                   H = Type III SSCP Matrix for group
                         E = Error SSCP Matrix

                      S=3  M=-0.5  N=18

Statistic                      Value    F Value   Num DF    Den DF   Pr > F

Wilks' Lambda             0.42249084       4.37        9    92.633   <.0001
Pillai's Trace            0.57947763       3.19        9   120       0.0017
Hotelling-Lawley Trace    1.36225938       5.65        9    56.616   <.0001
Roy's Greatest Root       1.35883727      18.12        3    40       <.0001

       NOTE: F Statistic for Roy's Greatest Root is an upper bound.
```

12.3 Multivariate Analysis of Variance—The Randomized Complete Block Experiment

In the usual one-way experimental layout, the units to which the treatments are applied are expected to be reasonably homogeneous. For instance, in studying the effect of different teaching methods on learning of students, it is usual to use students of the same class, who went through the same learning experience, who have about the same level of intelligence and are reasonably equally motivated to learn. Under such conditions, results would be free of influence of individual young students. When there is reason to believe that these conditions are not met by the subjects being used, the experimenter would be better advised to use blocking to eliminate the influence of individual students on the result. In that case, the primary interest of the experimenter should be in the effect of the treatments. The blocking effect should only be studied to ensure that it had been effective. Effective blocking would have taken place if the blocking effects are found significant in MANOVA or the means square of blocks is deemed to be large. If this is not the case, then blocking is ineffective, and the ordinary completely randomized experiment (one-way layout) would have been sufficient for the study.

As we pointed out earlier, blocking in this way was originally applied in agricultural experiments where the experimental portion of land may come in inhomogeneous parcels; those parcels of the land which are homogeneous are then considered as blocks for the experiment. In industrial research, batches of raw materials may constitute blocks in a study; shifts of workers may also be used as blocks if there is reason to believe that skills vary among the shifts. Apart from satisfying the need for the elimination of the differences in experimental units, blocking has found its use in the reduction of the number of experiments to be performed as in the balanced incomplete block designs (BIBD). Other useful applications are in the Latin square and balanced lattice designs.

In Chapter 3, we studied the univariate two-way experimental layout, which was essentially an extension of the one-way layout by the addition of an extra factor. The principle remains the same in the relationship between the one-way MANOVA and two-way MANOVA for the randomized complete block experiment. Although blocking is added as a factor, as we explained previously, it is not usually tested in ANOVA due to controversy arising from the fact that blocking represents a restriction on randomization. Here, we have the block, which is treated as the second factor in the two-way classification; the treatments form the levels of the first factor.

The MANOVA for RCBD involves the replacement of the univariate response variable y_{ij} with a vector-valued random variable; a p-dimensional vector $\boldsymbol{Y}_{ij}$, where $\boldsymbol{Y}'_{ij} = [y_{ij1} \quad y_{ij2} \quad \dots \quad y_{ijp}]$ relating to p-response variables obtained from each subject or unit used in the experiment. Further, $\boldsymbol{Y}_{ij}$ refers to the p responses obtained from the individual in the ith block who was subjected to jth treatment. We can rewrite the responses in terms of the differences between the responses of the grand centroid. Just as we indicated after subtracting the grand mean from the observations for the one-way MANOVA, this does not affect whatever conclusion we shall reach after RCBD MANOVA.

Suppose that there are c blocks and r treatments, then taking a cue from the model which was used in the univariate case, the appropriate model for the multivariate randomized block design is

$$\boldsymbol{Y}_{ij} = \boldsymbol{\mu} + \boldsymbol{\gamma}_i + \boldsymbol{\beta}_j + \boldsymbol{\varepsilon}_{ij} \quad i = 1, 2, \dots, r \quad j = 1, 2, \dots, c \tag{12.25}$$

$$\begin{aligned}
\mu &= [\mu_{11}, \mu_{12}, \dots, \mu_{1p}] \\
\gamma'_1 &= [\gamma_{11}, \gamma_{12}, \dots, \gamma_{1p}] \\
\gamma'_2 &= [\gamma_{21}, \gamma_{22}, \dots, \gamma_{2p}] \\
&\vdots \\
\gamma'_r &= [\gamma_{r1}, \gamma_{r2}, \dots, \gamma_{rp}] \\
\beta'_1 &= [\beta_{11}, \beta_{12}, \dots, \beta_{1p}] \\
\beta'_2 &= [\beta_{21}, \beta_{22}, \dots, \beta_{2p}] \\
&\vdots \\
\beta'_c &= [\beta_{c1}, \beta_{c2}, \dots, \beta_{cp}]
\end{aligned}$$

provided that the treatments are arranged in rows and the blocks in columns. For the multivariate application, given that each $\boldsymbol{Y}_{ij}$ is a p-variate vector, let

$$\overline{Y}_{i.} = \frac{\sum_{j=1}^{c} Y_{ij}}{c} \quad \overline{Y}_{.j} = \frac{\sum_{i=1}^{r} Y_{ij}}{r} \quad \overline{Y}_{..} = \frac{\sum_{i=1}^{r} \sum_{j=1}^{c} Y_{ij}}{rc}$$

These are the treatments, blocks, and grand centroids, respectively. If we use the definition (as in the previous section (Section 12.2) on the one-way MANOVA) in which $\boldsymbol{y}_{ij} = \boldsymbol{Y}_{ij} - \overline{\boldsymbol{Y}}$, then for the two-way MANOVA

$$\boldsymbol{y}_{ij} = (\boldsymbol{Y}_{i.} - \overline{\boldsymbol{Y}}_{..}) + (\boldsymbol{Y}_{.j} - \overline{\boldsymbol{Y}}_{..}) + (\boldsymbol{Y}_{ij} - \overline{\boldsymbol{Y}}_{.j} - \overline{\boldsymbol{Y}}_{i.} - \overline{\boldsymbol{Y}}_{..})$$

We can obtain the total sum of squares for all the observations by comparing with the univariate case described in Chapter 5 as

$$\sum_{i=1}^{r}\sum_{j=1}^{c} \boldsymbol{y}_{ij}\boldsymbol{y}_{ij}' = \sum_{i=1}^{r}\sum_{j=1}^{c}(\boldsymbol{Y}_{i.} - \overline{\boldsymbol{Y}}_{..})(\boldsymbol{Y}_{i.} - \overline{\boldsymbol{Y}}_{..})' + \sum_{i=1}^{r}\sum_{j=1}^{c}(\boldsymbol{Y}_{i.} - \overline{\boldsymbol{Y}}_{..})(\boldsymbol{Y}_{i.} - \overline{\boldsymbol{Y}}_{..})'$$
$$+ \sum_{i=1}^{r}\sum_{j=1}^{c}(\boldsymbol{Y}_{ij} - \overline{\boldsymbol{Y}}_{.j} - \overline{\boldsymbol{Y}}_{i.} - \overline{\boldsymbol{Y}}_{..})(\boldsymbol{Y}_{ij} - \overline{\boldsymbol{Y}}_{.j} - \overline{\boldsymbol{Y}}_{i.} - \overline{\boldsymbol{Y}}_{..})'$$

As usual, total sum of squares is denoted by $\boldsymbol{T}$, where

$$\boldsymbol{T} = \sum_{i=1}^{r}\sum_{j=1}^{c} \boldsymbol{y}_{ij}\boldsymbol{y}_{ij}' \tag{12.26}$$

The block sum of squares and cross products is

$$\boldsymbol{B} = \sum_{i=1}^{r}\sum_{j=1}^{c}(\boldsymbol{Y}_{.j} - \overline{\boldsymbol{Y}}_{..})(\boldsymbol{Y}_{.j} - \overline{\boldsymbol{Y}}_{..})' = r\sum_{j=1}^{c}(\boldsymbol{Y}_{.j} - \overline{\boldsymbol{Y}}_{..})(\boldsymbol{Y}_{.j} - \overline{\boldsymbol{Y}}_{..})' \tag{12.27}$$

while the treatment sum of squares is

$$\boldsymbol{G} = \sum_{i=1}^{r}\sum_{j=1}^{c}(\boldsymbol{Y}_{i.} - \overline{\boldsymbol{Y}}_{..})(\boldsymbol{Y}_{i.} - \overline{\boldsymbol{Y}}_{..})' = c\sum_{j=1}^{c}(\boldsymbol{Y}_{i.} - \overline{\boldsymbol{Y}}_{..})(\boldsymbol{Y}_{i.} - \overline{\boldsymbol{Y}}_{..})' \tag{12.28}$$

$$\boldsymbol{W} = \sum_{i=1}^{r}\sum_{j=1}^{c}(\boldsymbol{Y}_{ij} - \overline{\boldsymbol{Y}}_{.j} - \overline{\boldsymbol{Y}}_{i.} - \overline{\boldsymbol{Y}}_{..})(\boldsymbol{Y}_{ij} - \overline{\boldsymbol{Y}}_{.j} - \overline{\boldsymbol{Y}}_{i.} - \overline{\boldsymbol{Y}}_{..})' \tag{12.29}$$

In Equation 12.25, $\varepsilon_{ij} \sim IN_p(\boldsymbol{0}, \boldsymbol{\Sigma})$ so that $\boldsymbol{\varepsilon}_{ij}$ and $\boldsymbol{Y}_{ij}$ are p-variate random vectors of the responses and error obtained from each individual involved in the experiment. The hypothesis of major interest is about the treatments:

$$H_0\colon \gamma_1 = \gamma_2 = \cdots = \gamma_r = 0 \quad \text{vs} \quad H_1\colon \text{at least one } \gamma_i \neq 0 \tag{12.30}$$

Although generally the block is not tested because of the reasons we adduced in the univariate case (Chapter 5), sometimes a two-way ANOVA is carried out without replication if it is known that there is no interaction between the two factors. In that case, we adapt the above multivariate randomized block design to handle the analysis and the following hypothesis is therefore tested for the second factor:

$$H_0\colon \beta_1 = \beta_2 = \cdots = \beta_c = 0 \quad \text{vs} \quad H_1\colon \text{at least one } \beta_j \neq 0 \tag{12.31}$$

In the RCBD MANOVA, the tests depend on the solutions of two characteristic equations:

$$|\boldsymbol{B} - \lambda \boldsymbol{W}| = 0$$
$$|\boldsymbol{G} - \lambda \boldsymbol{W}| = 0 \qquad (12.32)$$

The solutions of these characteristic equations are used to obtain different criteria, which determine the significance or nonsignificance of the centroid of the vectors of treatment and block effects. In particular, if we use Roy's criterion, we need to obtain the parameters s, m, and n for testing hypotheses about treatments and blocks. Table 12.3 gives the formulas for determining these parameters.

The MANOVA for the analysis of data is presented in Table 12.4.

The tests for the hypotheses under different criteria are performed as follows:

Reject H_0 if

(i) *Wilks' criterion for blocks*

$$\Lambda = \frac{|\boldsymbol{W}|}{|\boldsymbol{W} + \boldsymbol{B}|} < U^{\alpha}(p, v_{\mathrm{B}}, v_{\mathrm{e}}) \qquad (12.33a)$$

where λ_i are the ordered nonzero roots of characteristic equation 12.10 and v_{B} and v_{e} the block and error degrees of freedom, respectively.

TABLE 12.3: Statistics for Factors in the Two-Way Layout

Treatments (r treatments)	**Blocks (c blocks)**
$s = \min(r-1, p)$	$s = \min(c-1, p)$
$m = \dfrac{\lvert r-p-1\rvert - 1}{2}$	$m = \dfrac{\lvert c-p-1\rvert - 1}{2}$
$n = \dfrac{N-r-c-p}{2}$	$n = \dfrac{N-r-c-p}{2}$

TABLE 12.4: MANOVA Table for the Two-Way Layout

Source	**Degrees of freedom**	**Sum of squares and cross products**
Blocks	$c-1$	$\boldsymbol{B}$
Treatment	$r-1$	$\boldsymbol{G}$
Within error	$(c-1)(r-1)$	$\boldsymbol{W}$
Total	$rc-1$	$\boldsymbol{T}$

Wilks' criterion for treatments

$$\Lambda = \frac{|\boldsymbol{W}|}{|\boldsymbol{W}+\boldsymbol{B}|} < U^{\alpha}(p, v_{\mathrm{G}}, v_{\mathrm{e}}) \tag{12.33b}$$

where λ_i are the ordered nonzero roots of characteristic equation 12.32 and v_{G}, and v_{e} are the treatment and error degrees of freedom, respectively.

(ii) *Roy's criterion*

$$\theta_s = \frac{\lambda_1}{1+\lambda_1} > \theta^{\alpha}(s, m, n)$$

The largest root of the characteristic equation 12.10 should be used for blocks, while the largest root of the characteristic equation 12.32 should be used for treatments. The appropriate values of s, m, and n should be obtained by referring to Table 12.3.

(iii) *Lawley–Hotelling criterion*

$$U^{(s)} = \sum_{i=1}^{s} \lambda_i > U_0^{\alpha}(s, m, n) \tag{12.34}$$

The roots of the characteristic equation 12.10 should be used for blocks, while the roots of the characteristic equation 12.32 should be used for treatments. The appropriate values of s, m, and n should be obtained by referring to Table 12.3. The same applies to Pillai's criterion below.

(iv) *Pillai's criterion*

$$\sum_{i=1}^{s} \left(\frac{\lambda_i}{1+\lambda_i}\right) > V^{\alpha}(s, m, n) \tag{12.35}$$

On the right-hand side of each criterion is the value of the critical value of the test statistic obtained from the tables, α being the level of significance. Whenever the degrees of freedom of the hypothesis being tested $v_{\mathrm{h}} = 1$ or $s = \min(n, p) = 1$, Wilks', Roy's, and Lawley–Hotelling criteria are equivalent. Since the tables for these criteria do not exist for this combination of parameters, the hypotheses could all be referred to F-distribution using the following formulas:

$$\frac{v_{\mathrm{e}}+p+1}{|p-v_{\mathrm{h}}|+1}\frac{1-\Lambda}{\Lambda} \sim F(|p-v_{\mathrm{h}}|+1, v_{\mathrm{e}}+p+1) \tag{12.36}$$

$$\frac{v_{\mathrm{e}}+p+1}{|p-v_{\mathrm{h}}|+1}\frac{\theta_s}{1-\theta_s} \sim F(|p-v_{\mathrm{h}}|+1, v_{\mathrm{e}}+p+1) \tag{12.37}$$

$$\frac{v_{\mathrm{e}}+p+1}{|p-v_{\mathrm{h}}|+1}U^{(s)} \sim F(|p-v_{\mathrm{h}}|+1, v_{\mathrm{e}}+p+1) \tag{12.38}$$

Alternatively, when $s=1$, we use the critical regions of the Wilks' criterion, that is, $U^{(\alpha)}(p,1,v_e)$ to obtain the critical region for the Roy's criterion and hence the Pillai's criterion in this case. It can be shown that $V^{(s)}=1-U^{(\alpha)}(p,1,v_e)$ for this combination of parameters.

Further, for the values of parameters outside the tabulated regions of the Lawley–Hotelling criterion, a suitable approximation is

$$F(v_1, v_2) = \frac{v_2}{v_1}\frac{U^{(s)}}{s} \tag{12.39}$$

where

$$v_1 = s(2m + s + 1) \quad v_2 = 2(sn + 1)$$

Example 12.2

A study was carried out to evaluate the performances of three grades of students: the students were divided according to the grade, sex, and to test whether grade and sex made a difference in their performances. The students were subjected to three tests s_1, s_2, and s_3. The data of the experiment are presented below according to grade, sex, and test. We carry out the analysis of the responses of the experiment as a RCBD with sex considered as block.

Grade I			**Grade II**			**Grade III**		
s_1	s_2	s_3	s_1	s_2	s_3	s_1	s_2	s_3
Boys								
120	90	140	80	60	125	95	75	95
130	100	150	85	65	110	96	70	96
70	30	100	89	78	95	100	76	100
80	60	110	92	82	90	110	80	110
70	85	20	101	50	110	123	68	105
Girls								
91	57	52	51	91	90	80	60	70
89	68	120	44	89	94	85	65	75
90	58	130	52	90	65	89	78	81
95	52	90	50	95	78	92	82	85
67	62	81	57	30	85	104	110	30

Analysis:

For manual computation, we reduce the figures we are dealing with by subtracting a constant from each of the scores. This is not going to affect the values of the dispersion matrices. For this particular analysis, we subtract 90 from each observation so that all the scores are presented as follows:

Grade I			**Grade II**			**Grade III**		
s_1	s_2	s_3	s_1	s_2	s_3	s_1	s_2	s_3
Boys								
30	0	50	−10	−30	35	5	−15	5
40	10	60	−5	−25	20	6	−20	6
−20	−60	10	−1	−12	5	10	−14	10
−10	−30	20	2	−8	0	20	−10	20
−20	−5	−70	11	−40	20	33	−22	15
Girls								
1	−33	−38	−39	1	0	−10	−30	−20
−1	−22	30	−46	−1	4	−5	−25	−15
0	−32	40	−38	0	−25	−1	−12	−9
5	−38	0	−40	5	−12	2	−8	−5
−23	−28	−9	−33	−60	−5	14	20	−60

To obtain the elements of the total dispersion matrix $\boldsymbol{T}$, we calculate the following quantities:

$$\sum s_1 = 20 - 3 + 74 - 18 - 196 + 0 = -123$$

$$\sum s_2 = -85 - 115 - 81 - 153 - 55 - 55 = 544$$

$$\sum s_3 = 70 + 80 + 56 + 23 - 38 - 109 = 82$$

$$\sum s_1^2 = 30^2 + 40^2 + \cdots + 2^2 + 14^2 = 13953$$

$$\sum s_2^2 = 0^2 + 10^2 + \cdots + (-8)^2 + (-60)^2 = 23502$$

$$\sum s_3^2 = 50^2 + 60^2 + \cdots + (-5)^2 + (-60)^2 = 19948$$

$$\sum s_1 s_2 = 30 \times 0 + 40 \times 10 + \cdots + 14 \times 20 = 3651$$

$$\sum s_1 s_3 = 30 \times 50 + 40 \times 60 + \cdots + 14 \times (-60) = 3905$$

$$\sum s_2 s_3 = 0 \times 50 + 10 \times 60 + \cdots + 20 \times (-60) = -3800$$

$$T_{11} = 13953 - \frac{(-123)^2}{30} = 13448.7$$

$$T_{22} = 19948 - \frac{(-544)^2}{30} = 10083.467$$

$$T_{11} = 23502 - \frac{82^2}{30} = 23277.867$$

$$T_{21} = T_{12} = 3651 - \frac{(-123)(-544)}{30} = 1420.60$$
$$T_{31} = T_{13} = 3905 - \frac{(-123)(82)}{30} = 7041.2$$
$$T_{23} = T_{32} = -3800 - \frac{(-544)(82)}{30} = -2313.067$$

The total matrix $\boldsymbol{T}$ is obtained following the above calculations as

$$\boldsymbol{T} = \begin{bmatrix} 13448.7 & 1420.60 & 7041.2 \\ 1420.60 & 10083.46 & -2313.067 \\ 7041.2 & -2313.067 & 23277.8 \end{bmatrix}$$

We consider the sex of the students to be the block in the design with two levels, namely, boys and girls. To obtain the elements of the matrix for the block factor, $\boldsymbol{B}$, we make the following calculations:

For Boys:

$$\sum s_1 = 20 - 3 + 74 = 91$$
$$\sum s_2 = -85 - 115 - 81 = -281$$
$$\sum s_3 = 70 + 80 + 56 = 206$$

For Girls:

$$\sum s_1 = -18 - 196 + 0 = -214$$
$$\sum s_2 = -153 - 55 - 55 = -263$$
$$\sum s_3 = 23 - 38 - 109 = -124$$

Using the above quantities, the elements of the matrix $\boldsymbol{B}$ are obtained as follows:

$$B_{11} = \frac{(86^2 + 214^2)}{15} - \frac{(-123)^2}{30} = 3100.83$$
$$B_{22} = \frac{(-281)^2 + 263^2}{15} - \frac{(-544)^2}{30} = 10.80$$
$$B_{33} = \frac{206^2 + (-124)^2}{15} - \frac{82^2}{30} = 3630$$
$$B_{12} = B_{21} = \frac{91(-281) + (-214)(-263)}{15} - \frac{(-544)(-123)}{30} = -183$$
$$B_{31} = B_{13} = \frac{91(206) + (-124)(-214)}{15} - \frac{82(-123)}{30} = 3355$$
$$B_{32} = B_{23} = \frac{(-281)(206) + (-124)(-214)}{15} - \frac{(-263)(-124)}{30} = -198$$

The matrix $\boldsymbol{B}$ for the factor sex (block) is obtained as

$$\boldsymbol{B} = \begin{bmatrix} 3100.83 & -183 & 3355 \\ -183 & 10.80 & -198 \\ 3355 & -198 & 3630 \end{bmatrix}$$

To obtain the dispersion matrix for the Grades $\boldsymbol{G}$, we calculate the following quantities:

For Grade I:

$$\sum s_1 = 20 - 18 = 2$$
$$\sum s_2 = -85 - 153 = -238$$
$$\sum s_3 = 70 + 23 = 93$$

For Grade II:

$$\sum s_1 = -3 - 196 = -199$$
$$\sum s_2 = -115 - 55 = -170$$
$$\sum s_3 = 80 - 38 = 42$$

For Grade III:

$$\sum s_1 = 74 + 0 = 74$$
$$\sum s_2 = -81 - 55 = -136$$
$$\sum s_3 = 56 - 109 = -53$$

The elements of the $\boldsymbol{G}$ matrix are obtained as

$$G_{11} = \frac{[2^2 + (-199)^2 + 74^2]}{10} - \frac{(-123)^2}{30} = 4003.8$$

$$G_{22} = \frac{[(-238)^2 + (-170)^2 + (-136)^2]}{10} - \frac{(-544)^2}{30} = 539.467$$

$$G_{33} = \frac{[(-238)^2 + 42^2 + (-53)^2]}{10} - \frac{82^2}{30} = 1098.067$$

$$G_{12} = G_{21} = \frac{2(-238) + (-199)(-170) + 74(-136)}{10} - \frac{(-123)(-544)}{30} = 98.60$$

$$G_{13} = G_{31} = \frac{2(93) + (-42)(-199) + 74(-136)}{10} - \frac{(-123)(82)}{30} = -873.2$$

$$G_{23} = G_{32} = \frac{(93)(-238) + (-42)(-170) + (-53)(-136)}{10} - \frac{(82)(-544)}{30}$$
$$= -719.667$$

so that the dispersion matrix is

$$G = \begin{bmatrix} 4003.80 & 98.60 & -873.20 \\ 98.60 & 539.467 & -719.667 \\ -873.20 & -719.667 & 1098.067 \end{bmatrix}$$

We obtain the error matrix, $\boldsymbol{W}$ as

$$\boldsymbol{W} = \boldsymbol{T} - \boldsymbol{G} - \boldsymbol{B} = \begin{bmatrix} 6344.07 & 1505 & 4559.4 \\ 1505 & 9533.2 & -1395.4 \\ 4559.4 & -1395.4 & 18549.8 \end{bmatrix}$$

To solve the Equations 12.10 and 12.32, we obtain

$$\boldsymbol{B}\boldsymbol{W}^{-1} = \begin{bmatrix} 3100.83 & -183 & 3355 \\ -183 & 10.8 & -198 \\ 3355 & -198 & 3630 \end{bmatrix} \begin{bmatrix} 6344.07 & 1505 & 4559.4 \\ 1505 & 9533.2 & -1395.467 \\ 4559.4 & -1395.4 & 18549.8 \end{bmatrix}^{-1}$$

$$= \begin{bmatrix} 0.46545 & -0.083859 & 0.060144 \\ -0.027469 & 0.004949 & -0.0035495 \\ 0.50360 & -0.090733 & 0.065073 \end{bmatrix}$$

We obtain the solutions of the $|\boldsymbol{B}\boldsymbol{W}^{-1} - \boldsymbol{\lambda I}| = 0$, which are the eigenvalues of $\boldsymbol{B}\boldsymbol{W}^{-1}$. The eigenvalues/solutions are

$$\lambda_1 = 0.53547 \quad \lambda_2 = 1.074 \times 10^{-7} \quad \lambda_3 = -5.9076 \times 10^{-7}$$

Similarly for Grades,

$$\boldsymbol{G}\boldsymbol{W}^{-1} = \begin{bmatrix} 4003.80 & 98.60 & -873.20 \\ 98.60 & 539.467 & -719.667 \\ -873.20 & -719.667 & 1098.067 \end{bmatrix} \begin{bmatrix} 6344.07 & 1505 & 4559.4 \\ 1505 & 9533.2 & -1395.467 \\ 4559.4 & -1395.4 & 18549.7 \end{bmatrix}^{-1}$$

$$= \begin{bmatrix} 0.86649 & -0.16641 & -0.27259 \\ 0.037101 & 0.044195 & -0.044586 \\ -0.20947 & -0.02649 & 0.10869 \end{bmatrix}$$

We obtain the solutions of $|\boldsymbol{G}\boldsymbol{W}^{-1} - \boldsymbol{\lambda I}| = 0$, which are the eigenvalues of $\boldsymbol{G}\boldsymbol{W}^{-1}$. The eigenvalues/solutions are

$$\lambda_1 = 0.92755 \quad \lambda_2 = 9.1821 \times 10^{-2} \quad \lambda_3 = -1.223 \times 10^{-7}$$

We use the two sets of solutions of the two characteristic equations above to construct different criteria for testing the hypotheses in this study. First, we consider the matrix $\boldsymbol{B}$, which contains the sum of squares and cross products for the blocks to test the hypothesis of Equation 12.30. For this hypothesis, $s = \min(v_{\mathrm{B}}, p) = \min(1, 3) = 1$. The degrees of freedom for the block v_{B}; p is the number of response variables measured for each individual in the designed experiment. Similarly, $m = \{|p - v_{\mathrm{B}}| - 1\}/2 = \{|1 - 2| - 1\}/2 = 1/2$; while $n = (v_{\mathrm{e}} - p - 1)/2 = (26 - 3 - 1)/2 = 11$. With the above values of s, m, and n, we obtain the critical values for different criteria (such as Roy's, Lawley–Hotelling, etc.). For $\boldsymbol{G}$, $s = \min(v_G, p) = 2$; $m = \{|n-p|-1\}/2 = 0$; $n = (v_e - p - 1)/2 = 11$.

For the factor B, we test the hypothesis of Equation 12.30 under

(i) *Wilks' criterion:*

$$\Lambda = \frac{|\boldsymbol{W}|}{|\boldsymbol{W}+\boldsymbol{B}|} = \frac{8.5012 \times 10^{11}}{1.3053 \times 10^{12}} = 0.65128$$

We compare this with $U^{0.05}(3,1,26) = 0.726681$. Clearly, $U^{0.05}(3,1,26) > \Lambda$, so we reject the null hypothesis of Equation 12.30 at $\alpha = 0.05$.

(ii) *Roy's criterion:*

For $\boldsymbol{B}$, the Roy's criterion does not exist since $s = 1$. We therefore use the table of Wilks' criterion to obtain the critical value through Roy's and Pillai's criteria. The test statistic is

$$\theta_s = \frac{\lambda_1}{1+\lambda_1} = \frac{0.53547}{1.53547} = 0.34873 = V^{(s)} \text{ (i.e., Pillai's)}$$

Using the tables, we obtain

$$\begin{aligned} \theta^{\alpha}(s,m,n) &= 1 - U^{\alpha}(p,1,v_e) \\ &\Rightarrow \theta^{0.05}\left(1,\frac{1}{2},11\right) = 1 - U^{0.05}(3,1,26) \\ &= 1 - 0.726681 = 0.273319 \end{aligned}$$

We see that

$$\theta_s, V^{(s)} > \theta^{0.05}\left(1,\frac{1}{2},11\right)$$

therefore we reject the null hypothesis of Equation 12.30.

(iii) *Lawley–Hotelling criterion:*

The criterion does not exist for $\boldsymbol{B}$ (since $s = 1$), we therefore refer the test to F-distribution. The test statistic is

$$U^{(s)} = \sum_{i=1}^{s} \lambda_i = \lambda_1 = 0.62628$$

From the tables, we note that $U_0^{0.05}(1,1/2,11) = F(0.05,3,24) = 3.01 \Rightarrow U^{(s)} = 0.37625$. We see that $U^{(s)} > U_0^{\alpha}(s,m,n)$ so we reject the null hypothesis of Equation 12.30. The overall conclusion using different criteria is that blocking has been effective. That is, there is significant difference between performances of boys and girls.

(iv) *Bartlett's χ^2 test*:

Bartlett devised a χ^2-based test for the above hypothesis. Applying the Bartlett's χ^2 approximate test for difference between the vectors of block effects, we note that $\Lambda = 0.65128$, $v_e = 26$, $v_h = 1$, $pv_h = 3$, and $p = 3$, so that

$$\chi_c^2 = -[26 + 1 - (3 + 1 + 1)/2 - (3 - 2)/2] \ln(0.65128) = 11.6711$$

From the table of χ^2 distribution, $\chi^2_{(0.05,3)} = 7.815$. It is clear that the vectors of block effects are significantly different from each other, confirming that blocking was effective. If we prefer to use the *Rao F-approximation*, then for this test, using Equations 12.21a through 12.21d, we note that $n_1 = 3$,

$$n_2 = [26 + 1 - (3 + 1 + 1)/2 - (3 - 2)/2] = 24 \quad s = 1, \text{ so that}$$
$$x = (0.65128)^{1/1} = 0.65128$$
$$F_c = \frac{1 - 0.65128}{0.65128}\left(\frac{24}{3}\right) = 4.2835$$

From the tables, $F(0.05, 3, 24) = 3.009$. We reject the hypothesis of equality of block effects to the null vector.

For the factor $\boldsymbol{G}$, we test the hypothesis of Equation 12.31 under

(i) *Wilks' criterion*:

$$\boldsymbol{\Lambda} = \frac{|\boldsymbol{W}|}{|\boldsymbol{W} + \boldsymbol{G}|} = \frac{8.5012 \times 10^{11}}{1.7892 \times 10^{12}} = 0.47514$$

From the tables, $U^{0.05}(3, 2, 26) = 0.603899 \Rightarrow \Lambda < U^{\alpha}(s, m, n)$. We therefore reject the null hypothesis.

(ii) *Roy's criterion*:

The test statistic is

$$\theta_s = \frac{\lambda_1}{1 + \lambda_1} = \frac{0.92755}{1.92755} = 0.48121$$

From the table, $\theta^{0.05}(2, 0, 11) = 0.3548$. With $\theta_s > \theta^{\alpha}(s, m, n)$, we reject the null hypothesis of Equation 12.31.

(iii) *Lawley–Hotelling criterion*

Here, $s = 2$, so that the test statistic is $U^{(s)} = \lambda_1 + \lambda_2 = 0.091821 + 0.92755 = 1.019371$. The tabulated value $U_0^{0.05}(2, 0, 11) = 0.14801$. Since for $s = 2$, $m = 0$, and $n = 11$, the table of the Lawley–Hotelling criterion does not exist, the value of $U_0^{\alpha}\ (s, m, n)$ is obtained from the F-distribution using the relationship in Equation 12.39. We can easily show by making all the necessary substitutions in Equation 12.39 that $U^{(s)} > U_0^{\alpha}(s, m, n)$ therefore, we reject the null hypothesis of Equation 12.31.

(iv) *Pillai's criterion*

The test statistic is

$$V^{(s)} = \frac{\lambda_1}{1+\lambda_1} + \frac{\lambda_2}{1+\lambda_2} = \frac{0.92755}{1+0.92755} + \frac{0.091821}{1+0.091821} = 0.56530566$$

From the tables, $V^{0.05}_{(2,0,11)} = 0.4274 < V^{(s)}$ so we reject the null hypothesis.

Using all the criteria, we come to the same conclusion that there is significant difference between the performances in different grades.

Next, we test the significant difference between the grades ($\boldsymbol{G}$) using the χ^2 approximation: for this test, $p=3$, $v_\text{h}=2$, $v_\text{e}=26$, and $\Lambda = 0.47514$, so that

$$\chi^2_f = -[26+2-(3+2+1)/2]\ln(0.47514) = 18.6036$$

From the tables, $\chi^2_{(0.05,6)} = 12.592$. The test is significant at 5% level of significance. We reject the null hypothesis of equality of vectors of all grade effects to a null vector.

Using the general form of Rao's approximation (Equation 12.21) to carry out the above test, we note that for grades ($\boldsymbol{G}$),

$$n_1 = 6 \quad n_2 = 2[26+2-(3+2+1)/2] - (6-2)/2 = 48$$

$$s = \sqrt{\frac{6^2-4}{3^2+2^2-5}} = 2 \quad \Lambda^{1/s} = \Lambda^{1/2} = \sqrt{0.47514} = 0.689304$$

$$F_c = \left(\frac{1-0.689304}{0.689304}\right)\left(\frac{48}{6}\right) = 3.6059$$

From the tables, $F(0.05, 6, 48) = 2.294601$. This again confirms the same conclusion we reached by other test criteria, the rejection of the hypothesis of equality of vector of grade effects to a null vector. □

12.3.1 SAS Solution for Example 12.2

Here, we present a SAS program for the analysis of the responses of Example 12.2.

SAS PROGRAM

```
data examp12_2;
do gender='boys', 'girls';
do k=1 to 5;
do grade=1 to 3;
```

```
input y1 y2 y3@@;y1=y1-90; y2=y2-90; y3=y3-90;
output;
end; end;end;
datalines;
120   90    140   80    60    125   95    75    95
130   100   150   85    65    110   96    70    96
70    30    100   89    78    95    100   76    100
80    60    110   92    82    90    110   80    110
70    85    20    101   50    110   123   68    105
91    57    52    51    91    90    80    60    70
89    68    120   44    89    94    85    65    75
90    58    130   52    90    65    89    78    81
95    52    90    50    95    78    92    82    85
67    62    81    57    30    85    104   110   30
;
proc print data=examp12_2;
proc glm data=examp12_2;
class gender grade;
model y1-y3=gender grade;
manova h=_all_ /printe printh; run;
```

SAS OUTPUT

Obs	gender	k	grade	y1	y2	y3
1	boys	1	1	30	0	50
2	boys	1	2	-10	-30	35
3	boys	1	3	5	-15	5
4	boys	2	1	40	10	60
5	boys	2	2	-5	-25	20
6	boys	2	3	6	-20	6
7	boys	3	1	-20	-60	10
8	boys	3	2	-1	-12	5
9	boys	3	3	10	-14	10
10	boys	4	1	-10	-30	20
11	boys	4	2	2	-8	0
12	boys	4	3	20	-10	20
13	boys	5	1	-20	-5	-70
14	boys	5	2	11	-40	20
15	boys	5	3	33	-22	15
16	girl	1	1	1	-33	-38
17	girl	1	2	-39	1	0
18	girl	1	3	-10	-30	-20
19	girl	2	1	-1	-22	30
20	girl	2	2	-46	-1	4
21	girl	2	3	-5	-25	-15
22	girl	3	1	0	-32	40
23	girl	3	2	-38	0	-25
24	girl	3	3	-1	-12	-9
25	girl	4	1	5	-38	0
26	girl	4	2	-40	5	-12
27	girl	4	3	2	-8	-5
28	girl	5	1	-23	-28	-9
29	girl	5	2	-33	-60	-5
30	girl	5	3	14	20	-60

```
                 The GLM Procedure

               Class Level Information

            Class      Levels        Values

            gender          2      boys girl

            grade           3      1 2 3

           Number of Observations Read      30
           Number of Observations Used      30

                          The GLM Procedure


Dependent Variable: y1

                                   Sum of
    Source               DF       Squares    Mean Square    F Value    Pr > F

    Model                 3    7104.63333     2368.21111       9.71    0.0002

    Error                26    6344.06667      244.00256

    Corrected Total      29   13448.70000

             R-Square    Coeff Var    Root MSE      y1 Mean

             0.528277    -380.9898    15.62058    -4.100000

    Source               DF     Type I SS    Mean Square    F Value    Pr > F

    gender                1   3100.833333    3100.833333      12.71    0.0014
    grade                 2   4003.800000    2001.900000       8.20    0.0017

    Source               DF   Type III SS    Mean Square    F Value    Pr > F

    gender                1   3100.833333    3100.833333      12.71    0.0014
    grade                 2   4003.800000    2001.900000       8.20    0.0017

                          The GLM Procedure

Dependent Variable: y2

                                   Sum of
    Source               DF       Squares    Mean Square    F Value    Pr > F

    Model                 3     550.26667      183.42222       0.50    0.6854

    Error                26    9533.20000      366.66154

    Corrected Total      29   10083.46667

             R-Square    Coeff Var    Root MSE      y2 Mean

             0.054571    -105.5978    19.14841    -18.13333
```

Source	DF	Type I SS	Mean Square	F Value	Pr > F
gender	1	10.8000000	10.8000000	0.03	0.8651
grade	2	539.4666667	269.7333333	0.74	0.4889

Source	DF	Type III SS	Mean Square	F Value	Pr > F
gender	1	10.8000000	10.8000000	0.03	0.8651
grade	2	539.4666667	269.7333333	0.74	0.4889

The GLM Procedure

Dependent Variable: y3

Source	DF	Sum of Squares	Mean Square	F Value	Pr > F
Model	3	4728.06667	1576.02222	2.21	0.1110
Error	26	18549.80000	713.45385		
Corrected Total	29	23277.86667			

R-Square	Coeff Var	Root MSE	y3 Mean
0.203114	977.2155	26.71056	2.733333

Source	DF	Type I SS	Mean Square	F Value	Pr > F
gender	1	3630.000000	3630.000000	5.09	0.0327
grade	2	1098.066667	549.033333	0.77	0.4735

Source	DF	Type III SS	Mean Square	F Value	Pr > F
gender	1	3630.000000	3630.000000	5.09	0.0327
grade	2	1098.066667	549.033333	0.77	0.4735

The GLM Procedure
Multivariate Analysis of Variance

E = Error SSCP Matrix

	y1	y2	y3
y1	6344.0666667	1505	4559.4
y2	1505	9533.2	-1395.4
y3	4559.4	-1395.4	18549.8

Partial Correlation Coefficients from the Error SSCP Matrix / Prob > |r|

DF = 26	y1	y2	y3
y1	1.000000	0.193523	0.420295
	0.3335	0.0290	

```
y2       0.193523    1.000000   -0.104932
         0.3335      0.6024

y3       0.420295   -0.104932    1.000000
         0.0290      0.6024

              The GLM Procedure
       Multivariate Analysis of Variance

      H = Type III SSCP Matrix for gender

              y1          y2          y3

y1    3100.8333333     -183         3355
y2    -183               10.8       -198
y3    3355             -198         3630

   Characteristic Roots and Vectors of: E Inverse * H, where
          H = Type III SSCP Matrix for gender
               E = Error SSCP Matrix

Characteristic        Characteristic Vector V'EV=1
         Root    Percent              y1            y2             y3

   0.53547562     100.00      0.01142238   -0.00205825     0.00147622
   0.00000000       0.00     -0.00865346    0.00264331     0.00814208
   0.00000000       0.00      0.00059821    0.01013627     0.00000000

MANOVA Test Criteria and Exact F Statistics for the Hypothesis of No Overall
                              gender Effect
             H = Type III SSCP Matrix for gender
                     E = Error SSCP Matrix

                      S=1   M=0.5   N=11

Statistic                      Value    F Value    Num DF    Den DF    Pr > F

Wilks' Lambda             0.65126400       4.28         3        24    0.0148
Pillai's Trace            0.34873600       4.28         3        24    0.0148
Hotelling-Lawley Trace    0.53547562       4.28         3        24    0.0148
Roy's Greatest Root       0.53547562       4.28         3        24    0.0148

             H = Type III SSCP Matrix for grade

                   y1              y2              y3

y1          4003.8       98.6             -873.2
y2            98.6      539.46666667      -719.6666667
y3          -873.2     -719.6666667       1098.0666667

                   The GLM Procedure
            Multivariate Analysis of Variance

   Characteristic Roots and Vectors of: E Inverse * H, where
              H = Type III SSCP Matrix for grade
                    E = Error SSCP Matrix
```

```
    Characteristic          Characteristic Vector V'EV=1
              Root   Percent              y1              y2              y3

        0.92750637     90.99      0.01422806     -0.00254132     -0.00459755
        0.09183223      9.01     -0.00149843      0.00757140     -0.00420026
        0.00000000      0.00      0.00101396      0.00708399      0.00544913

MANOVA Test Criteria and F Approximations for the Hypothesis of No Overall
                              grade Effect
                   H = Type III SSCP Matrix for grade
                          E = Error SSCP Matrix

                            S=2    M=0    N=11

  Statistic                       Value    F Value    Num DF    Den DF    Pr > F

  Wilks' Lambda              0.47516918       3.61         6        48    0.0050
  Pillai's Trace             0.56530334       3.28         6        50    0.0084
  Hotelling-Lawley Trace     1.01933860       4.00         6    30.286    0.0046
  Roy's Greatest Root        0.92750637       7.73         3    25        0.0008

          NOTE: F Statistic for Roy's Greatest Root is an upper bound.
                  NOTE: F Statistic for Wilks' Lambda is exact.
```

12.4 Multivariate Two-Way Experimental Layout with Interaction

Here, the performance of the experiments is similar to the two-factor factorial design (Chapter 3) except that p observations are elicited from each experimental unit. Grouping of responses according to the levels of the two factors lead to a two-way layout. The measurements or observations are replicated for each cell which means that interaction of factors is possible as discussed for similar situations in the univariate case (Chapter 3). None of the factors is treated as block as we discussed in the previous model. If two factors, A and B with levels r and c are being considered, then there should be rc samples, each containing n_s experimental units. These experimental units are part of the rcn_s units in the entire experiment which are assigned at random to the rc samples. By analogy with the univariate model of the two-way layout (Equation 3.1), we obtain the multivariate model. We replace the single response variable of the univariate model by a p-variate vector of responses. However, for each individual or experimental unit, there is at least a two-variate vector of measurements (representing repeated or simultaneous measurements under the combined levels of the factors). The resulting multivariate model is

$$Y_{ijk} = \mu + \gamma_i + \beta_j + \tau_{ij} + \varepsilon_{ijk} \tag{12.40}$$

where $\varepsilon_{ijk} \sim IN_p(0, \Sigma)$ and Y_{ijk} is a p-variate random vector; $N = rcn_s$; $i = 1, \ldots, r$; $j = 1, \ldots, c$; $k = 1, \ldots, n_s$

$$\overline{Y}_{ij.} = \frac{\sum_{k=1}^{n_s} Y_{ijk}}{n_s} \quad \overline{Y}_{i..} = \frac{\sum_{j=1}^{c}\sum_{k=1}^{n_s} Y_{ijk}}{cn_s} \quad \overline{Y}_{.j.} = \frac{\sum_{i=1}^{r}\sum_{k=1}^{n_s} Y_{ij}}{rn_s}$$

$$\overline{Y}_{...} = \frac{\sum_{i=1}^{r}\sum_{j=1}^{c}\sum_{k=1}^{n_s} y_{ijk}}{rcn_s} \tag{12.41}$$

By analogy with the univariate case, we define the side conditions, which govern the fixed effects model. With these side conditions, estimable functions of γ and τ are obtained. The conditions are as follows:

$$\beta = \frac{1}{c}\sum_{j=1}^{c} \beta_j = 0 \quad \text{for all } i$$

$$\gamma = \frac{1}{r}\sum_{i=1}^{r} \gamma_i = 0 \quad \text{for all } j$$

$$\tau_{i.} = \frac{1}{c}\sum_{j=1}^{c} \tau_{ij} = 0 \quad \text{for all } i$$

$$\tau_{.j} = \frac{1}{r}\sum_{i=1}^{r} \tau_{ij} = 0 \quad \text{for all } j$$

With these side conditions, $\gamma = 0$; $\beta = 0$; $\tau_{i.} = 0$ for all i and $\tau_{.j} = 0$ for all j. Once these conditions apply along with the model above (Equation 12.40), the individual parameters of this multivariate design become estimable as

$$\begin{aligned}
\hat{\mu} &= \overline{Y}_{...} \\
\hat{\gamma} &= \overline{Y}_{i..} - \overline{Y}_{...} \\
\hat{\beta} &= \overline{Y}_{.j.} - \overline{Y}_{...} \\
\hat{\tau}_{ij} &= \overline{Y}_{ij.} - \overline{Y}_{i..} - \overline{Y}_{.j.} + \overline{Y}_{...}
\end{aligned}$$

To test all the hypotheses in this design, we obtain the sum of squares and cross products to be used in MANOVA. The sum of squares are

$$\boldsymbol{A} = cn_s \sum_{i=1}^{r} (\overline{Y}_{i..} - \overline{Y}_{...})(\overline{Y}_{i..} - \overline{Y}_{...})' \tag{12.42}$$

$$\boldsymbol{B} = rn_s \sum_{j=1}^{c} (\overline{Y}_{.j.} - \overline{Y}_{...})(\overline{Y}_{.j.} - \overline{Y}_{...})' \tag{12.43}$$

$$\boldsymbol{AB} = n_s \sum_{i=1}^{r} \sum_{j=1}^{c} (\overline{Y}_{ij.} - \overline{Y}_{i..} - \overline{Y}_{.j.} + \overline{Y}_{...})(\overline{Y}_{ij.} - \overline{Y}_{i..} - \overline{Y}_{.j.} + \overline{Y}_{...})' \quad (12.44)$$

$$\boldsymbol{W} = \sum_{i=1}^{r} \sum_{j=1}^{c} \sum_{k=1}^{n_s} (Y_{ijk} - \overline{Y}_{ij.})(Y_{ijk} - \overline{Y}_{ij.})' \quad (12.45)$$

$$\boldsymbol{T} = \sum_{i=1}^{r} \sum_{j=1}^{c} \sum_{k=1}^{n_s} (Y_{ijk} - \overline{Y}_{...})(Y_{ijk} - \overline{Y}_{...})' \quad (12.46)$$

TABLE 12.5: MANOVA Table for the Two-Way Layout with Interaction

Source	Degrees of freedom	Sum of squares and cross products
Factor $\boldsymbol{A}$	$r-1$	$\boldsymbol{A}$
Factor $\boldsymbol{B}$	$c-1$	$\boldsymbol{B}$
Interaction ($\boldsymbol{AB}$)	$(c-1)(r-1)$	$\boldsymbol{AB}$
Error	$rc(n_s-1)$	$\boldsymbol{W}$
Total	rcn_s-1	$\boldsymbol{T}$

$\boldsymbol{A}$, $\boldsymbol{B}$, $\boldsymbol{AB}$, $\boldsymbol{W}$, and $\boldsymbol{T}$ have $r-1$, $c-1$, $(r-1)$ $(c-1)$, $rc(n-1)$, and $N-1$ degrees of freedom, respectively. $\boldsymbol{W}$ and $\boldsymbol{T}$ are the error and total sum of squares and cross products. Table 12.5 summarizes the sum of squares and cross products as well as their corresponding degrees of freedom in the design.

The null hypotheses and the alternatives to be tested under this design using factors A and B are

$$H_0\colon \gamma_1 = \gamma_2 = \cdots = \gamma_r = 0 \quad \text{vs} \quad H_1\colon \text{at least one } \gamma_i \text{ is different}$$

$$H_0\colon \beta_1 = \beta_2 = \cdots = \beta_c = 0 \quad \text{vs} \quad H_1\colon \text{at least one } \beta_j \text{ is different}$$

Usually, the interaction sum of squares and cross products is used for testing for parallelism or significant interactions. However, when the constraints which lead to fixed effects model (stated above) are included in the design, this hypothesis is equivalent to

$$H_{AB} : \tau_{ij} = 0 \quad (\text{for all } i, j) \quad \text{vs} \quad H'_{AB} : \tau_{ij} \neq 0 \quad (\text{for some } i, j)$$

To test these hypotheses, we need to calculate the values of s, m, and n with which we shall obtain the critical values under different test criteria discussed in Section 12.3. For the hypothesis involving factor $\boldsymbol{A}$,

$$s = \min(r-1, p) \quad m = \frac{|r-1-p|-1}{2} \quad n = \frac{rc(n_s-1)-p-1}{2}$$

For the hypothesis involving factor $\boldsymbol{B}$, we obtain

$$s = \min(c-1, p) \quad m = \frac{|c-1-p|-1}{2}$$

$$n = \frac{rc(n_s-1)-p-1}{2}$$

In testing the hypothesis about the significance of $\boldsymbol{AB}$ interaction effects, we use

$$s = \min[(r-1)(c-1), p]$$

$$m = \frac{|(r-1)(c-1)-p|-1}{2} \quad n = \frac{rc(n_s-1)-p-1}{2}$$

It is usual to carry out a test for significance of interaction before carrying out the test for the significance of the main effects, that is, effects of the rest of the factors, $\boldsymbol{A}$ and $\boldsymbol{B}$. If the interaction effect is significant, which means that there is the confounding of main effects, $\boldsymbol{A}$, $\boldsymbol{B}$ with interaction, then, there would be very little point testing for the significance of the main effects since they are not deemed to be acting independently but in consort with each other.

Example 12.3

A study was carried out to evaluate some methods of teaching. Three methods were used to teach pupils drawn from schools which are located in areas of the city occupied by low income families (LES), middle income families (MES), and high income families (HES). The three methods were used to teach pupils and before they were subjected to three tests in English (E), mathematics (M), and general knowledge (G). Apart from testing whether there are any differences in performances between pupils subjected to different methods of teaching, tests were carried out as well for differences in performances between pupils from different social backgrounds. The scores from the tests which were scored over 20 are presented as follows:

	Method I								
	HES			**MES**			**LES**		
	M	E	G	M	E	G	M	E	G
	11	12	10	11	9	10	9	11	10
	13	13	14	12	12	11	8	12	10
	10	10	12	10	12	10	11	13	9
	11	12	13	10	11	9	11	13	10
	13	14	12	12	12	11	9	14	9
	13	10	11	12	12	13	9	14	12
Total	71	71	72	67	68	64	57	77	60

Method II

	HES			MES			LES		
	M	**E**	**G**	**M**	**E**	**G**	**M**	**E**	**G**
	12	14	13	13	12	12	11	12	12
	13	18	10	12	14	10	12	13	10
	14	15	9	14	13	10	10	13	12
	14	17	12	13	10	12	9	14	11
	15	15	11	11	13	11	9	13	12
	14	12	9	10	14	12	11	12	13
Total	82	91	64	73	76	67	62	77	70

Method III

	HES			MES			LES		
	M	**E**	**G**	**M**	**E**	**G**	**M**	**E**	**G**
	9	12	14	11	10	12	9	13	9
	8	10	12	10	10	11	8	9	10
	11	13	11	9	12	9	10	8	11
	9	12	10	10	12	10	9	10	12
	8	14	10	11	11	13	8	10	10
	11	12	11	12	9	12	10	11	9
Total	56	73	68	63	64	67	54	61	61

Analysis:
We calculate the overall sum of all the scores and sum of squares of the scores (for all methods) for mathematics (M), English (E), and general knowledge (G).

Sum of scores:

$$\sum \mathrm{M} = 71 + 67 + 57 + 82 + 73 + 62 + 56 + 63 + 54 = 585$$

$$\sum \mathrm{E} = 71 + 68 + 77 + 91 + 76 + 77 + 73 + 64 + 61 = 658$$

$$\sum \mathrm{G} = 72 + 64 + 60 + 64 + 67 + 70 + 68 + 67 + 61 = 593$$

Sum of squares of scores:

$$\sum \mathrm{M}^2 = 11^2 + 13^2 + 10^2 + \cdots + 9^2 + 8^2 + 10^2 = 6513$$

$$\sum \mathrm{E}^2 = 12^2 + 13^2 + 10^2 + \cdots + 10^2 + 10^2 + 11^2 = 8216$$

$$\sum \mathrm{G}^2 = 10^2 + 14^2 + 12^2 + \cdots + 12^2 + 10^2 + 9^2 = 6611$$

We also calculate the sum of cross products for all the three tests as

$$\begin{aligned}\sum \text{ME} &= 11 \times 12 + 13 \times 13 + 10 \times 10 + \cdots \\ &\quad + 9 \times 10 + 8 \times 10 + 10 \times 11 = 7193 \\ \sum \text{MG} &= 11 \times 10 + 13 \times 14 + 10 \times 12 + \cdots \\ &\quad + 9 \times 12 + 8 \times 10 + 10 \times 9 = 6441 \\ \sum \text{EG} &= 12 \times 10 + 13 \times 14 + 10 \times 12 + \cdots \\ &\quad + 10 \times 12 + 10 \times 10 + 11 \times 9 = 7217\end{aligned}$$

We now calculate all the elements of the total matrix $\boldsymbol{T}$, first the diagonal elements:

$$\text{For M: } T_{11} = 6513 - \frac{585^2}{54} = 175.50$$

$$\text{For E: } T_{22} = 8216 - \frac{658^2}{54} = 198.15$$

$$\text{For G: } T_{33} = 6611 - \frac{593^2}{54} = 98.537$$

$$\text{For M-E: } T_{12} = T_{21} = 7193 - \frac{585(658)}{54} = 64.667$$

$$\text{For M-G: } T_{13} = T_{31} = 6441 - \frac{585(593)}{54} = 16.833$$

$$\text{For E-G: } T_{23} = T_{32} = 7217 - \frac{658(593)}{54} = -8.8152$$

We can now write down the matrix $\boldsymbol{T}$ as

$$\boldsymbol{T} = \begin{bmatrix} 175.50 & 64.667 & 16.833 \\ 64.667 & 198.15 & -8.8152 \\ 16.833 & -8.8152 & 98.537 \end{bmatrix}$$

We obtain the matrix of the treatment (method) sum of squares, $\boldsymbol{G}$ by considering the three methods, I, II, and III: First, we calculate the sum of the mathematics scores, M_1, M_2, and M_3 corresponding to the three methods.

$$\sum M_1 = 71 + 67 + 57 = 195$$

$$\sum M_2 = 82 + 73 + 62 = 217$$

$$\sum M_3 = 56 + 63 + 54 = 173$$

Next, we calculate the sums of English and General Knowledge scores corresponding to three methods.

$$\sum E_1 = 71 + 68 + 77 = 216$$

$$\sum E_2 = 91 + 76 + 77 = 244$$

$$\sum E_3 = 73 + 64 + 61 = 198$$

$$\sum G_1 = 72 + 64 + 60 = 196$$

$$\sum G_2 = 64 + 67 + 70 = 201$$

$$\sum G_3 = 68 + 67 + 62 = 196$$

The elements of the matrix $\boldsymbol{G}$, which are the sums of squares and cross products for the methods, are obtained as

$$G_{12} = G_{21} = \frac{195 \times 216 + 217 \times 244 + 173 \times 198}{18} - \frac{585(658)}{54} = 56.222$$

$$G_{13} = G_{31} = \frac{195 \times 196 + 217 \times 201 + 173 \times 196}{18} - \frac{585(593)}{54} = 6.11111$$

$$G_{23} = G_{32} = \frac{196 \times 216 + 201 \times 244 + 196 \times 198}{18} - \frac{593(658)}{54} = 6.8518$$

$$G_{11} = \frac{195^2 + 217^2 + 173^2}{18} - \frac{585^2}{54} = 53.778$$

$$G_{22} = \frac{216^2 + 244^2 + 198^2}{18} - \frac{658^2}{54} = 59.704$$

$$G_{33} = \frac{196^2 + 201^2 + 196^2}{18} - \frac{593^2}{54} = 0.92592$$

so that

$$\boldsymbol{G} = \begin{bmatrix} 53.778 & 56.222 & 6.1111 \\ 56.222 & 59.704 & 6.8518 \\ 6.1111 & 6.8518 & 0.92592 \end{bmatrix}$$

We obtain the sums of M, E, and G on all the social background of the children, HES, MES, and LES as

$$\sum M_{\mathrm{H}} = 71 + 82 + 56 = 209$$

$$\sum M_{\mathrm{M}} = 67 + 73 + 63 = 203$$

$$\sum M_{\mathrm{L}} = 57 + 62 + 54 = 173$$

$$\sum E_{\mathrm{H}} = 71 + 91 + 73 = 235$$

$$\sum E_{\mathrm{M}} = 68 + 76 + 64 = 208$$

$$\sum E_{\mathrm{L}} = 77 + 77 + 61 = 215$$

$$\sum G_{\mathrm{H}} = 72 + 64 + 68 = 204$$

$$\sum G_{\text{M}} = 64 + 67 + 67 = 198$$
$$\sum G_{\text{L}} = 60 + 70 + 61 = 191$$

We find the elements of the matrix $\boldsymbol{B}$ as

$$B_{11} = \frac{(209^2 + 203^2 + 173^2)}{18} - \frac{585^2}{54} = 41.333$$
$$B_{22} = \frac{(235^2 + 208^2 + 215^2)}{18} - \frac{658^2}{54} = 21.815$$
$$B_{33} = \frac{(204^2 + 198^2 + 187^2)}{18} - \frac{593^2}{54} = 4.7037$$
$$B_{12} = B_{21} = \frac{(209 \times 235 + 203 \times 208 + 173 \times 215)}{18} - \frac{(585 \times 658)}{54} = 12.444$$
$$B_{13} = B_{31} = \frac{(209 \times 204 + 203 \times 198 + 173 \times 191)}{18} - \frac{(585 \times 593)}{54} = 13.2222$$
$$B_{23} = B_{32} = \frac{(235 \times 204 + 208 \times 198 + 215 \times 191)}{18} - \frac{(658 \times 593)}{54} = 6.907$$

$$\boldsymbol{B} = \begin{bmatrix} 41.333 & 12.444 & 13.2222 \\ 12.444 & 21.815 & 6.907 \\ 13.2222 & 6.907 & 4.7037 \end{bmatrix}$$

We obtain the matrix of the interaction between teaching methods and socioeconomic background of children. The matrix is $\boldsymbol{BG}$:

$$(BG)_{11} = \frac{(71^2 + 67^2 + 57^2 + 82^2 + 73^2 + 62^2 + 56^2 + 63^2 + 54^2)}{6} - \frac{585^2}{54} - 53.778 - 41.333 = 16.889$$
$$(BG)_{22} = \frac{(71^2 + 68^2 + 77^2 + 91^2 + 76^2 + 77^2 + 73^2 + 64^2 + 612)}{6} - \frac{658^2}{54} - 59.704 - 21.815 = 21.629$$

$$(BG)_{33} = \frac{(72^2 + 64^2 + 60^2 + 64^2 + 67^2 + 70^2 + 68^2 + 67^2 + 61^2)}{6}$$

$$- \frac{593^2}{54} - 0.92592 - 4.7037 = 15.51853$$

$$(BG)_{12} = (BG)_{21} = \frac{(71 \times 71 + 67 \times 68 + 57 \times 77 + \cdots + 54 \times 61)}{6}$$

$$- \frac{658(585)}{54} - 56.222 - 12.444 = 0.3333$$

$$(BG)_{13} = (BG)_{31} = \frac{(71 \times 72 + 67 \times 64 + 57 \times 77 \cdots + 54 \times 61)}{6}$$

$$- \frac{585(593)}{54} - 6.111 - 13.2222 = -6.50$$

$$(BG)_{23} = (BG)_{32} = \frac{(71 \times 64 + 68 \times 67 + 57 \times 77 \cdots + 61 \times 61)}{6}$$

$$- \frac{593(658)}{54} - 6.8518 - 6.907 = -12.407$$

$$\boldsymbol{BG} = \begin{bmatrix} 16.8890 & 0.3333 & -6.5 \\ 0.3333 & 21.6290 & -12.407 \\ -6.5 & -12.407 & 15.5185 \end{bmatrix}$$

We obtain the matrix of the within error sum of squares, $\boldsymbol{W}$ as

$$\boldsymbol{W} = \boldsymbol{T} - \boldsymbol{G} - \boldsymbol{B} - \boldsymbol{BG} \tag{12.47}$$

so that

$$\boldsymbol{W} = \begin{bmatrix} 63.50 & -4.3330 & 4 \\ -4.3330 & 95.002 & -10.1667 \\ 4 & -10.1667 & 77.8333 \end{bmatrix}$$

To test the hypothesis of equality of centroids, we need the solutions of the three characteristic equations as follows:

$$\left|\boldsymbol{BW}^{-1} - \lambda \boldsymbol{I}\right| = 0 \quad \left|\boldsymbol{GW}^{-1} - \lambda \boldsymbol{I}\right| = 0 \quad \left|(\boldsymbol{BG})\boldsymbol{W}^{-1} - \lambda \boldsymbol{I}\right| = 0 \tag{12.48}$$

Using the matrices $\boldsymbol{W}$, $\boldsymbol{B}$, $\boldsymbol{G}$, and $\boldsymbol{BG}$, we obtain the following results which are required to obtain the solutions of the above characteristic equations:

$$\boldsymbol{BW}^{-1} = \begin{bmatrix} 0.653 & 0.17784 & 0.15955 \\ 0.2061 & 0.2509 & 0.11092 \\ 0.21043 & 8.8853 \times 10^{-2} & 6.1225 \times 10^{-2} \end{bmatrix}$$

$$\boldsymbol{GW}^{-1} = \begin{bmatrix} 0.8835 & 0.64465 & 0.11732 \\ 0.92391 & 0.6845 & 0.12996 \\ 0.10053 & 7.8523 \times 10^{-2} & 1.6987 \times 10^{-2} \end{bmatrix}$$

$$(\boldsymbol{BG})\,\boldsymbol{W}^{-1} = \begin{bmatrix} 0.27245 & 5.577\times10^{-3} & -9.6785\times10^{-2} \\ 0.028268 & 0.21475 & -0.13281 \\ -0.12227 & -0.11579 & 0.19054 \end{bmatrix}$$

We obtain the eigenvalues, solutions of $|\boldsymbol{BW}^{-1} - \boldsymbol{\lambda I}| = 0$ as

$$\lambda_1 = 0.78878 \quad \lambda_2 = 0.17634 \quad \lambda_3 = 1.3022\times10^{-6}$$

Similarly, the solutions of $|\boldsymbol{GW}^{-1} - \boldsymbol{\lambda I}| = 0$ are

$$\lambda_1 = 1.5762 \quad \lambda_2 = 0.0087464 \quad \lambda_3 = 0.0000043969$$

Also, the solutions for $|(\boldsymbol{BG})\,\boldsymbol{W}^{-1} - \boldsymbol{\lambda I}| = 0$ are

$$\lambda_1 = 0.39394 \quad \lambda_2 = 0.23114 \quad \lambda_3 = 0.052653$$

For the tests which follow, we confine ourselves to three of the four criteria previously described, which give consistent results. For $\boldsymbol{BG}$: $s = 3$; $m = 0$; $n = 20.5$. We test the hypothesis

$$H_0 : \tau_{11} = \tau_{12} = \cdots = \tau_{33} = 0 \quad \text{vs} \quad H_1 : \tau_{ij} \neq 0 \quad \text{for some}\, i, j$$

Criteria from the tables

$$\begin{aligned} \text{Wilks':}\ & \Lambda = 0.55356 > U^{0.01}(3, 4, 45) = 0.5505 \\ \text{Roy's:}\ & \theta_s = 0.2826 < \theta_{0.01}(3, 0, 20.5) = 0.36407 \\ \text{Pillai's:}\ & V^{(s)} = 0.520373 < V^{0.01}(s, m, n) = 0.4998 \\ \text{Lawley–Hotelling's:}\ & 0.67773 \end{aligned}$$

It should be noted that Roy's criterion is different from Roy's largest root criterion implemented in SAS.

Those are the values of the statistics for testing the hypothesis of no significant interaction effect using different criteria. The conclusions reached under the χ^2 and F-approximations are better.

To use Bartlett's and Rao's approximations in the test for significance of the interaction effect, we calculate the following using Equations 8.20 and 8.21:

$$\Lambda = 0.55356 \quad p = 3 \quad v_{\text{h}} = 4 \quad v_{\text{e}} = 45 \quad s = 2.6458 \quad n_1 = 12$$

$$n_2 = 114.0588 \quad \Lambda_1/s_1 = 0.79969 \quad f = 12$$

$$\text{Bartlett's: } \chi_f^2 = -\frac{[45 + 4 - (3 + 4 + 10)]}{2} \ln(0.55356)$$

$$\chi^2 = -\frac{[45 + 4 - (3 + 4 + 1)]}{2} \ln(0.55356) = 26.612$$

From the tables: $\chi^2_{(0.01,12)} = 26.22$

$$\text{Rao's } F\text{: } F_c = \frac{[(1 - 0.79969)]}{0.79969} \frac{(114.0558)}{12} = 2.3806$$

From the tables: $F(0.01, 12, 114.0558) = 2.3519$

So we confirm that the interaction effect is significant and reject the null hypothesis.

For G: $s=2$; $m=0$; $n=20.5$. We test the hypothesis

$$H_0: \gamma_1 = \gamma_2 = \gamma_3 = 0 \quad \text{vs} \quad H_1: \gamma_i \neq 0 \quad \text{for some } i$$

Criteria from the tables

$$\begin{aligned} \text{Wilks': } & \Lambda = 0.3848 < U^{0.01}(3,2,45) = 0.6752 \\ \text{Roy's: } & \theta_s = 0.6118 > \theta_{0.01}(s,m,n) = 0.3248 \\ \text{Pillai's: } & V^{(s)} = 0.6205 > V^{0.05}(s,m,n) = 0.3107 \end{aligned}$$

We reject the null hypothesis H_0 and accept H_1 which means that there is significant treatment (teaching method) effect.

Applying the Bartlett's and Rao's approximations to the test for significance of the main effect G, we calculate the following:

$$s = 2 \quad v_{\text{e}} = 45 \quad v_{\text{h}} = 2 \quad n_1 = 6 \quad n_2 = 86 \quad \Lambda = 0.3848$$
$$\Lambda_1/s_1 = 0.6203225 \quad p = 3$$

$$\text{Bartlett's: } \chi_f^2 = -[45 + 2 - (3 + 2 + 1)/2] \ln(0.3848) = 42.0213$$
$$\text{From the tables: } \chi^2_{(0.01,6)} = 16.81$$

$$\text{Rao's } F\text{: } F_c = \{(1 - 0.6203225)/0.3203225\}\{86/6\} = 8.7729$$
$$\text{From the tables: } F(0.01, 6, 86) = 3.0484$$

We therefore confirm (by both methods) the rejection of the null hypothesis and therefore accept that the treatment (i.e., teaching method) effect is significant at $\alpha=1\%$.

For $\boldsymbol{B}$: $s=2$; $n=0$; $m=20.5$. The hypothesis to be tested is

$$H_0\text{: } \beta_1 = \beta_2 = \beta_3 = 0 \quad \text{vs} \quad H_A\text{: } \beta_j\text{'s} \neq 0 \quad \text{for some } j$$

Criteria from the tables

$$\begin{aligned} \text{Wilks': } & \Lambda = 0.475236 < U^{0.01}(3,2,45) = 0.6752 \\ \text{Roy's: } & \theta_s = 0.44096 > \theta^{0.01}(s,m,n) = 0.2832 \\ \text{Pillai's: } & V^{(s)} = 0.590867 > V^{0.05}(s,m,n) = 0.3107 \end{aligned}$$

We reject the null hypothesis H_0 and conclude that social background has significant effect.

Just as in others, we retest the hypothesis on the $\boldsymbol{B}$ effect using both Bartlett's and Rao's approximations by obtaining the following quantities:

$$\Lambda = 0.475236 \quad p = 3 \quad v_e = 2 \quad v_h = 45 \quad s_1 = 2 \quad n_1 = 6$$
$$n_2 = 86 \ \Lambda^{1/s_1} = 0.689374$$

$$\text{Bartlett's:}\quad \chi_f^2 = -[45+2-(3+2+1)/2]\ln(0.475236) = 32.7335$$
$$\text{From the tables:}\quad \chi^2(0.01,6) = 16.81$$

$$\text{Rao's } F\text{:}\quad F_c = \{(1-0.0.689374)/0.0.689374\}\{86/6\} = 6.4585$$
$$\text{From the tables:}\quad F(0.01,6,86) = 3.0484$$

We have confirmed from both tests that social background effect is significant. □

12.4.1 SAS Program for Example 12.3

We present a SAS program for the analysis of the responses of Example 12.3:

SAS PROGRAM

```
data examp12_3;
do METHOD=1 TO 3;
do K=1 to 6;
do SOCIAL='HES', 'MES','LES';
input M E G @@;
output;
end; end;end;
datalines;
11   12  10 11  9  10  9   11 10
13  13 14   12    12 11 8  12   10
10   10 12  10   12   10 11   13  9
11 12 13    10 11   9   11 13 10
13 14         12   12   12 11  9   14  9
13   10   11    12          12 13   9   14   12
12   14  13 13    12    12   11  12      12
13  18        10    12    14   10      12    13    10
14     15   9     14    13    10 10      13      12
14    17     12    13    10      12   9     14      11
15    15     11    11    13 11 9   13      12
14    12      9    10    14      12    11   12    13
9   12   14        11    10   12       9     13    9
8  10  12 10       10 11 8      9  10
11  13 11   9      12 9   10     8 11
9  12   10   10      12      10    9      10 12
8 14    10 11        11       13  8       10 10
11      12 11        12     9  12 10     11  9
;
proc print data=examp12_3;
proc glm data=examp12_3;
class METHOD SOCIAL ;
model M E G=METHOD SOCIAL METHOD*SOCIAL;
manova h=_all_ /printe printh;
run;
```

SAS OUTPUT

Obs	METHOD	K	SOCIAL	M	E	G
1	1	1	HES	11	12	10
2	1	1	MES	11	9	10
3	1	1	LES	9	11	10
4	1	2	HES	13	13	14
5	1	2	MES	12	12	11
6	1	2	LES	8	12	10
7	1	3	HES	10	10	12
8	1	3	MES	10	12	10
9	1	3	LES	11	13	9
10	1	4	HES	11	12	13
11	1	4	MES	10	11	9
12	1	4	LES	11	13	10
13	1	5	HES	13	14	12
14	1	5	MES	12	12	11
15	1	5	LES	9	14	9
16	1	6	HES	13	10	11
17	1	6	MES	12	12	13
18	1	6	LES	9	14	12
19	2	1	HES	12	14	13
20	2	1	MES	13	12	12
21	2	1	LES	11	12	12
22	2	2	HES	13	18	10
23	2	2	MES	12	14	10
24	2	2	LES	12	13	10
25	2	3	HES	14	15	9
26	2	3	MES	14	13	10
27	2	3	LES	10	13	12
28	2	4	HES	14	17	12
29	2	4	MES	13	10	12
30	2	4	LES	9	14	11
31	2	5	HES	15	15	11
32	2	5	MES	11	13	11
33	2	5	LES	9	13	12
34	2	6	HES	14	12	9
35	2	6	MES	10	14	12
36	2	6	LES	11	12	13
37	3	1	HES	9	12	14
38	3	1	MES	11	10	12
39	3	1	LES	9	13	9
40	3	2	HES	8	10	12
41	3	2	MES	10	10	11
42	3	2	LES	8	9	10
43	3	3	HES	11	13	11
44	3	3	MES	9	12	9
45	3	3	LES	10	8	11
46	3	4	HES	9	12	10
47	3	4	MES	10	12	10
48	3	4	LES	9	10	12
49	3	5	HES	8	14	10
50	3	5	MES	11	11	13

Obs	METHOD	K	SOCIAL	M	E	G
51	3	5	LES	8	10	10
52	3	6	HES	11	12	11
53	3	6	MES	12	9	12
54	3	6	LES	10	11	9

The GLM Procedure
Class Level Information

Class	Levels	Values
METHOD	3	1 2 3
SOCIAL	3	HES LES MES

Number of Observations Read	54
Number of Observations Used	54

The GLM Procedure

Dependent Variable: M

Source	DF	Sum of Squares	Mean Square	F Value	Pr > F
Model	8	112.0000000	14.0000000	9.92	<.0001
Error	45	63.5000000	1.4111111		
Corrected Total	53	175.5000000			

R-Square	Coeff Var	Root MSE	M Mean
0.638177	10.96525	1.187902	10.83333

Source	DF	Type I SS	Mean Square	F Value	Pr > F
METHOD	2	53.77777778	26.88888889	19.06	<.0001
SOCIAL	2	41.33333333	20.66666667	14.65	<.0001
METHOD*SOCIAL	4	16.88888889	4.22222222	2.99	0.0284

Source	DF	Type III SS	Mean Square	F Value	Pr > F
METHOD	2	53.77777778	26.88888889	19.06	<.0001
SOCIAL	2	41.33333333	20.66666667	14.65	<.0001
METHOD*SOCIAL	4	16.88888889	4.22222222	2.99	0.0284

The GLM Procedure

Dependent Variable: E

Source	DF	Sum of Squares	Mean Square	F Value	Pr > F
Model	8	103.1481481	12.8935185	6.11	<.0001
Error	45	95.0000000	2.1111111		
Corrected Total	53	198.1481481			

R-Square	Coeff Var	Root MSE	E Mean
0.520561	11.92404	1.452966	12.18519

Source	DF	Type I SS	Mean Square	F Value	Pr > F
METHOD	2	59.70370370	29.85185185	14.14	<.0001
SOCIAL	2	21.81481481	10.90740741	5.17	0.0096
METHOD*SOCIAL	4	21.62962963	5.40740741	2.56	0.0512

Source	DF	Type III SS	Mean Square	F Value	Pr > F
METHOD	2	59.70370370	29.85185185	14.14	<.0001
SOCIAL	2	21.81481481	10.90740741	5.17	0.0096
METHOD*SOCIAL	4	21.62962963	5.40740741	2.56	0.0512

The GLM Procedure

Dependent Variable: G

Source	DF	Sum of Squares	Mean Square	F Value	Pr > F
Model	8	21.14814815	2.64351852	1.53	0.1744
Error	45	77.83333333	1.72962963		
Corrected Total	53	98.98148148			

R-Square	Coeff Var	Root MSE	G Mean
0.213658	11.97611	1.315154	10.98148

Source	DF	Type I SS	Mean Square	F Value	Pr > F
METHOD	2	0.92592593	0.46296296	0.27	0.7664
SOCIAL	2	4.70370370	2.35185185	1.36	0.2671
METHOD*SOCIAL	4	15.51851852	3.87962963	2.24	0.0793

Source	DF	Type III SS	Mean Square	F Value	Pr > F
METHOD	2	0.92592593	0.46296296	0.27	0.7664
SOCIAL	2	4.70370370	2.35185185	1.36	0.2671
METHOD*SOCIAL	4	15.51851852	3.87962963	2.24	0.0793

The GLM Procedure
Multivariate Analysis of Variance

E = Error SSCP Matrix

	M	E	G
M	63.5	-4.333333333	4
E	-4.333333333	95	-10.16666667
G	4	-10.16666667	77.833333333

Partial Correlation Coefficients from the Error SSCP Matrix / Prob > |r|

DF = 45	M	E	G
M	1.000000	-0.055792	0.056897
		0.7127	0.7072

```
DF = 45          M           E           G

E       -0.055792    1.000000   -0.118232
         0.7127      0.4339

G        0.056897   -0.118232    1.000000
         0.7072      0.4339

                 The GLM Procedure
          Multivariate Analysis of Variance

          H = Type III SSCP Matrix for METHOD

                      M                E              G

    M   53.777777778   56.222222222   6.1111111111
    E   56.222222222   59.703703704   6.8518518519
    G    6.1111111111   6.8518518519  0.9259259259

   Characteristic Roots and Vectors of: E Inverse * H, where
             H = Type III SSCP Matrix for METHOD
                    E = Error SSCP Matrix

          Characteristic      Characteristic Vector V'EV=1
         Root  Percent               M              E              G


   1.57625203    99.45    0.09580573     0.07041265    0.01307714
   0.00874427     0.55   -0.07311528     0.06149668    0.07102689
   0.00000000     0.00    0.03624071    -0.04429420    0.08858841

MANOVA Test Criteria and F Approximations for the Hypothesis of No Overall
                             METHOD Effect
                 H = Type III SSCP Matrix for METHOD
                        E = Error SSCP Matrix

                           S=2 M=0 N=20.5

    Statistic                     Value    F Value  Num DF    Den DF    Pr > F


    Wilks' Lambda            0.38479602      8.77        6        86    <.0001
    Pillai's Trace           0.62050769      6.60        6        88    <.0001
    Hotelling-Lawley Trace   1.58499630     11.24        6    55.591    <.0001
    Roy's Greatest Root      1.57625203     23.12        3        44    <.0001

         NOTE: F Statistic for Roy's Greatest Root is an upper bound.
                NOTE: F Statistic for Wilks' Lambda is exact.

                 H = Type III SSCP Matrix for SOCIAL

                     M                E                G

   M      41.333333333   12.444444444    13.222222222
   E      12.444444444   21.814814815     6.9074074074
   G      13.222222222    6.9074074074    4.7037037037

                     The GLM Procedure
             Multivariate Analysis of Variance
```

```
  Characteristic Roots and Vectors of: E Inverse * H, where
             H = Type III SSCP Matrix for SOCIAL
                    E = Error SSCP Matrix

   Characteristic          Characteristic Vector V'EV=1
             Root    Percent             M            E            G

       0.78879679      81.73    0.11197950   0.04214427   0.03098169
       0.17633633      18.27   -0.04940830   0.09284196   0.02098430
       0.00000000       0.00   -0.02928110  -0.01749261   0.10799783

MANOVA Test Criteria and F Approximations for the Hypothesis of No Overall
                              SOCIAL Effect
                H = Type III SSCP Matrix for SOCIAL
                       E = Error SSCP Matrix

                          S=2  M=0  N=20.5

  Statistic                   Value     F Value    Num DF   Den DF    Pr > F

 Wilks' Lambda           0.47523398        6.46         6       86    <.0001
 Pillai's Trace          0.59086800        6.15         6       88    <.0001
 Hotelling-Lawley Trace  0.96513312        6.84         6   55.591    <.0001
 Roy's Greatest Root     0.78879679       11.57         3       44    <.0001

       NOTE: F Statistic for Roy's Greatest Root is an upper bound.
              NOTE: F Statistic for Wilks' Lambda is exact.
                H = Type III SSCP Matrix for METHOD*SOCIAL

                      M               E                G

        M  16.888888889     0.3333333333       -6.5
      E   0.3333333333   21.62962963       -12.40740741
      G  -6.5           -12.40740741        15.518518519

         Characteristic Roots and Vectors of: E Inverse * H, where
             H = Type III SSCP Matrix for METHOD*SOCIAL
                        E = Error SSCP Matrix

        Characteristic      Characteristic Vector V'EV=1
                  Root     Percent             M            E            G

            0.39394396       58.13   0.08118764   0.04755617  -0.06968258
            0.23114105       34.10   0.08660986  -0.06730281   0.01369593
            0.05265609        7.77   0.04177644   0.06253401   0.08955722

                          The GLM Procedure
                    Multivariate Analysis of Variance

MANOVA Test Criteria and F Approximations for the Hypothesis of No Overall
                           METHOD*SOCIAL Effect
                H = Type III SSCP Matrix for METHOD*SOCIAL
                         E = Error SSCP Matrix

                           S=3 M=0 N=20.5

Statistic                  Value   F Value   Num DF    Den DF      Pr > F

Wilks' Lambda         0.55355448      2.38       12    114.06      0.0089
Pillai's Trace        0.52037855      2.36       12    135         0.0087
```

```
Statistic                       Value    F Value    Num DF    Den DF      Pr > F

Hotelling-Lawley Trace 0.67774110      2.38        12    71.053      0.0120
Roy's Greatest Root    0.39394396      4.43         4        45      0.0042

          NOTE: F Statistic for Roy's Greatest Root is an upper bound.
```

12.5 Two-Stage Multivariate Nested or Hierarchical Design

In Chapter 10, we discussed the univariate case of the nested or hierarchical design. The results of the least squares estimations in the univariate case can easily be generalized to give the multivariate results. For the multivariate two-stage design, instead of making a single scalar observation for each individual in the experiment, a vector of observations (reflecting different responses being studied) is obtained. If the number of response variables is p, then a p-vector is obtained for each individual in the multivariate design; N individuals are studied under a single level of $\boldsymbol{B}$, which is nested in $\boldsymbol{A}$. Suppose that there are a levels of $\boldsymbol{A}$, and $b_1 + b_2 + \cdots + b_a$ levels of $\boldsymbol{B}$, with b_1 levels of $\boldsymbol{B}$ nested within first level of $\boldsymbol{A}$, b_2 levels of $\boldsymbol{B}$ nested within second level of $\boldsymbol{A}$, ... and b_a levels of B nested within the ath level of $\boldsymbol{A}$, then the layout of the responses of such a multivariate two-stage nested design could be presented as shown in Table 12.6.

TABLE 12.6: Typical Layout for a Nested or Hierarchical Design

Factor $\boldsymbol{A}$							
$\boldsymbol{A}_1$			...	$\boldsymbol{A}_a$			
B_1	B_2 ...	B_{b_1}	...	B_1	B_2	...	B_{b_a}
y_{111}	y_{121}	y_{1b_11}		y_{a11}	$y_{a21...}$	...	y_{ab_a1}
y_{112}	y_{122}	y_{1b_12}		y_{a12}	$yy_{a22...}$	...	y_{ab_a2}
	⋮		⋮		⋮		
$y_{11N_{11}}$	$y_{12N_{12}}$	$y_{1b_1N_{b_1}}$	...	y_{a11}	$y_{a21...}$	...	y_{ab_a1}
$y_{11.}$	$y_{12.}$ ...	y_{1b_1}		$y_{a1.}$	$y_{a2.}$	...	y_{ab_1}

We generalize the results of the univariate model to the multivariate case by replacing the scalar observations y_{ijk} by the p-vector y_{ijk} in every formula. The formulas become

$$\overline{y}_{ij.} = \frac{\sum_{k=1}^{N_{ij}} y_{ijk}}{N_{ij}}$$

$$\overline{y}_{i..} = \frac{\sum_{j=1}^{b_i}\sum_{k=1}^{N_{ij}} y_{ijk}}{N_i} = \frac{\sum_{j=1}^{b_i} N_{ij}y_{ij.}}{N_i}$$

$$\overline{y}_{...} = \frac{\sum_{i=1}^{a}\sum_{j=1}^{b_i}\sum_{k=1}^{N_{ij}} y_{ijk}}{N} = \frac{\sum_{i=1}^{a} N_i y_{i..}}{N}$$

$$N_i = \sum_{j=1}^{b_i} N_{ij} \quad N = \sum_{i=1}^{a} N_i = \sum_{i=1}^{a}\sum_{j=1}^{b_i} N_{ij}$$

The general linear model for each vector observation is

$$y_{ijk} = \mu + \tau_i + \beta_{j(i)} + \varepsilon_{ijk} \quad i = 1, 2, \ldots, a \;\; j = 1, 2, \ldots, b_i \;\; k = 1, 2, \ldots, N_{ij}$$

$$\varepsilon_{ijk} \sim IN_p\left(0, \Sigma\right)$$

We can extend the restrictions imposed on the univariate to the multivariate case, so that

$$\sum_{i=1}^{a} N_i \tau_i = 0 \quad \sum_{j=1}^{b_i} N_{ij}\beta_{j(i)} = 0$$

Based on these restrictions using least squares analysis, it is possible to estimate the parameters of the model. The procedure though complicated, is straight forward. Based on these restrictions and other model assumptions stated above, the following hypotheses can be tested:

$H_{B(A)}$: $\beta_{1(i)} = \beta_{2(i)} = \cdots = \beta_{b_i(i)}$ vs $H'_{B(A)}$: at least one $\beta_{j(i)}$ is different

H_A: $\tau_1 = \tau_2 = \cdots = \tau_a$ vs H'_A: at least one τ_i is different

To test these hypotheses, we partition the sum of squares and cross products into the following:

$$\text{Total} = \boldsymbol{T} = \sum_{i=1}^{a}\sum_{j=1}^{b_i}\sum_{k=1}^{N_{ij}} (y_{ijk} - \overline{y}_{...})(y_{ijk} - \overline{y}_{...})'$$

With the total degrees of freedom,

$$N - 1 = \sum_{i=1}^{a}\sum_{j=1}^{b_i} N_{ij} - 1$$

For factor $\boldsymbol{A}$,

$$\boldsymbol{A} = \sum_{i=1}^{a} N_i(\overline{y}_{i..} - \overline{y}_{...})(\overline{y}_{i..} - \overline{y}_{...})'$$

$$\boldsymbol{A} = \sum_{i=1}^{a} N_i(\overline{y}_{i..} - \overline{y}_{...})(\overline{y}_{i..} - \overline{y}_{...})' \tag{12.49}$$

with $a - 1$ degrees of freedom

For factor $\boldsymbol{B}$ nested in $\boldsymbol{A}$,

$$\boldsymbol{B}(\boldsymbol{A}) = \sum_{i=1}^{a}\sum_{j=1}^{b_i} N_{ij}(\overline{y}_{ij.} - \overline{y}_{...})(\overline{y}_{i..} - \overline{y}_{...})'$$

$$\text{with} \sum_{i=1}^{a} (b_i - 1) \text{ degrees of freedom} \tag{12.50}$$

For error $\boldsymbol{W}$,

$$\boldsymbol{W} = \sum_{i=1}^{a}\sum_{j=1}^{b_i}\sum_{k=1}^{N_{ij}} (y_{ijk} - \overline{y}_{ij.})(\overline{y}_{i..} - \overline{y}_{ij.})'$$

$$\text{with} \sum_{i=1}^{a}\sum_{j=1}^{b_i} (N_{ij} - 1) \text{ degrees of freedom} \tag{12.51}$$

To test the hypotheses above, we use $\boldsymbol{A}$, $\boldsymbol{B}(\boldsymbol{A})$, and $\boldsymbol{W}$. First, we find the solutions of the determinants

$$|\boldsymbol{A}\boldsymbol{W}^{-1} - \boldsymbol{\lambda}\boldsymbol{I}| = 0 \tag{12.52a}$$

$$|\boldsymbol{B}(\boldsymbol{A}) - \boldsymbol{\lambda}\boldsymbol{W}| = 0 \tag{12.52b}$$

and use the solutions to obtain the test statistics.

As usual, to test the hypotheses, we need the parameters, s, m, and n. These parameters are different for the two hypotheses to be tested and the formulas for obtaining them are presented in Table 12.7.

TABLE 12.7: Parameters for Testing Hypothesis in Nested Designs

For H_A	For $H_{B(A)}$
$s = \min(a - 1, p)$	$s = \min(v_{B(A)}, p)$
$m = \dfrac{\lvert a - p - 1\rvert - 1}{2}$	$m = \dfrac{\lvert v_{B(A)} - p\rvert - 1}{2}$
$n = \dfrac{\lvert v_W - p - 1\rvert}{2}$	$n = \dfrac{\lvert v_W - p - 1\rvert}{2}$

As usual, the hypothesis $H_{\boldsymbol{A}}$ is rejected if

(i) *Wilks' criterion*

$$\Lambda = \frac{|\boldsymbol{W}|}{|\boldsymbol{W}+\boldsymbol{A}|} = \prod_{i=1}^{s}(1+\lambda_i)^{-1} < U^{\alpha}(p, a-1, v_{\boldsymbol{w}})$$

(ii) *Roy's criterion*

$$\theta_s = \frac{1}{1+\lambda_1} > \theta^{\alpha}(s, m, n)$$

(iii) *Lawley–Hotelling's criterion*

$$U^{(s)} = \sum_{i=1}^{s}\lambda_i > U_0^{\alpha}(s, m, n)$$

(iv) *Pillai's criterion*

$$V^{(s)} = \sum_{i=1}^{s}\left(\frac{\lambda_i}{1+\lambda_i}\right) > V^{\alpha}(s, m, n)$$

where λ_i is the solutions of $|\boldsymbol{A} - \boldsymbol{\lambda}\boldsymbol{W}| = 0$; that is, Equation 12.52a.

The hypothesis, $H_{B(A)}$ is rejected if

(i) *Wilks' criterion*

$$\Lambda = \frac{|\boldsymbol{W}|}{|\boldsymbol{W}+\boldsymbol{B}(\boldsymbol{A})|} = \prod_{i=1}^{s}(1+\lambda_i)^{-1} < U^{\alpha}(p, v_{B(A)}, v_w)$$

(ii) *Roy's criterion*

$$\theta_s = \frac{1}{1+\lambda_1} > \theta^{\alpha}(s, m, n)$$

(iii) *Lawley–Hotelling's criterion*

$$U^{(s)} = \sum_{i=1}^{s}\lambda_i > U_0^{\alpha}(s, m, n)$$

(iv) *Pillai's criterion*

$$V^{(s)} = \sum_{i=1}^{s}\left(\frac{\lambda_i}{1+\lambda_i}\right) > V^{\alpha}(s, m, n)$$

where λ_i is the solutions of $|\boldsymbol{B}(\boldsymbol{A}) - \boldsymbol{\lambda}\boldsymbol{W}| = 0$; that is, Equation 12.52b.

Example 12.4

In a study similar to that discussed in Section 12.4, the method of teaching was nested within socioeconomic background of the pupils. Tests in English (T_1) and Mathematics (T_2) only were administered to the pupils. The responses of the experiments are presented below as follows:

	Social background											
	HES (A_1)				**MES (A_2)**				**LES (A_3)**			
	M_1		M_2		M_3		M_4		M_5		M_6	
	T_1	T_2	T_1	T_2	T_1	T_2	T_1	T_2	T_1	T_2	T_1	T_2
	11	12	12	14	11	9	11	10	11	12	9	13
	13	13	13	18	12	12	10	10	12	13	8	9
	10	10	14	15	10	12	9	12	10	13	10	8
	11	12	14	17	10	11	10	12	9	14	9	10
	13	14	15	15	12	12	11	11	9	13	8	10
	13	10	14	12	12	12	12	9	11	12	10	11
Total	71	71	82	91	67	68	63	64	62	77	54	61

We illustrate the analysis of responses of the multivariate two-stage nested design using these responses; we test the appropriate hypothesis at 1% and 5% levels of significance.

Analysis:

First, we obtain the total sum of squares and cross products for all the responses as follows:

$$\sum T_1 = 11 + 13 + \cdots + 8 + 10 = 399$$

$$\sum T_2 = 12 + 13 + \cdots + 10 + 11 = 432$$

$$\sum T_1^2 = (11^2 + 13^2 + \cdots + 8^2 + 10^2) - \frac{3992}{36} = 110.75$$

$$\sum T_2^2 = (12^2 + 13^2 + \cdots + 10^2 + 11^2) - \frac{432^2}{36} = 166.00$$

$$\sum T_1 T_2 = (11 \times 12 + 13 \times 13 + \cdots + 8 \times 10 + 10 \times 11) - \frac{399 \times 432}{36}$$
$$= 70.00$$

We obtain the sum of squares and cross products for the social backgrounds (A) as:

$$A_{11} = (130^2 + 116^2 + 153^2)/12 - 399^2/36 \quad \text{i.e., } 4480 - 399^2/36 = 58.1667$$

$$A_{22} = (162^2 + 132^2 + 138^2)/12 - 432^2/36 = 5226.0 - 432^2/36 = 42.00$$

$$A_{12} = A_{21} = (130 \times 132 + 116 \times 138 + 162 \times 153)/12 - 432 \times 399/36$$
$$= 4829.5 - 399 \times 432/36 = 41.50$$

Finally, we obtain the sum of squares and cross products for the methods nested within social backgrounds $\boldsymbol{B}(\boldsymbol{A})$ as:

$$B(A)_{11} = \frac{71^2 + 82^2 + 67^2 + 63^2 + 62^2 + 54^2}{6} - \frac{130^2 + 116^2 + 153^2}{12}$$
$$= 4497.1666 - 4480.4166 = 16.75$$

$$B(A)_{22} = \frac{71^2 + 91^2 + \cdots + 77^2 + 61^2}{6} - \frac{162^2 + 132^2 + 138^2}{12}$$
$$= 5282 - 5226 = 56.0$$

$$B(A)_{12} = B(A)_{21} = \frac{71 \times 71 + 82 \times 91 + \cdots + 62 \times 77 + 54 \times 61}{6}$$
$$- \frac{130 \times 132 + 116 \times 138 + 162 \times 153}{12}$$
$$= 4859.8333 - 4829.5 = 30.3333$$

The matrices of sum of squares and cross products are

$$\boldsymbol{T} = \begin{bmatrix} 110.75 & 70.00 \\ 70.00 & 166.00 \end{bmatrix} \qquad \boldsymbol{A} = \begin{bmatrix} 58.1667 & 41.50 \\ 41.50 & 42.0 \end{bmatrix}$$

$$\boldsymbol{B}(\boldsymbol{A}) = \begin{bmatrix} 16.75 & 30.3333 \\ 30.3333 & 56.000 \end{bmatrix} \qquad \boldsymbol{W} = \begin{bmatrix} 35.8333 & -1.833 \\ -1.833 & 68.000 \end{bmatrix}$$

The matrix of sum of squares and cross products for error can be obtained by subtraction as follows:

$$\boldsymbol{W} = \boldsymbol{T} - \boldsymbol{A} - \boldsymbol{B}(\boldsymbol{A})$$

To test the hypotheses

$$H_{B(A)}\colon \beta_{1(i)} = \beta_{2(i)} \quad \text{vs} \quad H'_{B(A)}\colon \beta_{1(i)} \neq \beta_{2(i)} \text{ is different}$$

$$H_A\colon \tau_1 = \tau_2 = \tau_3 \quad \text{vs} \quad H'_A\colon \text{at least one } \tau_i \text{ is different } (i = 1, 2, 3)$$

Using Wilks' criterion for both hypotheses

$$\Lambda_{B(A)} = \frac{|\boldsymbol{W}|}{|\boldsymbol{W} + \boldsymbol{B}(\boldsymbol{A})|} = \frac{2433.3034}{5708.0792} = 0.4262914$$

$$\Lambda_A = \frac{|\boldsymbol{W}|}{|\boldsymbol{W} + \boldsymbol{A}|} = \frac{2433.3034}{8766.5529} = 0.277566$$

With $p = 2$; $v_e = 30$; $v_A = 2$; $v_{B(A)} = 3$, then we obtain the tabulated values for the Wilks' test for the test of the hypotheses, H_A and $H_{B(A)}$ because $U^{0.05}(2,\ 2,\ 27) = 0.724899$ and $U^{0.05}(2,\ 3,\ 30) = 0.656962$. We reject the null hypotheses in both cases at 5% level of significance because $\Lambda_{B(A)} < U^{0.05}(2,\ 2,\ 30)$ and $\Lambda_A < U^{0.05}(2,\ 3,\ 30)$.

To use other test criteria to these hypotheses, we solve the following characteristic equations:

$$|\boldsymbol{A} - \boldsymbol{\lambda} \boldsymbol{W}| = 0 \qquad |\boldsymbol{B}(\boldsymbol{A}) - \boldsymbol{\lambda} \boldsymbol{W}| = 0 \tag{12.53}$$

The characteristic roots for the first equation are 2.1700 and 0.13649 while the roots of the second equation are 1.3329 and 0.00551.

Under the Roy's criterion:

For $\boldsymbol{A}$: $\theta_s = l/(1+\lambda) = 2.1700/3.1700 = 0.6845$. $s=2$; $m=-1/2$; $n=12$. From the tables (using the Heck's chart), $\theta^{0.05}$ $(2, -1/2, 12) = 0.1868$. So we reject the null hypothesis $H_{0(A)}$ and accept the alternative $H_{\boldsymbol{A}}$.

For $\boldsymbol{B}(\boldsymbol{A})$: $\theta_s = 1.3329/2.3329 = 0.5713$. $s=2$; $m=0$; $n=12$ so that from the tables $\theta^{0.05}(2,\ 0,\ 12) = 0.2329$ so we again reject the null hypothesis, $H_{\mathbf{0}\boldsymbol{B}(\boldsymbol{A})}$ and accept the alternative $H_{\boldsymbol{B}(\boldsymbol{A})}$.

As we did in previous sections (Sections 12.2.1 and 12.2.2), we can test the hypotheses using both the Bartlett's χ^2 and Rao's approximate F tests. Applying these to the nested design, using Equations 12.20 and 12.21, we obtain the following: $p=2$; $v_e=30$; $v_A=2$; $\Lambda=0.277566$; $s_1=2$; $\Lambda^{1/s_1}=0.5268$; $n_1=4$; $n_2=58$

For $\boldsymbol{A}$:

$$\text{Bartlett's: } \chi_f^2 = -\left[30 + 2 - \frac{(2+2+1)}{2}\right] \ln(0.277566) = 37.8100$$

From the tables: $\chi^2(0.01, 4) = 13.28$

$$\text{Rao's } F\text{: } F_c = \left[\frac{(1-0.5268)}{0.5268}\right]\left(\frac{58}{4}\right) = 13.0247$$

From the tables: $F(0.01, 4, 58) = 3.6680$

Both tests confirm that the $\boldsymbol{A}$ effect is significant at 1%.

For $\boldsymbol{B}(\boldsymbol{A})$:

$$p=2 \quad v_e=30 \quad v_{\boldsymbol{B}(\boldsymbol{A})}=3 \quad \Lambda=0.4262914 \quad n_1=6 \quad s_1=2$$
$$n_2=58 \quad \Lambda^{1/s_1}=0.6529$$

$$\text{Bartlett's: } \boldsymbol{\chi}_f^2 = -\left[30 + 3 - \frac{(2+3+1)}{2}\right] \ln(0.4262914) = 25.5790$$

From the tables: $\chi^2_{(0.01.6)} = 16.81$

$$\text{Rao's } F\text{: } \quad F_c = \left[\frac{(1-0.6529)}{0.6529}\right]\left(\frac{58}{6}\right) = 5.14$$

From the tables: $F(0.01, 6, 58) = 3.1370$.

Once again both tests confirm that $\boldsymbol{B}(\boldsymbol{A})$ effect is significant at 1%. □

12.5.1 SAS Program for Example 12.4

Next, we present the SAS program that would analyze the responses of Example 12.4.

SAS PROGRAM

```
data examp12_4;
SOCIAL='HES';
DO K=1 TO 6;
DO METHOD='M1', 'M2';
input Y1-Y2 @@;
output;
end;END;
SOCIAL='MES';
DO K=1 TO 6;
DO METHOD='M3', 'M4';
input Y1-Y2@@ ;
output;
end;END;
SOCIAL='LES';
DO K=1 TO 6;
DO METHOD='M5', 'M6';
input Y1-Y2@@;
output;
end; END;
datalines;
11     12     12     14
 13 13 13    18
 10    10     14  15
 11    12     14     17
 13    14     15  15
 13 10 14 12
 11    9 11  10
 12    12 10 10
10  12  9 12
 10 11       10     12
12     12  11       11
12     12     12     9
11     12  9  13
12     13  8  9
10     13     10  8
9      14     9 10
9      13  8 10
11     12     10     11
;
proc print data=examp12_4;
proc glm data=examp12_4;
class METHOD SOCIAL ;
model Y1-Y2=SOCIAL METHOD(SOCIAL);
manova h=_all_ /printe printh;run;
```

SAS OUTPUT

Obs	SOCIAL	K	METHOD	Y1	Y2
1	HES	1	M1	11	12
2	HES	1	M2	12	14
3	HES	2	M1	13	13
4	HES	2	M2	13	18

Obs	SOCIAL	K	METHOD	Y1	Y2
5	HES	3	M1	10	10
6	HES	3	M2	14	15
7	HES	4	M1	11	12
8	HES	4	M2	14	17
9	HES	5	M1	13	14
10	HES	5	M2	15	15
11	HES	6	M1	13	10
12	HES	6	M2	14	12
13	MES	1	M3	11	9
14	MES	1	M4	11	10
15	MES	2	M3	12	12
16	MES	2	M4	10	10
17	MES	3	M3	10	12
18	MES	3	M4	9	12
19	MES	4	M3	10	11
20	MES	4	M4	10	12
21	MES	5	M3	12	12
22	MES	5	M4	11	11
23	MES	6	M3	12	12
24	MES	6	M4	12	9
25	LES	1	M5	11	12
26	LES	1	M6	9	13
27	LES	2	M5	12	13
28	LES	2	M6	8	9
29	LES	3	M5	10	13
30	LES	3	M6	10	8
31	LES	4	M5	9	14
32	LES	4	M6	9	10
33	LES	5	M5	9	13
34	LES	5	M6	8	10
35	LES	6	M5	11	12
36	LES	6	M6	10	11

The GLM Procedure

Class Level Information

Class	Levels	Values
METHOD	6	M1 M2 M3 M4 M5 M6
SOCIAL	3	HES LES MES

Number of Observations Read 36
Number of Observations Used 36

The GLM Procedure

Dependent Variable: Y1

Source	DF	Sum of Squares	Mean Square	F Value	Pr > F
Model	5	74.9166667	14.9833333	12.54	<.0001
Error	30	35.8333333	1.1944444		
Corrected Total	35	110.7500000			

R-Square	Coeff Var	Root MSE	Y1 Mean
0.676448	9.860810	1.092906	11.08333

Source	DF	Type I SS	Mean Square	F Value	Pr > F
SOCIAL	2	58.16666667	29.08333333	24.35	<.0001
METHOD(SOCIAL)	3	16.75000000	5.58333333	4.67	0.0085

Source	DF	Type III SS	Mean Square	F Value	Pr > F
SOCIAL	2	58.16666667	29.08333333	24.35	<.0001
METHOD(SOCIAL)	3	16.75000000	5.58333333	4.67	0.0085

The GLM Procedure

Dependent Variable: Y2

Source	DF	Sum of Squares	Mean Square	F Value	Pr > F
Model	5	98.0000000	19.6000000	8.65	<.0001
Error	30	68.0000000	2.2666667		
Corrected Total	35	166.0000000			

R-Square	Coeff Var	Root MSE	Y2 Mean
0.590361	12.54621	1.505545	12.00000

Source	DF	Type I SS	Mean Square	F Value	Pr > F
SOCIAL	2	42.00000000	21.00000000	9.26	0.0007
METHOD(SOCIAL)	3	56.00000000	18.66666667	8.24	0.0004

Source	DF	Type III SS	Mean Square	F Value	Pr > F
SOCIAL	2	42.00000000	21.00000000	9.26	0.0007
METHOD(SOCIAL)	3	56.00000000	18.66666667	8.24	0.0004

The GLM Procedure
Multivariate Analysis of Variance

E = Error SSCP Matrix

	Y1	Y2
Y1	35.833333333	-1.833333333
Y2	-1.833333333	68

Partial Correlation Coefficients from the Error SSCP Matrix / Prob > |r|

```
DF = 30            Y1            Y2
Y1           1.000000     -0.037140
               0.8428
Y2          -0.037140      1.000000
               0.8428

              The GLM Procedure
          Multivariate Analysis of Variance

        H = Type III SSCP Matrix for SOCIAL

                          Y1          Y2

        Y1      58.166666667       41.5
        Y2      41.5               42

Characteristic Roots and Vectors of: E Inverse * H, where
          H = Type III SSCP Matrix for SOCIAL
                     E = Error SSCP Matrix
Characteristic          Characteristic  Vector V'EV=1
          Root    Percent              Y1              Y2

    2.17003688      94.08      0.14605960      0.06292534
    0.13649630       5.92     -0.08131493      0.10376221

MANOVA Test Criteria and F Approximations for the Hypothesis of No Overall
                              SOCIAL Effect
           H = Type III SSCP Matrix for SOCIAL
                       E = Error SSCP Matrix

                       S=2    M=-0.5   N=13.5

Statistic                    Value  F Value    Num DF    Den DF    Pr > F


Wilks' Lambda           0.27756689    13.02         4        58    <.0001
Pillai's Trace          0.80464898    10.10         4        60    <.0001
Hotelling-Lawley Trace  2.30653318   16.55          4    33.787    <.0001
Roy's Greatest Root     2.17003688    32.55         2    30        <.0001

      NOTE: F Statistic for Roy's Greatest Root is an upper bound.
           NOTE: F Statistic for Wilks' Lambda is exact.
            H = Type III SSCP Matrix for METHOD(SOCIAL)

                         Y1                      Y2

             Y1      16.75              30.333333333
             Y2     30.333333333      56
                        The GLM Procedure
                 Multivariate Analysis of Variance

   Characteristic Roots and Vectors of: E Inverse * H, where
        H = Type III SSCP Matrix for METHOD(SOCIAL)
                  E = Error SSCP Matrix
```

Characteristic Root	Percent	Characteristic Vector V'EV=1 Y1	Y2
1.33294741	99.59	0.10355883	0.09798826
0.00551536	0.41	-0.13122916	0.07158558

MANOVA Test Criteria and F Approximations for the Hypothesis of No Overall METHOD(SOCIAL) Effect
H = Type III SSCP Matrix for METHOD(SOCIAL)
E = Error SSCP Matrix

S=2 M=0 N=13.5

Statistic	Value	F Value	Num DF	Den DF	Pr > F
Wilks' Lambda	0.42629118	5.14	6	58	0.0003
Pillai's Trace	0.57684278	4.05	6	60	0.0018
Hotelling-Lawley Trace	1.33846277	6.37	6	36.941	0.0001
Roy's Greatest Root	1.33294741	13.33	3	30	<.0001

NOTE: F Statistic for Roy's Greatest Root is an upper bound.
NOTE: F Statistic for Wilks' Lambda is exact.

12.6 Multivariate Latin Square Design

In Chapter 5, we discussed the Latin square and related designs such as the Graeco-Latin squares. The Latin square design is a very useful way of eliminating extraneous error by treating their sources as factors, thereby isolating the factor(s) of interest for proper study. The extraneous factors are treated as blocks in which interest is in eliminating their effects on the conclusions of the experiment. Sometimes, especially in the behavioral sciences, it is used as a building block for very complex designs.

The model for the univariate Latin square design is

$$y_{ijk} = \mu + \gamma_i + \beta_j + \tau_k + \varepsilon_{ijk} \tag{12.54}$$

$$\varepsilon_{ijk} \sim N(0, \sigma^2) \quad i, j, k = 1, 2, \ldots, r$$

where y_{ijk} represents the observation in the ith row, jth column, which receives the treatment k. It is assumed for this model that no interaction occurs between factors.

The multivariate model is simply an extension of this univariate design on the factors of primary interest. In the multivariate case, however, the entire r observations are vector-valued; each contains p observations (representing p different responses being studied in the experiment). The least squares linear model for this design is

$$y_{ijk} = \mu + \gamma_i + \beta_j + \tau_k + \varepsilon_{ijk} \tag{12.55}$$

where $\varepsilon_{ijk} \sim IN_p(0, \Sigma)$; that is, ε_{ijk} is assumed to have a multivariate normal distribution with zero centroid and dispersion matrix, Σ. Since the basic theories underlying the univariate case apply, we shall not discuss them further. We rather concentrate on the types of hypotheses that may be of interest under the multivariate design. We may be interested in testing the following hypotheses:

$$H_\beta : \beta_1 = \beta_2 = \cdots = \beta_r \quad \text{vs} \quad H'_\beta\text{: at least one } \beta_i \text{ is different}$$
$$H_\tau : \tau_1 = \tau_2 = \cdots = \tau_r \quad \text{vs} \quad H'_\tau\text{: at least one } \tau_j \text{ is different}$$
$$H_\gamma : \gamma_1 = \gamma_2 = \cdots = \gamma_r \quad \text{vs} \quad H'_\beta\text{: at least one } \gamma_k \text{ is different}$$

By simple analogy with the univariate case, we present the partitions of the sum of squares and cross products of the responses as follows:

$$\boldsymbol{R}(\text{rows}) = r\sum_{i=1}^{r} (\overline{\boldsymbol{y}}_{i..} - \overline{\boldsymbol{y}}_{...})(\overline{\boldsymbol{y}}_{i..} - \overline{\boldsymbol{y}}_{...})' \tag{12.56}$$

$$\boldsymbol{C}(\text{columns}) = r\sum_{j=1}^{r} (\overline{\boldsymbol{y}}_{.j.} - \overline{\boldsymbol{y}}_{...})(\overline{\boldsymbol{y}}_{.j.} - \overline{\boldsymbol{y}}_{...})' \tag{12.57}$$

$$\boldsymbol{L}(\text{Latin or Treatment}) = r\sum_{k=1}^{r} (\overline{\boldsymbol{y}}_{..k} - \overline{\boldsymbol{y}}_{...})(\overline{\boldsymbol{y}}_{..k} - \overline{\boldsymbol{y}}_{...})' \tag{12.58}$$

$$\begin{aligned}\boldsymbol{W}(\text{error}) = \sum_{i=1}^{r}\sum_{j=1}^{r}\sum_{k-1}^{r} &(\boldsymbol{y}_{ijk} - \overline{\boldsymbol{y}}_{i..} - \overline{\boldsymbol{y}}_{.j.} - \overline{\boldsymbol{y}}_{..k} + 2\overline{\boldsymbol{y}}_{...}) \\ &\times (\boldsymbol{y}_{ijk} - \overline{\boldsymbol{y}}_{i..} - \overline{\boldsymbol{y}}_{.j.} - \overline{\boldsymbol{y}}_{..k} + 2\overline{\boldsymbol{y}}_{...})'\end{aligned} \tag{12.59}$$

$$\boldsymbol{T}(\text{total}) = \sum_{i=1}^{r}\sum_{j=1}^{r}\sum_{k=1}^{r} (\boldsymbol{y}_{ijk} - \overline{\boldsymbol{y}}_{...})(\boldsymbol{y}_{ijk} - \overline{\boldsymbol{y}}_{...})'$$

$\boldsymbol{W}$ can also be obtained by subtraction from $\boldsymbol{T}$ as

$$\boldsymbol{W} = \boldsymbol{T} - \boldsymbol{R} - \boldsymbol{C} - \boldsymbol{L} \tag{12.60}$$

$\boldsymbol{R}$, $\boldsymbol{C}$, $\boldsymbol{L}$, each has $r - 1$ degrees of freedom, while $\boldsymbol{T}$ and $\boldsymbol{W}$ have $r^2 - 1$ and $(r-1)(r-1)$ degrees of freedom, respectively.

Example 12.5

In a study of mathematics, five pupils were chosen from each of five different grades (columns); they were taught mathematics using five different methods (A, B, C, D and E) at five different hours of the day (rows) and subjected to two tests (mental [E_1] and written [E_2]), 20 min after the session. The design most suited for this investigation was a 5×5 multivariate Latin square design and the responses from the experiments are as follows:

E_1,E_2	E_1,E_2	E_1,E_2	E_1,E_2	E_1,E_2
C 9,16	E 19,9	D 6,4	B 9,5	A 7,13
B 6,8	C 2,7	E 19,10	A 7,20	D 10,24
E 15,9	B 5,6	A 8,9	D 6,12	C 5,18
A 14,18	D 2,6	C 5,14	E 12,10	B 8,9
D 10,14	A 10,15	B 4,8	E 17,24	C 3,11

Analyze the responses and test for significance of the factors in the design.

Solution:
First, we calculate the response totals for row, column, and treatments as follows:

	Row totals		**Column totals**		**Treatment totals**	
	$y_{1..} = 50$	47	$y_{.1} = 54$	65	$y_{..1} = 46$	75
	$y_{2..} = 44$	69	$y_{.2.} = 38$	43	$y_{..2} = 32$	36
	$y_{3..} = 39$	54	$y_{.3.} = 42$	45	$y_{..3} = 24$	66
	$y_{4..} = 41$	57	$y_{.4.} = 37$	58	$y_{..4} = 34$	60
	$y_{5..} = 44$	72	$y_{.5.} = 47$	88	$y_{..5} = 82$	62
Total	218	299	218	299	218	299

Then, we calculate the total sum of squares and cross products for all the responses, T as follows:

$$\text{For } E_1\text{: } (9^2 + 19^2 + \cdots + 3^3 + 17^2) - \frac{218^2}{25} = 579.04$$

$$\text{For } E_2\text{: } (16^2 + 9^2 + \cdots + 11^2 + 24^2) - \frac{299^2}{25} = 744.96$$

$$\text{For } E_1E_2\text{: } (9 \times 16 + 19 \times 9 + \cdots + 3 \times 11 + 17 \times 24) - \frac{(218 \times 299)}{25} = 187.72$$

We calculate the sum of squares and cross products for the rows, R as follows:

$$\text{For } E_1\text{: } \frac{(50^2 + 44^2 + 39^2 + 41^2 + 44^2)}{5} - \frac{218^2}{25} = 13.84$$

$$\text{For } E_2\text{: } \frac{(47^2 + 69^2 + 54^2 + 57^2 + 72^2)}{5} - \frac{299^2}{25} = 87.76$$

$$\text{For } E_1E_2\text{: } \frac{(50 \times 47 + 44 \times 69 + 39 \times 54 + 41 \times 57 + 44 \times 72)}{5} - \frac{(218 \times 299)}{25} = -7.88$$

We obtain the sum of squares and cross products for the Columns, C as follows:

$$\text{For } E_1\text{: } \frac{(54^2 + 38^2 + 42^2 + 37^2 + 47^2)}{5} - \frac{218^2}{25} = 39.44$$

$$\text{For } E_2\text{: } \frac{(65^2 + 43^2 + 45^2 + 58^2 + 88^2)}{5} - \frac{299^2}{25} = 265.36$$

$$\text{For } E_1E_2\text{: } \frac{(54 \times 65 + 38 \times 43 + 42 \times 45 + 37 \times 58 + 47 \times 88)}{5}$$

$$- \frac{(218 \times 299)}{25} = 55.92$$

We calculate the sum of squares and cross products for the treatments (represented by A, B, C, D, and E), L:

$$\text{For } E_1\text{: } \frac{(46^2 + 32^2 + 24^2 + 34^2 + 82^2)}{5} - \frac{218}{25} = 418.24$$

$$\text{For } E_2\text{: } \frac{(75^2 + 36^2 + 66^2 + 60^2 + 62^2)}{5} - \frac{299}{25} = 168.16$$

$$\text{For } E_1E_2\text{: } \frac{(46 \times 75 + 32 \times 36 + 24 \times 66 + 34 \times 60 + 82 \times 62)}{5}$$

$$- \frac{(218 \times 299)}{25} = 54.72$$

The sum of squares and cross products for the residual or error, $\boldsymbol{W}$ is obtained by subtraction so that the matrices of sums of squares and cross products are as follows:

$$\boldsymbol{T} = \begin{bmatrix} 579.04 & 187.72 \\ 187.72 & 744.96 \end{bmatrix} \quad \boldsymbol{R} = \begin{bmatrix} 13.84 & -7.88 \\ -7.88 & 87.76 \end{bmatrix} \quad \boldsymbol{C} = \begin{bmatrix} 39.44 & 55.92 \\ 55.92 & 265.36 \end{bmatrix}$$

$$\boldsymbol{L} = \begin{bmatrix} 418.24 & 54.72 \\ 54.72 & 168.16 \end{bmatrix} \quad \boldsymbol{W} = \begin{bmatrix} 107.52 & 84.96 \\ 84.96 & 223.68 \end{bmatrix}$$

We test the following three hypotheses:

$$H_R\text{: } \tau_1 = \tau_2 = \tau_3 = \tau_4 = \tau_5 \quad \text{vs} \quad H_R'\text{: at least one } \tau_i \text{ is different}$$
$$H_C\text{: } \beta_1 = \beta_2 = \beta_3 = \beta_4 = \beta_5 \quad \text{vs} \quad H_C'\text{: at least one } \beta_j \text{ is different}$$
$$H_L\text{: } \gamma_1 = \gamma_2 = \gamma_3 = \gamma_4 = \gamma_5 \quad \text{vs} \quad H_L'\text{: at least one } \gamma_k \text{ is different}$$

To test these hypotheses, we find the solutions of the following characteristic equations:

$$|\boldsymbol{R} - \lambda \boldsymbol{W}| = 0 \qquad (12.61a)$$

$$|\boldsymbol{C} - \lambda \boldsymbol{W}| = 0 \qquad (12.61b)$$

$$|\boldsymbol{L} - \lambda \boldsymbol{W}| = 0 \qquad (12.61c)$$

The roots of Equation 12.61a are used for testing the hypothesis on the vector of row effects (hours) are

$$\lambda_1 = 0.73031 \quad \lambda_2 = 0.093765$$

The roots of the second characteristic equation (Equation 12.61b), which are used for testing the hypothesis on the vector of column effects (grades) are

$$\lambda_1 = 1.3235 \quad \lambda_2 = 0.32885$$

The roots of the third characteristic equation (Equation 12.61c), which are used for testing the hypothesis on vector of treatment effects (Latin) are

$$\lambda_1 = 5.3291 \quad \lambda_2 = 0.75070$$

We use the Wilks' Lambda to test the hypotheses of significant effects as follows:

$$H_R\colon \Lambda = \frac{|\boldsymbol{W}|}{|\boldsymbol{W}+\boldsymbol{R}|} = \frac{\begin{vmatrix} 107.52 & 84.96 \\ 84.96 & 223.68 \end{vmatrix}}{\begin{vmatrix} 121.36 & 77.08 \\ 77.08 & 311.44 \end{vmatrix}} = 0.528394$$

$$H_C\colon \Lambda = \frac{|\boldsymbol{W}|}{|\boldsymbol{W}+\boldsymbol{C}|} = \frac{\begin{vmatrix} 107.52 & 84.96 \\ 84.96 & 223.68 \end{vmatrix}}{\begin{vmatrix} 146.96 & 140.88 \\ 140.88 & 489.04 \end{vmatrix}} = 0.323555$$

$$H_L\colon \Lambda = \frac{|\boldsymbol{W}|}{|\boldsymbol{W}+\boldsymbol{L}|} = \frac{\begin{vmatrix} 107.52 & 84.96 \\ 84.96 & 223.68 \end{vmatrix}}{\begin{vmatrix} 525.76 & 139.68 \\ 139.68 & 391.84 \end{vmatrix}} = 0.090252$$

We use the Bartlett's χ^2 and Rao's F to test the hypotheses regarding the rows, columns, and treatment effects in the above designed experiment. We obtain the quantities required to test the hypotheses as

$$v_{\text{row}} = v_{\text{col}} = v_{\text{treat}} = 4 \quad \Lambda_{\text{row}} = 0.528394 \quad \Lambda_{\text{col}} = 0.323555$$

$$\Lambda_{\text{treat}} = 0.090252 \quad s = 2 \;\; p = 2 \;\; v_{\text{e}} = 12 \;\; n_1 = 8 \;\; n_2 = 22$$

$$\Lambda_{\text{row}}^{1/s} = \sqrt{0.528394} = 0.726907 \quad \Lambda_{\text{methods}}^{1/s} = \sqrt{0.0.090252} = 0.30042$$

$$\Lambda_{\text{columns}}^{1/s} = \sqrt{0.528394} = 0.568819$$

$$\text{Row: } \chi_f^2 = 7.97391 \quad F_c = 1.033151$$

$$\text{Column: } \chi_f^2 = 14.1048 \quad F_c = 2.084576$$

$$\text{Treatment: } \chi_f^2 = 30.0644 \quad F_c = 6.40386$$

From the tables, $\chi^2(0.01, 8) = 20.09$ and $F(0.01, 8, 22) = 3.4500$. From these, we conclude that the day and grades effects are not significant at $\alpha = 1\%$ but the method effect is significant.

12.6.1 SAS Program

Next, we present a SAS program for analyzing the responses of the design in Example 12.5.

SAS PROGRAM

```
options nodate nonumber ls=76 ps=67;
data latin;
input grades day method$ m1 m2 @@;
cards;
1 1 C 9 16 2 1 E 19 9 3 1 D 6 4 4 1 B 9 5 5 1 A 7 13
1 2 B 6 8 2 2 C 2 7 3 2 E 19 10 4 2 A 7 20 5 2 D 10 24 1 3 E 15 9
2 3 B 5 6 3 3 A 8 9 4 3 D 6 12 5 3 C 5 18 1 4 A 14 18 2 4 D 2 6 3 4
C 5 14 4 4 E 12 10 5 4 B 8 9 1 5 D 10 14 2 5 A 10 15 3 5 B 4 8 4 5
C 3 11 5 5 E 17 24
;
PROC GLM DATA=LATIN;
CLASS GRADES DAY METHOD;
MODEL M1 M2=GRADES DAY METHOD;
MANOVA H=_ALL_ /PRINTH PRINTE; RUN;
```

SAS OUTPUT

```
                          The SAS System

                        The GLM Procedure

                     Class Level Information

                    Class      Levels     Values

                   grades           5    1 2 3 4 5

                   day              5    1 2 3 4 5

                   method           5    A B C D E

                  Number of Observations Read     25
                  Number of Observations Used     25

                        The SAS System

                       The GLM Procedure

Dependent Variable: m1
                                    Sum of
 Source                 DF         Squares    Mean Square   F Value   Pr > F

 Model                  12     471.5200000     39.2933333      4.39   0.0080

 Error                  12     107.5200000      8.9600000

 Corrected Total        24     579.0400000
```

```
              R-Square     Coeff Var  Root MSE    m1 Mean

                0.814313    34.32713   2.993326   8.720000

Source             DF      Type I SS   Mean Square   F Value    Pr > F

grades              4     39.4400000     9.8600000      1.10    0.4003
day                 4     13.8400000     3.4600000      0.39    0.8145
method              4    418.2400000   104.5600000     11.67    0.0004

Source             DF    Type III SS   Mean Square   F Value    Pr > F

grades              4     39.4400000     9.8600000      1.10    0.4003
day                 4     13.8400000     3.4600000      0.39    0.8145
method              4    418.2400000   104.5600000     11.67    0.0004

                           The SAS System

                          The GLM Procedure
Dependent Variable: m2

                                Sum of
Source             DF          Squares   Mean Square   F Value    Pr > F

Model              12    521.2800000    43.4400000      2.33    0.0785

Error              12    223.6800000    18.6400000

Corrected Total    24    744.9600000

              R-Square    Coeff Var    Root MSE    m2 Mean

              0.699742    36.09872    4.317407    11.96000

Source             DF      Type I SS   Mean Square   F Value    Pr > F

grades              4    265.3600000    66.3400000      3.56    0.0390
day                 4     87.7600000    21.9400000      1.18    0.3693
method              4    168.1600000    42.0400000      2.26    0.1237

Source             DF    Type III SS   Mean Square   F Value    Pr > F

grades              4    265.3600000    66.3400000      3.56    0.0390
day                 4     87.7600000    21.9400000      1.18    0.3693
method              4    168.1600000    42.0400000      2.26    0.1237

                              The SAS System

                            The GLM Procedure
                     Multivariate Analysis of Variance

                          E = Error SSCP Matrix

                                  m1            m2

                    m1        107.52         84.96
                    m2         84.96        223.68

 Partial Correlation Coefficients from the Error SSCP Matrix / Prob > |r|

                     DF = 12         m1            m2

                     m1        1.000000      0.547843
                                             0.0526
```

```
             DF = 12        m1           m2

             m2        0.547843     1.000000
                       0.0526

                            The SAS System

                           The GLM Procedure
                    Multivariate Analysis of Variance

             H = Type III SSCP Matrix for grades

                            m1           m2

             m1          39.44        55.92
             m2          55.92       265.36

       Characteristic Roots and Vectors of: E Inverse * H, where
               H = Type III SSCP Matrix for grades
                     E = Error SSCP Matrix

     Characteristic        Characteristic Vector V'EV=1
               Root   Percent              m1             m2

         1.32583835     80.13   -0.04312120     0.07838773
         0.32885104     19.87    0.10690948    -0.01559632

        MANOVA Test Criteria and F Approximations for
         the Hypothesis of No Overall grades Effect
             H = Type III SSCP Matrix for grades
                 E = Error SSCP Matrix

                       S=2  M=0.5  N=4.5

   Statistic                    Value   F Value   Num DF   Den DF   Pr > F

Wilks' Lambda            0.32355206       2.08        8   22        0.0826
Pillai's Trace           0.81751771       2.07        8   24        0.0799
Hotelling-Lawley Trace   1.65468939       2.18        8   13.589    0.0984
Roy's Greatest Root      1.32583835       3.98        4   12        0.0279

   NOTE: F Statistic for Roy's Greatest Root is an upper bound.
          NOTE: F Statistic for Wilks' Lambda is exact.

          H = Type III SSCP Matrix for day

                      m1           m2


         m1        13.84        -7.88
         m2        -7.88        87.76

     Characteristic Roots and Vectors of: E Inverse * H, where
          H = Type III SSCP Matrix for day
             E = Error SSCP Matrix

     Characteristic        Characteristic Vector V'EV=1
               Root   Percent              m1             m2

         0.73031471     88.62   -0.08399749    0.07769824
         0.09375619     11.38    0.07895249    0.01873139
```

```
                         The SAS System

                       The GLM Procedure
                Multivariate Analysis of Variance

           MANOVA Test Criteria and F Approximations for
             the Hypothesis of No Overall day Effect
               H = Type III SSCP Matrix for day
                  E = Error SSCP Matrix

                      S=2  M=0.5  N=4.5
Statistic                        Value  F Value  Num DF  Den DF    Pr > F

Wilks' Lambda               0.52838974     1.03       8  22        0.4417
Pillai's Trace              0.50778992     1.02       8  24        0.4473
Hotelling-Lawley Trace      0.82407091     1.09       8  13.589    0.4267
Roy's Greatest Root         0.73031471     2.19       4  12        0.1316

      NOTE: F Statistic for Roy's Greatest Root is an upper bound.
              NOTE: F Statistic for Wilks' Lambda is exact.

            H = Type III SSCP Matrix for method

                        m1           m2

            m1      418.24        54.72
            m2       54.72       168.16

       Characteristic Roots and Vectors of: E Inverse * H, where
            H = Type III SSCP Matrix for method
                E = Error SSCP Matrix

       Characteristic        Characteristic Vector V'EV=1
                 Root   Percent             m1            m2

           5.32910175     87.65     0.11526456   -0.04481106
           0.75070114     12.35     0.00177636    0.06618044

           MANOVA Test Criteria and F Approximations for
             the Hypothesis of No Overall method Effect
               H = Type III SSCP Matrix for method
                     E = Error SSCP Matrix

                           S=2  M=0.5  N=4.5

Statistic                        Value  F Value  Num DF  Den DF   Pr > F

Wilks' Lambda               0.09024973     6.40       8  22       0.0002
Pillai's Trace              1.27079998     5.23       8  24       0.0007
Hotelling-Lawley Trace      6.07980289     8.02       8  13.589   0.0005
Roy's Greatest Root         5.32910175    15.99       4  12       <.0001

        NOTE: F Statistic for Roy's Greatest Root is an upper bound.
               NOTE: F Statistic for Wilks' Lambda is exact.
```

□

12.7 Exercises

Exercise 12.1

In devising a test for interviewing engineers for company scholarship, an oil company studied three grades of students to see whether engineering aptitude differed between them. They were also interested in finding out whether there was any difference in the performances of female and male students. Sixty students were used for the test, equally split among the sexes. The grades of students used were (a) secondary school graduates (SSG), (b) lower sixth students (LSS), and (c) upper sixth students (USS). The students were subjected to two tests (a) engineering structures aptitude (ESA) and (b) mathematics aptitude test (MAT). The scores of the students in the study are presented as follows:

Females					
SSG		**LSS**		**USS**	
ESA	**MAT**	**ESA**	**MAT**	**ESA**	**MAT**
20	24	25	40	27	40
21	28	27	36	29	45
29	29	28	35	32	46
26	17	25	27	32	45
28	26	25	36	29	41
27	24	22	35	36	41
19	23	24	31	33	39
19	23	27	31	31	40
22	29	33	33	31	35
26	25	31	36	33	39

Males					
SSG		**LSS**		**USS**	
ESA	**MAT**	**ESA**	**MAT**	**ESA**	**MAT**
20	22	25	40	41	41
22	23	29	36	40	50
24	25	27	32	36	53
29	28	28	41	35	42
24	24	26	35	38	41
28	25	24	31	31	46
26	27	22	29	42	42
22	24	24	38	37	44
20	24	26	35	36	46
29	24	26	31	31	46

□

Exercise 12.2

Three compounds (C_1, C_2, C_3) useful for the drug production could be extracted from three plants (P_1, P_2, P_3), which grow in different parts of the world. Four technicians were chosen at random from each of three different laboratories in those parts of the world to determine the percentages of the compounds in each of those plants. The responses of the investigations are presented below as follows:

	Plant											
Compound	P_1				P_2				P_3			
	1	**2**	**3**	**4**	**5**	**6**	**7**	**8**	**9**	**10**	**11**	**12**
C_1	8.7	9.3	8.5	7.8	10.2	9.8	10.6	9.9	11.2	9.7	10.5	9.4
C_2	6.3	7.1	5.8	6.1	8.1	8.4	7.3	6.9	7.5	8.5	6.8	7.7
C_3	3.9	5.3	4.7	4.9	4.2	4.8	4.0	4.7	6.0	4.1	4.3	5.1

Determine whether any of the plants yields more or less of the chemicals. Is there any significant difference in the skills of the technicians? ▯

Exercise 12.3

An experiment was performed to study three different methods of teaching pupil's arithmetical skills of addition (A), multiplication (M), and subtraction (S). The scores obtained from different groups of pupils used in the experiment are presented below as follows:

Group I			Group II			Group III		
A	M	S	A	M	S	A	M	S
87	56	76	66	39	79	56	35	96
84	51	69	60	32	64	50	33	80
86	42	77	74	38	80	50	47	89
74	40	69	76	43	83	52	44	87
80	42	70	60	42	82	50	38	84
78	45	71	60	37	72	42	38	84
72	40	62	63	39	67	55	42	86
55	55	58	75	33	77	60	39	79
85	54	69	51	35	68	52	43	87
80	40	72	60	40	70	58	50	70

Can it be said that any of the methods used is better for teaching these mathematical skills? ▯

Exercise 12.4

Ten students from each of three disciplines, geography, sociology, and history were selected at random to participate in a study whose aim was to find out whether a person's major field of study affects his or her intellectual ability to perceive some basic concepts. They were subjected to three basic tests in mathematics (T_1), algebra (T_2), and computing (T_3), and their scores are presented in the table below:

Geography majors				Sociology majors				History majors			
S.No.	T_1	T_2	T_3	S.No.	T_1	T_2	T_3	S.No.	T_1	T_2	T_3
1	19	17	18	11	14	22	26	21	27	29	18
2	26	22	16	12	22	16	22	22	21	25	20
3	19	28	22	13	18	21	25	22	20	24	20
4	22	24	19	14	24	19	22	21	18	20	18
5	29	23	24	15	23	22	23	25	15	21	17
6	20	32	28	16	32	18	22	24	27	28	21
7	25	22	18	17	22	18	22	23	29	30	18
8	17	29	21	18	29	21	29	28	15	20	16
9	18	19	16	19	19	16	19	19	20	20	19
10	26	24	18	20	24	18	23	20	17	22	18

Analyze the data and find out if any of the groups is intellectually superior to the other. □

Exercise 12.5

A research was carried out into the yield of paw-paw under nine fertilizer treatments in three different experimental farms. However, only three of the fertilizers were used exclusively in each of the three farms, so that fertilizer treatments were nested within farms. The following data represent the yields in pounds of paw-paw for early harvest and late harvests from the same paw-paw tree.

Farm I						Farm II					
F1		F2		F3		F4		F5		F6	
E	L	E	L	E	L	E	L	E	L	E	L
54.1	78.6	46.2	65	44.7	66	44.1	58.6	36.4	45	34.7	51
44.6	68.9	46.8	63	55.4	63	48.5	58.9	36.8	43	35.5	53
55.7	67.5	47.8	58	47.2	61.7	45.7	57.5	38.9	44	37.3	46
34.6	46.7	43.2	55.5	47.3	65.7	44.6	48.7	32	45.5	37	65.7
47.8	56.9	46.4	58.9	46.6	61	41.8	56.9	31.6	46.7	36.7	61

Farm III					
F7		F8		F9	
E	L	E	L	E	L
54	58.6	46.4	45	39.8	53
58	67	44.8	43	40.5	56
43	57.8	43.4	44	39.7	56
54.4	58.7	38	45.5	37	63.5
41.8	57	40.6	46.7	45	62.2

Analyze the responses and determine whether there are significant differences between farms and fertlizers embedded in farms. □

Exercise 12.6
Write a SAS program to carry out the analysis of the responses of the experiment in Exercise 12.5. □

Exercise 12.7
Six methods were available to be used to extract Oleoresin (OLE) and a byproduct (BYP), from ginger. Three researchers, each uses two of the methods to make determinations of the ginger content of OLE and BYP leading to nested design. The percentages of OLE and BYP obtained are presented in the table below:

Researcher 1				Researcher 2				Researcher 3			
M1		M2		M3		M4		M5		M6	
OLE	**BYP**	**OLE**	**BYP**	**OLE**	**BYP**	**OLE**	**BYP**	**OLE**	**BYP**	**OLE**	**BYP**
27	13	24	9.2	20.5	11.1	21	9	17.9	10	19.2	9
25.8	10.6	24.7	9.6	20.8	12	20.7	9.6	16.4	10.1	18.4	8.7
26	12.1	25	9.8	20.1	10	19	8	16.6	9.6	19	9.1
20.5	10	24.6	9	17.8	9	18.5	8.7	16.5	9.1	17.9	10
27.6	9.5	25.4	11	17.1	9	17	10	16	9.5	16.9	8.4

Analyze the responses and determine whether there are significant differences between farms and fertlizers nested within farms. □

Exercise 12.8
Write a SAS program to carry out the analysis of responses of the experiment in Exercise 12.3. □

Exercise 12.9
Write a SAS program to carry out the analysis of responses of the experiment in Exercise 12.1. □

Exercise 12.10
Write a SAS program to carry out the analysis of responses of the experiment in Exercise 12.2. □

Exercise 12.11
Write a SAS program to carry out the analysis of responses of the experiment in Exercise 12.7. □

Exercise 12.12

A car manufacturer concerned about two issues: rising cost of fuel and pollution from car emissions, decided to produce more fuel-efficient cars. He carried out an experiment to test a set of five prototypes of burners he designed, which were to be fitted to the new generation of cars he wished to produce. These burners were designated (A, B, C, D). He then added the old burner (E) which had been in use, to compare them to see whether the new burners had improved fuel efficiency, and reduced the emission of pollutants. In his experiments, he planned to use five different drivers and five different cars. He realized that the skills of the drivers could be different and suspected that cars could have effect on the fuel consumption and therefore decided to use a 5×5 Latin square design to study the efficiency of the burners. Randomization led to the following layout. Experiments were performed, changing the burners in each car in the order shown. Cars were driven on the same road, at the same speed for the same distance before fuel consumption was measured in miles per gallon. He recorded the consumption and the units of emission/gallon released.

	Driver				
Car	**1**	**2**	**3**	**4**	**5**
1	E 54,10	A 76,6	C 60,6	D 59,6	B 60,7
2	D 57,6	E 50,12	B 69,8	C 54,6	A 70,6
3	C 53,5	D 66,8	A 75,5	B 59,8	E 50,13
4	B 60,6	C 60,5	E 53,12	A 64,5	D 60,6
5	A 64,6	B 70,7	D 69,8	E 51,13	C 56,5

Analyze the responses and determine whether the new burners have significant effect on the fuel. □

Exercise 12.13

Use SAS to carry out the analysis required for Exercise 12.12. □

References

Bartlett, M.S. (1947) Multivariate analysis, *Journal Royal Statistical Society Supplement, Series B*, 9, 176–197.

Davis, A.W. (1970) Exact distributions of Hotelling's generalized T_o^2, *Biometrika*, 57, 187–191.

Hotelling, H. (1951) A generalized T-test and measure of multivariate dispersion, Proceeding 2nd Berkeley Symposium, 23–41.

Kres, H. (1983) *Statistical Tables for Multivariate Analysis*, Springer-Verlag, New York.

Lawley, D.N. (1938) A generalization of Fisher's Z-test, *Biometrika*, 30, 180–187.

Lohnes, P.R. (1961) Test space and discriminant space classification models and related significant test, *Educational Psychological Measurement*, 21, 559–574.

Pillai, K.C.S. (1960) *Statistical Tables for Tests of Multivariate Hypothesis*, The Statistics Center, University of the Philippines Manila.

Pillai, K.C.S. (1965) On the distribution of the largest root of a matrix in multivariate analysis, *Biometrics*, 52, 405–414.

Pillai, K.C.S. (1967) Upper percentage of the largest root of a matrix in multivariate analysis, *Biometrics*, 54, 189–194.

Press, W.H., Flannery, B.P., Teukolsky, S.A., and Vettering, W.T. (1988) *Numerical Recipes: The Art of Scientific Computing*, Cambridge University Press, New York.

Rao, C.R. (1952) *Advanced Statistical Methods in Biometric Research*, Wiley, New York.

Roy, S.N. (1957) *Some Aspects of Multivariate Analysis*, Wiley, New York.

Roy, S.N. and Bose, R.C. (1953) Simultaneous confidence interval estimates, *Annals of Mathematical Statistics*, 24, 220–238.

Timm, N.H. (1975) *Multivariate Analysis with Applications in Education and Psychology*, Brooks/Cole.

Wilks, S.S. (1932) Certain generalizations in the analysis of variance, *Biometrika*, 24, 471–494.

Appendix

Statistical Tables

TABLE A1: Cumulative Probabilities of the Standard Normal Distribution

Z	**0.00**	**−0.01**	**−0.02**	**−0.03**	**−0.04**	**−0.05**	**−0.06**	**−0.07**	**−0.08**	**−0.09**
−4.0	0.00003	0.00003	0.00003	0.00003	0.00003	0.00003	0.00002	0.00002	0.00002	0.00002
−3.9	0.000048	0.000046	0.000044	0.000042	0.000041	0.000039	0.000037	0.000036	0.000034	0.000033
−3.8	0.000072	0.000069	0.000067	0.000064	0.000062	0.000059	0.000057	0.000054	0.000052	0.000050
−3.7	0.000108	0.000104	0.000100	0.000096	0.000092	0.000088	0.000085	0.000082	0.000078	0.000075
−3.6	0.00015911	0.000153	0.000147	0.000142	0.000136	0.000131	0.000126	0.000121	0.000117	0.000112
−3.5	0.00023263	0.000224	0.000216	0.000208	0.0002	0.000193	0.000185	0.000178	0.000172	0.000165
−3.4	0.00033693	0.000325	0.000313	0.000302	0.000291	0.00028	0.00027	0.00026	0.000251	0.000242
−3.3	0.00048342	0.000466	0.00045	0.000434	0.000419	0.000404	0.00039	0.000376	0.000362	0.000349
−3.2	0.00068714	0.000664	0.000641	0.000619	0.000598	0.000577	0.000557	0.000538	0.000519	0.000501
−3.1	0.0009676	0.000935	0.000904	0.000874	0.000845	0.000816	0.000789	0.000762	0.000736	0.000711
−3.0	0.0013499	0.001306	0.001264	0.001223	0.001183	0.001144	0.001107	0.00107	0.001035	0.001001
−2.9	0.00186581	0.001807	0.00175	0.001695	0.001641	0.001589	0.001538	0.001489	0.001441	0.001395
−2.8	0.00255513	0.002477	0.002401	0.002327	0.002256	0.002186	0.002118	0.002052	0.001988	0.001926
−2.7	0.00346697	0.003364	0.003264	0.003167	0.003072	0.00298	0.00289	0.002803	0.002718	0.002635
−2.6	0.00466119	0.004527	0.004396	0.004269	0.004145	0.004025	0.003907	0.003793	0.003681	0.003573
−2.5	0.00620967	0.006037	0.005868	0.005703	0.005543	0.005386	0.005234	0.005085	0.00494	0.004799
−2.4	0.00819754	0.007976	0.00776	0.007549	0.007344	0.007143	0.006947	0.006756	0.006569	0.006387
−2.3	0.01072411	0.010444	0.01017	0.009903	0.009642	0.009387	0.009137	0.008894	0.008656	0.008424
−2.2	0.01390345	0.013553	0.013209	0.012874	0.012545	0.012224	0.011911	0.011604	0.011304	0.011011
−2.1	0.01786442	0.017429	0.017003	0.016586	0.016177	0.015778	0.015386	0.015003	0.014629	0.014262

−2.0	0.02275013	0.022216	0.021692	0.021178	0.020675	0.020182	0.019699	0.019226	0.018763	0.018309
−1.9	0.02871656	0.028067	0.027429	0.026803	0.02619	0.025588	0.024998	0.024419	0.023852	0.023295
−1.8	0.03593032	0.035148	0.03438	0.033625	0.032884	0.032157	0.031443	0.030742	0.030054	0.029379
−1.7	0.04456546	0.043633	0.042716	0.041815	0.04093	0.040059	0.039204	0.038364	0.037538	0.036727
−1.6	0.05479929	0.053699	0.052616	0.051551	0.050503	0.049471	0.048457	0.04746	0.046479	0.045514
−1.5	0.0668072	0.065522	0.064255	0.063008	0.06178	0.060571	0.05938	0.058208	0.057053	0.055917
−1.4	0.08075666	0.07927	0.077804	0.076359	0.074934	0.073529	0.072145	0.070781	0.069437	0.068112
−1.3	0.09680048	0.095098	0.093418	0.091759	0.090123	0.088508	0.086915	0.085343	0.083793	0.082264
−1.2	0.11506967	0.113139	0.111232	0.109349	0.107488	0.10565	0.103835	0.102042	0.100273	0.098525
−1.1	0.13566606	0.1335	0.131357	0.129238	0.127143	0.125072	0.123024	0.121	0.119	0.117023
−1.0	0.15865525	0.156248	0.153864	0.151505	0.14917	0.146859	0.144572	0.14231	0.140071	0.137857
−0.9	0.18406013	0.181411	0.178786	0.176186	0.173609	0.171056	0.168528	0.166023	0.163543	0.161087
−0.8	0.2118554	0.20897	0.206108	0.203269	0.200454	0.197663	0.194895	0.19215	0.18943	0.186733
−0.7	0.24196365	0.238852	0.235762	0.232695	0.22965	0.226627	0.223627	0.22065	0.217695	0.214764
−0.6	0.27425312	0.270931	0.267629	0.264347	0.261086	0.257846	0.254627	0.251429	0.248252	0.245097
−0.5	0.30853754	0.305026	0.301532	0.298056	0.294599	0.29116	0.28774	0.284339	0.280957	0.277595
−0.4	0.34457826	0.340903	0.337243	0.333598	0.329969	0.326355	0.322758	0.319178	0.315614	0.312067
−0.3	0.38208858	0.37828	0.374484	0.3707	0.366928	0.363169	0.359424	0.355691	0.351973	0.348268
−0.2	0.42074029	0.416834	0.412936	0.409046	0.405165	0.401294	0.397432	0.39358	0.389739	0.385908
−0.1	0.46017216	0.456205	0.452242	0.448283	0.44433	0.440382	0.436441	0.432505	0.428576	0.424655
−0.0	0.5000000	0.4960106	0.492022	0.48803	0.48405	0.480061	0.47608	0.472097	0.46812	0.464144

TABLE A2: Cumulative Probabilities of the Standard Normal Distribution

Z	**0.00**	**0.01**	**0.02**	**0.03**	**0.04**	**0.05**	**0.06**	**0.07**	**0.08**	**0.09**
0	0.500000	0.503989	0.507978	0.511966	0.515953	0.519939	0.523922	0.527903	0.531881	0.535856
0.1	0.539828	0.503989	0.507978	0.511966	0.515953	0.519939	0.523922	0.527903	0.531881	0.575345
0.2	0.579260	0.543795	0.547758	0.551717	0.555670	0.559618	0.563559	0.567495	0.571424	0.614092
0.3	0.617911	0.583166	0.587064	0.590954	0.594835	0.598706	0.602568	0.606420	0.610261	0.651732
0.4	0.655422	0.621720	0.625516	0.629300	0.633072	0.636831	0.640576	0.644309	0.648027	0.687933
0.5	0.691462	0.659097	0.662757	0.666402	0.670031	0.673645	0.677242	0.680822	0.684386	0.722405
0.6	0.725747	0.694974	0.698468	0.701944	0.705401	0.708840	0.712260	0.715661	0.719043	0.754903
0.7	0.758036	0.729069	0.732371	0.735653	0.738914	0.742154	0.745373	0.748571	0.751748	0.785236
0.8	0.788145	0.761148	0.764238	0.767305	0.770350	0.773373	0.776373	0.779350	0.782305	0.813267
0.9	0.815940	0.791030	0.793892	0.796731	0.799546	0.802337	0.805105	0.807850	0.810570	0.838913
1.0	0.841345	0.818589	0.821214	0.823814	0.826391	0.828944	0.831472	0.833977	0.836457	0.862143
1.1	0.864334	0.843752	0.846136	0.848495	0.850830	0.853141	0.855428	0.857690	0.859929	0.882977
1.2	0.884930	0.866500	0.868643	0.870762	0.872857	0.874928	0.876976	0.879000	0.881000	0.901475
1.3	0.903200	0.886861	0.888768	0.890651	0.892512	0.894350	0.896165	0.897958	0.899727	0.917736
1.4	0.919243	0.904902	0.906582	0.908241	0.909877	0.911492	0.913085	0.914657	0.916207	0.931888
1.5	0.933193	0.920730	0.922196	0.923641	0.925066	0.926471	0.927855	0.929219	0.930563	0.944083
1.6	0.945201	0.934478	0.935745	0.936992	0.938220	0.939429	0.940620	0.941792	0.942947	0.954486
1.7	0.955435	0.946301	0.947384	0.948449	0.949497	0.950529	0.951543	0.952540	0.953521	0.963273
1.8	0.964070	0.956367	0.957284	0.958185	0.959070	0.959941	0.960796	0.961636	0.962462	0.970621

1.9	0.971283	0.964852	0.965620	0.966375	0.967116	0.967843	0.968557	0.969258	0.969946	0.976705
2.0	0.977250	0.971933	0.972571	0.973197	0.973810	0.974412	0.975002	0.975581	0.976148	0.981691
2.1	0.982136	0.977784	0.978308	0.978822	0.979325	0.979818	0.980301	0.980774	0.981237	0.985738
2.2	0.986097	0.982571	0.982997	0.983414	0.983823	0.984222	0.984614	0.984997	0.985371	0.988989
2.3	0.989276	0.986447	0.986791	0.987126	0.987455	0.987776	0.988089	0.988396	0.988696	0.991576
2.4	0.991802	0.989556	0.989830	0.990097	0.990358	0.990613	0.990863	0.991106	0.991344	0.993613
2.5	0.993790	0.992024	0.992240	0.992451	0.992656	0.992857	0.993053	0.993244	0.993431	0.995201
2.6	0.995339	0.993963	0.994132	0.994297	0.994457	0.994614	0.994766	0.994915	0.995060	0.996427
2.7	0.996533	0.995473	0.995604	0.995731	0.995855	0.995975	0.996093	0.996207	0.996319	0.997365
2.8	0.997445	0.996636	0.996736	0.996833	0.996928	0.997020	0.997110	0.997197	0.997282	0.998074
2.9	0.998134	0.997523	0.997599	0.997673	0.997744	0.997814	0.997882	0.997948	0.998012	0.998605
3.0	0.998650	0.998193	0.998250	0.998305	0.998359	0.998411	0.998462	0.998511	0.998559	0.998999
3.1	0.999032	0.998694	0.998736	0.998777	0.998817	0.998856	0.998893	0.998930	0.998965	0.999289
3.2	0.999313	0.999065	0.999096	0.999126	0.999155	0.999184	0.999211	0.999238	0.999264	0.999499
3.3	0.999517	0.999336	0.999359	0.999381	0.999402	0.999423	0.999443	0.999462	0.999481	0.999651
3.4	0.999663	0.999534	0.999550	0.999566	0.999581	0.999596	0.999610	0.999624	0.999638	0.999758
3.5	0.999767	0.999675	0.999687	0.999698	0.999709	0.999720	0.999730	0.999740	0.999749	0.999835
3.6	0.999841	0.999776	0.999784	0.999792	0.999800	0.999807	0.999815	0.999822	0.999828	0.999888
3.7	0.999892	0.999847	0.999853	0.999858	0.999864	0.999869	0.999874	0.999879	0.999883	0.999925
3.8	0.999928	0.999896	0.999900	0.999904	0.999908	0.999912	0.999915	0.999918	0.999922	0.999950
3.9	0.999952	0.999931	0.999933	0.999936	0.999938	0.999941	0.999943	0.999946	0.999948	0.999967
4.0	0.999968	0.999954	0.999956	0.999958	0.999959	0.999961	0.999963	0.999964	0.999966	0.999978

TABLE A3: Upper Percentage Points (α) of the t-Distribution

Degrees of freedom	0.4	0.25	0.2	0.1	0.05	0.025	0.005	0.0025
1	0.32492	1	1.376382	3.077684	6.313752	12.7062	63.65674	127.3213
2	0.288675	0.816497	1.06066	1.885618	2.919986	4.302653	9.924843	14.08905
3	0.276671	0.764892	0.978472	1.637744	2.353363	3.182446	5.840909	7.453319
4	0.270722	0.740697	0.940965	1.533206	2.131847	2.776445	4.604095	5.597568
5	0.267181	0.726687	0.919544	1.475884	2.015048	2.570582	4.032143	4.773341
6	0.264835	0.717558	0.905703	1.439756	1.94318	2.446912	3.707428	4.316827
7	0.263167	0.711142	0.89603	1.414924	1.894579	2.364624	3.499483	4.029337
8	0.261921	0.706387	0.88889	1.396815	1.859548	2.306004	3.355387	3.832519
9	0.260955	0.702722	0.883404	1.383029	1.833113	2.262157	3.249836	3.689662
10	0.260185	0.699812	0.879058	1.372184	1.812461	2.228139	3.169273	3.581406
11	0.259556	0.697445	0.87553	1.36343	1.795885	2.200985	3.105807	3.496614
12	0.259033	0.695483	0.872609	1.356217	1.782288	2.178813	3.05454	3.428444
13	0.258591	0.693829	0.870152	1.350171	1.770933	2.160369	3.012276	3.372468
14	0.258213	0.692417	0.868055	1.34503	1.76131	2.144787	2.976843	3.325696
15	0.257885	0.691197	0.866245	1.340606	1.75305	2.13145	2.946713	3.286039
16	0.257599	0.690132	0.864667	1.336757	1.745884	2.119905	2.920782	3.251993
17	0.257347	0.689195	0.863279	1.333379	1.739607	2.109816	2.898231	3.22245
18	0.257123	0.688364	0.862049	1.330391	1.734064	2.100922	2.87844	3.196574
19	0.256923	0.687621	0.860951	1.327728	1.729133	2.093024	2.860935	3.173725
20	0.256743	0.686954	0.859964	1.325341	1.724718	2.085963	2.84534	3.153401
21	0.25658	0.686352	0.859074	1.323188	1.720743	2.079614	2.83136	3.135206
22	0.256432	0.685805	0.858266	1.321237	1.717144	2.073873	2.818756	3.118824
23	0.256297	0.685306	0.85753	1.31946	1.713872	2.068658	2.807336	3.103997

24	0.256173	0.68485	0.856855	1.317836	1.710882	2.063899	2.796939	3.090514
25	0.25606	0.68443	0.856236	1.316345	1.708141	2.059539	2.787436	3.078199
26	0.255955	0.684043	0.855665	1.314972	1.705618	2.055529	2.778715	3.066909
27	0.255858	0.683685	0.855137	1.313703	1.703288	2.05183	2.770683	3.05652
28	0.255768	0.683353	0.854647	1.312527	1.701131	2.048407	2.763262	3.046929
29	0.255684	0.683044	0.854192	1.311434	1.699127	2.04523	2.756386	3.038047
30	0.255605	0.682756	0.853767	1.310415	1.697261	2.042272	2.749996	3.029798
32	0.255464	0.682234	0.852998	1.308573	1.693889	2.036933	2.738481	3.014949
34	0.255339	0.681774	0.852321	1.306952	1.690924	2.032244	2.728394	3.001954
36	0.255227	0.681366	0.85172	1.305514	1.688298	2.028094	2.719485	2.990487
38	0.255128	0.681001	0.851183	1.30423	1.685954	2.024394	2.711558	2.980293
40	0.255039	0.680673	0.8507	1.303077	1.683851	2.021075	2.704459	2.971171
44	0.254884	0.680107	0.849867	1.30109	1.68023	2.015368	2.692278	2.955534
48	0.254756	0.679635	0.849174	1.299439	1.677224	2.010635	2.682204	2.942616
50	0.254699	0.679428	0.848869	1.298714	1.675905	2.008559	2.677793	2.936964
55	0.254576	0.678977	0.848205	1.297134	1.673034	2.004045	2.668216	2.924701
60	0.254473	0.678601	0.847653	1.295821	1.670649	2.000298	2.660283	2.914553
65	0.254387	0.678283	0.847186	1.294712	1.668636	1.997138	2.653604	2.906015
70	0.254312	0.678011	0.846786	1.293763	1.666914	1.994437	2.647905	2.898734
75	0.254248	0.677775	0.84644	1.292941	1.665425	1.992102	2.642983	2.89245
80	0.254191	0.677569	0.846137	1.292224	1.664125	1.990063	2.638691	2.886972
85	0.254142	0.677387	0.84587	1.291591	1.662979	1.988268	2.634914	2.882154
90	0.254097	0.677225	0.845633	1.291029	1.661961	1.986674	2.631565	2.877884
95	0.254058	0.677081	0.845421	1.290527	1.661052	1.985251	2.628576	2.874073
100	0.254022	0.676951	0.84523	1.290075	1.660234	1.983971	2.625891	2.870652
105	0.25399	0.676833	0.845058	1.289666	1.659495	1.982815	2.623465	2.867562
∞	1.150349	1.281552	1.644854	1.959964	2.241403	2.807034	3.023342	1.150349

TABLE A4: Upper Percentage Points (α) of the χ^2 Distribution

Degrees of freedom	0.995	0.99	0.975	0.95	0.5	0.025	0.02	0.01	0.005
1	3.92704 E − 05	0.000157	0.000982	0.003932	0.454936	5.023886	5.411895	6.634897	7.879439
2	0.010025084	0.020101	0.050636	0.102587	1.386294	7.377759	7.824046	9.21034	10.59663
3	0.071721775	0.114832	0.215795	0.351846	2.365974	9.348404	9.837409	11.34487	12.83816
4	0.206989093	0.297109	0.484419	0.710723	3.356694	11.14329	11.66784	13.2767	14.86026
5	0.411741904	0.554298	0.831212	1.145476	4.35146	12.8325	13.38822	15.08627	16.7496
6	0.675726778	0.87209	1.237344	1.635383	5.348121	14.44938	15.03321	16.81189	18.54758
7	0.989255685	1.239042	1.689869	2.16735	6.345811	16.01276	16.62242	18.47531	20.27774
8	1.344413088	1.646497	2.179731	2.732637	7.344122	17.53455	18.16823	20.09024	21.95495
9	1.734932909	2.087901	2.70039	3.325113	8.342833	19.02277	19.67902	21.66599	23.58935
10	2.155856482	2.558212	3.246973	3.940299	9.341818	20.48318	21.16077	23.20925	25.18818
11	2.603221895	3.053484	3.815748	4.574813	10.341	21.92005	22.61794	24.72497	26.75685
12	3.073823653	3.570569	4.403789	5.226029	11.34032	23.33666	24.05396	26.21697	28.29952
13	3.565034584	4.106915	5.008751	5.891864	12.33976	24.7356	25.47151	27.68825	29.81947
14	4.074674969	4.660425	5.628726	6.570631	13.33927	26.11895	26.87276	29.14124	31.31935
15	4.600915599	5.229349	6.262138	7.260944	14.33886	27.48839	28.2595	30.57791	32.80132
16	5.142205451	5.812213	6.907664	7.961646	15.3385	28.84535	29.63318	31.99993	34.26719
17	5.697217119	6.40776	7.564186	8.67176	16.33818	30.19101	30.99505	33.40866	35.71847
18	6.264804719	7.014911	8.230746	9.390455	17.3379	31.52638	32.34616	34.80531	37.15645
19	6.843971456	7.63273	8.906517	10.11701	18.33765	32.85233	33.68743	36.19087	38.58226
20	7.433844283	8.260398	9.590778	10.85081	19.33743	34.16961	35.01963	37.56623	39.99685
21	8.033653456	8.897198	10.2829	11.59131	20.33723	35.47888	36.34345	38.93217	41.40106
22	8.642716463	9.542492	10.98232	12.33801	21.33705	36.78071	37.6595	40.28936	42.79565

23	9.260424795	10.19572	11.68855	13.09051	22.33688	38.07563	38.96831	41.6384	44.18128
24	9.886233535	10.85636	12.40115	13.84843	23.33673	39.36408	40.27036	42.97982	45.55851
25	10.51965217	11.52398	13.11972	14.61141	24.33659	40.64647	41.56607	44.3141	46.92789
26	11.16023749	12.19815	13.84391	15.37916	25.33646	41.92317	42.85583	45.64168	48.28988
27	11.80758738	12.8785	14.57338	16.1514	26.33634	43.19451	44.13999	46.96294	49.64492
28	12.46133599	13.56471	15.30786	16.92788	27.33623	44.46079	45.41885	48.27824	50.99338
29	13.12114895	14.25645	16.04707	17.70837	28.33613	45.72229	46.6927	49.58788	52.33562
30	13.78671996	14.95346	16.79077	18.49266	29.33603	46.97924	47.9618	50.89218	53.67196
32	15.13403215	16.36222	18.29076	20.07191	31.33586	49.48044	50.4867	53.48577	56.32811
34	16.50127257	17.78915	19.80625	21.66428	33.33571	51.966	52.99524	56.06091	58.96393
36	17.88672669	19.23268	21.33588	23.26861	35.33557	54.43729	55.48886	58.61921	61.58118
38	19.28891165	20.69144	22.87848	24.8839	37.33545	56.89552	57.9688	61.16209	64.18141
40	20.70653548	22.16426	24.43304	26.5093	39.33535	59.34171	60.43613	63.69074	66.76596
44	23.58369335	25.14803	27.57457	29.78748	43.33516	64.20146	65.33667	68.70951	71.89255
48	26.51059112	28.17701	30.75451	33.09808	47.33501	69.02259	70.19676	73.68264	76.96877
50	27.99074904	29.70668	32.35736	34.76425	49.33494	71.4202	72.61325	76.15389	79.48998
55	31.73475764	33.57048	36.39811	38.95803	54.33479	77.38047	78.61914	82.29212	85.74895
60	35.53449152	37.48485	40.48175	43.18796	59.33467	83.29768	84.57995	88.37942	91.9517
65	39.3831412	41.44361	44.60299	47.44958	64.33456	89.17715	90.50124	94.42208	98.10514
70	43.27517986	45.44172	48.75757	51.73928	69.33448	95.02318	96.38754	100.4252	104.2149
75	47.20604798	49.47503	52.94194	56.05407	74.3344	100.8393	102.2425	106.3929	110.2856
80	51.17193241	53.54008	57.15317	60.39148	79.33433	106.6286	108.0693	112.3288	116.3211
85	55.16960469	57.63393	61.38878	64.7494	84.33427	112.3934	113.8706	118.2357	122.3246
90	59.19630494	61.75408	65.64662	69.12603	89.33422	118.1359	119.6485	124.1163	128.2989
95	63.2496491	65.89836	69.92487	73.51984	94.33417	123.858	125.4049	129.9727	134.2465
100	67.32756375	70.0649	74.22193	77.92947	99.33414	129.5612	131.1417	135.8067	140.1695

TABLE A5: Upper Percentage of the F-Distribution for $\alpha = 0.05$

Degrees of freedom of denominator	Degrees of freedom of numerator							
	1	**2**	**3**	**4**	**5**	**6**	**7**	**8**
1	161.44764	199.5	215.70735	224.58324	230.16188	233.986	236.7684	238.88269
2	18.512821	19	19.164292	19.246794	19.29641	19.329534	19.353218	19.370993
3	10.127964	9.5520945	9.2766282	9.1171823	9.0134552	8.9406451	8.886743	8.8452385
4	7.7086474	6.9442719	6.5913821	6.3882329	6.2560565	6.1631323	6.0942109	6.0410445
5	6.607891	5.786135	5.4094513	5.1921678	5.0503291	4.9502881	4.8758717	4.8183195
6	5.9873776	5.1432528	4.7570627	4.533677	4.3873742	4.2838657	4.2066585	4.1468042
7	5.5914478	4.7374141	4.3468314	4.1203117	3.9715232	3.8659689	3.7870435	3.7257253
8	5.3176551	4.4589701	4.0661806	3.8378534	3.6874987	3.5805803	3.5004639	3.4381012
9	5.117355	4.2564947	3.8625484	3.6330885	3.4816587	3.3737536	3.2927458	3.2295826
10	4.9646027	4.102821	3.7082648	3.4780497	3.3258345	3.2171745	3.1354648	3.0716584
11	4.8443357	3.982298	3.5874337	3.35669	3.2038743	3.0946129	3.0123303	2.9479903
12	4.7472253	3.8852938	3.4902948	3.2591667	3.1058752	2.9961204	2.9133582	2.8485651
13	4.6671927	3.8055653	3.4105336	3.1791171	3.0254383	2.9152692	2.8320975	2.7669132
14	4.6001099	3.7388918	3.3438887	3.1122498	2.9582489	2.847726	2.7641993	2.6986724
15	4.5430771	3.6823203	3.2873821	3.0555683	2.9012945	2.790465	2.7066268	2.6407969
16	4.4939984	3.6337235	3.2388715	3.0069173	2.8524092	2.7413108	2.6571966	2.5910962
17	4.4513217	3.5915306	3.1967768	2.9647081	2.8099962	2.6986599	2.614299	2.5479554

18	4.4138734	3.5545571	3.1599076	2.9277442	2.7728532	2.6613045	2.5767217	2.5101579
19	4.3807497	3.5218933	3.12735	2.8951073	2.7400575	2.628318	2.5435343	2.4767701
20	4.3512435	3.4928285	3.0983912	2.8660814	2.7108898	2.5989777	2.5140111	2.4470637
21	4.3247937	3.4668001	3.072467	2.8400998	2.6847807	2.5727116	2.4875777	2.4204622
22	4.3009495	3.4433568	3.049125	2.8167083	2.6612739	2.5490614	2.4637738	2.3965033
23	4.2793443	3.4221322	3.0279984	2.7955387	2.6399994	2.5276553	2.4422261	2.3748121
24	4.2596772	3.4028261	3.0087866	2.7762893	2.6206541	2.5081888	2.4226285	2.3550815
25	4.241699	3.38519	2.9912409	2.7587105	2.6029874	2.49041	2.4047281	2.3370572
26	4.2252012	3.3690164	2.975154	2.7425941	2.5867901	2.4741088	2.3883137	2.3205272
27	4.2100084	3.3541308	2.9603513	2.7277653	2.5718864	2.4591084	2.3732077	2.3053132
28	4.1959717	3.3403856	2.9466853	2.7140758	2.5581275	2.4452594	2.3592599	2.291264
29	4.1829642	3.3276545	2.9340299	2.7013993	2.5453865	2.4324341	2.3463419	2.2782508
30	4.1708768	3.3158295	2.9222772	2.6896276	2.5335545	2.4205232	2.334344	2.2661633
40	4.0847457	3.231727	2.8387454	2.6059749	2.4494664	2.3358524	2.2490243	2.1801705
50	4.0343095	3.1826099	2.7900084	2.5571792	2.4004091	2.2864359	2.1992021	2.1299228
60	4.0011913	3.1504113	2.7580783	2.5252151	2.3682702	2.254053	2.1665412	2.0969683
70	3.9777793	3.1276756	2.7355415	2.5026565	2.3455863	2.2311924	2.143478	2.0736904
80	3.9603523	3.1107662	2.718785	2.4858849	2.3287206	2.2141928	2.1263243	2.0563726
90	3.9468756	3.097698	2.7058381	2.472927	2.3156892	2.2010565	2.1130667	2.0429857
100	3.9361428	3.0872959	2.6955343	2.4626149	2.3053182	2.1906009	2.1025133	2.0323276
∞	3.8414621	2.995731	2.604913	2.371935	2.214098	2.098594	2.009589	1.938411

TABLE A5: Continued

Degrees of freedom of denominator	Degrees of freedom of numerator							
	9	**10**	**11**	**12**	**13**	**14**	**15**	**16**
1	240.54325	241.88175	242.98346	243.90604	244.68985	245.36398	245.94993	246.46392
2	19.384826	19.395897	19.404958	19.412511	19.418904	19.424384	19.429135	19.433293
3	8.8122996	8.7855247	8.7633328	8.7446407	8.7286812	8.7148964	8.7028701	8.6922863
4	5.998779	5.9643706	5.9358127	5.9117291	5.891144	5.8733463	5.8578054	5.8441174
5	4.7724656	4.7350631	4.7039672	4.6777038	4.6552255	4.6357677	4.6187591	4.603764
6	4.0990155	4.0599628	4.027442	3.9999354	3.9763627	3.9559339	3.938058	3.9222834
7	3.6766747	3.6365231	3.6030373	3.5746764	3.5503426	3.5292314	3.5107402	3.4944081
8	3.3881302	3.3471631	3.3129507	3.283939	3.2590192	3.2373781	3.2184055	3.2016343
9	3.1788931	3.1372801	3.1024854	3.0729471	3.0475493	3.0254727	3.006102	2.9889656
10	3.0203829	2.978237	2.9429573	2.9129767	2.8871747	2.8647277	2.8450165	2.8275664
11	2.8962228	2.8536249	2.8179305	2.7875693	2.7614174	2.7386482	2.7186396	2.7009144
12	2.7963755	2.7533868	2.7173314	2.6866371	2.6601775	2.6371236	2.6168512	2.5988812
13	2.7143558	2.6710242	2.6346505	2.6036607	2.5769271	2.5536188	2.53311	2.5149197
14	2.6457907	2.6021551	2.5654974	2.5342433	2.5072634	2.4837257	2.4630031	2.4446132
15	2.5876264	2.5437185	2.5068057	2.475313	2.4481102	2.4243644	2.4034471	2.384875
16	2.5376665	2.4935132	2.4563694	2.42466	2.3972542	2.3733182	2.3522228	2.3334836
17	2.4942915	2.4499155	2.4125614	2.3806542	2.3530625	2.328952	2.3076927	2.2887995

18	2.4562811	2.411702	2.3741556	2.3420668	2.3143042	2.2900329	2.2686222	2.2495866
19	2.4226989	2.3779337	2.3402104	2.3079544	2.2800341	2.2556139	2.2340629	2.214895
20	2.3928141	2.3478776	2.3099912	2.2775806	2.249514	2.2249557	2.2032743	2.1839832
21	2.3660482	2.3209534	2.2829161	2.250362	2.2221595	2.1974726	2.1756696	2.1562634
22	2.3419373	2.296696	2.2585184	2.2258308	2.1975016	2.1726947	2.1507779	2.131264
23	2.3201052	2.2747276	2.2364194	2.2036073	2.1751597	2.1502404	2.128217	2.108602
24	2.3002435	2.2547388	2.2163086	2.1833801	2.1548216	2.1297969	2.1076734	2.0879633
25	2.282097	2.2364736	2.1979292	2.1648915	2.1362289	2.111105	2.0888873	2.0690876
26	2.2654527	2.2197181	2.1810666	2.1479262	2.1191657	2.0939485	2.0716419	2.0517577
27	2.2501315	2.2042925	2.1655403	2.1323034	2.1034505	2.0781452	2.0557547	2.0357904
28	2.2359817	2.1900445	2.1511975	2.1178694	2.0889293	2.0635408	2.0410708	2.0210308
29	2.2228738	2.1768441	2.1379076	2.1044935	2.0754709	2.0500036	2.0274583	2.0073464
30	2.210697	2.1645799	2.1255588	2.0920632	2.0629626	2.0374204	2.0148037	1.9946236
40	2.1240293	2.077248	2.0375803	2.0034594	1.9737563	1.9476352	1.9244628	1.9037498
50	2.0733512	2.026143	1.9860565	1.9515277	1.9214291	1.8949256	1.871384	1.8503149
60	2.0400981	1.992592	1.9522119	1.9173959	1.8870176	1.8602423	1.8364374	1.8151134
70	2.0166007	1.9688749	1.9282776	1.8932482	1.8626615	1.8356832	1.8116808	1.7901651
80	1.9991148	1.9512203	1.9104556	1.8752616	1.8445135	1.8173776	1.7932218	1.7715567
90	1.985595	1.9375668	1.8966693	1.8613442	1.8304675	1.8032059	1.7789272	1.7571425
100	1.9748292	1.9266925	1.8856869	1.8502551	1.8192735	1.7919092	1.7675301	1.7456472
∞	1.8798823	1.830704	1.788646	1.752173	1.720156	1.691774	1.666382	1.643514

TABLE A5: Continued

Degrees of freedom of denominator	Degrees of freedom of numerator							
	17	**18**	**19**	**20**	**21**	**22**	**23**	**24**
1	246.91844	247.32324	247.68605	248.01308	248.30937	248.57906	248.82557	249.05177
2	19.436961	19.440223	19.443142	19.445768	19.448145	19.450307	19.45228	19.454089
3	8.6829	8.6745191	8.6669903	8.6601898	8.654017	8.6483887	8.643236	8.638501
4	5.8319696	5.8211156	5.8113592	5.8025419	5.7945342	5.7872295	5.7805391	5.7743887
5	4.5904445	4.5785342	4.5678205	4.5581315	4.5493268	4.5412906	4.5339264	4.5271531
6	3.9082593	3.8957093	3.884412	3.8741886	3.8648926	3.8564031	3.8486194	3.8414569
7	3.4798767	3.4668628	3.4551401	3.4445248	3.4348669	3.4260422	3.417947	3.4104944
8	3.1867007	3.1733174	3.161254	3.1503238	3.1403737	3.1312774	3.122929	3.1152398
9	2.973696	2.9600025	2.947652	2.9364554	2.9262575	2.9169299	2.9083654	2.9004738
10	2.812007	2.7980451	2.7854452	2.7740164	2.7636019	2.7540717	2.7453175	2.7372477
11	2.6850999	2.6709008	2.6580801	2.6464452	2.6358378	2.626127	2.617203	2.6089736
12	2.5828389	2.5684276	2.5554087	2.5435883	2.5328072	2.5229331	2.5138556	2.5054815
13	2.4986721	2.484069	2.4708705	2.4588818	2.4479425	2.4379197	2.428702	2.4201957
14	2.428179	2.4134011	2.4000387	2.3878961	2.3768119	2.3666526	2.3573062	2.3486781
15	2.3682701	2.3533321	2.3398193	2.327535	2.3163175	2.3060322	2.2965668	2.2878261
16	2.3167218	2.3016363	2.2879847	2.2755696	2.2642286	2.2538266	2.2442508	2.2354054
17	2.2718929	2.256671	2.2428908	2.2303543	2.2188985	2.2083881	2.1987094	2.1897665
18	2.2325457	2.2171971	2.2032974	2.1906479	2.1790853	2.1684735	2.1586987	2.1496645

19	2.1977293	2.1822628	2.1682516	2.1554966	2.143834	2.1331274	2.1232626	2.1141429
20	2.166701	2.1511244	2.137009	2.1241552	2.1123989	2.1016034	2.0916539	2.0824537
21	2.1388723	2.1231926	2.1089794	2.096033	2.0841886	2.0733094	2.0632804	2.0540043
22	2.1137709	2.0979944	2.0836893	2.0706557	2.0587284	2.0477703	2.0376662	2.0283185
23	2.091013	2.0751455	2.0607539	2.047638	2.0356326	2.0246	2.0144248	2.0050095
24	2.070284	2.0543306	2.0398575	2.026664	2.0145846	2.0034815	1.9932391	1.9837596
25	2.0513231	2.0352887	2.0207385	2.0074715	1.9953221	1.9841522	1.9738461	1.9643056
26	2.0339126	2.0178016	2.0031783	1.9898418	1.9776259	1.9663927	1.956026	1.9464277
27	2.0178692	2.0016855	1.986993	1.9735904	1.9613115	1.9500181	1.9395938	1.9299403
28	2.0030373	1.9867848	1.9720265	1.9585611	1.9462222	1.9348714	1.9243923	1.9146863
29	1.9892843	1.9729663	1.9581455	1.9446204	1.9322244	1.9208189	1.9102875	1.9005313
30	1.9764962	1.9601159	1.9452356	1.9316535	1.919203	1.9077454	1.897164	1.88736
40	1.8851117	1.8682417	1.8528918	1.8388593	1.8259767	1.8141042	1.8031242	1.792937
50	1.8313337	1.8141329	1.7984644	1.7841248	1.7709461	1.7587881	1.7475326	1.7370796
60	1.7958854	1.7784461	1.7625468	1.7479841	1.7345896	1.7222228	1.7107653	1.7001167
65	1.7823329	1.7647976	1.7488053	1.7341524	1.7206705	1.7082189	1.6966793	1.685951
70	1.7707512	1.7531316	1.7370575	1.7223252	1.7087663	1.6962401	1.6846279	1.6738293
80	1.7519976	1.7342368	1.7180255	1.7031601	1.6894718	1.6768198	1.6650854	1.6541679
90	1.7374667	1.7195924	1.7032708	1.6882978	1.6745049	1.6617511	1.6499177	1.6389038
100	1.7258758	1.7079083	1.6914957	1.6764343	1.6625551	1.6497172	1.6378019	1.6267081
∞	1.6227694	1.603852	1.5865	1.570524	1.555739	1.54202	1.529236	1.517294

TABLE A5: Continued

Degrees of freedom of denominator	Degrees of freedom of numerator							
	25	**26**	**27**	**28**	**29**	**30**	**31**	**32**
1	249.26008	249.45252	249.63085	249.79657	249.95096	250.09515	250.23012	250.35672
2	19.455753	19.45729	19.458712	19.460033	19.461263	19.462411	19.463485	19.464492
3	8.634135	8.6300964	8.6263498	8.6228645	8.6196142	8.6165759	8.6137294	8.6110573
4	5.7687153	5.7634655	5.7585936	5.7540602	5.7498312	5.745877	5.7421715	5.7386921
5	4.5209024	4.515116	4.509744	4.5047434	4.500077	4.4957123	4.4916209	4.487778
6	3.834844	3.8287197	3.8230318	3.8177351	3.8127906	3.8081643	3.8038263	3.7997506
7	3.4036106	3.3972328	3.3913071	3.385787	3.3806322	3.3758075	3.3712822	3.3670292
8	3.1081345	3.1015488	3.0954277	3.0897236	3.0843952	3.0794065	3.0747259	3.0703258
9	2.8931784	2.8864141	2.8801247	2.8742617	2.8687832	2.8636523	2.8588371	2.8543091
10	2.7297847	2.7228625	2.7164241	2.7104203	2.7048084	2.6995512	2.694616	2.689974
11	2.6013603	2.5942962	2.5877236	2.5815928	2.5758606	2.5704891	2.5654453	2.5606999
12	2.4977317	2.4905385	2.4838438	2.4775972	2.4717551	2.4662791	2.4611359	2.4562959
13	2.4123208	2.4050093	2.3982024	2.3918494	2.3859061	2.3803339	2.375099	2.3701716
14	2.340688	2.3332673	2.3263568	2.3199053	2.3138684	2.308207	2.3028871	2.2978785
15	2.2797293	2.2722073	2.2652006	2.2586576	2.2525336	2.2467892	2.24139	2.2363057
16	2.2272094	2.2195931	2.2124967	2.2058684	2.199663	2.1938409	2.1883676	2.1832124
17	2.1814778	2.1737734	2.1665931	2.1598849	2.1536032	2.1477084	2.1421654	2.1369436

18	2.1412891	2.1335021	2.1262432	2.1194599	2.1131066	2.1071433	2.1015349	2.0962504
19	2.1056859	2.0978214	2.0904885	2.0836346	2.0772137	2.0711859	2.0655157	2.060172
20	2.0739202	2.0659825	2.0585799	2.0516593	2.0451748	2.0390859	2.0333573	2.0279575
21	2.0453985	2.0373919	2.0299233	2.0229397	2.0163949	2.0102483	2.0044643	1.9990115
22	2.0196443	2.0115725	2.0040416	1.9969983	1.9903964	1.984195	1.9783585	1.9728552
23	1.9962706	1.988137	1.9805469	1.973447	1.9667908	1.9605375	1.954651	1.9490998
24	1.9749594	1.9667671	1.9591209	1.9519672	1.9452594	1.9389565	1.9330226	1.9274257
25	1.9554472	1.9471991	1.9394995	1.9322947	1.9255378	1.9191877	1.9132085	1.907568
26	1.9375138	1.9292127	1.9214622	1.9142086	1.9074048	1.9010098	1.8949873	1.8893052
27	1.9209737	1.912622	1.904823	1.8975228	1.8906744	1.8842364	1.8781725	1.8724507
28	1.9056693	1.8972694	1.8894241	1.8820794	1.8751882	1.8687092	1.8626057	1.8568458
29	1.8914663	1.8830202	1.8751307	1.8677435	1.8608114	1.854293	1.8481517	1.8423553
30	1.8782491	1.8697589	1.861827	1.8543991	1.8474278	1.8408717	1.8346941	1.8288626
40	1.7834576	1.7746129	1.7663395	1.7585826	1.7512941	1.744432	1.737959	1.7318423
50	1.7273435	1.7182507	1.7097376	1.7017489	1.6942362	1.6871569	1.6804737	1.6741533
60	1.6901911	1.6809148	1.6722239	1.6640627	1.6563826	1.649141	1.6423002	1.6358266
70	1.6637581	1.6543403	1.645512	1.6372173	1.6294075	1.6220398	1.6150762	1.6084832
80	1.6439811	1.634451	1.6255134	1.6171123	1.6091989	1.6017302	1.5946683	1.5879795
90	1.6286231	1.6190015	1.6099747	1.6014868	1.5934887	1.5859375	1.578795	1.5720276
100	1.6163496	1.6066521	1.5975513	1.5889911	1.5809225	1.5733023	1.5660927	1.5592596
∞	1.5061002	1.495583	1.485676	1.476325	1.467482	1.459098	1.451139	1.44357

TABLE A6: Upper Percentage Points of the F-Distribution for $\alpha = 0.025$

Degrees of freedom of denominator	Degrees of freedom of numerator							
	1	**2**	**3**	**4**	**5**	**6**	**7**	**8**
1	647.78901	799.5	864.16297	899.58331	921.8479	937.11108	948.21689	956.65622
2	38.506329	39	39.165495	39.248418	39.298228	39.331458	39.355205	39.373022
3	17.443443	16.044106	15.439182	15.100979	14.884823	14.734718	14.624395	14.539887
4	12.217863	10.649111	9.9791985	9.6045299	9.3644708	9.1973111	9.0741411	8.9795804
5	10.006982	8.4336207	7.7635895	7.3878858	7.1463818	6.9777019	6.8530756	6.757172
6	8.8131006	7.2598557	6.5987985	6.2271612	5.9875651	5.8197566	5.6954705	5.599623
7	8.0726689	6.5415203	5.8898192	5.5225943	5.2852369	5.1185966	4.9949092	4.8993406
8	7.5708821	6.0594674	5.4159623	5.0526322	4.8172756	4.6516955	4.5285621	4.4332599
9	7.2092832	5.7147054	5.0781187	4.7180785	4.4844113	4.3197218	4.1970466	4.1019557
10	6.9367282	5.4563955	4.8256215	4.4683416	4.2360857	4.0721313	3.9498241	3.8548909
11	6.7241297	5.2558893	4.630025	4.2750716	4.0439982	3.8806512	3.7586379	3.663819
12	6.5537687	5.0958672	4.4741848	4.1212086	3.8911339	3.7282921	3.6065146	3.5117767
13	6.4142543	4.9652657	4.3471781	3.9958976	3.7666741	3.6042564	3.4826693	3.3879873
14	6.2979386	4.8566979	4.2417276	3.8919144	3.6634231	3.5013649	3.3799329	3.285288
15	6.1995009	4.7650483	4.152804	3.8042713	3.5764153	3.4146647	3.2933598	3.1987381
16	6.1151271	4.6866654	4.0768231	3.7294165	3.5021163	3.3406309	3.2194313	3.1248222
17	6.0420133	4.6188743	4.0111631	3.6647541	3.4379437	3.276689	3.1555771	3.0609728

18	5.9780525	4.5596717	3.9538634	3.6083436	3.3819678	3.2209153	3.0998769	3.0052714
19	5.9216312	4.507528	3.9034285	3.5587061	3.3327184	3.1718442	3.0508679	2.9562569
20	5.8714937	4.4612555	3.8586987	3.5146952	3.2890558	3.12834	3.0074163	2.9127965
21	5.8266477	4.4199182	3.8187607	3.4754085	3.2500836	3.089509	2.9686303	2.8739993
22	5.7862991	4.3827684	3.7828859	3.4401263	3.2150866	3.0546388	2.9337987	2.8391546
23	5.7498048	4.3492022	3.7504858	3.4082678	3.1834878	3.0231543	2.9023474	2.807689
24	5.7166386	4.3187258	3.7210802	3.379359	3.1548163	2.9945864	2.8738082	2.7791346
25	5.6863658	4.2909324	3.6942732	3.3530092	3.1286845	2.9685487	2.8477954	2.753106
26	5.658624	4.2654832	3.6697357	3.3288939	3.1047698	2.94472	2.8239883	2.7292828
27	5.6331091	4.2420941	3.6471917	3.306741	3.0828022	2.9228312	2.8021184	2.7073965
28	5.6095636	4.2205252	3.6264083	3.2863207	3.0625537	2.902655	2.7819588	2.6872204
29	5.5877682	4.2005723	3.6071873	3.2674379	3.0438303	2.8839984	2.7633166	2.6685619
30	5.5675349	4.1820606	3.5893591	3.2499254	3.0264664	2.8666962	2.7460272	2.6512563
40	5.4239371	4.0509921	3.4632597	3.1261142	2.9037223	2.7443816	2.623781	2.5288635
50	5.3403231	3.9749309	3.3901888	3.054415	2.8326541	2.673555	2.5529737	2.457942
60	5.2856105	3.9252654	3.3425197	3.0076594	2.7863148	2.6273696	2.5067915	2.4116718
70	5.2470253	3.8902905	3.3089718	2.9747633	2.7537139	2.5948749	2.4742942	2.3791055
80	5.2183536	3.8643291	3.2840813	2.9503612	2.7295319	2.5707705	2.4501849	2.3549411
90	5.1962103	3.8442952	3.2648805	2.93154	2.7108811	2.5521789	2.4315878	2.336299
100	5.1785938	3.8283669	3.2496189	2.916582	2.696059	2.5374032	2.4168067	2.3214805
∞	5.0238848	3.688872	3.116134	2.785824	2.566503	2.40823	2.287537	2.191815

TABLE A6: Continued

Degrees of freedom of denominator	Degrees of freedom of numerator							
	9	10	11	12	13	14	15	16
1	963.28458	968.62744	973.0252	976.70795	979.83678	982.52781	984.86684	986.91866
2	39.386883	39.397975	39.407051	39.414615	39.421017	39.426505	39.431261	39.435423
3	14.473081	14.418942	14.37418	14.336552	14.30448	14.276816	14.252711	14.23152
4	8.9046816	8.843881	8.7935355	8.7511589	8.7149963	8.6837731	8.6565412	8.6325808
5	6.6810543	6.6191543	6.5678186	6.5245492	6.4875797	6.4556251	6.4277282	6.4031611
6	5.5234066	5.4613237	5.409761	5.3662439	5.3290197	5.2968115	5.2686668	5.2438605
7	4.8232171	4.7611164	4.7094699	4.6658297	4.62846	4.5960944	4.5677873	4.5428178
8	4.3572331	4.295127	4.2434128	4.1996675	4.1621704	4.1296653	4.1012127	4.0760959
9	4.0259942	3.9638652	3.9120745	3.8682203	3.8305956	3.7979525	3.7693573	3.7440969
10	3.7789626	3.7167919	3.664914	3.6209455	3.5831908	3.5504097	3.5216732	3.4962714
11	3.5878987	3.5256717	3.4736991	3.429613	3.391728	3.3588102	3.3299348	3.3043946
12	3.4358456	3.3735528	3.3214813	3.2772771	3.2392633	3.2062117	3.1772011	3.1515267
13	3.3120324	3.249668	3.1974962	3.1531752	3.1150357	3.0818544	3.0527132	3.0269095
14	3.2093003	3.1468612	3.0945898	3.0501548	3.0118937	2.9785875	2.9493211	2.9233936
15	3.1227117	3.0601969	3.0078277	2.9632824	2.9249044	2.8914787	2.8620925	2.8360467
16	3.0487535	2.9861632	2.9336991	2.8890476	2.8505577	2.8170178	2.7875176	2.7613591
17	2.9848594	2.922195	2.8696391	2.824886	2.7862893	2.7526407	2.7230318	2.6967662

18	2.9291125	2.8663757	2.8137316	2.7688813	2.730183	2.6964309	2.6667188	2.6403513
19	2.880052	2.8172451	2.7645165	2.7195735	2.6807783	2.6469279	2.6171177	2.5906536
20	2.8365461	2.7736714	2.7208619	2.6758306	2.6369433	2.6029995	2.5730961	2.54654
21	2.7977039	2.734764	2.6818773	2.6367618	2.597787	2.5637544	2.5337625	2.507119
22	2.7628152	2.6998127	2.6468521	2.6016566	2.5625985	2.5284816	2.4984054	2.4716789
23	2.7313068	2.6682441	2.6152131	2.5699414	2.5308041	2.4966068	2.4664506	2.439645
24	2.7027108	2.6395904	2.5864923	2.5411479	2.5019352	2.4676615	2.4374291	2.4105483
25	2.6766418	2.6134662	2.560304	2.5148903	2.4756058	2.4412595	2.4109545	2.384002
26	2.6527796	2.5895511	2.5363276	2.490848	2.4514951	2.4170794	2.3867052	2.3596844
27	2.6308556	2.5675764	2.5142945	2.4687519	2.4293337	2.394852	2.3644118	2.3373258
28	2.6106432	2.5473155	2.4939776	2.448375	2.4088945	2.3743497	2.3438466	2.3166983
29	2.5919496	2.5285754	2.4751841	2.4295241	2.3899842	2.3553791	2.3248159	2.2976081
30	2.5746101	2.5111913	2.4577489	2.412034	2.3724373	2.3377746	2.3071539	2.2798892
40	2.4519392	2.3881611	2.3343099	2.288157	2.2481068	2.2129842	2.1819033	2.1541825
50	2.3808209	2.3167942	2.2626622	2.2162092	2.175848	2.1404094	2.1090116	2.0809756
60	2.3344059	2.2701983	2.2158627	2.1691922	2.1286054	2.0929372	2.0613084	2.0330423
70	2.301729	2.2373843	2.1828948	2.1360597	2.095302	2.0594596	2.0276554	1.999214
80	2.2774777	2.2130257	2.1584157	2.1114518	2.0705604	2.0345814	2.0026395	1.9740603
90	2.2587656	2.1942273	2.1395206	2.0924533	2.0514543	2.0153655	1.9833126	1.9546223
100	2.2438894	2.1792804	2.1244944	2.0773422	2.0362549	2.0000757	1.9679316	1.9391497
∞	2.1136387	2.048316	1.99273	1.944723	1.902737	1.86564	1.832559	1.802835

TABLE A6: Continued

Degrees of freedom of denominator	Degrees of freedom of numerator							
	17	**18**	**19**	**20**	**21**	**22**	**23**	**24**
1	988.73307	990.34901	991.79732	993.1028	994.28558	995.36217	996.34625	997.24925
2	39.439096	39.442361	39.445282	39.447911	39.45029	39.452453	39.454428	39.456238
3	14.212744	14.195993	14.180955	14.167381	14.155068	14.143846	14.133578	14.124146
4	8.6113354	8.592368	8.5753308	8.5599432	8.5459766	8.5332428	8.5215854	8.5108735
5	6.3813604	6.3618832	6.3443765	6.3285552	6.3141871	6.3010804	6.2890759	6.2780401
6	5.2218305	5.2021345	5.1844197	5.1684009	5.1538454	5.140561	5.128388	5.1171924
7	4.5206266	4.5007732	4.4829057	4.4667396	4.4520424	4.4386222	4.4263191	4.4149991
8	4.0537585	4.0337616	4.0157544	3.999453	3.9846253	3.9710796	3.958656	3.9472203
9	3.7216172	3.7014809	3.6833381	3.6669055	3.6519513	3.6382839	3.6257435	3.6141957
10	3.4736521	3.4533794	3.4351041	3.4185435	3.403466	3.38968	3.3770257	3.3653687
11	3.2816392	3.2612337	3.2428297	3.2261448	3.2109475	3.1970464	3.1842817	3.1725188
12	3.1286395	3.1081057	3.0895774	3.0727725	3.0574598	3.0434477	3.0305763	3.0187112
13	3.0038957	2.9832386	2.9645909	2.9476708	2.9322472	2.9181284	2.9051547	2.8931913
14	2.9002585	2.8794834	2.8607215	2.8436912	2.8281614	2.8139406	2.8008689	2.7888114
15	2.8127959	2.7919083	2.7730374	2.755902	2.7402708	2.7259526	2.7127872	2.7006397
16	2.7379981	2.7170033	2.6980287	2.680793	2.6650652	2.6506541	2.6373993	2.6251659
17	2.6733004	2.6522037	2.6331303	2.6157991	2.5999794	2.5854797	2.5721397	2.5598244

18	2.6167859	2.5955922	2.5764249	2.559003	2.5430957	2.5285115	2.5150903	2.5026969
19	2.5669934	2.5457076	2.5264509	2.5089426	2.4929519	2.4782874	2.4647888	2.4523208
20	2.5227899	2.5014165	2.4820748	2.4644843	2.4484142	2.4336731	2.4201006	2.4075616
21	2.4832833	2.4618266	2.4424041	2.4247352	2.4085895	2.3937754	2.3801325	2.3675256
22	2.4477618	2.4262258	2.4067263	2.3889829	2.372765	2.3578812	2.3441712	2.3314995
23	2.4156505	2.394039	2.3744661	2.3566516	2.340365	2.3254148	2.3116406	2.298907
24	2.3864801	2.3647966	2.3451538	2.3272714	2.3109192	2.2959055	2.2820701	2.2692773
25	2.3598635	2.3381114	2.3184018	2.3004548	2.2840398	2.2689655	2.2550715	2.2422221
26	2.3354789	2.3136613	2.293888	2.2758791	2.2594043	2.244272	2.2303218	2.2174183
27	2.3130563	2.2911762	2.271342	2.253274	2.2367418	2.2215541	2.2075502	2.1945947
28	2.2923676	2.2704278	2.2505354	2.2324108	2.2158238	2.2005829	2.1865276	2.1735223
29	2.2732189	2.2512219	2.2312738	2.2130951	2.1964555	2.1811636	2.167059	2.154006
30	2.2554441	2.2333923	2.2133909	2.1951603	2.1784703	2.1631295	2.1489776	2.1358787
40	2.1292882	2.106796	2.0863642	2.067714	2.0506158	2.0348783	2.020341	2.0068683
50	2.0557697	2.0329711	2.0122387	1.9932945	1.975909	1.9598914	1.9450812	1.9313428
60	2.0076083	1.9845847	1.9636311	1.9444698	1.9268718	1.9106463	1.895633	1.8816963
70	1.9736061	1.9504105	1.9292874	1.9099594	1.8921979	1.8758121	1.8606419	1.8465517
80	1.9483154	1.9249841	1.903727	1.8842672	1.866376	1.8498629	1.8345679	1.8203553
90	1.9287666	1.9053256	1.88396	1.8643932	1.8463967	1.8297802	1.8143835	1.8000711
100	1.9132028	1.8896712	1.868216	1.8485607	1.830477	1.8137746	1.7982936	1.7838983
∞	1.7759395	1.751466	1.729067	1.70848	1.689469	1.671852	1.655461	1.64017

TABLE A6: Continued

Degrees of freedom of denominator	Degrees of freedom of numerator							
	25	**26**	**27**	**28**	**29**	**30**	**31**	**32**
1	998.08079	998.84903	999.56093	1000.2225	1000.8388	1001.4144	1001.9532	1002.4586
2	39.457904	39.459441	39.460865	39.462187	39.463417	39.464566	39.465641	39.466648
3	14.115452	14.107414	14.099959	14.093027	14.086563	14.080523	14.074866	14.069557
4	8.5009963	8.4918601	8.4833844	8.4755001	8.4681473	8.461274	8.4548349	8.4487899
5	6.2678603	6.2584404	6.2496984	6.2415636	6.2339748	6.2268789	6.2202293	6.2139851
6	5.1068609	5.097297	5.0884182	5.0801533	5.0724407	5.0652268	5.0584648	5.0521134
7	4.4045485	4.3948708	4.3858832	4.3775142	4.3697021	4.362393	4.3555398	4.3491011
8	3.936659	3.9268752	3.917786	3.9093197	3.9014144	3.8940159	3.887077	3.8805561
9	3.603527	3.5936402	3.5844523	3.5758915	3.5678955	3.5604102	3.553388	3.5467871
10	3.3545952	3.3446081	3.335324	3.326671	3.3185867	3.3110167	3.3039132	3.2972343
11	3.1616437	3.1515592	3.1421818	3.1334393	3.1252692	3.1176168	3.1104343	3.1036796
12	3.007738	2.9975595	2.988092	2.979263	2.97101	2.963278	2.9560191	2.949191
13	2.882124	2.871855	2.8623007	2.8533886	2.8450557	2.837247	2.8299145	2.8230156
14	2.7776537	2.7672981	2.7576606	2.7486686	2.740259	2.7323767	2.7249734	2.7180065
15	2.6893956	2.6789571	2.6692399	2.6601714	2.6516884	2.6437355	2.6362643	2.6292321
16	2.6138393	2.6033215	2.5935282	2.5843865	2.5758331	2.5678126	2.5602764	2.5531816
17	2.548419	2.5378255	2.5279594	2.5187478	2.5101271	2.5020418	2.4944433	2.4872885

18	2.4912163	2.4805505	2.4706149	2.4613364	2.4526514	2.4445042	2.436846	2.4296338
19	2.4407685	2.4300337	2.4200317	2.4106893	2.4019427	2.3937362	2.386021	2.3787537
20	2.3959408	2.3851401	2.3750746	2.3656711	2.3568656	2.3486024	2.3408325	2.3335126
21	2.3558395	2.3449758	2.3348496	2.3253876	2.3165258	2.3082083	2.300386	2.2930156
22	2.319751	2.3088271	2.2986429	2.2891249	2.2802092	2.2718397	2.2639672	2.2565483
23	2.2870988	2.2761173	2.2658776	2.2563061	2.2473387	2.2389193	2.2309987	2.2235333
24	2.2574119	2.2463754	2.2360825	2.2264597	2.2174427	2.2089756	2.2010088	2.1934989
25	2.2303021	2.2192127	2.2088689	2.1991969	2.1901325	2.1816195	2.1736084	2.1660556
26	2.2054459	2.194306	2.1839134	2.1741942	2.1650842	2.1565271	2.1484735	2.1408796
27	2.1825722	2.1713838	2.1609443	2.1511798	2.1420261	2.1334267	2.1253321	2.1176987
28	2.1614517	2.1502167	2.1397322	2.1299243	2.1207285	2.1120884	2.1039545	2.096283
29	2.1418892	2.1306095	2.1200818	2.1102321	2.1009958	2.0923166	2.0841449	2.0764367
30	2.1237175	2.112395	2.1018257	2.0919357	2.0826606	2.0739438	2.0657355	2.057992
40	1.9943445	1.9826702	1.97176	1.9615394	1.9519437	1.942916	1.9344063	1.9263703
50	1.9185602	1.9066343	1.8954792	1.8850205	1.8751932	1.8659402	1.8572113	1.8489622
60	1.8687202	1.8566056	1.8452665	1.8346283	1.824626	1.8152024	1.8063073	1.7978959
70	1.8334255	1.8211642	1.8096817	1.7989036	1.7887645	1.7792073	1.7701816	1.7616428
80	1.8071094	1.7947308	1.7831336	1.7722431	1.7619942	1.7523295	1.7431987	1.7345571
90	1.7867274	1.7742529	1.7625618	1.7515793	1.7412402	1.7314872	1.72227	1.7135436
100	1.7704732	1.7579189	1.7461494	1.7350901	1.7246757	1.7148489	1.7055592	1.6967619
∞	1.6258581	1.612429	1.599796	1.587885	1.57663	1.565975	1.555867	1.546264

TABLE A7: Upper Percentage of the F-Distribution for $\alpha = 0.01$

Degrees of freedom of denominator	Degrees of freedom of numerator							
	1	2	3	4	5	6	7	8
1	4052.1807	4999.5	5403.352	5624.5833	5763.6496	5858.9861	5928.3557	5981.0703
2	98.502513	99	99.166201	99.249372	99.299296	99.332589	99.356374	99.374215
3	34.116222	30.81652	29.456695	28.709898	28.237081	27.910657	27.671696	27.489177
4	21.19769	18	16.694369	15.977025	15.521858	15.206865	14.975758	14.798889
5	16.258177	13.273934	12.059954	11.391928	10.967021	10.672255	10.455511	10.289311
6	13.745023	10.924767	9.7795382	9.148301	8.7458953	8.4661253	8.2599953	8.1016514
7	12.246383	9.546578	8.4512851	7.8466451	7.4604355	7.1914048	6.9928328	6.8400491
8	11.258624	8.6491106	7.5909919	7.0060766	6.6318252	6.3706807	6.1776243	6.0288701
9	10.561431	8.0215173	6.9919172	6.4220855	6.0569407	5.8017703	5.6128655	5.4671225
10	10.044289	7.5594322	6.5523126	5.9943387	5.6363262	5.385811	5.2001213	5.0566931
11	9.6460341	7.2057134	6.2167298	5.6683002	5.3160089	5.0692104	4.886072	4.7444676
12	9.3302121	6.9266081	5.9525447	5.4119514	5.0643431	4.8205735	4.6395024	4.4993653
13	9.0738057	6.7009645	5.7393803	5.2053302	4.8616212	4.6203634	4.4409974	4.302062
14	8.8615927	6.5148841	5.5638858	5.035378	4.6949636	4.45582	4.2778819	4.1399461
15	8.6831168	6.3588735	5.4169649	4.8932096	4.555614	4.3182731	4.1415463	4.0044532
16	8.5309653	6.2262353	5.2922141	4.772578	4.4374205	4.2016337	4.0259466	3.8895721
17	8.3997401	6.1121137	5.1849999	4.6689676	4.3359391	4.1015053	3.9267194	3.7909642

18	8.2854195	6.0129048	5.0918895	4.579036	4.2478821	4.0146365	3.8406387	3.7054219
19	8.1849468	5.925879	5.0102868	4.5002577	4.170767	3.9385726	3.7652694	3.6305246
20	8.095958	5.8489319	4.9381934	4.4306902	4.1026846	3.8714268	3.6987402	3.5644121
21	8.0165969	5.7804157	4.8740462	4.3688152	4.0421439	3.8117255	3.6395896	3.5056318
22	7.9453857	5.7190219	4.8166058	4.3134295	3.9879632	3.7583014	3.5866602	3.4530335
23	7.8811336	5.6636988	4.7648768	4.2635675	3.9391949	3.7102184	3.5390239	3.4056947
24	7.8228706	5.6135912	4.7180508	4.2184453	3.8950697	3.6667167	3.4959275	3.3628671
25	7.7697984	5.5679971	4.6754648	4.1774202	3.8549572	3.627174	3.456754	3.3239375
26	7.7212544	5.5263347	4.6365696	4.1399605	3.8183358	3.5910751	3.420993	3.2883985
27	7.676684	5.4881178	4.6009069	4.1056221	3.7847702	3.5579905	3.3882185	3.2558272
28	7.6356193	5.4529369	4.5680909	4.0740318	3.7538945	3.527559	3.3580727	3.2258677
29	7.5976632	5.420445	4.5377947	4.0448732	3.7253988	3.4994746	3.3302522	3.1982188
30	7.5624761	5.3903459	4.5097396	4.0178768	3.6990188	3.4734766	3.3044989	3.172624
40	7.3140999	5.1785082	4.3125692	3.8282935	3.5138398	3.2910124	3.1237571	2.9929809
50	7.1705767	5.0566109	4.1993435	3.7195452	3.4076795	3.1864342	3.0201683	2.8900077
60	7.0771057	4.977432	4.1258919	3.6490475	3.3388844	3.1186743	2.9530492	2.8232802
70	7.0113988	4.9218723	4.0743968	3.5996471	3.290689	3.0712086	2.906032	2.7765332
80	6.962688	4.8807382	4.0362967	3.5631096	3.2550493	3.0361109	2.8712655	2.7419641
90	6.9251352	4.8490583	4.0069681	3.5349915	3.2276258	3.0091058	2.8445149	2.715364
100	6.895301	4.8239098	3.9836953	3.5126841	3.2058718	2.9876845	2.8232953	2.6942627
∞	6.6349001	4.605175	3.781622	3.319174	3.017251	2.80198	2.639329	2.511276

TABLE A7: Continued

Degrees of freedom of denominator	Degrees of freedom of numerator							
	9	**10**	**11**	**12**	**13**	**14**	**15**	**16**
1	6022.4732	6055.8467	6083.3168	6106.3207	6125.8647	6142.674	6157.2846	6170.1012
2	99.388093	99.399196	99.408281	99.415852	99.422259	99.427751	99.432511	99.436676
3	27.345206	27.228734	27.132567	27.051819	26.983057	26.923797	26.872195	26.826857
4	14.659134	14.545901	14.452284	14.373587	14.306502	14.248633	14.198202	14.15386
5	10.157762	10.051017	9.9626484	9.8882755	9.8248106	9.7700137	9.7222195	9.6801643
6	7.9761214	7.8741185	7.7895697	7.7183327	7.6574832	7.6048973	7.5589944	7.5185738
7	6.7187525	6.6200627	6.5381656	6.4690913	6.410034	6.3589538	6.3143309	6.2750098
8	5.9106188	5.8142939	5.7342746	5.6667193	5.6089105	5.5588706	5.5151248	5.4765511
9	5.3511289	5.256542	5.1778904	5.111431	5.0545143	5.0052101	4.9620784	4.9240223
10	4.9424207	4.8491468	4.7715181	4.7058697	4.6496055	4.6008331	4.5581396	4.5204482
11	4.6315397	4.5392818	4.462436	4.3974011	4.3416241	4.2932431	4.2508673	4.2134358
12	4.38751	4.2960544	4.21982	4.1552578	4.0998508	4.0517622	4.0096191	3.9723742
13	4.1910778	4.1002673	4.0245184	3.9603264	3.9052044	3.8573366	3.8153655	3.7782545
14	4.0296803	3.9393964	3.864039	3.8001408	3.7452408	3.6975412	3.6556973	3.6186822
15	3.8947881	3.8049397	3.7299019	3.6662398	3.6115143	3.5639435	3.5221937	3.4852461
16	3.7804152	3.6909314	3.6161574	3.5526867	3.4980996	3.4506276	3.4089469	3.3720456
17	3.6822415	3.5930661	3.5185122	3.4551981	3.4007212	3.3533251	3.3116943	3.2748234

18	3.5970739	3.5081617	3.4337929	3.3706079	3.3162192	3.268881	3.2272855	3.1904328
19	3.5225025	3.4338169	3.3596049	3.296527	3.2422091	3.1949149	3.1533433	3.1164993
20	3.4566756	3.3681864	3.2941084	3.2311198	3.1768588	3.1295973	3.0880407	3.0511984
21	3.3981474	3.3098296	3.2358667	3.172953	3.1187375	3.0715001	3.0299515	2.9931052
22	3.3457728	3.2576056	3.1837422	3.1208914	3.0667124	3.0194919	2.9779458	2.9410913
23	3.2986336	3.2105994	3.1368225	3.0740248	3.0198748	2.9726656	2.9311177	2.8942516
24	3.2559851	3.168069	3.0943674	3.0316148	2.9774876	2.9302852	2.888732	2.8518518
25	3.2172168	3.129406	3.0557706	2.9930561	2.9389468	2.8917474	2.8501862	2.8132899
26	3.181824	3.0941076	3.0205303	2.9578482	2.9037525	2.8565531	2.8149819	2.778068
27	3.1493854	3.0617539	2.988228	2.9255733	2.8714879	2.8242861	2.7827031	2.7457706
28	3.119547	3.0319921	2.9585118	2.8958805	2.8418024	2.7945961	2.7530001	2.7160482
29	3.092009	3.0045236	2.9310837	2.8684723	2.8143991	2.7671868	2.7255767	2.6886048
30	3.0665159	2.9790936	2.9056898	2.8430952	2.7890249	2.7418052	2.7001803	2.6631881
40	2.8875604	2.8005451	2.7273519	2.6648274	2.6107256	2.5634005	2.5216157	2.4844237
50	2.7849557	2.6981394	2.6250263	2.5624968	2.5083283	2.4608912	2.4189614	2.3816001
60	2.7184544	2.6317508	2.5586703	2.4961159	2.4418809	2.3943466	2.352297	2.3147993
70	2.6718592	2.5852258	2.5121579	2.4495746	2.3952804	2.3476647	2.3055173	2.2679097
80	2.6373984	2.5508119	2.4777473	2.4151361	2.3607909	2.3131072	2.2708793	2.2331818
90	2.6108793	2.5243257	2.45126	2.388623	2.334234	2.2864933	2.2441983	2.2064264
100	2.5898406	2.5033111	2.4302422	2.3675821	2.3131555	2.2653662	2.2230147	2.1851803
∞	2.4073307	2.320922	2.247723	2.184749	2.129862	2.08152	2.038524	1.999996

TABLE A7: Continued

Degrees of freedom of denominator	Degrees of freedom of numerator							
	17	**18**	**19**	**20**	**21**	**22**	**23**	**24**
1	6181.4348	6191.5287	6200.5756	6208.7302	6216.1184	6222.8433	6228.9903	6234.6309
2	99.440351	99.443617	99.44654	99.449171	99.451551	99.453715	99.455691	99.457502
3	26.786708	26.750905	26.718779	26.689791	26.663502	26.639552	26.617642	26.597523
4	14.114566	14.079505	14.048027	14.019609	13.993825	13.970326	13.94882	13.929064
5	9.6428716	9.6095749	9.5796637	9.5526462	9.5281214	9.5057593	9.4852857	9.4664708
6	7.4827065	7.4506627	7.4218609	7.3958319	7.3721931	7.3506294	7.3308785	7.3127208
7	6.2400957	6.2088852	6.180817	6.1554384	6.1323795	6.1113355	6.092053	6.0743193
8	5.4422798	5.4116266	5.3840454	5.3590949	5.336415	5.3157082	5.2967272	5.2792644
9	4.8901916	4.8599163	4.8326616	4.8079952	4.7855639	4.7650759	4.7462884	4.7289976
10	4.4869234	4.4569069	4.4298725	4.4053948	4.3831259	4.3627785	4.3441133	4.3269292
11	4.1801252	4.1502863	4.1234	4.0990462	4.0768816	4.056622	4.038031	4.0209096
12	3.9392137	3.909496	3.8827076	3.8584331	3.8363323	3.8161242	3.7975743	3.7804855
13	3.7451985	3.7155618	3.688836	3.6646091	3.642544	3.6223617	3.6038298	3.5867525
14	3.5856975	3.556113	3.5294242	3.5052223	3.4831727	3.4629982	3.4444679	3.4273874
15	3.4523084	3.4227549	3.3960847	3.3718916	3.3498429	3.3296633	3.311123	3.2940286
16	3.3391369	3.3095989	3.2829336	3.2587374	3.2366791	3.2164851	3.1979265	3.1808108
17	3.24193	3.2123959	3.1857256	3.1615175	3.1394422	3.119227	3.1006441	3.0835018

18	3.1575446	3.1280057	3.101323	3.0770967	3.0549988	3.0347576	3.0161463	2.9989737
19	3.0836085	3.0540584	3.0273579	3.0031088	2.9809843	2.9607138	2.942071	2.9248656
20	3.0182991	2.9887328	2.9620105	2.9377353	2.9155814	2.8952792	2.8766032	2.8593633
21	2.9601929	2.930607	2.9038599	2.8795562	2.8573711	2.8370357	2.818325	2.8010496
22	2.9081627	2.8785544	2.8517806	2.8274467	2.8052291	2.7848595	2.7661134	2.7488019
23	2.8613042	2.8316716	2.8048695	2.7805044	2.7582536	2.7378492	2.7190675	2.7017197
24	2.8188836	2.7892254	2.7623941	2.7379972	2.7157127	2.6952734	2.676456	2.6590721
25	2.7802997	2.7506149	2.7237538	2.6993248	2.6770066	2.6565324	2.6376795	2.6202597
26	2.7450547	2.7153429	2.6884517	2.6639905	2.6416386	2.6211299	2.6022418	2.5847866
27	2.7127336	2.6829943	2.656073	2.6315798	2.6091946	2.5886517	2.5697289	2.5522389
28	2.682987	2.6532202	2.6262689	2.601744	2.5793258	2.5587493	2.5397925	2.5222681
29	2.6555193	2.6257251	2.598744	2.5741878	2.5517372	2.5311275	2.5121373	2.4945794
30	2.6300783	2.6002568	2.5732463	2.5486592	2.5261768	2.5055347	2.4865117	2.468921
40	2.451085	2.4210135	2.3937382	2.3688761	2.3461122	2.3251849	2.3058751	2.2879977
50	2.3480744	2.3178032	2.2903194	2.2652428	2.2422607	2.2211132	2.2015826	2.183485
60	2.2811252	2.2506969	2.2230498	2.1978059	2.1746539	2.1533353	2.1336331	2.1153643
70	2.2341167	2.2035631	2.1757863	2.1504095	2.1271227	2.1056683	2.08583	2.0674253
80	2.1992922	2.168637	2.140755	2.1152707	2.0918748	2.0703104	2.0503619	2.0318471
90	2.1724569	2.1417179	2.1137493	2.0881763	2.0646904	2.0430354	2.0229959	2.0043901
100	2.1511439	2.1203346	2.0922933	2.0666461	2.043085	2.0213539	2.0012382	1.9825562
∞	1.9652139	1.933629	1.90478	1.878312	1.853911	1.831336	1.810363	1.790825

TABLE A7: Continued

Degrees of freedom of denominator	Degrees of freedom of numerator							
	25	**26**	**27**	**28**	**29**	**30**	**31**	**32**
1	6239.8251	6244.6239	6249.0708	6253.2031	6257.053	6260.6486	6264.0142	6267.1711
2	99.459168	99.460706	99.46213	99.463452	99.464684	99.465833	99.466908	99.467916
3	26.578983	26.561844	26.545952	26.531176	26.517403	26.504534	26.492482	26.481172
4	13.910852	13.894011	13.878391	13.863864	13.850319	13.83766	13.825803	13.814673
5	9.4491208	9.4330709	9.4181801	9.4043273	9.3914074	9.3793292	9.3680131	9.3573891
6	7.2959708	7.2804708	7.2660857	7.2526992	7.2402109	7.2285331	7.2175893	7.2073125
7	6.0579546	6.042806	6.0287428	6.015652	6.0034361	5.9920102	5.9812998	5.9712398
8	5.2631442	5.2482172	5.2343554	5.2214485	5.209401	5.1981295	5.1875615	5.1776329
9	4.7130308	4.6982414	4.6845032	4.6717079	4.6597614	4.6485817	4.6380972	4.6282448
10	4.3110559	4.2963486	4.282683	4.2699517	4.258062	4.2469328	4.2364932	4.226681
11	4.0050895	3.9904272	3.9767997	3.9641007	3.9522381	3.9411318	3.9307113	3.920915
12	3.7646909	3.7500482	3.7364352	3.7237467	3.7118911	3.7007888	3.6903699	3.680573
13	3.5709642	3.5563234	3.5427088	3.5300157	3.5181531	3.5070418	3.4966123	3.4868035
14	3.4115917	3.3969405	3.3833128	3.3706046	3.3587254	3.3475961	3.3371477	3.3273192
15	3.2782161	3.2635457	3.249897	3.2371664	3.2252637	3.2141102	3.203637	3.1937834
16	3.1649747	3.1502789	3.1366037	3.1238456	3.1119147	3.1007326	3.0902307	3.0803482
17	3.0676374	3.0529121	3.0392064	3.0264172	3.014455	3.0032415	2.992708	2.9827943

18	2.9830777	2.9683198	2.954581	2.9417585	2.9297628	2.9185159	2.9079493	2.8980026
19	2.9089358	2.8941435	2.88037	2.8675125	2.855482	2.8442005	2.8335996	2.8236191
20	2.8433982	2.8285703	2.8147609	2.8018677	2.7898017	2.7784849	2.7678492	2.7578343
21	2.7850485	2.7701842	2.7563385	2.7434091	2.7313072	2.7199549	2.7092841	2.6992348
22	2.7327644	2.7178635	2.7039813	2.6910157	2.6788778	2.6674901	2.6567844	2.6467008
23	2.6856457	2.6707083	2.6567897	2.643788	2.6316144	2.6201915	2.6094512	2.5993335
24	2.6429617	2.627988	2.6140334	2.6009959	2.5887871	2.5773294	2.5665549	2.5564036
25	2.6041134	2.5891038	2.5751136	2.5620409	2.5497973	2.5383053	2.5274972	2.5173128
26	2.5686048	2.55356	2.5395347	2.5264274	2.5141496	2.502624	2.4917828	2.4815661
27	2.5360222	2.5209427	2.506883	2.4937417	2.4814305	2.469872	2.4589984	2.4487499
28	2.5060173	2.4909038	2.4768104	2.4636359	2.4512919	2.4397012	2.428796	2.4185165
29	2.4782951	2.4631483	2.4490221	2.435815	2.423439	2.4118168	2.4008807	2.3905709
30	2.4526039	2.4374247	2.4232663	2.4100275	2.3976203	2.3859674	2.3750011	2.3646616
40	2.2713954	2.2559333	2.2414953	2.2279808	2.2153019	2.203382	2.1921536	2.181557
50	2.1666638	2.1509847	2.1363322	2.1226059	2.1097185	2.0975934	2.0861633	2.0753685
60	2.0983728	2.082525	2.0677054	2.0538141	2.040764	2.0284785	2.0168905	2.0059405
70	2.0502986	2.0343165	2.019364	2.0053412	1.9921612	1.9797477	1.9680335	1.9569593
80	2.0146106	1.9985195	1.983459	1.9693293	1.9560436	1.9435257	1.9317086	1.9205329
90	1.9870631	1.9708819	1.9557319	1.9415137	1.9281404	1.915536	1.9036333	1.8923732
100	1.9651531	1.9488962	1.9336712	1.9193785	1.9059314	1.893254	1.8812793	1.8699481
∞	1.7725649	1.75545	1.739368	1.724224	1.709928	1.696406	1.683594	1.67143

TABLE A8: Significant Ranges for the Duncan's Multiple Range Test for $\alpha = 0.01$

$r_{0.01}\ (p, f)$

f	p = 2	3	4	5	6	7	8	9	10	20	50	100
1	90	90	90	90	90	90	90	90	90	90	90	90
2	14	14	14	14	14	14	14	14	14	14	14	14
3	8.26	8.5	8.6	8.7	8.8	8.9	8.9	9	9	9.3	9.3	9.3
4	6.51	6.8	6.9	7	7.1	7.1	7.2	7.2	7.3	7.5	7.5	7.5
5	5.7	5.96	6.11	6.18	6.26	6.33	6.4	6.44	6.5	6.8	6.8	6.8
6	5.24	5.51	5.65	5.73	5.81	5.88	5.95	6	6	6.3	6.3	6.3
7	4.95	5.22	5.37	5.45	5.53	5.61	5.69	5.73	5.8	6	6	6
8	4.74	5	5.14	5.23	5.32	5.4	5.47	5.51	5.5	5.8	5.8	5.8
9	4.6	4.86	4.99	5.08	5.2	5.25	5.32	5.36	5.4	5.7	5.7	5.7
10	4.48	4.73	4.88	4.96	5.06	5.13	5.2	5.24	5.28	5.55	5.55	5.55
11	4.39	4.63	4.77	4.86	4.94	5.01	5.06	5.12	5.15	5.39	5.39	5.39
12	4.32	4.55	4.68	4.76	4.84	4.92	4.96	5.02	5.07	5.26	5.26	5.26
13	4.26	4.48	4.62	4.69	4.74	4.84	4.88	4.94	4.98	5.15	5.15	5.15
14	4.21	4.42	4.55	4.63	4.7	4.78	4.83	4.87	4.91	5.07	5.07	5.07
15	4.17	4.37	4.5	4.58	4.64	4.72	4.77	4.81	4.84	5	5	5
16	4.13	4.34	4.45	4.54	4.6	4.67	4.72	4.76	4.79	4.94	4.94	4.94
17	4.1	4.3	4.41	4.5	4.56	4.63	4.68	4.73	4.75	4.89	4.89	4.89
18	4.07	4.27	4.38	4.46	4.46	4.59	4.64	4.68	4.71	4.85	4.85	4.85
19	4.05	4.24	4.35	4.43	4.43	4.56	4.61	4.64	4.67	4.82	4.82	4.82
20	4.02	4.22	4.33	4.4	4.4	4.53	4.58	4.61	4.65	4.79	4.79	4.79
30	3.89	4.06	4.16	4.22	4.32	4.36	4.41	4.45	4.48	4.65	4.71	4.71
40	3.82	3.99	4.1	4.17	4.24	4.3	4.34	4.37	4.41	4.59	4.69	4.69
60	3.76	3.92	4.03	4.12	4.17	4.23	4.27	4.31	4.34	4.53	4.66	4.66
100	3.71	3.86	3.98	4.06	4.11	4.17	4.21	4.25	4.29	4.48	4.64	4.65
∞	3.64	3.8	3.9	3.98	4.04	4.09	4.14	4.17	4.2	4.41	4.6	4.68

TABLE A9: Significant Ranges for the Duncan's Multiple Range Test for $\alpha = 0.05$

$r_{0.05}\ (p, f)$

	p											
f	**2**	**3**	**4**	**5**	**6**	**7**	**8**	**9**	**10**	**20**	**50**	**100**
1	18	18	18	18	18	18	18	18	18	18	18	18
2	6.09	6.09	6.09	6.09	6.09	6.09	6.09	6.09	6.09	6.09	6.09	6.09
3	4.5	4.5	4.5	4.5	4.5	4.5	4.5	4.5	4.5	4.5	4.5	4.5
4	3.93	4.01	4.02	4.02	4.02	4.02	4.02	4.02	4.02	4.02	4.02	4.02
5	3.64	3.74	3.79	3.83	3.83	3.83	3.83	3.83	3.83	3.83	3.83	3.83
6	3.46	3.58	3.64	3.68	3.68	3.68	3.68	3.68	3.68	3.68	3.68	3.68
7	3.35	3.47	3.54	3.58	5.53	3.61	3.61	3.61	3.61	3.61	3.61	3.61
8	3.26	3.39	3.47	3.52	3.58	3.56	3.56	3.56	3.56	3.56	3.56	3.56
9	3.2	3.34	3.41	3.47	3.5	3.52	3.52	3.52	3.52	3.52	3.52	3.52
10	3.15	3.3	3.37	3.43	3.46	3.47	3.47	3.47	3.47	3.48	3.48	3.48
11	3.11	3.27	3.35	3.39	3.43	3.44	3.46	3.46	3.46	3.48	3.48	3.48
12	3.08	3.23	3.33	3.36	3.4	3.42	3.44	3.44	3.46	3.48	3.48	3.48
13	3.06	3.21	3.3	3.35	3.38	3.41	3.44	3.44	3.45	3.47	3.47	3.47
14	3.03	3.18	3.27	3.33	3.37	3.39	3.42	3.42	3.44	3.47	3.47	3.47
15	3.01	3.16	3.25	3.31	3.36	3.38	3.42	3.42	3.43	3.47	3.47	3.47
16	3	3.15	3.23	3.3	3.34	3.37	3.41	3.41	3.43	3.47	3.47	3.47
17	2.98	3.13	3.22	3.28	3.33	3.36	3.4	3.4	3.42	3.47	3.47	3.47
18	2.97	3.12	3.21	3.27	3.32	3.35	3.39	3.39	3.41	3.47	3.47	3.47
19	2.96	3.11	3.19	3.26	3.31	3.35	3.39	3.39	3.41	3.47	3.47	3.47
20	2.95	3.1	3.18	3.25	3.3	3.34	3.38	3.38	3.4	3.47	3.47	3.47
30	2.89	3.04	3.12	3.2	3.25	3.29	3.35	3.35	3.37	3.47	3.47	3.47
40	2.86	3.01	3.1	3.17	3.22	3.27	3.33	3.33	3.35	3.47	3.47	3.47
60	2.83	2.98	3.08	3.14	3.2	3.24	3.31	3.31	3.33	3.47	3.48	3.48
100	2.8	2.95	3.05	3.12	3.18	3.22	3.26	3.29	3.32	3.47	3.53	3.53
∞	2.77	2.92	3.02	3.09	3.15	3.19	3.23	3.26	3.29	3.47	3.61	3.67

TABLE A10: Table of Orthogonal Polynomial Coefficients

Order of polynomial (n)	X	1	2	3	4	5	6	7	8	9	10	$\sum [g_k(x)]^2$	λ
3	Linear	−1	0	1								2	1
	Quadratic	1	−2	1								6	3
4	Linear	−3	−1	1	3							20	2
	Quadratic	1	−1	−1	1							4	1
	Cubic	−1	3	−3	1							20	10/3
5	Linear	−2	−1	0	1	2						10	1
	Quadratic	2	−1	−2	−1	2						14	1
	Cubic	−1	2	0	−2	1						10	5/6
	Quartic	1	−4	6	−4	1						70	35/12
6	Linear	−5	−3	−1	1	3	5					70	2
	Quadratic	5	−1	−4	−4	−1	5					84	3/2
	Cubic	−5	7	4	−4	−7	5					180	5/3
	Quartic	1	−3	2	2	−3	1					28	7/12
	Quintic	−1	5	−10	10	−5	1					252	21/10
7	Linear	−3	−2	−1	0	1	2	3				28	1
	Quadratic	5	0	−3	−4	−3	0	5				84	1
	Cubic	−1	1	1	0	−1	−1	1				6	1/6
	Quartic	3	−7	1	6	1	−7	3				154	7/12
	Quintic	−1	4	−5	0	5	−4	1				84	7/20
	Sextic	1	−6	15	−20	15	−6	1				924	77/60

8	Linear	−7	−5	−3	−1	1	3	5	7			168	2
	Quadratic	7	1	−3	−5	−5	−3	1	7			168	1
	Cubic	−7	5	7	3	−3	−7	−5	7			264	2/3
	Quartic	7	−13	−3	9	9	−3	−13	7			616	7/12
	Quintic	−7	23	−17	−15	15	17	−23	7			2184	7/10
	Sextic	1	−5	9	−5	−5	9	−5	1			264	11/60
	Septic	−1	7	−21	35	−35	21	−7	1			3432	
9	Linear	−4	−3	−2	−1	0	1	2	3	4		60	1
	Quadratic	28	7	−8	−17	−20	−17	−8	7	28		2772	3
	Cubic	−14	7	13	9	0	−9	−13	−7	14		990	5/6
	Quartic	14	−21	−11	9	18	9	−11	−21	14		2002	7/12
	Quintic	−4	11	−4	−9	0	9	4	−11	4		468	3/20
	Sextic	4	−17	22	1	−20	1	22	−17	4		1980	11/60
	Septic	−1	6	−14	14	0	−14	14	−6	1		858	
	Octic	1	−8	28	−56	70	−56	28	−8	1		12870	
10	Linear	−9	−7	−5	−3	−1	1	3	5	7	9	330	2
	Quadratic	6	2	−1	−3	−4	−4	−3	−1	2	6	132	1/2
	Cubic	−42	14	35	31	12	−12	−31	−35	−14	42	8580	5/3
	Quartic	18	−22	−17	3	18	18	3	−17	−22	18	2860	5/12
	Quintic	−6	14	−1	−11	−6	6	11	1	−14	6	780	1/10
	Sextic	3	−11	10	6	−8	−8	6	10	−11	3	660	11/240
	Septic	−9	47	−86	42	56	−56	−42	86	−47	9	29172	
	Octic	1	−7	20	−28	14	14	−28	20	−7	1	2860	
	Novic	−1	9	−36	84	−126	126	−84	36	−9	1	48620	

Note: X refers to the columns whereas (n) refers to the degrees of the polynomial, that is, 3 4 5 6 7 8 9 10. The arrow on X indicates the values of the first row of Table A10, that is, 1 2 3 4 5 .....

TABLE A11: Critical Values of the Studentized Range Statistic for $\alpha = 0.05$

	Number of means (k)											
Degrees of freedom	**2**	**3**	**4**	**5**	**6**	**7**	**8**	**9**	**10**	**12**	**15**	**20**
2	6.08	8.33	9.80	10.88	11.74	12.44	13.03	13.54	13.99	14.75	15.65	16.77
3	4.50	5.91	6.82	7.50	8.04	8.48	8.85	9.18	9.46	9.95	10.52	11.24
4	3.93	5.04	5.76	6.29	6.71	7.05	7.35	7.60	7.83	8.21	8.66	9.23
5	3.64	4.60	5.22	5.67	6.03	6.33	6.58	6.80	6.99	7.32	7.72	8.21
6	3.46	4.34	4.90	5.30	5.63	5.90	6.12	6.32	6.49	6.79	7.14	7.59
7	3.34	4.16	4.68	5.06	5.36	5.61	5.82	6.00	6.16	6.43	6.76	7.17
8	3.26	4.04	4.53	4.89	5.17	5.40	5.60	5.77	5.92	6.18	6.48	6.87
9	3.20	3.95	4.41	4.76	5.02	5.24	5.43	5.59	5.74	5.98	6.28	6.64
10	3.15	3.88	4.33	4.65	4.91	5.12	5.30	5.46	5.60	5.83	6.11	6.47
12	3.08	3.77	4.20	4.51	4.75	4.95	5.12	5.27	5.39	5.61	5.88	6.21
14	3.03	3.70	4.11	4.41	4.64	4.83	4.99	5.13	5.25	5.46	5.71	6.03
16	3.00	3.65	4.05	4.33	4.56	4.74	4.90	5.03	5.15	5.35	5.59	5.50
18	2.97	3.61	4.00	4.28	4.49	4.67	4.82	4.96	5.07	5.27	5.50	5.79
20	2.95	3.58	3.96	4.23	4.45	4.62	4.77	4.90	5.01	5.20	5.43	5.71
24	2.92	3.53	3.90	4.17	4.37	4.54	4.68	4.81	4.92	5.10	5.32	5.59
30	2.89	3.49	3.85	4.10	4.30	4.46	4.60	4.72	4.82	5.00	5.21	5.47
40	2.86	3.44	3.79	4.04	4.23	4.39	4.52	4.63	4.73	4.90	5.11	5.36
60	2.83	3.40	3.74	3.98	4.16	4.31	4.44	4.55	4.65	4.81	5.00	5.24
120	2.80	3.86	3.68	3.92	3.10	4.24	4.36	4.47	4.56	4.71	4.90	5.13
∞	2.77	3.31	3.63	3.86	4.03	4.17	4.29	4.39	4.47	4.62	4.80	5.10

TABLE A12: Critical Values of the Studentized Range Statistic for $\alpha = 0.10$

Degrees of freedom	Number of means (k)											
	2	**3**	**4**	**5**	**6**	**7**	**8**	**9**	**10**	**12**	**15**	**20**
2	4.14	5.74	6.78	7.54	8.14	8.63	9.05	9.41	9.73	10.26	10.89	11.68
3	3.33	4.47	5.20	5.74	6.16	6.51	6.81	7.06	7.29	7.67	8.12	8.68
4	3.02	3.98	4.59	5.04	5.39	5.68	5.93	6.14	6.33	6.65	7.03	7.50
5	2.85	3.72	4.26	4.67	4.98	5.24	5.46	5.65	5.82	6.10	6.44	6.86
6	2.75	3.56	4.07	4.44	4.73	4.97	5.17	5.34	5.50	5.76	6.08	6.47
7	2.68	3.45	3.93	4.28	4.56	4.78	4.97	5.14	5.28	5.53	5.83	6.19
8	2.63	3.38	3.83	4.17	4.43	4.65	4.83	4.99	5.13	5.36	5.64	6.00
9	2.59	3.32	3.76	4.08	4.34	4.55	4.72	4.87	5.01	5.24	5.51	5.85
10	2.56	3.27	3.70	4.02	4.26	4.47	4.64	4.78	4.91	5.13	5.40	5.73
12	2.52	3.20	3.62	3.92	4.16	4.35	4.51	4.65	4.78	4.99	5.24	5.55
14	2.49	3.16	3.56	3.85	4.08	4.27	4.42	4.56	4.68	4.88	5.12	5.43
16	2.47	3.12	3.52	3.81	4.03	4.21	4.36	4.49	4.61	4.81	5.04	5.33
18	2.45	3.10	3.49	3.77	3.98	4.16	4.31	4.44	4.55	4.75	4.98	5.26
20	2.44	3.08	3.46	3.74	3.95	4.12	4.27	4.40	4.51	4.70	4.92	5.20
24	2.42	3.05	3.42	3.69	3.90	4.07	4.21	4.34	4.45	4.63	4.85	5.12
30	2.40	3.02	3.39	3.65	3.85	4.02	4.16	4.28	4.38	4.56	4.77	5.03
40	2.38	2.99	3.35	3.61	3.80	3.96	4.10	4.22	4.32	4.49	4.69	4.95
60	2.36	2.96	3.31	3.56	3.76	3.91	4.04	4.16	4.25	4.42	4.62	4.86
120	2.35	2.93	3.28	3.52	3.71	3.86	3.99	4.10	4.19	4.35	4.54	4.78
∞	2.33	2.90	3.24	3.48	3.66	3.81	3.93	4.04	4.13	4.28	4.47	4.69

TABLE A13: Critical Values of the Studentized Range Statistic for $\alpha = 0.01$

Degrees of freedom	Number of means (k)											
	2	**3**	**4**	**5**	**6**	**7**	**8**	**9**	**10**	**12**	**15**	**20**
2	14.04	19.02	22.29	24.72	26.63	28.20	29.53	30.68	31.69	33.40	35.43	37.95
3	8.26	10.62	12.17	13.33	14.24	15.00	15.64	16.20	16.69	17.53	18.52	19.77
4	6.51	8.12	9.17	9.96	10.58	11.10	11.55	11.93	12.27	12.84	13.53	14.40
5	5.70	6.85	7.80	8.42	8.91	9.32	9.67	9.97	10.24	10.70	11.24	11.93
6	5.24	6.33	7.03	7.56	7.97	8.32	8.61	8.87	9.10	9.48	9.95	10.54
7	4.95	5.92	6.54	7.00	7.37	7.68	7.94	8.17	8.37	8.71	9.12	9.65
8	4.75	5.64	6.20	6.62	6.96	7.24	7.47	7.68	7.86	8.18	8.55	9.03
9	4.60	5.43	5.96	6.35	6.66	6.91	7.13	7.33	7.49	7.78	8.13	8.57
10	4.48	5.27	5.77	6.14	6.43	6.67	6.87	7.05	7.21	7.49	7.81	8.23
12	4.32	5.05	5.50	5.84	6.10	6.32	6.51	6.67	6.81	7.06	7.36	7.73
14	4.21	4.89	5.32	5.63	5.88	6.08	6.26	6.41	6.54	6.77	7.05	7.39
16	4.13	4.79	5.19	5.49	5.72	5.92	6.08	6.22	6.35	6.56	6.82	7.15
18	4.07	4.70	5.09	5.38	5.60	5.79	5.94	6.08	6.20	6.41	6.65	6.97
20	4.02	4.64	5.02	5.29	5.55	5.69	5.84	5.97	6.09	6.28	6.52	6.82
24	3.96	4.55	4.91	5.17	5.37	5.54	5.69	5.81	5.92	6.11	6.33	6.61
30	3.89	4.45	4.80	5.05	5.24	5.40	5.54	5.65	5.76	5.93	6.14	6.41
40	3.82	4.37	4.70	4.93	5.11	5.26	5.39	5.50	5.60	5.76	5.96	6.21
60	3.76	4.28	4.59	4.82	4.99	5.13	5.25	5.36	5.45	5.60	5.70	6.01
120	3.70	4.20	4.50	4.71	4.87	5.01	5.12	5.21	5.30	5.44	5.61	5.83
∞	3.64	4.12	4.40	4.60	4.76	4.88	4.99	5.08	5.16	5.29	5.45	5.65

Index